MXene Reinforced Polymer Composites

Scrivener Publishing
100 Cummings Center, Suite 541J
Beverly, MA 01915-6106

Publishers at Scrivener
Martin Scrivener (martin@scrivenerpublishing.com)
Phillip Carmical (pcarmical@scrivenerpublishing.com)

MXene Reinforced Polymer Composites

Fabrication, Characterization and Applications

Edited by

Kalim Deshmukh

New Technologies-Research Centre, University of West Bohemia, Plzeň, Czech Republic

Mayank Pandey

Department of Electronics, Kristu Jayanti College, Bengaluru, India

and

Chaudhery Mustansar Hussain

Department of Chemistry & Environmental Sciences, New Jersey Institute of Technology, Newark, New Jersey, United States

Scrivener Publishing

WILEY

This edition first published 2024 by John Wiley & Sons, Inc., 111 River Street, Hoboken, NJ 07030, USA and Scrivener Publishing LLC, 100 Cummings Center, Suite 541J, Beverly, MA 01915, USA

For more information about Scrivener publications please visit www.scrivenerpublishing.com.

Wiley Global Headquarters
111 River Street, Hoboken, NJ 07030, USA

For details of our global editorial offices, customer services, and more information about Wiley products visit us at www.wiley.com.

Library of Congress Cataloging-in-Publication Data

ISBN 978-1-119-90104-4

Cover images: Pixabay.Com
Cover design by Russell Richardson

Set in size of 11pt and Minion Pro by Manila Typesetting Company, Makati, Philippines

Printed in the USA

10 9 8 7 6 5 4 3 2 1

Contents

Preface

Recently, MXenes, a relatively new and exciting class of two-dimensional (2D) materials, have attracted much attention in various research disciplines. MXenes, also called transition metal carbides, nitrides and carbonitrides, exhibit various compositions and remarkable properties, such as easy dispersibility, high surface-to-volume ratio, metallic conductivity and exceptional mechanical and structural characteristics. These properties make them promising candidates to be used as nanofillers in polymer composites. Polymer/MXene composites are benefitted from the attractive physicochemical properties of MXenes and the flexibility and facile processability of polymer matrices.

This book provides a detailed discussion of fundamental characteristics, synthesis and processing methods, structure, properties and characterizations of MXenes. Furthermore, it discusses surface chemistry and functionalization strategies of MXene and their incorporation into various polymer matrices to form high-performance polymer composites, followed by a systematic review of various strategies employed to design and synthesize advanced polymer/MXene composites comprising different polymers and different types of MXenes. The book further summarizes various applications of polymer/MXene composites as dielectrics, microwave absorption and electromagnetic interference (EMI) shielding, supercapacitors and electrochemical double layer capacitors, gas and volatile organic compounds sensing, flexible wearable sensors, biomedical engineering and biomedicine, water desalination and wastewater treatment, as well as pervaporation and gas separation. This book serves as a unique resource that critically describes the important research accomplishments and findings in the area of MXene based polymer composites, putting forth key technical challenges and future research perspectives in this field.

This book covers a comprehensive discussion on various promising aspects ranging from fundamental characteristics, synthesis, exfoliation and delamination techniques, surface chemistry, surface functionalization, and various properties of MXenes to fabrication, processing, characterizations, and numerous applications of MXene-reinforced polymer composites. The book comprises 15 chapters, which are summarized as follows.

Chapter 1 introduces 2D MXenes, their fundamental characteristics, processing, compositions, structure and various applications. Chapter 2 gives state-of-the-art recent progress in different chemical exfoliation and delamination methods of MXenes. Chapter 3 describes surface terminations, surface chemistry, and different functionalization strategies of MXenes. Chapter 4 discusses the electronic, electrical, and optical properties of MXener, while the magnetic, mechanical and thermal properties of MXenes are discussed in Chapter 5.

Chapter 6 provides information about different fabrication and processing methods and properties of MXene-reinforced polymer composites. Chapter 7 discusses the structural, morphological and tribological properties of polymer/MXene composites. Chapters 8-15 discuss various applications of MXene-reinforced polymer composites including dielectrics, microwave absorption and EMI shielding, supercapacitors and electrochemical double layer capacitors, gas and volatile organic compounds sensing, flexible wearable sensors, biomedical engineering and biomedicine, water desalination and wastewater treatment, as well as pervaporation and gas separation. Overall, this book will benefit all academic and industrial researchers who work in the emerging field of 2D MXenes and their polymer composites.

We are deeply grateful to all authors for their excellent contributions to this book. We also highly appreciate the dedicated support and valuable assistance rendered by Martin Scrivener and the Scrivener Publishing team during the publication of this book.

Dr. Kalim Deshmukh
Dr. Mayank Pandey
Prof. Chaudhery Mustansar Hussain
December 2023

1

Two-Dimensional MXenes: Fundamentals, Characteristics, Synthesis Methods, Processing, Compositions, Structure, and Applications

Sudipta Goswami and Chandan Kumar Ghosh*

School of Materials Science and Nanotechnology, Jadavpur University, Jadavpur, Kolkata, India

Abstract

During the last decade, among various two-dimensional materials, the carbides, nitrides or carbonitrides of transition metal ions (MXenes), defined by $M_{n+1}X_nT_x$ (n = 1 – 4), where M, X, and T_x stand for, respectively, the transition metal ions (e.g. Ti, V, Nb, Mo, etc.), carbon and/or nitrogen, and the surface terminated groups of different population, have drawn a lot of interest of the scientists as they exhibit unique features such as large conductivity, possibility of processing in the form of solution, large aspect ratio of the structure, and tunability of the properties. In this chapter, fundamental properties and classification of MXenes are discussed in detail along with different synthesis strategies and applications. Emphasis is given on discussing MXene hydrogel as they are widely being used in flexible electronics. Since surface functionalization plays a prominent role in this class of materials, controlling surface functionalization is discussed thoroughly. Correlation between applications of MXene and their structure is also discussed here.

***Keywords*:** MXene, MAX phase, 2D materials, surface termination, exfoliation, etching, energy storage, biomedical application

1.1 Introduction

In this chapter we discuss certain salient features of a class of two-dimensional compounds which are known by the acronym MXenes. They are composed of two-dimensional transition metal ion layers interleaved with carbon and nitrogen layers (thereby forming carbide or nitrides). The hydroxyl or oxy-halide groups are attached onto the top layers. During the last fifteen years or so, intense research is being pursued on this class of low-dimensional systems for their tremendous use in storage and harvesting of energy as well as in biomedical and sensor sectors. Among a wide range of oxide/non-oxide compounds including the lower-dimensional van der Waals solids, the MXenes (classified based on the type of transition metal ion and the number of layers) have carved a niche for themselves for their many unique features. We shall highlight those features while discussing the crystallographic and electronic structures, their synthesis and

**Corresponding author*: chandan.kghosh@jadavpuruniversity.in properties, and, finally, use in a score of applications. We first describe the fundamentals of these compounds along with their classification. Next, we discuss the electronic, physical, and chemical properties. The different techniques being used for synthesizing the MXenes are presented then and, different application potential of MXenes in various sectors have been discussed.

1.2 Fundamentals

The carbides or nitrides of transition metal ions in two-dimensional (2D) form are described by the general chemical formula $M_{n+1}X_nT_x$ where M, X, and T, respectively, designate the transition metal ion, carbon or nitrogen, and the hydroxyl (OH) or oxy-halide (OCl, OF, OBretc) ions which terminate the surface; 'n' defines the number of layers [1]. There are, primarily, three types of MXenes – M_2XT_x, $M_3X_2T_x$, and $M_4X_3T_x$. (Figure 1.1) [2] Interest

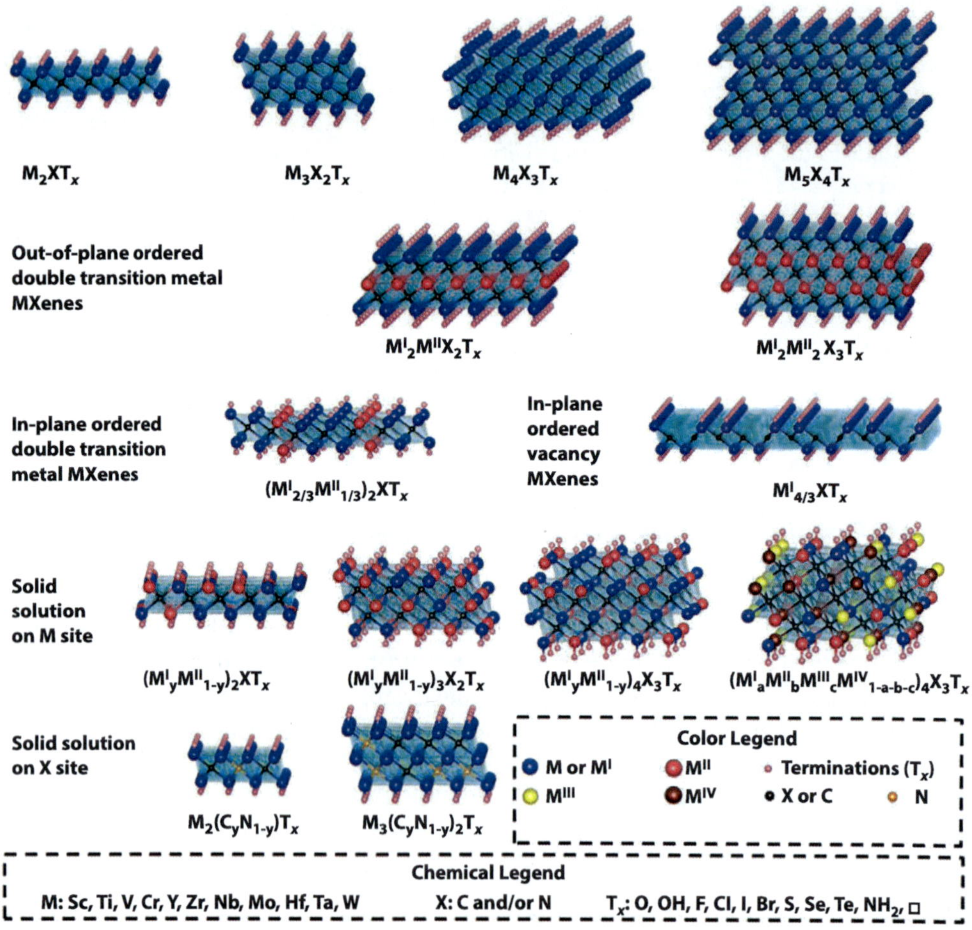

Figure 1.1 The crystallographic structure of single-, double-, and triple-layered MXene compounds $M_{n+1}X_nT_x$, (n = 1 to 4) where MX, and T represent, respectively, early transition metals, C or N, and F, O, or OH ions. More than one transition metal atoms may occupy the M sites forming solid solutions or ordered structures. Reproduced with permission from Ref. [34]. Copyright 2021, Wiley-VCH.

has been generated due to their application potential in areas such as harvesting and/or storage of energy (electrochemical supercapacitors, Li-ion, Na-ion batteries), catalysis (hydrogen or oxygen evolution reaction, CO_2 reduction), electronics/spintronics (memories, sensors), environment (membranes for clean air or water), structural, biomedical (biosensors, cancer treatment), sensors (gas, humidity, strain) etc. Unique features such as large electrical and thermal conductivity (multilayered MXenes exhibit higher conductivity than multilayered graphene), tunability of the band gap via surface terminated ions (from metallic to semiconducting), mechanical strength, etc., have made this class of compounds quite attractive for a variety of applications [3]. Here, we shall first examine their crystallographic and electronic structures for understanding the origin of the uniqueness of the properties of MXenes.

1.2.1 Crystallographic Structure

MXenes are derivatives of MAX phases where the A ions belong to the III–IV groups of the periodic table (i.e., the Al, Ga, In, Tl, Si, Ge, Sn, Pb, P, Bi, As, B, Te, S, etc.). The crystallographic structure of the MAX phase is shown in Figure 1.2 [4]. They assume hexagonal $P6_3/mmm$ (No. 194) structure [5]. By selectively leaching out the 'A' ions, it is possible to synthesize the MXenes. Therefore, the two-dimensional layers of MXenes to assume hexagonal structure (space group $P6_3/mmm$). Apart from the pure systems, alloy MXenes can also be synthesized where two different M ions – M′ and M″ – coexist and form a solid solution. In such alloy systems, ordering of the *M′* and *M″* ions have been noticed. The *M′* and *M″* ions could be ordered along the outer direction or within the planar structure [6]. In the case of ordering along the outer direction, the M′(M″) ions could be on the outermost layers while the M″(M′) ions occupy the inner layers. In the case of in-plane ordering,

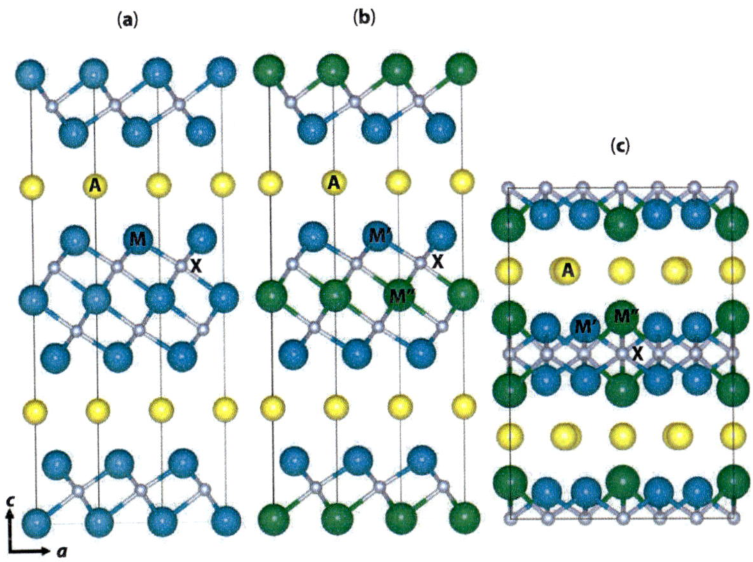

Figure 1.2 The crystal structures of the (a) MAX $[M_3AX_2]$ phase and (b) out-of- and (c) in-plane ordered $M'_2M''AX_2$ phases. Reproduced with permission from Ref. [83]. Copyright 2019, Elsevier.

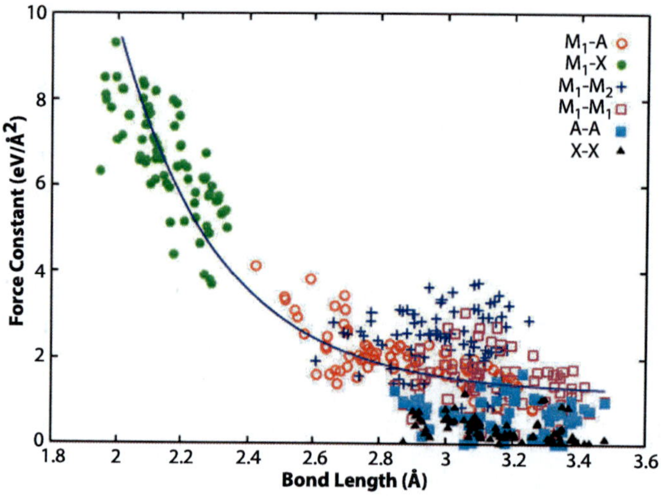

Figure 1.3 Variation of the force constant with bond length in various MAX phases. Reproduced with permission from Ref. [9]. Copyright 2018, The Royal Society of Chemistry.

the M′ and M″ ions are ordered within each layer– inner and outer. However, the in-plane ordering of M′ and M″ ions can be different from the hexagonal structure and close to the Kagome structure [7] when the size of M′ and M″ ions differ and the size of M′ ion turns out to be smaller than that of M″ ions. The space group in such cases becomes C2/c (monoclinic). The derivation of MXenes from the MAX phases requires mechanical or chemical exfoliation. The HF treatment, for example, yields formation of AF_2 and H_2 as the longer M-A bonds are relatively weak than the shortest M-X bonds. Theoretical simulation [8, 9] of the bond strengths (Figure 1.3) and the exfoliation processes offers insights behind the exfoliation processes. The MXenes also contain the F, OH, O surface ions. During the exfoliation process, vacancies are generated at the X-sites below the M layers. As a result, two sites – one with an X ion and another without (i.e., with a vacancy) – are generated. The F, OH, O groups are adsorbed at the vacancy sites and, thereby, form the stable structure with octahedral field of transition metal M ions.

1.2.2 Electronic Structure

Most of the MXenes are either metals or semimetals or semiconductors where spin-orbit coupling (SOC) does not have any significant effect on the electronic bands. In fact, when the functionalization of the surface is absent, the pristine MXenes are metallic. In these cases, the Fermi energy lies on the d-bands of the M ions. Surface functionalization by F, OH, O ions leads to the formation of new bands with hybridization of M d bands. The Fermi energy then shifts to the gap between M d bands and X p bands and the compound becomes semiconducting [10]. Such an observation has been made in the cases of Sc_2CT_2 (T = O, OH, F) and M_2CO_2 (M = Ti, Zr, Hf). In the cases of (M′, M″)XT_x systems, Fermi energy shifts to the gap generated from d band splitting due to octahedral crystal field and,

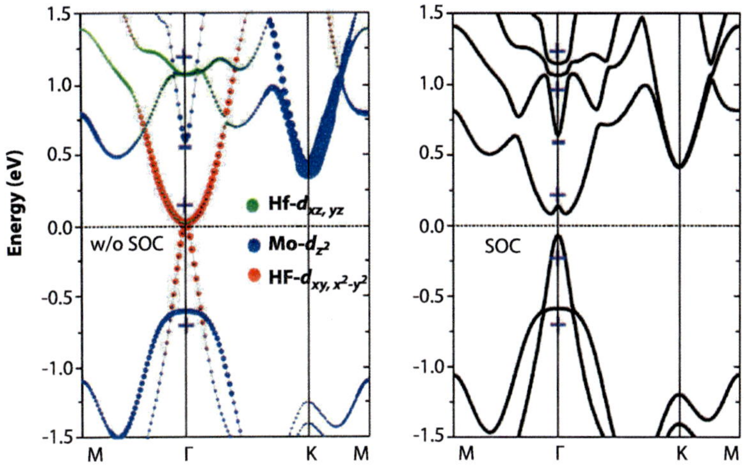

Figure 1.4 The electronic band structure of $Mo_2HfC_2O_2$ with and without spin-orbit coupling (SOC). Reproduced with permission from Ref. [11]. Copyright 2016, The American Chemical Society.

as a result, the compounds exhibit semiconducting behavior with finite energy gap. In general, single layer MXenes could exhibit semiconducting property whereas the double- or triple-layer ones are primarily metallic. However, in some cases, the SOC plays a significant role. When SOC is not present, the valence and the conduction bands (comprising of the d levels of the M ions) touch at the Γ point and give rise to semimetallic behavior. SOC splits the bands and opens a gap at the Γ point (Figure 1.4) [11]. The compounds M_2CO_2 (M = W, Mo) and $M_2'M''CO_2$ (M′ = Mo, W; M″ = Ti, Zr, Hf) exhibit such two-dimensional topological semimetallic or insulator behavior with topologically protected states with conducting edges which remain robust against nonmagnetic impurities and disorder [11].

1.2.3 Magnetic Structure

The MXenes exhibit finite magnetism and magnetic order depending on the electron states, band splitting, spin-orbit coupling, etc. For example, most of the compounds containing Cr and/or Mn exhibit magnetic order [12]. If the nonbonding d orbital resides in between bonding and anti-bonding orbitals and the Fermi level passes through these levels, finite magnetism is predicted. Theoretical calculations [13] predicted magnetism in a score MXene compounds such as M_2X (M = Ti, V, Cr, Ni, Mn; X = C, N), M_2MnC_2 (M = Ti, Hf), M_2TiX_2 (M = V, Cr, Mn), Hf_2VC_2, and $Mo_3N_2F_2$. The crystal field of the surrounding ligands of the transition metal ions determines the splitting of the bands including that of the spin bands and thus gives rise to the formation of majority and minority carriers. Accordingly, the compound exhibits metallic, half-metallic, or semiconducting properties. Most of the compounds exhibit antiferromagnetic ground state while the ones containing Mn ions are ferromagnetic. Presence of finite spin-orbit coupling in $Hf_2VC_2F_2$, on the other hand, induces noncollinear ordering with 120° rotation among the near-neighbor spins [14].

1.3 General Characteristics of the MXenes

1.3.1 Physical Properties

Some of the MXenes such as Sc_2CO_2 exhibit finite ferroelectric polarization [15] both along in- and out-of-plane directions because of asymmetric O ion cage structure. The O ions occupy the vacant sites below which C ions are present in one side and absent in another side and thus create structural noncentrosymmetry. The $(M'_{2/3}M''_{1/3})_2CO_2$ (M' = Mo, W; M'' = Sc, Y) compounds exhibit finite piezoelectric coefficient d_{11} ranging from 4 to ~25 pVm^{-1} [16] which is comparable to that of the dichalcogenides such as MoS_2, $MoSe_2$. Therefore, these compounds may find application in transducer device industry.

Some of the semiconducting compounds such as Sc_2CT_2 (T = O, OH, F) and M_2CO_2 (M = Ti, Zr, Hf) exhibit large thermoelectric power [17] because of reasonably high electrical conductivity (σ) and large Seebeck coefficient (S). The thermal conductivity (κ), on the other hand, is in the 10–60 W.m^{-1}.K^{-1} range which is comparable to that of the dichalcogenides. As a result, the thermoelectric figure of merit ZT ($=S^2\sigma T/\kappa$) for these compounds approaches a high value of ~1.1.

Very interestingly, the Mo_2C is found to be superconducting with $T_c < 4$ K [18]. This compound is derived from the MAX phase Mo_2GaC which also exhibits superconductivity at T_c ~4 K [19]. Influence of the surface terminated ions such as O, F, OH, H, Br, S, Se, etc. on the superconducting transition temperature has been theoretically estimated. The compound with H and Br were predicted to exhibit a higher T_c of ~13 K [20]. These systems appear to fall in the category of Bardeen-Cooper-Schrieffer (BCS) superconductors.

The functionalization of the MXenes by different surface terminated ions (e.g., by F, O, OH etc.) gives rise to change in the optical properties because of the formation of the defect states at close to the Fermi energy. The absorption, transmittance, reflectivity characteristics change and, as a result, the compounds become suitable for various optical applications. For example, O terminated Ti_3C_2 exhibits more prominent peaks in the real part of the dielectric permittivity $\varepsilon'(\omega)$ at the frequencies below the visible light range in comparison to the F or OH terminated compounds [21]. The O-functionalized M_2C (M = Ti, Zr, Hf) also exhibits distinct absorption peaks [i.e., peaks in the imaginary part of the dielectric permittivity $\varepsilon''(\omega)$] in the visible light range. This makes them suitable for optical devices. Figure 1.5 shows the dielectric permittivity (real and imaginary) versus energy (frequency) for pristine, O, F, and OH functionalized Ti_3C [21]. The complex dielectric permittivity exhibits anisotropy [22] as well – along in- and out-of-plane directions [Figure 1.5(b)]. MXenes such as V_2CT_x (T = O, OH, F) with large absorption in the 500–2700 nm range and high conductivity are also very good candidates for applications as conductive transport electrodes.

1.3.2 Chemical Properties

The surface functionalization changes the work function (difference in Fermi energy level and the vacuum) and also the dipole moments at the surface. The changes in the work function and surface dipole moment (ΔW and ΔP, respectively) are found to maintain linear correlation. Because of such changes, the interface developed between two-dimensional MXenes and other such two-dimensional systems could give rise to Schottky-barrier-free regions [23]. By varying the surface terminated ions – O, F, OH – it is possible to reconstruct

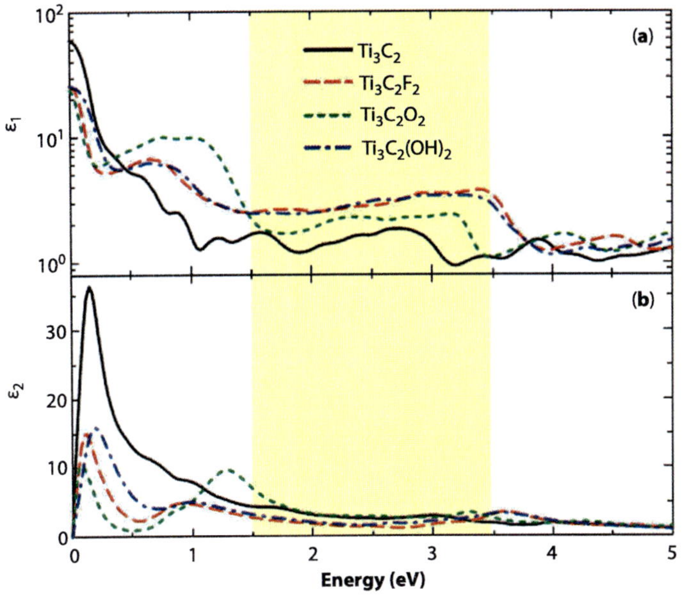

Figure 1.5 (a) The dielectric permittivity (real and imaginary) versus energy (frequency) for pristine, O, F, and OH functionalized Ti_3C_2; (b) in- and out-of-plane imaginary part of the dielectric permittivity for O-functionalized Ti, Hf, and Zr MXenes. Reproduced with permission from Ref. [21]. Copyright 2016, AIP Publishing.

the work function and thus the corresponding MXene can be made suitable for gas sensor applications [24] irrespective of their thickness (n = 1–3).

Since the surface area is large which results in enhanced surface activities, the MXenes are suitable for catalytic applications, especially, water splitting as well as hydrogen and oxygen evolution reactions (HER and OER) at the cathode and anode, respectively. The Gibbs free energy corresponding to the hydrogen absorption is close to zero for O-terminated Ti_2CO_2 and W_2CO_2. This has made them suitable for HER [25]. A theoretical calculation shows that among all the MXenes, the Mo_2CT_x possesses the most suitable surface activity for HER. Introducing metal ions (i.e., less electronegative) such as Fe it is possible to enhance the HER even further as the charge transfer to the O ions weakens the O-H bonds in water and thus facilitates the HER. The hybrid structure Ti_3C_2/g-C_3N_4 improves the OER [26], on the contrary, as the charge transfer between Ti and carbon nitride in graphitic form facilitates the oxygen evolution.

MXenes are potential candidates for energy storage (e.g., as capacitors and batteries) because of their large surface area and electrical conductivity. The diffusion barrier for selective ions can be engineered suitably by functionalizing the surface by O, F, or OH ions in various MXene compounds. For example, the diffusion barrier for Li ion can be changed [27] by introducing defect states in Ti_2C system and/or by developing hybrid structure with graphene. The Ti_2CT_2 (T = O, OH)/graphene heterostructure offers higher barrier for the diffusion of Li ions than what is observed in the pristine and monolayered MXene. The diffusion of alkali ions such as Na and K has also been studied in such structures [28]. It has been found that these monovalent ions diffuse more effectively than the multivalent Mg, Ca, Al ions. Monolayer MXene such as Ti_2C is found to be suitable for hydrogen adsorption as well [29]. The hydrogen adsorbed form hydrides whereas in the cases of double or triple layered compounds, molecular hydrogen was found to form.

1.4 Synthesis Methods

MXenes was first prepared from its corresponding MAX phase by removing "A". It produces two-dimensional (2D) flakes of general formula $M_{n+1}X_nT_x$ (n = 1, 2, 3). After the etching of "A", the surface terminations, T_x, cover the produced sheet. M stands for ions such as Ti, Nb, Cr, Mo, etc., X represents carbon and/or nitrogen which are connected with layers of group IIIA or IVA atoms and T stands for fluorine (F), hydroxyl (OH), or oxygen (O) ions

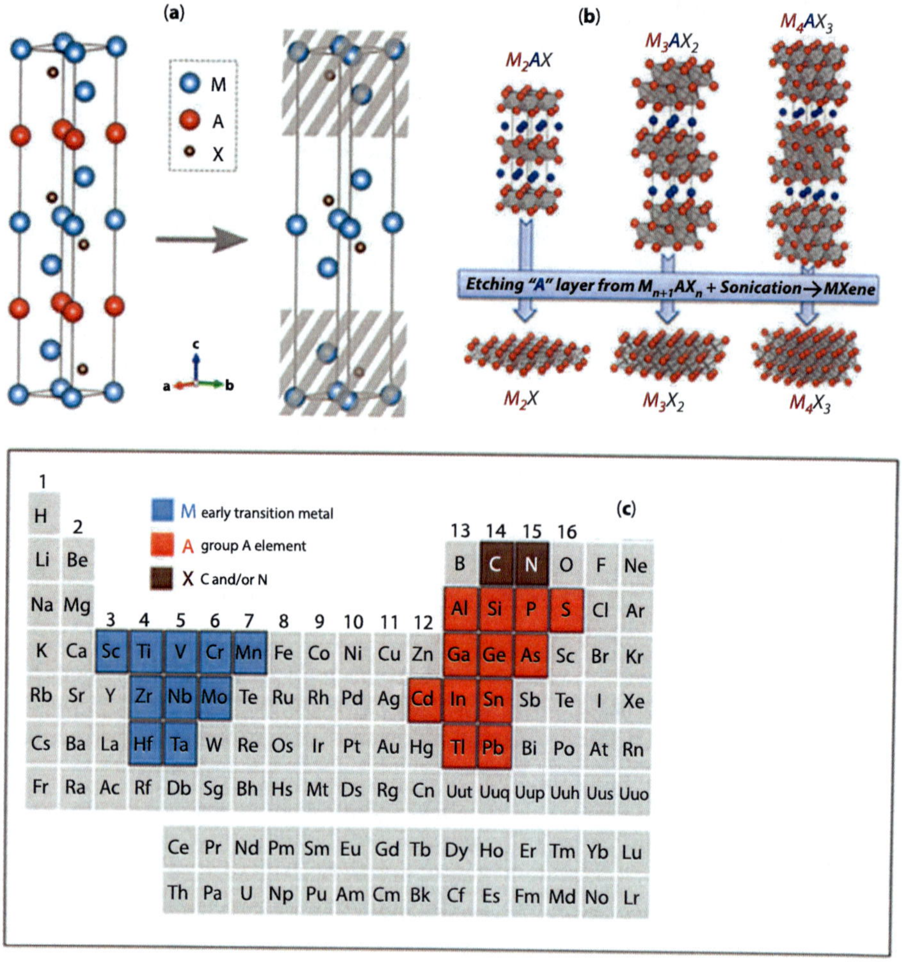

Figure 1.6 (a) The M_3AC_2 MAX phase primitive cell (left panel), and the resulting M_3C_2 MXene primitive cell (right panel). The blue spheres stand for the M atoms, the red for A atoms, and the brown for C atoms; (b) Structure of MAX phases from the corresponding MXenes after etching; (c) the location of the components of a MAX phase in the periodic table. Reproduced with permission from Ref. [30]. Copyright 2020, AIP Publishing.

which terminate the surface [30] (Figure 1.6). The selective elimination of element "A" can be achieved by two different ways – either top-down or bottom-up techniques. The top-down method basically includes liquid based etching process and is widely utilized. The bottom-up method, on the other hand, forces assembling of small atoms/molecules into 2D structures of different patterns. They include the CVD, carbonization etc. Different kinds of synthesis strategies for making 2D MXenes have been reported so far; e.g., wet selective etching, chemical vapor deposition (CVD), plasma-enhanced pulsed laser deposition (PEPLD), template method, intercalation etc. and discussed in details in following sections.

1.4.1 Wet Chemical Etching

The strategy for the preparation of 2D MXenes via wet selective etching involves the engraving of atomic layers of "A" from a multilayered MAX phase at room temperature. The higher reactivity of "A" and the weaker M-A bonds (than the layer-to-layer M-X bonds) are the key driving forces in this reaction [31]. In this process (Figure 1.7a) [32] etchant like aqueous HF of specific concentration is mixed with MAX phase powder by vigorous stirring for a specified time. HF breaks down the weak M-A bonds easily. Upon the removal of A-layer, interaction between the 2D sheets is weakened. Therefore, one can separate the layers very easily by ultrasonication or centrifugation and/or filtration [33–35]. The solid precipitate containing MXene phase is thoroughly washed with deionized water (DI) and the final pH of the suspension is maintained in between 4 to 6. The etching reaction converts the dense solid MAX phase into a loosely packed accordian-like MXenes phase; which looks alike exfoliated graphite [36]. The mechanism of etching process for Al-based MAX phase can be represented as follows:

$$M_{n+1}AlX_n + HF \rightarrow M_n+1X_n + AlF_3 + 1.5H_2 \tag{1.1}$$

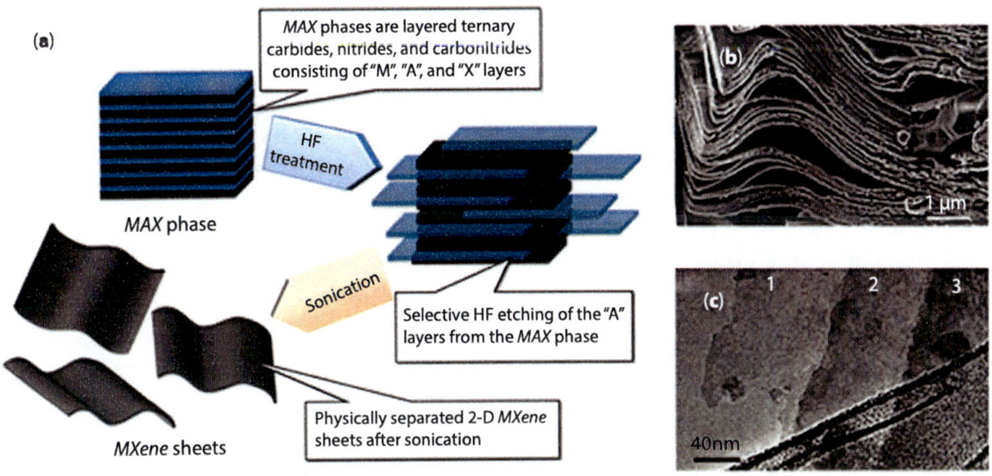

Figure 1.7 (a) Schematic for the formation of MXenes from of MAX phases by exfoliation process and; (b) The stacking of the multi-layer and (c) single layer MXenes. Reproduced with permission from Ref. [32]. Copyright 2012, The American Chemical Society.

$$M_n+1X_n+ 2H_2O \rightarrow M_n+1X_n(OH)_2+ H_2 \tag{1.2}$$

$$M_{n+1}X_n+ 2HF \rightarrow M_n+1X_nF_2+ H_2 \tag{1.3}$$

Numerous reports are available in literature for preparation of MXenes using the above principal; for example, Naguib *et al.* reported the replacement of Al atom with hydroxyl, Oxygen or Fluorine terminated surfaces from the bulk Ti_3AlC_2 by using aqueous HF. The resulting $M_{n+1}X_n$ shows a graphene like single sheet [37]. The etching process ultimately controls the defects concentration and surface termination reaction resulting in different morphology of the product; the concentration of etching solution, etching time, ultrasonic time etc. can be varied to obtain desirable microstructure like single layer or multilayer MXenes with high aspect ratio (Figures 1.7b and c). MXenes with more than 20 different compositions and stacking can be achieved by varying these process conditions. Different MXenes needs different etching time to complete the etching reaction because the etching process is dynamically controlled; with the increase in chain length (n) of MXenes a stronger and larger etching time is necessary for the complete conversion. In general, stability of the Mxene enhances with the rise in 'n'. Anasori *et al.* [38] have shown that with the increase in chain length in $Mo_2Ti_2AlC_2$/$Mo_2Ti_2AlC_3$ (from n = 2 to n = 3), the etching time doubles keeping all other parameter constant and Ti_2AlC requires 50% HF to yield Ti_3C_2, whereas, Ti_2AlC requires 10% HF to yield Ti_2C [32]. The combined effect of HF concentration and reaction time on the exfoliation of Al-containing MAX phases [e.g., Ti_2AlC, Ta_4AlC_3, $(Ti_{0.5}Nb_{0.5})_2AlC$, $(V_{0.5}Cr_{0.5})3AlC_2$, and Ti_3AlCN] has been studied in detail [39]. The MAX powder was immersed in different concentration of HF at room temperature. The reaction time was varied between 10 to 72 hrs. The etching reaction details are tabulated in Table 1.1. Importantly, it is noted that even after 65 h in a 50% HF solution, the reaction of the $(V_{0.5}Cr_{0.5})_3AlC_2$ powder was not complete; for all other cases the yields are quite high. The wide variation of etching parameters to remove the same element, here it is "Al" may also lie in the difference of M-Al bond energies of the corresponding MAX phase. For example, the Ti–Al bong energy (0.98 eV) in Ti_2AlC is much higher than the Nb-Al bond energy (1.21 eV) in Nb_2AlC [42]. This difference has been reflected in the longer etching time and higher HF concentration requirement for etching Al from Nb_2AlC compared to Ti_2AlC.

The SEM images of MAX phases and MXenes produced after exfoliation are shown in Figure 1.8. The figures reveal that the exfoliation of individual particles is quite successful. The importance of etching process parameters has also been reflected in an atomically laminated i-MAX phase, $(Mo_{2/3}Y_{1/3})_2AlC$, in which two different replaceable transition metal ions (Al and Y) are present in the basal plane. $(Mo_{2/3}Y_{1/3})_2AlC$ MAX phase when treated with different etching protocols gives two different types of MXene phases. Removal of Al atoms, selectively, by dissolving 1 gm $(Mo_{2/3}Y_{1/3})_2AlC_2$ into20 ml of 48%HF at room temperature with a stirring time of 12 h produces $(Mo_{2/3}Y_{1/3})_2C$ phase with in-plane elemental ordering. In another approach, removal of both Al and Y atoms have been achieved by dissolving 1 g of $(Mo_{2/3}Y_{1/3})_2AlC_2$ in 20 ml of 10 wt% HF at room temperature with a stirring time of 72 h, producing $Mo_{1.33}C$ with ordered vacancies [43]. Using liquid exfoliation via intercalation of molecules it is possible to prepare Mxenes. Introduction of appropriate molecules expands the interlayer space and weakens the interaction between layers. It eventually splits the multilayers into single sheets (Figure 1.9) [44].

Table 1.1 Reaction conditions and unit cell c-axis parameter of the of MXenes synthesized from corresponding MAX phases.

			RT etching conditions		C lattice parameter (Å)		Domain size (nm)	Yield (wt %)	Ref.
MAX structure	MAX	MXene	HF conc. (%)	Time, h	MAX	MXene			
211	Ti_2AlC	Ti_2CT_x	10	10	13.6	15.04	6	60	[32]
	V_2AlC	V_2CT_x	50	8	13.13	23.96			[40]
				90		19.73			
	Nb_2AlC	Nb_2CT_x	50	90	13.88	22.34			[40]
	$(Ti_{0.5}Nb_{0.5})_2AlC$	$(Ti_{0.5}Nb_{0.5})_2CT_x$	50	28	13.79	14.88	5	80	[32]
312	Ti_3AlC_2	$Ti_3C_2T_x$	50	2	18.42	20.51	11	100	[33, 32]
			40	20	18.62	20.89			[41]
	$(V_{0.5}Cr_{0.5})_3AlC_2$	$(V_{0.5}Cr_{0.5})_3C_2T_x$	50	69	17.73	24.26	28	NA	[32]
	Ti_3AlCN	Ti_3CNT_x	30	18	18.41	22.28	7	80	[32]
413	Ta_4AlC_3	$Ta_4C_3T_x$	50	72	24.08	30.34	38	90	[32]
	Nb_4AlC_3	$Nb_4C_3T_x$	50	90	24.19	30.47			[40]

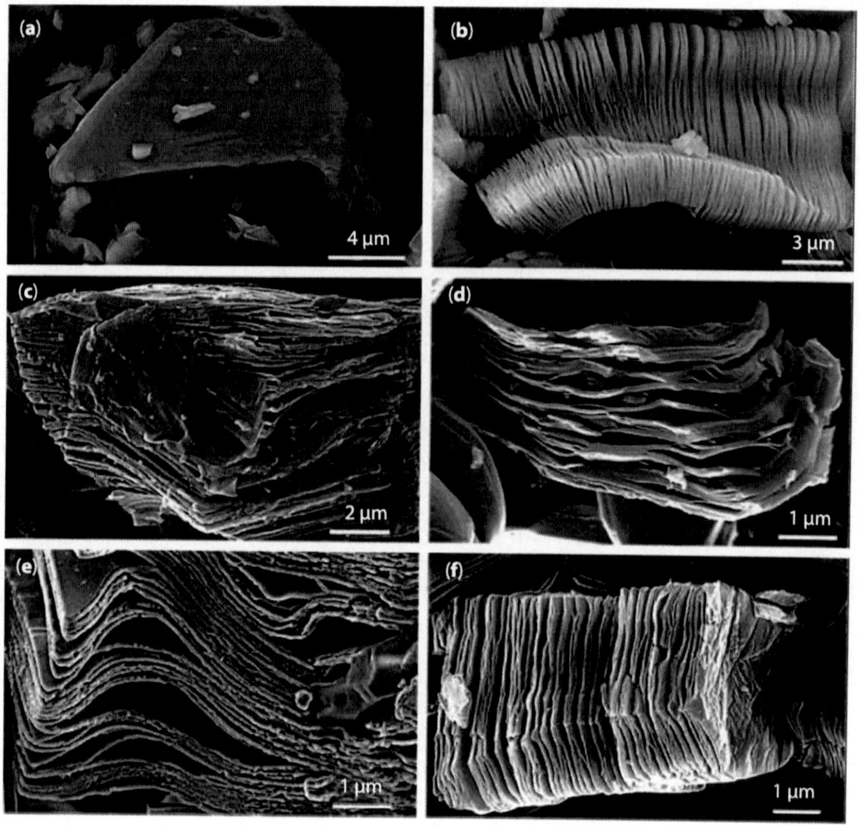

Figure 1.8 SEM images of (a) a typical unreacted Ti_3AlC_2 MAX phases before HF treatment, the exfoliation is obvious in the reaction thus producing (b) Ti_3AlC_2 (c) Ti_2AlC (d) Ta_4AlC_3 (e) TiNbAlC and (f) Ti_3AlCN "MXenes". Reproduced with permission from Ref. [32]. Copyright 2012, The American Chemical Society.

The exfoliation and intercalation can be done by using proper ultrasonication technique [45]. The etching time of Ti_3AlC_2 powders by using 49% HF could be reduced from 24 h to 4 h by using ultrasonic irradiation [46]. Ultrasonic vibration of high intensity breaks the MXene flakes into smaller sheets and facilitates the etching process. Zhang *et al.* [47] reported ultrasonication driven exfoliation of $Ti_3Si_{0.75}Al_{0.25}C_2$ using different solvents. They observed that the exfoliation is possible only if the A layer is composed of different atoms. This technique, however, cannot exfoliate pure Ti_3SiC_2. Thickness of the exfoliated ultrathin sheets is found to be ≈ 4 nm while the lateral dimensions are 100–200 nm. Obviously, intercalation increases the lattice parameter c (Δc), and the increment depends on the size of molecule of the solvent used; for $Ti_3C_2T_x$, it rises from 0.7 Å (for sodium sulphate) to 15.4 Å [for dimethyl-sulfoxide (DMSO)] [44, 48]. If ambient moisture co-intercalates spontaneously in DMSO intercalated $Ti_3C_2T_x$, it increases the c-axis parameter to a large extent and it can be further increased by storing the DMSO intercalated sample in ambient condition for long time. In general, the strategy for the preparation of $Ti_3C_2T_x$ at room temperature with HF or in-situ HF to produce delamination of multi-layered MXenes using different kinds of intercalants are tabulated in Figure 1.10 [37].

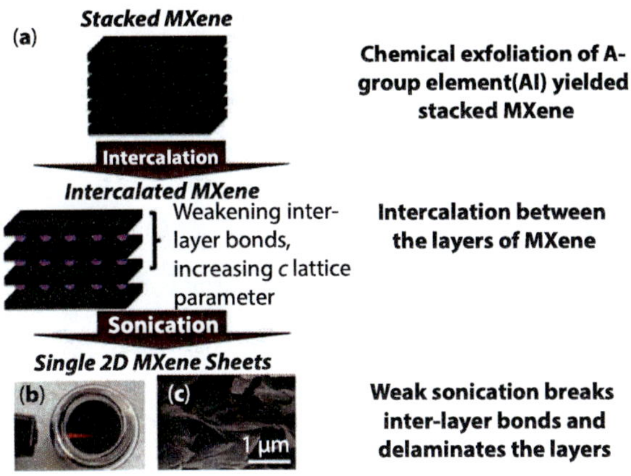

Figure 1.9 (a) The intercalation and delamination process for producing $Ti_3C_2T_x$; (b) the Tyndall scattering effect from the aqueous suspension of $Ti_3C_2T_x$ flakes with laser beam, and (c) the SEM image of delaminated flakes separated from the aqueous suspension by filtration. Reproduced with permission from Ref. [36, 44].

MXenes, prepared by wet chemical etching process, exhibit higher electrical conductivity and have fewer atomic defects. The main drawback of this process is the use of carcinogenic HF which is not at all environment friendly. Different approaches to replace HF by other etching medium like a mixture of hydrochloric acid and Lithium Fluoride which form in situ HF [49] had been tried. Other fluoride salts like NaF, KF, and NH_4F have also been used [50, 51], along with different (LIF)/HCl molar ratio. The electrochemical etching

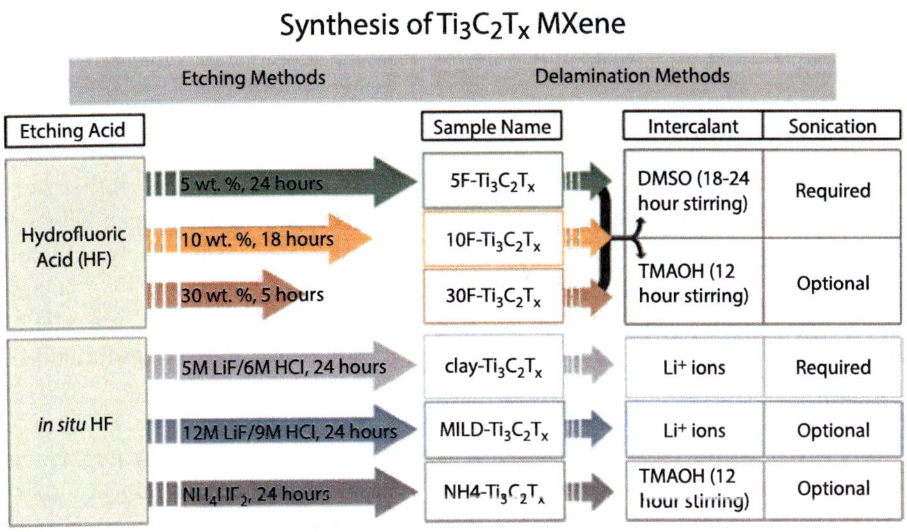

Figure 1.10 General strategy for $Ti_3C_2T_x$ synthesis from Ti_3AlC_2. Reproduced with permission from Ref. [37]. Copyright 2017, The American Chemical Society.

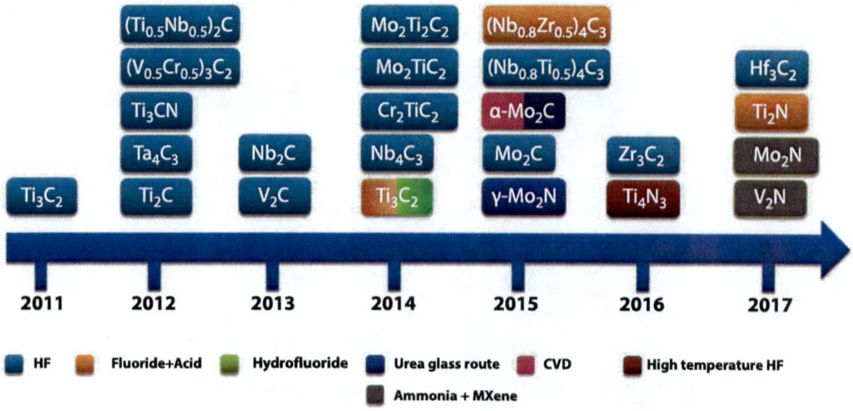

Figure 1.11 From the first discovery in 2011 the progress of the etching methods used to produce MXenes up to 2017. Reproduced with permission from Ref. [53]. Copyright 2019, Elsevier.

of Ti_2AlC in diluted HCl also produces MXenes without F [52]. It is possible to prepare biocompatible wearable sensors by using these synthesis techniques. The timeline of all the etching technique used so far for the preparation of different types of MXenes until 2017 is tabulated in Figure 1.11 [53].

1.4.2 Urea Glass Route

Phase pure nano-crystalline Mo_2C and Mo_2N MXenes were synthesized from the metal precursor $MoCl_5$ by using the urea glass process. The molar ratio of the urea/metal precursor [54] was changed for the synthesis. Mo-orthoesters formed when the precursor salts are dissolved into ethanol followed by gelation after reaction with urea. Silver black precipitate obtained after calcination at 800°C in inert atmosphere. Urea plays the dual role of C/N source and stabilizing agent [55]. The process is environment friendly. The large surface area of the produced MXene phases is quite effective for their use as catalysts.

1.4.3 Chemical Vapor Deposition

The technique of chemical vapor deposition (CVD) is quite useful for the synthesis of high quality 2D materials possessing large surface area. Xu *et al.* [19, 56] developed 2D ultrathin α-Mo_2C crystals by this technique using methane as the source for C. The Cu foil above a Mo foil was used as the substrate above 1085°C. At this high temperature Cu foil melts to form Mo-Cu alloy at the liquid Cu/Mo interface and Mo atoms diffuse from the interface to the surface of the liquid Cu. Decomposition of methane produces C which eventually reacts with Mo to form Mo_2C crystals (Figure 1.12).

It is important to control the concentration of methane which will otherwise produce graphene instead of Mo_2C. The thickness of the resulting crystals is in nanometer range. Their size is over 100 μm and they are very stable under ambient conditions. Other carbide crystals (for example, the ultrathin tungsten and tantalum carbides) can also be

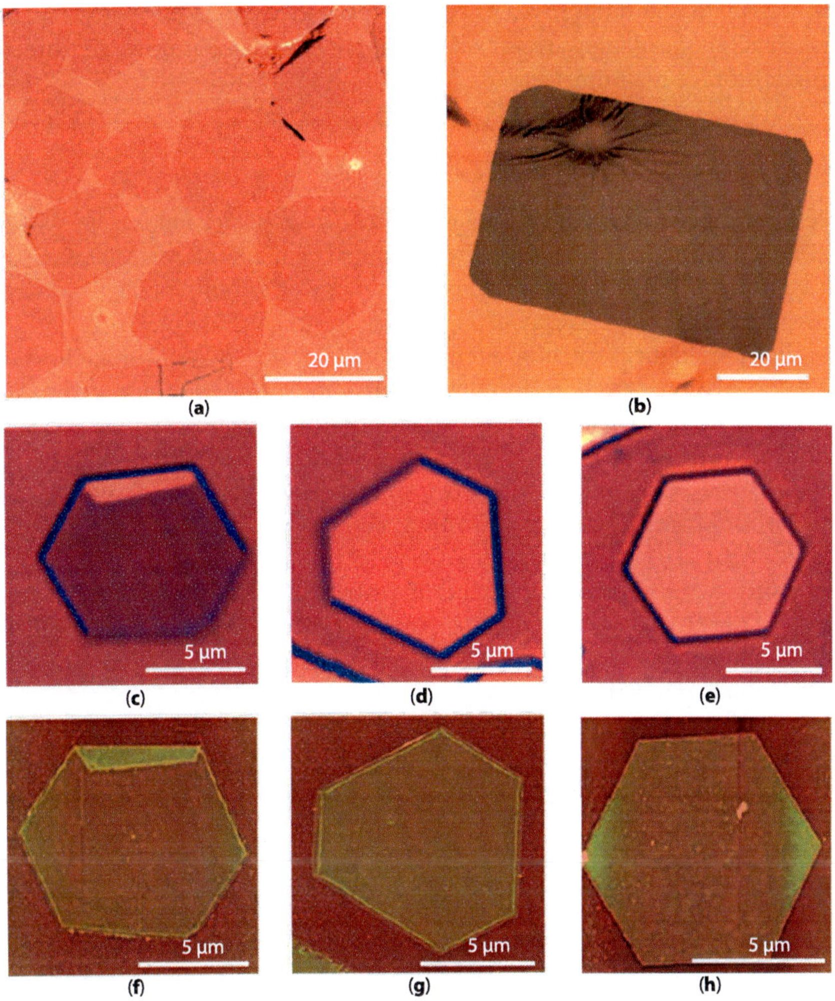

Figure 1.12 (a) Optical image of large-area 2D ultrathin α-Mo_2C crystals on Cu/Mo substrate prepared by CVD, showing different regular shapes; (b) Octagonal shaped crystal with average size of ~100 μm; Hexagonal shaped crystals with thicknesses of (c) 6.7nm (d) 8.2nm and (e) 11.2nm transferred onto a 290-nm-thick SiO_2/Si substrate, different colors arise from different thickness; (f–h) AFM images of the hexagonal crystals as shown in Figure 1.12c–e, respectively. Reproduced with permission from Ref. [19]. Copyright 2015, Springer Nature.

grown by this technique. The resulting crystals possess high crystallinity and very low defect concentration. Wang *et al.* [57] prepared ultrathin tantalum carbide, nitride, and boride using the CVD technique. A diffusion coupled Tantalum-Copper bilayer was heated at 1077°C for 30 min and 1 h in inert atmosphere with corresponding precursor (acetylene for C, ammonia for N and boron powder for B) for producing uniformly distributed sheets of TaC, TaB, and TaN on Cu. Here, diffusion of Ta takes place through thin Cu sheet to the opposite end of the atomically thin coating. Thus, it can protect the Cu matrix from oxidation and improves its scratch resistance. Since the electrical conductivity of these TMCs is good, they are suitable for applications in batteries, sensors, catalysis, and hydrogen storage medium.

1.4.4 Molten Salt Etching

It is interesting to note here that the etching of "Al" from corresponding MAX phases are much easier than etching of "N" from corresponding N containing MAX phases. There may be two reasons behind this observation; one is the lower stability of $Ti_{n+1}N_n$ compared to $Ti_{n+1}C_n$, because the cohesive energy of $Ti_{n+1}N_n$ are less than those of $Ti_{n+1}C_n$ [58]. Another reason may lie in the higher formation energy of $Ti_{n+1}N_n$ from $Ti_{n+1}AlN_n$ than those of $Ti_{n+1}C_n$ from $Ti_{n+1}AlC_n$. Al atoms are bonded more strongly in $Ti_{n+1}AlN_n$ compared to $Ti_{n+1}AlC_n$ and thus require more energy for their extraction. But nitride MXenes have quite a few advantages over the carbide systems. From the theoretical calculation, higher density of states (DOS) at the Fermi level $N(E_F)$ has been predicted for Ti_4N_3 (containing O, F, or OH at the surface) compared to what is observed in either Ti_4C_3 or Ti_3C_2 containing the same ions at the surface [59]. The higher DOS in nitrides counters the electron withdrawal tendency by surface terminated groups and increases the metallic character and electronic conductivity of the nitride MXene [60, 61]. Nitride based MXenes are thus promising candidates for electrodes in electrochemical capacitors, plasmonic material, transformation optics and metamaterial devices [62, 63].

The molten salt synthesis process successfully prepared $Ti_4N_3T_x$ MXenes which was otherwise impossible to prepare by wet chemical etching in acidic medium. Instead of etching the MAX phase precursor, selectively, in aqueous HF solutions, Urbankowski *et al.* [64]

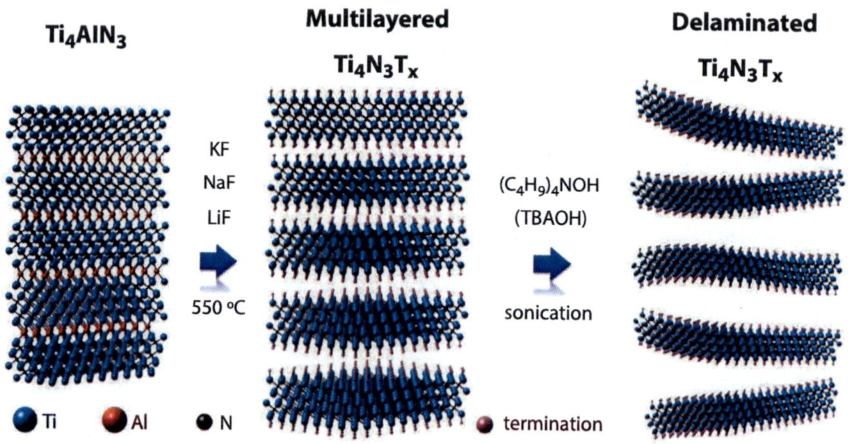

Figure 1.13 Schematic illustration of the synthesis of $Ti_4N_3T_x$ by molten salt treatment from Ti_4AlN_3 and delamination of the produced multilayer MXene by tetrabutylammonium hydroxide. Reproduced with permission from Ref. [64]. Copyright 2016, The Royal Society of Chemistry.

showed that Al can also be etched from a Ti_4AlN_3 MAX powder precursor by a molten fluoride salt (1:1 mass ratio) at 550 °C for 30 min. in an inert atmosphere. The fluoride salt mixture contains 59 wt% KF, 29 wt% LiF, and 12 wt% NaF. It corresponds to the eutectic point. The resulting MXene can be further delaminated to produce few and single-layered flakes of $Ti_4N_3T_x$, where T represents F, O, or OH ions (Figure 1.13), by mixing with tetrabutylammonium hydroxide (TBAOH) followed by centrifugation. The resulting Ti_4N_3 showed typical accordion-like structures under scanning electron microscopy.

1.4.5 Hydrothermal Synthesis

Use of corrosive hydrofluoric acid for etching of MAX phases in wet chemical process may affect the performance of lithium-ion batteries and supercapacitors using MXenes. It is theoretically feasible to etch Ti_3AlC_2 by using alkali because of its binding ability with amphoteric Al (stability constant of $Al(OH)_4^-$, lgβ4 is 33.3 and that of AlF_6^{3-}, lgβ6 is 19.84 [65]. The main problem in preparing MXene by using alkali etching is a kinetic one rather than thermodynamic which could be attributed to the formation of some oxide/hydroxide layers on the Ti_3AlC_2 surface [66] formed during alkali treatment of MAX phases. In Bayer process

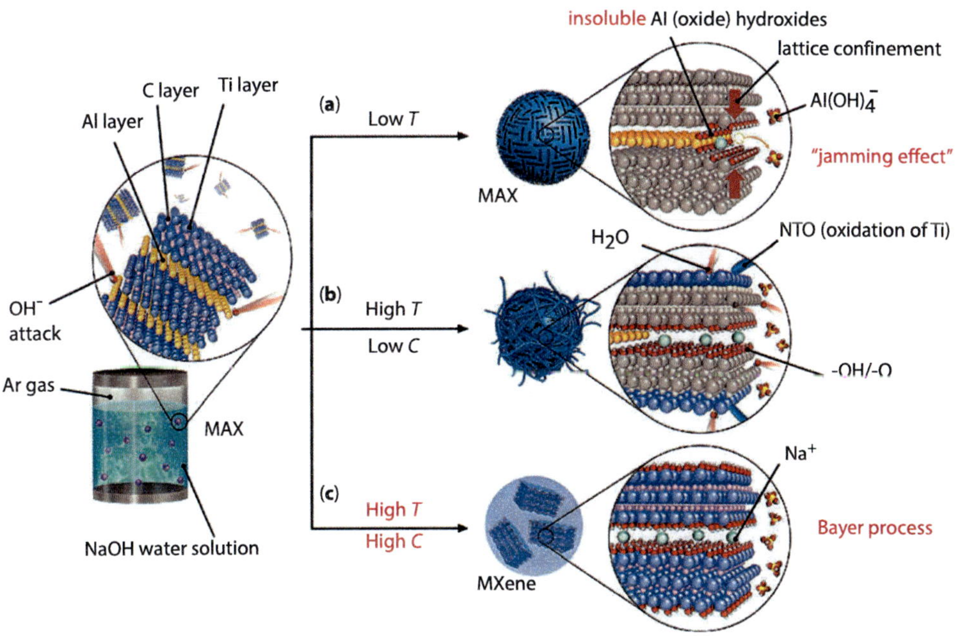

Figure 1.14 Schematic of the reaction between Ti_3AlC_2 and NaOH aqueous solution under different reaction conditions. (a) formation of Al (oxide) hydroxides blocks the Al extraction process at low temperatures; (b) at high temperatures and low concentration of NaOH some amount of Al (oxide) hydroxides dissolves; but NTO forms due to oxidation of "MXenes" and high-water content; (c) Bayer process, showing the effect of high temperatures and high NaOH concentrations to dissolve the Al (oxide) hydroxides. Reproduced with permission from Ref. [67]. Copyright 2018, Wiley-VCH.

this problem was solved by using higher concentration of alkali and high reaction temperature which eventually dissolve the Al (oxide) hydroxides at the surface thus improves the etching reaction (Figure 1.14) [67]. The hydrothermal process was indeed inspired by Bayer process where the elevated temperature improves the $Ti_3C_2T_x$ yield and the alkali concentration affects the purity of the product. The fluorine-free MXenes are prepared by treating the Ti_3AlC_2 powder, hydrothermally, using 27.5 (M) NaOH, at 270 °C [68]. This treatment produces 92 wt% $Ti_3C_2T_x$.

1.4.6 Electrochemical Synthesis at Room Temperature

Another efficient process to prepare Fluoride free MXenes is electrochemical synthesis. Mono and bilayer $Ti_3C_2T_x$ (T = O, OH....) has been prepared successfully by anodic corrosion of Ti_3AlC_2in a binary aqueous electrolyte with high yield (over 90 %) [69]. Chemical

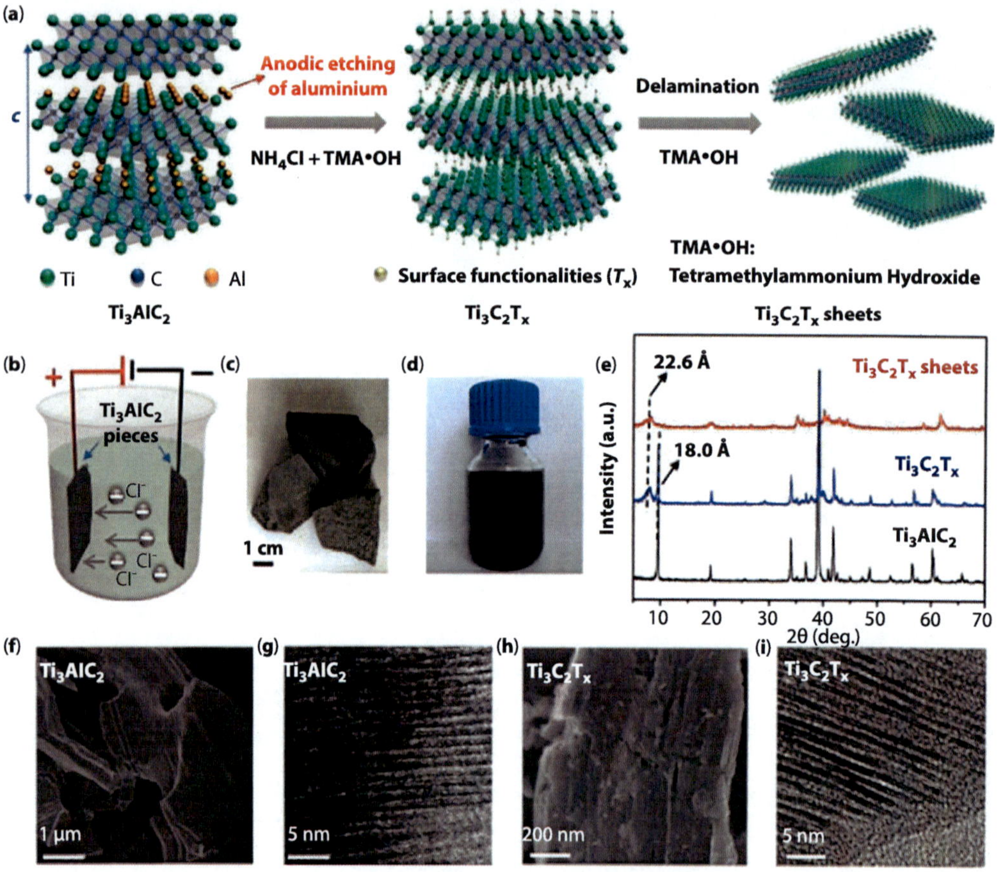

Figure 1.15 Anodic etching of bulk Ti_3AlC_2 in a binary aqueous electrolyte. (a) Schematic of the etching and delamination process; (b) The configuration of electrochemical cell. Reproduced with permission from Ref. [69]. Copyright 2018, Wiley-VCH.

etching is in fact an electrochemical process itself, so it can be a good alternative to prepare fluoride free MXenes [70]. The design of electrolyte is the main challenge in this process so that Al ions can be removed selectively leaving behind 1D structure of Ti_3C_2 (Figure 1.15). In the electrochemical process where two pieces of bulk Ti_3AlC_2 were used as anode and cathode; anode takes part in etching reactions, whereas the cathode was used as a counter electrode; the electrolyte used was a mixture of 1 M NH_4Cl and 0.2 M tetramethylammonium hydroxide (TMA·OH) with a pH value >9. Chloride ions helps etching the anodic Al, rapidly, and breaking the Ti-Al bonds. It is followed by ammonium hydroxide (NH_4OH) intercalation. It results in opening up the edges and underlying surface of the etched anode leading to efficient electrochemical reaction. A low voltage (+5V) is sufficient to delaminate the anode in a short period of time (~5hrs.) at room temperature and the produced colloidal dispersion of $Ti_3C_2T_x$ sheets were separated by filtering. $Ti_3C_2T_x$ powders thus obtained were ground thoroughly. They were then transferred into 25% w/w TMAOH to finally produce the individual sheets of $Ti_3C_2T_x$. As the product does not contain any fluoride termination, the additive-free films exhibit very good areal capacitance (220 mF cm^{-2}) when used as a supercapacitor. This technique then was extended to produce other MXenes successfully (e.g., V_2CT_x and Cr_2CT_x), thus proving the effectiveness of this process to prepare fluoride free MXenes [71]. The possible chemical reactions which take place during the electrochemical etching are:

$$Ti_3AlC_2 + 3e^- + 3Cl^- \Rightarrow Ti_3C_2 + AlCl_3 \quad (1.4)$$

$$Ti_3C_2 + 2OH^- - 2e^- \Rightarrow Ti_3C_2(OH)_2 \quad (1.5)$$

$$Ti_3C_2 + 2H_2O \Rightarrow Ti_3C_2(OH)_2 + H_2 \quad (1.6)$$

1.5 Applications

With the development of technologies, energy requirement is increasing day by day and at the same time, limited fossil fuels and global pollution issue have forced us to think about alternative green technology and renewable energy sources for our sustainability. During recent past, huge efforts have been devoted to generate and store energy from different renewable sources like solar, wind, hydro, biomass etc. and the tidal reduce energy deficiencies and decrease environment issues. Herein, it may be stated that energy generation and storage appear as big challenge as there are several local environmental issues those limit the scope along with many sporadic imitations. In this context, it may be stated that various advanced electrochemical energy conversion protocols have been developed for energy generation, while several energy storage technologies like fuel cells, metal – air batteries, metal – ion batteries, supercapacitor etc. have been developed to overcome the crisis. Herein, catalysts lead a promising role in converting earth-abundant resources like H_2O, CO_2, CO, N_2 to clean fuels and different chemical feedstocks e.g. H_2, hydrocarbons, NH_3 etc. It has been observed that they are promising, stable as required to facilitate long-term energy storage in H_2 fuel cells. However, in recent time it has been identified that catalysts, particularly electrode catalysts play a crucial role to deal the technological issues where electrochemistry-based energy conversion and storage depending on chemical reaction, physical adsorption facilitates the development. Earlier, Pt metal and related materials such PtIr, IrO_2 etc. were mostly used in these applications, specifically in H_2 evolution reaction (HER), O_2 reduction reaction (ORR), O_2 evolution reaction (OER) etc. Although few of them have already been commercialized, however due to scarcity their usage is limited. Hence, researchers working in this field are highly engaged to explore alternatives. In this context, it may be stated that a copious series of MXene materials have shown great promise as electrocatalyst due to their suitable physical, chemical properties, metallic electrical conductivity, superlative character with high surface area, high mechanical strength etc. as required for energy conversion and storage. Briefly, metallic conductivity facilitates charge transport across electrode – electrolyte interface when it is coupled with other material having high active reaction sites, but is electrically conductive. In this context, it may be stated that such combination helps to overcome the limitation of carbon-based support like graphene whose hydrophobic nature limits its use in aqueous medium. Herein, it may be stated that conductivity of graphene is highly reduced due to formation of –OH in between graphene and MXenes. It has been examined that conversion and storage efficiencies highly depend on type of MXene electrocatalyst as well as their architecture. Herein, it may be stated that approximately 30 different compositions of MXenes have been predicted to have application in energy storage and conversion. It has been well-experimented fact that when size of a material gets reduced around 10–20 nm, called quantum dots (QD), it exhibits various interesting activities, sometimes novel features due to confinement and edge effect depending on their size and shape. As an example, 2D basal planes of MXenes are found to be HER active, while in case of other 2D systems like MoS_2, HER activity mainly originates from surface edges. In fact, it has been theoretically studied that this important series of materials have excellent photocatalytic, electrocatalytic, and photoelectrocatalytic activities leading various reactions like HER, OER, ORR, CO_2 reduction (CO_2RR) etc. (Figure 1.16). Like other materials, MXene QDs also show these types of improved properties, specifically in energy generation and storage applications. Herein, it may be stated that research on MXenes had been started few years back and there are plenty of research papers on different electrocatalytic aspects, especially for catalytic H_2O splitting, using variety of MXenes. Its hydrophilic nature makes it robust and efficient to split acidic, basic even if sea water. In comparison, MXene QD is new hot topic of research where improved energy generation and storage performance have been reported. In general, due to smaller size, ion diffusion path length is very low, hence catalytic sites become easy accessible to electrolytes giving higher efficiency. Sometime, it has also been explored that MXenes composites show much better performance over bare MXenes as reaction kinetic becomes faster, more catalytic sites get exposed in composite. In these subsequent sections, we have provided a brief overview of the performance of variety of MXenes and their composite in energy generation and storage applications such as oxygen and nitrogen reduction reaction, supercapacitor, batteries etc.

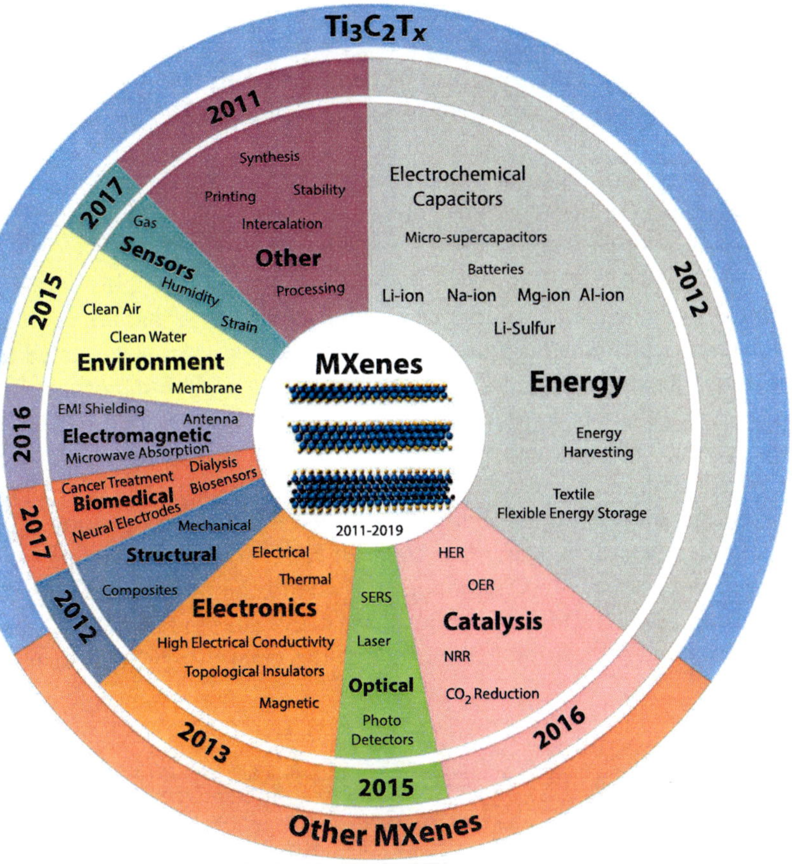

Figure 1.16 Different application potential of MXenes. Reproduced with permission from Ref. [3]. Copyright 2019, The American Chemical Society.

1.5.1 Nitrogen Reduction Reaction (NRR)

The requirement of an electrocatalyst which will efficiently produce NH_3 selectively is very high, as day to day usage of NH_3 in fuel cell for industry and various agricultural purposes is increasingly rapidly. Though there are few commercial electrocatalysts those produce NH_3 through electrocatalytic NRR technique, but their performance is limited due to various energy requirement and environment pollution related issues. Hence alternative green electrocatalysts for NRR is high demand for sustainability. In this context, it may be stated that there are few experimental and theoretic reports on MXene QDs showing its great promise as efficient NRR catalysts. It has been revealed that surface atoms, exposed to electrolyte and active edges of MXenes play most crucial role in NRR activity, the effect appears to be more pronounced in case of MXene QDs and active research stands with developing the MXene QD with various morphologies. As an example, exposed Ti-edge site atom of $Ti_3C_2T_x$ QD facilitates adsorption and activation of N_2. In this context, it may be stated that

Jin *et al.* noted that Ti-edge terminated and –OH containing $Ti_3C_2T_x$ QD possesses high NRR yielding NH_3 due to high Faradaic reaction [72]. As per density functional based theoretical studies by the same group, –OH helps in N_2 adsorption, followed by activation of the reaction. In a comparative study among bare and –OH (Ti_3C_2OH) and –F (Ti_3C_2F) functionalized $Ti_3C_2T_x$ QDs illustrates that Ti_3C_2OH shows highest NRR activity giving 62.94 μg h^{-1} mg^{-1} yield of NH_3 at potential of –0.5V and long term durability up to 25 h in the presence of 0.1M of HCl. Most importantly, no product other than NH_3 has been obtained. Herein, another comparative study among $Ti_3C_2T_x$, alk-$Ti_3C_2T_x$ and Ti_3C_2F reveal NH_3 production of 16.8, 22.3 and 48.3 μg h^{-1} mg^{-1} respectively. This research group has demonstrated that fluorination takes place at the Ti-edge sites that improves conductivity, thus NRR becomes facilitated (Figure 1.17). In this context, it may be stated that Cheng *et al.* examined NRR activity of Mo_2C implanted ultrathin carbon nanosheet (Mo_2C/C) and studied that Mo_2C/C catalyzes to produce NH_3 under ambient conditions with yield ~11.3 μg h^{-1} mg^{-1} having 7.8% Faradaic efficiency at –0.3V in the presence 0.5 M Li_2SO_4 solution [73]. They have identified from MNR spectroscopy minute formation of N_2H_4 as byproduct, however it corresponds good durability (81.6% after six cycles of operation). It was investigated by the same research group that superior catalytic activity originates from electrostatic interaction between Mo_2C nanodots and due to the presence of oxygen containing functional group. Actually, Mo_2C nanodots have active sites to facilitate N_2 adsorption and activation, however a calculation using density functional theory N_2 adsorption sites have

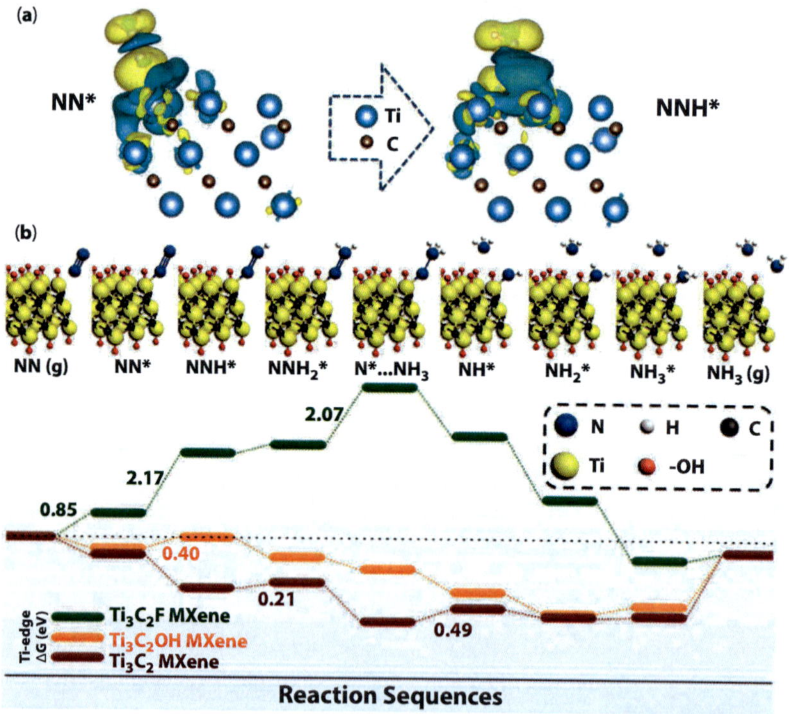

Figure 1.17 (a) Modified charge density distribution of the N_2-adsorbed Ti_3C_2OH MXene surface. (b) Reaction mechanism and free energy diagram of Nitrogen Reduction Reaction on the edge Ti of Ti_3C_2MXene. Reproduced with permission from Ref. [72]. Copyright 2020, Wiley-VCH.

special feature towards N≡N bond cleavage and hydrogenation which in consequence facilitates high NRR performance.

1.5.2 Oxygen Evolution Reaction (OER)/Oxygen Reduction Reaction (ORR)

These are the techniques to have potentiality to overcome energy crisis and environmental pollution issues as required for our sustainability. Previously, nanostructure of Pt, Ru, Ir, etc. were vastly used as electrocatalyst; however, their practical usages are limited due to scarcity, liberation of CO, low oxidation, slow kinetics etc. After them, researchers were very much engaged to examine scope of transition metals, but they are not promising due to insufficient catalytic sites, low conductivity, unstable operational potential etc. In this context, it has been identified that MXene QDs have good potentiality, but their main disadvantage is self-aggregation that makes working potential unstable. To overcome the issue, MXene QDs are often functionalized with other materials in which OER/ORR occurs synergistically. As an example, Han *et al.* have prepared NiFe-layered di-hydride (LDH)/Ti_3C_2/nitrogen doped graphene heterostructure (Figure 1.18) and studied ORR and OER activities [74]. Their investigations reveal that the catalyst has the very low overpotential ~1.50V at 10 mA/cm^2, faster OER kinetics (Tafel slope ~57 mV/dec) in comparison with bare NiFe-LDH and nitrogen doped graphene. It has been also examined that the above material also exhibits high ORR activity with half-wave potential ~ +0.69V (on RHE scale). Herein, analysis of the Koutecky – Levich plot indicates first order reaction kinetics with average number of transferred electrons ~3.94. Presently, the difference between OER overpotential and ORR

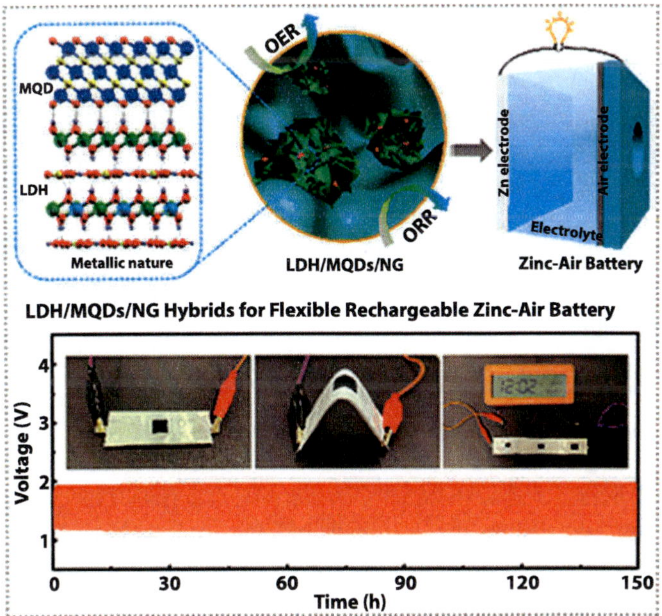

Figure 1.18 Electronically coupled metallic hybrids of NiFe layered double hydroxide nanosheet/Ti_3C_2 MXene quantum dots deposited on a nitrogen-doped graphene surface (LDH/MQD/NG) for ORR and QER. The digital photographs of different electronics powered by such devices. Reproduced with permission from Ref. [74]. Copyright 2021, John Wiley & Sons.

half-wave potential is measured to be ~0.81V which is lower than that of LDH/nitrogen doped graphene (~0.92 V) and previously reported novel metal supported electrocatalyst (Pt/C ~1.01 V). As per theoretical studies, MXene QD contributes to density of states and partial density of states at the Fermi energy level of LDH resulting increase of conductivity and consequently it's ORR and OER activities get enhanced. Although, LDHs have excellent OER and ORR activity in alkaline solution, however they have poor electrical conductivity and get aggregated easily. Initially, LDHs were combined with C support such as graphene to improve charge transport. Later these C supports were functionalized with graphene oxide/reduced graphene oxide for better structural integrity with LDH, hence MXenes were mostly composites with graphene oxide/reduced graphene oxide. In a separate study, Yang *et al.* have studied ORR activity of multi-walled carbon nanotubes (MWCNT) inflated with MoS_2 QDs and Ti_3C_2Tx QDs ($MoS_2QD@Ti_3C_2TxQD@MWCNT$) and observed excellent ORR performance having onset potential 0.87V, half-potential ~0.75V, Tafel slope ~90 mV/dec involving four electrons [75]. In this context, it may be stated that the material corresponds less onset potential in comparison with that of commercially available electrocatalyst (Pt/C ~0.94 V) and exhibits stability up to 1000 cycles for more than 2 h. This excellent ORR activity is attributed to the synergistic effects from higher conductivity, surface area, number of active edges of the heterostructures, where confinement affect plays a prominent role.

1.5.3 Hydrogen Evolution Reaction (HER)

After theoretical prediction by several research groups, MXenes have gained wide attention due to HER activity. Unlike 2D MOS_2, MXene exhibit HER activity from basal plane rather than edges. Hence, more advantageous is that active site can be increased significantly. In this context, it may be stated that HER activity was first identified in Mo_2CT_x and Ti_2CT_x, where $-OT_x$ surface group primarily gives the HER active sites on their basal plane. It has been identified that the performance can be significantly improved either by increasing density electrochemical active sites by making their nanostructures or by modifying MXenes to have more number of thermoneutral *H adsorption sites. Briefly the later technique is generally achieved by controlling T_x as well by doping with other transitional metals as this effect was identified on $Ti_4N_3T_x$ using scanning electrochemical microscopy (SECM). Though V doping didn't increase conductivity of $Ti_4N_3T_x$, however it increases HER activity significantly. In contrast, $VOSO_4$ enhances conductivity of $Ti_4N_3T_x$, but correspondences very little change of HER activity. In this context, it may be stated that earlier electrical conductivity and HER activity were considered to be correlated phenomenon; however, researchers have identified few experimental evidences indicating that they are decoupled. It has been identified that HER activity of MXene can be further improved by preparing MXene hybridized with other 2D materials such as TM chalcogenides, phosphides, carbides, metallic alloys etc. in which MXenes due to large area conducting and mechanical support helps charge transfer, while counterpart improves H^+ adsorption centers, hence HER activity gets improved significantly (Figure 1.19) [76]. As cumulative HER activity is higher in comparison with their summative activity, hence it is believed that hybrids have PGM-like property. As an example, $MoS_2/Ti_3C_2T_x$ @ C architecture, prepared by Wu *et al.*, exhibits charge transfer resistance ~1.28 Ω, over potential ~135 mV, hence this composite has most profound HER activity. Similarly, coupling of MXene with other

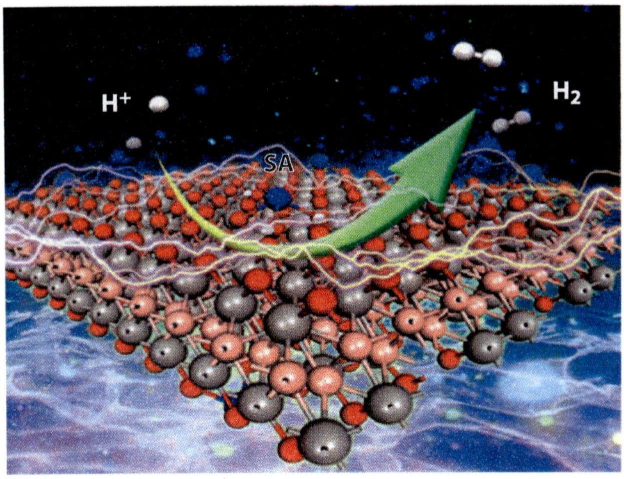

Figure 1.19 Schematic illustration of hydrogen evolution of TM-doped MXenes. Adapted from Ref. [76]. Copyright 2021, Springer Nature.

TM chalcogenide gives similar phenomenon. In this context, a hybrid of Co-doped Mo_2C, $Ti_3C_2T_x$ and N-doped carbon is found to have high electrical conductivity and exhibits very high HER activity over wide range of pH even if in the presence of seawater. Herein, a theoretical calculation illustrates that MXene $Co_xMo_{2-x}C$/NC composite decreases H^+ absorption barrier and gives much lower water dissociation barrier (~0.78 V) in comparison with that of Pt (111) (~0.89 V). An compositional optimization reveals that Co0.31Mo1.49C/NC has the highest efficiency with Tafel slope ~75 mV in the presence of 1.0M KOH. It has been examined that QDs anchored MXene significantly improves adsorption kinetics; herein Du *et al.* have synthesized Nb doped $Ti_3C_2T_x$/Ni, Co alloy composite to examine their HER activity and they have found that $Ti_{2.5}Nb_{0.25}C_2T_x$ gives highest HER activity due to 43 mV over potential which was 516 mV of bare $Ti_{2.5}Nb_{0.25}C_2T_x$. TM phosphides, integrated with MXene, have been examined to exhibit HER and OER activities. In this context, Xiu *et al.* have synthesized CoP@3D $Ti_3C_2T_x$MXene architectures and have examined electrochemical catalytic activity and water splitting capability [77]. Functionalization of MXenes with TM phosphides in general shifts d-band center to higher energy side which in consequence reduces occupancy of the anti-bonding from TM and adsorbed O states and enhances –OH adsorption and OER activity. Herein, it has been noted that 1D CoPnanorods, grown on top of MXene $Ti_3C_2T_x$nanosheets, induces tensile stress and bending those in consequence expose more electrochemically active surface area for HER to take place.

1.5.4 Energy Storages

In the present scenario of energy crisis and other environmental pollution related issues, search of efficient, reliable, reversible energy storage devices (EED) appears to be very attractive research topic in recent time. Among various EED, metal – air battery, metal – ion battery and supercapacitor show great promise along with sustainability. However, it has been noted that the efficiency of EED is very much dependent on electrode material and their architectures. In recent time, various 2D materials e.g. graphene, phosphorene

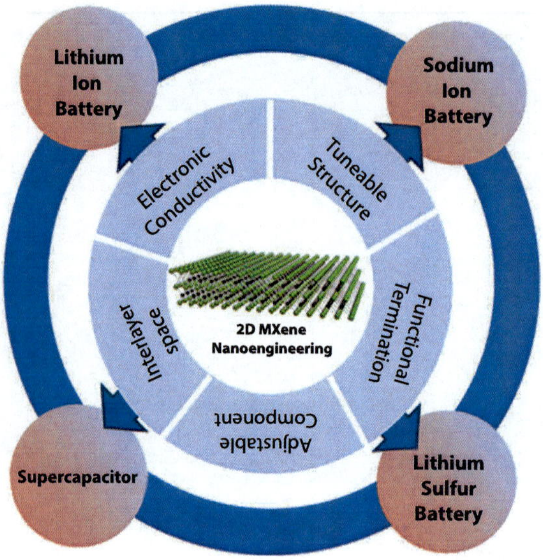

Figure 1.20 2D MXenes for different Energy storage application. Reproduced with permission from Ref. [78]. Copyright 2019, John Wiley & Sons.

etc. exhibit excellent efficiency where MXenes and MXene QDs have shown great scope attracting researchers to develop MXene based EED (Figure 1.20) [78].

1.5.4.1 Battery

It is the most efficient EED in recent time which is being commercially used in various electrical and electronic devices. Among various types, Li^+-ion battery has come into market as it has overcome barrier of threshold limit of energy density, durability, power and environmental friendliness. Anode material is the heart of these devices and plays significant role in the performance of the battery. Commonly, graphite is used as anode, however due to low capacity (~372 $mAhg^{-1}$), researchers are actively engaged to develop alternative anode materials with high capacity. In this context, it may be stated that MXenes has shown it potentiality to replace graphite as steady state capacity can easily be achieved during intercalation of Li^+ into MXenes. Such easy intercalation increases the storage capacity and durability. In addition, it has been identified that MXenes can easily be modified with various functional groups those acts as trapping centers to facilitate adsorption electrolyte and in consequence shuttle effects gets minimized. As the advantage of nanosized, MXene QDs exhibit higher efficiency in comparison with their bulk counterpart and in this background lots of different types of investigations have been carried out so far. As an example, Ti_3C_2MXenenanodot-interspersed nanosheet based electrode have been prepared by Xiao *et al.* for Li – S battery [79] and their study reveals that the electrode has high energy density, discharge capacity ~1081 $mAhg^{-1}$, stability up to 3000 cycles with average Coulombic efficiency ~98.5%. The same research group has further shown that the capacity is dependent of S content and the highest capacity which they have achieved is 1957 $mAhcm^{-3}$. In this connection, the capacity $Ti_3C_2T_x$ electrode has been examined by Zhang *et al.* for Na^+ batteries and the electrode's capacity is measured ~600 mAg^{-1} with stability up to 150 cycles [80]. In this context, it may be stated that Li – S batteries have gained wide attention as next generation energy storage devices due to unique conversion principle and very high theoretical energy density. Mechanistically, discharge process involves oxidation of Li anode into Li^+ that travels to the S electrode, reacts with it to form Li_2S through several intermediate steps involving various intermediate species. It has been noticed that intermediate species diffuse to the Li anode where chemical reduction takes place before diffusing back to the cathode for re-oxidation in an internal short circuit/self-discharge condition. In this process, ionic and electronic conductivity of the S electrode, enabled with high volume expansion ratio ~80%, imparts structural stress and limits electrochemical reaction. The barrier is overcome S electrode is hybridized with C materials that increase electrical conductivity. Simultaneously, electrode is functionalized with polymeric binders, TM oxides etc. to enhance adsorption of intermediate products. Due to unique feature of T_x group, MXenes are being adopted for cathode as well as separator as they easily facilitate transportation of intermediate products. Here a series of MXene type materials such as Ti_2CT_x, $Ti_3C_2T_x$, Ti_3CNT_x, V_2CT_x have been investigate for cathode materials. Mechanistically, it has been established that intermediate phases are strongly bonded with Ti atoms of Ti_2CT_x through Ti – S bonding which has been confirmed by theoretical study using density functional theory. The study indicates that –O and –F of T_x group leads immobilizations than –F in conventional electrolytes. N-doping further improves the charge transport capacity and Ti – S binding strength. In addition, it has been realized that MXene are very robust to adjust volumetric changes of S cathode during charging/discharging. However, it has been identified that MXene has the tendency of self-restack that limits electrode access of Li^+ at the cathode. Herein, several strategies have been taken to integrate different secondary materials while maintaining high electrical conductivity and affinity of the intermediate phase towards S electrode. In this context, Zhang *et al.* have prepared self-assembled 3D $S/MnO_2@Ti_3C_2T_x$ and found strong chemical affinity of intermediate phase towards MnO_2, while the platform has 70% S loading capacity and have the ability to adjust volumetric changes during cycle of operation [81]. Cathode exhibits discharge capacity ~1140 $mAhg^{-1}$ and durability up to 500 cycle of operation. In another work, it has been noticed that TiO_2, MoS_2 etc. have similar features of attracting intermediate phases. After careful analysis, it has been identified that incorporation of MoS_2 or TiO_2 increase conductivity, hence facilitates storage capacity. MXene and carbon nanotube composite is also found to play an important role in Li – S battery.

Higher energy storage, durability (240 cycles at current density of 200 mAg^{-1}) and discharge capability (~16022 $mAhg^{-1}$ at current density of 200 mAg^{-1}) have been obtained from MXene quantum dot clusters anchored N-doped Carbon nanosheet electrode in case of Li – O_2 batteries [82]. It has been revealed that the composite contains very high concentration of grain boundary and edge defects those in consequence modifies coordination number and local atomic arrangement those facilitate catalytic activity. A density function based theoretical study, carried out by the same research group, indicates that local structure modifies Li_xO_y intermediate's adsorption mechanism, followed by lowering of energy barrier and enhancing the reaction kinetics. MXene QDs are also investigated to be used as electrode material of Zn – air battery (74). Herein, a hybrid material consisting of MXene QDs, NiFe-LDH nanosheet and nitrogen doped graphene with power density ~113.8 $mWcm^{-2}$ has been identified to have good scope. In recent time, due to their compactness, light weight and high theoretical energy capacity metal – air batteries have gained wide attention. Commonly, metal – air battery comprises of the architecture of metal anode based traditional battery and cathode exposed fuel cell. In this context, it may be stated that the performance of bare MXenes is limited due to complex multistep, four-electron mediated redox reaction. The problem is overcome by hybridization of MXenes with conductive supports that modulates electronic structure facilitating OER and ORR activities. In this context, Fe – N – C materials appears to be most promising in which highly active FeN_4 motifs plays most crucial role. Later, Fe phthalocyanine/$Ti_3C_2T_x$ hybridized structure shows ORR activity which is 5 times higher than conventional Pt/C composite. It has been studied that negatively charged –OH/–F T_x groups of $Ti_3C_2T_x$ strongly interact with Fe^{2+} through van der Waals interaction and reduce electron density on Fe^{2+} giving spin transition that optimizes ORR activity. In a follow-up work, Fe(phen)$_2$/$Ti_3C_2T_x$, prepared pyrolysis process, shows ORR activity. A 3D N-$CoSe_2$/MXene has been synthesized using poly(vinyl alcohol) as cross-linker, followed by annealing in ammonia atmosphere, possesses high concentration of electrochemically active sites, hence mass transport gets facilitated significantly. $Ti_3C_2T_x$/trimetallic NiFeMN nitride thin sheet composite has been investigated to have bifunctionality of OER and ORR activities. Herein a theoretical calculation reveals interfacial coupling between nitride and $Ti_3C_2T_x$ acts as ORR active motifs.

1.5.4.2 Supercapacitor

In modern technology, there are plenty of electronic and electrical devices like drilling machine requiring high power density and very fast charging – discharging capacity. For these types of devices, energy storage device having the capability of high charging and discharging rate and high power density is highly essential. In this context, it may be stated that conventional batteries are good choice for these application as batteries have very low power density. In recent time, a completely new technology has been developed for high power density devices and it has been developed in the form of supercapacitor. Like batteries, the performance of the supercapacitor-based energy storage devices also depends on electrode materials as well their architectures. Researchers working in this field are highly engaged to develop electrode materials with higher capacity, durability etc. Among various 2D materials, MXenes also exhibit excellent potentiality due to their unique chemical and physical features suitable supercapacitor application such as 2D lamellar structure enabling fast ion transport, high electrical and metallic conductivity, high redox activity, easy functionality with various elements like –O, –OH, –F those make them hydrophilic which in consequence significantly alters charge storage mechanism in aqueous medium. In this connection plenty of reports are available in literature mentioning different aspects of MXenes as active electrode and based on the information various strategies have been adopted to engineer electrode materials and architectures. Like batteries, here also MXene QDs exhibit high performances, ascribed less ion diffusion path length. Other advantage of MXene QDs is that they can easily be functionalized giving easy tuneability of voltage window that has direct impact on the capacitance [83]. Later, it has been explored that asymmetric configuration leads higher capacitance. In this context, several studies have also been carried out, particularly in flexible energy storage devices. As an example, a prototype supercapacitor, developed using $Ti_3C_2T_x$ nanosheets/$Ti_3C_2T_x$ QDs/reduced graphene oxide and $Ti_3C_2T_x$ nanosheets/PEDOT:PSS positive and negative electrode respectively (Zhou *et al.*), exhibits volumetric capacitance ~542 F cm^{-3} at current density of 0.25 A cm^{-3} and voltage window

~1.5 V, stability up to 5000 cycles. Other advantage of this configuration is that it is flexible maintaining 97.6% retention up to 500 times bending at 90° angle as it has very good tensile strength ~40 MPa.

1.5.4.3 Electromagnetic Interference Shielding

In the recent era of technological development in all around us, we are adversely exposed to different types of interfering electromagnetic signals from various electronic gadgets leading health hazardous like nausea, eye problem, cancer, even if causing detrimental effects on development of infant's brain. In addition, it may also be noted that electromagnetic waves facilitates malfunctions in the modern warfare systems. Primarily the effect of MXenes is being realized as electromagnetic interface shielding when inherent charge carriers of MXenes interact with electromagnetic fields and consequently weakens the detrimental effect of the radiation field. As thumb rule, it can be stated that high electrical conductivity of these MXenes is highly essential in order to get benefit of MXenes as shield materials. In this context, it may be revealed that $Ti_3C_2T_x$ had emerged as first 2D shield material in the year 2016, till then a series of MXenes have been proposed as potential shield candidate due to high surface area, low density, high electrical conductivity, lightweight etc. However, it may be expressed here that not only virgin MXenes, but also few MXenes based composites, hydrogels etc. have been developed to have high shielding performance. Among various architectures, layer-by-layer (LBL) assembly formation may be considered as most useful key structural formulation methodology to provide electromagnetic shielding where two distinct phases of the MXenes of different impedance causes major internal scattering at the internal surface that leads attenuation of the incident electromagnetic radiation. In general, a LBL structure of $(MAX)_n$ formula is deposited on both sides of a conductive textile, made of silk fabric, by an atomizer in which $Ti_3C_2T_x$ layer and silver nanowire get stacked alternatively, where M, A x and n denote MXene, silver nanowire, concentration and number of deposited cycles. Very briefly, a leaf-like nanocomposite gets formed on the porous textile composites which consist of conductive lamina of $Ti_3C_2T_x$ and conductive veins of silver nanowires. Importantly, such fabrication is found be very uniform and exhibits good mechanical integrity with respect to their bulk counterpart. Although increase of silver nanowire ratio enhances electrical conductivity which in consequence facilitates shielding, however beyond critical concentration silver nanowires results poor air permeability as compactness of the film gets increased with increasing silver nanowire content. Hence, silver nanowires ration needs optimization. Higher the spraying cycles, shielding performance gets increased; hence thicknesses of the films have to be optimized in order get highest shielding efficiency. It has been investigated that composite textile of approximately 480 μm thick MXene shows shielding ~90 dB at 12.4 GHz. Another scope of this MXene is their ability in wearable electronics which in general gets synthesized by repeated dipping and drying of poly ethylene terephthalate (PET) fibers. Often, MXenes are functionalized with different polymers to facilitate their applications. For example, polypyrole functionalized $Ti_3C_2T_x$MXenes sheets render high electrical conductivity (~1000 S/m) as well as maintains air permeability [84]. It has been realized this functionalization yields very high shielding performance. Very briefly, it has been observed that $Ti_3C_2T_x$ sheets exhibits shielding of 25 dB, when its electrical conductivity is 150 S/m, however shielding effects gets increased up to 66 dB at electrical conductivity ~4000 S/m. Later, it has been realized that lamination of these MXene sheets further enhances shielding effect up to ~90 dB. For durability, silicon which increases shielding efficiency > 30 dB is mostly adopted. For commercial uses, these 2D MXene sheets are often deposited on cotton fabric vertically that gives interconnected conductive network and helps shielding performance. In this context, it may be stated that 6 wt% MXene functionalized fabric has shown great promise for commercial use as it maintains flexibilities. Herein, it has been observed that highest MXene loading shows shielding ~36 dB, hence it gives a tuneability of achieving the performance. It has been also observed that nacre induced ultrathin $Ti_3C_2T_x$/carbon nanofiber composite (known as brick-and-mortar) thin film (thickness ~47 μm), prepared on vacuum filtration unit, provides specific shielding performance ~2647 dB cm^2/g, while its electrical conductivity is measured to be 739.4 S/m. It has been identified that mechanical property gets significantly increased due to inclusion of the nanofibers, however it depends on wt% of mixing. At 50 wt%, tensile strength ~135.4 MPa, strain fracture ~16.7% have been obtained. Later, it has been examined that MXene has good structural integrity with cellular nanofiber leading possibility of preparing compact, good-stacked 2D structures which shows great promise in flexible wearable electronics. In a similar fashion, a multilayered MXene/carbon nanofibers that plays crucial role as reinforcement component has been prepared with high specific shielding capacity ~7029 dB cm^2/g, herein this multilayer exhibits mechanical strength ~112.5 MPa, toughness ~2.7 MJ/m^3. Followed by MXene/nanofiber multilayer, a more complex thin film structure that consists of MXene, carbon nanofiber and silver nanoparticles have prepared. It has been identified that silver nanoparticles and nanofiber synergistically increases shielding performance ~507 dB, as well as its mechanical strength (~34.2 MPa), toughness get increased. In this context, it may be stated that few research groups working in this field have prepared freestanding $Ti_3C_2T_x$ MXene and aramid nanofibers based paper with high commercial opportunity as it exhibits high mechanical, electrical and shielding properties. It has been examine that aramid plays an important role as the intermolecular cross-linker to enhance stability of the film. Most importantly, aramid creates an interconnected fibrous network that helps in releasing stress. However, it has been observed that higher $Ti_3C_2T_x$ content reduces mechanical stability. Thus, it may be stated that MXene and polymer nanocomposite shows great promise of shielding performance for electromagnetic radiation, however lightweight efficient shielding devices is still lacking and it may be stated that active research is going in this field.

1.5.5 Biomedical Applications

MXene QDs have also gained attention in the field of biomedical applications, particularly in cancer diagnosis and therapy. Primary advantage is that they are biocompatible consisting if Ti and C. It has been explored that MXene QDs can be a potential candidate for photo thermal therapy to cancer where light induced heat is being employed to kill cancer cells (Figure 1.21) [85].

In this context, pioneering work includes functionalized MXene QDs with Al oxoanions that results high absorption of near infrared light (extinction coefficient ~52.8 Lg^{-1} cm^{-1} at 808 nm) and high proficiency ~52.5%. In addition, it can act as contrast agent for photoacoustic imaging. After noticing insignificant toxicity (*in vivo* and *in vitro*), it has been proposed as a potential material for biomedical purposes. In another study, 2D Ti_3C_2 QDs have been functionalized with Fe_3O_4 and glucose oxidase and their efficacy has been tested in cancer treatment. It has been identified that glucose oxidase through catalytic process

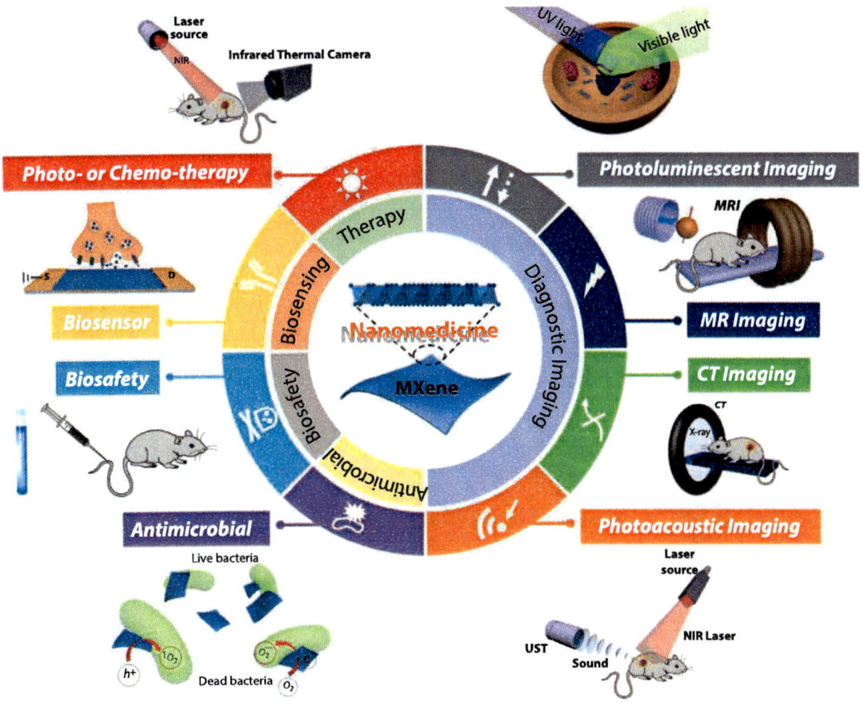

Figure 1.21 Different bio-medical application of MXenes. Adapted from Ref. [85]. Copyright 2018, Wiley-VCH.

produces H_2O_2 of substantial quantity which gets converted in hydroxyl radicals by Fe_3O_4, which in consequence obstruct cancerous cell. Herein, photothermal therapeutic property of MXene QDs synergistically makes the procedure more effective. In this context, Liu *et al.* had also prepared Ti_3C_2/Fe_3O_4 and their study demonstrated that the composite has excellent magnetic resonance (MR) imaging capability with T_2relaxivity ~394 $mM^{-1}s^{-1}$ and photothermal translation proficiency ~48.6% [86]. An *in vivo* study, carried out using this matrix, illustrates that it as the good potentiality to treat cancer. Due to porous nature, Ti_3C_2 can be also be used as drug carrier, hence therapeutic performance can be further improved. In fact, it has been identified that Ti_3C_2 has ~211.8% drug loading capacity, while drug release can be controlled by pH of the medium or it can be infrared light assisted. Biosafety as examined by *in vivo* studies reveal that it has the potentiality as theranostic agent. In another study, it has been revealed that Ti_3C_2 can be used carrier of Al that can be utilized to minimize Al loss during treatment. In this regard, ultrathin Ti_3C_2 having extinction coefficient ~67 Lg^{-1} cm^{-1} at 808 nm, 58.3% photothermal conversion efficiency and capacity to produce single oxygen ($^1O_2$), is generally adopted. As Ti_3C_2 MXene is easily functionalized with hyaluronic acid and doxorubicin through layer-by-layer deposition technique, hence synergistic chemo-effect can easily be obtained. Another study reveals that surface of Ti_3C_2 MXenes can easily be modified via an exterior-nanopore engineering protocol where in consequence silica can be incorporated for further biosafety. It has been examined that hydrogels often exhibit high performances in various biomedical applications. In this context, a Ti_3C_2 based hydrogel has been synthesized with cellulose.

Importantly, hydrogel is highly infrared sensitive with photothermal efficacy where release of drug is highly dependent on infrared irradiation. After through study, it may be stated that this hydrogel through intratumoral injection may destruct tumor instantly with less relapse chance. Later, it has been revealed that a core – shell structure can be prepared with Ti_3C_2 nanosheet and Au nanoparticle to club photoacoustic and computed tomographic (CT) imaging techniques into a single platform. This platform has shown tremendous increase of infrared absorption of Ti_3C_2 that may be ascribed to presence of Au nanoparticles along with improved tumor oxygenation and radiotherapy. Herein, soybean phospholipid amended Ti_3C_2nanosheet, prepared through double-stage MAX phase Ti_3AlC_2 exfoliation technique, followed by TPAOH merging, and has exhibited excellent photothermal translation proficiency during intravenous dispersion. A Ti_3C_2/MnO_x nanocomposite has been synthesized has shown potentiality of multifunctional theranostic platform with capability of magnetic resonance and photoacoustic imaging for diagnosis part and photothermal therapy in therapeutic part. In this context, polyethylene coated Ti_2C has been examined to have excellent photothermal efficacy to cancerous cells, while they are highly compatible towards normal cells. According to tumor xenograft model, Ti^{3+} of Ti_2C reacts with H_2O_2 in tumor microenvironment and produce extremely toxic reactive oxygen species that kills cancer cells. Ti_3C_2, functionalized with GdW_{10}-centered polyoxometalates, shows hyperthermia affect along with magnetic resonance/CT imaging capability. In this context, it may be stated that few other MXene QDs such as Ta_4C_3, Nb_2C TiCN, V_2C have also been reported for their potentiality in these applications. As an example, Ta_4C_3/MnO_x nanocomposite exhibits contrast efficiency of dual mode i.e. CT and T_1 weighted MR imaging, where CT originates from Ta and while MR is ascribed to presence of Mn. Due to excellent photothermal efficiency, this nanocomposite shows good potentiality for tumor ablation. Nb_2C is another MXene that has drawn attention. It has been researched that Nb_2C corresponds good photothermal conversion efficiency ~45.6% and 36.4% in near infrared II and I region respectively along with high photothermal stability. Herein, Tin *et al.* have prepared Nb_2C – mesoporous silica nanoparticles (MSNs) – S core – shell nanostructure having potentiality of photonic thermos-gaseous cancer therapy. It has been examined that this core – shell structure have infrared triggered nitric oxide release and photothermal activity those synergistically improves therapeutic activity, while diagnostic mostly involves photoacoustic imaging. TiCN being an important candidate of MXene series has also been adopted for various biomedical applications, particularly in cancer treatment due to its inherent fluorescence property that makes it excellent materials for bioluminescence imaging. In this context, Cao *et al.* have identified photothermal activity of V_2C MXene in near infrared II region [87]. In addition, they have observed that V_2C QDs have long circulation time, excellent biocompatibility, and complete destruction ability of cancer cells, followed by endosomal escape, hence it may be a potential candidate to treat cancer.

1.6 Conclusion and Future Scope

As discussed above, the class of two-dimensional compounds called MXenes is quite attractive both for their fundamental science and application potential in a variety of devices. In recent time, lower-dimensional systems such as two-dimensional ferromagnets, ferroelectrics (ferroelectricity has been observed in one-two unit cell thick platforms), topological materials, van der Waals solids (where two-dimensional layers are stacked along the third dimension and coupled by van der Waals force) etc. have started becoming extremely important. For example, in two-dimensional magnets, the Rashba spin-orbit coupling and antisymmetric Dzyloashinskii-Moriya exchange interaction appear to have given rise to novel magnetic and ferroelectric phases (even at room temperature) which are not at all realizable in bulk scale. The advent of MXenes should be examined within this backdrop. It has been found that the surface termination ions, the number of layers, and the in-plane and out-of-plane ordering of the transition metal ions influence the properties of the MXenes significantly. Therefore, it is a challenge to synthesize them with appropriate structure for optimizing the properties. In future, attempts should be made to find out the permissible extent of variation in the surface terminated ions for observing stabilization of the properties. This would ease out the synthesis techniques. It is also important to identify the applications which do not depend on the property change induced by the variation of surface terminated ions so greatly. It is possible alongside to introduce different extent of ordering of the magnetic transition metal ions and thereby stabilizing esoteric magnetic and even multiferroic phases. Some work has been done but more such work needs to be done in future. Observing multiferroicity in lower-dimensional compounds (including Anomalous Hall effect) is quite an important development in the present context. Likewise, very little work has so far been done on observation of superconductivity in a class of MXenes. The Fermi surface, its nesting, and electron-phonon coupling need to be examined in a variety of MXene compounds in order to identify the potential superconductors. Lower-dimensional superconductivity in MXene compounds would open a new avenue of research. Another very important area of research is development of heterostructures with other systems such as graphene or graphene-like two-dimensional systems. Already, efficacy of heterostructures in oxygen evolution reaction has been demonstrated. Heterostructures should also be designed to inject electron or hole across the interfaces for developing novel electronic devices. Yet another class of systems could be designed by engineering the vacancies. By controlling the extent of point defects and also by inducing their ordering, it is possible to enhance the charge storage and gas storage characteristics. This area still requires more effort in near future. Scope is also there for theoretical investigation of the charge transfer mechanism across the interfaces for enhancing the charge and gas storage and transfer properties. Only limited work has so far been done on these issues. Therefore, in short, exciting days are ahead for the MXenes.

Acknowledgement

The authors acknowledge the IMPRINT-II Program of the MHRD, Government of India, during this work.

References

1. Anasori, B., Lukatskaya, M.R., Gogotsi, Y., 2D metal carbides and nitrides (MXenes) for energy storage. *Nat. Rev. Mater.*, 2, 16098, 2017.

2. Naguib, M., Barsoum, M.W., Gogotsi, Y., Ten years of progress in the synthesis and development of MXenes. *Adv. Mater.* 33, 2103393, 2021.
3. Gogotsi, Y. and Anasori, B., The rise of MXenes. *ACS Nano*, 13, 8, 8491, 2019.
4. Khazaeia, M., Mishrab, A., Venkataramananc, N.S., Singhb, A.K., Yunoki, S., Recent advances in MXenes: From fundamentals to applications. *Curr. Opin. Solid State Mater. Sci.*, 23, 3, 164–178, 2019.
5. Barsoum, M.W., The $M_{N+1}AX_N$ phases: A new class of solids: Thermodynamically stable nanolaminates. *Prog. Solid State Chem.*, 28, 1–4, 201, 2000.
6. Anasori, B., Dahlqvist, M., Halim, J., Moon, E.J., Lu, J., Hosler, B.C., Caspi, E.N., May, S.J., Hultman, L., Eklund, P., Rosén, J., Barsoum, M.W., Experimental and theoretical characterization of ordered MAX phases Mo_2TiAlC_2 and $Mo_2Ti_2AlC_3$. *J. Appl. Phys.*, 118, 83, 94304, 2015.
7. Dahlqvist, M., Petruhins, A., Lu, J., Hultmanand, L., Rosen, J., Origin of chemically ordered atomic laminates (*i*-MAX): Expanding the elemental space by a theoretical/experimental approach. *ACS Nano*, 12, 8, 7761, 2018.
8. Srivastava, P., Mishra, A., Mizuseki, H., Lee, K.-R., Singh, A.K., Mechanistic insight into the chemical exfoliation and functionalization of Ti_3C_2 MXene. *ACS Appl. Mater. Interfaces*, 8, 36, 24256, 2016.
9. Khazaei, M., Ranjbar, A., Esfarjani, K., Bogdanovski, D., Dronskowski, R., Yunoki, S., Insights into exfoliation possibility of MAX phases to MXenes. *Phys. Chem. Chem. Phys.*, 20, 8579, 2018.
10. Khazaei, M., Arai, M., Sasaki, T., Chung, C.-Y., Venkataramanan, N.S., Novel electronic and magnetic properties of two-dimensional transition metal carbides and nitrides. *Adv. Funct. Mater.*, 23, 2185, 2013.
11. Hasan, M.Z. and Kane, C.L., *Colloquium*: Topological insulators. *Rev. Mod. Phys.*, 82, 3045, 2010.
12. Ingason, A.S., Dahlqvist, M., Rosen, J., Magnetic MAX phases from theory and experiments; a review. *J. Phys. Condens. Matter*, 28, 433003, 2016.
13. Kumar, H., Frey, N.C., Dong, L., Anasori, B., Gogotsi, Y., Shenoy, V.B., Tunable magnetism and transport properties in nitride MXenes. *ACS Nano*, 11, 8, 7648, 2017.
14. Zhang, J.-J., Lin, L., Zhang, Y., Wu, M., Yakobsonand, B.I., Dong, S., Type-II multiferroic $Hf_2VC_2F_2$ MXene monolayer with high transition temperature. *J. Am. Chem. Soc.*, 140, 30, 9768, 2018.
15. Chandrasekaran, A., Mishraand, A., Singh, A.K., Ferroelectricity, antiferroelectricity, and ultrathin 2D electron/hole gas in multifunctional monolayer MXene. *Nano Lett.*, 17, 5, 3290, 2017.
16. Khazaei, M., Wang, V., Sevik, C., Ranjbar, A., Arai, M., Yunoki, S., Electronic structures of iMAX phases and their two-dimensional derivatives: A family of piezoelectric materials. *Phys. Rev. Mater.*, 2, 074002, 2018.
17. Kumar, S. and Schwingenschlögl, U., Thermoelectric performance of functionalized Sc_2C MXenes. *Phys. Rev. B*, 94, 035405, 2016.
18. Jeitschko, W., Nowotny, H., Benesovsky, F. Kohlenstoffhaltige ternäre Verbindungen (H-Phase), *Monatsh. Chem.*, 94, 672, 1963.
19. Xu, C., Wang, L., Liu, Z., Chen, L., Guo, J., Kang, N., Ma, X.-L., Cheng, H.-M., Ren, W., Large-area high-quality 2D ultrathin Mo_2C superconducting crystals. *Nat. Mater.*, 14, 1135, 2015.
20. Lei, J., Kutana, A., Yakobson, B.I., Predicting stable phase monolayer Mo_2C (MXene), a superconductor with chemically-tunable critical temperature. *J. Phys. Chem. C*, 5, 3438, 2017.
21. Berdiyorov, G.R., Optical properties of functionalized $Ti_3C_2T_2$ (T = F, O, OH) MXene: First-principles calculations. *AIP Adv.*, 6, 055105, 2016.
22. Zhang, H., Yang, G., Zuo, X., Tang, H., Yang, Q., Li, G., Computational studies on the structural, electronic and optical properties of graphene-like MXenes (M_2CT_2, M = Ti, Zr, Hf; T = O,

F, OH) and their potential applications as visible-light driven photocatalysts. *J. Mater. Chem. A*, 4, 12913, 2016.
23. Liu, Y., Xiao, H., Goddard, W.A., Schottky-barrier-free contacts with two-dimensional semiconductors by surface-engineered MXenes. *J. Am. Chem. Soc.*, 138, 15853, 2016.
24. Khazaei, M., Ranjbar, A., G-Asl, M., Arai, M., Sasaki, T., Liang, Y., Yunoki, S., Nearly free electron states in MXenes. *Phys. Rev. B*, 93, 205125, 2016.
25. Ling, C., Shi, L., Ouyang, Y., Wang, J., Searching for highly active catalysts for hydrogen evolution reactionbased on O-terminated MXenes through a simple descriptor. *Chem. Mater.*, 28, 9026, 2016.
26. An, X., Wang, W., Wang, J., Duan, H., Shi, J., Yu, X., The synergeticeffects of Ti_3C_2 MXene and Pt as co-catalysts for highly efficient photocatalytic hydrogen evolution over g-C_3N_4. *Phys. Chem. Chem. Phys.*, 20, 11405, 2018.
27. Wan, Q., Li, S., Liu, J.-B., First-principle study of Li-ion storage of functionalized Ti2C monolayer with vacancies. *ACS Appl. Mater. Interfaces*, 10, 7, 6369, 2018.
28. Xie, Y., Dall'Agnese, Y., Naguib, M., Gogotsi, Y., Barsoum, M.W., Zhuang, H.L., Kent, P.R.C., Prediction and characterization of MXene nanosheet anodes for non-lithium-ion batteries. *ACS Nano*, 8, 9, 9606, 2014.
29. Hu, Q., Sun, D., Wu, Q., Wang, H., Wang, L., Liu, B., Zhou, A., He, J., MXene: A new family of promising hydrogen storage medium. *J. Phys. Chem. A*, 117, 51, 14253, 2013.
30. Papadopoulou, K.A., Chroneos, A., Parfitt, D., Christopoulos, S.-R.G., A perspective on MXenes: Their synthesis, properties, and recent applications. *J. Appl. Phys.*, 128, 170902, 2020.
31. Xin, M., Li, J., Ma, Z., Pan, L., Shi, Y., MXenes and their applications in wearable sensors. *Front. Chem.*, 8, 297, 2020.
32. Naguib, M., Mashtalir, O., Carle, J., Presser, V., Lu, J., Hultman, L., Gogotsi, Y., Barsoum, M.W., Two-dimensional transition metal carbides. *ACS Nano*, 6, 2, 1322, 2012.
33. Naguib, M., Kurtoglu, M., Presser, V., Lu, J., Niu, J., Heon, M., Hultman, L., Gogotsi, Y., Barsoum, M.W., Two-dimensional nanocrystals produced by exfoliation of Ti_3AlC_2. *Adv. Mater.*, 23, 37, 4248–4253, 2011.
34. Baig, R.N. and Varma, R.S., Alternative energy input: Mechanochemical, microwave and ultrasound-assisted organic synthesis. *Chem. Soc. Rev.*, 41, 1559–1584, 2012.
35. Varma, R.S., Chemical activation by mechanochemical mixing, microwave and ultrasonic irradiation. *Green Chem.*, 10, 1129–1130, 2008.
36. Naguib, M., Mochalin, V.N., Barsoum, M.W., Gogotsi, Y., 25th Anniversary article: MXenes: A new family of two-dimensional materials. *Adv. Mater.*, 26, 7, 992–1005, 2014.
37. Alhabeb, M., Maleski, K., Anasori, B., Lelyukh, P., Clark, L., Sin, S., Gogotsi, Y., Guidelines for synthesis and processing of two-dimensional titanium carbide ($Ti_3C_2T_x$MXene). *Chem. Mater.*, 29, 18, 7633–7644, 2017.
38. Anasori, B., Xie, Y., Beidaghi, M., Lu, J., Hosler, B.C., Hultman, L., Kent, P.R.C., Gogotsi, Y., Barsoum, M.W., Two-dimensional, ordered, double transition metals carbides (MXenes). *ACS Nano*, 9, 10, 9507–9516, 2015.
39. Anasori, B. and Gogotsi, Y., The rise of MXenes. *ACS Nano*, 13, 8491, 2019.
40. Naguib, M., Halim, J., Lu, J., Hultman, L., Gogotsi, Y., Barsoum, M.W., New two-dimensional niobium and vanadium carbides as promising materials for Li-ion batteries. *J. Am. Chem. Soc.*, 135, 15966, 2013.
41. Chang, F., Li, C., Yang, J., Tang, H., Xue, M., Synthesis of a new graphene-like transition metal carbide by de-intercalating Ti_3AlC_2. *Mater. Lett.*, 109, 295, 2013.
42. Sun, Z., Li, S., Ahuja, R., Schneider, J.M., Calculated elastic properties of M_2AlC (M = Ti, V, Cr, Nb and Ta). *Solid State Commun.*, 129, 589, 2004.

43. Persson, I., el Ghazaly, A., Tao, Q., Halim, J., Kota, S., Darakchieva, V., Palisaitis, J., Barsoum, M.W., Rosen, J., Persson, P. O. Å., Tailoring structure, composition, and energy storage properties of MXenes from selective etching of in-plane, chemically ordered MAX Phases. *Small*, 14, 17, 1703676, 2018.
44. Mashtalir, O., Naguib, M., Mochalin, V.N., Dall'Agnese, Y., Heon, M., Barsoum, M.W., Gogotsi, Y., Intercalation and delamination of layered carbides and carbonitrides. *Nat. Commun.*, *4*, 1716, 2013.
45. Malaki, M., Maleki, A., Varma, R.S., MXenes and ultrasonication. *J. Mater. Chem. A*, 7, 10843–10857, 2019.
46. Feng, W., Luo, H., Wang, Y., Zeng, S., Tan, Y., Zhang, H., Peng, S., Ultrasonic assisted etching and delaminating of Ti3C2 MXene. *Ceram. Int.*, 44, 6, 7084–7087, 2018.
47. Zhang, X., Xu, J., Wang, H., Zhang, J., Yan, H., Pan, B., Zhou, J., Xie, Y., Ultrathin nanosheets of MAX phases with enhanced thermal and mechanical properties in polymeric compositions: $Ti_3Si_{(0.75)}Al_{(0.25)}C_2$. *Angew. Chem. Int. Ed.*, *52*, 4361, 2013.
48. Lukatskaya, M.R., Mashtalir, O., Ren, C.E., Dall'Agnese, Y., Rozier, P., Taberna, P.L., Naguib, M., Simon, P., Barsoum, M.W., Gogotsi, Y., Cation intercalation and high volumetric capacitance of two-dimensional titanium carbide. *Science*, *341*, 1502, 2013.
49. Ghidiu, M., Lukatskaya, M.R., Zhao, M.-Q., Gogotsi, Y., Barsoum, M.W., Conductive two-dimensional titanium carbide 'clay' with high volumetric capacitance. *Nature*, 516, 78, 2014.
50. Liu, F., Zhou, A., Chen, J., Jia, J., Zhou, W., Wang, L., Hu, Q., Preparation of Ti_3C_2 and Ti_2C MXenes by fluoride salts etching and methane adsorptive properties. *Appl. Surf. Sci.*, 416, 781–789, 2017.
51. Huang, S. and Mochalin, V.N., Combination of high pH and an antioxidant improves chemical stability of two-dimensional transition-metalcarbides and carbonitrides (MXenes) in aqueous colloidal solutions. *Inorg. Chem.*, 61, 9877–9887, 26, 2022.
52. Sun, W., Shah, S.A., Chen, Y., Tan, Z., Gao, H., Habib, T., Radovic, M., Green, M.J., Electrochemical etching of Ti_2AlC to Ti_2CTx (MXene) in low-concentration hydrochloric acid solution. *J. Mater. Chem. A*, 5, 21663–21668, 2017.
53. Peng, J., Chen, X., Ong, W.-J., Zhao, X., Li, N., Surface and heterointerface engineering of 2D MXenes and their nanocomposites: Insights into electro- and photocatalysis. *Chem*, 5, 1, 18–50, 10 2019.
54. Ma, L., Ting, L.R.L., Molinari, V., Giordano, C., Yeo, B.S., Efficient hydrogen evolution reaction catalyzed by molybdenum carbide and molybdenum nitride nanocatalysts synthesized via the urea glass route. *J. Mater. Chem. A*, 3, 8361, 2015.
55. Giordano, C., Erpen, C., Yao, W.T., Antoniett, M., Synthesis of Mo and W carbide and nitride nanoparticles via a simple "urea glass" route. *Nano Lett.*, 8, 4659–4663, 2008.
56. Gogotsi, Y., Transition metal carbides go 2D. *Nat. Mater.*, 14, 1079–1080, 2015.
57. Wang, Z., Kochat, V., Pandey, P., Kashyap, S., Chattopadhyay, S., Samanta, A., Sarkar, S., Manimunda, P., Zhang, X., Asif, S., Singh, A.K., Chattopadhyay, K., Tiwary, C.S., Ajayan, P.M., Metal immiscibility route to synthesis of ultrathin carbides, borides, and nitrides. *Adv. Mater. Weinheim*, 29, 1700364, 2017.
58. Shein, I.R. and Ivanovskii, A.L., Graphene-like titanium carbides and nitrides Tin+1Cn, Tin+1Nn (n = 1, 2, and 3) from de-intercalated MAX phases: First-principles probing of their structural, electronic properties and relative stability. *Comput. Mater. Sci.*, *65*, 104, 2012.
59. Xie, Y. and Kent, P.R.C., Hybrid density functional study of structural and electronic properties of functionalized $Ti_{n+1}X_n$ (X = C, N) monolayers. *Phys. Rev. B: Condens. Matter*, 87, 235441, 2013.
60. Zhong, Y., Xia, X., Shi, F., Zhan, J., Tu, J., Fan, H.J., Transition metal carbides and nitrides in energy storage and conversion. *Adv. Sci.*, 3, 1500286, 2016.

61. Wang, L., Wen, J., Jiang, Y., Ou, Q., Yu, L., Xiong, B.-S., Yang, B., Zhang, C., Tong, Y., Electrical conduction characteristic of a 2D MXene device with Cu/Cr_2C/TiN structure based on density functional theory. *Materials*, 13, 17, 3671, 2020.
62. Morel, A., Borjon-Piron, Y., Porto, R.L., Brousse, T., Bélanger, D., Vanadium nitride electrodes: Limitations and practical use in electrochemical capacitors. *J. Electrochem. Soc.*, 163, A1077–A1082, 2016.
63. Naik, G.V., Schroeder, J.L., Ni, X., Kildishev, A.V., Sands, T.D., Boltasseva, A., Titanium nitride as a plasmonic material for visible and near-infrared wavelengths. *Opt. Mater. Express*, 2, 478–489, 2012.
64. Urbankowski, P., Anasori, B., Makaryan, T., Er, D., Kota, S., Walsh, P.L., Zhao, M., Shenoy, V.B., Barsouma, M.W., Gogotsi, Y., Synthesis of two-dimensional titanium nitride Ti_4N_3 (MXene). *Nanoscale*, 8, 11385, 2016.
65. Hogfeldt, E., *Stability constants of metal-ion complexes: Part A: Inorganic ligands*, vol. 21, Pergamon Press, Oxford, 1982.
66. Xuan, J., Wang, Z., Chen, Y., Liang, D., Cheng, L., Yang, X., Liu, Z., Ma, R., Sasaki, T., Geng, F., Organic-base-driven intercalation and delamination for the production of functionalized titanium carbide nanosheets with superior photothermal therapeutic performance. *Angew. Chem.*, 128, 47, 14789–14794, 2016.
67. Li, T., Yao, L., Liu, Q., Gu, J., Luo, R., Li, J., Yan, X., Wang, W., Liu, P., Chen, B., Zhang, W., Abbas, W., Naz, R., Zhang, D., Fluorine-free synthesis of high purity $Ti_3C_2T_x$ (T = -OH, -O) via alkali treatment. *Angew. Chem.*, 57, 21, 6115–6119, 2018.
68. Li, T., Yao, L., Liu, Q., Gu, J., Luo, R., Li, J., Yan, X., Wang, W., Liu, P., Chen, B., Zhang, W., Abbas, W., Naz, R., Zhang, D., Fluorine-free synthesis of high-purity $Ti_3C_2T_x$(T = OH, O) via alkali treatment. *Angew. Chem. Int. Ed.*, 57, 6115–6119, 2018.
69. Yang, S., Zhang, P., Wang, F., Ricciardulli, A.G., Lohe, M.R., Blom, P.W., Feng, X., Fluoride-free synthesis of two-dimensional titanium carbide (MXene) using a binary aqueous system. *Angew. Chem.*, 130, 15717, 2018.
70. Yang, S., Ricciardulli, A.G., Liu, S., Dong, R., Lohe, M.R., Becker, A., Squillaci, M.A., Samorì, P., Müllen, K., Feng, X., Ultrafast delamination of graphite into high-quality graphene using alternating currents. *Angew. Chem. Int. Ed.*, *56*, 23, 6669–6675, 2017.
71. Pang, S.-Y., Wong, Y.-T., Yuan, S., Liu, Y., Tsang, M.-K., Yang, Z., Huang, H., Wong, W.-T., Hao, J., Universal strategy for HF-free facile and rapid synthesis of two-dimensional MXenes as multifunctional energy materials. *J. Am. Chem. Soc.*, 141, 9610–9616, 2019.
72. Jin, Z., Liu, C., Liu, Z., Han, J., Fang, Y., Han, Y., Niu, Y., Wu, Y., Sun, C., Xu, Y., Rational design of hydroxyl-rich Ti3C2Tx MXene quantum dots for high-performance electrochemical N2 reduction. *Adv. Energy Mater.*, 10, 22, 2000797, 2020.
73. Cheng, H., Ding, L.-X., Chen, G.-F., Zhang, L., Xue, J., Wang, H., Molybdenum carbide nanodots enable efficient electrocatalytic nitrogen fixation under ambient conditions. *Adv. Mater.*, 30, 46, 1803694, 2018.
74. Han, X., Li, N., Xiong, P., Jung, M.G., Kang, Y., Dou, Q., Liu, Q., Lee, J.Y., Park, H.S., Electronically coupled layered double hydroxide/MXene quantum dot metallic hybrids for high-performance flexible zinc–air batteries. *InfoMat*, 1–11, 2021.
75. Yang, X., Jia, Q., Duan, F., Hu, B., Wang, M., He, L., Song, Y., Zhang, Z., Multiwall carbon nanotubes loaded with MoS2 quantum dots and MXene quantum dots: Non–Pt bifunctional catalyst for the methanol oxidation and oxygen reduction reactions in alkaline solution. *Appl. Surf. Sci.*, 464, 78–87, 2019.
76. Bai, S., Yang, M., Jiang, J., He, X., Zou, J., Xiong, Z., Liao, G., Liu, S., Recent advances of MXenes as electrocatalysts for hydrogen evolution reaction. *NPJ 2D Mater. Appl.*, 5, 78, 2021.

77. Xiu, L., Wang, Z., Yu, M., Wu, X., Qiu, J., Aggregation-resistant 3D MXene-based architecture as efficient bifunctionalelectrocatalyst for overall water splitting. *ACS Nano*, 12, 8017–8028, 2018.
78. Nan, J., Guo, X., Xiao, J., Li, X., Chen, W., Wu, W., Liu, H., Wang, Y., Wu, M., Wang, G., Nanoengineering of 2D MXene-based materials for energy storage applications. *Small*, 1902085, 17, 9, 2021.
79. Xiao, Z., Li, Z., Li, P., Meng, X., Wang, R., Ultrafine Ti3C2 MXene nanodots-interspersed nanosheet for high-energy-density lithium–sulfur batteries. *ACS Nano*, 13, 3, 3608–3617, 2019.
80. Zhang, T., Jiang, X., Li, G., Yao, Q., Lee, J.Y., A red-phosphorous-assisted ball-milling synthesis of few-layered Ti_3C_2Tx (MXene) nanodot composite. *ChemNanoMat*, 4, 1, 56–60, 2018.
81. Zhang, H., Qi, Q., Zhang, P., Zheng, W., Chen, J., Zhou, A., Tian, W., Zhang, W., Sun, Z., Self-assembled 3D MnO2 nanosheets@ delaminated-Ti_3C_2 aerogel as sulfur host for lithium–sulfur battery cathodes. *ACS Appl. Energy Mater.*, 2, 705–714, 2019.
82. Wang, Z., Yu, K., Feng, Y., Qi, R., Ren, J., Zhu, Z., VO_2(p)-V_2C(MXene) Grid structure as a lithium polysulfide catalytic host for high-performance. *ACS Appl. Mater. Interfaces*, 11, 47, 44282–44292, 2019.
83. Mohanty, B., Giri, L., Jena, B.K., MXene-derived quantum dots for energy conversion and storage applications. *Energy Fuels*, 35, 14304–14324, 2021.
84. Oliveira, F.M. and Gusmao, R., Recent advances in the electromagnetic interference shielding of 2D materials beyond graphene. *ACS Appl. Electron. Mater.*, 2, 3048–3071, 2020.
85. Lin, A.H., Chen, Y., Shi, J., Insights into 2D MXenes for versatile biomedical applications: Current advances and challenges. *Adv. Sci.*, 5, 1800518, 2018.
86. Liu, Z., Zhao, M., Lin, H., Dai, C., Ren, C., Zhang, S., Peng, W., Chen, Y., 2D magnetic titanium carbide MXene for cancer theranostics. *J. Mater. Chem. B*, 6, 3541–3548, 2018.
87. Cao, Y., Wu, T., Zhang, K., Meng, X., Dai, W., Wang, D., Dong, H., Zhang, X., Engineered exosome-mediated near-infrared-II region V2C quantum dot delivery for nucleus-target low-temperature photothermal therapy. *ACS Nano*, 13, 1499–1510, 2019.

2

Chemical Exfoliation and Delamination Methods of MXenes

Kaili Gong, Lian Yin and Keqing Zhou*

Faculty of Engineering, China University of Geosciences (Wuhan), Wuhan, Hubei, PR China

Abstract

Since MXene was prepared by HF in 2011, its outstanding physical/chemical properties had shown great potential on multiple fronts, which also promoted to the exploration of its synthesized methods. And safe and efficient synthesis routes are gradually being proposed. Because of their milder and safer than HF, *in-situ* HF-forming etching methods are widely studied firstly, such as fluoride salts (e.g., LiF, KF, and NH_4F) in HCl or H_2SO_4 and hydrothermal route with low toxicity etching agents (e.g., $NaBF_4$ in HCl). However, the releases of volatile HF gas formed during mixing are still risky. Bifluoride etchants (e.g., $NaHF_2$, KHF_2, and NH_4HF_2), molten fluorinated salts etchants (e.g., NH_3F, NaF) and Lewis acidic molten salts etchants (e.g., $CuCl_2$, $CoCl_2$) have also been used in the preparation of MXene with good results. Thereafter, non-hazardous synthesis routes or fluorine-free etching methods are gradually being developed to explore efficient preparation methods and enrich MXene performance. Considering the process of electrons transfer for chemical etching, electrochemical etching methods were used to prepare MXene. In addition, alkali etching methods (e.g., KOH, NaOH), halogen-based etching methods (e.g., molecular chlorine, bromine, and iodine), and ionic liquid-based etching methods (e.g., $EMIMBF_4$, $BMIMPF_6$) have also been explored with the successful preparation of MXene, also revealing the safe, efficient methods. Besides, diverse attempts have also been applied to organic-soluble etching methods for the preparation and storage of MXene to make up for MXene's weak stability in water. Although the preparation methods of MXenes are being explored extensively, a mature and scalable method is still a challenge.

Keywords: Transition metal carbides and/or nitrides, MXenes, top-down etching methods, chemical exfoliation, intercalation, delamination, two-dimensional (2D) materials, scale-up production

2.1 Introduction

For the bulk MAX precursor, it is 3D crystal structure made up of 2D layered structures stacked by bondings, in which the M-A layer is linked with ionic/metallic/covalent bonding, and only metallic bonds exist between the M and A atoms, which is the key to etching [1, 2]. The weak bonding of the M-A layer can be broken using etching methods to

**Corresponding author*: zhoukq@cug.edu.cn

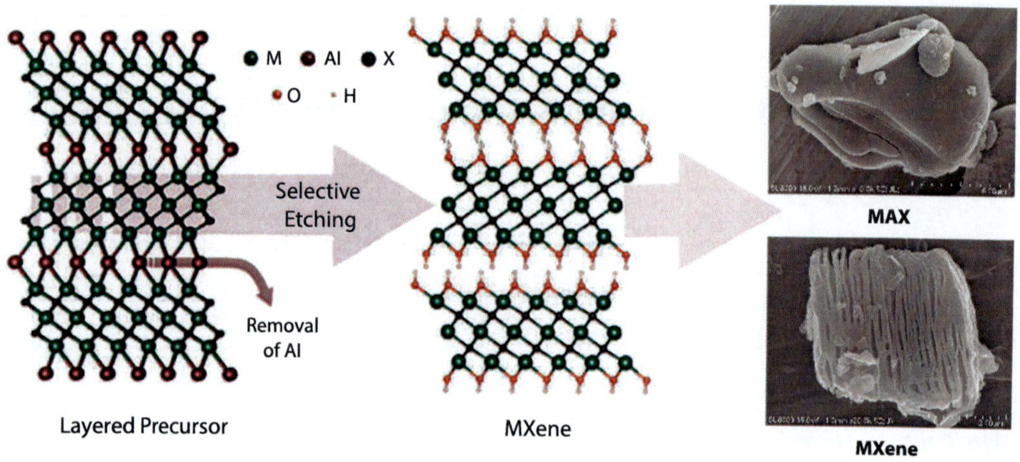

Figure 2.1 Schematic of the etching process of MAX [3].

obtain MXenes with more stable M-X layers, as shown in Figure 2.1. The unique structure of MXene makes it shine, but its practical use depends on an effective synthesis process, which is an area that many researchers are currently working on. Since the discovery of MXene in 2011, a top-down etching procedure from the MAX phase toward safe, effective, large-scale synthesis routes has been extensively investigated.

2.2 HF Etching Method

In 2011, Gogotsi *et al.* immersed Ti_3AlC_2 MAX into 50% HF at room temperature and successfully removed the Al layer from MAX due to the selectively etching reaction of F-ions to obtain accordion-like MXene structures displayed in Figure 2.2b [4]. Moreover, Figure 2.2a shows the XRD patterns before and after the reaction, from which it can be seen that the strong peak at 39° disappears after completion of etching and the (002) characteristic peak is shifted towards a lower angle compared to that of MAX phase, with the shift being more apparent after sonication, indicating the rise in layer spacing. During reaction, HF can react with Al layer to form AlF_3 (Equation 2.1), while the aqueous etching method allows abundant surface functional groups (-OH, =O) to grow on the surface of the lamellae, presented in Equations 2.2–2.4, which endows them with better properties.

$$Ti_3AlC_2 + HF = AlF_3 + 3/2\ H_2 + Ti_3C_2 \tag{2.1}$$

$$Ti_3C_2 + 2H_2O = Ti_3C_2(OH)_2 + H_2 \tag{2.2}$$

$$Ti_3C_2 + 2H_2O = Ti_3C_2(O)_2 + 2H_2 \tag{2.3}$$

$$Ti_3C_2 + 2HF = Ti_3C_2(F)_2 + H_2 \tag{2.4}$$

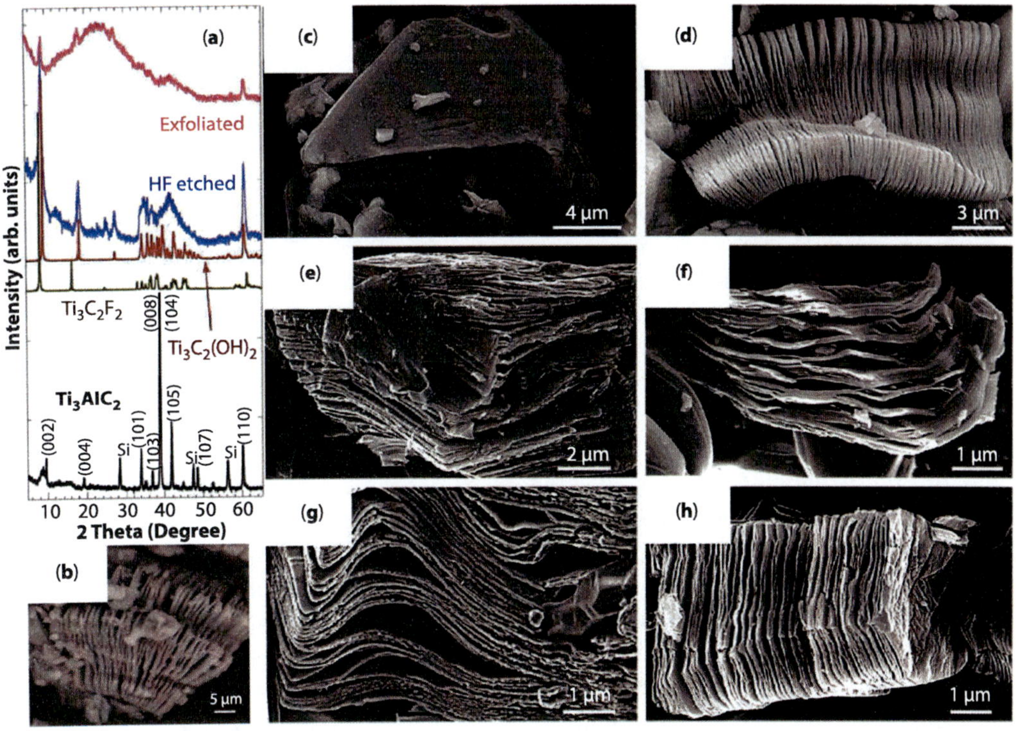

Figure 2.2 (a) XRD patterns for Ti_3AlC_2 before and after HF treatment (in order from bottom to top, MAX precursor, $Ti_3C_2F_2$ (simulated) and $Ti_3C_2(OH)_2$ (simulated), accordion-like MXene and exfoliated MXene nanosheets); (b) SEM image for synthesized accordion-like Ti_3C_2 MXene [4]. Copyright 2011. Reproduced with permission from Wiley. SEM images of (c) Ti_3AlC_2 MAX, (d) Ti_3C_2 MXene, (e) Ti_2C MXene, (f) Ta_4C_3 MXene, (g) TiNbC MXene, and (h) Ti_3CN MXene [5]. Copyright 2012. Reproduced with permission from American Chemical Society.

The successfully prepared Ti_3C_2 MXene opened the "door" to the exploration of MXenes, and subsequently, different components of MXene were synthesized based on the HF etching method. In order to verify that Al-containing MAX phases with diverse chemical characters can be etched, Naguib *et al.* chose a variety of MAX (Ti_2AlC, Ta_4AlC_3, Ti_3AlCN, $(V_{0.5}, Cr_{0.5})_3AlC_2$ and TiNbAlC) to investigate the effectiveness of the HF etching method with successes [5]. Figures 2.2c–h show the many accordion-like multilayer MXenes, which also confirms the suitability of the HF etching method. Since then, a diverse range of MXene types have been synthesized, including Nb_2C/V_2C [6], Zr_3C_2 [7], V_4C_3 [8], Ti_2N/Mo_2C [9], Nb_4C_3 [10] and Hf_3C_2 [11]. Besides, 2D ordered, carbides MXenes ($M'_2M''C_2$ and $M'_2M''_2C_3$) also enriches the number of MXene families with HF as etchant, such as $Mo_2TiC_2T_x$, $Mo_2Ti_2C_3T_x$, and $Cr_2TiC_2T_x$ [12, 13]. According to the etching process, the reaction using HF as etchant can be summarized as follows:

$$M_{n+1}AX_n + 3HF = M_{n+1}X_n + AlF_3 + 3/2H_2 \tag{2.5}$$

$$M_{n+1}X_n + 2H_2O = M_{n+1}X_n(OH)_2 + H_2 \tag{2.6}$$

$$M_{n+1}X_n + 2HF = M_{n+1}X_nF_2 + H_2 \tag{2.7}$$

Despite the acceptable etching results of the HF-etching methods, the etching conditions (acid concentration, temperature, time) are the key to ensuring high yields and low defects of MXenes due to the different interactions between M-A layer in MAXs, also in relation to the application area [14]. The etching time and etching temperature were tailored by Mashtalir and Gogotsi *et al.* to investigate their effect on the production of MXene at a concentration of 50% HF [15]. As shown in XRD patterns in Figures 2.3a-b, (104) characteristic peaks of Ti_3AlC_2 MAX fade with increasing etching temperature and time, accompanied by (002) diffraction peak moving towards a lower angle, demonstrating the growing of layer spacing and the removal of aluminum layers. Moreover, HF etching strategies with different reaction concentrations (Figures 2.4a–c) were explored and the results showed that higher concentrations allowed for a more pronounced accordion-like structure of the multilayer MXene (Figures 2.4d–f) [16]. By establishing three different HF etching protocols (e.g., 5 wt% for 24 h, 10 wt% for 18 h, and 30 wt% for 5 h), Mohamed *et al.* [17] demonstrated that higher concentrations of HF could achieve shorter reaction times. Furthermore, increasing etching time and temperature would resulted in increased defect sites and reduced flake sizes, which can be of benefit in catalytic and gas-phase sensor applications, but not for optics, electronics, and electromagnetism [1, 18]. The HF-etching methods have simple operation and easily available reaction conditions, and it provides good etching results for most Al-based MAX. However, its high corrosiveness and toxicity pose operational and environmental risks that cannot be ignored. It is imperative to optimize the preparation of MXene towards safe, efficient, large-scale synthetic routes.

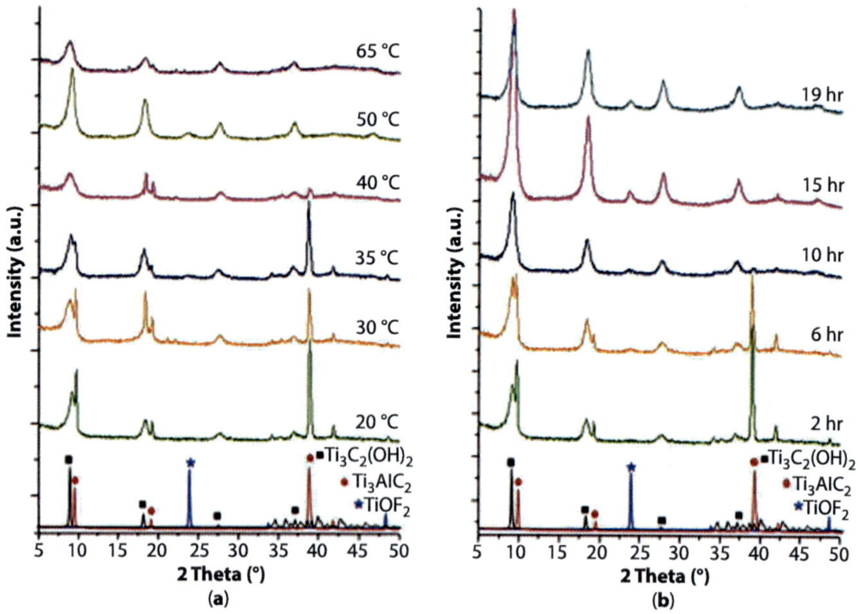

Figure 2.3 XRD patterns of Ti_3AlC_2 after etching with 50% HF solution: (a) varying reaction temperatures for 2 h; (b) varying reaction times at room temperature [15]. Copyright 2013. Reproduced with permission from Elsevier.

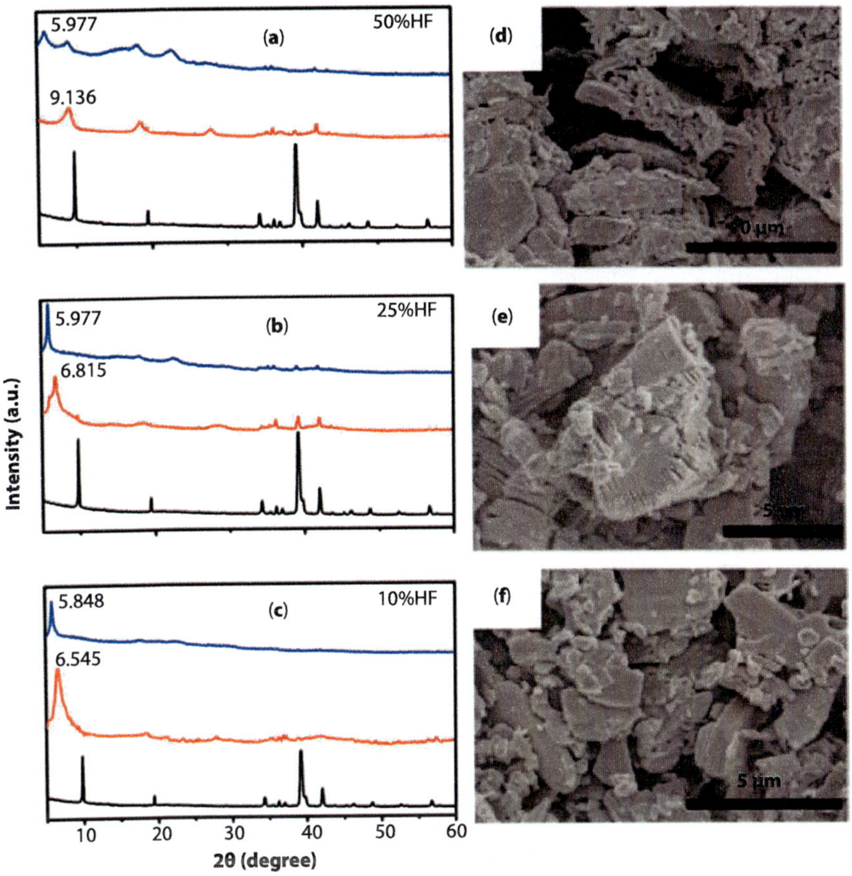

Figure 2.4 XRD patterns and SEM images of the Ti_3AlC_2 precursor after treating with different concentration of HF solution: 50 wt% (a, d), 25 wt% (b, e), 10 wt% (c, f) [16]. Copyright 2015. Reproduced with permission from American Chemical Society.

2.3 *In-Situ* HF-Forming Etching Method

The preparation of MXenes is extremely dangerous because of the high risk of HF, so it is vital to figure out how to avoid using HF solutions directly during the preparations. From the etching of Al-containing MAX, it is clear that the efficient binding effect of F ions to Al layer is the key to the etching results. As a safer etching strategy, the *in-situ* HF-forming etching methods are widely explored and also shown excellent etching results. The class of etchants can be roughly divided into two categories: fluoride salts/acids etching method and bifluoride salts etching method.

2.3.1 Fluoride Salts/Acids Etching Method

In 2014, to address the high risk of concentrated HF during synthesized MXene, Ghidiu *et al.* [19] combined fluoride salts with more benign HCl to obtain etchants with the etching capable for some of the MAX phases (Figure 2.5a). After HCl-LiF etching, the resulting products display clay-like morphology. Moreover, the presence of Li^+ ions and water molecules between the layers allows MXene to exhibit enlarged interlayer spacing and swollen states (Figure 2.5b), also to acquire a higher shapeability [14]. Here, when the milder LiF is dissolved in HCl solutions, it produces F ions that combine with H^+ ions to form HF, which avoids the direct use of HF solution with higher risk and is safer. Upon this, Lipatov *et al.* [20] explored the optimal ratio of LiF based on the LiF-HCl etching method to obtain high-quality MXene flakes, corresponding to two etching schemes shown in Figure 2.5c. In comparison, Route 2 offers better etching results than Route 1, which is reflected by the

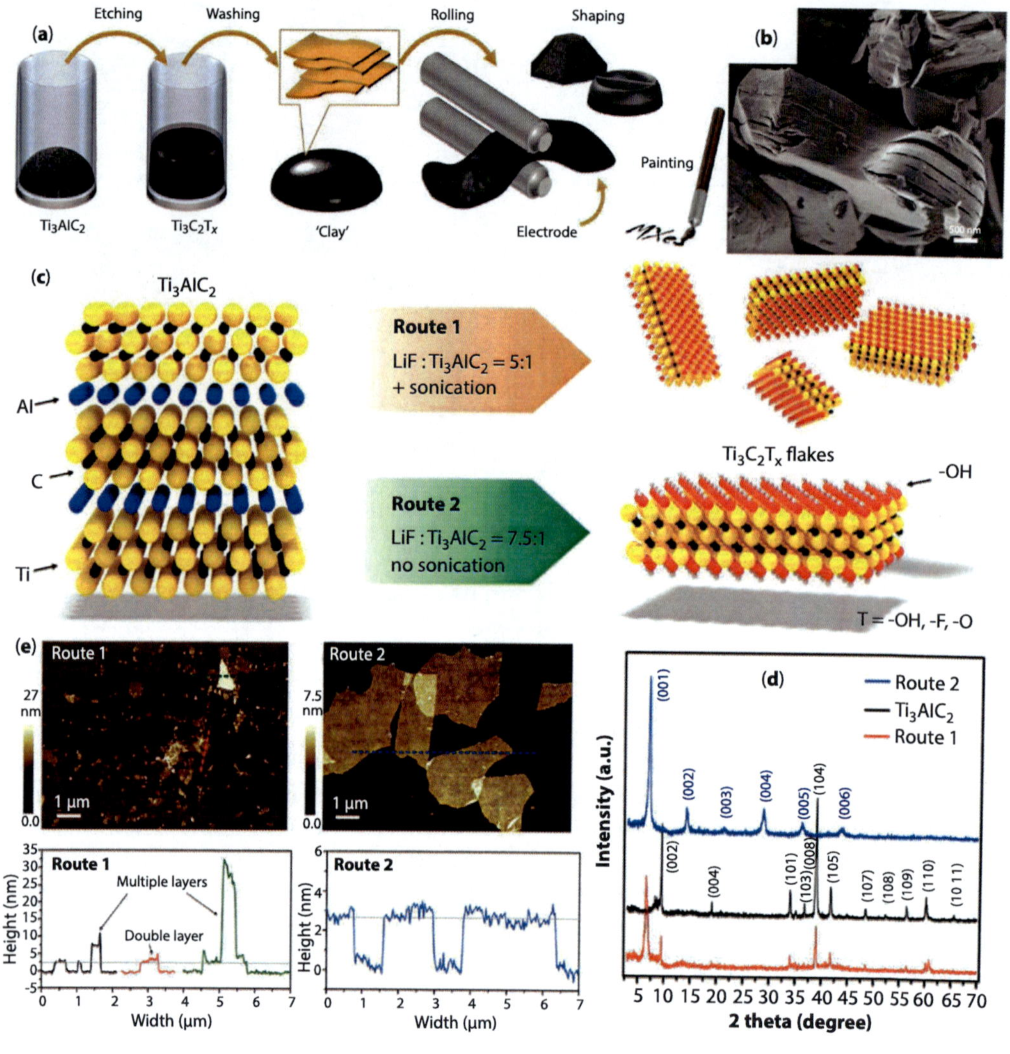

Figure 2.5 (a) Etching of the MAX phase, washing, rolling and molding of the products and (b) SEM image of multilayer Ti_3C_2 MXene [19]. Copyright 2013. Reproduced with permission from Springer Nature. (c) Schematic structures of the conversion of Ti_3AlC_2 MAX to Ti_3C_2 MXene; (d) XRD patterns of Ti_3AlC_2 phase and Ti_3C_2 MXene and (e) AFM height profiles of Ti_3C_2 MXene nanosheets produced by Route 1 and Route 2 [20]. Copyright 2016. Reproduced with permission from Wiley.

disappearing (104) characteristic peaks in XRD patterns (Figure 2.5d), visually defect-free surfaces, sizes and clean edges of flakes from AFM images (Figure 2.5e).

Except for the LiF-HCl etching system, various fluorine salts (e.g. NaF, KF [21], NH_4F, CaF_2, FeF_3 [22]) have been added to HCl/H_2SO_4 to study their ability of etching for MAX phase [19, 23]. To explore the influences of temperature, time and types of fluoride salts/acids etching system on the synthesis of MXenes, Liu *et al.* [24] designed the synthesis methods of Ti_3C_2 and Ti_2C MXenes through the etching routes of $LiF/NaF/KF/NH_4F$ in HCl (Figures 2.6a–d). The optimum etching conditions of MAX phases were obtained for the synthesis of Ti_3C_2 and Ti_2C MXenes. Taking KF-HCl etching system for Ti_2AlN MAX as an example [9], a possible etching reaction for fluoride salts/acids etching method is shown in Equations 2.8–2.9. In addition, considering both steric and electrostatic effects of terminations on electrolyte ion infiltration, Guo *et al.* [25] used H_2SO_4 instead of HCl to engineer $Ti_3C_2T_x$ MXene with $-SO_3$ termination (Figure 2.7a). From Figure 2.7b, the synthesized Ti_3C_2 MXene using LiF-H_2SO_4 etchant displayed lower angle of (002) characteristic peaks in contrast to MXene with HF and LiF-HCl as etchants, which was also presented by SEM images of accordion-like multilayered MXene in Figures 2.7c–f.

$$2Ti_2AlN + 6KF + 6HCl = 2Ti_2NT_x + K_3AlF_6 + AlCl_3 + 3KCl + 3H_2 \tag{2.8}$$

$$2Ti_2AlN + 6KF + 6HCl = 2Ti_2NT_x + 2AlF_3 + AlCl_3 + 6KCl + 3H_2 \tag{2.9}$$

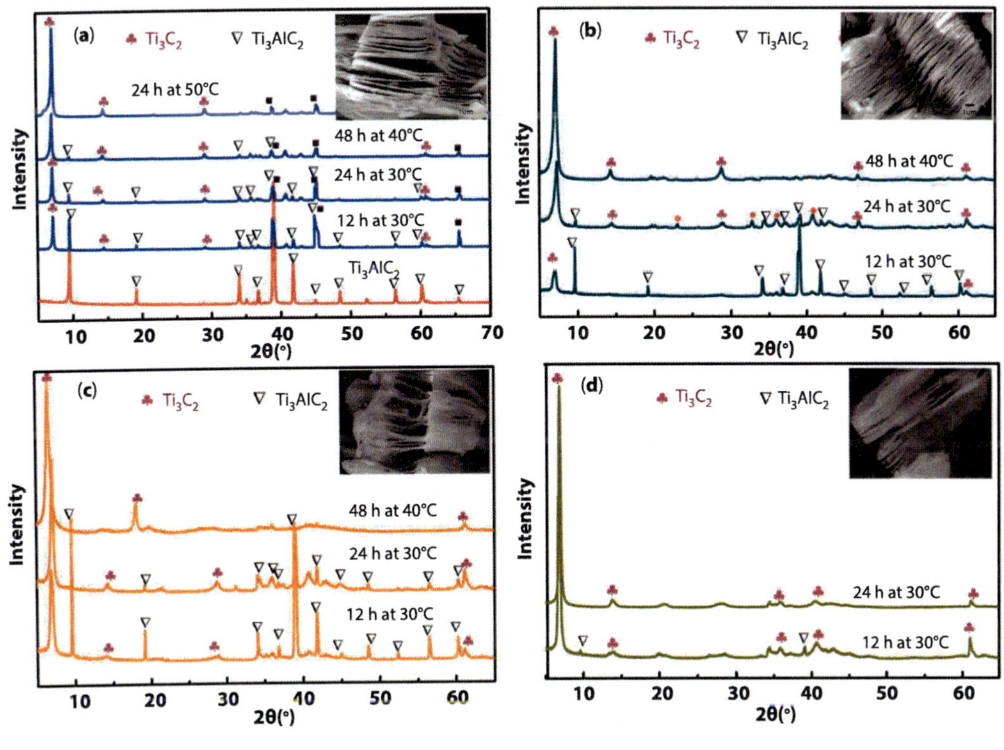

Figure 2.6 XRD patterns of exfoliating Ti_3AlC_2 by (a) LiF-HCl etchant, (b) NaF-HCl etchant, (c) KF-HCl etchant, (d) NH_4F-HCl etchant [24]. Copyright 2017. Reproduced with permission from Elsevier.

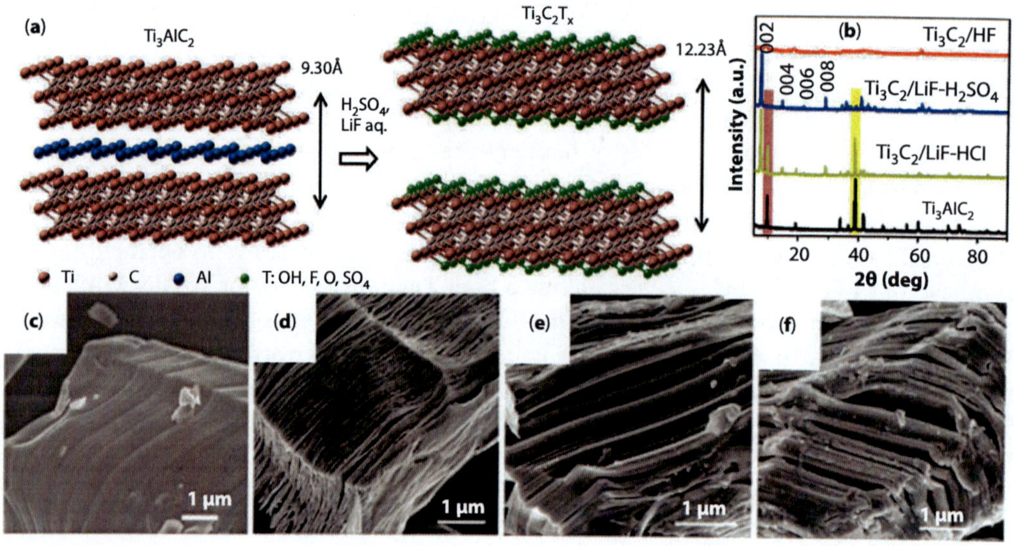

Figure 2.7 (a) Schematic structures of synthesis for Ti_3C_2 through LiF-H_2SO_4 etchant; (b) XRD patterns of MXene and SEM images of (c) Ti_3AlC_2 and Ti_3C_2 MXene etched by (d) HF etchant, (e) LiF-HCl etchant, and (f) LiF-H_2SO_4 etchant [25]. Copyright 2020. Reproduced with permission from American Chemical Society.

In-situ forming HF through the combination of fluorine salts and strong acids produces the same etching effect as concentrated HF, and the difference is that the metal cations from the former can be intercalated into the interlayer of MXene flakes, which raises the interlayer spacing and effectively prevents the self-agglomeration of MXene layers due to van der Waals or hydrogen bonding [26]. These facilitate the obtaining of accordion-like structures and thus enable more efficient MXene nanosheet stripping. The fluoride salts/acids etching method simplifies operations and makes the etching process more accessible to the laboratories, which also boosts the application and development of MXene materials. However, unetched MAX phase residues are still remaining and further improvements in yield are an area that needs to be pursued in the future.

2.3.2 Bifluoride Salts Etching Method

Another approach to tackle the challenge of high risk for HF etching method is bifluoride salts etching method. In 2014, Halim *et al.* [27] firstly applied the NH_4HF_2 etchant in the etching of Ti_3AlC_2 MAX to remove Al layers (Figure 2.8a). As shown in Figure 2.8b, the atom layers of Ti and C can be seen to be regularly aligned with wide layer spacing from STEM image, confirming the successful extraction of Al atom layers. Moreover, Feng *et al.* [28] compared the etching effect of NH_4HF_2-etched Ti_3C_2 MXene with HF-etched Ti_3C_2. As can be seen from the XRD results in Figure 2.8c, the (002) peak for NH_4HF_2-etched Ti_3C_2 displayed lower angle than HF-etched Ti_3C_2, and the c lattice parameter of NH_4HF_2-$Ti_3C_2T_x$ is from 18.4 Å for Ti_3AlC_2 to 24.8 Å that is above 19.8 Å for HF-$Ti_3C_2T_x$. During etching, ammonium species (NH_3 and NH^{4+}) can act as an intercalator to facilitate the etching of MAX, and the etching of the aluminum layer is synchronized with the intercalation process.

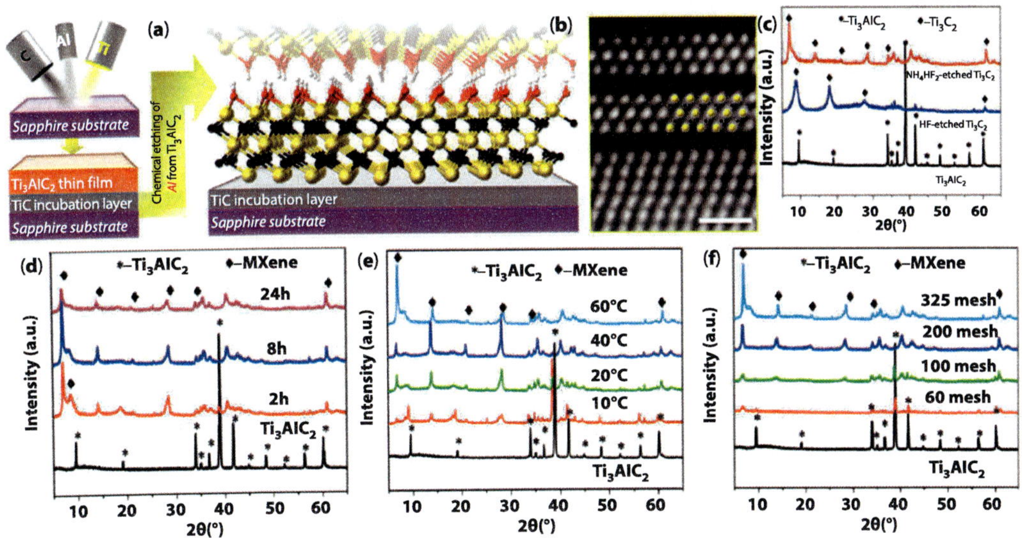

Figure 2.8 (a) The synthesis process of Ti_3C_2 MXene by bifluoride salts; (b) STEM image of Ti_3C_2 MXene [27]. Copyright 2014. Reproduced with permission from American Chemical Society. (c) XRD patterns of Ti_3AlC_2 and Ti_3C_2 MXene etched by HF and NH_4HF_2 [28]. Copyright 2018. Reproduced with permission from Springer Nature. XRD patterns of Ti_3AlC_2 and multilayered MXene exfoliated by NH_4HF_2: (d) variable etching time at 60°C; (e) variable etching temperature for 8 h; (f) variable diameter of Ti_3AlC_2 powder [29]. Copyright 2017. Reproduced with permission from Elsevier.

Furthermore, compared to HF-etched Ti_3C_2, Ti_3C_2 MXene etching by NH_4HF_2 possesses the better intercalation pseudo-capacity and the specific capacitance. Against the HF etching process, the etching process of the bifluoride salts is shown in Equations 2.10–2.11.

$$Ti_3AlC_2 + 3NH_4HF_2 = Ti_3C_2 + (NH_4)_3AlF_6 + 3/2H_2 \tag{2.10}$$

$$Ti_3C_2 + aNH_4HF_2 + bH_2O = (NH_3)_b(NH_4)dTi_3C_2(OH)_xF_y \tag{2.11}$$

To investigate the optimal reaction conditions for NH_4HF_2 etchant, Li *et al.* [29] designed experiments in terms of reaction temperature, reaction time and size of MAX. As shown in Figures 2.8d–f, when etching MAX using NH_4HF_2, an optimized reaction conditions was demonstrated: 325 mesh MAX at 60°C for 8 h. In addition, the stability of the obtained MXene was also explored by knowing that the MXene obtained from the bifluoride salt had higher thermal stability than that obtained from HF-etching. In addition to NH_4HF_2, $NaHF_2$ and KHF_2 were also used as etchants to explore the etching effect on MAX [30]. At 60°C, MAX was added to 1M bifluoride salts solution and reacted at different times, followed by centrifugation and drying to obtain MXene powder (Figure 2.9a). From XRD patterns (Figures 2.9b, d and f), more than 8 h reaction time was required for the beginning of Ti_3AlC_2 exfoliation, possibly longer for $NaHF_2$ and KHF_2, which can be judged by the change of characteristic peaks, such as the shift of (002) peak (from 18.4Å of Ti_3AlC_2 MAX phase to 21.4 Å of $NaHF_2$ etchant, 24.8Å of KHF_2 etchant and 24.8Å of NH_4HF_2 etchant) and the disappearance of (104) peak. After the reaction, the obvious accordion-like structure can be seen

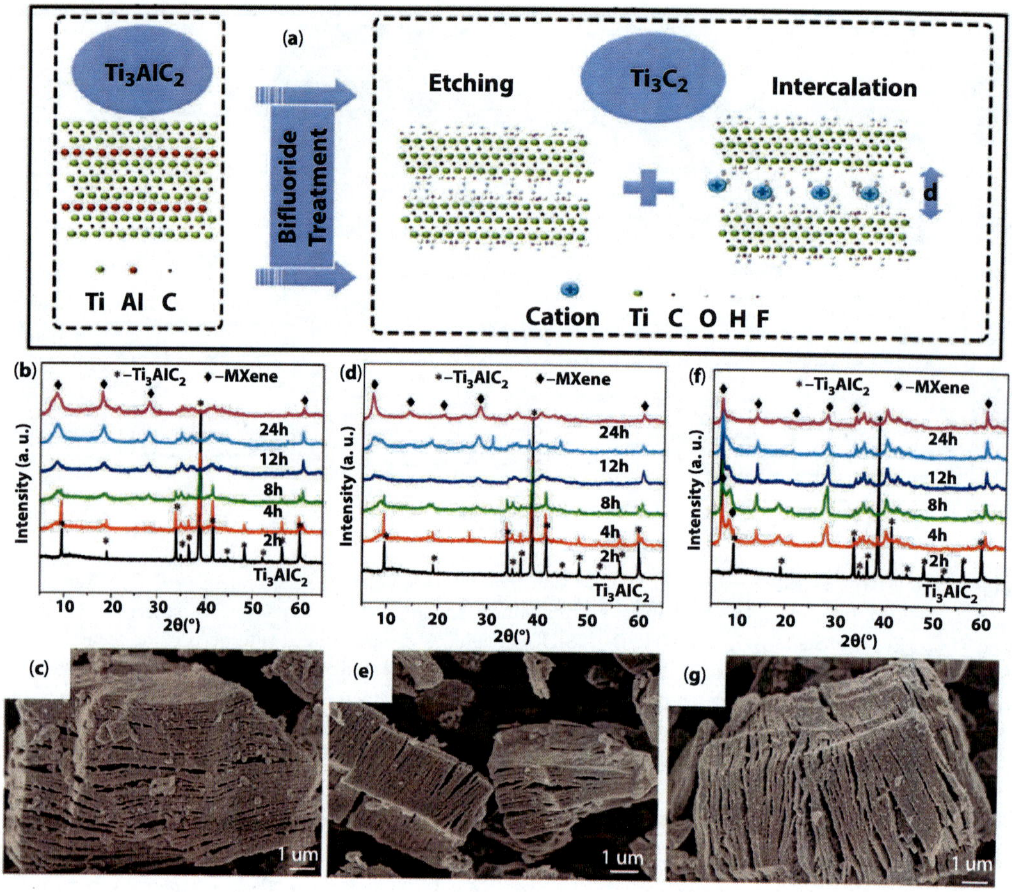

Figure 2.9 (a) Schematic illustration of reaction between Ti_3AlC_2 and bifluorides; XRD patterns of Ti_3AlC_2 and the products etching by (b) $NaHF_2$, (d) KHF_2, (f) NH_4HF_2; SEM images of multilayered MXene etching by (c) $NaHF_2$, (e) KHF_2, (g) NH_4HF_2 [30]. Copyright 2017. Reproduced with permission from Elsevier.

(Figures 2.9c, e and g), which also proves the success of the etching. And the reaction process of bifluoride salts etching method can be summarized as Equations 2.12–12.13.

$$Ti_3AlC_2 + XHF_2 \rightarrow Ti_3C_2 + X_aAlF_b + AlF_3 + H_2\uparrow \tag{2.12}$$

$$Ti_3C_2 + XHF_2 + bH_2O \rightarrow Ti_3C_2(OH)_xF_yX_z \tag{2.13}$$

Based on the above analysis, the bifluoride etchant has a higher safety profile and the etched MXene product is superior in some aspects to the MXene etched by HF. However, the bifluoride etching method is far from being applied and further exploration is needed.

2.4 Molten Salt Etching Method

2.4.1 Fluorine-Containing Molten Salt Etching Route

For Al-MAX, HF-etching method and *in-situ* HF-forming etching method exhibit unparalleled results. However, for some transition metal nitrides and non-Al-MAX phase, these etching methods do not work well [1, 31, 32]. In 2011, Naguib *et al.* [33] explored the topological transformation of the Ti_2AlC MAX phase in molten LiF under air. From Figure 2.10a,

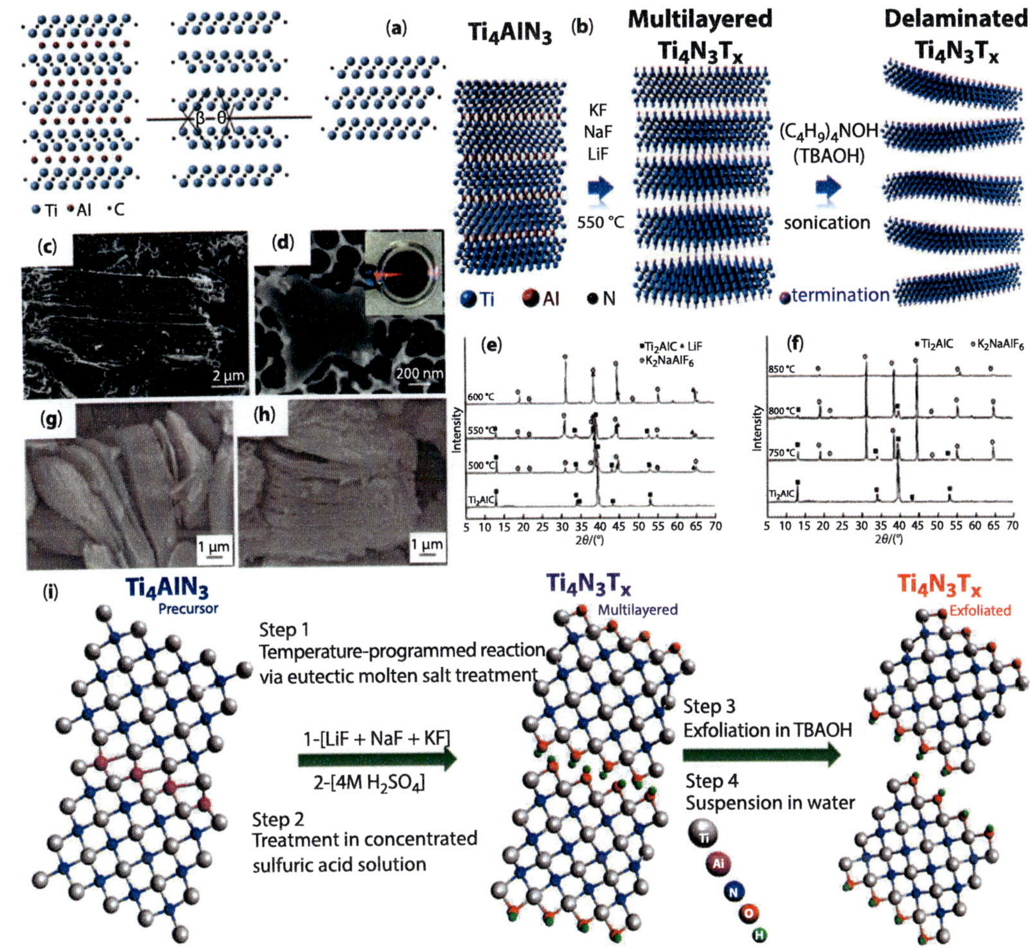

Figure 2.10 (a) Schematic for the transforming of Ti_2AlC into $TiC_{0.5}$ [33]. Copyright 2011. Reproduced with permission from American Ceramic Society. (b) Schematic for the preparation of $Ti_4N_3T_x$ MXene via molten salt etching methods under Ar atmosphere; SEM image of (c) molten salt treated Ti_4AlN_3 and (d) few-layered $Ti_4N_3T_x$ MXene [34]. Copyright 2016. Reproduced with permission from Royal Society of Chemistry. XRD patterns of etching Ti_2AlC by (e) LiF-NaF-KF and (f) NaF-KF etching systems; SEM images for multilayer MXene of (g) LiF-NaF-KF etching system at 600°C; (h) NaF-KF etching system at 800°C [35]. Copyright 2019. Reproduced with permission from Springer Nature. (i) Schematic for $Ti_4N_3T_x$ MXene obtained by eutectic molten salt treatment at 550°C for 7 h [36]. Copyright 2019. Reproduced with permission from American Chemical Society.

the results showed that Al-layer would diffuse out of the structure to react with LiF to form Li_3AlF_6, remaining Ti-C layers, which illustrated the possibility of molten salt as etchants. Afterwards, molten fluorinated salts were explored as etchants for MAX, such as LiF, NaF, NH_3F. Urbankowski *et al.* [34] had proved the molten fluoride-salt-etching method to synthesize Ti_4N_3-based MXene at 550°C under argon atmospheres. As shown in Figure 2.10b, Ti_4AlN_3 precursor powder is added to molten LiF-NaF-KF mixture heated to 550°C under argon atmospheres for 0.5 h, and subsequently ultrasonically exfoliated to obtain delaminated MXene. After the treatment, the accordion-like structure of the multilayer MXene (Figure 2.10c) is obtained, while the delaminated MXene also exhibits a lamellar morphology (Figure 2.10d).

In order to better probe the optimal reaction conditions of molten salts on Ti_2AlC MAX, Yan *et al.* [35] investigated the etching effect of different molten fluorine salt etching systems. LiF-NaF-KF and NaF-KF etching systems were used at different etching conditions. During the etching reaction, LiF was not involved in the reaction, but acted as a role to reduce the eutectic temperature of the mixture. Moreover, XRD patterns (Figures 2.10e–f) confirm the absence of Ti-containing fluorides, which indicates the selective etching effects of the etching systems. The work reports that for LiF-NaF-KF etching system, 600°C is the optimal reaction temperature and 850°C for NaF-KF system, and the obtained accordion-like structure for Ti_2C MXene is shown in Figures 2.10g–h. A possible reaction formula of LiF-NaF-KF etching and NaF-KF etching systems is shown in Equation 2.14. In addition, to investigate that transition metal nitrides have electrical and optical properties similar to those of carbides, Li *et al.* similarly used the LiF-NaF-KF ternary hybrid etching system to obtain $Ti_4N_3T_x$ MXene, as shown by the reaction mechanism in Figure 2.10i. It was shown that $Ti_4N_3T_x$ MXene not only possessed semiconductor performance, but had intriguing optical, (photo) electrochemical, and electrical properties [36].

$$Ti_2AlC + Na^+ + 2K^+ + 6F^- = Ti_2CT_x + K_2NaAlF_6 \tag{2.14}$$

2.4.2 Fluorine-Free Molten Salt Etching Route

It is well known that surface terminations of MXene can have an important influence on its properties. The application of aqueous etching methods can limit its direct use of MXenes in water-sensitive applications. In addition, when MXene materials are used as supercapacitor electrodes, the inert fluorine-containing terminations can have negative impacts on electrochemical performance. To solve such problems, fluorine-free molten salt etching methods have been explored. Li *et al.* [37] proposed a redox-controlled A-site etching of MAX phases in Lewis acidic molten salts, which was validated by etching unconventional MAX-phase precursors with A elements (Zn, Si, Zn and Ga) to synthesized different MXenes. As shown in Figure 2.11a, as Ti_3SiC_2 phase is submerged into molten salt of $CuCl_2$, the element Si, which has a weak interaction with Ti, is oxidized to Si^{4+} by Cu^{2+} ions from the Lewis acid, and then Cu^{2+} is reduced to Cu, accompanied by the production of the volatile $SiCl_4$ phase, which is reflected by Equations 2.15–2.16. The resulting copper is then removed by ammonium persulphate solution, while producing O-containing terminations onto MXene. Figure 2.11b displays the change of characteristic peaks with the (002)

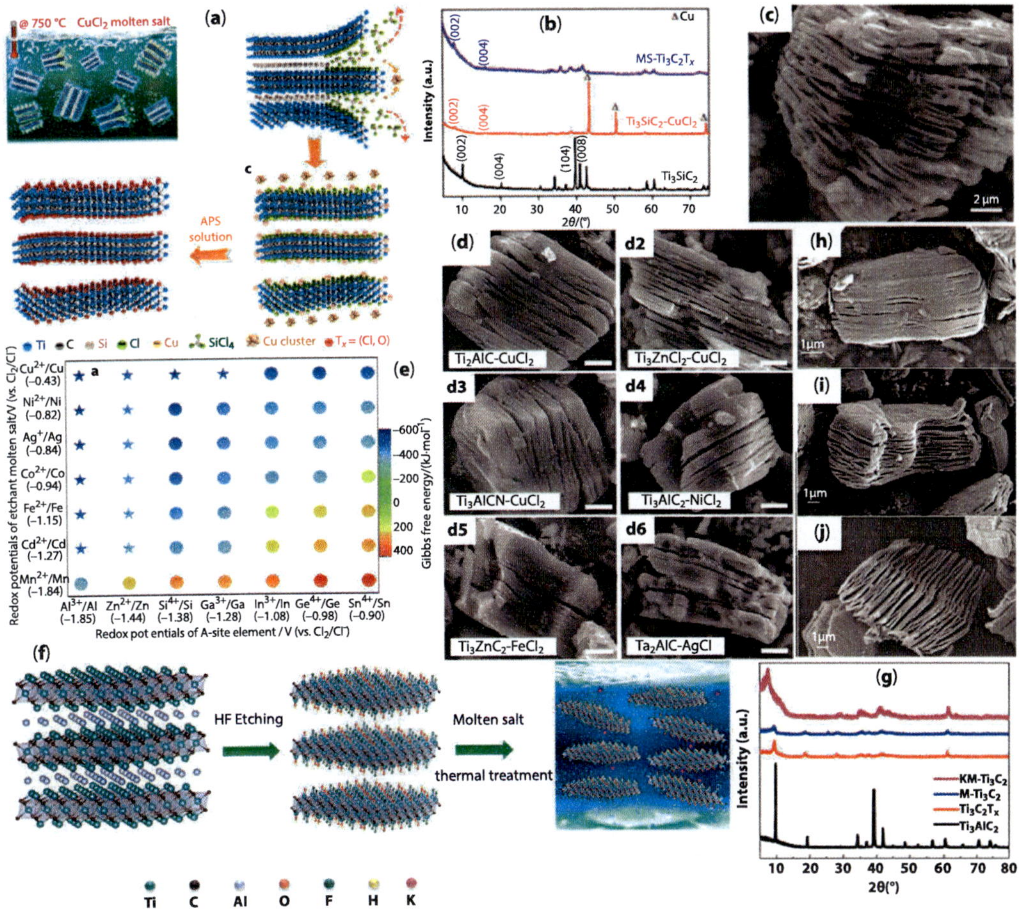

Figure 2.11 (a) Illustration for the synthesis of $Ti_3C_2T_x$ MXene by Lewis acidic molten salt etching route; (b) XRD patterns of Ti_3SiC_2 MAX phase and MS-$Ti_3C_2T_x$ MXene obtained via molten $CuCl_2$ etching route; (c) SEM image of multilayer Ti_3C_2 MXene [37]. Copyright 2020. Reproduced with permission from Springer Nature. (d) SEM images of accordion-like MXenes obtained by various molten Lewis acid Cl salts; (e) Gibbs free energy mapping (700°C) to guide the choosing of appropriate Lewis acid Cl salts [38]. Copyright 2020. Reproduced with permission from Springer Nature. (f) Schematic of molten salt modified $Ti_3C_2T_x$ (KM-$Ti_3C_2T_x$); (g) XRD patterns of MAX precursors and resulted MXene; SEM images of (h) HF-etched $Ti_3C_2T_x$, (i) M-$Ti_3C_2T_x$ and (j) KM-$Ti_3C_2T_x$ [39]. Copyright 2021. Reproduced with permission from Elsevier.

characteristic peak changing from 10.13° to 8.07° after the reaction, while other characteristic peaks disappear and shift, and the accordion-like structure of the multilayer MXene can be seen in the SEM image (Figure 2.11c), which all proves the success of the Lewis acidic molten salt etching route. The study also found that MXene obtained by molten salt etching route had excellent performance in the field of energy storage.

$$Ti_3SiC_2 + 2CuCl_2 = Ti_3C_2 + SiCl_4(g) + 2Cu \quad (2.15)$$

$$Ti_3C_2 + CuCl_2 = Ti_3C_2Cl_2 + Cu \quad (2.16)$$

In addition to $CuCl_2$, another late transition-metal halides (e.g., $ZnCl_2$, $FeCl_2$, AgCl) were also be utilized to explore selectively etching effects on MAX in terms of their high electrochemical redox potential. Figure 2.11d illustrates the effects of etching with different Lewis acid molten salts etching routes, and the accordion-like structure can be clearly seen, indicating their acceptable etching effect for different etchants. According to the redox potentials of A elements in MAX phases and the cations in molten salt melts, a Gibbs free energy mapping was constructed to predict the redox coupling between the two, as shown in Figure 2.11e, which provided guidance for the rational selection of suitable Lewis acid melts for the preparation of MXene [38]. In addition, in view of the controllability of the molten salt etching route on the terminations of MXene, Guo *et al.* [39] placed the multilayer MXene etched with HF into the molten salt to remove the fluorine-containing functional groups (Figure 2.11f). The LiCl-KCl (M-$Ti_3C_2T_x$) and LiCl-KCl-K_2CO_3 (KM-$Ti_3C_2T_x$) molten salt routes were used to replace -F by -O surface terminations for MXene. In Figure 2.11g, the XRD patterns demonstrate that the former has little shifts for (002) peak, while the latter has a significant variation from 9.7° to 7.45°, which is also confirmed by the SEM image in Figures 2.11h–j. This can be attributed to various potassium oxygenated anions facilitating the intercalation for LiCl-KCl-K_2CO_3 molten salt routes. The results show that MXene obtained by LiCl-KCl-K_2CO_3 route has higher conductivity and more electrochemical activity-specific surface area as electrode materials than these for MXene etched with LiCl-KCl route, which may be because of the replacement of F-containing terminations by O-terminated functional groups to form more $Ti_3C_2O_2$ for LiCl-KCl-K_2CO_3 molten salt routes.

2.5 Electrochemical Etching Method

The etching process of MAX is an electrochemical reaction accompanied by electron transfer. Thus to selectively removal of A layers from MAX phase, electrochemical etching method is also a plausible alternative approach to synthesis MXene. Sun *et al.* [40] adopted an etching route to obtain Ti_2CT_x MXene with only -Cl, -O, and -OH terminal groups by electrochemical etching of Ti_2AlC MAX using HCl as the electrolyte. During the etching, two different concentrations of HCl electrolyte (1M and 2M) were used, into which Ti_2AlC MAX was placed and reacted for different times at an working voltage of 0.6V. Although it is possible to obtain MXene without -F functional groups in this way, over-etched can also occur. As shown in Table 2.1, as the reaction proceeds, not only is the Al removed, but also the Ti atoms in the surface layer are removed. In Figure 2.12a, there are two stages of electrochemical etching. In the first stage (Equation 2.17), the Al layer is selectively etched. But as the reaction progresses, the outer layer of MXene is etched, finally producing carbide-derived carbon (CDC). Moreover, Yang *et al.* [41] presented a binary aqueous electrolyte system (ammonium chloride (NH_4Cl) and 0.2 M tetramethylammonium hydroxide (TMAOH)) for the electrochemical etching method to synthesize $Ti_3C_2T_x$ MXene (T= -OH, =O), as shown in Figures 2.12b and c. During the etching, Cl-ions enable rapid etching of anodic Al, resulting in the break of Ti-Al bonds and followed by the intercalation of ammonium hydroxide (NH_4OH), in which the anode (Ti_3AlC_2) is completely etched in 5 h under ambient conditions with a yield of 90%. In XRD patterns (Figure 2.12d), the (002) peak of Ti_3AlC_2 ($2\theta = 9.8°$) shifts to lower angle ($2\theta = 7.8°$) after the completion of etching, which verifies the successful preparation of the MXene, as well as the SEM (Figure 2.12e) and

Table 2.1 EDS results before and after MAX etching under different conditions, using chemical etching (CE-MXene) as a comparison [40]. Copyright 2017. Reproduced with permission from Royal Society of Chemistry.

	MAX	EE–1 M–0.6 V–1 day	EE–2 M–0.6 V–5 days	EE–2 M–0.6 V–14 days	CE-MXene
Ti atomic%	30.57	35.15	22.10	0.55	38.66
Al atomic%	15.29	14.58	9.30	0.25	0.82
C atomic%	54.14	41.97	58.10	73.68	32.48
O atomic%	0.00	8.30	9.40	24.68	7.63
Cl atomic%	0.00	0.00	1.10	0.84	12.31
F atomic%	0.00	0.00	0.00	0.00	8.10
Ti/Al	2.00	2.40	2.37	2.20	47.15

TEM (Figure 2.12f) images. Moreover, the mechanism of the etching process is proposed in Equations 2.17–2.20.

$$Ti_2AlC + yCl + (2x + z)H_2O \rightarrow Ti_2C(OH)_{2x}Cl_yO_z + Al^{3+} + (x + z)H_2 + (y + 3)e^- \tag{2.17}$$

$$Ti_3AlC_2 - 3e^- + 3Cl^- = Ti_3C_2 + AlCl_3 \tag{2.18}$$

$$Ti_3C_2 + 2OH^- + 2e^- = Ti_3C_2(OH)_2 \tag{2.19}$$

$$Ti_3C_2 + 2H_2O = Ti_3C_2(OH)_2 + H_2 \tag{2.20}$$

During the chemical etching process, the lithium ions have an important intercalation role to enable the obtained multilayer MXene to be well delaminated. Based on the electrochemical etching method, Sun *et al.* [42] proposed a mechanism for the selective lithiation-expansion-microexplosion synthesis of fluoride-free $Ti_3C_2T_x$ MXene nanosheet using 1M LiTFSI as the electrolyte. As shown in Figure 2.12g, during the etching, the Li^+ from the electrolyte would intercalate between Ti-C layers to form Li_xAl_y alloys, a process called lithiation. The alloys would undergo 100% volume expansion, which can expand the interlayer spacing. Finally, the formed Al–Li alloys can react with H_2O to produce $Al(OH)_3$, $Li(OH)_2$ and H_2 gas, followed by sonication to obtain MXene nanosheets (Figures 2.12h-i).

To exclude the hazards posed by fluoride ions, electrochemical etching methods are used to prepare biocompatible Nb_2CT_x MXene materials to fabricate an enzymatic electrochemical biosensor for ultrasensitive pesticide phosmet detection [43]. As shown in Figure 2.13, multilayer Nb_2CT_x MXene is obtained by electrochemical etching methods at conduction of 1M HCl electrolyte with 40°C under 1V for 4 h, followed by exfoliation to get single/few layers MXene nanosheets. The synthesized fluorine-free Nb_2CT_x MXene not only possesses the advantages of low cytotoxicity, high porosity, excellent conductivity and good chemical

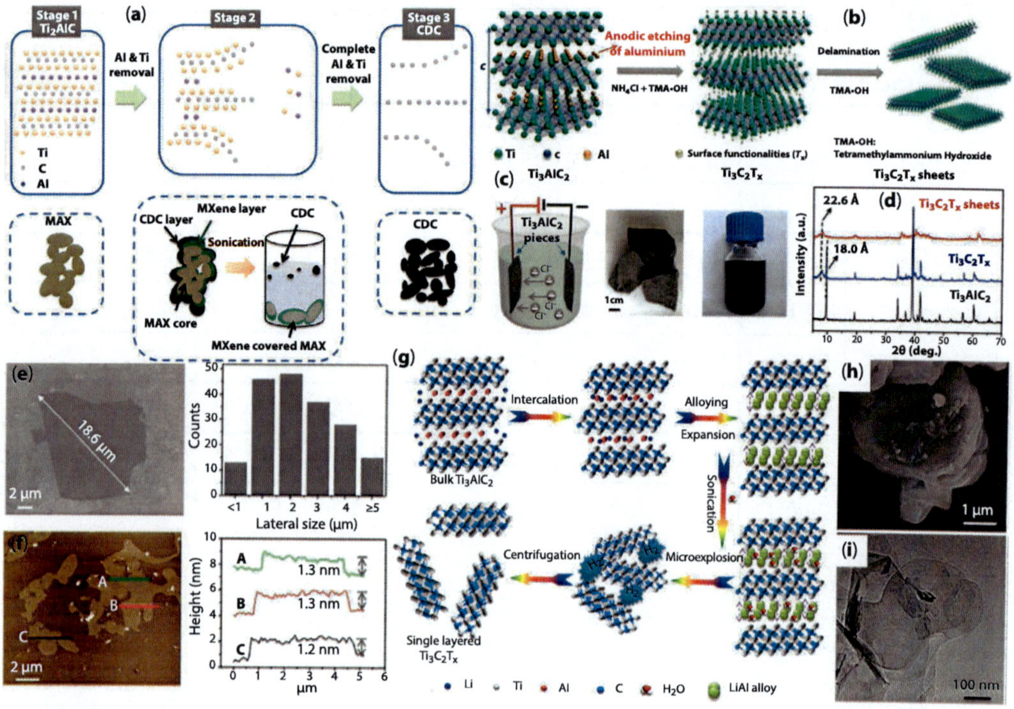

Figure 2.12 (a) Mechanism for the preparation of MXene by electrochemical etching in hydrochloric acid aqueous electrolyte [40]. Copyright 2017. Reproduced with permission from Royal Society of Chemistry. (b) Schematic for the etching reaction of Ti_3AlC_2 MAX and delamination process; (c) Schematic for the structure of electrochemical cell, MAX precursor and the aqueous dispersion of delaminated $Ti_3C_2T_x$ MXene; (d) XRD patterns of Ti_3AlC_2, $Ti_3C_2T_x$, and $Ti_3C_2T_x$ film; SEM (e) and AFM (f) image of $Ti_3C_2T_x$ powder [41]. Copyright 2018. Reproduced with permission from Wiley. (g) Schematic of the lithiation intercalation-expansion-microexplosion process; (h) SEM images of Ti_3AlC_2 and (i) HRTEM images of Ti_3C_2Tx [42]. Copyright 2019. Reproduced with permission from American Chemical Society.

stability, but also the MXene based biosensor has good stability and enhanced enzymatic activity and electrochemical properties. In addition, Chen *et al.* [44] employed the mixture of lithium fluoride and lithium hydroxide as the electrolyte and synthesized $Ti_3C_2T_x$ MXene that was stable in aqueous solution for 15 days by electrochemical etching (Figure 2.14a). In addition, the XRD patterns (Figure 2.14c) of MXene obtained using different etching routes (Figure 2.14b) are evaluated and the results show that the MXene layers obtained by electrochemical etching combined with ultrasound are comparable to those obtained by organic intercalation by other methods. The filtrated $Ti_3C_2T_x$ film not only achieves a tensile strength of 20.5 MPa, electrical conductivity of 1663 S/cm and Young's modulus of 13.4 GPa, but also obtains 323.7 F/g, 1.39 F/cm and 1160 F/cm in specific capacitances for supercapacitors. As shown in Figure 2.14d, MXene obtained by electrochemical etching can be used for several applications such as smart devices due to its excellent electrochemical properties.

The electrochemical etching method is a safe, green etching technique, and the synthesized MXene has tunable surface functional groups, such as improved electrochemical properties for Cl-terminated MXene and stronger mechanical properties for oxygen-terminated MXene than fluorine- and hydroxyl-functionalized MXenes. In addition, the good

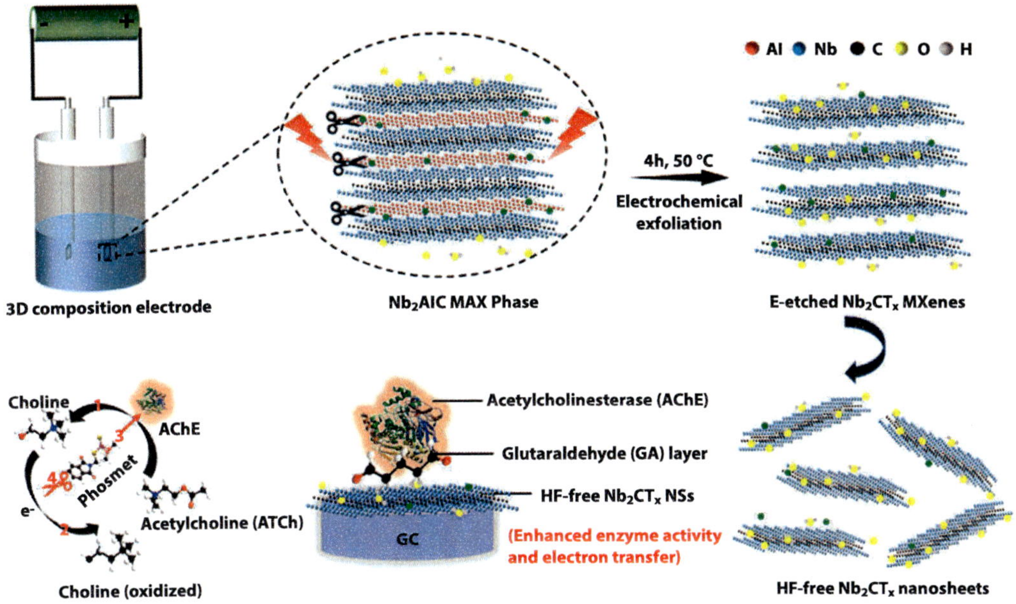

Figure 2.13 (a) Illustration for etching, exfoliation and delamination process of Nb_2AlC phase and HF-free Nb_2CT_x/AChE based biosensor showing the enzyme inhibition effect for phosmet detection [43]. Copyright 2020. Reproduced with permission from Wiley.

electrical properties of the fluorine-free MXene prepared by this method make it promising for the preparation of MXene for electrical applications, such as supercapacitors, batteries, etc. Nevertheless, the manipulation of the reaction conditions is a major pain point for this method, or else over-etching can occur. In view of the etching conditions, the scale-up preparation is also an aspect that needs to be further explored.

2.6 Hydrothermal Etching Method

Whether it is the HF-etching method or the *in-situ* HF-forming etching method, the risk of HF is always present in the liquid/gas phase. Also, improving the yield of MXene etching is also a challenge for the synthesis of MXene. Therefore, a mild and simple route for the preparation of MXene using on-pot hydrothermal methods has been proposed. Wang *et al.* [45] used NH_4F as etchant and obtained MXene by hydrothermal etching route. As shown in Figure 2.15a, the effects of different amounts of etchants, etching temperatures and etching times on the etching effect are explored, and different states of the accordion-like structure can be observed. The comparisons show that the volume of etchant and etching temperature is proportional to the etching effect of MXene, but that high temperatures lead to oxidation of MXene. Peng *et al.* [46] used *in-situ* HF-forming ($NaBF_4$, HCl) route coupled with hydrothermal methods to etch MAX phase (Ti_3AlC_2, Nb_2AlC) and explored the effects between different reaction times and etching results. Compared to the XRD patterns (Figure 2.15b) of MXene obtained by HF-etching method, the multilayer MXene obtained by hydrothermal method has a larger interlayer distance, as shown in Figure 2.15c.

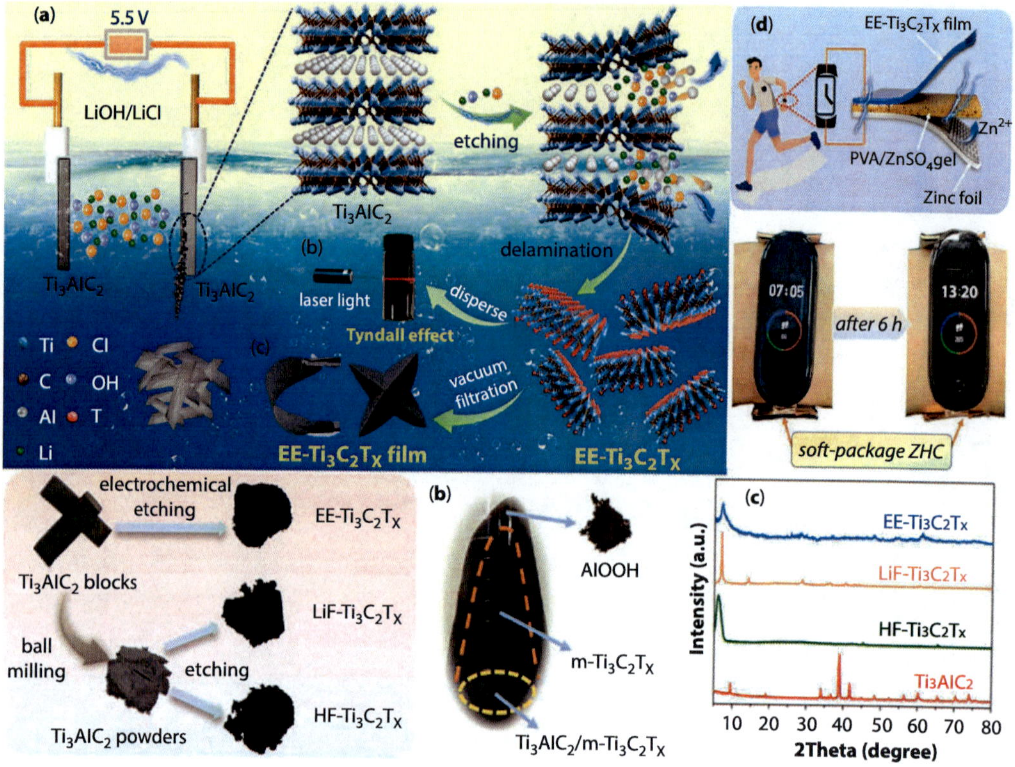

Figure 2.14 (a) Schematic for the preparation of EE-$Ti_3C_2T_x$; (b) The etching routes of Ti_3AlC_2 via different methods and d-$Ti_3C_2T_x$ powder products; (c) XRD patterns of Ti_3AlC_2, EE-$Ti_3C_2T_x$, LiF-$Ti_3C_2T_x$, and HF-$Ti_3C_2T_x$; (d) Schematic for the application of quasi-solid-state EE-$Ti_3C_2T_x$ ZHC [44]. Copyright 2022. Reproduced with permission from American Chemical Society.

Figure 2.15d also demonstrates the excellent results of the hydrothermal etching route in terms of Nb_2AlC MAX, as the evidence of SEM image in Figure 2.15e Nb_2AlC MAX. Furthermore, based on hydrothermal etching routes, other fluorine salts (LiF, NaF, KF, etc.) have also been used to explore the etching effect on V_2AlC MAX with success [47]. During the hydrothermal process, the fluoride ions react more slowly with the hydrogen ions, which results in better A-layer removal and larger layer spacing of MXenes (Figure 2.15f). For Mo_2CT_x MXene, excellent electrochemical and thermoelectric properties have led to its wide range of applications, but its preparation is a major challenge. Mo_2Ga_2C MAX, as a precursor for the preparation of Mo_2CT_x MXene, has a double A-layer structure that makes it difficult for etching [48]. The hydrothermal etching method have been applied to synthesize Mo_2CT_x MXene. Guo *et al.* [49] investigated the etching effect of etchants obtained by mixing different fluorine salts (NaF, KF, LiF and NH_4F) with HCl on Mo_2Ga_2C MAX under hydrothermal conditions. In XRD patterns (Figures 2.16a-d), the NH_4F-HCl etching route offers a lower temperature of 140°C than other etching methods for above 140°C, in which the characteristic peak of Mo_2Ga_2C MAX disappears and the (002) characteristic peak appears and shifts towards a lower angle. As can be seen from the SEM images in Figure 2.16e and f, the accordion-like structure of the etched MAX can be significantly

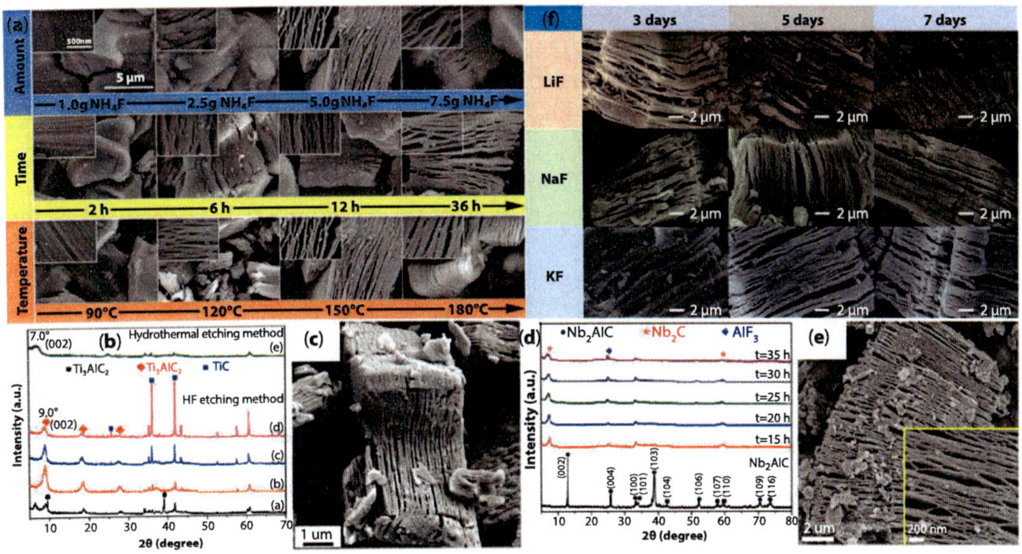

Figure 2.15 (a) SEM images of $Ti_3C_2T_x$ obtained under different conditions [45]. Copyright 2016. Reproduced with permission from Springer Nature. (b) XRD patterns of Ti_3C_2 with HF etching for different times (4h, 12h, 24h, and 36h) and Ti_3C_2 obtained by hydrothermal reaction for 16 h; (c) SEM image of Ti_3C_2 prepared with hydrothermal etching route for 16h; (d) XRD patterns of Nb_2AlC and Nb_2C prepared by hydrothermal route; (e) SEM image of Nb_2C prepared with hydrothermal reaction for 25 h [46]. Copyright 2018. Reproduced with permission from Elsevier. (f) SEM images of V_2CT_x obtained under different conditions at 90°C [47]. Copyright 2020. Reproduced with permission from Elsevier.

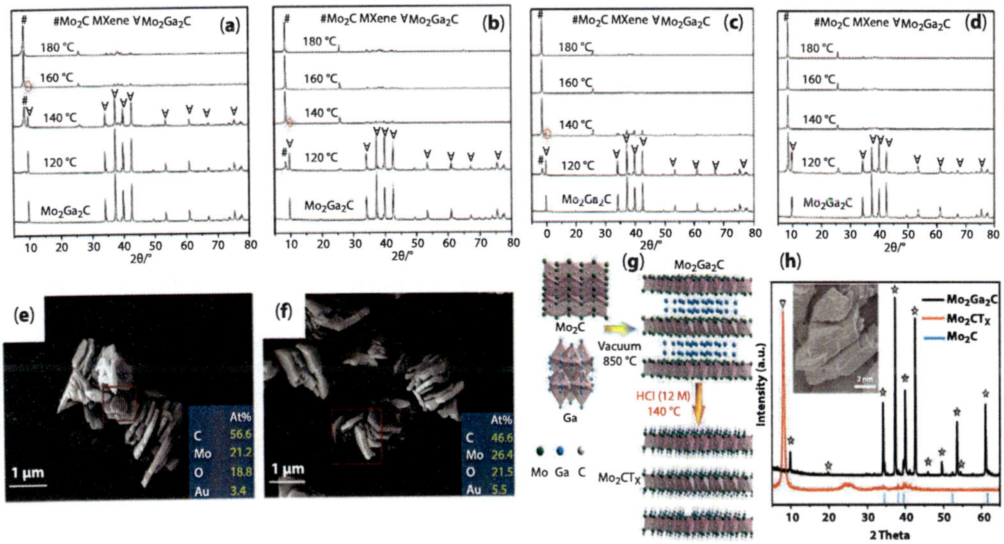

Figure 2.16 XRD patterns of etching Mo_2Ga_2C with different etching systems: (a) LiF/HCl, (b) NaF/HCl, (c) KF/HCl and (d) NH_4F/HCl; SEM images of samples etched by (e) LiF/HCl and (f) NH_4F/HCl [49]. Copyright 2020. Reproduced with permission from Elsevier. (g) Schematic for the synthesis of fluoride-free Mo_2CT_x via hydrothermal etching strategy with the sole addition of hydrochloric acid; (h) XRD patterns of Mo_2Ga_2C and Mo_2CT_x MXenes; (Insert picture of figure h) SEM image of Mo_2CT_x MXene [50]. Copyright 2020. Reproduced with permission from Wiley.

observed in the reaction at 180°C for 24 h. Wang *et al.* [50] verified the etching feasibility of HCl on MAX by density functional theory and experimentally proved the effect of the HCl-etching method under hydrothermal conditions (Figure 2.16g). By reacting at 140°C for 24 h, the Ga layer is removed, demonstrating an accordion-like structure, while the XRD patterns exhibit the appearance of the (002) peak and the absence of the characteristic peaks of Mo_2Ga_2C MAX (Figure 2.16h). The synthesized MXene not only exhibits high electrochemical performance in supercapacitors and sodium-ion batteries, but also achieves a 98% yield.

Hydrothermal etching methods produce better etching results and higher yields because of their higher reaction temperatures. However, high temperatures can also damage the MXene with poor stability, leading to oxidation and deterioration, which may also have a negative impact on the dimensions. At present, hydrothermal etching methods need further exploration, including the choice of etchant, etching temperature and solvent, etc.

2.7 Alkali Etching Method

In addition to acidic solutions, alkali solutions have also been explored for etching MAX. Li *et al.* [51] mixed the mixture containing KOH and a small amount of water with Ti_3AlC_2 MAX, followed by hydrothermal reaction at 180°C for 24 h and washing with water to obtain MXene (Figures 2.17a, b). Under high temperature conditions, the Al layer in MAX reacts with KOH and water to form $KAlO_2$, which are removed by aqueous washing, as shown in Equation 2.21–2.22. In order to investigate the optimum conditions for etching MAX with alkaline solutions, several etching strategies were established to investigate the effect on etching results [52]. As shown in Figure 2.17c, the Al layer would react with OH^- to form Al(oxide)hydroxides in alkali solution, and with further dissolution in alkaline solution to complete the etching of the Al layer. At low temperatures, however, the Al(oxide)hydroxides cannot be dissolved in alkali solution, creating a jamming effect that inhibits the etching of the Al layer. For high temperature and low NaOH concentrations, the presence of high content of water causes the oxidation of Ti layer to form Na/K-Ti-O compounds (NTOs), which would result in the destruction of the MXene nanosheets. According to the Bayer process, the Al(oxide)hydroxides can dissolve in NaOH solution under the condition of high temperatures and high NaOH concentrations, contributing to the removal of Al layer. Figures 2.17d–f exhibit the microstructure characterizations of Ti_3C_2Tx flakes, demonstrating the successful preparation of MXene. There is no fluorine involved in the process and the multilayer $Ti_3C_2T_x$ with OH and O terminated groups achieve 92 wt% purity (using 27.5 M NaOH, 270°C). In addition, through the small amounts of water-assisted KOH etching method, Zhang *et al.* [53] produced novel two-dimensional $Ti_3C_2(OH)_2$-MXene nanoribbons ($Ti_3C_2(OH)_2$ NRs) under hydrothermal conditions and also proposed a reaction mechanism (Equations 2.21–2.22) for the etching of MXene by potassium hydroxide. In Figure 2.17g, under high temperature and confined conditions, high-pressure water vapor with KOH would form, which acts like polar wedges hit and etch the Ti_3AlC_2 bulks. The sugarcane-like etching products would form, expand along the axial slits, and then be torn to ribbons because of the existing surface tension, also resulted in by-product $KAlO_2$. As the reaction proceeds, the thick band of the etched product continues to be stripped into thin nanoribbons (Figures 2.17h–l).

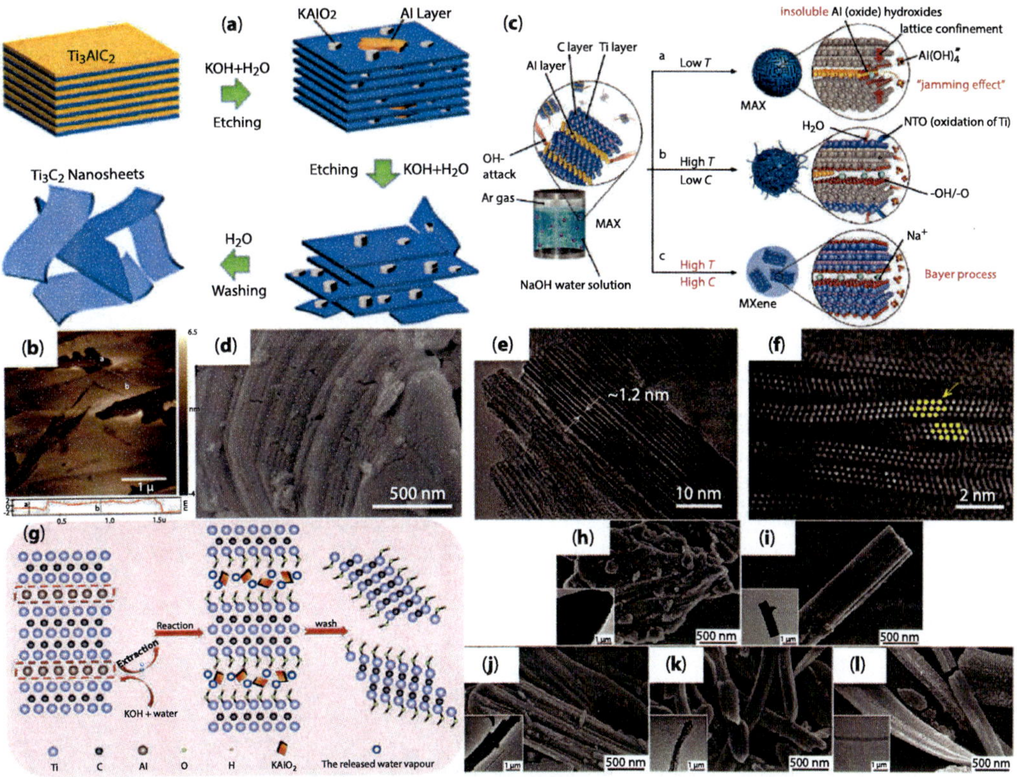

Figure 2.17 (a) Schematic for the etching and delamination of Ti_3AlC_2 via Alkali etching method; (b) AFM image of $Ti_3C_2(OH)_2$ nanosheets [51]. Copyright 2017. Reproduced with permission from American Chemical Society. (c) The reaction between Ti_3AlC_2 and NaOH aqueous solution; (d) SEM, (e) TEM and (f) HAADF-STEM images of $Ti_3C_2T_x$ flakes [52]. Copyright 2018. Reproduced with permission from Wiley. (g) Schematic illustration for $Ti_3C_2(OH)_2$ NRs obtained by the small amounts of water-assisted KOH etching method under hydrothermal conditions; SEM images of (h) Ti_3AlC_2 precursors and (i) Sugarcane-like etching products, (j) Slab-like etching products, (k) Thick ribbon-like etching products and (l) Thin ribbon-like etching products [53]. Copyright 2019. Reproduced with permission from Elsevier.

$$Ti_3AlC_2 + KOH + H_2O = Ti_3C_2 + KAlO_2 + 2/3H_2 \tag{2.21}$$

$$Ti_3C_2 + 2H_2O = Ti_3C_2(OH)_2 + H_2 \tag{2.22}$$

2.8 Other Etching Methods

With the gradual progress of MXene research, their preparation methods were gradually improved and extended. In 2020, Husmann *et al.* [54] proposed a novel etching method using ionic liquids. This method uses ionic liquids (ILs) as etchant and intercalant, discarding the previous high-risk strong acid and alkali etchants. The work explored the reaction atmospheres and times on the etching effect using the mixture of 1-Ethyl-3-methylimidazolium tetrafluoroborate ($EMIMBF_4$) ionic liquid and water. As shown in Figure 2.18a, adjusting the reaction conditions can achieve material exfoliation. In addition, the study also explored the etching effect of different ionic liquid etchants ($EMIMBF_4$ and 1-butyl-3-methylimidazolium hexafluorophosphate (BMIMPF6)) on Ti_2AlC MAX, and the good etching effect (accordion-like structure) can be seen in the SEM images in Figure 2.18b. Due to the unique characteristics of ILs such as non-toxicity, non-volatility, high thermal stability, good solubility, safety and potential for solvent recovery, the ionic liquid etching method was also used by Lu *et al.* [55] to prepare bio-applicable MXene (Figure 2.18c). The Ti_3AlC_2 MAX phase was placed into imidazole-based ILs ($[C_{n+1}mim]Br$) for 24 h at 80°C to obtain IL-MXene. As shown in Figure 2.18d, the (002) characteristic peaks are shifted towards a lower angle compared to that of MAX, while the accordion-like structure is shown in Figure 2.18f, also showing the lamellar structure in the TEM image (Figure 2.18e). It has been shown that the IL-MXene has not only high photothermal conversion efficiency, but also high drug loading capacity as chemotherapeutic drug (DOX). Evidently, the ionic liquid etching method has good potential, but it is still under-explored.

Moreover, halogen materials are used as etchants to synthesize MXenes. Based on DFT theoretical calculations, Li *et al.* [56] obtained Ti_3C_2 MXene with unitary, binary, and ternary

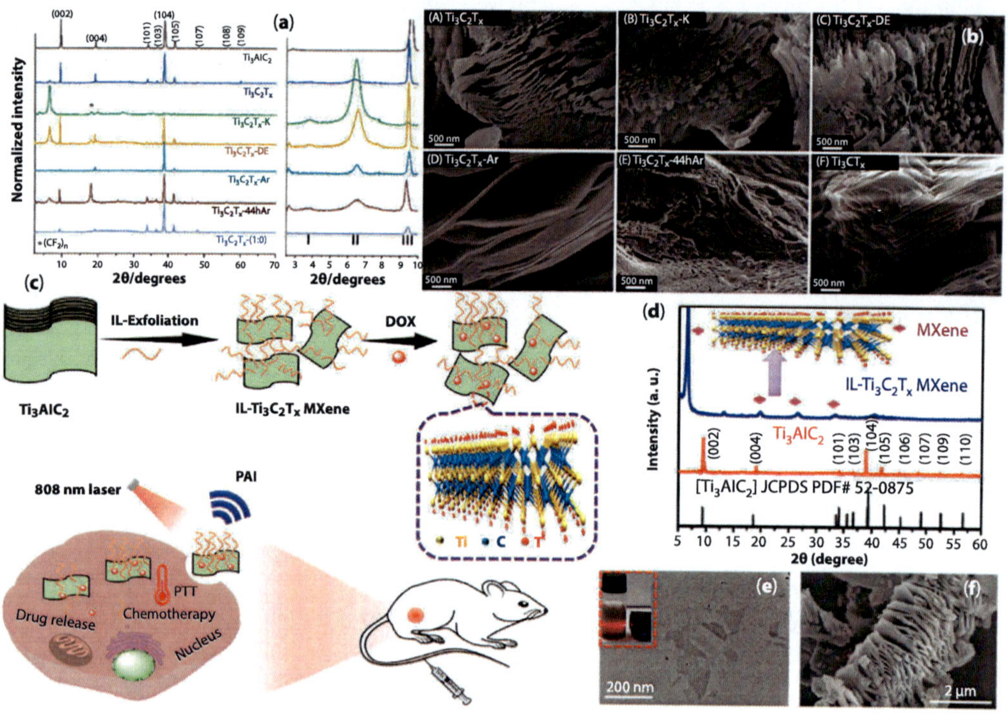

Figure 2.18 (a) XRD patterns of MAX phases treated with ionic liquid under different conditions and (b) SEM images of MAX phases etched with $EMIMBF_4$ in the efficient synthetic conditions [54]. Copyright 2021. Reproduced with permission from American Chemical Society. (c) Schematic for the synthesis of $Ti_3C_2T_x$ MXene nanosheets obtained by IL-assist etching method; (d) XRD patterns, (e) TEM image and (f) SEM image of IL-$Ti_3C_2T_x$ MXene [55]. Copyright 2022. Reproduced with permission from Royal Society of Chemistry.

halogen terminals, e.g., –Br, –Cl, –I, –BrI, and –ClBrI, using molten halogen salts as etching agents. As shown in Equations 2.23–2.24, at high temperatures, the halogen ions can react with the Al layer to form halogenated aluminum salts, while halogen terminated groups can be anchored on the surface of MXene. The effectiveness of the synthetic routes is verified by XRD patterns (Figure 2.19a) and SEM (Figures 2.19b–g), while the synthesized MXene has different applications in electrochemistry due to the different surface end groups. Jawaid *et al.* [57] used liquid halogen as etchants to remove the Al layer by the high reactivity between the halogen element and Al elements. As shown in Figure 2.19i, when Ti_3AlC_2 MAX is mixed with liquid halogen etchants under inert gas atmosphere, the halogen ions would react with the Al layer to form halogenated aluminum salts and halogen functional groups on MXene surface, enabling the etching and delamination of MXene. Figures 2.19h and j show the XRD patterns and SEM image of $Ti_3C_2Br_x$ MXene obtained by liquid Br_2, showing good etching results. Besides, I_2, ICl and IBr are also utilized as etchants to synthesize MXenes from Ti_3AlC_2, indicating that the absolute concentration of the halogen and the solvent, as well as molar ratio of the halogen to MAX phase and temperature, are closely related to the etching effect.

$$Ti_3AlC_2 + 2.5CuBr \rightarrow Ti_3C_2Br_2 + 2.5Cu + AlBr_3\uparrow \quad (2.23)$$

$$Ti_3AlC_2 + 5CuI = Ti_3C_2I_2 + 5Cu + AlI_3\uparrow \quad (2.24)$$

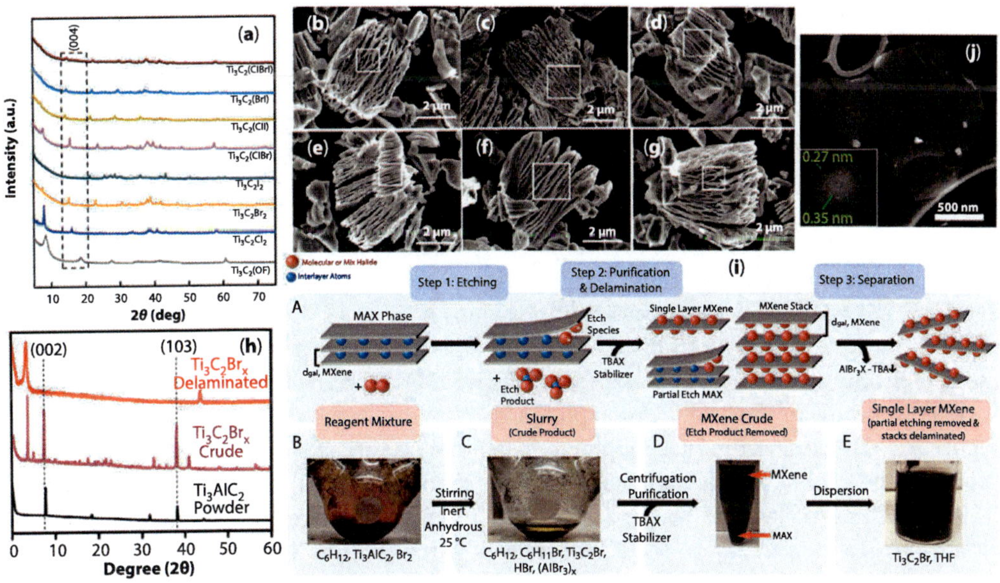

Figure 2.19 (a) XRD patterns of the MXenes obtained by different halogen salts etching routes; SEM of (b) $Ti_3C_2Cl_2$, (c) $Ti_3C_2Br_2$, (d) $Ti_3C_2(ClBr)$, (e) $T_{i3}C_2(ClI)$, (f) $Ti_3C_2(BrI)$ and (g) $Ti_3C_2(ClBrI)$ [56]. Copyright 2021. Reproduced with permission from American Chemical Society. (h) XRD patterns of Ti_3AlC_2 bulks, resulting products and $Ti_3C_2Br_x$; (i) Schematic illustration for the preparation of delaminated MXenes with halogen functional groups; (j) STEM image of $Ti_3C_2Br_x$ MXene nanosheet [57]. Copyright 2021. Reproduced with permission from American Chemical Society.

Apart from the above methods, other diverse preparation methods are being explored and gradually adapted to the needs of the application, such as thermal reduction strategy [58], UV-induced selective etching [59], surface acoustic waves that facilitated derivation of MXene [60], which further extends the performance and application range of MXene. Although the preparation methods of MXene are gradually expanding, most of them are still limited to laboratory preparation and cannot be carried out on a large scale, which is a problem that needs to be solved for the etching methods of MAX. In addition, environmentally friendly, safe, efficient and simple etching methods are also an area that researchers are currently working on.

2.9 Exfoliation Strategies of Multilayered MXene

MXene nanosheets exhibit excellent properties due to their rich functional groups, tunable layer spacing, large specific surface area and other characteristics, and are widely used. During preparing MXene by chemical etching of MAX, intercalation and exfoliation are necessary to obtain multilayer or mono/few-layer MXene nanosheets. After the completion of chemical etching, the A layer is removed from the MAX phase, resulting in an accordion-like structure MXene. However, the strong interaction between the layers makes it difficult to separate them by simple ultrasonication. Therefore, widening the spacing between layers by molecular intercalation and stopping the re-stacking of layers is also an important aspect for the preparation of MXene nanosheets.

After etching and washing to neutral, the multi-layer MXene would form a clay-like state, so a variety of intercalators are applied to intercalate MXene by analogy to clay intercalation [1]. In 2013, Mashtalir *et al.* [61] experimented with numerous compounds for the intercalation of multilayer MXene, including urea, hydrazine monohydrate $N_2H_4 \cdot H_2O$ (HM), HM dissolved in DMF and dimethyl sulphoxide (DMSO). As shown in Figure 2.20a, on the basis of stacked MXene after chemical exfoliation of A-group element, organic molecules are intercalated between the layers of MXene to weaken the van der Waals forces between the layers, followed by sonication to obtain MXene nanosheets. In XRD patterns (Figures 2.20b, c), the organic molecules mentioned above can be intercalated between MXene layers to raise the layer spacing. Thus, diverse organic molecular intercalation routes have been explored and applied to the preparation of various MXene nanosheets. However, as exploration progressed, it was found that some organic molecules did not act significantly on molecules other than $Ti_3C_2T_x$ MXene, such as DMSO. Moreover, other organic molecules such as isopropylamine have been explored and have demonstrated excellent results [62]. Figure 2.20d illustrates the successful intercalation of mulitilayer Nb_2CT_x MXene using isopropylamine to obtain a homogeneous aqueous solution after sonication, and XRD (Figure 2.20e) and TEM (Figure 2.20f) results verify the effect of delamination. In 2015, Naguib *et al.* [63] explored the large-scale preparation of MXene by means of organic intercalation in Figure 2.21a. The work compared the intercalation effect of choline hydroxide, n-butylamine and tetrabutylammonium hydroxide (TBAOH) and found that TBAOH not only displayed better intercalation effect, but also demonstrated an intercalation effect on different MXene (V_2CT_x and Ti_3CNT_x) (Figure 2.21b). Moreover, tetramethylammonium hydroxide (TMAOH) was also used as intercalator to assist the delamination of MXene nanosheets (Figure 2.21c) [17, 64]. In addition to this, liquid-phase exfoliation has been

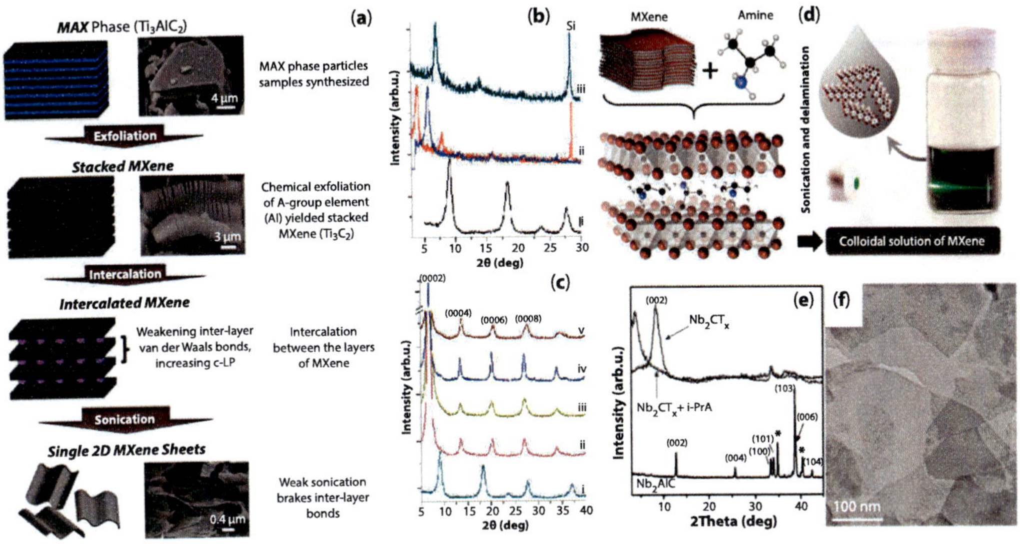

Figure 2.20 (a) Illustration for the delamination of MXene flakes; XRD patterns of f-Ti_3C_2 before and after (b) DMSO/urea treatment and (c) HM/DMF treatment [61]. Copyright 2013. Reproduced with permission from Springer Nature. (d) Schematic for isopropylamine intercalation Nb_2CT_x and delamination process; (e) XRD patterns of Nb_2AlC powder and exfoliated Nb_2CT_x; (f) TEM image of the d-Nb_2CT_x flakes [62]. Copyright 2013. Reproduced with permission from Wiley.

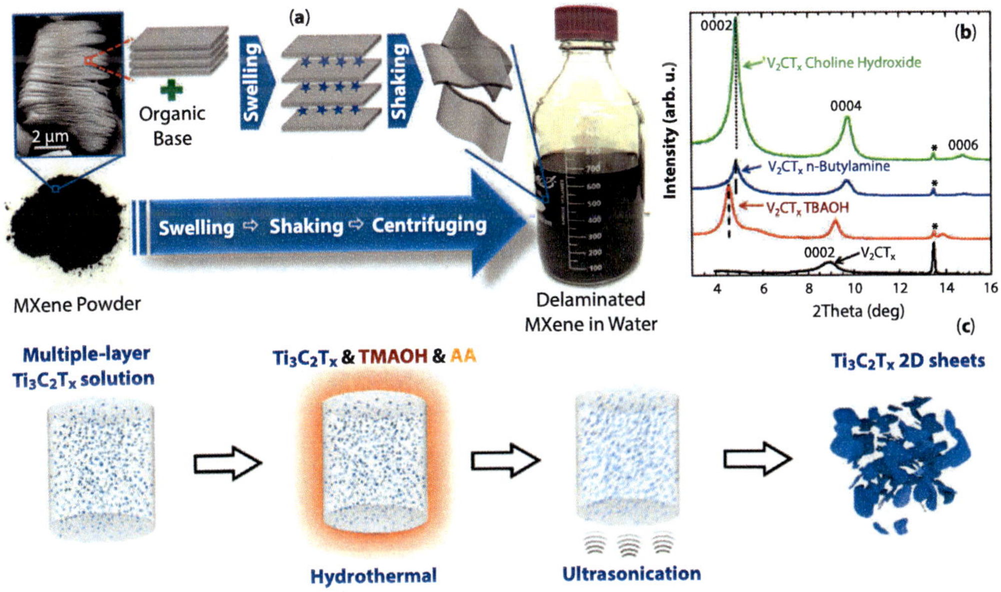

Figure 2.21 (a) XRD patterns for V_2CT_x samples before and after treatments; (b) illustration of the delamination process for MXene based on the reaction between an organic base and MXene [63]. Copyright 2015. Reproduced with permission from Royal Society of Chemistry. (c) The synthesize process of 2D $Ti_3C_2T_x$ sheets via designed HAI strategy [64]. Copyright 2019. Reproduced with permission from American Chemical Society.

used to the delamination of various 2D materials including MXene using 1-methyl-2-pyrrolidinone (NMP), 1-cyclohexyl-2-pyrrolidinone (CHP) and N,N-dimethylformamide (DMF), and two ILs, 1-butyl-3-methylimidazolium hexafluoro phosphate ($BMIMPF_6$) and 1-hexyl-3-methylimidazolium hexafluorophosphate ($HMIMPF_6$), which has led to new routes for the delamination of multilayer MXene materials [65].

Since multilayer MXene is negatively charged after etching, it has an adsorption effect on cations [66]. For example, *in-situ* HF-forming etching methods, there is the intercalation for metal ions or ammonium ions appearing in the etching solution, which promotes an increase in the space of the MXene layer. During etching of MAX using the LiF-HCl etching method, the Li^+ in the solution would spontaneously intercalate between the MXene layers,

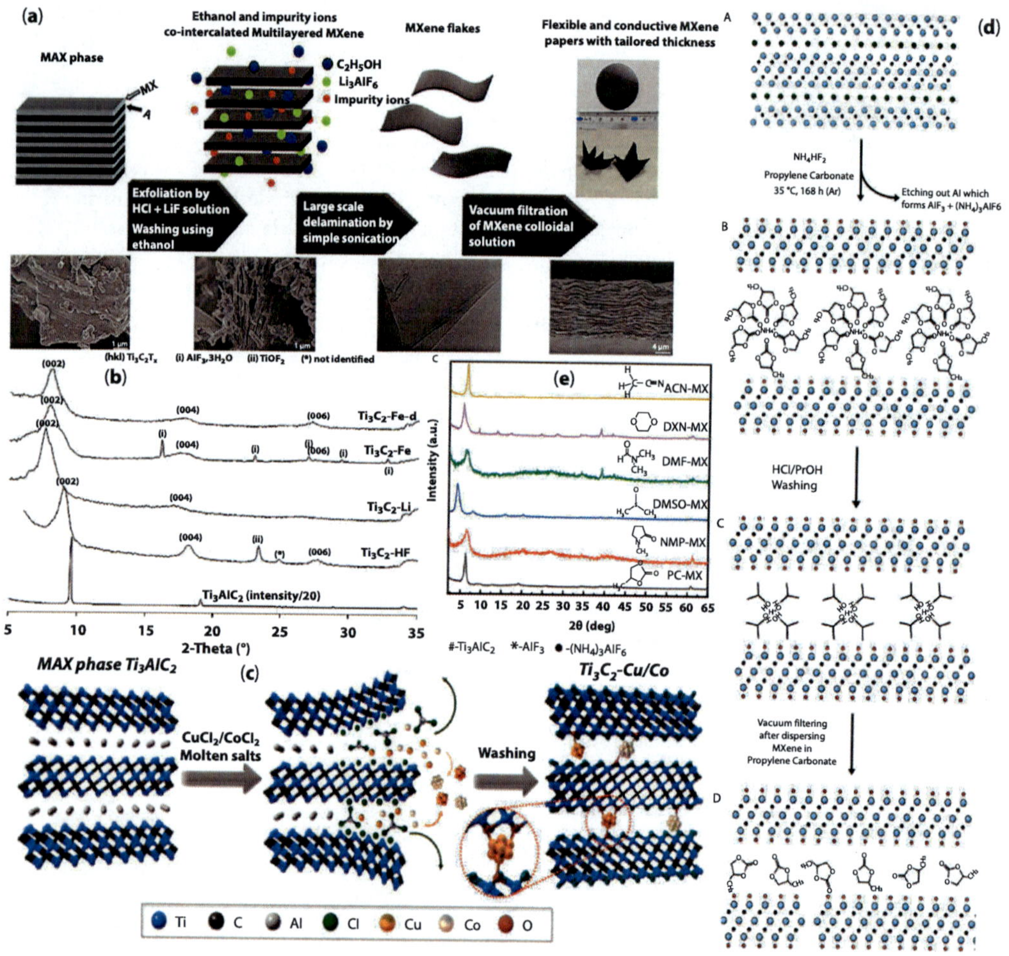

Figure 2.22 (a) Schematic of etching and delamination of MXene using LiF/HCl etching route; (b) Preparation of Ti_3C_2-Cu/Co hybrids [68]. Copyright 2021. Reproduced with permission from Wiley. (c) XRD patterns of Ti_3AlC_2 bulk, Ti_3C_2–HF, Ti_3C_2–Li, Ti_3C_2–Fe and Ti_3C_2–Fe-d [22]. Copyright 2017. Reproduced with permission from Royal Society of Chemistry. (d) Schematic of etching and washing steps using NH_4HF_2 etchants in organic solvent and (e) XRD patterns of $Ti_3C_2T_z$ obtained in organic solvents [69]. Copyright 2020. Reproduced with permission from Elsevier.

increasing the MXene layer spacing [67]. Alternatively, washing with organic solvents such as alcohol can be used as an intercalator to facilitate the exfoliation of MXene layers. Another example is the FeF_3/HCl etching route, where the presence of Fe^{3+} had a similar intercalation effect with Li^+ (Figure 2.22a) [22]. As shown in Figure 2.22c, by comparing the XRD patterns of the HF, FeF_3/HCl and LiF/HCl etching systems, it can be seen that the (002) characteristic peak of FeF_3/HCl has lower angle and shows a larger MXene layer spacing. For the molten salt etching method (Figure 2.22b), metal particles such as Cu and Co formed by catalytic reduction during the reaction would intercalate between the MXene layers and promote the delamination of MXene [68]. In addition to water as a solvent for the reaction, organic solvents are also important reaction solvents. During etching, organic molecules can be used as intercalators to facilitate MXene delamination during the etching process [69, 70], as shown in Figures 2.22d and e.

2.10 Conclusion

In general, since the discovery of MXene in 2011, there has been a gradual increase of etching methods available for MAX and a rise in the number of MXene types, but the exfoliation, intercalation and delamination of preparation methods for MXene are still awaiting in-depth investigation. And there are also many areas of the prepared MXene that need further exploration, such as stability, storage strategies and scale-up preparation methods etc. i) The varieties of structural flexibility and surface tunable chemistry of MXenes are the basis for its application. Although MXene with =O, -OH, -F and halogen terminations can already be prepared, but the properties of MXene obtained by different etching methods have not been studied enough. ii) MXene can be obtained as a homogeneous aqueous solution, but has a short storage life and is prone to oxidation. Organic solutions are an effective way for storing MXene but many MXenes are not well dispersed in it. iii) At present, The scaleable preparation methods available are also only a few tens of grams and cannot be used industrially. And green, safe and efficient preparation methods are also a major challenge currently. Evidently, the preparation of MXene still has a long way to go and further exploration is needed to achieve its better development and application.

Acknowledgement

This work was supported by the National Natural Science Foundation of China (No. 22005277 and 52074247), the Young Top-notch Talent Cultivation Program of Hubei Province, Experimental Technology Research Project of China University of Geosciences (Wuhan) (No. SJ202118).

References

1. Wei, Y., Zhang, P., Soomro, R.A., Zhu, Q., Xu, B., Advances in the synthesis of 2D MXenes. *Adv. Mater.*, 33, 39, 2103148, 2021.

2. Gong, K., Zhou, K., Qian, X., Shi, C., Yu, B., MXene as emerging nanofillers for high-performance polymer composites: A review. *Compos. Part B: Eng.*, 217, 108867, 2021.
3. Maleski, K. and Alhabeb, M., *Top-down MXene synthesis (selective etching)*, pp. 69–87, Springer International Publishing, Cham, 2019.
4. Naguib, M., Kurtoglu, M., Presser, V., Lu, J., Niu, J., Heon, M., Hultman, L., Gogotsi, Y., Barsoum, M.W., Two-dimensional nanocrystals produced by exfoliation of Ti_3AlC_2. *Adv. Mater.*, 23, 37, 4248–4253, 2011.
5. Naguib, M., Mashtalir, O., Carle, J., Presser, V., Lu, J., Hultman, L., Gogotsi, Y., Barsoum, M.W., Two-dimensional transition metal carbides. *ACS Nano*, 6, 2, 1322–1331, 2012.
6. Naguib, M., Halim, J., Lu, J., Cook, K.M., Hultman, L., Gogotsi, Y., Barsoum, M.W., New two-dimensional niobium and vanadium carbides as promising materials for Li-ion batteries. *J. Am. Chem. Soc.*, 135, 43, 15966–15969, 2013.
7. Zhou, J., Zha, X., Chen, F.Y., Ye, Q., Eklund, P., Du, S., Huang, Q., A two-dimensional zirconium carbide by selective etching of Al_3C_3 from nanolaminated $Zr_3Al_3C_5$. *Angew. Chem. Int. Ed.*, 55, 16, 5008–5013, 2016.
8. Tran, M.H., Schäfer, T., Shahraei, A., Dürrschnabel, M., Molina-Luna, L., Kramm, U.I., Birkel, C.S., Adding a new member to the MXene family: Synthesis, structure, and electrocatalytic activity for the hydrogen evolution reaction of $V_4C_3T_x$. *ACS Appl. Energy Mater.*, 1, 8, 3908–3914, 2018.
9. Soundiraraju, B. and George, B.K., Two-dimensional titanium nitride (Ti_2N) MXene: Synthesis, characterization, and potential application as surface-enhanced raman scattering substrate. *ACS Nano*, 11, 9, 8892–8900, 2017.
10. Ghidiu, M., Naguib, M., Shi, C., Mashtalir, O., Pan, L.M., Zhang, B., Yang, J., Gogotsi, Y., Billinge, S.J.L., Barsoum, M.W., Synthesis and characterization of two-dimensional Nb_4C_3 (MXene). *Chem. Commun.*, 50, 67, 9517–9520, 2014.
11. Zhou, J., Zha, X., Zhou, X., Chen, F., Gao, G., Wang, S., Shen, C., Chen, T., Zhi, C., Eklund, P., Du, S., Xue, J., Shi, W., Chai, Z., Huang, Q., Synthesis and electrochemical properties of two-dimensional hafnium carbide. *ACS Nano*, 11, 4, 3841–3850, 2017.
12. Anasori, B., Xie, Y., Beidaghi, M., Lu, J., Hosler, B.C., Hultman, L., Kent, P.R.C., Gogotsi, Y., Barsoum, M.W., Two-dimensional, ordered, double transition metals carbides (MXenes). *ACS Nano*, 9, 10, 9507–9516, 2015.
13. Li, G., Natu, V., Shi, T., Barsoum, M.W., Titova, L.V., Two-dimensional MXenes $Mo_2Ti_2C_3T_z$ and $Mo_2TiC_2T_z$: Microscopic conductivity and dynamics of photoexcited carriers. *ACS Appl. Energy Mater.*, 3, 2, 1530–1539, 2020.
14. Naguib, M., Barsoum, M.W., Gogotsi, Y., Ten years of progress in the synthesis and development of MXenes. *Adv. Mater.*, 33, 39, 2103393, 2021.
15. Mashtalir, O., Naguib, M., Dyatkin, B., Gogotsi, Y., Barsoum, M.W., Kinetics of aluminum extraction from Ti_3AlC_2 in hydrofluoric acid. *Mater. Chem. Phys.*, 139, 1, 147–152, 2013.
16. Ying, Y., Liu, Y., Wang, X., Mao, Y., Cao, W., Hu, P., Peng, X., Two-dimensional titanium carbide for efficiently reductive removal of highly toxic chromium(VI) from water. *ACS Appl. Mater. Interfaces*, 7, 3, 1795–1803, 2015.
17. Alhabeb, M., Maleski, K., Anasori, B., Lelyukh, P., Clark, L., Sin, S., Gogotsi, Y., Guidelines for synthesis and processing of two-dimensional titanium carbide ($Ti_3C_2T_x$) MXene. *Chem. Mater.*, 29, 18, 7633–7644, 2017.
18. Sang, X., Xie, Y., Lin, M.-W., Alhabeb, M., Van Aken, K.L., Gogotsi, Y., Kent, P.R.C., Xiao, K., Unocic, R.R., Atomic defects in monolayer titanium carbide ($Ti_3C_2T_x$) MXene. *ACS Nano*, 10, 10, 9193–9200, 2016.

19. Ghidiu, M., Lukatskaya, M.R., Zhao, M.-Q., Gogotsi, Y., Barsoum, M.W., Conductive two-dimensional titanium carbide 'clay' with high volumetric capacitance. *Nature*, 516, 7529, 78–81, 2014.
20. Lipatov, A., Alhabeb, M., Lukatskaya, M.R., Boson, A., Gogotsi, Y., Sinitskii, A., Effect of synthesis on quality, electronic properties and environmental stability of individual monolayer Ti_3C_2 MXene flakes. *Adv. Electron. Mater.*, 2, 12, 1600255, 2016.
21. Wang, C., Xu, J., Wang, Y., Song, Y., Guo, J., Huang, W., Ge, Y., Hu, L., Liu, J., Zhang, H., MXene (Ti_2NT_x): Synthesis, characteristics and application as a thermo-optical switcher for all-optical wavelength tuning laser. *Sci. China Mater.*, 64, 1, 259–265, 2021.
22. Wang, X., Garnero, C., Rochard, G., Magne, D., Morisset, S., Hurand, S., Chartier, P., Rousseau, J., Cabioc'H, T., Coutanceau, C., Mauchamp, V., Célérier, S., A new etching environment (FeF_3/HCl) for the synthesis of two-dimensional titanium carbide MXenes: A route towards selective reactivity vs. water. *J. Mater. Chem. A*, 5, 41, 22012–22023, 2017.
23. Zou, X., Liu, H., Xu, H., Wu, X., Han, X., Kang, J., Reddy, K.M., A simple approach to synthesis Cr_2CT_x MXene for efficient hydrogen evolution reaction. *Mater. Today Energy*, 20, 100668, 2021.
24. Liu, F., Zhou, A., Chen, J., Jia, J., Zhou, W., Wang, L., Hu, Q., Preparation of Ti_3C_2 and Ti_2C MXenes by fluoride salts etching and methane adsorptive properties. *Appl. Surf. Sci.*, 416, 781–789, 2017.
25. Guo, M., Geng, W.-C., Liu, C., Gu, J., Zhang, Z., Tang, Y., Ultrahigh areal capacitance of flexible MXene electrodes: Electrostatic and steric effects of terminations. *Chem. Mater.*, 32, 19, 8257–8265, 2020.
26. Wang, H.-W., Naguib, M., Page, K., Wesolowski, D.J., Gogotsi, Y., Resolving the structure of $Ti_3C_2T_x$ MXenes through multilevel structural modeling of the atomic pair distribution function. *Chem. Mater.*, 28, 1, 349–359, 2016.
27. Halim, J., Lukatskaya, M.R., Cook, K.M., Lu, J., Smith, C.R., Näslund, L.-Å., May, S.J., Hultman, L., Gogotsi, Y., Eklund, P., Barsoum, M.W., Transparent conductive two-dimensional titanium carbide epitaxial thin films. *Chem. Mater.*, 26, 7, 2374–2381, 2014.
28. Feng, A., Yu, Y., Mi, L., Yu, Y., Song, L., Comparative study on electrosorptive behavior of NH_4HF_2-etched Ti_3C_2 and HF-etched Ti_3C_2 for capacitive deionization. *Ionics*, 25, 2, 727–735, 2019.
29. Feng, A., Yu, Y., Jiang, F., Wang, Y., Mi, L., Yu, Y., Song, L., Fabrication and thermal stability of NH_4HF_2-Etched Ti_3C_2 MXene. *Ceram. Int.*, 43, 8, 6322–6328, 2017.
30. Feng, A., Yu, Y., Wang, Y., Jiang, F., Yu, Y., Mi, L., Song, L., Two-dimensional MXene Ti_3C_2 produced by exfoliation of Ti_3AlC_2. *Mater. Des.*, 114, 161–166, 2017.
31. Abdelmalak, M.N., *MXenes: A new family of two-dimensional materials and its application as electrodes for li-ion batteries*, p. 176, Drexel University, Ann Arbor, 2014.
32. Cheng, Y., Lyu, W., Wang, Z., Ouyang, H., Zhang, A., Sun, J., Yang, T., Fu, B., He, B., MXenes: Synthesis, incorporation, and applications in ultrafast lasers. *Nanotechnology*, 32, 39, 392003, 2021.
33. Naguib, M., Presser, V., Tallman, D., Lu, J., Hultman, L., Gogotsi, Y., Barsoum, M.W., On the topotactic transformation of Ti_2AlC into a Ti-C-O-F cubic phase by heating in molten lithium fluoride in air. *J. Am. Ceram. Soc.*, 94, 12, 4556–4561, 2011.
34. Urbankowski, P., Anasori, B., Makaryan, T., Er, D., Kota, S., Walsh, P.L., Zhao, M., Shenoy, V.B., Barsoum, M.W., Gogotsi, Y., Synthesis of two-dimensional titanium nitride Ti_4N_3 (MXene). *Nanoscale*, 8, 22, 11385–11391, 2016.
35. Yan, M., Yang, L., Li, C., Zou, Y., Preparation of two-dimensional Ti_2CT_x by molten fluorinated salt method. *J. Wuhan Univ. Technol.-Mater. Sci. Ed.*, 34, 2, 299–302, 2019.

36. Djire, A., Zhang, H., Liu, J., Miller, E.M., Neale, N.R., Electrocatalytic and optoelectronic characteristics of the two-dimensional titanium nitride Ti4N3Tx MXene. *ACS Appl. Mater. Interfaces*, 11, 12, 11812–11823, 2019.
37. Li, Y., Shao, H., Lin, Z., Lu, J., Liu, L., Duployer, B., Persson, P.O.Å., Eklund, P., Hultman, L., Li, M., Chen, K., Zha, X.-H., Du, S., Rozier, P., Chai, Z., Raymundo-Piñero, E., Taberna, P.-L., Simon, P., Huang, Q., A general lewis acidic etching route for preparing MXenes with enhanced electrochemical performance in non-aqueous electrolyte. *Nat. Mater.*, 19, 8, 894–899, 2020.
38. Xiu, L.-Y., Wang, Z.-Y., Qiu, J.-S., General synthesis of MXene by green etching chemistry of fluoride-free lewis acidic melts. *Rare Met.*, 39, 11, 1237–1238, 2020.
39. Guo, L., Jiang, W.-Y., Shen, M., Xu, C., Ding, C.-X., Zhao, S.-F., Yuan, T.-T., Wang, C.-Y., Zhang, X.-Q., Wang, J.-Q., High capacitance of MXene ($Ti_3C_2T_x$) through intercalation and surface modification in molten salt. *Electrochim. Acta*, 401, 139476, 2022.
40. Sun, W., Shah, S.A., Chen, Y., Tan, Z., Gao, H., Habib, T., Radovic, M., Green, M.J., Electrochemical etching of Ti_2AlC to Ti_2CT_x (MXene) in low-concentration hydrochloric acid solution. *J. Mater. Chem. A*, 5, 41, 21663–21668, 2017.
41. Yang, S., Zhang, P., Wang, F., Ricciardulli, A.G., Lohe, M.R., Blom, P.W.M., Feng, X., Fluoride-free synthesis of two-dimensional titanium carbide (MXene) using a binary aqueous system. *Angew. Chem. Int. Ed.*, 57, 47, 15491–15495, 2018.
42. Sun, Z., Yuan, M., Lin, L., Yang, H., Nan, C., Li, H., Sun, G., Yang, X., Selective lithiation–expansion–microexplosion synthesis of two-dimensional fluoride-free Mxene. *ACS Mater. Lett.*, 1, 6, 628–632, 2019.
43. Song, M., Pang, S.Y., Guo, F., Wong, M.C., Hao, J., Fluoride-free 2D niobium carbide MXenes as stable and biocompatible nanoplatforms for electrochemical biosensors with ultrahigh sensitivity. *Adv. Sci.*, 7, 24, 2001546, 2020.
44. Chen, J., Chen, M., Zhou, W., Xu, X., Liu, B., Zhang, W., Wong, C., Simplified synthesis of fluoride-free $Ti_3C_2T_x$ via electrochemical etching toward high-performance electrochemical capacitors. *ACS Nano*, 16, 2, 2461–2470, 2022.
45. Wang, L., Zhang, H., Wang, B., Shen, C., Zhang, C., Hu, Q., Zhou, A., Liu, B., Synthesis and electrochemical performance of Ti3C2Tx with hydrothermal process. *Electron. Mater. Lett.*, 12, 5, 702–710, 2016.
46. Peng, C., Wei, P., Chen, X., Zhang, Y., Zhu, F., Cao, Y., Wang, H., Yu, H., Peng, F., A hydrothermal etching route to synthesis of 2D MXene (Ti_3C_2, Nb_2C): Enhanced exfoliation and improved adsorption performance. *Ceram. Int.*, 44, 15, 18886–18893, 2018.
47. Wang, L., Liu, D., Lian, W., Hu, Q., Liu, X., Zhou, A., The preparation of V_2CT_x by facile hydrothermal-assisted etching processing and its performance in lithium-ion battery. *J. Mater. Res. Technol.*, 9, 1, 984–993, 2020.
48. He, H., Wang, J., Xia, Q., Wang, L., Hu, Q., Zhou, A., Effect of electrolyte on supercapacitor performance of two-dimensional molybdenum carbide (Mo_2CT_x) MXene prepared by hydrothermal etching. *Appl. Surf. Sci.*, 568, 150971, 2021.
49. Guo, Y., Jin, S., Wang, L., He, P., Hu, Q., Fan, L.-Z., Zhou, A., Synthesis of two-dimensional carbide Mo_2CT_x MXene by hydrothermal etching with fluorides and its thermal stability. *Ceram. Int.*, 46, 11, 19550–19556, 2020.
50. Wang, C., Shou, H., Chen, S., Wei, S., Lin, Y., Zhang, P., Liu, Z., Zhu, K., Guo, X., Wu, X., Ajayan, P.M., Song, L., HCl-based hydrothermal etching strategy toward fluoride-free MXenes. *Adv. Mater.*, 33, 27, 2101015, 2021.
51. Li, G., Tan, L., Zhang, Y., Wu, B., Li, L., Highly efficiently delaminated single-layered MXene nanosheets with large lateral size. *Langmuir*, 33, 36, 9000–9006, 2017.

52. Li, T., Yao, L., Liu, Q., Gu, J., Luo, R., Li, J., Yan, X., Wang, W., Liu, P., Chen, B., Zhang, W., Abbas, W., Naz, R., Zhang, D., Fluorine-free synthesis of high-purity $Ti_3C_2T_x$ (T=OH, O) via alkali treatment. *Angew. Chem. Int. Ed.*, 57, 21, 6115–6119, 2018.
53. Zhang, B., Zhu, J., Shi, P., Wu, W., Wang, F., Fluoride-free synthesis and microstructure evolution of novel two-dimensional $Ti_3C_2(OH)_2$ nanoribbons as high-performance anode materials for lithium-ion batteries. *Ceram. Int.*, 45, 7, 8395–8405, 2019.
54. Husmann, S., Budak, Ö., Shim, H., Liang, K., Aslan, M., Kruth, A., Quade, A., Naguib, M., Presser, V., Ionic liquid-based synthesis of MXene. *Chem. Commun.*, 56, 75, 11082–11085, 2020.
55. Lu, B., Hu, S., Wu, D., Wu, C., Zhu, Z., Hu, L., Zhang, J., Ionic liquid exfoliated $Ti_3C_2T_x$ MXene nanosheets for photoacoustic imaging and synergistic photothermal/chemotherapy of cancer. *J. Mater. Chem. B*, 10, 8, 1226–1235, 2022.
56. Li, M., Li, X., Qin, G., Luo, K., Lu, J., Li, Y., Liang, G., Huang, Z., Zhou, J., Hultman, L., Eklund, P., Persson, P.O.Å., Du, S., Chai, Z., Zhi, C., Huang, Q., Halogenated Ti_3C_2 MXenes with electrochemically active terminals for high-performance zinc ion batteries. *ACS Nano*, 15, 1, 1077–1085, 2021.
57. Jawaid, A., Hassan, A., Neher, G., Nepal, D., Pachter, R., Joshua Kennedy, W., Ramakrishnan, S., Vaia, R.A., Halogen etch of Ti_3AlC_2 MAX phase for MXene fabrication. *ACS Nano*, 15, 2, 2771–2777, 2021.
58. Mei, J., Ayoko, G.A., Hu, C., Sun, Z., Thermal reduction of sulfur-containing MAX phase for MXene production. *Chem. Eng. J.*, 395, 125111, 2020.
59. Mei, J., Ayoko, G.A., Hu, C., Bell, J.M., Sun, Z., Two-dimensional fluorine-free mesoporous Mo_2C MXene via UV-induced selective etching of Mo_2Ga_2C for energy storage. *Sustain. Mater. Technol.*, 25, e156, 2020.
60. El Ghazaly, A., Ahmed, H., Rezk, A.R., Halim, J., Persson, P.O.Å., Yeo, L.Y., Rosen, J., Ultrafast, one-step, salt-solution-based acoustic synthesis of Ti_3C_2 MXene. *ACS Nano*, 15, 3, 4287–4293, 2021.
61. Mashtalir, O., Naguib, M., Mochalin, V.N., Agnese, Y.D., Heon, M., Barsoum, M.W., Gogotsi, Y., Intercalation and delamination of layered carbides and carbonitrides. *Nat. Commun.*, 4, 1, 1716, 2013.
62. Mashtalir, O., Lukatskaya, M.R., Zhao, M.-Q., Barsoum, M.W., Gogotsi, Y., Amine-assisted delamination of Nb_2C MXene for li-ion energy storage devices. *Adv. Mater.*, 27, 23, 3501–3506, 2015.
63. Naguib, M., Unocic, R.R., Armstrong, B.L., Nanda, J., Large-scale delamination of multi-layers transition metal carbides and carbonitrides "MXenes". *Dalton Trans.*, 44, 20, 9353–9358, 2015.
64. Han, F., Luo, S., Xie, L., Zhu, J., Wei, W., Chen, X., Liu, F., Chen, W., Zhao, J., Dong, L., Yu, K., Zeng, X., Rao, F., Wang, L., Huang, Y., Boosting the yield of MXene 2D sheets via a facile hydrothermal-assisted intercalation. *ACS Appl. Mater. Interfaces*, 11, 8, 8443–8452, 2019.
65. Wu, W., Xu, J., Tang, X., Xie, P., Liu, X., Xu, J., Zhou, H., Zhang, D., Fan, T., Two-dimensional nanosheets by rapid and efficient microwave exfoliation of layered materials. *Chem. Mater.*, 30, 17, 5932–5940, 2018.
66. Lukatskaya, M.R., Mashtalir, O., Ren, C.E., Dall'Agnese, Y., Rozier, P., Taberna, P.L., Naguib, M., Simon, P., Barsoum, M.W., Gogotsi, Y., Cation intercalation and high volumetric capacitance of two-dimensional titanium carbide. *Science*, 341, 6153, 1502–1505, 2013.
67. Zhang, T., Pan, L., Tang, H., Du, F., Guo, Y., Qiu, T., Yang, J., Synthesis of two-dimensional $Ti_3C_2T_x$ MXene using HCl+LiF etchant: Enhanced exfoliation and delamination. *J. Alloys Compd.*, 695, 818–826, 2017.

68. Bai, Y., Liu, C., Chen, T., Li, W., Zheng, S., Pi, Y., Luo, Y., Pang, H., MXene-copper/cobalt hybrids via lewis acidic molten salts etching for high performance symmetric supercapacitors. *Angew. Chem. Int. Ed.*, 60, 48, 25318–25322, 2021.
69. Natu, V., Pai, R., Sokol, M., Carey, M., Kalra, V., Barsoum, M.W., 2D $Ti_3C_2T_z$ MXene synthesized by water-free etching of Ti_3AlC_2 in polar organic solvents. *Chem*, 6, 3, 616–630, 2020.
70. Zhao, X., Radovic, M., Green, M.J., Synthesizing MXene nanosheets by water-free etching. *Chem*, 6, 3, 544–546, 2020.

3

Surface Terminations and Surface Functionalization Strategies of MXenes

Lekshmi A. G., Rejithamol. R., Santhy A., Akhila Raman, Asok Aparna and Appukuttan Saritha*

Department of Chemistry, Amrita Vishwa Vidyapeetham, Amritapuri, Kollam, Kerala, India

Abstract

MXenes, a novel group of two-dimensional (2D) transition metal nitrides and carbides, have attained substantial attention in research field. The exceptional physicochemical characteristics and diverse chemical compositions of MXenes make them attractive materials for manifold applications; catalysis, energy conversion and storage, biomedicine, sensing etc. Modifications in MXenes via surface termination, surface functionalization, and interlayer engineering can colossally influence the properties of MXenes by increasing the number of electrochemical reactive sites and enhancement in the electronic structure. The surface engineering of MXene surfaces using imido group, halogen, selenium, tellurium, oxygen, sulphur etc. have shown noteworthy electronic and configurational properties. Over the last few years, various surface functionalization and termination strategies were developed to broaden the functional applications of MXenes in various fields. This chapter throws light on various surface termination and surface functionalization strategies involved in developing modified MXenes and their employment in diverse fields.

Keywords: MXenes, surface functionalization, surface terminations, etching, catalysis, antibacterial, biomedicine, energy storage

3.1 Introduction

Over the last few years, 2D nanomaterials including transition metal dichalcogenides (TMDs) [1], graphene [2], black phosphorus nanosheets [3], MXenes [4] etc. have attained remarkable recognition because of their appealing physical and chemical characteristics and distinctive structure. They have functional applications in different fields like energy storage [5], biomedicine [6], sensing [7], catalysis [8] etc. Specifically, MXenes have been accepted worldwide as they are potentially and economically applicable in the above-mentioned fields. MXene was initially synthesized via selective etching of aluminum (Al) utilizing hydrofluoric acid and was formulated as Ti_3C_2 [9]. Generally, MXenes are designated as $M_{n+1}X_nT_x$ (n = 1–4), with M, X, and T are transition metals, carbon or nitrogen, and

*Corresponding author: sarithatvla@gmail.com

surface functional groups sequentially. Number of functional moieties present in MXenes are represented by the integer x [10].

Even though, there are many proposed compositions for MXenes, only a few of them have been fabricated. The general chemical formula Mn+1AXn describes the hexagonally stacked structure of the reactants (MAX phase) that give rise to MXenes. They are formed through covalent bonding between M and X atoms and the A atom's layers are alternately placed between M–X layers. M and A atoms possess metallic bonds between them. Moreover, M–X bonds are stronger compared to M–A bonds [11]. The existence of various functional groups like fluoride (-F), chloride (-Cl), hydroxyl (-OH) etc. on MXene surfaces enables their functionalization on surfaces and thereby, increases the adaptability of MXene materials [12]. Apart from that, their exceptional biocompatibility, make them potentially applicable in biomedical fields [13].

MXenes can be synthesized by various methods including bottom-up methods like salt-templated growth [14] and chemical vapor deposition [15], top-down methods such as selective etching [16] of non-MAX and MAX phases and chemical transformation [17] methods. Nanostructured MXenes with both single layer and multilayers can be developed utilizing these methods. Although MXenes own several attractive properties, poor mechanical flexibility, ease of restacking, low stability in the vicinity of molecular oxygen, poor water dispersibility limits the application of pure MXenes in various industries. Therefore, surface modifications and functionalizations can significantly improve their stability and performance. The surface modifications on MXenes can be either non-covalent or covalent in nature [18]. The covalent type of surface functionalization can be of three types; i) involving surface-initiated polymerization via macromolecules [19, 20], ii) involving small molecules like acid halides, alkali metal hydroxides, organic amine and epoxy compounds [21, 22], iii) involving single heteroatoms [23]. The non-covalent functionalization includes hydrogen bonding, electrostatic attraction and van der Waals forces [24]. This chapter includes various surface termination and modification strategies for MXenes and the prospective significances of modified MXenes in different fields including energy storage and conversion, biomedicine, catalysis, sensing etc.

3.2 Surface Termination Strategies in MXenes

MXene's properties are greatly governed by changing the surface chemistry of these materials and analysis of the surface properties of MXene is an interesting area of research. MXene formation involves the etching of the element A (main group elements) from the MAX phases where M stands for the transition metals, and X is N or C [10]. During the etching process, the surface of the MXenes may be terminated with different groups such as =O, -F, -Cl, -I, =S, -NH, -OH etc. [25] and these groups are commonly designated as T_x. Thus, the surface terminated MXenes are entirely different in structure and properties compared to the parent MXene. Literature studies established that changing the surface groups in the termination process alters the optical, electronic and magnetic properties of MXenes which is rarely seen in the unmodified MXene [26]. Different strategies are involved in the surface termination properties of MXenes such as hydrofluoric acid-based etching, alkali-based etching, electrochemically assisted etching, molten salt etching etc.

3.2.1 Hydrofluoric Acid-Based Etching

HF-based etching is the most common method adopted for the synthesis of MXenes for various applications including, hydrogen storage, sensors, catalysis etc. [27, 28]. The surface termination of MAX phase by hydrofluoric acid can be done in direct HF etching and *in-situ* HF etching method. The direct HF etching method involves the exfoliation of the MAX phase by treatment with HF directly to form surface terminated MXenes [27]. The *in-situ* method of HF etching is performed by treating the MAX phase by a fluoride salt in HCl solution [28]. The bonding in the M-A layers of MAX phase is weaker and can be easily removed by etching with fluoride solutions to form 2-dimensional MXenes [29]. A 2-dimensional TiVC alloy or solid solution MXene was synthesized via HF based etching and their surface properties was studied by Sanaz Yazdanparast and co-workers in 2020 [30]. The $TiVCT_x$ MXene was prepared from the TiVAlC MAX phase according to previously reported literature [28, 31]. The MAX phase was treated with a solution of LiF in HCl for the etching process. The resultant $TiVCT_x$ MXene was delaminated using tetrabutylammonium hydroxide solution to form a colloidal suspension which was further made into sheets by drop casting over a glass plate. The graphic representation of the etching and delamination process is given in Figure 3.1 [30].

The results were also compared by synthesizing MXene with direct etching by treating with HF for the same MAX phase. The nature of the etchant also performs a crucial part in the properties of the MXenes synthesized. The MXenes synthesized by both methods showed terminations of =O and -F according to the elemental analysis. But the $TiVCT_x$ formed by HF showed termination with -Cl also. According to density functional theory (DFT) calculations, the MXenes with =O terminations were more stable compared to other terminations. The elemental studies proved that percentage of =O termination increased,

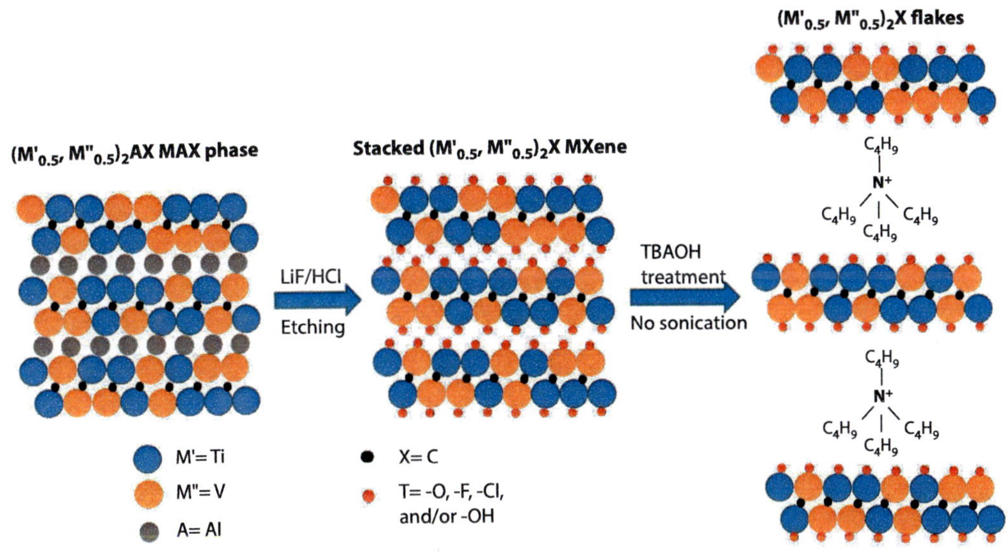

Figure 3.1 The pictorial illustration of the TiVAlC MAX phase etching and delamination processes via HF method [30]. Copyright 2020, Reproduced with permission from the American Chemical Society.

and -F terminations decreased in the case of using LiF in HCl as the etchant compared to HF alone. Thus, the etching with LiF in HCl method was found to be a better choice for the surface termination of $TiVCT_x$ and also it could avoid toxicity of HF as etchant in the termination process [32].

In other research work, Halim *et al.* described the synthesis of titanium carbide MXene thin film by HF based etching [33]. The Ti_3AlC_2 MAX phase was surface terminated with -OH and -F groups by two different processes, with HF and NH_4HF_2 as the etchants respectively and the corresponding MXenes obtained may be represented as $Ti_3C_2T_x$ and $Ti_3C_2T_y$. The elemental analysis of both MXenes confirmed the etching process and the termination of the surface with -OH and -F groups. However, intercalated MXenes are formed in the case of NH_4HF_2 with NH_3 and NH_4^+ as intercalating species. The optical properties such as transmittance were compared for the parent MAX phase, $Ti_3C_2T_x$ and $Ti_3C_2T_y$. The results showed that the transmittance is higher for, $Ti_3C_2T_x$ and least for Ti_3AlC_2. Moreover, the electrical resistivity of three species were studied and confirmed that resistivity was in the order of $Ti_3C_2T_y > Ti_3C_2T_x > Ti_3AlC_2$. The higher resistivity of $Ti_3C_2T_y$ may be due to the intercalation process and this agreed with the previously reported work [34]. Both $Ti_3C_2T_y$ and $Ti_3C_2T_x$ showed metallic character from 300 K to 100 K and the resistivity increased with increase in temperature below 100 K. Hence the synthesized MXenes may be a good candidate for a range of applications including electrochemical sensors, optoelectronics, photonics etc.

3.2.2 Molten Salt Etching

Molten salt etching provides an alternative route for the surface termination of MXenes without using the hazardous HF in the etching process. Also, this process requires fewer exposure to the chemicals during the reaction and not perilous as in that of HF based etching [35]. In this process, molten salts of transition metals are used for the selective etching of element A from the MAX phase. Literature studies revealed that numerous works reported the synthesis of carbide MXenes [36, 37], whereas only a few works reported the synthesis of nitride MXenes. In 2016, Urbankowsk *et al.* synthesized the first nitride MXene, $Ti_4N_3T_x$ using molten salt etching process [38]. To synthesize $Ti_4N_3T_x$, Ti_4AlN_3 was heated with the mixture of molten fluoride salts of KF, LiF, and NaF at 500°C in the inert atmosphere. After this reaction, Al was selectively etched from the MAX phase and the product was washed with tetrabutylammonium hydroxide solution to form delaminated $Ti_4N_3T_x$, where Tx = -OH, F and =O. The etching process was confirmed by SEM, XRD, EDX, Raman spectra etc. DFT calculations proved that $Ti_4N_3T_x$ terminated with =O is more energetically feasible than other terminations and a reduction in the magnetic properties was found to be in the case of terminated MXene. $Ti_4N_3T_x$ may be a potential scaffold for various electrochemical applications than analogous carbides [39].

Recently, a transition metal carbide was synthesized by molten salt etching method by Liu *et al.* [40]. The Ti_3AlC_2 was mixed with $CuCl_2$ and NaCl–KCl mixture and heated at 680°C to form a molten state. The Al in the MAX phase gets oxidized and simultaneously the Cu^{2+} gets reduced and deposited as metallic Cu on the surface of MXene. The Al is thus removed from the MXene and the formed Ti_3C_2 again reacted with the molten mixture containing $CuCl_2$ to form $Ti_3C_2Cl_2$. Further Cu is eliminated by the treatment with ammonium persulphate. Thus, the MXenes gets terminated with -Cl group rather than -F or -OH

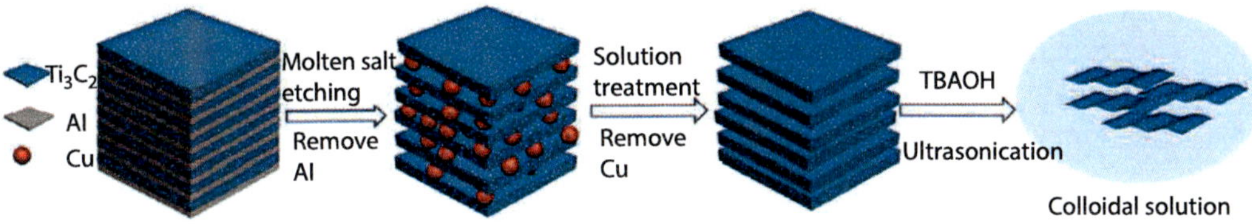

Figure 3.2 Diagrammatic depiction of $Ti_3C_2Cl_2$ synthesis [40]. Copyright 2022, Reproduced with permission from the American Chemical Society.

terminations. Further the obtained MXene gets delaminated with ammonium solutions to form flakes of MXenes as colloidal suspensions. The schematic representation of synthesis of $Ti_3C_2Cl_2$ is given in Figure 3.2 [40].

3.2.3 Alkali-Based Etching

Literature survey revealed that only a few reports have been published based on etching of MXenes with alkali compared to other methods. The use of alkali etchants for MXenes are theoretically possible as the elemental Al in the MAX phase can form strong bond with alkali [41]. However, the reaction of alkali with MXenes may result in the formation of hydroxides on the MXene surface [42]. This could be one of the limitations of alkali based etching process. The synthesis of $Ti_3C_2T_x$ has been reported by Li *et al.* by etching of Ti_3AlC_2 with NaOH. This method selectively etched Al from the MAX phase to form -OH and =O terminated MXene. The above hydrothermal etching process produced multi-layered MXenes with fluorine free terminations and high purity. In a different work, Li *et al.* synthesized a single layered MXene with alkali treatment for the delamination of Ti_3AlC_2 [43]. Al in the MAX phase is effectively removed by etching with KOH and water as the etching medium to form $Ti_3C_2(OH)_2$. The graphical representation of the etching process followed by exfoliation of MXene is given in the Figure 3.3 [43]. Restacking of the delaminated $Ti_3C_2(OH)_2$

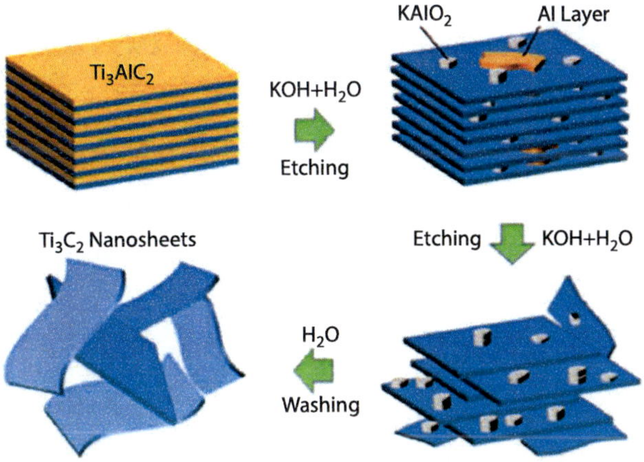

Figure 3.3 A symbolic representation of alkali-based etching for Ti_3C_2 nanosheets synthesis [43]. Copyright 2017, Reproduced with permission from the American Chemical Society.

to form layered nanosheets was confirmed by XRD analysis. Also, this restacked MXene showed easy delamination again when dispersed in water for more than a few hours with shaking. This property combined with its metallic conductivity and hydrophilicity makes it an excellent material for a wide variety of applications.

3.2.4 Electrochemically-Assisted Etching

Electrochemical etching are another type of etching process for MXenes where the etching results are not significantly affected by the etchant used for the process [44]. Yin *et al.* synthesized a titanium carbide MXene with fluoride terminations by the electrochemical etching method for the battery applications [45]. The fluorinated MXenes are good candidates for cathodes in high energy lithium batteries and provides much storage capacity for lithium ions in battery applications [46]. The synthesis of MXene was done in an electrochemical device consisting of three electrode system, in a non-aqueous medium with ionic liquid containing fluorine as the electrolyte. Under inert conditions, a constant potential of approximately 7V vs. Ag wire was applied to the working electrode (Ti_3AlC_2). Pt wire was the counter electrode used for this process. The graphical representation of the etching process is given in Figure 3.4 [45]. The Al atoms in the MXenes were exfoliated during this process and the resultant product was centrifuged and dried to get pure Ti_3C_2F. An environmental friendly electrochemical etching process with molten salts was reported for the preparation of MXenes in 2021 [47]. This process generated a chloride terminated MXene, $Ti_3C_2Cl_2$. The terminated groups could be easily tailored with other groups such as =O, -OH etc. and this method provides a route for the synthesis of a vast diversity of MXenes with different surface groups. $Ti_3C_2T_x$, especially the oxygen terminated MXene was found to be a fascinating material for electrodes for supercapacitors compared to the MXenes generated by other etching procedures.

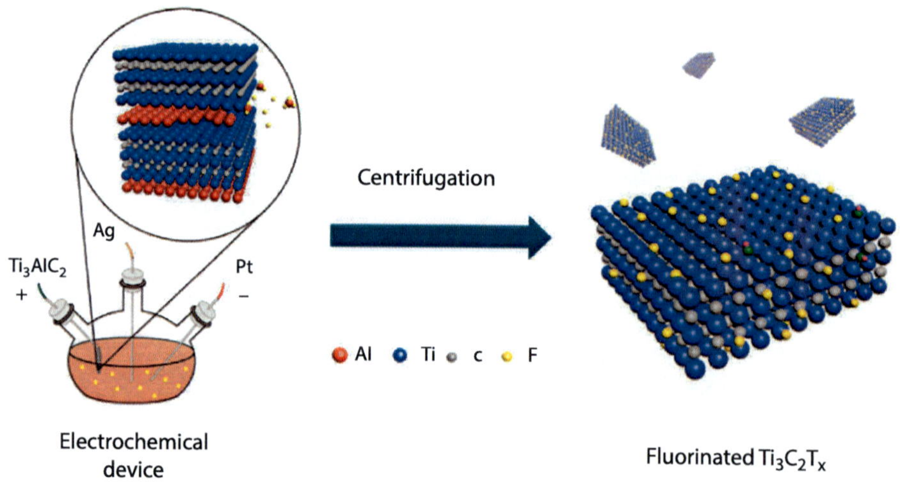

Figure 3.4 The illustrative overview of Ti_3AlC_2 electrochemical etching [45]. Copyright 2021, Reproduced with permission from Elsevier.

3.2.5 Manipulation of Terminations: Surface Modification and Doping in MXenes

MXenes are notable for their adaptable surface terminations (e.g., F, O, OH). It has been found in numerous studies that the composition of the surface termination has a major control on MXene's electronic and electrochemical properties, making them attractive candidates for batteries, supercapacitors, hybrid capacitors etc. [48]. MXenes' intrinsic physical, chemical, and mechanical properties, including work functions, electronic band structures, and hydrogen evolution reactivity are clearly influenced by surface functional terminations. MXenes show wide range of applications because of their excellent electrical conductivity, unique optical response and other properties. However, various problems have hampered the development of pure MXenes, including inescapable restacking, low flexibility, high contact resistance and poor oxygen stability. Elemental doping procedures have been manifested to be successful ways to optimize virgin MXene's properties including electrical, magnetic, and optical properties, as motivated by various doping tactics in graphene. In general, doped materials have a very low heteroatom content (less than 1%), whereas substituted materials have a greater heteroatom content. Due to the incorporated heteroatoms, substituted and doped MXenes can be considered as hetero-MXenes. During the production of MXenes, defects may arise, allowing for elemental doping and substitution. In general, multiple elements can be substituted and doped for both M and X elements of MXenes during the MAX phase or after exfoliation, producing stoichiometric or non-stoichiometric MXenes with distinct characteristics. Theoretical computations of hetero MXenes typically focuses on three MXene locations (M, X, and T), which are referred to as M-doped, X-doped, and T-substituted MXenes, respectively [49]. According to Ling *et al.*, $TiVCO_2$, an M-doped MXene show stronger performance in hydrogen evolution reaction (HER) performance than Ti_2CO_2 and W_2CO_2 [50]. The electronic properties of X-doped Sc_2CF_2 are explored by substitution of C atoms with B, F, Si, Sn, Ge, S, N, and B+N [51]. Experimentally substituting C atoms by S, N, F, and B+N is favorable based on calculated total energies, however doping of Sn, Si or Ge element is less favorable. Yang and colleagues studied the stability and electrical characteristics of 20 different M_2XS_2 compounds (M = Sc, Ti, V, Cr, Zr, Nb, Mo, Hf, Ta, and W; X = C and N) [52]. V_2CS_2, Cr_2NS_2, Nb_2CS_2, Hf_2CS_2, and Ta_2CS_2 are projected to be the most stable structures based on molecular dynamic simulations and phonon frequency analyses. Hetero MXenes have diversity of applications like energy storage in supercapacitors, rechargeable batteries, as electrocatalysts, optoelectronic devices, sensors, catalysts, seawater desalination etc.

3.3 Methods of Surface Functionalization in MXenes

The characteristics of chemically modified MXenes, and their applicability in purification, healthcare, sensory, battery storage, chemical transformation, and catalytic reactivity is increasingly noticed by the research community. This chapter aims to provide substantial innovative ideas to the readers based on the comparison of published works on different MXene surface functionalization techniques (Figure 3.5). Surface-active initiators, small molecules, and polymers can be utilized to control surface terminations of MXenes. The copious hydroxyl groups and oxygen on the MXenes surface act as reactive centers for

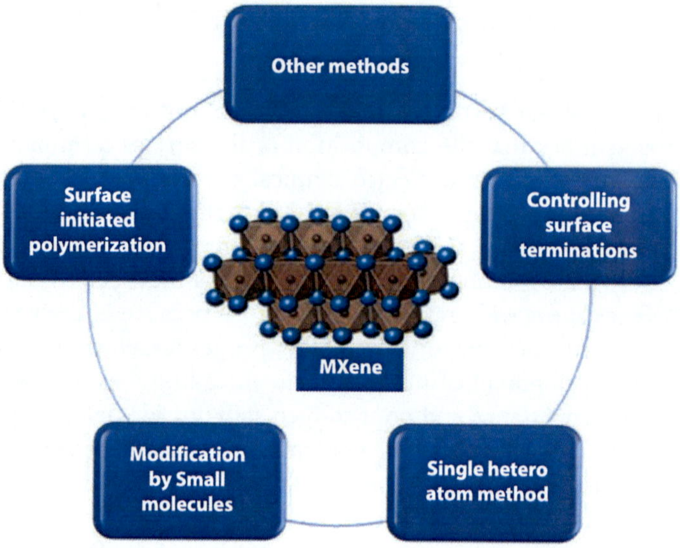

Figure 3.5 Surface functionalization methods of MXene.

covalent bonding [53–57]. MXene properties can be easily enhanced by adding appropriate functional groups due to their hydrophilic nature and hence chemical surface functionalization of MXene has been extensively studied [18].

3.3.1 Controlling Surface Terminations

The characteristic features of MXenes can be adjusted by manipulating the T groups on their surfaces. Traditional etching methods, such as HF or salts containing fluoride, can easily create –O and –F surface terminations. More crucially, the types and concentrations of moieties can be customized by varying the parameters of etching, as well as employing several post-treatments including in hydrazine and annealing in varied environments [59, 60].

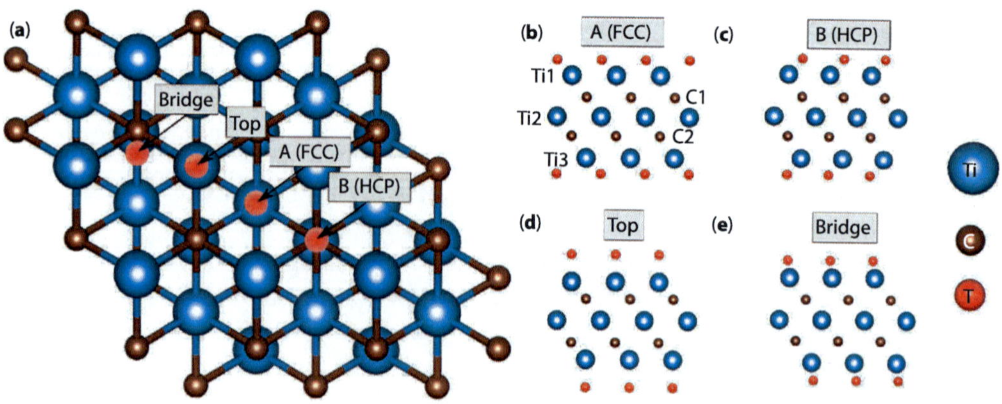

Figure 3.6 Diagrammatic depiction of various adsorption sites in Ti3C2Tx chosen for the DFT calculations [58]. Copyright 2019, Reproduced with permission from the American Chemical Society.

The physical and chemical characteristics of surface controlled Mxenes and their applications were studied by different groups. Schultz and co-workers reported the influence of surface modifications on the electronic conductivity of Ti_3C_2 using density functional theory (DFT) [58]. Different Ti3C2Tx adsorption sites that were taken into account for the DFT calculations in this work are shown schematically in Figure 3.6 [58]. Ti_3C_2 has a high dispersion due to the existence of bonds between Titanium and Oxygen and subjecting to an annealing temperature of 3800°C which improves the work function due to the removal of water, OH species and impurities. But higher temperatures, from 500 to 750°C reduced the work function due to removal of fluorine [18].

3.3.2 Single Heteroatom Method

The introduction of single hetero atoms such as P, N, and S, into the interlayer space of MXenes, can improve their electrochemical characteristics. Significant improvements in the electrochemical performance of MXenes have been achieved by doping MXenes with heteroatoms [62–65]. There are reports available on the enhanced electrochemical properties of MXenes using single heteroatoms [59]. Liu *et al.* studied the doping of N on Nb_2C MXene produced by hydrothermal method [60]. The morphology of Nb_2C is maintained after N-doping, indicating the consistent distribution of O, Nb and C elements. SEM image

Figure 3.7 SEM image of the (a) Nb2AlC, (b) Nb2CTx MXene, (c) N doped Nb2CTx [60]. Copyright 2019, Reproduced with permission from Elsevier.

of the Nb2AlC, Nb2CTx MXene and N doped Nb2CTx are depicted in Figure 3.7 [60]. Later, P-doped, and N-doped V_2C were synthesized by Yoon and co-workers via heat treatment [61]. Large interlayer spacing (1.03 nm) was attained for p-doped V_2C via triphenylphosphine (TPP) heating. The N-doped V_2C was developed by annealing in the presence of ammonia, produced a uniform distribution of all the elements. To enhance the performance of MXenes, there were reports in which several other atoms such as Ge, Si, B, Pt, Pd, and other cations like NH_4^+, Al^{3+}, Mg^{2+}, Ca^{2+}, and K^+ doped into MXenes [67–69].

3.3.3 Small Molecules

Surface functionalization of MXenes using a range of small compounds, which are economical and ease to prepare, can improve performance attributes such as solution stability, electrical properties, and mechanical stability [23, 70]. Xia *et al.* illustrated the introduction of hexaethylene glycol monododecyl ether ($C_{12}E_6$), a non-ionic surfactant on the surface of Ti_3C_2 MXenes ($C_{12}E_6$@Ti_3C_2) to improve molecular interactions [62]. Lei *et al.* reported a single step reaction of Ti_3C_2 with diazonium salt to obtain a sulfonic functionalized MXene (Ti_3C_2–SO_3H). The sulfonic functionalized MXene showed granular structures rather than the lamellar layered structures of the pristine Ti_3C_2 [63]. Jin *et al.* reported the development of electromagnetic wave (EMW) absorbing performance on the surface of the Nb_2C MXenes by a solvothermal method [64]. The use of ethanol produced superior EMW performance for Nb_2C MXenes than deionized water because of amplified interlayer spacing and increased surface functional groups making them better dielectric loss.

Hexaethylene glycol monododecyl ether as a non-ionic surfactant was used to improve the molecular interactions of MXene nanolayers by the research group of Yang *et al.* [65]. There are reports of surface functionalization of MXenes utilizing other small molecules like bismuth ferrite ($BiFeO_3$) [75, 76] and sodium aluminum hydride ($NaAlH_4$) [66]. The incorporation of biomolecules on the surface of amine modified MXenes can be obtained by chemical coupling method (Figure 3.8) [67]. Wang and co-workers devised a way for large-scale MXene delamination, a method for the expansion of the interlayer spacing by

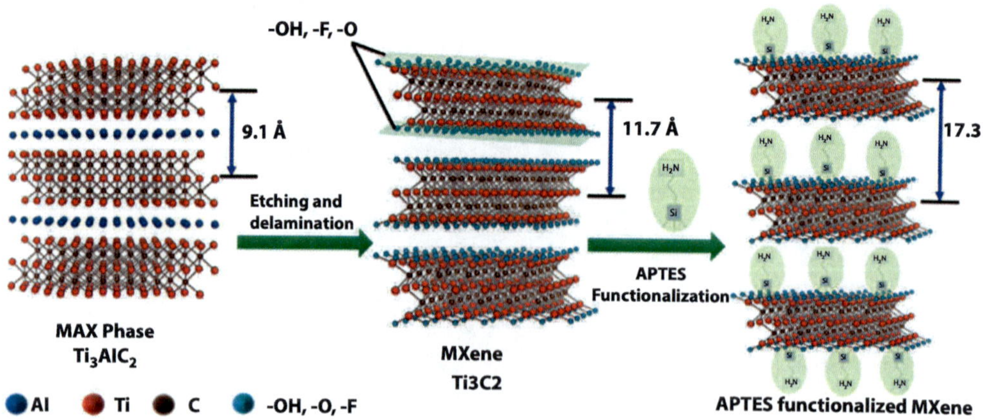

Figure 3.8 Functionalization of Ti3C2-MXene using aminosilane is shown schematically [67]. Copyright 2018, Reproduced with permission from Elsevier.

weakening the M-X bond of MXenes [21]. This was formulated by introducing aryl-surface linkage using diazonium and sulphanilic salts. The covalent bonding of $C_6H_4SO_3H$ moieties on Ti_3C_2 improves the electrical properties. The surface of MXene can easily form covalent bond with crosslinkers such as trimesoyl chloride, epichlorohydrin, and silane agents via the free hydroxyl functional groups on them.

3.3.4 Surface-Initiated Polymerization

In situ or *ex situ* blended polymerization on the Mxene surface is highly appreciable due to the availability of numerous active functional groups on MXene surface [19, 79–85]. To facilitate the efficient interaction between polymers and MXenes, *ex situ* blending approach can be carried out which involves hydrogen bonding [68] and electrostatic interactions [87, 88]. Sun *et al.* introduced compression molding to produce nanocomposites of MXene with polystyrene to get highly conductive MXene polystyrene (Ti3C2Tx@PS) nanocomposites [69]. $Ti_3C_2T_x$@PS nanocomposites are shown in a schematic throughout their creation is shown in Figure 3.9 [69]. The active groups present on the surface of MXene allow them to make strong electrostatic interactions, weak van der Waals forces, and hydrogen bonding. It has been reported that electrostatic interactions occur at smaller contents of Ti_3C_2 MXenes, whereas hydrogen bonds and van der Waals forces favor higher contents.

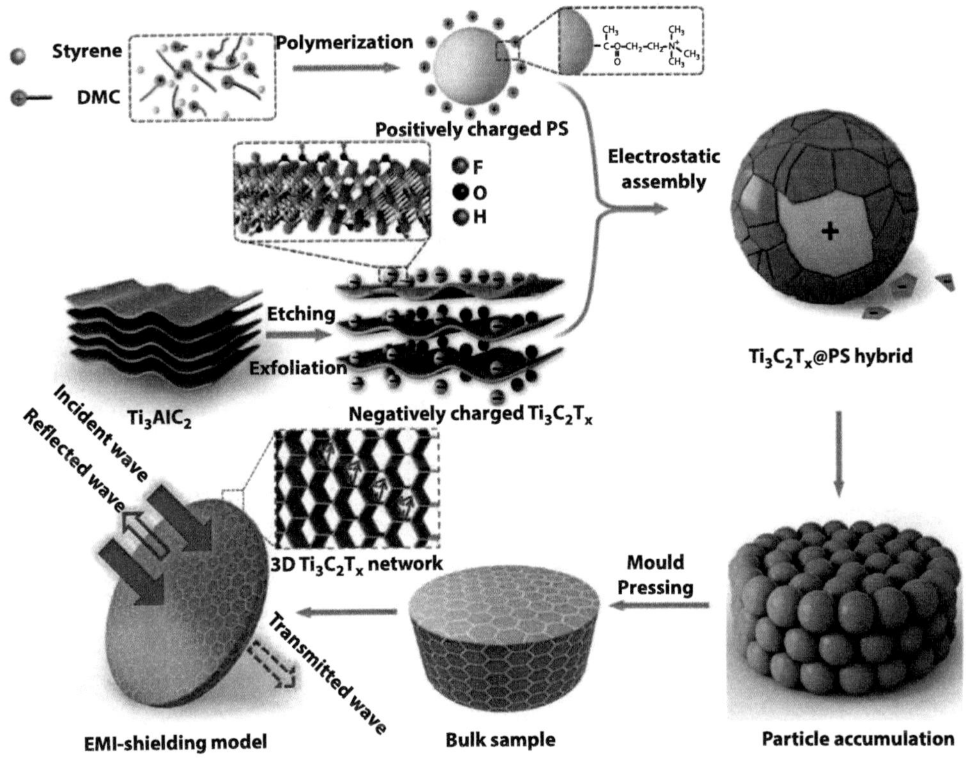

Figure 3.9 Diagrammatic depiction of Ti3C2Tx@PS nanocomposites synthesis [69]. Copyright 2017, Reproduced with permission from Wiley-VCH.

Ling *et al.* discussed the preparation of nanocomposites of Ti_3C_2 with polyvinyl alcohol (PVA) and poly-diallyl dimethylammonium (PDDA) by vacuum-assisted filtration [70]. A strong electrostatic interaction arises between PDDA/Ti_3C_2 composite and PVA/Ti_3C_2 composite can be obtained by hydrogen bonding. Boota *et al.* reported *in situ* polymerization with monomers like caprolactone and pyrrole by strong covalent bonds between polymers and MXenes. Hydrogen bonding between the oxygen atoms on the Ti_3C_2 and nitrogen and hydrogen atoms of the intercalated polypyrrole (PPy) chains stabilize the PPy/Ti_3C_2 composites [19].

Lin and co-workers detailed the surface functionalization of Ti_3C_2 by biocompatible soybean phospholipid (SP), which was reported to improve tumor blood flow and biodistribution in a physiological environment [71]. In addition, the use of poly (ethylene glycol) (PEG) was observed to improve the stability of MXene in saline [72]. Polydopamine (PDA) is another useful drug for modifying the surface of MXene due to its rich catechol and amino moieties in the backbone and can create non-shared and shared interactions with inorganic and organic surfaces [73]. The adsorption of cobalt on Ti_3C_2 surface forming CNT in the PDA layer leads to the generation of Ti_3C_2 @ carbon nanotube (CNT) composite [74]. Shamsabadi *et al.* [75] reported the amino silane modification of Ti_3C_2 by the incorporation of TiO_2 nanoparticles and the grafting of 2-aminoethylamino propyl trimethoxysilane (AEAPTMS) onto the surface of Ti_3C_2 by a chemical reaction between the -OH group of Ti_3C_2 and AEAPTMS. The terminal of the amine group causes an electrostatic interaction, and the terminal of the silanol group is a bond to a covalent bond. MXene materials that have been amino functionalized have a wide range of potential outcomes for future reactions. The presence of free amine groups that can act as initiators for atom transfer radical polymerization and epoxy hardeners. This makes MXene materials apt for the sensing of biomolecules, fabrication of energy storage devices and fuel cells, and act as catalysts and dye adsorbents [76–80].

3.3.5 Other Methods

Self-assembled monolayers (SAMs) are obtained by the covalent bonding of hydroxyl groups to diverse metal and oxide particles [81, 82]. The main application of superhydrophobic SAM on 2D materials makes them used for diffusion barriers, antibacterial coatings, water purification or corrosion inhibitors [83]. Because the surface of MXene is completely coated with surface functional groups e.g., -O and OH, molecular engineering of its surface using SAMs could be a simple and effective way to reduce its environmental deterioration. Stanciu *et al.* developed a surface functionalization approach for $Ti_3C_2T_x$ using fluoroalkyl silane (FOTS) molecules, resulting in a superhydrophobic surface, with improved mechanical and environmental stability, for ethanol and acetone sensing applications. The fabricated system showed high sensitivity, selectivity, repeatability, and fast response time with long-term stability. The significant energy of adsorption of ethanol on $Ti_3C_2T_x$- F and the local structural distortion generated by adsorption of ethanol, was calculated by density functional theory simulations. The developed $Ti_3C_2T_x$-F sensor contributed a concentration range for ethanol from 5 to 80 percent relative humidity range [84].

Catechols are used to functionalize exfoliated MXenes ($Ti_3C_2T_x$) in a colloidal suspension by James *et al.* [85]. Catechols react spontaneously with $Ti_3C_2T_x$ surfaces, with binding originating from a charge-transfer complex, as established by UV-vis and DFT. X-ray diffraction (XRD) measurements show that catechol functionalization increases $Ti_3C_2T_x$ sheet interlayer spacing, while atomic force microscopy-infrared spectroscopy (AFM-IR) and Raman spectral data show that catechols bind to the $Ti_3C_2T_x$ surface via metal-oxygen bonds, confirmed by DFT measurements. The group provides a method for modifying MXene surfaces through aqueous functionalization using catechols, in which colloidal stability may be altered and additional functionality can be added, potentially providing ideal anchoring places for growth [85].

Pazniak *et al.* [86] showed how low-energy ion implantation may be used to modify two-dimensional MXenes, resulting in the insertion of various ions in $Ti_3C_2T_x$. They reported the implantation with manganese ions to modify Ti3C2Tx MXenes. Recent theoretical predictions on manipulating MXene magnetic characteristics via strain, doping, or defect engineering piqued the curiosity of the spintronics and quantum information fields, prompting the study. In comparison to the C atoms in the MXene layer, ion implantation causes preferred Ti atom sputtering. This type of sputtering creates new crystallographic sites for functionalization groups (oxygen), resulting in a significant rise in the O content. Spectroscopic studies, combining high resolution and high resolution, are used to study the structural changes and damage caused by ion implantation.

Shenoy *et al.* [87] introduce the influence of chemical dopants on the surface of MXenes to increase the electron density on them, enabling controllable magnetic order. Electrostatic doping and external electric fields can be used to tune the exchange mechanisms of MXenes with chemical dopants. The research group provides a strategy for exploiting functionalization of MXenes to achieve magnetism under ambient conditions. The introduction of chemical dopants such as Cr_2C, Mn_2N, V_2C, and Ti_2C to MXene was found to be robustly antiferromagnetic.

3.4 Application of Surface Modified MXenes

3.4.1 Energy Generation and Storage

MXenes on account of their copious active sites, high conductivity, adjustable surface chemistry and layered structure, attract significant attention in research on Li-S batteries. Li-S batteries suffer short life span because of factors like:

a. In the cell, the kinetics of sulphur redox is limited by the sulphur's insulative nature and Li_2 Swhich is the final discharge product.
b. The solubility of LiPSs, the intermediate lithium polysulphides in the electrolyte based on ether, aids them to migrate from the cathode to the metal anode due to the electric field and concentration gradient. It results in the rapid decay of reversible capacity and loss of active sulphur.
c. Sulphur gets reduced to Li_2S during the discharge process. This results in volume enhancement of 80%, that harm the stability of the electrode, which in turn results in the collapsing of the whole structure and a significantly cause a sharp reduction in reversible capacity.

Many studies have been conducted in this field to address these issues. The results concluded that for a room temperature Li–S battery, an ideal Sulphur host material require to be the one with sufficient physical confinement, high electronic conductivity and the ability to upgrade slack kinetics of sulphur redox reactions and to efficaciously hamper the shuttling effect of LiPSs intermediate [88, 89]. It was found that MXenes are competitive candidates for the host because of its graphene-like 2D structure, metal metallic conductivity and strong interaction with Li_2S_n (1 <_ n <_8) [90]. Surface functionalization of MXenes can overcome the poor propensity and capacity of restacking of bare MXenes, which limits their applicability in storage of energy and its conversion [83].

The first application of MXenes as a sulphur host was done by Zhao *et al.* They prepared layered Ti_3C_2 (L-Ti_3C_2) with accordion-like structure by HF etching method. It was utilized as a sulphur host [91]. The S/L-Ti_3C_2 composite displayed excellent cycle performance and high capacity, with a sulfur content of 57.6% due to its reduced resistivity (0.05 Ω m) and accordion-like structure. Xu *et al* prepared a small sulfur/MXene electrode which was flexible, without binder and conductive agent, where the backbone and conductive network was 2D MXene with metallic conductivity [92]. The self-assembly process resulted in small sulphur molecules to be fully in contact with the MXene, thereby exhibiting good electrochemical performances and outstanding flexibility of bendable Li–S cells. MXenes also promote the nucleation of Li_2S [93].

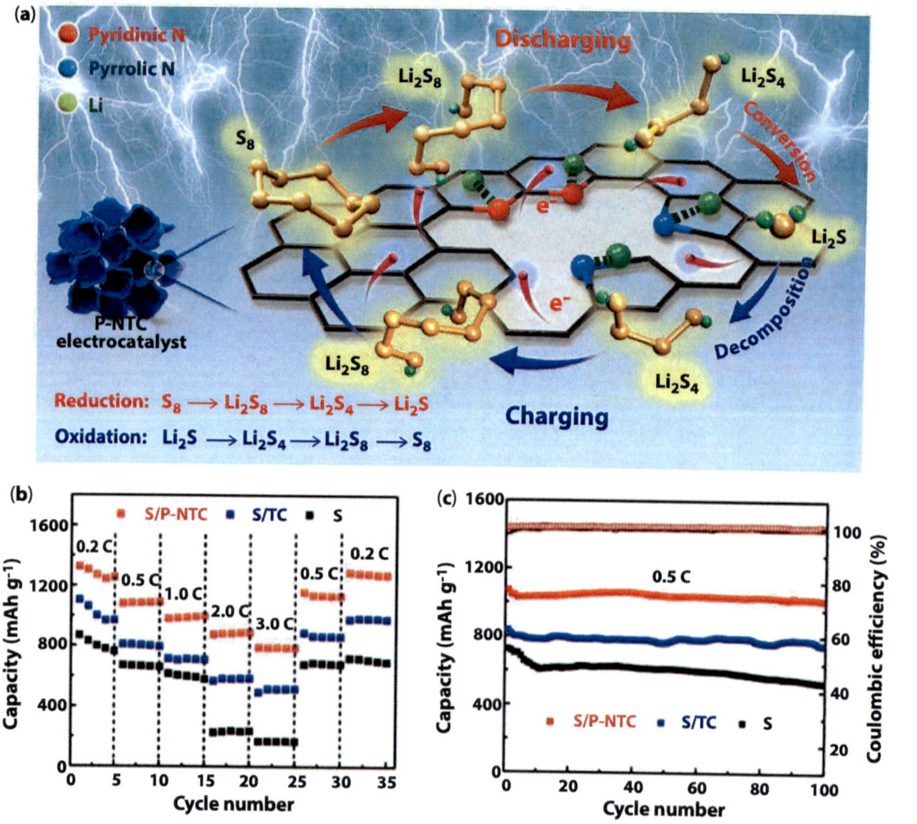

Figure 3.10 (a) The graphical representation of the impact of N-doped Ti_3C_2 in the Li-S cell (b) N-doped Ti_3C_2, pristine Ti_3C_2 and bare S cathode performances. (c) S/N-doped Ti_3C_2's cycle performance at 0.5 °C [94]. Copyright 2020, Reproduced with permission from Elsevier.

N-doped Ti_3C_2 is a highly efficacious electrocatalyst. The process by which porous N-doped Ti_3C_2 MXenes convert and capture lithium polysulfides (LiPSs) has been depicted in Figure 3.10a [94]. As MXenes have ample surface area, porous structure and high electrical conductivity, several active sites are available for LiPSs adsorption as well as conversion. On comparison between S-porous N-doped Ti_3C_2 cathode with S/Ti_3C_2 cathode and pristine S cathode, the former exhibited higher capacities (Figure 3.10b) [94]. Figure 3.10c shows the coulombic efficiency and initial capacity, where S-porous N-doped Ti_3C_2 MXenes displayed exceptional performance at 0.5 C when compared to those of S/Ti_3C_2 and pristine S cathode [94].

MXenes are widely used as electrochemical capacitors owing to their excellent intercalation effects, superior electrical conductivity and electrochemical energy storage characteristics [95, 96]. The performance of a capacitive device can be significantly impacted by surface modification. A N-doped Ti_2C MXene was created by C3N4 breakdown followed by intercalation, exfoliation and heat treatment [97]. Upon analyzing the capacitance, it was found that for the original Ti_2C and N-doped Ti_2C when the temperature rises, the surface area of the cyclic voltammetry (CV) curve also rises (Figure 3.11a) [97]. The scan rate has a nearly linear connection with the N-doped Ti_2C's discharge current densities, showing that the whole surface of the electrode was subjected to quick electrochemical reaction

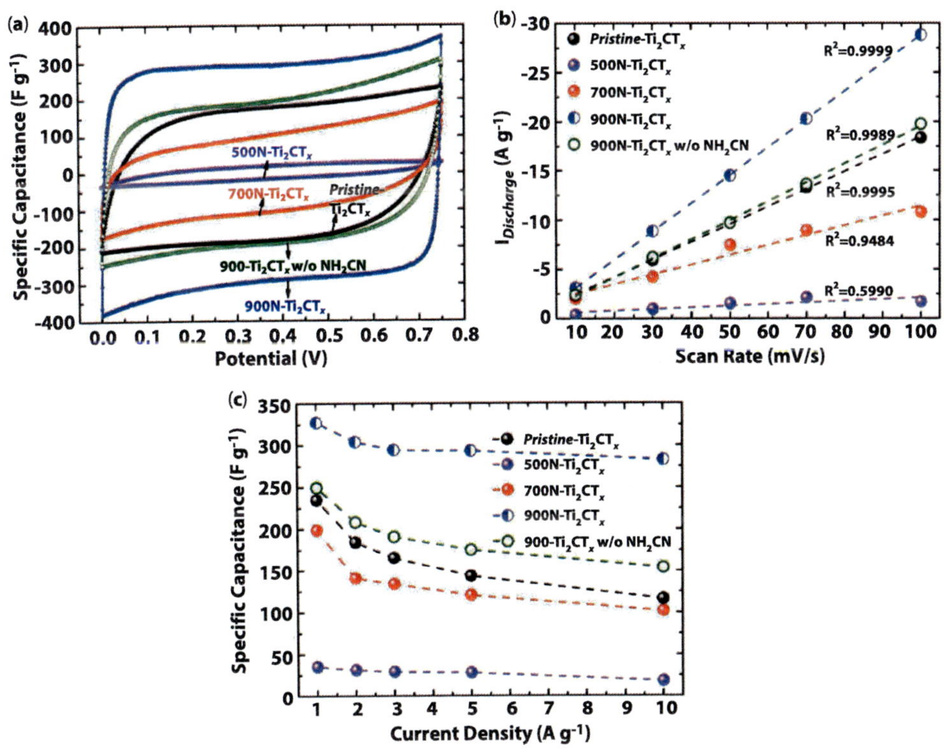

Figure 3.11 CV of pristine Ti_3C_2, 900N-Ti_3C_2, 700N-Ti_3C_2, 500N-Ti_3C_2, and 900N-Ti_3C_2 without NH_2CN electrodes in 6M KOH electrolyte; (a) Cyclic voltammetric graphs at 100 mVS^{-1}. (b) Discharge current densities v/s scan rate. (c) The plot of gravimetric specific capacitance v/s current densities [97]. Copyright 2018, Reproduced with permission from Wiley-VCH.

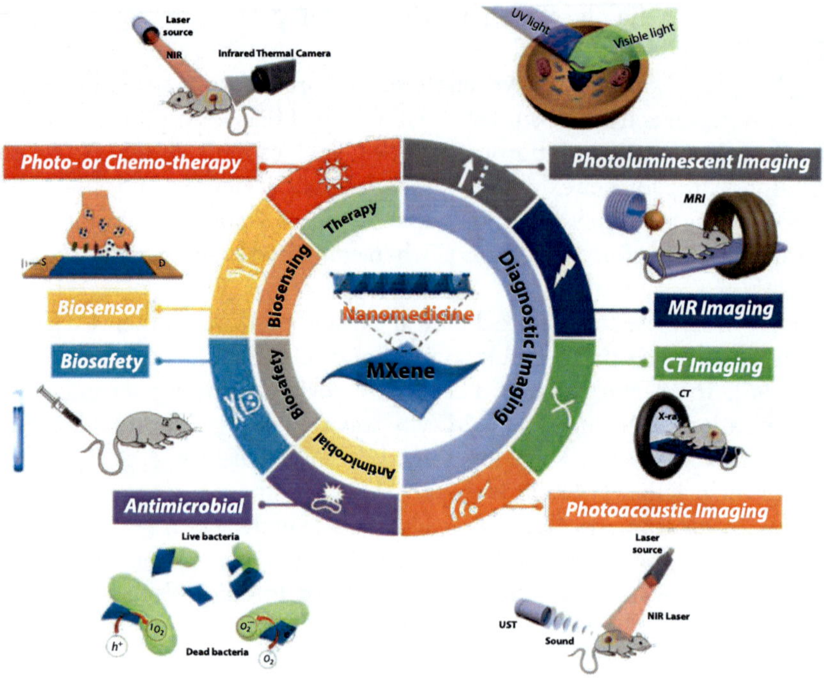

Figure 3.12 An overview of applications of advancing 2D MXenes in nanomedicines Adapted from Ref. [102]. Copyright 2018, John Wiley & Sons.

(Figure 3.11b) [97]. The N-doped Ti_2C electrode has the maximum specific capacitance at 900 °C, 327 ± 1Fg $^{-1}$, which is greater than that of original Ti_2C (Figure 3.11c) [97].

Chen and co-workers examined the theoretical and experimental effect of functionalization of $Ti_3C_2O_2$ MXenes by glycine on storage of charges as organic molecules can hinder restacking by enhancing interlayer space [65]. The robust chemical reaction between $Ti_3C_2O_2$ MXenes and glycine was due to the bond formed between titanium and oxygen via sharing of electrons. The interlayer spacing of bare $Ti_3C_2O_2$ was augmented from 10.16 to 14.06 Å by surface functionalization which led to improved charging rate, enhanced cycling performance and better ion availability. An improvement of the rate capability and specific capacity was observed when the surface of MXene was modified by TiO_2 [114–121]. When -O terminations enhance the specific capacity, -OH and -F terminations should be avoided as they lead to a relatively lower capacity. Moreover, it was also found that MXene materials have the tendency to accommodate a variety of cations e.g., Na+ as they have large surface area and high electrical conductivity [98]. Hence, Wang *et al.* published that the Ti_2CT_x MXene can be used as an anode in Na-ion hybrid capacitors [99]. To promote the integrated capacity, hierarchical MXenes with 3D porous structure can be designed so that it will enhance the mass loading of sulfur in cathode so that the integrated capacity can be greatly promoted. Besides separators and cathodes, MXenes can improve the properties of lithium metal anodes for lithium–sulfur batteries by suppressing the growth of lithium dendrites [100].

3.4.2 Biomedicine

Surface functionalized MXenes show substantial performances in biomedical applications and they are emerging candidates in nanomedicines because of their synergistic properties when compared to pristine MXenes. MXene-based materials possess high potential for diverse biomedical applications. MXene-based materials are extensively applied for bio-imaging, biosensing, drug delivery system, photothermal therapy, antibacterial agents and theranostics nano-platforms [101]. Figure 3.12 gives an overview of applications of advancing 2D MXenes in nanomedicines [102].

MXenes prepared via traditional methods have excessively large sideward size of sheets that restrict their biomedical application [103, 104]. Furthermore, pristine MXene has low dispersibility in water and has the tendency to form cluster and even precipitate which prevent their easy transportation through circulation system [105]. To overcome these problems, MXene surface was decorated by various hydrophilic polymers through physical adsorption or non-covalent interaction [72, 106]. For example, MXene-based composites exhibited stable circulation, enhanced retention ability and permeability when natural polymer soyabean phospholipid (SP) was immobilized on MXene-based composite surface through physical adsorption (Figure 3.13) [71]. PEGylation of MXenes was also found to be an efficient and economical refinement via electrostatic adsorption for boosting the dispersibility of MXenes in water [72]. However, physical adsorption and non-covalent interaction cannot keep up enough rapidness in some practical applications [107, 108]. Moreover, through physical adsorption common conjugating drugs cannot achieve controlled release. Thus, it necessitates the development of dynamic approaches of surface modification to realize full potential of MXene-based materials for biomedical applications.

3.4.2.1 *Biosensing and Bioimaging*

Because of their huge specific surface area, superior electrical characteristics, and excellent biocompatibility, MXene-based materials have proven to be attractive nanomaterials for biosensing. Scanning Electron Microscopic (SEM) and Tunneling Electron Microscopic (TEM) images of Ag@Ti3C2Tx nanocomposites are shown in Figure 3.14 [109]. A new organ-like MXene-based nanomaterial was developed by Wang *et al.* for H_2O_2 [110].

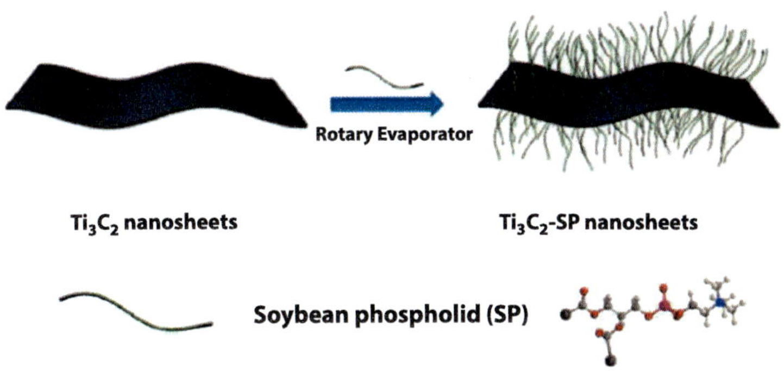

Figure 3.13 Physical adsorption [71]. Copyright 2017, Reproduced with permission from the American Chemical Society.

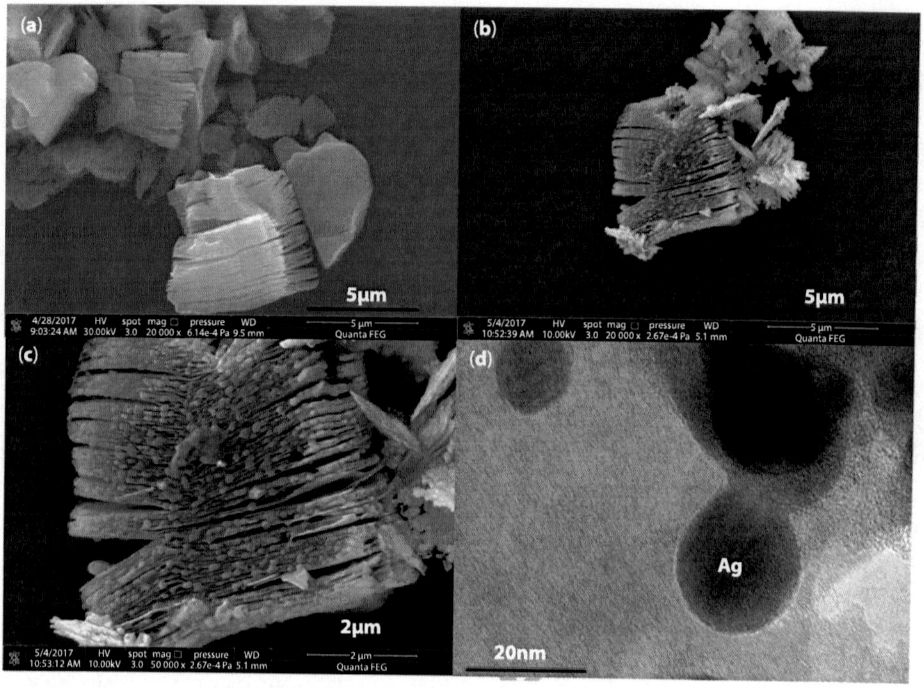

Figure 3.14 SEM images of (a) layered $Ti_3C_2T_x$ nanosheets, (b) Ag@ $Ti_3C_2T_x$ nanocomposites, (c) high-magnification SEM image of (b d) TEM image of Ag@Ti3C2Tx nanocomposites [109]. Copyright 2018, Reproduced with permission from Elsevier.

As hemoglobin has an effective collision with the substrate, the surface of Ti_3C_2 MXene-based materials was modified by Hemoglobin as a model protein. The electro-inactive state of Nafion/MXenes/a glassy carbon (GC) electrode in the potential window is indicated by the absence of redox peak. A couple of well-defined and stable redox peaks is shown by Nafion/Hb/MXenes/GC electrode. The results confirm that Nafion/Hb/MXene/GC electrode significantly enhances the direct transfer of electrons between GC electrode and Hb molecules. MXenes find application in imaging methods, like computed tomography (CT), photoacoustic (PA), magnetic resonance (MR) imaging [13, 111] and contrast agents for CT imaging [111]. In addition, MXene-based materials' excellent photothermal-conversion capability make them suitable for PA imaging [13].

3.4.2.2 *Photothermal Therapy*

MXene-based materials serve as potential substitutes to conventional nanomaterials for photothermal therapy due to their outstanding photothermal property. They show powerful absorbance and preserve high effectiveness for photothermal-conversion in the NIR bio-windows (Figures 3.15a, b) [112]. MXene nanosheets were prepared by room temperature etching of the precursor ceramic powder in acid [112]. The surface of the MXene was modified by poly (vinyl pyrrolidine) (PVP) through physical adsorption. This improved their biocompatibility and water dispersibility. They also looked into the potential PVP modified Nb_2C for ablation of mice tumor xenografts via photothermal method. The tumor volumes in 6 groups were evaluated in alternate days using a digital caliper, and it was found that even after two weeks of various treatments, the tumor locations remained enormous (Figures 3.15c-e). In contrast, the mice in certain groups fully lost their tumor locations (Figures 3.15d and e). The TEM image of the composites is shown in Figure 3.15f indicating the particle size of about 100 nm. The time-dependent tumor growth curves after different treatments are shown in Figure 3.15g. The curves showing the survival of mice is depicted in Figure 3.15h. MXene based nanosheets conjugated with SP also exhibited excellent photothermal property for tumor-bearing mice [71].

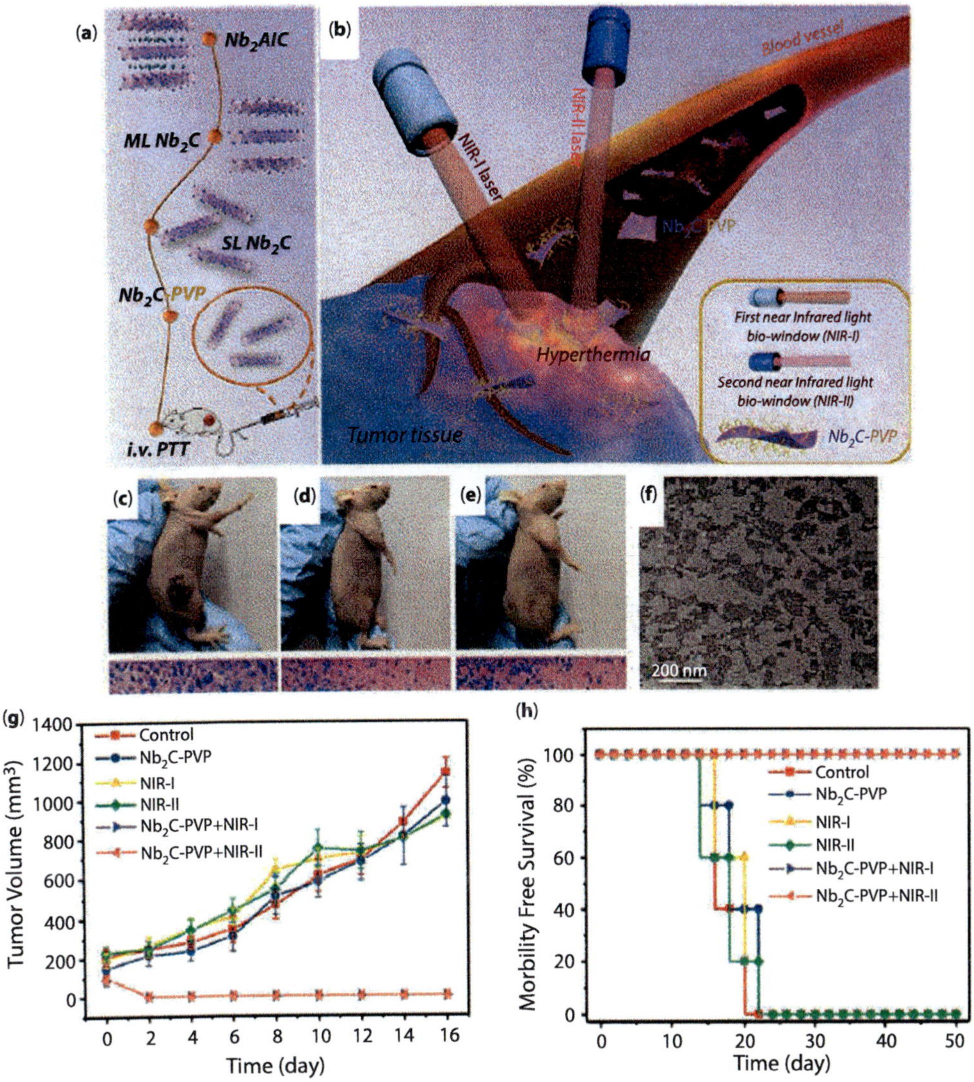

Figure 3.15 (a, b) Diagrammatic illustration of 2D biodegradable Nb_2C for NIR-I and NIR-II bio-window photothermal tumor ablation. (c-e) Photographs of 4T1-tumor-bearing mice and their tumor regions after 16 days. (f) TEM image of Nb_2C-PVP/PBS. (g) Time-dependent tumor growth curves after different treatments. (h) Survival curves of mice after various treatments [112]. Copyright 2017, Reproduced with permission from American Chemical Society.

3.4.2.3 *Drug Delivery*

MXene-based materials possess multifarious merits for developing drug delivery systems. For example, the different 2D structure provides large surface area for MXene-based materials which in turn render ample sites for anchoring of the molecules that are involved in therapy. Moreover, the nanoscale size contributes to an effective accumulation of the materials into tumor sites during cancer therapy and its intravenous administration. Negatively charged fluorine or hydroxyl groups on the surface of MXene make it simple for positively charged drug molecules to engage electrostatically with them. When these nanocomposites flow in the bloodstream, the negatively charged multilayer polymeric coating protects the medications [113]. These drug-loaded MXenes include two sensors to track the release of chemotherapeutic drugs: pH sensitivity and temperature sensitivity [114]. MXene's innate ability of high pH-sensitivity and photothermal conversion insure the synergistic effect of photothermal ablation. It also aids targeted drug release. A MXene-based highly efficient drug delivery system for chemotherapy and phototherapy has been reported. Figures 3.16a, b illustrates the creation and modification of MXene-based materials for the creation of drug delivery systems. Ti3C2, a material based on MXene, was immobilized with soybean phospholipid to provide excellent water dispersibility and transit inside blood arteries [115]. The anticancer drug doxorubicin (Dox), which features NIR-accelerated and pH-responsive drug release, was attached by electrostatic adsorption to the surface of Ti3C2-SP. Dox could quickly release in an acidic environment if H+ interferes with the electrostatic connection between Ti_3C_2-SP and Dox. The moieties of Dox@Ti_3C_2-SP alone and Ti_3C_2-SP + laser, however, continue to retain better cell viability (Figures 3.16c, d). The most apoptosis was clearly produced by the laser irradiation and Dox@Ti_3C_2-SP combination, as seen by the mostly entire red color (Figure 3.16e). In the *in vivo* testing, the Dox@Ti_3C_2-SP + laser group performed the best. This could eliminate the tumor and prevent recurrence (Figures 3.16f, g). MXene-based material when used for novel drug delivery system exhibited high synergistic efficiency of chemotherapy and PTT. In another work, Ti_3C_2 surface was modified by mesoporous-silica shell which was then loaded with Dox for constructing drug delivery systems [13].

3.4.2.4 *Antibacterial Activity*

MXenes always show higher antibacterial activity than GO due to its higher conductivity. The interaction with the bacterial cell membrane is augmented by the anion nature and hydrophilicity of the MXene surface [117]. The hydrogen bonding between MXene functional groups and the cell membrane lipopolysaccharide molecules hamper the nutrients-intake resulting in cell inactivation and thereby prevent the growth of bacteria. A conductive bridge is formed over the lipid bilayer from the cell of bacteria to the external environment. It aids the transfer of electrons, thus enabling cell death [118]. Mayerberg *et al.* developed electrospun MXene ($Ti_3C_2T_z$) incorporated chitosan (CS) nanofibers which showed antibacterial properties for biodegradable medical bandage. The positively charged nitrogen-containing groups of chitosan and the negatively charged MXene functional groups are bound by mostly electrostatic interactions or hydrogen bonding [119].

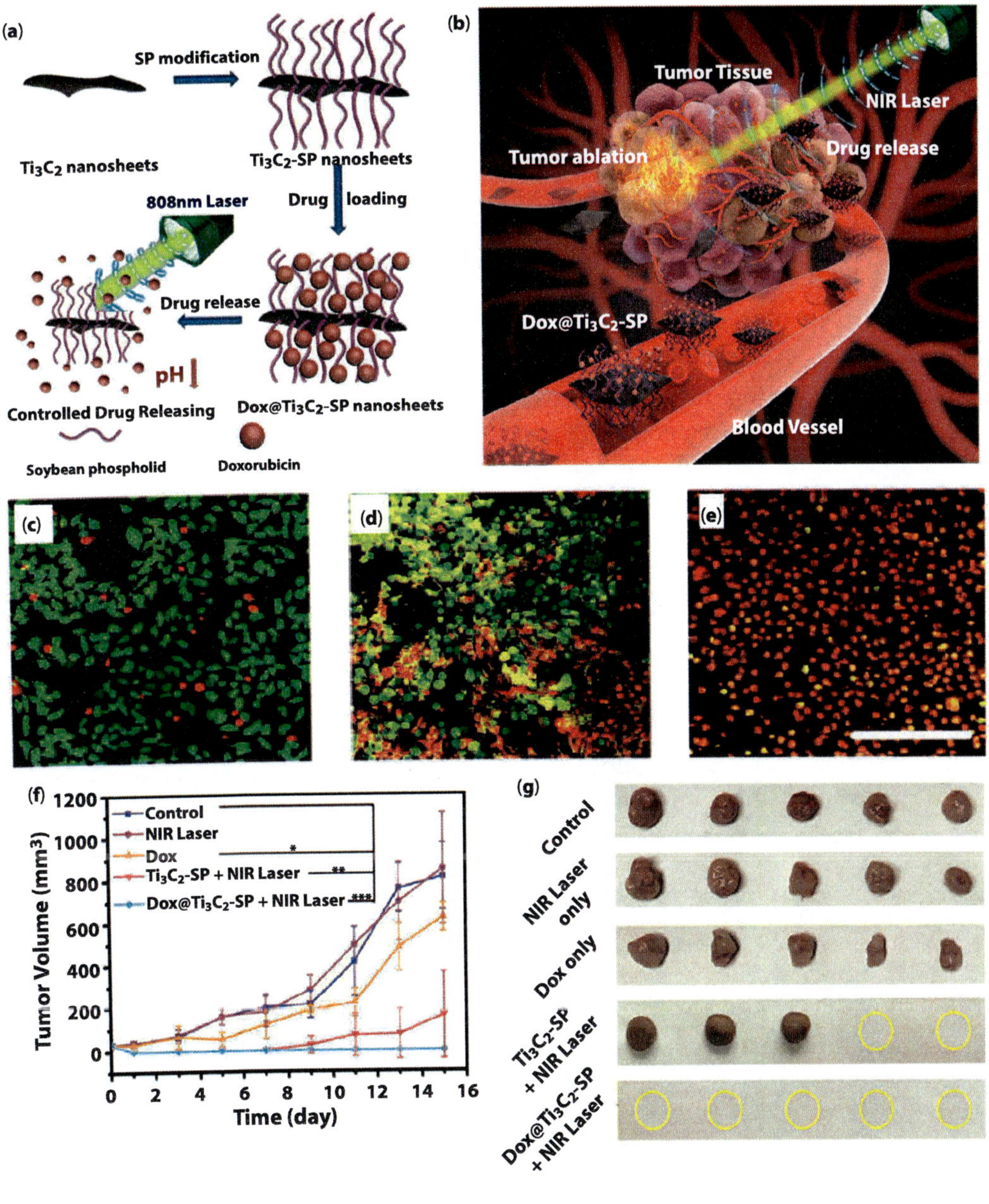

Figure 3.16 (a) The drug loading, external or internal irradiation, and stimuli-responsive drug release by surface alteration of Ti_3C_2 nanosheets by SP. (b) A diagram illustrating an *in vivo* synergistic chemotherapy and photothermal therapy for cancer using a Ti_3C_2-based drug delivery system. (c-e) Images taken using confocal laser scanning microscopy of 4T1 cells following Dox@Ti_3C_2-SP, Ti_3C_2-SP + laser, and Dox@Ti_3C_2-SP + laser treatments. Green fluorescence indicates living cells, while red fluorescence indicates dead cells. (f) Curves of tumor volume over time following various treatments. (g) Images of the tumors removed at the end of therapy from sacrificed mice in all treatment groups [116]. Copyright 2018, Reproduced with permission from John Wiley & Sons.

3.4.3 Catalysis

Experimental evidence and computational predictions reported that surface functionalized MXene materials exhibit outstanding performance in the field of heterogeneous catalytic processes such as activation and conversion of CO_2 (Figure 3.17b), CO oxidation (Figure 3.17a), Water-gas shift (WGS) (Figure 3.17c), N2 fixation (Figure 3.17d), hydrodeoxygenation (HDO) reaction (Figure 3.17e) and direct dehydrogenation (DDH) (Figure 3.7f) [120].

3.4.3.1 CO Oxidation

This is a key reaction in proton-exchange membrane fuel cells for removing CO contaminants from H_2 fuel, preventing Pt catalyst poisoning, pollution control, and as a probe for the rational design of heterogeneous catalysts [121]. Three distinct pathways for the interaction of CO and O_2 have been postulated for MXenes: (i) I the Langmuir Hinshelwood (LH) mechanism, the traditional Eley-Rideal (ER) mechanism and the trimolecular Eley-Rideal (TER) mechanism [122, 123]. Ti@Ti_2CO_2 was found to have significant activity in studies for low-temperature CO oxidation [124]. Later, Cheng and co-workers examined the efficiency of single Pd atoms supported by Mo_2CO_2 MXene, showing unusual CO oxidation activity via the TER mechanism [125]. Cheng and co-workers studied Zn, Fe, Ag, Ru, Co, Rh, Ni, Cu, Zn, Au, Pt and Ir atoms beared on the $Mo_2CO_{2-\delta}$ MXene [126].

3.4.3.2 Activation and Conversion of CO_2

As CO_2 has good chemical stability and so reacts poorly with most substrates, active CO_2 adsorption substrates are difficult to come by. To put it another way, CO_2 capture is

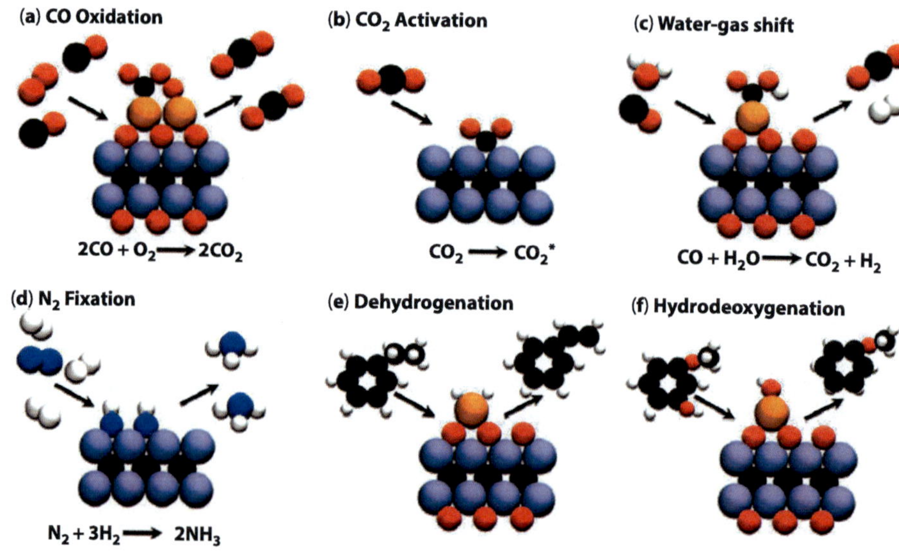

Figure 3.17 Graphical illustration of the different catalytic heterogeneous reactions carried out on MXenes: Color code: black for carbon, white for hydrogen, light purple for an early transition-metal element, blue for nitrogen, orange for a single-atom transition metal, red for oxygen [120]. Copyright 2020, Reproduced with permission from The American Chemical Society.

straightforward conceptually, but putting it into practice on a large enough scale is a major difficulty [127–129]. Mo-, W-and Cr- derived MXenes are potential for CO_2 catalytic conversion because of their low energy of activation. Furthermore, MXene's anticipated CO_2 uptake is close to 8 mol of CO2 kg^{-1}, making it competitive with existing materials like I3X zeolites and Ca-X, bulk MgO nano-powders and graphene derivatives [128]. Surprisingly, the theoretical predictions for the Ti_3C_2 MXene were later validated by tests [130].

3.4.3.3 *Water-Gas Shift (WGS)*

MXene has a high catalytic activity, as well as good selectivity and stability toward CO_2 and H_2, with a selectivity of over 99 percent. As MXene sinters to bulk Mo_2C at higher temperatures, partly due to defunctionalization of the MXene surface, the reaction was conducted below 550 °C. Li's group previously reported a comprehensive investigation of the MXene-based $Pt@Nb_2CT_x$ system for the WGS reaction [131]. The $Pt@Nb_2CT_x$ catalyst was more stable under the WGS conditions than $Pt@Al_2O_3$, which had a same particle size as determined by platinum dispersion. The improved performance was due to the MXene catalyst's greater contacts with H_2O or hydroxyl groups.

MXenes also catalyze a wide range of other reactions. The main reasons why MXene-based materials received significant attention in catalysis application because of their appealing properties such as: (i) their tunable band gap obtained by the alignment of surface chemistries. (ii) the presence of rich surface groups and abundant active sites and (iii) their extraordinary carrier mobility, leading to improved migration and dissociation of photogenerated electron-hole pairs [132–134]. The coordination interactions between ruthenium atoms and Ti_3C_2 via the N and S heteroatom doping technique were reported by Ramalingam and his co-workers. A comparison study on Hydrogen Evolution Reaction's (HER) polarization curves of RuSA-N-S-Ti_3C_2, RuSA-Ti_3C_2, N-STi_3C_2, Ti_3C_2 and carbon paper was conducted. It was observed that the HER baseline of RuSA, N, and S co-doped Ti_3C_2 and RuSA-doped Ti_3C_2 are not at zero potential, which can be attributed to the capacitance impact of the nanocarbons of Ti_3C_2. It was evident that RuSA-N-S-Ti_3C_2 has pronounced potentiality for electrocatalytic HER performance [86]. The reaction kinetics in the HER process was evaluated by calculating Tafel plots' slope. It was observed that the electrocatalysts followed Volmer–Heyrovsky mechanism as the slope of Tafel plot for RuSA-N-S-Ti3C2 was about 90mV dec^{-1}. The mechanism involves two steps [135, 136]:

(i) fast initial discharge reaction: $H_3O^+ + e^- \rightarrow H_{ads} + H_2O$
(ii) slow electrochemical desorption reaction: $H_{ads} + H_3O^+ + e^- \rightarrow H_{ads} + H_2O$

RuSA, N and S co-doped showed exceptional performance compared to other RuSA based electrocatalysts which was confirmed by turnover frequency measurements (ToF). Highly active catalysts can be formed by combining better flexible nature of the two phases and the large exposed basal plane formed by coupling other nano-materials with the interfacial engineering of MXene-based systems. The change in electronic structure can also improve the reaction kinetics [137]. For example, the impact of interfacial structure of Ti_3C_2 MXene and TiO_2 nanoparticles on the photocatalytic reactions have been studied [138–140]. The TiO_2/Ti_3C_2 was developed by growing TiO_2 *in situ* on Ti_3C_2 surface by a hydrothermal strategy [141]. TiO_2/Ti_3C_2 displayed predominant photocatalytic degradation of approximately 98 % in comparison to 42 % of pristine TiO_2 in 30 minutes. An oxidation method which is mild was used to grow Nb_2O_5 homogeneously on the surface of Nb_2C MXenes for HER [142]. The contact between conductive Nb_2O_5 and Nb_2C generated charge carriers at Nb_2O_5/Nb_2C interface. The HER activity of Nb-doped Ti_3C_2 was intensified by surface modification with an Ni/Co alloy, enfeebling the M–H affinity [143]. It was observed from the DFT calculations that a superior catalyst can be prepared by Ti replacement with Ni or Co. That can be ascribed to the synergistic effect of Ni/Co@Nb-doped Ti_3C_2. MXene-based materials are utilized in clean energy conversion technology, photocatalytic water splitting as they have the suitable hydrophilicity, outstanding electronic conductivity and abundant exposed metal sites [144]. The oxygen evolution reaction (OER) permit the yield of immensely pure O_2, making it crucial step in water splitting process. The geometric properties can be adjusted to obtain higher catalytic performance [137]. N-doped Ti_3C_2 which is denoted as $Ti_3C_{1.6}N_{0.4}$ possess high electrical conductivity compared to pristine Ti_3C_2 [145]. The suitability of $Ti_3C_{1.6}N_{0.4}$ for the OER can be confirmed from the fact that it makes smaller contact angle with KOH electrolyte. The onset potential and the Tafel slope of $Ti_3C_{1.6}N_{0.4}$ film were 245.8 mV and 216.4 mV respectively. It indicates that it possesses finer electrocatalytic properties than pristine Ti_3C_2.

3.4.4 Other Applications of Surface Modified MXenes

3.4.4.1 Sensors

Attributable to their thermoelectric and piezoelectric feedback, large surface area, excellent anisotropy of conductivity and rich surface groups, MXenes appear to possess propitious functionality for sensing applications like electrochemical sensors [146], strain sensors [147, 148], biosensors [149–151] and gas sensors [152, 153]. Shi *et al.* used a microscale "brick-and-mortar" design to create layered Ti_3C_2–Ag nanowire-PDA/Ni^{2+} nanocomposite film-based strain sensors [148]. They made Ti_3C_2–Ag nanowires with either Ni^{2+} or PDA. The Ti_3C_2–Ag nanowire-PDA/Ni^{2+} nanocomposite outperformed the pure Ti_3C_2–Ag nanowire (32 percent), Ti_3C_2–Ag nanowire-Ni^{2+} (47 percent), and Ti_3C_2–Ag nanowire-PDA nanocomposite (83 percent) in terms of stretchability (75 percent). The output signals of Ti_3C_2–Ag nanowire-PDA/Ni^{2+} sensors were examined at different strains and the sensor's stable electrical responses produced a 1.2–8 mm S^{-1} strain rate at a strain of 80 percent.

Kumar *et al.* used APTES to detect cancer using a biofunctionalized Ti_3C_2 (f-Ti_3C_2) (Figure 3.18) [67]. Furthermore, by depositing TiO_2 NPs on MXenes may provide additional benefits for detecting H_2O_2 [148]. As a result, by loading the surface of MXenes with remarkable biocompatible TiO_2, a sheltering microenvironment for improving enzyme activity and stability can be created. When TiO_2–Ti_3C_2 was compared to pristine Ti_3C_2, the active sites of enzyme adsorption increased considerably. Furthermore, Chen *et al.* observed that changing the temperature and CO_2 concentration alters the conductivity and transmittance of V_2C, where PDMAEMA was employed to amend the surface of V_2C through self-initiated photopolymerization and photo grafting [154]. The hydrophilic character of PDMAEMA transforms to hydrophobic as the solution temperature increases over the lower critical solution temperature (LCST), causing an increase in the transmittance of V_2C@PDMAEMA. Furthermore, due of V_2C@ PDMAEMA's CO_2 responsive property,

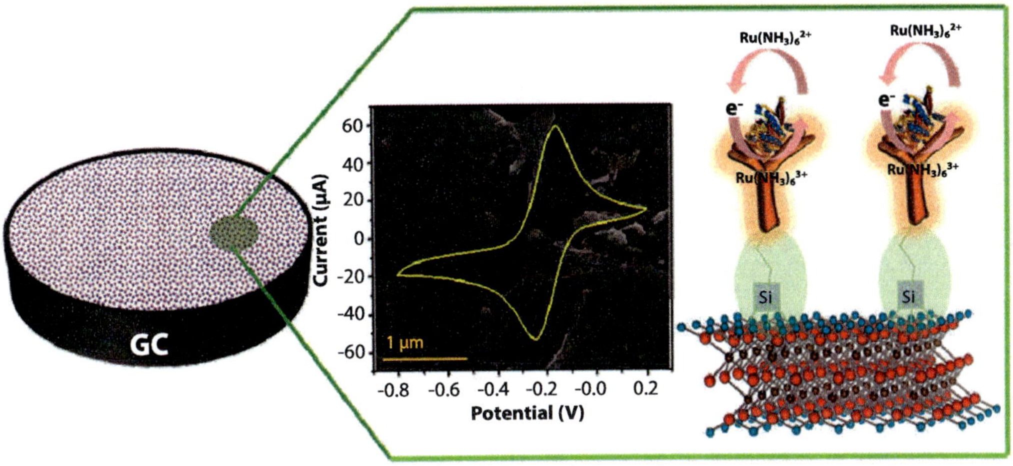

Figure 3.18 Mechanism of electrochemical carcinoembryonic antigen detection [67]. Copyright 2018, Reproduced with permission from Elsevier.

the conductivity of V_2C@PDMAEMA enhanced on addition of CO_2 to water to create charged ammonium bicarbonate, revealing the sensitive nature of V_2C@PDMAEMA to CO_2 and temperature.

3.4.4.2 Membrane-Based Separation

3.4.4.2.1 Pervaporation

Ascribed to their high metallic conductivity and mechanical strength, exceptional separation capabilities are being demonstrated by MXene-based materials [18, 155, 156]. For pervaporation dehydration of isopropanol, Liu and co-workers created an ordered laminar structure of Ti_3C_2 MXene membrane by adding positively charged polyelectrolytes (like polyethylene (PEI), polyallylamine hydrochloride (PAH), and Poly Diallyldimethylammonium chloride (PDDA) to develop electrostatic interactions with negatively charged Ti_3C_2 nanosheets [157]. When compared to pure MXene, MXene–PEI, and MXene-PAH membranes, MXene–PDDA has the lowest water contact angle, indicating more absorption of water molecules and facilitating the water transport channel, resulting in improved separation performance. The water content of MXene-PDDA membrane was ~99.5 wt percent, while the pure MXene, MXene–PEI, and MXene–PAH membranes had ~18.9, ~63, and ~57 wt. percent, respectively.

3.4.4.2.2 Nanofiltration

To increase flux and selectivity for alcohol-based solvents, functionalized Ti_3C_2 MXenes were used in thin-film nanocomposite membranes (TFNs) [158]. To create functionalized Ti_3C_2–$C_{12}H_{26}$, Ti_3C_2–NH_2, Ti_3C_2–COOR and Ti_3C_2–C_6H_6, four modifiers were used respectively: (dodecyl)triethoxysilane (DCTES), APTES, (g methacryloxypropyl) trimethoxysilane (MPTMS) and (methyl aniline) triethoxysilane (MATES). The solvent uptake and area swelling of the produced TFN membranes were investigated using polar solvents (isopropanol

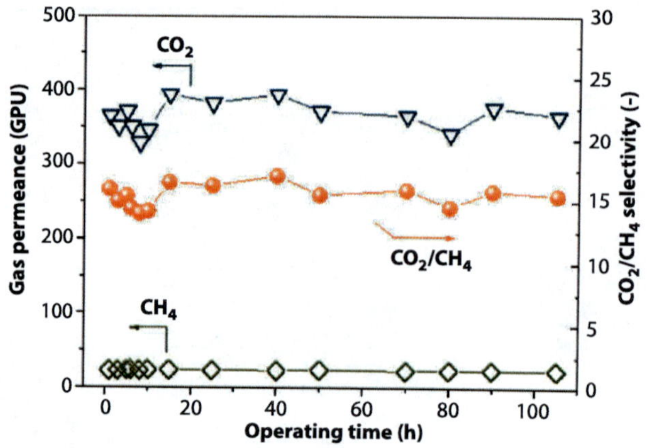

Figure 3.19 CO_2/CH_4 long-term operation test of gad permeation [160]. Copyright 2018, Reproduced with permission from John Wiley & Sons.

and ethyl acetate) and non-polar solvents (toluene and n-heptane). Consequently, the addition of Ti_3C_2 MXene reduced area swellings and solvent uptakes, with the exception of isopropanol, which has no swelling properties. The –OH groups on Ti_3C_2, on the other hand, increased the adsorption and storage of isopropanol by H-interaction. Ti_3C_2–COOR and Ti_3C_2–NH_2 embedded TFN membranes showed greater solvent uptakes and area swelling for polar solvents, while Ti_3C_2–C_6H_6 and Ti_3C_2–$C_{12}H_{26}$ showed more solvent uptakes and area swelling for non-polar solvents. Depending on the polarity properties of functional groups and solvents, all Ti_3C_2-M added TFN membranes enhanced affinity for a given solvent. Ti_3C_2–NH_2 and Ti_3C_2–COOR incorporated membrane's affinity was boosted towards polar solvents and embedding Ti_3C_2–$C_{12}H_{26}$ and Ti_3C_2–C_6H_6 enhanced the transport of non-polar solvents.

3.4.4.2.3 Gas Separation

The interlayer spacing of MXene-based materials can be tuned with appropriate modification for effective gas separation performance with good reproducibility and mechanical stability, caused by the presence of well-ordered channels in lamellar MXene membranes [159]. The interlayer space of MXene nanosheets can be boosted by surface modification, resulting in improved selective molecule separation. For H_2/CO_2 separation, Shen and co-workers created a nanofilm based on MXene on an anodic aluminum oxide (AAO) matrix. They used borate crosslink with the –O moieties on MXenes to reduce the interlayer distance, boosting the selective nature of CO_2/N_2 and CO_2/CH_4. With the potential reaction of amine groups on PEI and Ti suboxides, robust electrostatic interactions could be produced between the negatively charged MXene and positively charged PEI [160]. Properties of CO_2-selective MXene nanofilms for gas transfer are shown in Figure 3.19 [160].

3.5 Conclusion and Future Perspectives

MXenes comprises an extraordinary family of materials based on their unparalleled potential to tune the properties by controlling surface terminations and functionalizing their surface by various groups which further leads to the synthesis of various structures and chemistries. MXenes possess astonishing optical, thermoelectric and electronic properties which can be enhanced by functionalizing the surfaces that are already composed by -O, -OH and -F groups. This review showcases the exceptional opportunities present in the surface termination of MXenes along with the challenges which can be helpful in identifying the stability of surface termination as well as to improve the properties of MXenes. We have outlined the current strategies of surface terminations including the choice of etchant and the manipulation of surface terminations by modification and doping. In addition, we have discussed the methods for surface functionalization followed by the wide variety of applications including energy storage, biomedicine, biosensor, drug delivery and catalysis. MXenes will have a substantial influence on two-dimensional materials research in future which will result in the generation of materials with exceptional characteristics and applications.

References

1. Choi, W., Choudhary, N., Han, G.H., Park, J., Akinwande, D., Lee, Y.H., Recent development of two-dimensional transition metal dichalcogenides and their applications. *Mater. Today*, 20, 116–130, 2017.
2. Geim, Andre K., and Konstantin S. Novoselov. "The rise of graphene." *Nat. Mater.*, 6, no. 3, 183-191, 2007.
3. Eswaraiah, V., Zeng, Q., Long, Y., Liu, Z., Black phosphorus nanosheets: Synthesis, characterization and applications. *Small*, 12, 3480–3502, 2016.
4. Li, X., Wang, C., Cao, Y., Wang, G., Functional MXene materials: Progress of their applications. *Chem. Asian J.*, 13, 2742–2757, 2018.
5. Zhang, X., Zhang, Z., Zhou, Z., MXene-based materials for electrochemical energy storage. *J. Energy Chem.*, 27, 73–85, 2018.
6. Soleymaniha, M., Shahbazi, M., Rafieerad, A.R., Maleki, A., Amiri, A., Promoting role of MXene nanosheets in biomedical sciences: Therapeutic and biosensing innovations. *Adv. Healthcare Mater.*, 8, 1801137, 2019.
7. Pei, Y., Zhang, X., Hui, Z., Zhou, J., Huang, X., Sun, G., Huang, W., Ti3C2TX MXene for sensing applications: Recent progress, design principles, and future perspectives. *ACS Nano*, 15, 3996–4017, 2021.
8. Li, Z. and Wu, Y., 2D early transition metal carbides (MXenes) for catalysis. *Small*, 15, 1804736, 2019.
9. Naguib, M., Kurtoglu, M., Presser, V., Lu, J., Niu, J., Heon, M., Hultman, L., Gogotsi, Y., Barsoum, M.W., Two-dimensional nanocrystals produced by exfoliation of Ti3AlC2. *Adv. Mater.*, *23*, 4248–4253, 2011.
10. Naguib, M., Mochalin, V.N., Barsoum, M.W., Gogotsi, Y., 25th anniversary article: MXenes: A new family of two-dimensional materials. *Adv. Mater.*, 26, 992–1005, 2014.
11. Magnuson, M., Halim, J., Näslund, L.-Å., Chemical bonding in carbide MXene nanosheets. *J. Electron Spectros. Relat. Phenomena*, 224, 27–32, 2013.
12. Alli, U., Hettiarachchi, S.J., Kellici, S., Chemical functionalisation of 2D materials by batch and continuous hydrothermal flow synthesis. *Chem. Eur. J.*, 26, 6447–6460, 2020.
13. Huang, H., Jiang, R., Feng, Y., Ouyang, H., Zhou, N., Zhang, X., Wei, Y., Recent development and prospects of surface modification and biomedical applications of MXenes. *Nanoscale*, 12, 1325–1338, 2020.

14. Jin, H., Liu, X., Vasileff, A., Jiao, Y., Zhao, Y., Zheng, Y., Qiao, S., Single-crystal nitrogen-rich two-dimensional Mo5N6 nanosheets for efficient and stable seawater splitting. *ACS Nano*, 12, 12761–12769, 2018.
15. Xu, C., Wang, L., Liu, Z., Chen, L., Guo, J., Kang, N., Ma, X., Cheng, H., Ren, W., Large-area high-quality 2D ultrathin Mo2C superconducting crystals. *Nat. Mater.*, 14, 1135–1141, 2015.
16. Alhabeb, M., Maleski, K., Mathis, T.S., Sarycheva, A., Hatter, C.B., Uzun, S., Levitt, A., Gogotsi, Y., Selective etching of silicon from Ti3SiC2 (MAX) to obtain 2D titanium carbide (MXene). *Angew. Chem.*, 130, 19, 5542–5546, 2018.
17. Gao, X., Zheng, Y., Chang, J., Xu, H., Hui, Z., Dai, H., Wang, H., Xia, Z., Zhou, J., Sun, G., Universal strategy for preparing highly stable PBA/Ti3C2T x MXene toward lithium-ion batteries via chemical transformation. *ACS Appl. Mater. Interfaces*, 14, 15298–15306, 2022.
18. Mozafari, M. and Soroush, M., Surface functionalization of MXenes. *Mater. Adv.*, 2, 7277–7307, 2021.
19. Boota, M., Anasori, B., Voigt, C., Zhao, M., Barsoum, M.W., Gogotsi, Y., Pseudocapacitive electrodes produced by oxidant-free polymerization of pyrrole between the layers of 2D titanium carbide (MXene). *Adv. Mater.*, 28, no. 7: 1517–1522, 2016.
20. Qin, L., Tao, Q., Liu, X., Fahlman, M., Halim, J., OÅ Persson, P., Rosen, J., Zhang, F., Polymer-MXene composite films formed by MXene-facilitated electrochemical polymerization for flexible solid-state microsupercapacitors. *Nano Energy*, 60, 734–742, 2019.
21. Wang, H., Zhang, J., Wu, Y., Huang, H., Li, G., Zhang, X., Wang, Z., Surface modified MXene Ti3C2 multilayers by aryl diazonium salts leading to large-scale delamination. *Appl. Surf. Sci.*, 384, 287–293, 2016.
22. Zhu, X., Liu, B., Hou, H., Huang, Z., Zeinu, K.M., Huang, L., Yuan, X., Guo, D., Hu, J., Yang, J., Alkaline intercalation of Ti3C2 MXene for simultaneous electrochemical detection of Cd (II), Pb (II), Cu (II) and Hg (II). *Electrochim. Acta*, 248, 46–57, 2017.
23. Tian, Y., Que, W., Luo, Y., Yang, C., Yin, X., Kong, L.B., Surface nitrogen-modified 2D titanium carbide (MXene) with high energy density for aqueous supercapacitor applications. *J. Mater. Chem. A*, 7, 5416–5425, 2019.
24. Thurakkal, S. and Zhang, X., Noncovalent functionalization of $Ti3C2T_x$ using cationic porphyrins with enhanced stability against oxidation. *Mater. Chem. Front.*, 6, 561–569, 2022.
25. Björk, J. and Rosen, J., Functionalizing MXenes by tailoring surface terminations in different chemical environments. *Chem. Mater.*, 33, 9108–9118, 2021.
26. Kamysbayev, V., Filatov, A.S., Hu, H., Rui, X., Lagunas, F., Wang, D., Klie, R.F., Talapin, D.V., Covalent surface modifications and superconductivity of two-dimensional metal carbide MXenes. *Science*, 369, 979–983, 2020.
27. Naguib, M., Mashtalir, O., Carle, J., Presser, V., Lu, J., Hultman, L., Gogotsi, Y., Barsoum, M.W., Two-dimensional transition metal carbides. *ACS Nano*, 6, 1322, 2012.
28. Wang, Y., Zheng, W., Zhang, P., Tian, W., Chen, J., Sun, Z.M., Preparation of (Vx, Ti1−x)2C MXenes and their performance as anode materials for LIBs. *J. Mater. Sci.*, 54, 11991–11999, 2019.
29. Khazaei, M., Mishra, A., Venkataramanan, N.S., Singh, A.K., Yunoki, S., Recent advances in MXenes: From fundamentals to applications. *Curr. Opin. Solid State Mater. Sci.*, 23, 164–178, 2019.
30. Yazdanparast, S., Soltanmohammad, S., Fash-White, A., Tucker, G.J., Brennecka, G.L., Synthesis and surface chemistry of 2D TiVC solid-solution MXenes. *ACS Appl. Mater. Interfaces*, 12, 20129–20137, 2020.
31. Naguib, M., Gogotsi, Y., Barsoum, M.W., Mxenes: A new family of two-dimensional materials and its application as electrodes for Li and Na-ion batteries. *ECS Meet. Abstr.*, *227*, 849–849, 2015.

32. Shuck, C.E., Ventura-Martinez, K., Goad, A., Uzun, S., Shekhirev, M., Gogotsi, Y., Safe synthesis of MAX and MXene: Guidelines to reduce risk during synthesis. *ACS Chem. Health Saf.*, 28, 326–338, 2021.
33. Halim, J., Lukatskaya, M.R., Cook, K.M., Lu, J., Smith, C.R., Näslund, L.A., May, S.J., Hultman, L., Gogotsi, Y., Eklund, P., Barsoum, M.W., Transparent conductive two-dimensional titanium carbide epitaxial thin films. *Chem. Mater.*, *26*, 2374–2381, 20142014.
34. Mashtalir, O., Naguib, M., Mochalin, V.N., Dall'Agnese, Y., Heon, M., Barsoum, M.W., Gogotsi, Y., Intercalation and delamination of layered carbides and carbonitrides. *Nat. Commun.*, 4, 1716, 2013.
35. Arole, K., Blivin, J.W., Saha, S., Holta, D.E., Zhao, X., Sarmah, A., Cao, H., Radovic, M., Lutkenhaus, J.L., Green, M.J., Water-dispersible Ti3C2Tz MXene nanosheets by molten salt etching. *Iscience*, 24, 103403, 2021.
36. Aslam, M.K., Niu, Y., Xu, M., MXenes for non-lithium-ion (Na, K, Ca, Mg, and Al) batteries and supercapacitors. *Adv. Energy Mater.*, 11, 2000681, 2021.
37. Naguib, M., Barsoum, M.W., Gogotsi, Y., Ten years of progress in the synthesis and development of MXenes. *Adv. Mater.*, 33, 2103393, 2021.
38. Urbankowski, P., Anasori, B., Makaryan, T., Er, D., Kota, S., Walsh, P.L., Zhao, M., Shenoy, V.B., Barsoum, M.W., Gogotsi, Y., Synthesis of two-dimensional titanium nitride Ti 4 N 3 (MXene). *Nanoscale*, 8, 11385–11391, 2016.
39. Brousse, T., Morel, A., Bélanger, D., Borjon-Piron, Y., Vanadium nitride electrodes: Limitations and practical use in electrochemical capacitors. *ECS Meet. Abstr.*, 230, 932–932, 2016.
40. Liu, L., Orbay, M., Luo, S., Duluard, S., Shao, H., Harmel, J., Rozier, P., Taberna, P., Simon, P., Exfoliation and delamination of Ti3C2T x MXene prepared via molten salt etching route. *ACS Nano*, 16, 111–118, 2022.
41. Luo, J., Matios, E., Wang, H., Tao, X., Li, W., Interfacial structure design of MXene-based nanomaterials for electrochemical energy storage and conversion. *InfoMat*, 2, 1057–1076, 2020.
42. Bai, S., Yang, M., Jiang, J., He, X., Zou, J., Xiong, Z., Liao, G., Liu, S., Recent advances of MXenes as electrocatalysts for hydrogen evolution reaction. *NPJ 2D Mater. Appl.*, 5, 1–15, 2021.
43. Li, L., Li, G., Tan, L., Zhang, Y., Wu, B., Highly efficiently delaminated single-layered MXene nanosheets with large lateral size. *Langmuir*, 33, 9000–9006, 2017.
44. Zhou, C., Zhao, X., Xiong, Y., Tang, Y., Ma, X., Tao, Q., Sun, C., Xu, W., A review of etching methods of MXene and applications of MXene conductive hydrogels. *Eur. Polym. J.*, *167*, 111063, 2022.
45. Yin, T., Li, Y., Wang, R., Al-Hartomy, O.A., Al-Ghamdi, A., Wageh, S., Luo, X., Tang, X., Zhang, H., Synthesis of Ti3C2Fx MXene with controllable fluorination by electrochemical etching for lithium-ion batteries applications. *Ceram. Int.*, 47, 28642–28649, 2021.
46. Feng, W., Long, P., Feng, Y.Y., Li, Y., Two-dimensional fluorinated graphene: Synthesis, structures, properties and applications. *Adv. Sci.*, 3, 1500413, 2016.
47. Shen, M., Jiang, W., Liang, K., Zhao, S., Tang, R., Zhang, L., Wang, J., One-pot green process to synthesize MXene with controllable surface terminations using molten salts. *Angew. Chem.*, 133, 27219–27224, 2021.
48. Bao, Z., Lu, C., Cao, X., Zhang, P., Yang, L., Zhang, H., Sha, D., He, W., Zhang, W., Pan, L., Sun, Z., Role of MXene surface terminations in electrochemical energy storage: A review. *Chin. Chem. Lett.*, *32*, 2648–2658, 2021.
49. Gao, L., Bao, W., Kuklin, A.V., Mei, S., Zhang, H., Ågren, H., Hetero-MXenes: Theory, synthesis, and emerging applications. *Adv. Mater.*, 33, 1–43, 2021.
50. Ling, C., Shi, L., Ouyang, Y., Wang, J., Searching for highly active catalysts for hydrogen evolution reaction based on O-terminated MXenes through a simple descriptor. *Chem. Mater.*, 28, 9026–9032, 2016.

51. Balcı, E., Akkuş, Ü. Ö., Berber, S., Band gap modification in doped MXene: Sc2CF2. *J. Mater. Chem. C*, 5, 5956–5961, 2017.
52. Yang, J., Wang, A., Zhang, S., Wu, H., Chen, L., Stability and electronic properties of sulfur terminated two-dimensional early transition metal carbides and nitrides (MXene). *Comput. Mater. Sci.*, 153, 303–308, 2018.
53. Chen, J., Huang, Q., Huang, H., Mao, L., Liu, M., Zhang, X., Wei. Recent progress, Y., and advances in the environmental applications of MXene related materials. *Nanoscale*, 12, 3574–3592, 2020.
54. Long, Z., Liu, M., Jiang, R., Wan, Q., Mao, L., Wan, Y., Deng, F., Zhang, X., Wei, Y., Preparation of water soluble and biocompatible AIE-active fluorescent organic nanoparticles via multicomponent reaction and their biological imaging capability. *J. Chem. Eng.*, 308, 527–534, 2017.
55. Wan, Q., Huang, Q., Liu, M., Xu, D., Huang, H., Zhang, X., Wei, Y., Aggregation-induced emission active luminescent polymeric nanoparticles: Non-covalent fabrication methodologies and biomedical applications. *Appl. Mater. Today*, 9, 145–160, 2017.
56. Jiang, R., Liu, M., Li, C., Huang, Q., Huang, H., Wan, Q., Wen, Y., Cao, Q., Zhang, X., Wei, Y., Facile fabrication of luminescent polymeric nanoparticles containing dynamic linkages via a one-pot multicomponent reaction: Synthesis, aggregation-induced emission and biological imaging. *Mater. Sci. Eng. C*, 80, 708–714, 2017.
57. Huang, L., Yang, S., Chen, J., Tian, J., Huang, Q., Huang, H., Wen, Y., Deng, F., Zhang, X., Wei, Y., A facile surface modification strategy for fabrication of fluorescent silica nanoparticles with the aggregation-induced emission dye through surface-initiated cationic ring opening polymerization. *Mater. Sci. Eng. C*, 94, 270–278, 2019.
58. Schultz, T., Frey, N.C., Hantanasirisakul, K., Park, S., May, S.J., Shenoy, V.B., Gogotsi, Y., Koch, N., Surface termination dependent work function and electronic properties of Ti3C2T x MXene. *Chem. Mater.*, 31, 6590–6597, 2019.
59. Lu, C., Yang, L., Yan, B., Sun, L., Zhang, P., Zhang, W., Sun, Z., Nitrogen-doped Ti3C2 MXene: Mechanism investigation and electrochemical analysis. *Adv. Funct. Mater.*, 30, 2000852, 2020.
60. Liu, R., Cao, W., Han, D., Mo, Y., Zeng, H., Yang, H., Li, W., Nitrogen-doped Nb2CTx MXene as anode materials for lithium ion batteries. *J. Alloys Compd.*, 793, 505–511, 2019.
61. Yoon, Y., Lee, M., Kim, S.K., Bae, G., Song, W., Myung, S., Lim, J., Lee, S.S., Zyung, T., An, K., A strategy for synthesis of carbon nitride induced chemically doped 2D MXene for high-performance supercapacitor electrodes. *Adv. Energy Mater.*, 8, 1703173, 2018.
62. Xia, Y., Mathis, T.S., Zhao, M., Anasori, B., Dang, A., Zhou, Z., Cho, H., Gogotsi, Y., Yang, S., Thickness-independent capacitance of vertically aligned liquid-crystalline MXenes. *Nature*, 557, 409–412, 2018.
63. Lei, Y., Cui, Y., Huang, Q., Dou, J., Gan, D., Deng, F., Liu, M., Li, X., Zhang, X., Wei, Y., Facile preparation of sulfonic groups functionalized Mxenes for efficient removal of methylene blue. *Ceram. Int.*, 45, 17653–17661, 2019.
64. Jin, Z., Fang, Y., Wang, X., Xu, G., Wei, S., Zhou, C., Zhang, Y., Xu, Y., Ultra-efficient electromagnetic wave absorption with ethanol-thermally treated two-dimensional Nb2CTx nanosheets. *J. Colloid Interface Sci.*, 537, 306–315, 2019.
65. Li, Z., Zhuang, Z., Lv, F., Zhu, H., Zhou, L., Luo, M., Zhu, J., Lang, Z., Feng, S., Chen, W., Mai, L., The marriage of the FeN4 moiety and MXene boosts oxygen reduction catalysis: Fe 3D electron delocalization matters. *Adv. Mater.*, *30*, 1803220, 2018.
66. Wu, R., Du, H., Wang, Z., Gao, M., Pan, H., Liu, Y., Remarkably improved hydrogen storage properties of NaAlH4 doped with 2D titanium carbide. *J. Power Sources*, 327, 519–525, 2016.
67. Kumar, S., Lei, Y., Alshareef, N.H., Quevedo-Lopez, M.A., Salama, K.N., Biofunctionalized two-dimensional Ti3C2 MXenes for ultrasensitive detection of cancer biomarker. *Biosens. Bioelectron.*, 121, 243–249, 2018.

68. Tu, S., Jiang, Q., Zhang, X., Alshareef, H.N., Large dielectric constant enhancement in MXene percolative polymer composites. *ACS Nano*, 12, 3369–3377, 2018.
69. Sun, R., Zhang, H., Liu, J., Xie, X., Yang, R., Li, Y., Hong, S., Yu, Z., Highly conductive transition metal carbide/carbonitride (MXene)@ polystyrene nanocomposites fabricated by electrostatic assembly for highly efficient electromagnetic interference shielding. *Adv. Funct. Mater.*, 27, 1702807, 2017.
70. Ling, Z., Ren, C.E., Zhao, M., Yang, J., Giammarco, J.M., Qiu, J., Barsoum, M.W., Gogotsi, Y., Flexible and conductive MXene films and nanocomposites with high capacitance. *Proc. Natl. Acad. Sci. U.S.A.*, 111, 16676–16681, 2014.
71. Lin, H., Wang, X., Yu, L., Chen, Y., Shi, J., Two-dimensional ultrathin MXene ceramic nanosheets for photothermal conversion. *Nano Lett.*, 17, 384–391, 2017.
72. Xuan, J., Wang, Z., Chen, Y., Liang, D., Cheng, L., Yang, X., Liu, Z., Ma, R., Sasaki, T., Geng, F., Organic-base-driven intercalation and delamination for the production of functionalized titanium carbide nanosheets with superior photothermal therapeutic performance. *Angew. Chem.*, 128, 14789–14794, 2016.
73. Liu, M., Zeng, G., Wang, K., Wan, Q., Tao, L., Zhang, X., Wei, Y., Recent developments in polydopamine: An emerging soft matter for surface modification and biomedical applications. *Nanoscale*, 8, 16819–16840, 2016.
74. Li, X., Zhu, J., Wang, L., Wu, W., Fang, Y., *In-situ* growth of carbon nanotubes on two-dimensional titanium carbide for enhanced electrochemical performance. *Electrochim. Acta*, 258, 291–301, 2017.
75. Shamsabadi, A.A., Seidi, F., Salehi, E., Nozari, M., Rahimpour, A., Soroush, M., Efficient CO_2-removal using novel mixed-matrix membranes with modified TiO 2 nanoparticles. *J. Mater. Chem. A*, 5, 4011–4025, 2017.
76. Nie, Z., Petukhova, A., Kumacheva, E., Properties and emerging applications of self-assembled structures made from inorganic nanoparticles. *Nat. Nanotechnol.*, 5, 15–25, 2010.
77. Fu, L., Yan, Z., Zhao, Q., Yang, H., Novel 2D nanosheets with potential applications in heavy metal purification: A review. *Adv. Mater. Interfaces*, 5, 1801094, 2018.
78. Ambade, R.B., Lee, H., Lee, K.H., Lee, H., Veerasubramani, G.K., Kim, Y.-B., Han, T.H., Ultrafast flashlight sintered mesoporous NiO nanosheets for stable asymmetric supercapacitors. *Chem. Eng. J.*, 436, 135041, 2022.
79. Ali, A., Hantanasirisakul, K., Abdala, A., Urbankowski, P., Zhao, M.-Q., Anasori, B., Gogotsi, Y., Aïssa, B., Mahmoud, K.A., Effect of synthesis on performance of MXene/iron oxide anode material for lithium-ion batteries. *Langmuir*, 34, 11325–11334, 2018.
80. VahidMohammadi, A., Moncada, J., Chen, H., Kayali, E., Orangi, J., Carrero, C.A., Beidaghi, M., Thick and freestanding MXene/PANI pseudocapacitive electrodes with ultrahigh specific capacitance. *J. Mater. Chem. A*, 6, 22123–22133, 2018.
81. Al-Hamadani, Y.A., Jun, B.M., Yoon, M., Taheri-Qazvini, N., Snyder, S.A., Jang, M., Heo, J., Yoon, Y., Applications of MXene-based membranes in water purification: A review. *Chemosphere*, 254, 126821, 2020.
82. Gao, L., Li, C., Huang, W., Mei, S., Lin, H., Ou, Q., Zhang, Y., Guo, J., Zhang, F., Xu, S., Zhang, H., MXene/polymer membranes: Synthesis, properties, and emerging applications. *Chem. Mater.*, 32, 1703–1747, 2020.
83. Huang, W., Hu, L., Tang, Y., Xie, Z., Zhang, H., Recent advances in functional 2D MXene-based nanostructures for next-generation devices. *Adv. Funct. Mater.*, 30, 2005223, 2020.
84. Chen, W.Y., Lai, S.N., Yen, C.C., Jiang, X., Peroulis, D., Stanciu, L.A., Surface functionalization of Ti3C2T x MXene with highly reliable superhydrophobic protection for volatile organic compounds sensing. *ACS Nano*, 14, 11490–11501, 2020.

85. Heckler, J.E., Neher, G.R., Mehmood, F., Lioi, D.B., Pachter, R., Vaia, R., Kennedy, W.J., Nepal, D., Surface functionalization of Ti3C2T x MXene nanosheets with catechols: Implication for colloidal processing. *Langmuir*, 37, 5447–5456, 2021.
86. Pazniak, H., Benchakar, M., Bilyk, T., Liedl, A., Busby, Y., Noël, C., Chartier, P., Hurand, S., Marteau, M., Houssiau, L., Larciprete, R., Ion implantation as an approach for structural modifications and functionalization of Ti3C2T x MXenes. *ACS Nano*, 15, 4245–4255, 2021.
87. Frey, N.C., Bandyopadhyay, A., Kumar, H., Anasori, B., Gogotsi, Y., Shenoy, V.B., Surface-engineered MXenes: Electric field control of magnetism and enhanced magnetic anisotropy. *ACS Nano*, 13, 2831–2839, 2019.
88. Manthiram, A., Fu, Y., Su, Y.S., Challenges and prospects of lithium–sulfur batteries. *Acc. Chem. Res.*, 46, 1125–1134, 2013.
89. Gu, L.L., Gao, J., Wang, C., Qiu, S.Y., Wang, K.X., Gao, X.T., Sun, K.N., Zuo, P.J., Zhu, X.D., Thin-carbon-layer-enveloped cobalt–iron oxide nanocages as a high-efficiency sulfur container for Li–S batteries. *J. Mater. Chem. A*, 8, 20604–20611, 2020.
90. Zhang, Y., Ma, C., He, W., Zhang, C., Zhou, L., Wang, G., Wei, W., MXene and MXene-based materials for lithium-sulfur batteries. *Prog. Nat. Sci.*, 31, 501–513, 2021.
91. Zhao, X., Liu, M., Chen, Y., Hou, B., Zhang, N., Chen, B., Yang, N., Chen, K., Li, J., An, L., Fabrication of layered Ti3C2 with an accordion-like structure as a potential cathode material for high performance lithium–sulfur batteries. *J. Mater. Chem. A*, 3, 7870–7876, 2015.
92. Zhao, Q., Zhu, Q., Miao, J., Zhang, P., Xu, B., 2D MXene nanosheets enable small-sulfur electrodes to be flexible for lithium–sulfur batteries. *Nanoscale*, 11, 8442–8448, 2019.
93. Nan, J., Guo, X., Xiao, J., Li, X., Chen, W., Wu, W., Liu, H., Wang, Y., Wu, M., Wang, G., Nanoengineering of 2D MXene-based materials for energy storage applications. *Small, 17*, 1902085, 2021.
94. Song, Y., Sun, Z., Fan, Z., Cai, W., Shao, Y., Sheng, G., Wang, M., Song, L., Liu, Z., Zhang, Q., Sun, J., Rational design of porous nitrogen-doped Ti3C2 MXene as a multifunctional electrocatalyst for Li–S chemistry. *Nano Energy*, 70, 104555, 2020.
95. Hantanasirisakul, K., Zhao, M.Q., Urbankowski, P., Halim, J., Anasori, B., Kota, S., Ren, C.E., Barsoum, M.W., Gogotsi, Y., Fabrication of Ti3C2Tx MXene transparent thin films with tunable optoelectronic properties. *Adv. Electron. Mater.*, *2*, 1600050, 2016.
96. Ghidiu, M., Halim, J., Kota, S., Bish, D., Gogotsi, Y., Barsoum, M.W., Ion-exchange and cation solvation reactions in Ti3C2 MXene. *Chem. Mater.*, 28, 3507–3514, 2016.
97. Yoon, Y., Lee, M., Kim, S.K., Bae, G., Song, W., Myung, S., Lim, J., Lee, S.S., Zyung, T., An, K.S., A strategy for synthesis of carbon nitride induced chemically doped 2D MXene for high-performance supercapacitor electrodes. *Adv. Energy Mater.*, *8*, 1703173, 2018.
98. Tang, Q., Zhou, Z., Shen, P., Are MXenes promising anode materials for Li ion batteries? Computational studies on electronic properties and Li storage capability of Ti3C2 and Ti3C2X2 (X= F, OH) monolayer. *J. Am. Chem. Soc.*, 134, 16909–16916, 2012.
99. Wang, X., Kajiyama, S., Iinuma, H., Hosono, E., Oro, S., Moriguchi, I., Okubo, M., Yamada, A., Pseudocapacitance of MXene nanosheets for high-power sodium-ion hybrid capacitors. *Nat. Commun.*, 6, 1–6, 2015.
100. Zhang, X., Lv, R., Wang, A., Guo, W., Liu, X., Luo, J., MXene aerogel scaffolds for high-rate lithium metal anodes. *Angew. Chem. Int. Ed.*, 57, 15028–15033, 2018.
101. George, S.M. and Kandasubramanian, B., Advancements in MXene-polymer composites for various biomedical applications. *Ceram. Int.*, 46, 8522–8535, 2020.
102. Lin, H., Chen, Y., Shi, J., Insights into 2D MXenes for versatile biomedical applications: Current advances and challenges ahead. *Adv. Sci.*, 5, 1800518, 2018.
103. Kurapati, R., Kostarelos, K., Prato, M., Bianco, A., Biomedical uses for 2D materials beyond graphene: Current advances and challenges ahead. *Adv. Mater.*, 28, 6052–6074, 2016.

104. Chimene, D., Alge, D.L., Gaharwar, A.K., Two-dimensional nanomaterials for biomedical applications: Emerging trends and future prospects. *Adv. Mater.*, 27, 7261–7284, 2015.
105. Friedman, Z., Marks, K.H., Maisels, M.J., Thorson, R., Naeye, R., Effect of parenteral fat emulsion on the pulmonary and reticuloendothelial systems in the newborn infant. *Pediatrics*, 61, 694–698, 1978.
106. Iwata, T., Hoshi, S., Takehisa, F., Tsutsumi, K., Furukawa, Y., Kimura, S., The effect of dietary safflower phospholipid and soybean phospholipid on plasma and liver lipids in rats fed a hypercholesterolemic diet. *J. Nutr. Sci. Vitaminol. Tokyo*, 38, 5, 471–479, 1992.
107. Huang, H., Liu, M., Jiang, R., Chen, J., Mao, L., Wen, Y., Tian, J., Zhou, N., Zhang, X., Wei, Y., Facile modification of nanodiamonds with hyperbranched polymers based on supramolecular chemistry and their potential for drug delivery. *J. Colloid Interface Sci.*, 513, 198–204, 2018.
108. Huang, H., Liu, M., Jiang, R., Chen, J., Huang, Q., Wen, Y., Tian, J., Zhou, N., Zhang, X., Wei, Y., Fabrication and characterization of hyperbranched polyglycerol modified carbon nanotubes through the host-guest interactions. *Mater. Sci. Eng. C*, 91, 458–465, 2018.
109. Jiang, Y., Zhang, X., Pei, L., Yue, S., Ma, L., Zhou, L., Huang, Z., He, Y., Gao, J., Silver nanoparticles modified two-dimensional transition metal carbides as nanocarriers to fabricate acetycholinesterase-based electrochemical biosensor. *J. Chem. Eng.*, 339, 547–556, 2018.
110. Wang, F., Yang, C., Duan, C., Xiao, D., Tang, Y., Zhu, J., An organ-like titanium carbide material (MXene) with multilayer structure encapsulating hemoglobin for a mediator-free biosensor. *J. Electrochem. Soc.*, 162, B16, 2014.
111. Dai, C., Chen, Y., Jing, X., Xiang, L., Yang, D., Lin, H., Liu, Z., Han, X., Wu, R., Two-dimensional tantalum carbide (MXenes) composite nanosheets for multiple imaging-guided photothermal tumor ablation. *ACS Nano*, 11, 12696–12712, 2017.
112. Lin, H., Gao, S., Dai, C., Chen, Y., Shi, J., A two-dimensional biodegradable niobium carbide (MXene) for photothermal tumor eradication in NIR-I and NIR-II biowindows. *J. Am. Chem. Soc.*, 139, 16235–16247, 2017.
113. Liu, G., Zou, J., Tang, Q., Yang, X., Zhang, Y., Zhang, Q., Huang, W., Chen, P., Shao, J., Dong, X., Surface modified Ti3C2 MXene nanosheets for tumor targeting photothermal/photodynamic/chemo synergistic therapy. *ACS Appl. Mater. Interfaces*, 9, 40077–40086, 2017.
114. Sur, S., Rathore, A., Dave, V., Reddy, K.R., Chouhan, R.S., Sadhu, V., Recent developments in functionalized polymer nanoparticles for efficient drug delivery system. *Nano-Struct. Nano-Objects*, 20, 100397, 2019.
115. Ide, T., Hirabayashi, S., Kano, S., Sugano, M., Soybean phospholipid dependent reductions in triacylglycerol concentration and synthesis in the liver of fasted-refed rats. *Biochim. Biophys. Acta (BBA)-Lipids Lipid Metab.*, 1124, 163–170, 1992.
116. Han, X., Huang, J., Lin, H., Wang, Z., Li, P., Chen, Y., 2D Ultrathin MXene-based drug-delivery nanoplatform for synergistic photothermal ablation and chemotherapy of cancer. *Adv. Healthc. Mater.*, 7, 1701394, 2018.
117. Gao, G., O'Mullane, A.P., Du, A., 2D MXenes: A new family of promising catalysts for the hydrogen evolution reaction. *ACS Catal.*, 7, 494–500, 2017.
118. Rasool, K., Helal, M., Ali, A., Ren, C.E., Gogotsi, Y., Mahmoud, K.A., Antibacterial activity of Ti3C2T x MXene. *ACS Nano*, 10, 3674–3684, 2016.
119. Mayerberger, E.A., Street, R.M., McDaniel, R.M., Barsoum, M.W., Schauer, C.L., Antibacterial properties of electrospun $Ti_3C_2T_z$ (MXene)/chitosan nanofibers. *RSC Adv.*, 8, 35386–35394, 2018.
120. Morales-Garcia, A., Calle-Vallejo, F., Illas, F., MXenes: New horizons in catalysis. *ACS Catal.*, 10, 13487–13503, 2020.
121. Freund, H., Meijer, G., Scheffler, M., Schlögl, R., Wolf, M., CO oxidation as a prototypical reaction for heterogeneous processes. *Angew. Chem. Int. Ed.*, 50, 10064–10094, 2011.

122. Campbell, C.T., Ertl, G., Kuipers, H., Segner, J., A molecular beam study of the catalytic oxidation of CO on a Pt (111) surface. *J. Chem. Phys.*, 73, 5862–5873, 1980.
123. Neugebohren, J., Borodin, D., Hahn, H.W., Altschäffel, J., Kandratsenka, A., Auerbach, D.J., Campbell, C.T., Schwarzer, D., Harding, D.J., Wodtke, A.M., Kitsopoulos, T.N., Velocity-resolved kinetics of site-specific carbon monoxide oxidation on platinum surfaces. *Nature*, 558, 280–283, 20182018.
124. Qiao, B., Wang, A., Yang, X., Allard, L.F., Jiang, Z., Cui, Y., Liu, J., Li, J., Zhang, T., Single-atom catalysis of CO oxidation using Pt1/FeOx. *Nat. Chem.*, 3, 634–641, 2011.
125. Cheng, C., Zhang, X., Wang, M., Wang, S., Yang, Z., Single Pd atomic catalyst on Mo2CO2 monolayer (MXene): Unusual activity for CO oxidation by trimolecular Eley–Rideal mechanism. *Phys. Chem. Chem. Phys.*, 20, 3504–3513, 2018.
126. Cheng, C., Zhang, X., Yang, Z., Hermansson, K., Identification of high-performance single-atom mxenes catalysts for low-temperature CO oxidation. *Adv. Theory Simul.*, 2, 1900006, 2019.
127. Burghaus, U. "Surface science studies of carbon dioxide chemistry." In *New and future developments in catalysis*, pp. 27–47. Amsterdam: Elsevier, 2013.
128. Morales-García, Á., Fernández-Fernández, A., Viñes, F., Illas, F., CO2 abatement using two-dimensional MXene carbides. *J. Mater. Chem. A*, 6, 3381–3385, 2018.
129. Morales-Salvador, R., Morales-Garcia, A., Vines, F., Illas, F., Two-dimensional nitrides as highly efficient potential candidates for CO2 capture and activation. *Phys. Chem. Chem. Phys.*, 20, 24490–24493, 2018.
130. Persson, I., Halim, J., Lind, H., Hansen, T.W., Wagner, J.B., Näslund, L.Å., Darakchieva, V., Palisaitis, J., Rosen, J., Persson, P.O., 2D transition metal carbides (MXenes) for carbon capture. *Adv. Mater.*, 31, 1805472, 2019.
131. Li, Z., Cui, Y., Wu, Z., Milligan, C., Zhou, L., Mitchell, G., Xu, B., Shi, E., Miller, J.T., Ribeiro, F.H., Wu, Y., Reactive metal–support interactions at moderate temperature in two-dimensional niobium-carbide-supported platinum catalysts. *Nat. Catal.*, 1, 349–355, 2018.
132. Ran, J., Gao, G., Li, F.-T., Ma, T.-Y., Du, A., Qiao, S.-Z., Ti3C2 MXene co-catalyst on metal sulfide photo-absorbers for enhanced visible-light photocatalytic hydrogen production. *Nat. Commun.*, 8, 13907, 2017.
133. Shao, B., Wang, J., Liu, Z., Zeng, G., Tang, L., Liang, Q., He, Q., Wu, T., Liu, Y., Yuan, X., Ti3C2 T x MXene decorated black phosphorus nanosheets with improved visible-light photocatalytic activity: Experimental and theoretical studies. *J. Mater. Chem. A*, 8, 5171–5185, 2020.
134. Wang, H., Wu, Y., Xiao, T., Yuan, X., Zeng, G., Tu, W., Wu, S., Lee, H.Y., Tan, Y.Z., Chew, J.W., Formation of quasi-core-shell In2S3/anatase TiO2@ metallic Ti3C2Tx hybrids with favorable charge transfer channels for excellent visible-light-photocatalytic performance. *Appl. Catal.*, 233, 213–225, 2018.
135. Kibsgaard, J., Jaramillo, T.F., Besenbacher, F., Building an appropriate active-site motif into a hydrogen-evolution catalyst with thiomolybdate [Mo3S13] 2–clusters. *Nat. Chem.*, 6, 248–253, 2014.
136. Yuan, W., Cheng, L., An, Y., Wu, H., Yao, N., Fan, X., Guo, X., MXene nanofibers as highly active catalysts for hydrogen evolution reaction. *ACS Sustain. Chem. Eng.*, 6, 8976–8982, 2018.
137. Zhu, Q., Cui, Y., Zhang, Y., Cao, Z., Shi, Y., Gu, J., Du, Z., Li, B., Yang, S., Strategies for engineering the MXenes toward highly active catalysts. *Mater. Today Nano*, 13, 100104, 2021.
138. Zhang, D., Wang, S., Hu, R., Gu, J., Cui, Y., Li, B., Chen, W., Liu, C., Shang, J., Yang, S., Catalytic conversion of polysulfides on single atom zinc implanted MXene toward high-rate lithium–sulfur batteries. *Adv. Funct. Mater.*, 30, 2002471, 2020.
139. Zhang, J., Yang, L., Wang, H., Zhu, G., Wen, H., Feng, H., Sun, X., Guan, X., Wen, J., Yao, Y., *In situ* hydrothermal growth of TiO2 nanoparticles on a conductive Ti3C2T x MXene nanosheet:

A synergistically active Ti-based nanohybrid electrocatalyst for enhanced N2 reduction to NH3 at ambient conditions. *Inorg. Chem.*, 58, 5414–5418, 2019.
140. Low, J., Zhang, L., Tong, T., Shen, B., Yu, J., TiO2/MXene Ti3C2 composite with excellent photocatalytic CO2 reduction activity. *J. Catal.*, 361, 255–266, 2018.
141. Gao, Y., Wang, L., Zhou, A., Li, Z., Chen, J., Bala, H., Hu, Q., Cao, X., Hydrothermal synthesis of TiO2/Ti3C2 nanocomposites with enhanced photocatalytic activity. *Mater. Lett.*, 150, 62–64, 2015.
142. Su, T., Peng, R., Hood, Z.D., Naguib, M., Ivanov, I.N., Keum, J.K., Qin, Z., Guo, Z., Wu, Z., Onestep synthesis of Nb2O5/C/Nb2C (MXene) composites and their use as photocatalysts for hydrogen evolution. *ChemSusChem*, 11, 688–699, 2018.
143. Du, C.F., Sun, X., Yu, H., Liang, Q., Dinh, K.N., Zheng, Y., Luo, Y., Wang, Z., Yan, Q., Synergy of Nb doping and surface alloy enhanced on water–alkali electrocatalytic hydrogen generation performance in Ti-based MXene. *Adv. Sci.*, 6, 1900116, 2019.
144. Ouyang, H., Zhou, M., Guo, Y., He, M., Huang, H., Ye, X., Feng, Y., Zhou, X., Yang, S., Metabolites profiling of Pulsatilla saponin D in rat by ultra performance liquid chromatography–quadrupole time-of-flight mass spectrometry (UPLC/Q-TOF-MS/MS). *Fitoterapia*, 96, 152–158, 2014.
145. Tang, Y., Yang, C., Tian, Y., Luo, Y., Yin, X., Que, W., The effect of *in situ* nitrogen doping on the oxygen evolution reaction of MXenes. *Nanoscale Adv.*, 2, 1187–1194, 2020.
146. Shahzad, F., Iqbal, A., Zaidi, S.A., Hwang, S.-W., Koo, C.M., Nafion-stabilized two-dimensional transition metal carbide (Ti3C2Tx MXene) as a high-performance electrochemical sensor for neurotransmitter. *J. Ind. Eng. Chem.*, 79, 338–344, 2019.
147. Zhang, Y.Z., Lee, K.H., Anjum, D.H., Sougrat, R., Jiang, Q., Kim, H., Alshareef, H.N., MXenes stretch hydrogel sensor performance to new limits. *Sci. Adv.*, 4, eaat0098, 2018.
148. Shi, X., Wang, H., Xie, X., Xue, Q., Zhang, J., Kang, S., Wang, C., Liang, J., Chen, Y., Bioinspired ultrasensitive and stretchable MXene-based strain sensor via nacre-mimetic microscale brick-and-mortar architecture. *ACS Nano*, 13, 649–659, 2018.
149. Lorencova, L., Gajdosova, V., Hroncekova, S., Bertok, T., Blahutova, J., Vikartovska, A., Parrakova, L., Gemeiner, P., Kasak, P., Tkac, J., 2D MXenes as perspective immobilization platforms for design of electrochemical nanobiosensors. *Electroanalysis*, 31, 1833–1844, 2019.
150. Chen, X., Li, J., Pan, G., Xu, W., Zhu, J., Zhou, D., Li, D., Chen, C., Lu, G., Song, H., Ti3C2 MXene quantum dots/TiO2 inverse opal heterojunction electrode platform for superior photoelectrochemical biosensing. *Sens. Actuators B Chem.*, 289, 131–137, 2019.
151. Liu, J., Jiang, X., Zhang, R., Zhang, Y., Wu, L., Lu, W., Li, J., Li, Y., Zhang, H., MXene-enabled electrochemical microfluidic biosensor: Applications toward multicomponent continuous monitoring in whole blood. *Adv. Funct. Mater.*, 29, 1807326, 2019.
152. Kim, S.J., Koh, H.J., Ren, C.E., Kwon, O., Maleski, K., Cho, S.Y., Anasori, B., Kim, C.K., Choi, Y.K., Kim, J., Gogotsi, Y., Metallic Ti3C2T x MXene gas sensors with ultrahigh signal-to-noise ratio. *ACS Nano*, 12, 986–993, 2018.
153. Yuan, W., Yang, K., Peng, H., Li, F., Yin, F., A flexible VOCs sensor based on a 3D Mxene framework with a high sensing performance. *J. Mater. Chem. A*, 6, 18116–18124, 2018.
154. Chen, J., Chen, K., Tong, D., Huang, Y., Zhang, J., Xue, J., Huang, Q., Chen, T., CO2 and temperature dual responsive smart MXene phases. *ChemComm*, 51, 314–317, 2015.
155. Ding, L., Wei, Y., Wang, Y., Chen, H., Caro, J., Wang, H., A two-dimensional lamellar membrane: MXene nanosheet stacks. *Angew. Chem. - Int. Ed.*, 56, 1825–1829, 2017.
156. Coleman, J.N., Lotya, M., O'Neill, A., Bergin, S.D., King, P.J., Khan, U., Young, K., Gaucher, A., De, S., Smith, R.J., Shvets, I.V., Two-dimensional nanosheets produced by liquid exfoliation of layered materials. *Science*, 331, 568–571, 2011.

157. Liu, G., Liu, S., Ma, K., Wang, H., Wang, X., Liu, G., Jin, W., Polyelectrolyte functionalized Ti2CTx mxene membranes for pervaporation dehydration of isopropanol/water mixtures. *Ind. Eng. Chem. Res.*, 59, 4732–4741, 2020.
158. Hao, L., Zhang, H., Wu, X., Zhang, J., Wang, J., Li, Y., Novel thin-film nanocomposite membranes filled with multi-functional Ti3C2Tx nanosheets for task-specific solvent transport. *Compos. Part A Appl. Sci. Manuf.*, 100, 139–149, 2017.
159. Ding, L., Wei, Y., Li, L., Zhang, T., Wang, H., Xue, J., Ding, L.X., Wang, S., Caro, J., Gogotsi, Y., MXene molecular sieving membranes for highly efficient gas separation. *Nat. Commun.*, 9, 1–7, 2018.
160. Shen, J., Liu, G., Ji, Y., Liu, Q., Cheng, L., Guan, K., Zhang, M., Liu, G., Xiong, J., Yang, J., Jin, W., 2D MXene nanofilms with tunable gas transport channels. *Adv. Funct. Mater.*, 28, no. 31: 1801511, 2018.

4

Electronic, Electrical and Optical Properties of MXenes

Deepthi Jayan K.* **and Ragin Ramdas M.**

Department of Basic Sciences & Humanities, Rajagiri School of Engineering & Technology (Autonomous), Kochi, Kerala, India

Abstract
MXenes represent a broad category of two-dimensional (2D) materials formed from the nitrides, carbides, or carbonitrides of transition metals, showing profound applications in diverse fields including photonics, photovoltaics, functional metamaterials-based device fabrication, ultrafast laser technology, biomedical and materials engineering. In this respect, tailoring various properties of MXenes, viz. optical, electronic, and electrical properties have a remarkable impact in identifying the most suitable MXene for a specific application. These materials usually exhibit a strong non-linear behavior arising from the phenomenon of plasmonic resonances together with exceptional optical transparency, large transmission coefficient, and other surface-dependent optical interactions. These properties are highly correlated with the structural and electronic properties of them. The estimation of electronic properties including band gap and analysis of band structure of functionalized MXenes reveal that most of them are metallic or semi-metallic. First principle density functional theory (DFT) studies furnish a very detailed and precise estimation of the electronic and optical properties of MXenes. The electrical properties of MXenes also have a high dependence on the structure and composition of the crystal and the nature of functional group attached to surface of structures. Cautious controlling of the synthetic route is indispensable for lowering the adverse effects of defects, which in turn can unfavorably impact the ultimate electrical properties of MXenes. This review covers the basic structure and different synthetic procedures of MXenes, examines their various optoelectronic and electrical properties, which have been identified through computational and experimental studies and the experimentally proven methods employed for the reinforcement of these properties.

Keywords: MXene, optoelectronic properties, electrical properties, density functional theory, surface functionalization, surface plasmon resonance

**Corresponding author*: deepthij@rajagiritech.edu.in

4.1 Introduction

Alloying, representing the mixing of chemical elements to form a solid solution, has been considered as a suitable method to enhance the properties of metals from the ancient days itself. It can also be applied to the class of certain inorganic compounds such as carbides, nitrides or dichalcogenides containing transition metals as the constituents, thus forming a solid solution [1, 2]. This method helps to optimize the electronic, optical, and mechanical properties of metal-based compounds without changing their crystal structure. Such an optimization of band structure, band gap, absorption and catalytic properties and other transport properties such as enhanced mobility and effective mass of carriers can be attained solely through a control of composition of constituent elements, which in no way alters the existing structure of the crystal [3–5]. In this context, two dimensional (2-D) materials play a significant role, essentially because of their atomically thin nature, thus providing several unique properties including tunable photoluminescence [6]. Recent research studies show that several 2-D transition metal carbide and nitride compounds are capable of exhibiting the formation of continuous solid solutions, which in turn helps in optimizing their physical and chemical properties by means of suitable composition engineering techniques [7, 8]. One such 2-D material called MXene, which corresponds to a class of 2-D transition metal carbide and nitride have been showed to exhibit exceptionally unique mechanical, elastic, optoelectronic and electrical properties when employed in electrochemical devices since its invention in 2018.

Several 2-D materials have been studied so far for employing in various optoelectronic and photovoltaic devices. Graphene, the widely employed 2-D material for many electronic applications possesses a zero band gap, thus making it unsuitable for optical applications. Hence, studies have been focused on semiconducting materials primarily and the 2-D compounds, namely transition metal dichalcogenide materials viz. MoS_2 and WS_2 were identified as suitable semiconductors for optoelectronic applications also [9–13]. Recently, a few research studies focusing on the method of chemical exfoliation of layered materials, caused the development of a new class of 2-D material. To exfoliate 2-D layered structures called MXenes, containing transition metal carbides and nitrides from members of MAX phase, the research group utilized hydrofluoric acid solution and ultrasonic techniques [14, 15]. A majority of the members of the MXene family are metals and the remaining exhibit the properties of semiconductors [16, 17]. MXenes exhibit several distinguishing structural, electronic, optical, mechanical, elastic, thermoelectric, thermodynamic, electrical, and magnetic properties, which can be utilized for several practical applications. However, for employing them in various devices, their properties need to be optimized accordingly. The estimation of the above-mentioned properties and their subsequent optimization can be performed though computational modeling studies, prior to the real experimental fabrication of device.

For estimating optoelectronic properties of MXenes, computational techniques of density functional theory (DFT) may be used. This method involves estimation of band gap and band structure, electron density, density of states (DOS), absorption spectrum, absorption coefficient, energy loss spectrum, reflectivity, index of refraction, dielectric permittivity, conduction band minimum (CBM), and valence band maximum (VBM) by employing various software such as WIEN2k, VASP, Quantum EXPRESSO, ABINIT etc. [18–21].

Various methods for enhancing the properties of these 2-D compounds have been employed including the addition of suitable amounts of dopants, the application of an electric field or gate voltages and external strains. The application of strain is considered as the most suitable method for altering the properties of MXenes as they possess excellent mechanical flexibility. It is possible to apply a biaxial strain between the 2-D material and the substrate under real experimental conditions by maintaining a lattice mismatch between them. On the other hand, while employing computational methods to assess the structural stability of MXenes, compressive and tensile strains of suitable amount can be provided and the structure can be optimized accordingly. In this chapter, various electrical and optoelectronic properties of MXenes are explained in depth, specifically in the context of experimental and computational methods to estimate and enhance the properties, as MXenes can be employed for diverse optoelectronic and photovoltaic applications. The subsequent sections summarize the structure, methods of synthesis, various optoelectronic and electrical properties of these 2D materials.

4.2 Structure of MXenes

The general chemical structure for MXenes corresponds to $M_{n+1}AX_n$, with n = 1, 2, 3. Here, M stands for transition metals like Sc, Ti, Y, Hf, V, Nb, Mo etc., A denotes a group 13 or 14 element and X denotes carbon or nitrogen. Based on chemical environment, a MXene structure occasionally consists of a termination group T_z including -OH, -F, =O or -Cl. Due to the availability of a number of A, M and X site elements, the chemical variety of MAX phases is also exceptional as indicated by the existence of more than 150 reported chemical compositions of MXenes so far [22–25]. Figure 4.1 indicates the chemical versatility of MXene materials [25]. Even though most of the MXenes are derived by selective etching of the Al planes from the MAX phase, the MXene Ti_3SiC_2 was synthesized via selectively etching non-Al containing MAX phases [24]. However, it is also possible to synthesis MXene via etching certain planes selectively from non-MAX phases also. One such example is the MXene Mo_2CT_z, which was synthesized by the selective etching of Ga layers from the non-MAX phase precursor, Mo_2Ga_2C [24, 25]. The pristine MXene structure can be obtained from parent MAX phase $M_{n+1}AX_n$ on removing A element, thus forming $M_{n+1}X_n$ (with n = 1, 2, 3) structure. Alike the parent MAX phases, MXene also exhibits a hexagonal crystal structure by forming a bilayer consisting of an X layer situated between two M layers. The isostructural out of the plane ordered MXenes have a chemical structure represented as M_2'M"X_2 or M_2'M_2"X_3, the outer layers are composed of M' element and inner layers composed of the M" element, thus forming a hexagonal symmetry. The second variety of MXenes obtained by exfoliating in plane MAX phases generally creates 2-D MXene having chemical structure $(M'_{2/3}M''_{1/3})_2X$ or $M_{1.33}X$, exhibiting a monoclinic symmetry. In these structures, the larger M" element will be shifted out of its usual position, by the existence of two different transition metals at M" position (Figure 4.1) [25].

Theoretical studies have been extremely useful in the hunt for novel MXenes and estimating the stability of existing MXenes. For estimating the formation and the structural stability of a class of MXenes, M_2C and M_2N with the surface terminal groups -F, -OH, or -O, researchers employed first-principle density functional theory (DFT) studies together with an analysis of the dynamical stability of structure by checking the presence

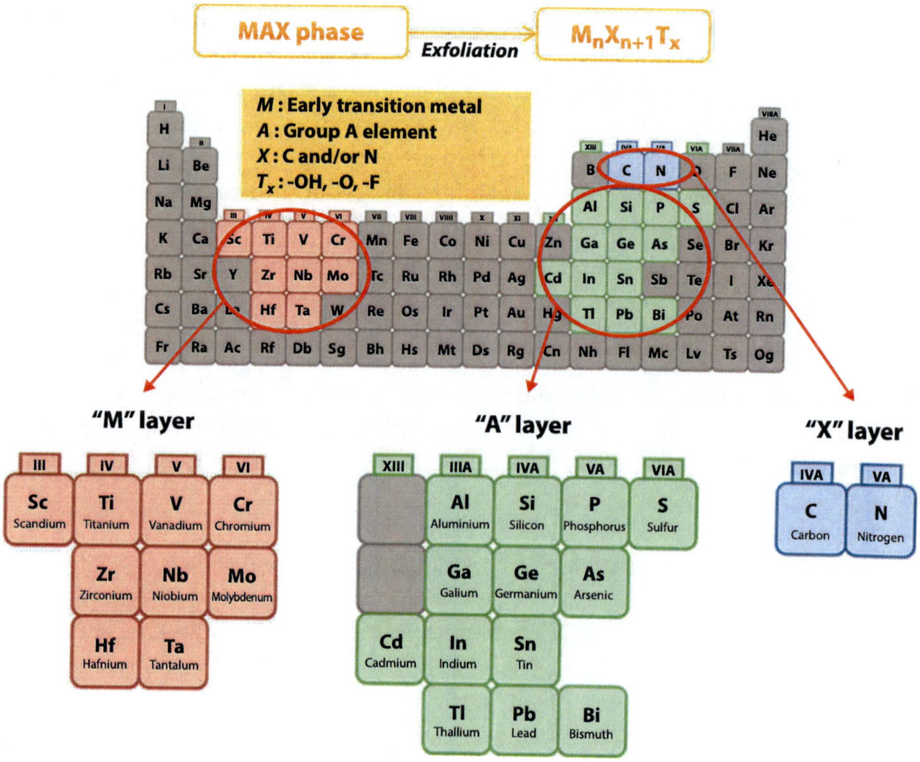

Figure 4.1 General composition of MXenes [25]. Copyright 2021. Reproduced with permission from Springer Nature.

of imaginary frequencies in phonon dispersion spectrum. After examining the structural stability of a number of o-MXenes, it was found that the ordered MXenes possess better stability than MAX phase counterpart, as well as the stability is found to be completely controlled by the composition of transition metals in them. To speculate its structural stability and ordering of predicted o-MXene and i-MXene, a two-step theoretical approach was employed. The first step involves an estimation of the formation enthalpy of the predicted structures using the first principle DFT studies. If its formation enthalpy becomes negative, then phase stability of structure can be assured. Second step involves a linear optimization to check the relative stability of all known competing phases for identifying the decay of the current phase into its competing ones, by performing a decomposition pathway analysis.

4.3 An Overview of Various Methods of Synthesis of MXenes

Several suitable methods are available for the synthesis of MXenes. However, two distinct methods that are mainly employed for their synthesis, namely aqueous acid etching (AAE) method as well as chemical vapor deposition (CVD) method are discussed in brief here.

4.3.1 Aqueous Acid Etching (AAE) Method

MXenes are developed from the precursor MAX phases by etching selected atomic planes of it. The method of mechanical exfoliation can be adopted during the stage of etching. However, in the case of $M_{n+1}X_n$ MXenes, M-A chemical link is strong metallic or covalent and hence cannot be exfoliated by mechanical methods. However, it is easy to etch A element from MAX phases since M-X chemical link exhibit stronger reactivity in comparison with M-A link. For the etching of A atomic layer and to prepare MXenes from its corresponding MAX phases, Fluoride acid etching method is usually employed. In this method, the fluoride acid is employed which is simply hydrofluoric acid (HF) or acid formed by reacting some fluorides with hydrochloric acid [26, 27]. The MAX phase etching with HF results in the formation of a few surface functional groups for generated MXenes that can vary properties of MXene. The intercalation of tetrapropylammonium hydroxide (TPAOH) with the etched MXene powder is employed to delaminate the MAX phase. Subsequently, the centrifugation process will result in the extraction of a few layered MXene nanosheets and removes large sheets of MXenes.

High temperature etching of MAX phases can also lead to the formation of MXenes. By adding molten fluoride mixture to the MAX phase, Ti_4AlN_3 at 550°C, it is possible to etch Al atomic layer and this results in the formation of the Ti_4N_3 MXene. The entire high temperature etching process is to be carried out in an argon rich environment [28]. Earlier, the etching of MAX phase was done in presence of some gaseous etchants. However, this led to the generation of carbide derived carbon, as these etchants were not capable of performing etching the atomic layers selectively [29]. Also, the high temperature etching process performed by keeping the temperature around 800°C results in the formation of bulk compounds rather than 2D nanolayers of MXenes. The etching process is to be carried out at an optimum temperature, the failing of which may result in the formation of the initial stereoscopic phase itself [30].

4.3.2 Chemical Vapor Deposition (CVD) Method

Chemical vapor deposition method, basically a bottom-up synthesis procedure, is employed for fabricating high quality and large surface area materials which can be employed for large scale applications and in diverse fields including integrated electronic equipment and flexible optoelectronic and photovoltaic devices. MXenes can also be synthesized from their raw materials by chemical vapor deposition method. In this method, the 2D sheets are synthesized by keeping the raw materials in a reaction chamber under optimum conditions. By employing chemical vapor deposition method, Xu *et al.* was able to synthesis Mo_2C in 2015 [31]. By employing chemical vapor deposition method, a heterostructure consisting of Mo_2C and grapheme was synthesized, which was able to exhibit a smaller onset voltage for evolution reaction of hydrogen in comparison to pure Mo_2C electrode [32].

One of the advantages of employing aqueous acid etching method is that a huge quantity of nanolayered MXene sheets can be formed within a very short span of time. However, this method cannot be employed for the formation of a huge quantity of monolayer MXene nanosheets. Another drawback of AAE method is related to the safety issues as the HF employed in this method is a strong and highly corrosive acid which may damage the structure. On the other hand, the chemical vapor deposition method results in generating bulk transverse sized materials with lesser defects as it is basically a bottom-up method for the fabrication of nanostructures. CVD also has some drawbacks such as high cost and bulk deposition of materials, which hinders the application of this method when large scale commercialization of the product is needed. Surface functionalization and termination control during the stage of synthesis of MXene are the two requirements for the effective optimization of their properties. The protection of oxidation of MXene during the stage of synthesis can also improve the properties of these materials [33].

4.4 Electronic Properties

MXenes possess high electronic conductivity in two dimensions and hence can be employed in diverse applications of electromagnetic interference shielding and energy storage. They can also be employed in energy conversion devices such as photovoltaic cells, which utilize their excellent conduction and absorption properties. The electronic conductivity of MXenes need to be optimized to enhance its properties further. In this regard, the tailoring of the surface chemistry of MXenes has a remarkable role as results of density functional theory calculations predict a strong dependence of Fermi level density of states on MXene surface terminations. The estimation of the electronic properties provides information about partial and total density of state, band structure along with band gap and electronic distribution in material under study. Both theoretical and experimental studies on the electronic structure of MXenes have been done and the results indicate that a majority of MXenes are either metallic or semi metallic and a minority of them shows the properties of semiconductors.

There are three main approaches that are utilized for the first principle estimation of the electronic properties of materials. The first approach utilizes a fixed basis set and the wave function is expanded in terms of linear superposition of the chosen fixed basis set that include linear combinations of atomic orbitals, Gaussian orbitals, plane waves etc. The second approach relies on a partial wave method, which employs partial waves as basis set for expanding the wave function and are completely energy and potential dependent. Some examples include the augmented plane wave method and the Korringa-Kohn-Rostoker method. The third approach, which is basically a combination of the above two approaches involves the Korringa-Kohn-Rostoker approach along with utilizing the benefits of fixed basis set or partial wave basis set depending on the nature of the problem under consideration. In this approach, the potential has a spherical symmetry inside nuclei and flat nature in interstitial region between the nuclei [34].

4.4.1 Density of States and Electronic Distribution

The Density of States (DOS) represents a popular and extensively used term in quantum mechanical systems in materials science and solid-state physics. The term represents the number of different energy states in which the electrons can occupy at a particular energy level. In other words, it represents the number of allowed electron states per unit volume per unit energy. Many important bulk properties of the materials such as viz. specific heat capacity, susceptibility and related magnetic properties, the conductive properties of the materials including the thermodynamic properties are estimated by the value of density of states via the DOS of electron in a material. The DOS calculations provide accurate information about the spacing between energy bands in semiconductors and a clear understanding of the distribution of electronic states [35–37]. The general equation for the DOS in 2-D is given as,

$$g(\omega)d\omega = \frac{L^2}{\pi}\frac{\omega}{v_s^2} \tag{4.1}$$

where v_s represents the velocity of sound and L represents the length of the region under consideration.

In the case of MXenes, which represent a class of 2-D material, the density of states in the energy ranges dE is given by the following equation,

$$g(E) = \frac{1}{2\pi^2}\left(\frac{2m^*E}{\hbar^2}\right)^{\frac{3}{2}} E^{\frac{1}{2}} \tag{4.2}$$

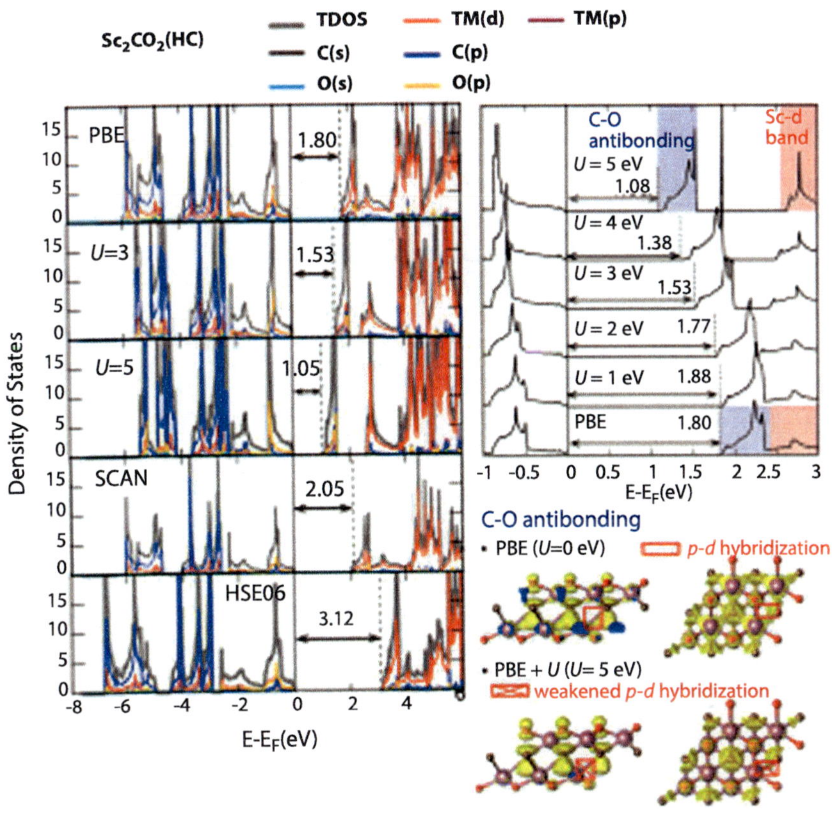

Figure 4.2 Total density of states (TDOS) and partial density of states (PDOS) of Sc_2CO_2. for different exchange-correlation functional [38]. Copyright 2020. Reproduced with permission from Elsevier.

This equation is valid for the region in the k-space, where the dispersion relation $E = \frac{\hbar^2 k^2}{2m}$ holds good.

Bae *et al.* performed DFT calculations on 22 transition metal based MXene compounds, M_2CT_2 with M representing various transition metal as well as T as O, F. They were able to estimate both the partial DOS and total DOS of these MXenes by employing diverse exchange-correlation functionals such as PBE, SCAN, HSE06 and PBE+U [38]. Figure 4.2 and Figure 4.3 represent the PDOS and TDOS of the two MXenes Sc_2CO_2 and Ti_2CO_2 respectively, estimated by employing DFT methods. The calculations were performed using the VASP package which relies on projector augmented wave (PAW). The PBE exchange correlation potential is extremely simple; however, it is prone to self-interaction error. In contrast, the PBE+U imposes the localization of electrons into specific orbitals, governed by non-local Hubbard *U*–type potential parameters [39]. In SCAN method, the Laplacian of the local density is incorporated along with gradient correlations to the local density function and hence is considered as a meta GGA. This approach helps to minimize the self-interaction errors in DFT calculations [40, 41]. The HSE06 provides the highest level of accuracy to the calculation results in DFT based estimations, as it mixes the exact exchange to density functional thus, reducing self-interaction error in calculations [42]. Under experimental conditions, the nature of the acid employed for the exfoliation process determines

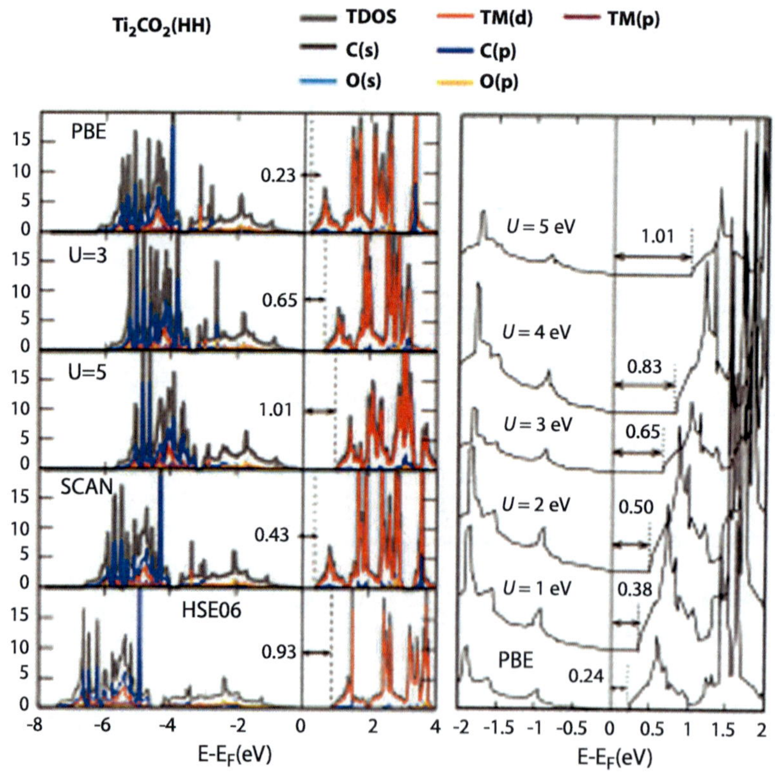

Figure 4.3 Total density of states (TDOS) and partial density of states (PDOS) of Ti_2CO_2 for different exchange-correlation functional [38]. Copyright 2020. Reproduced with permission from Elsevier.

the nature of the termination functional group in a MXene. Since the type of the functional group in a MXene plays a prominent role in extracting its electronic structure and the dynamical stability, a detailed structural stability analysis by performing a volume optimization is to be done to identify its volume for which energy becomes minimum and that structure represents the most stable structure of the analyzed MXene compound. The dynamical stability of MXenes can also be assessed by plotting the phonon dispersion spectrum. Meng *et al.* have done a DFT estimation of electronic structure and dynamical stability of Ti_3C_2 MXene and concluded that the structure possesses a metallic nature [43]. A DFT study performed on $Ti_3C_2X_2$ (where X = S, O, F, and Cl) indicates that the compounds $Ti_3C_2S_2$ and $Ti_3C_2O_2$ possess excellent anchoring behavior and the density of states (DOS) estimation shows that the compounds possess good conductivity and has the ability to possess structural integrity without undergoing chemical decomposition [44]. In the computational study performed by Bae *et al.*, five different structures have been identified for each of the chosen MXene materials based on the position of termination group, i.e. whether terminal group is attached to H site (hollow) or the C site (carbon top) or the T site (transition metal top). Thus, the identified structures include HH, CC, TT, CT and HC. They have performed a phonon stability analysis for the MXene structures, Ta_2CF_2 (HH), Ta_2CF_2 (CC), Hf_2CF_2 (CC) and Hf_2CF_2 (CC) [38]. The study indicates a negative phonon frequency for Ta_2CF_2 (HH), showing that the structure is unstable while all the other MXene structures are found to be stable with positive phonon frequencies only, as shown in Figure 4.4. The band structures of these MXene structures are included in Figure 4.4. The same study also includes an estimation of dynamical stability and electronic properties of vanadium based MXene, V_2CO_2 (HH) using the PBE GGA exchange correlation functional of DFT. The undistorted V_2CO_2 (HH) structure is identified as a non-magnetic (NM) symmetric phase

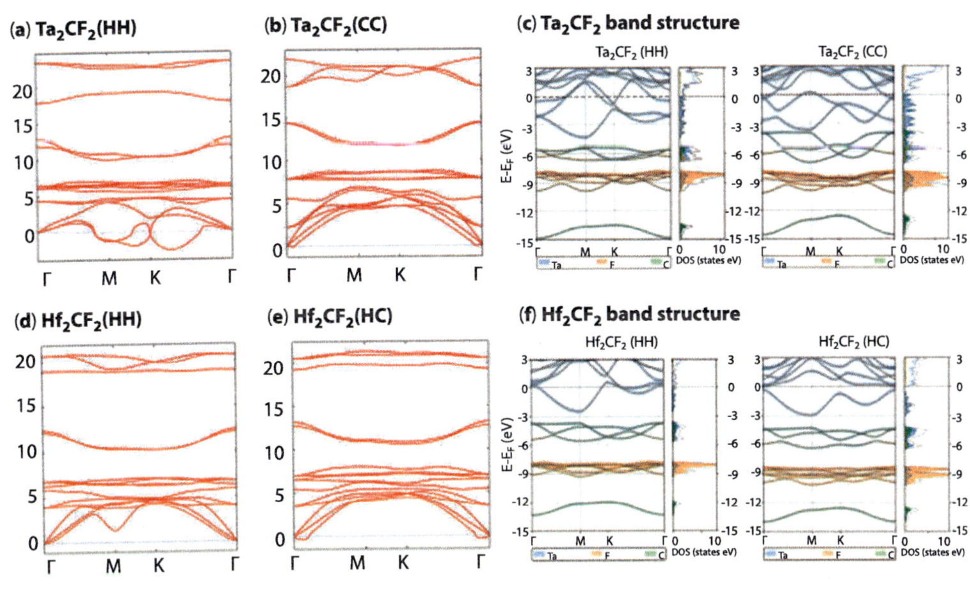

Figure 4.4 The phonon dispersion spectrum of the MXene structures (a) Ta_2CF_2 (HH) (b) Ta_2CF_2 (CC), (d) Hf_2CF_2 (HH) and (e) Hf_2CF_2 (HH) along with the band structure of (c) Ta_2CF_2 (HH) and Ta_2CF_2 (CC) and (f) Hf_2CF_2 (HH) and Hf_2CF_2 (HH) [38]. Copyright 2020. Reproduced with permission from Elsevier.

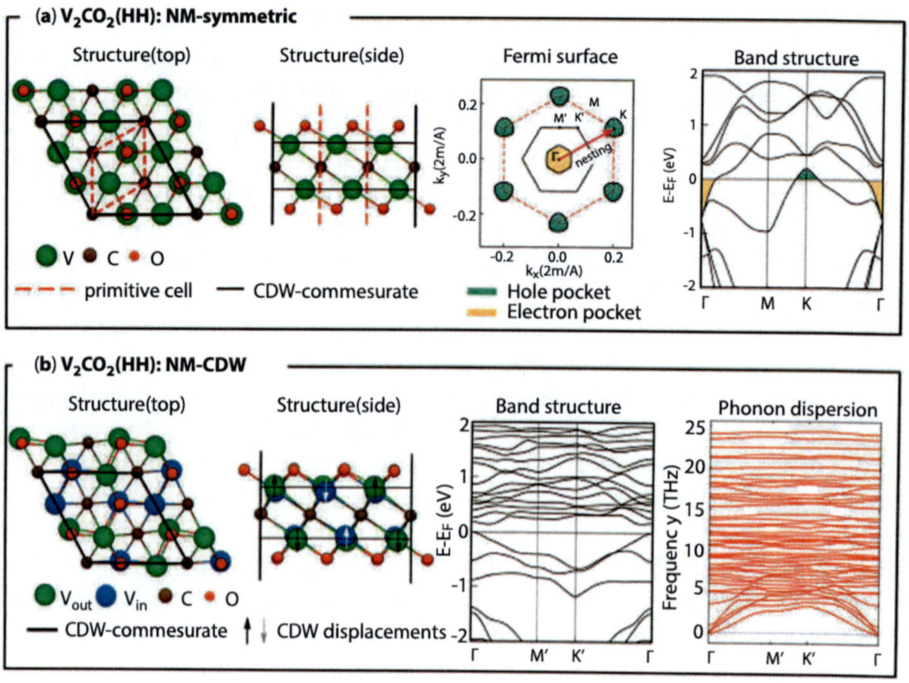

Figure 4.5 (a) Illustration of the dynamical stability attainment of the V_2CO_2 (HH) structure via CDW displacements induced by the vanadium atoms. (a) the NM symmetric phase of V_2CO_2 (HH) structure with the Fermi surface and the band structure and (b) the NM-CDW phase of the V_2CO_2 (HH) with the band structure and the phonon dispersion spectrum indicating dynamical stability [38]. Copyright 2020. Reproduced with permission from Elsevier.

and the corresponding band structure along with the Fermi surface are plotted in Figure 4.5(a) [38]. A displacement of vanadium atom arising from the charge density wave (CDW) results in the opening up of band gap and elimination of the dynamical instability of the structure via a metal – insulator transition. This is indicated via the phonon dispersion spectrum of the NM-CDW V_2CO_2 (HH) structure in Figure 4.5(b) [38].

4.4.2 Band Structure and Bandgap Estimation

The band structure and band gap represent two important parameters that determine the properties of materials and can be specifically employed for categorizing the material into conductor, semiconductor and insulator. A suitable computational code which employs either the pseudopotential method or the plane wave method can be employed for the estimation of the band structure and thereby the band gap of the material. If DFT based methods are employed for the estimation of the band structure of compounds, then most of their ground state properties such as charge density, total energy, electronic distribution etc. can be assessed from a mere knowledge of the structure of the material itself. In contrast, if the material exists in its excited state, then the properties of the material can be estimated by employing the time dependent density functional theory and the many body perturbation theory [45]. GW approximation using Green function is a many body perturbation method and is suitable for estimating the excited state properties of the material under study. It is executed by employing the random phase approximation (RPA), which utilizes a second coulomb interaction to estimate the vortex function in the reciprocal lattice [46]. The product of one electron Green function (G) and a screened Coulomb interaction (W) gives self-energy operator in GW approximation and is carried out by employing the perturbation theory from LDA or GGA results. It is difficult to compute a full self-consistent GW method for practical system and hence the best possible approximations to the G and W functions are initially taken to estimate the other related properties of the system [47].

Zhang *et al.* performed a GW quasiparticle calculation to estimate the electronic properties of the 2D MXene, M_2CO_2 (where M stands for Hf, Zr, and Ti) by employing the recently invented accelerated GW methods [48]. They predicted the band gaps of Hf_2CO_2, Zr_2CO_2, and Ti_2CO_2 as 2.45 eV, 2.13 eV, and 1.15 eV respectively and are identified as fully in agreement with the results of experimental study. The electronic distribution and total and partial DOS estimation indicate that the low band gap of Ti_2CO_2 arise from low energy of Ti 3d state in comparison to d states of Hf and Zr. The study also indicates that the structure $Hf_{2-2x}Ti_{2x}CO_2$ is chemically stable and it is anticipated that they can be fabricated with least complications. By employing DFT, W. Sun *et al.* predicted the existence of novel wide band gap for ordered MXene structures consisting of carbides and nitrides of Ti-V, Ti-Cr or Ti-Mn [49]. For the MXene TiM_2X_2T (M for V, Cr or Mn, X for C or N as well as T for H, F, O or OH), the study predicted surface termination-based metals to insulators transition in Cr-N or Mn-N series. The Hybrid exchange correlation functional HSE06 was employed for estimating the band gap of both the terminated $TiMn_2N_2F_2$ MXene and the unterminated $TiMn_2N_2$ MXene. The study predicted a wide band gap of 1 eV for the terminated MXene, $TiMn_2N_2F_2$ and the unterminated $TiMn_2N_2$ MXene was found to exhibit metallic nature. On the other hand, $TiCr_2C_2T$ MXene compounds were predicted to exhibit semiconductor properties.

The first principle calculations assisted by considering the many body effects was employed for estimating the optoelectronic properties of a Ti_2CO_2 monolayer. The study indicated that the pure Ti_2CO_2 possesses an indirect band gap of 1.32 eV [50]. The electronic distribution and the DOS calculation indicated the existence of conduction band minimum (CBM) and valence band maximum (VBM) at M and G symmetry point respectively. Without any applied strain, an estimation of band gap and binding energy for electron-hole pair provided values 1.260 eV and 0.560 eV respectively for the pristine Ti_2CO_2 monolayer. However, the application of biaxial strains shifted the lowest unoccupied band corresponding to G point slightly downwards while shifting the lowest unoccupied band of M point slightly upwards, thus resulting in a direct band gap when the pristine Ti_2CO_2 monolayer was 6% strained. The increase in strain also resulted in an increase of the lattice parameter and an exciton wave function expansion, which in turn caused a lowering of optical band gap and binding energy of first bright exciton. In a similar way, first principle DFT calculations were done on the MXene compounds Nd_2N as well as Nd_2NT_2 by research groups from China and the results indicated that Nd_2NT_2 MXene could exhibit half-metallic nature, with a band gap slightly above 1.70 eV and a work function value ranging from 1.830 eV to 6.500 eV as shown in Figure 4.6 [51]. An estimation of electronic properties of varying terminations of Nd_2NT_x with T = OH, O, or F and $x = 0.5$, 1 or 1.5, showed semiconductor nature for $Nd_2NO_{1.5}$, while $Nd_2NO(II)$ exhibits half-metallic semiconductor properties, which can be used for optoelectronic and spintronic applications in future [51]. The effect

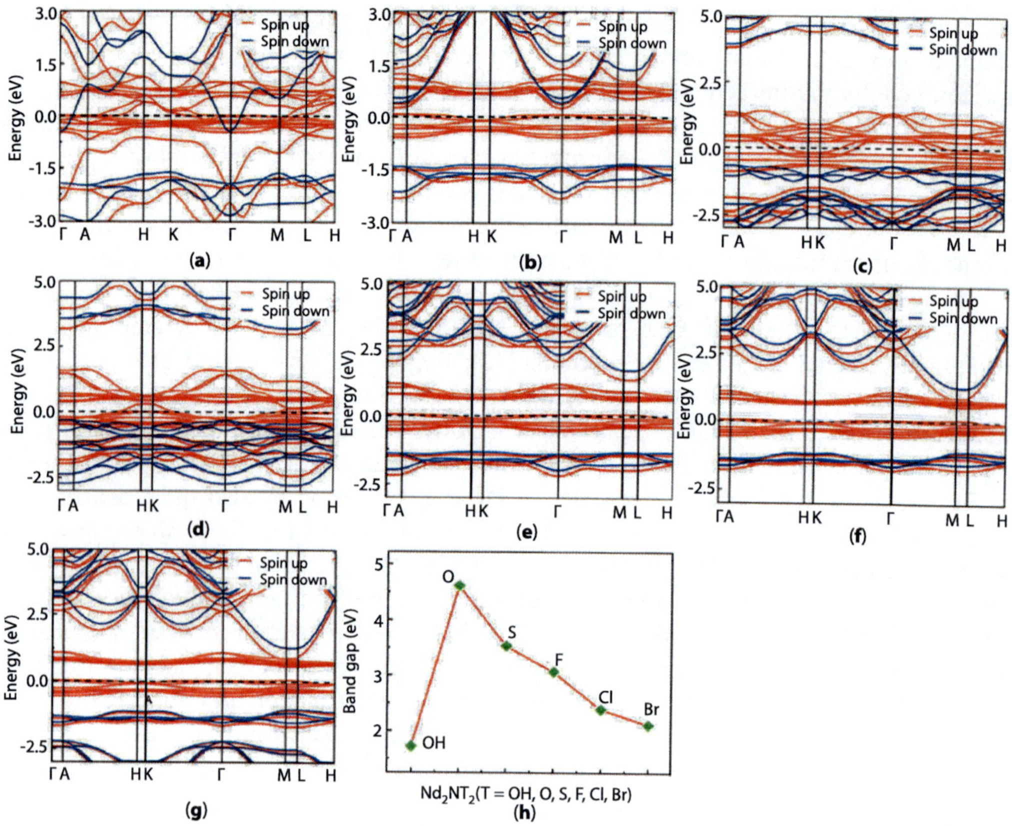

Figure 4.6 The estimated band structure of (a) Nd_2N and (b), (c), (d), (e) and (g) Nd_2NT_X MXenes (with T = OH, O, S, F, Cl, Br respectively) and (h) the band gap variation with the incorporation of different functional groups [51]. Copyright 2021. Reproduced with permission from Frontiers.

of various terminal groups on the total DOS is studied by plotting the TDOS of the chosen MXenes, Nd_2NT_x as represented in Figure 4.7 [51].

Machine learning methods can also be employed to estimate the band gap of existing MXenes and to predict the band gap of hitherto unexplored MXene structures. A.C. Rajan *et al.*, estimated electronic properties and band gap for a class of MXenes accurately by employing statistical learning [52]. The estimation of band gap was done by providing the readily available properties of MXenes such as the bond lengths, phases, atomic radii and boiling or melting points as input parameters to the regression algorithms using a machine learning approach. The models were developed with the help of kernel ridge (KRR), support vector (SVR), Gaussian process (GPR), and bootstrap aggregating regression algorithms. The predicted band gap was found to be accurate with the GPR algorithm with the least value of RMS error of 0.14 eV only and the computation was found to be fast unlike the DFT methods (Figure 4.8) [52]. While most of the DFT based approaches such as GW or GGA are time consuming and demands huge computational power, machine learning based models provide accurate results within a few seconds itself, indicating that they may outperform DFT estimations in near future.

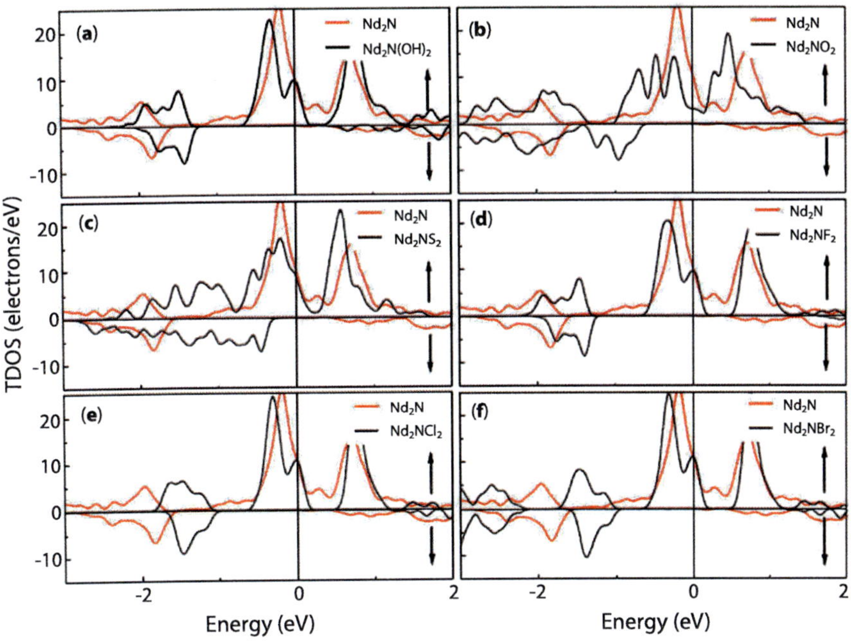

Figure 4.7 The effect of varying the functional group on the TDOS of various MXene structures, Nd_2NT_x with the TDOS of the Nd_2N as the reference [51]. Copyright 2021. Reproduced with permission from Frontiers.

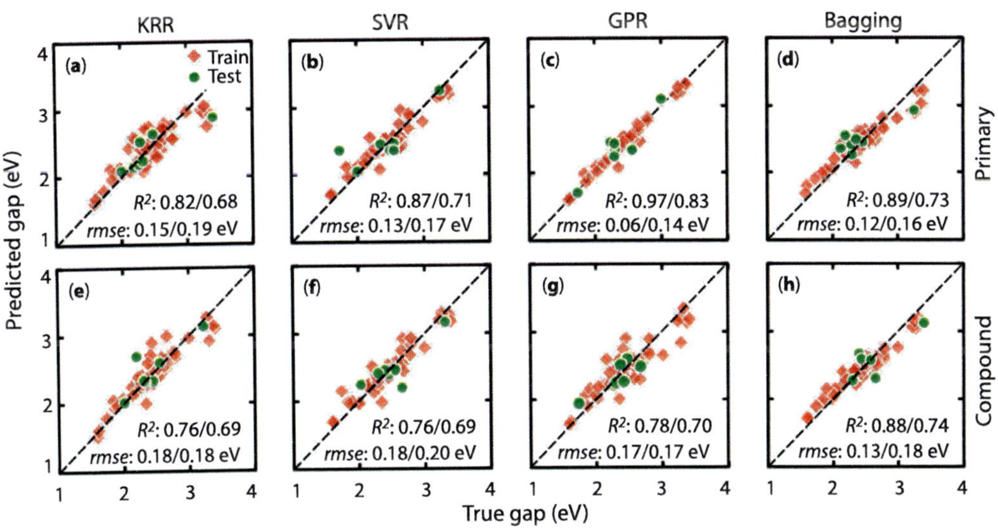

Figure 4.8 Predicted band gaps (indicated with scatter plots) of MXene using machine learning techniques R2 and rmse represent accuracy metrics of best models of (a/e) KRR, (b/f) SVR, (c/g) GPR and (d/h) bootstrap aggregating regression with 90% training (red) and 10% testing (green) data [52]. Copyright 2018. Reproduced with permission from American Chemical Society.

4.4.3 Methods to Enhance Electronic Properties

Since MXenes form a category of 2D layered transition metal carbide/nitride compounds, they possess diverse advantages such as excellent carrier transport properties, optimizable band gap, optimum mechanical properties and sufficiently favorable stable structure. Researchers have focused their attention to these materials with the aim to improve the existing properties of MXenes and to identify new MXene materials which can be apt for several applications. Y. Zhang *et al.* showed that band gap of $Hf_{2-2x}Ti_{2x}CO_2$ ($0 \leq x \leq 1$) MXene based semiconductor can be effectively optimized from 2.450 to 1.150 eV, thus making it suitable for future optoelectronics devices [48]. The nature of the T groups in functionalized MXenes determines the properties of the MXenes largely and they exhibit properties quite different from that of their corresponding MAX phase [53]. Hence, the structural and the chemical stability of these functionalized MXenes can be optimized to a large extent via surface functionalization of the structure [54]. Surface functionalization also affects the electronic, thermal, elastic, optical and magnetic properties of MXenes largely. New categories of MXenes with enhanced thermal and elastic properties can be achieved by means of surface functionalization and density functional theory-based approaches provide a great insight towards this direction [55]. Hence, the foundation to the practical applications of MXene in optoelectronic and photovoltaic devices is to identify the influence of surface functionalization on the properties of them.

Pure MXenes exhibit metallic properties without a band gap. Recently, researchers have focused their attention on engineering the band gap of MXenes, inspired mainly from the positive results of band gap engineering of graphene. Such studies have shed light on the properties of 2D MXenes that can be upgraded to a large extent via surface functionalization and strain engineering [56]. A group of researchers succeeded in converting the initially metallic Sc_2C monolayer to a semiconducting MXene having band gap in range of 0.450–1.800 eV by means of OH, F or O functionalization [57]. It was also identified that the initially metallic Sc_2CO_2 MXene can be converted to a direct band gap semiconducting MXene by the application of an external electric field or strain [58]. The electronic properties and thereby the band structure of semiconducting Sc_2CT_2 MXene can be enhanced by adding dopant transition metals such as Ti, Cr, V or Mn to the structure, thus, enabling the material to act as suitable for nanophotonic and optoelectronic application. The DFT studies on monolayers of WCrC and $WCrCO_2$ indicated that they are dynamically stable. These studies revealed the metallic and semi metallic nature of WCrC and $WCrCO_2$ respectively. The semi-metallic nature of $WCrCO_2$ with low band gap, gets optimized via strain engineering with an applied external electric field. Semi-metallic $WCrCO_2$ monolayers possess two band gap that gets preserved via applied electric field. The study also showed that the material can be changed to metallic nature by providing a tensile strain. This arises due to a gradual increase in band gap of the $WCrCO_2$ monolayer under the influence of a uniaxial strain thus, resulting in the formation of an exotic 2D double semi-metallic characteristics. Thus, strain engineering along with application of electric field can make $WCrCO_2$ monolayers as new platform for studies on high-speed low-power dissipation systems [59]. The application of a strain of 4%, results in a maximum band gap of 0.449 eV for the MXene Ti_2CO_2, while the material makes a transition from a semiconductor to a metal when the applied strain is reduced to -3% (Figure 4.9) [60]. When the strain is increased above 4%, a direct band gap between the symmetry points arises for the MXene material as shown in Figure 4.10 [60]. A similar study on the variation of work function of Ti_2CO_2 MXene as a function of biaxial strain is shown in Figure 4.11 [60]. As the biaxial strain applied to the material increases, the work function of the material decreases. A study conducted on the same MXene material Ti_2CO_2 indicated a variation in energy and band gap of the material under pressure. The energy under pressure increases and band gap lowers with applied pressure as illustrated in Figure 4.12 [61]. The tuning of electronic properties of materials are highly correlated to their enhanced optical properties also.

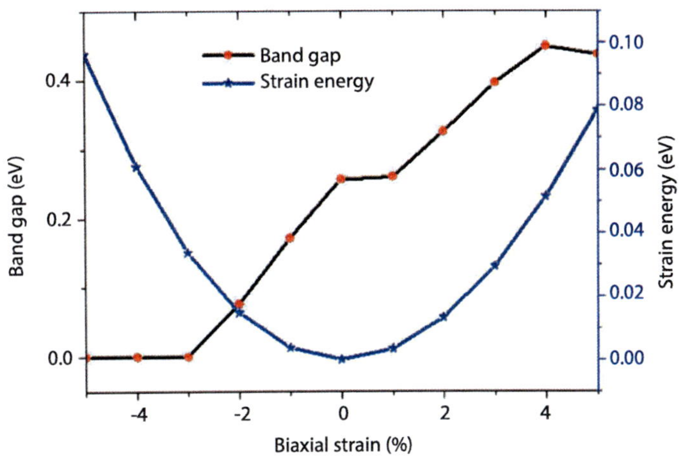

Figure 4.9 Variation of band gap and strain energy of Ti_2CO_2 MXene under different values of applied strain [60]. Copyright 2020. Reproduced with permission from American Chemical Society.

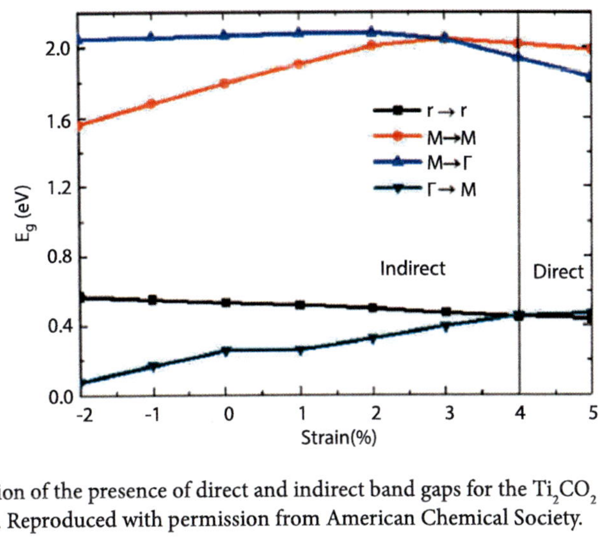

Figure 4.10 Illustration of the presence of direct and indirect band gaps for the Ti_2CO_2 MXene under strain [60]. Copyright 2020. Reproduced with permission from American Chemical Society.

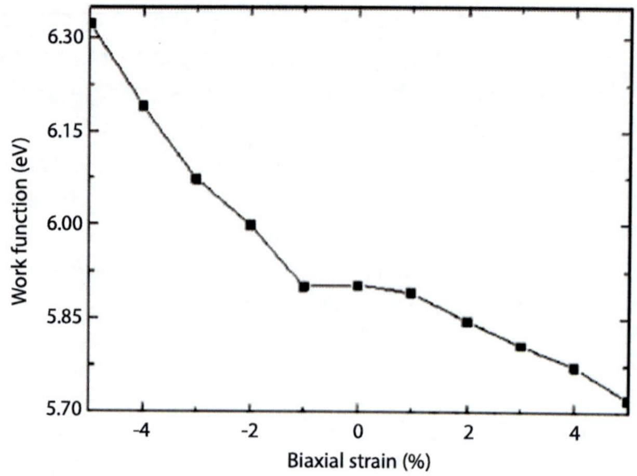

Figure 4.11 Work function of Ti_2CO_2 MXene as a function of the applied biaxial strain [60]. Copyright 2020. Reproduced with permission from American Chemical Society.

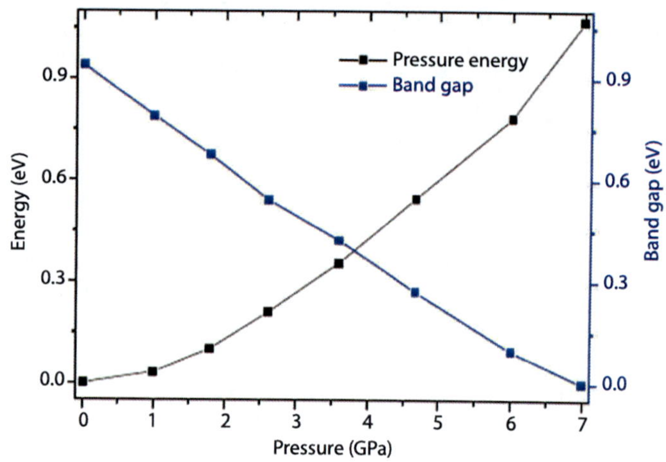

Figure 4.12 Variation of energy and band gap of Ti_2CO_2 as a function of pressure applied to the material [61]. Copyright 2020. Reproduced with permission from The American Chemical Society.

4.5 Electrical Properties

MXenes have drawn interest owing to the extraordinary electrical conductivity, low weight, flexibility, transparency, tunability of surface chemistry, and simplicity of solution processing since their discovery in 2011 [61, 62]. Here, both practical and theoretical perspectives on the possibilities of 2D MXenes for the development of conductive materials are reviewed, including electromagnetic interference shielding, flexible optoelectronics, sensors, and thermal heaters. One of the main factors contributing to MXene materials' quick growth in research is their superior electrical conductivity, greater in comparison with reduced graphene oxide (rGO) sheets by a factor of above 10. Graphene, the most well-known 2D material, has less potential than MXenes in applications requiring strong electrical conductivity and solution processing.

Ti_3C_2, which exhibits elegant performance in supercapacitors, batteries, water purification, flexible electronics composites, and sensors, is the most researched MXene. Electrically conductive [63] MXenes have galleries that can accommodate diverse ionic and molecular species. Because O, OH, and F terminate at their surfaces [64]. MXenes are hydrophilic [65, 66]. Recently, it was discovered that film of 100 cm x 10 cm MXene had exceptional tensile strength, electrical conductivity and EMI shielding capabilities, thereby outperforming the 2D compounds invented till date [67].

The electrical conductivity of MXene can be affected by extrinsic factors as its preparation conditions, MAX source material, post-etching treatments, ultrasonication etc. [68–73]. Depending on these features, the conductivity of a multi-layered MXene disc can range from 1 S/cm to 15000 S/cm for transparent MXene film. As a result, when creating MXenes for applications needing significant electrical conductivity, cautious management of some factors is necessary [74]. Shuck *et al.* showed the impact of source materials on electrical conductivity of $Ti_3C_2T_x$ MXene (Figure 4.13) [75]. A graphite-based carbon precursor had a higher electrically conducting nature (4400.0 $S.cm^{-1}$) as well as excellent stability in comparison to TiC-based carbon precursors of $Ti_3C_2T_x$ that exhibited an equivalent electrical conductivity (3480.0 $S.cm^{-1}$) but poor colloidal stability. The electrical conductivity (1020 S/cm) and chemical stability of Ti_3C_2Tx MXene made from carbon lampblack are the lowest (5.1 d) [75].

The outcomes of numerous studies clearly show the benefits of developing electrically conductive materials. In comparison to graphene (having conductivity 200 $S.cm^{-1}$) along with various man made material of equivalent thicknesses, a $Ti_3C_2T_x$ MXene film having conductivity 4600 $S.cm^{-1}$ (at 45 m thickness) obtained an EMI SE of 92 dB [76]. Additionally, MXenes have an excellent electrical conductivity and form dispersions and can be included within polymer matrices [77]. A recent study that looked at 16 different types of MXenes showed the impact of conductivity on EMI shielding effectiveness; the highest conductivity ($Ti_3C_2T_x$) leads to excellent EMI shielding capabilities. The very transparent conductivity of MXenes [78] has also been studied, with a lowered sheet resistance of 437 Ohm/sqnm and

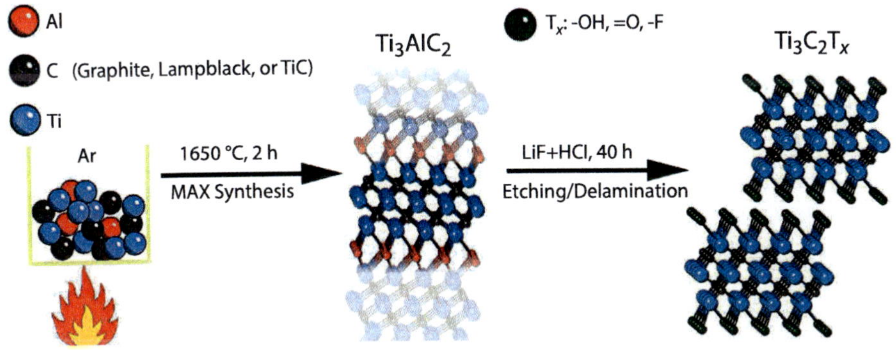

Figure 4.13 Preparation of Ti_3AlC_2 from different carbon sources (TiC, graphite, and carbon lampblack) followed by $Ti_3C_2T_x$ MXene synthesis [75]. Copyright 2019. Reproduced with permission from the American Chemical Society.

77% transmittance at 550 nm [79]. These exceptional features were attributed to excellent density of states (DOS) at Fermi level, huge flake size, and horizontally oriented 2D MXene sheets, which prevented percolation-related problems. Due to its better conductivity as well as sheet-like shape, MXene surface is now ideal as an immobilization platform for biological sensors in improving sensitivity and selectively detecting biological species. The acetylcholinesterase (AChE) enzyme was immobilized on Au/MXene composite substrate and then employed to identify the herbicide methamidophos (ATCI) [80]. MXenes have recently been studied for non-traditional applications including flexible and wearable THs due to their superior electrical conductivity.

Electrical features of MXene are governed by its crystal form, chemical makeup, as well as surface functionalization. Its structure-property relationship must be investigated in order to comprehend why the majority of MXenes have a metallic quality. Theoretical models have been extensively used to understand the underlying mechanism, and they are now supported by experimental results. It's crucial to properly control the synthesis process to minimize the possibility of errors that could negatively affect the electrical properties of the final product. Here, crystal structures of diverse MXenes along with structural and electrical properties are elaborated upon.

4.5.1 MXene Structure and Composition

MXene is prepared via selectively etching parent carbide MAX phase material, $M_{n+1}AX_n$ phase (211, 312 or 413 with n = 1). MAX phase exhibits layered hexagonal structure with P63/mmc space group. In MAX phase, "X" atoms fill octahedral positions while "M" layers become tightly packed. The "A" layer atoms in $M_{n+1}X_n$ layer are connected to "M" atoms via metallic link (M-A), however M-X link is often stronger than the M-A link owing to mixed ionic/covalent/metallic nature. Layers of the "A" group components are interspersed with M_6X octahedra in the unit cells [81]. It is believed that "M" group components within MXene generate 6 chemical links to nearby "X" site atom as well as surface terminations (T_x: OH, O, and F), resulting in $M_{n+1}X_nT_x$, such as M_2XF_2, $M_2X(OH)_2$, and M_2XO_2. There are MAX phases that may be created utilizing different M, A, and X element combinations and solid solutions [82]. Different MXene forms, viz. single metal, double metal, or double X solid solution MXene, are generated from various MAX phases (C and N, also called carbonitrides). The two types of double-metal MXenes are as follows: 1) a solid solution where M and M metals randomly occupy the M-atom sites, and 2) an ordered form where M elements can be organized either in plane (i MAX) and out of plane (o MAX). Many MXene family members, such as Ti_3AlCN and $Ti_3(Si,Al)C_2$, are created by mixing a 4th element to MAX phase for developing quaternary structure. The ordered double metal MXene (for example, $Mo_2Ti_2C_3T_x$, $Cr_2TiC_2T_x$, and $Mo_2TiC_2T_x$) created from matched o-MAX phases show noticeably different electrical properties when compared to normal MAX generated MXene materials like $Ti_3C_2T_x$.

4.5.2 Electrical Conductivity

MXenes are used in places where electrical conductivity is important because of their outstanding metallic conductivity, which ranges from 5000 to over 15000 S/cm [83]. In 5G network era, modern communication equipment and circuits work at tremendous speed than before, creating unwanted electromagnetic wave pollution namely electromagnetic interference (EMI). This ensuing EMI can reduce device performance, and long-term exposure to it may even lead to cancer in people. As a result, it is now a global priority to prevent and/or reduce dangerous electromagnetic (EM) radiations. The shield of excellent electrical conductivity frequently and substantially reflects incoming EM wave due to significant impedance mismatching at shield-air interfaces. These EM waves are significantly attenuated inside the conducting shield due to the ultra-strong eddy current along with ohmic loss [84].

Since they are the most conductive materials, traditional metals (Ag, Cu, Al, and Ni) have been used for decades to shield against EMI. However, their increased density and susceptibility to corrosion prevent them from being employed in lightweight mobile devices. As a result, several conducting materials have been researched as heavy metal substitutes, such as MoS_2, expanded graphite, reduced graphene oxide (rGO), graphene, carbon nanotubes (CNTs), and carbon nanofibers (1D) (2D materials). Although these materials have advantages including reduced weight, enhanced environmental stability, and greater corrosion resistance, they don't have enough electrical conductivity, requiring the usage of larger amounts of the materials [85–88]. In a pioneering attempt, Shahzad *et al.* explored EMI SE of independently synthesized mono-metal $Ti_3C_2T_x$, ordered double-metal $Mo_2TiC_2T_x$, and $Mo_2Ti_2C_3T_x$ MXenes [89].

4.5.3 Surface Functionalization

MXene can be specifically modified in order to obtain a particular type of features. On the other side, surface oxidation damages electrical properties and can be seriously affected by prolonged contact with air or aqueous dispersion. Regarding a specific stacking configuration of the next layer, MXenes can be stacked in either one of two ways: simple hexagonal (SH) or Bernal stacks. In addition, compared to pristine MXene, presence of surface functional groups on MXenes reduces interlayer coupling between the MXene sheets. However, it outperforms secondary interaction which is frequently up to six times more powerful than carbon based materials especially graphite or MoS_2 [90].

4.5.4 Methods to Enhance Electrical Properties

In the years that followed its discovery, a sizable portion of MXenes was found. The majority of the study for energy storage has focused on their electrochemical properties. As a result of an improved knowledge of the configuration-feature connection, new applications have developed in almost every area of materials research. The electronic properties of MXenes have gained attention since conducting electrodes have numerous applications. Like their MAX parent phases, MXenes are primarily metallic in nature. On the other side, surface modification alters electrical properties of MXenes.

DFT studies indicate that electrical properties of MXene are correlated to its composition together with surface termination. Despite the constant evolution of new MXene synthesis techniques, bare MXenes have not yet been identified. However, mixed termination like -F, -O, and -OH locate at MXene surface, producing localization of electronic state, thereby varying electrical and related characteristics. Hence, focused control of electrical characteristics can be achieved by altering composition along with termination of MXene surface. But, owing to the macroscopic arrangement of connected nanosheet layers of MXene film, intercalation of layers affects their electrical properties. Khazaei *et al.* performed a theoretical prediction using DFT and discovered a hydroxyl moiety containing $Sc_2C(OH)_2$ MXene exhibiting direct band gap while other semiconductors had indirect band gap [91]. The side and top views of the functionalized MXenes are given in Figure 4.14 [91]. This is because the electron numbers in the outermost, partially filled atomic orbitals of Titanium, Hafnium, and Zirconium are alike and they all belong to the same group. Surface similarities exist. Terminal groups impact electronic structure (F and OH). Due to the fact that neither of these can accept two electrons, whereas the O group can, their underlying electrical properties are encouraged to change. The electron transport characteristics of MXenes have been predicted by DFT calculations in a number of studies, although the majority, do not properly corroborate this experimental evidence. Schultz *et al.* obtained 3.9 eV work function for $Ti_3C_2T_x$ MXene in the as-loaded state that incremented to 4.8 eV when it was annealed at 380°C. In addition to the expected work function of pristine $Ti_3C_2T_x$ MXene, DFT predicts work function of 6.0 eV for $Ti_3C_2O_2$, which is significantly higher than the loaded samples (3.9 eV) [92]. The lack of consideration for adsorbed contaminations and terminal groups in DFT calculations represents the main reason for discrepancy between actual and theoretical work functions. Additionally, metallic (like $Ti_3C_2T_x$) or semiconducting characteristics of MXenes have been demonstrated (e.g., Ti_2Co_x, Mo_2CT_x). However, there have only been a few publications in recent years that provide experimental evidence for semiconducting properties ($Mo_2TiC_2T_x$ and Ti_3CNT_x MXenes) [93, 94].

It is unclear whether the behavior of semiconductors is influenced by their inherent stoichiometric makeup or by outside factors like the intercalant effect, terminal groups,

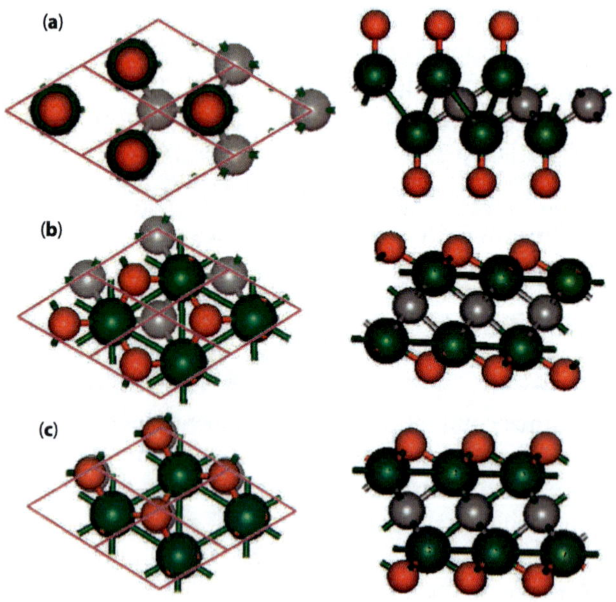

Figure 4.14 Models of the functionalized MXene systems in the top and side views: (a) two functional groups on top of two transition metals, (b) two functional groups on top of hollow sites A, (c) two functional groups located on top of hollow sites B [91]. Copyright 2013. Reproduced with permission from Advanced Functional Materials, Wiley.

measurement techniques, and so on. The experimental outcomes differ from the theoretical predictions because the synthetic conditions, fabricating parameters, surface terminations, and interlayer gaps between the MXene sheets are not under our control. The lattice parameters and layer thickness of MXenes have been the focus of some DFT research, although these parameters are insufficient to forecast the features of surface and stability of the structure. The theoretical calculations based on end terminals on the energy of cohesion and surface adsorption capacity result in a more accurate number that is equivalent to the actual data. The metallic electrical conductivity of single-layer MXene $Ti_3C_2T_x$ flakes was first discovered by Miranda *et al.* After applying gate voltage and observing the field-effect and perform calculations on density as well as field-effect mobility. Results showed that $Ti_3C_2T_x$ MXene kept its metallic properties even after being reduced to a single layer from a multi-layered structure [95]. The charge-carrier density ($3.10x10^{22}$ cm^{-3}) with mobility (0.90 $cm^2/V.s$) of an interconnected dense spin-cast $Ti_3C_2T_x$ MXene film were measured by Dillon *et al.* using the Hall effect [96].

It was demonstrated that maintaining a thin film thickness did not affect in-plane conductivity and that the proper alignment of the MXene flakes increased in-plane electrical conductivity. The difference in optical and DC conductivities between MXenes and pure metals indicates that inter-flake resistances, which obstruct electron transport, are most likely to blame for the lower electron mobilities in MXenes. At low temperatures, Halim *et al.* revealed that the variable-range hopping (VRH) between MXenes layer represent a dominant conduction process in various Mo containing MXene, but thermal excitation represents a dominant mechanism for carrier conduction under extreme values of temperature (Figure 4.15) [97]. Electronic transport is induced near the Fermi level by phonon aided hopping. A steady state between tunneling probability, active term as well as DOS available to hop determines electron hopping rate. The delocalized electrons, thermally stimulated over the mobility edge, are primarily responsible for electronic transport at higher temperatures (EC). Thermally induced transport and VRH together add to the overall conductivity of Mo-based MXenes. Hart *et al.* made similar observations with Ti_3CNT_x

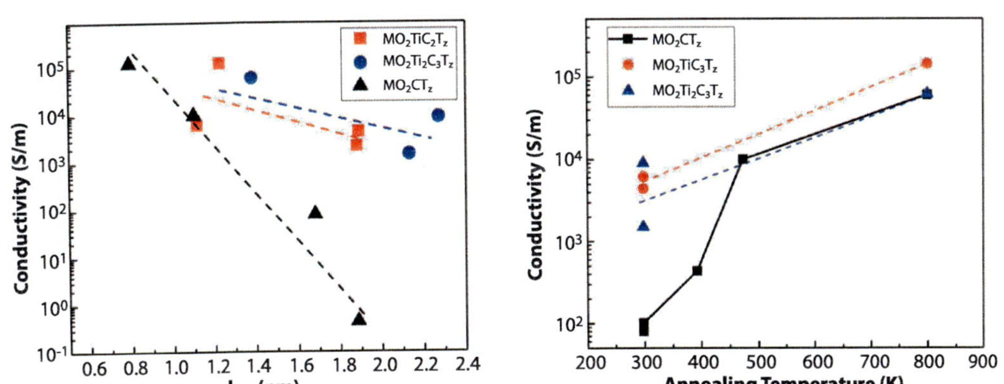

Figure 4.15 Conductivity of Mo_2T2C_3Tz, Mo_2TiC_2Tz, and Mo_2CTz freestanding thin films plotted against (a) interlayer distance $d_{c/2}$ and (b) annealing temperatures [97]. Copyright 2018. Reproduced with permission from Physical Review B, APS Physics.

MXene, indicating negatively changes resistance as a function of temperature, mostly due to an inter-flake electron hopping mechanism.

Recent research has shown the electrical conductivity for 16 different MXenes, like single metal, ordered double-metal as well as random solid solution MXene [98]. Ti-based MXenes (Ti_2CT_x, $Ti_3C_2T_x$, Ti_3CNT_x, and $Ti_{1.6}Nb_{0.4}CT_x$) and V_2CT_x MXenes have conductivity above 1000 $S.cm^{-1}$, but Nb containing MXene have low conductivity. The electrical conductivity of Nb containing MXene decremented as Nb-element content increased implying that the M and X elements' nature and composition can be altered to customize the conductivity of MXenes. As a result, novel spectrum for electrical conductivity distribution can emerge, though the synthesis conditions in this work were generally comparable, with just small variations in etching conditions, etchants and precursors, and other external considerations. Based on film thicknesses and volume of trapped ion or water molecule, MXenes' electrical properties can vary even more. It is commonly known that MXenes

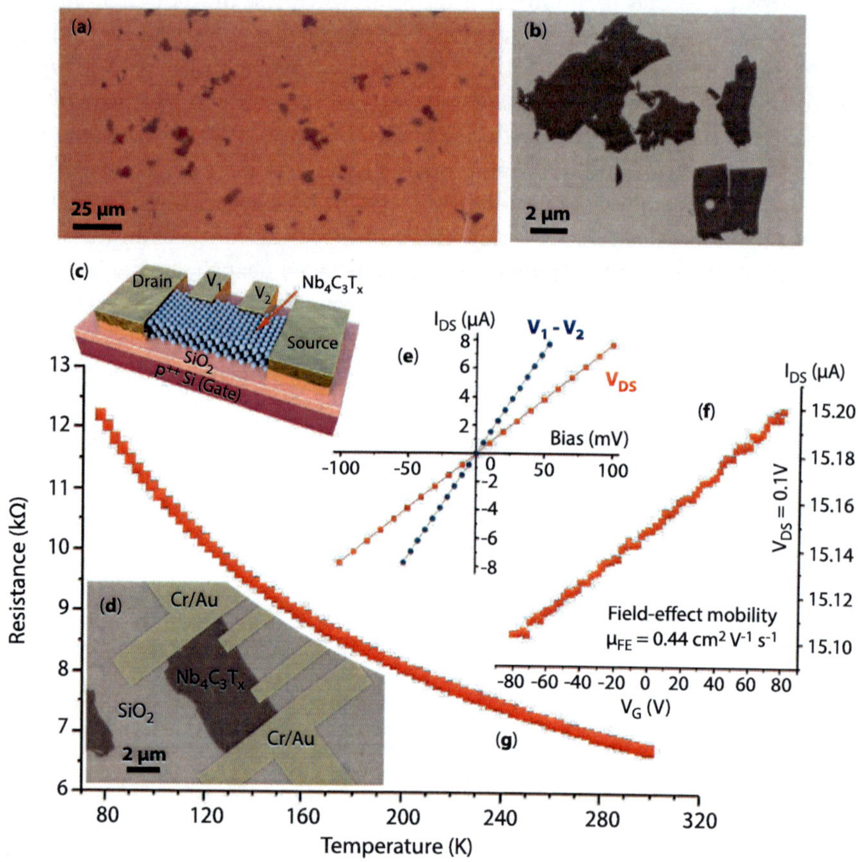

Figure 4.16 Electrical property of $Nb_4C_3T_x$ flake, (a) Optical image, (b) SEM image, (c) schematic representation of channel of $Nb_4C_3T_x$ on a four-terminal device, (d) monolayer $Nb_4C_3T_x$ MXene flake based 4-terminal equipment is depicted in a false-colored SEM picture, (e) I-V curves measured at $Nb_4C_3T_x$ device's four terminals are displayed in panel, (f) features of a monolayer $Nb_4C_3T_x$ FET's two-terminal transfer, (g) monolayer $Nb_4C_3T_x$ device resistance in relation to temperature [99]. Copyright 2020. Reproduced with permission from Advanced Electronic Materials, Wiley.

are very promising to diverse applications. Despite being helpful in revealing the physical characteristics of MXenes, individual monolayer measurements of these characteristics are still scarce. The first mechanical and electrical analyses of individual single-layer flakes of $Nb_4C_3T_x$ MXene made utilizing a more effective synthetic technique were published by Lipatov *et al.* [99]. The conductivity of $Nb_4C_3T_x$ monolayer was estimated as 1024 ∓ 165 S/cm, and is greater than conductivity of bulk $Nb_4C_3T_x$ assembly (Figure 4.16) [99]. In order to develop biomimetic core-shell MXene@graphene oxide (MX@GO) fiber that mimics the structure of wood, GO was employed as a wrapping layer for MXene. This layer made it simple to combine MXene particles into large-scale molecules while also shielding them from oxidation. The result is that the composite fiber has excellent tensile strength (290.0 MPa) along with high conductivity (2400.0 $S.m^{-1}$) in its as-fabricated state. The conductivity of composite fiber reduced by 30% only and became 1800 $S.m^{-1}$ at 100°C. The creation of MXene-based fibers with arbitrary functionalities and morphologies that can be utilized in a number of high-value fabric applications is made possible by the findings of this study [100].

Electrical synapses of MXene/SiO_2 resistive random-access memory (RRAM) equipment were constructed via utilizing characteristics of SiO_2 and MXenes. Ag/MXene/SiO_2/Pt, the suggested RRAM materials, exhibit resistance switching capabilities, allowing them to flip between volatile and non-volatile states. Ag/MXene/SiO_2/Pt devices are better suitable for imitating biological synaptic plasticity because of their intriguing properties. The electrical traits of Ag/MXene/SiO_2/Pt RRAM equipment can be achieved with both volatile and non-volatile behavior in a single device by adjusting a compliance current limit. When the compliance current limit is low (between 1 and 500 nA), the volatile switching behavior develops, however for high compliance current limit , a non-volatile switching behavior appears (between 10 A and 5 mA) [101]. Following post-treatments of alkalization and calcination for removing functional group from layered Ti_3AlC_2, which exfoliated the material in an aqueous HF solution, the process produced highly conductive MXene-Ti_3C_2 nanosheets (Figure 4.17) [101].

The impacts of post-treatment procedure on electrical properties and microstructures of 2D nanosheets as-prepared were investigated. It has been demonstrated that the alkaline treatment can increase the layer gap between 2D Ti_3C_2 layers and remove functional groups from the Ti_3C_2 surface. After one hour of calcination at 600°C, the conductivity of Ti_3C_2

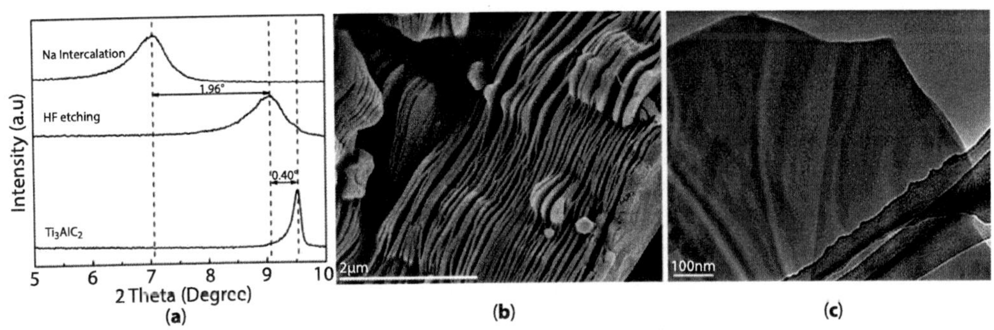

Figure 4.17 (a) MXene Ti_3C_2 nanosheets' XRD profiles before and after alkalization, Images of the after-alkalization in (b) SEM and (c) TEM [101]. Copyright 2020. Adapted from MDPI.

nanosheets improved from 850 S/cm to 2410 S/cm because of its smaller number of OH functional groups and short conduction paths [102]. $Ti_3C_2T_x$, MXene, and CNC reinforced electro spun PVA nanofibers all had excellent electrical conductivity that increased with $Ti_3C_2T_x$ content, especially for the C_0M_2 sample (0.140 wt. percentage $Ti_3C_2T_x$), which had conductivity value, 0.80 mS.cm^{-1} and narrow band-gap semiconducting properties, making it a good candidate for electrode materials in Li-ion batteries [103–106]. The secret to tailoring the properties of MXenes is the exact manipulation of intrinsic functional groups by post-treatment processing. It is also difficult to investigate the equivalent exterior termination grafting, which may complete the necessary MXenes characteristics and so meet the demands of numerous applications. Finally, the properties of MXenes will unavoidably be diminished when combined with other substances. It is crucial to create a hybrid system that may significantly enhance MXene properties with a small amount of additives while also interacting strongly with nanosheets.

4.6 Optical Properties

The novel 2D MXenes have high demand for diverse applications in various fields including optoelectronics, supercapacitors, solar cells and biomedicine. They exhibit exceptional electronic, optical, mechanical, elastic and transport properties, which make them a potential candidate for various applications. The exceptional non-linear optical response shown by MXenes have made them a potential candidate for applications in ultrafast photonics. The exceptional optical properties of MXenes are highly correlated to their electronic and structural properties. From a knowledge of the dielectric constant of material, it is possible to extract its optical properties. The dielectric constant of the material, $\varepsilon(\omega)$ consists of a real part ε_r and an imaginary part ε_i, which can provide valuable information about the optical behaviour and characteristics of the material under study.

$$\varepsilon(\omega) = \varepsilon_r(\omega) + i\varepsilon_i(\omega) \tag{4.3}$$

The dielectric constant is related to the electric susceptibility, $\chi(\omega)$ and is given by the following equation,

$$\varepsilon(\omega) = 1 + \chi(\omega) \tag{4.4}$$

The optical conductivity of the material is given as,

$$\sigma(\omega) = -i\omega\varepsilon_0\chi(\omega) \tag{4.5}$$

The refractive index of the material is related to complex dielectric constant as shown below.

$$n + ik = \sqrt{\varepsilon} \tag{4.6}$$

In this equation, k is the extinction coefficient and is correlated to absorption coefficient by the following relation,

$$\alpha_0 = \frac{2\omega}{c}k \tag{4.7}$$

In 2D MXenes, the presence of the functional group has a remarkable role in controlling the electronic structure and band structure of MXenes and thereby controls most of the optical properties exhibited by the them. The imaginary part of the dielectric constant is usually affected by the intraband and interband electronic transitions, specifically in materials consisting of transition metals as one of the constituent elements. The absorption spectrum and hence the absorption coefficient of the material in turn is determined mainly by total interband electronic transitions in materials. MXenes with hydroxylated or fluorinated terminations tend to accept a single electron and hence exhibit similar optical performance in visible regime of electromagnetic spectrum [107]. On the other hand, MXenes with oxygen terminations tend to accept two electrons and exhibit quite different behavior in the presence of visible light. In the case of Ti_3C_2, if the surface functionalization is done with oxygen, then the sample exhibits larger absorption in the visible region, while fluorinated samples exhibit weaker absorption characteristics in the visible region. This difference arises from the contribution of oxygen atoms towards the PDOS and TDOS and hence can be estimated and analyzed by performing a computational study employing DFT. Some of the interesting optical properties of 2D MXenes are discussed in the subsequent sections.

4.6.1 Photoluminescence Response

MXenes exhibit metallic character generally with excellent conductivity. Their surfaces consist of some hydrophilic groups and hence MXenes are capable of exhibiting surface redox capacitance and intercalation with very high conductivity values as large as 2400 cm^{-1}. This is the reason why MXenes exhibit excellent conduction properties as equivalent to that of graphene. Owing to the presence of these excellent properties, MXenes can act as a budding candidate for energy storage, energy conversion and other environmental applications. The 2D MXenes can be transformed to 0D quantum dot MXenes via quantum confinement and can be employed for energy conversion applications such as sensors, photovoltaic cells and photocatalysis. They exhibit exceptionally good optical properties as a result of the combined effects of quantum confinement, increased surface to volume ratio and band gap tuning. It is possible to reduce the lateral size of MXenes further to make them exhibit optimum optical properties. For example, single layered Ti_3C_2 MXene quantum dots having a size of 2.9 nm have been synthesized by a top-down hydrothermal approach and were found to exhibit photoluminescence providing a quantum yield of 10% approximately, and in turn can be employed in optoelectronic and biomedical fields [108]. Simultaneous interlayer cutting and lamination can lead to the formation of ultra-small Ti_3C_2 MXene nanosheets, which possesses a lateral size of 2 nm to 8 nm and exhibits fluorescence in the blue-green region of the electromagnetic spectrum [109]. Similar hydrothermal method can be employed to synthesis functionalized Ti_3C_2@PEG quantum dots, which exhibit pH-dependent photoluminescence in the blue region of the visible spectrum, showing a quantum yield of 7.13% [110]. Due to the influence of strong quantum confinement effects, the doped Ti_3C_2 quantum dot MXene synthesized by means of hydrothermal method was also showed to exhibit excellent photoluminescence that are excitation dependent providing a quantum yield of 18.7% [111]. So far, quantum dot MXene exhibiting photoluminescence in the blue region have been fabricated and the lack of suitable and effective procedures for preparation and blur luminescence mechanisms hinder the production of MXenes with photoluminescence in the white region of the electromagnetic spectrum.

4.6.2 Absorption Properties

MXenes were employed earlier for photocatalytic applications since they are capable of light absorption of wavelengths ranging from 300 nm to 600 nm of visible spectrum. The absorption properties of MXenes arise from their high conductivity. An important method of energy absorption in MXenes is photothermal conversion where the light energy is directly converted to heat. Efficient photothermal conversion and excellent absorption in UV-vis range of electromagnetic spectrum represents the characteristics of an ideal material. Since MXenes have the ability to convert electromagnetic radiation to heat, they find applications in photovoltaic conversion and storage devices, where maintaining an optimum temperature is a major concern specifically when employed for commercial applications. Figure 4.18 shows the UV-vis absorption spectrum of Ti_3AlC_2 and the corresponding pristine MXene $Ti_3C_2T_x$, indicating that the exfoliation of Al layer and the functionalization of MAX phase increases the UV-vis absorption properties of the corresponding MXene [112].

Both Ti_3C_2 and composites made of Ti_3C_2 can be employed for solar energy conversion devices due to their excellent ability to harvest sunlight. Metallic MXenes cannot favor photocatalytic reactions due to the existence of zero band gap in them and hence cannot absorb

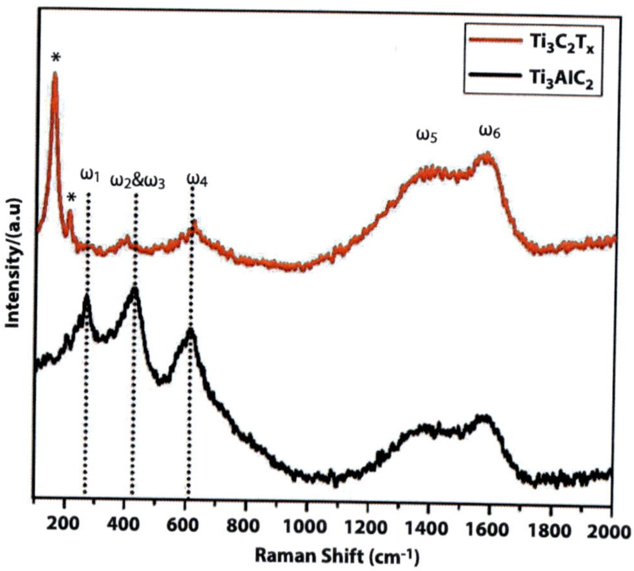

Figure 4.18 UV- vis absorption spectrum of the MAX phase, Ti_3AlC_2 and the corresponding MXene, Ti_3C_2TX [112]. Copyright 2021. Adapted from MDPI.

light energy. Some MXenes can be converted to semiconducting by means of doping or termination functionalization or composition alteration and can be made to absorb light. The DFT calculations on 48 different MXenes showed that the compounds Zr_2CO_2 and Hf_2CO_2 with suitable electronic structures could act as a potential candidate for photocatalysis in water splitting applications. It was also demonstrated that certain organic pollutants such as cationic methylene blue (MB) and anionic acid undergo degradation when exposed to blue wavelength under UV light [113]. The MXene materials can be tailored optically in a non-linear fashion to achieve absorption properties spanning in the ultraviolet to near infrared region of the electromagnetic spectrum. The enhanced optical absorption results directly from the enhanced conductivity of these compounds due to temperature and composition changes induced in them. Thus, the electrical conductivity can be altered under control above three times at room temperature and six times for the temperature range from 10 to 300 K [114]. In a similar way, the MXenes with oxidated termination groups tend to exhibit strong absorption in the UV region whereas sulphated or fluorinated MXenes show weak UV absorption properties. The type of surface termination of MXene also alters the absorption peaks and is visible if the absorption is plotted as a function of photon energy.

4.6.3 Dielectric Properties

The dielectric properties of MXenes together with the existence of a hydrophilic surface, excellent physiochemical properties, make them suitable for sensors and photodetectors, fillers in polymer composites, electromagnetic shielding interfaces, Li-ion batteries, supercapacitors and photovoltaic devices. One important application of MXenes include their ability to perform as ideal conductive fillers, which can be employed with flexible polymers to optimize dielectric properties together with the requirement of establishing high flexibility. The application of MXenes as a nanofiller in polymer composites results from their excellent metallic conductivity, leading to enhanced interfacial polarization arising from existence of a variation in conductivity of MXenes filler material and polymer matrices. MXenes are able to disperse in polymer matrices very easily and hence finds application in the field of supercapacitor networks. The dipole polarization at the interfacial spaces of the materials largely controls the dielectric loss in an oxidated MXene, even though it is not possible to measure the interfacial dipole polarization experimentally. The best method to measure the dipole polarization and thereby the dielectric loss is to perform a DFT analysis with the help of a simulation tool. The reinforcement of polymer dielectric materials by MXene nanofillers results in the formation of nanocomposites with excellent dielectric energy storage capacity by optimizing the morphology and the other related dielectric properties. $MoS_2@Ti_3C_2$ MXene formed by hydrothermal method can be incorporated into poly (vinylidene fluoride) (PVDF) to develop the PVDF based dielectric composites (MoS_2@MXene/PVDF) via suitable methods such as solution casting. The composite thus formed exhibits excellent elastic properties along with sufficiently good flexibility. The composite material (MoS_2@MXene/PVDF) possesses an excellent value of dielectric constant of 24.3 at room temperature. The study also showed that its dielectric constant also shows good frequency stability and lesser dielectric loss when compared to PVDF and MXene blend [115]. It is also evident that the enhanced dielectric properties and dielectric permittivity are related to charge accumulation on MXene-polymer matrix interface when electric field of suitable strength is applied to the composite material. The dielectric as well as

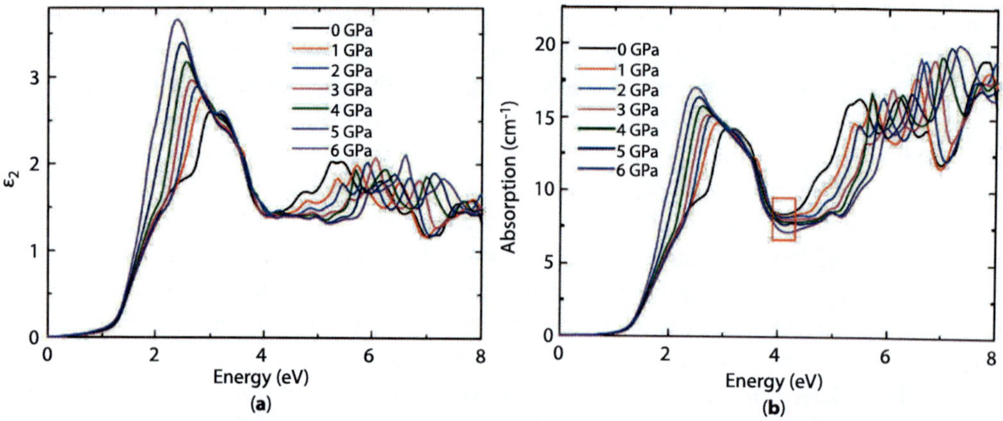

Figure 4.19 The variation of (a) dielectric constant's imaginary part, (b) absorption coefficient for Ti_2CO_2 MXene as a function of pressure [61]. Copyright 2020. Reproduced with permission from The American Chemical Society.

absorption characteristics of pristine Ti_2CO_2 can be enhanced by the application of pressure after the process of exfoliation from the corresponding MAX phase, Ti_2AlC as shown in Figure 4.19 [61]. On the other hand, percolation is the major factor for achieving ultrahigh dielectric constants in polymer/MXene composites [116]. However, the dielectric loss in such composites creates a major concern and it arises from the presence of huge leakage current. A study shows that a composite formed of Ti_2C MXene and ZrO_2 results in lesser leakage losses and a higher dielectric constant due to the interfacial effect of the Ti_2C/ZrO_2. The DFT studies on the same composite shows that the material exhibits a high dielectric constant of 40 and a low dielectric loss of 0.09 at room temperature and a frequency of 100 Hz [117].

4.6.4 Non-Linear Optical Properties

Materials capable of exhibiting excellent non-linear optical (NLO) response are of significance in certain applications such as data processing and storage, optical communication, light harvesting and conversion devices. When light interacts with such NLO materials, absorption, reverse saturable absorption (RSA) and multi-photon absorption (MPA) occur resulting in the generation of diverse and excellent optical properties for the material [118]. Owing to their excellent electronic and optical properties, certain transition metal based 2-D MXenes have captured the focus of researchers recently. The popular MXene Ti_3C_2 has been employed in various fields such as Q-switching and mode locking, due to its excellent non-linear optical properties. The same material can be employed for photonic and photovoltaic applications due to its ability to absorb two photons having wavelengths 780 nm and 1100 nm [119]. When excited with radiation of different possible wavelengths, the material is found to exhibit strong intensity dependent saturable non-linear absorption properties. Another significant property of MXenes known as the non-linear susceptibility can be obtained directly from the non-linear refractive index as well as absorption coefficient. Ti_3C_2 MXene possesses excellent non-linear susceptibility also [120]. Owing to their excellent non-linear optical properties, $Ti_3C_2T_X$ MXene and C_{60} module can be employed

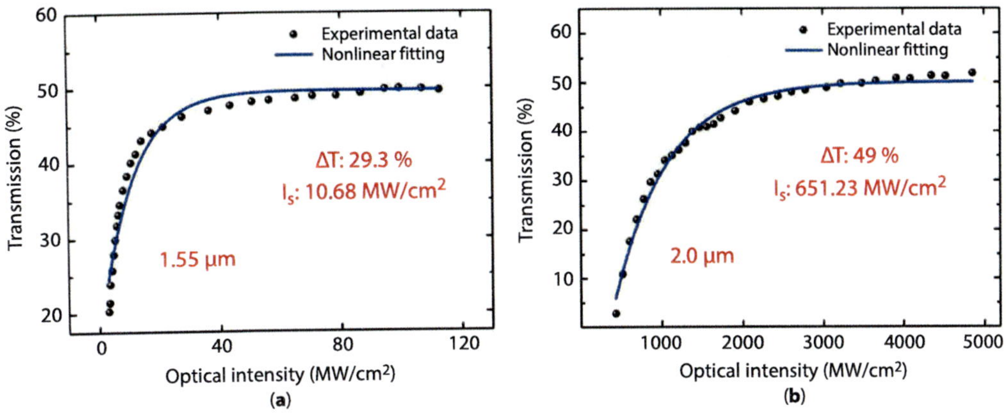

Figure 4.20 The non-linear optical transmittance of the 2D MXene, $Ti_3C_2T_X$ as a function of optical intensity of the incident radiation – (a) with erbium doped fiber pulsed laser of wavelength 1.5 μm and (b) thulium doped fiber pulsed laser of wavelength 2 μm as the sources [122]. Copyright 2020. Reproduced with permission from American Chemical Society.

for non-reciprocal transmission of laser pulses in photonic rectification process [121]. The non-linear optical transmittance properties of $Ti_3C_2T_X$ MXene indicates that it can be employed for ultrafast non-linear photonic equipment that can operate in the Near and Mid infrared wavelength regimes (Figure 4.20) [122]. Since surface termination of MXenes influences the electron distribution and other linear optical properties, it becomes obvious that surface termination of MXene may affect the non-linear optical responses of the materials also. One such research study by H. Li *et al.* have shown that the non-linear optical properties of $Ti_3C_2T_x$ can be optimized by a surface termination modulation. The study also showed that the magnitude and the sign of the non-linear absorption coefficient also changes when the surface termination of the MXene is modulated [123].

4.6.5 Plasmonic Properties

The plasmonic properties of MXenes include surface plasmon resonance and electric field confinement, and its estimation can be carried out via time dependent density functional theory. Various nanoclusters ($Ti_{48}C_{32}$, $Ti_{48}C_{32}F_{32}$, and $Ti_{48}C_{32}O_{32}$) were analyzed to exhibit broadband photo absorption spectrum along with localized surface plasmon resonance for IR wavelengths, which is considered as the wavelength region suitable for molecular sensing. These nanoclusters of $Ti_{48}C_{32}$, $Ti_{48}C_{32}F_{32}$, and $Ti_{48}C_{32}O_{32}$ produce a sizable confinement of developed electric field at their surface and its magnitude is found to vary with the F or O surface termination [124]. Functionalized MXenes are largely employed in the field of hydrogen storage, thermoelectric materials etc. With the help of first principle DFT studies by employing the hybrid functional HSE06, the optical properties of the Ti_2CO_2 MXene was studied under the action of pressure. The study showed that the MXene material exhibits excellent absorption in visible regime of electromagnetic spectrum [125]. MXene arrays formed of nanodisks of $Ti_3C_2T_x$ showed enhanced surface plasmon resonance that is found to be localized due to the quantum confinement effects at the near infra-red frequencies. It is possible to enhance the absorption of radiation in the

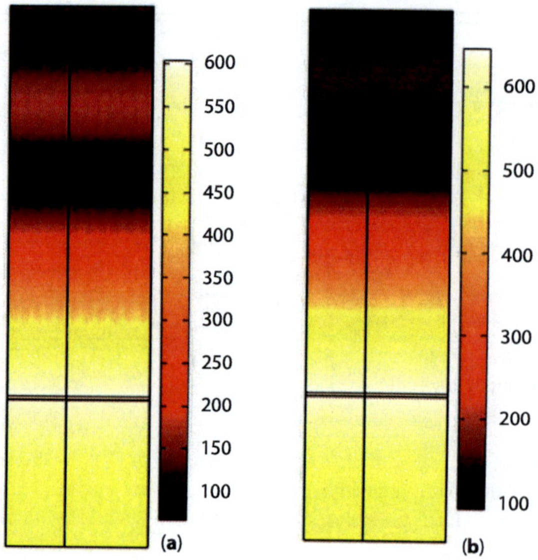

Figure 4.21 The enhanced confinement of electric field due to the enhanced SPPs in MXene sensors (a) without FG and (b) with FG [127]. Copyright 2021. Reproduced with permission from IEEE.

wide infra-red wavelength regime of the electromagnetic spectrum by utilizing the scattering enhancement at resonance in the case of these MXenes [126]. Figure 4.21 shows the enhanced electric field distribution in $Ti_3C_2T_X$ MXene based sensors, when a single layer of fluorinated graphene (FG) is added to it [127]. Without the presence of a single layer of FG, the electric filed is confined within 600 a.u. of the MXene-sensing layer interface while with FG the electric field distribution has been extended to 640 a.u., thus resulting in exciting the surface plasmon polaritons (SPPs) by utilizing the beneficial properties of MXene and FG [127]. Thus, the introduction of MXene and FG enhances the sensing capabilities of sensors via the enhanced excitation of SPPs through the enhanced electric field confinement with MXene as the sensing element.

4.6.6 Methods to Improve the Optical Properties

The electromagnetic wave responses arise due to the interaction of electromagnetic waves with the electrons of the materials. Hence, the electron distribution is important in materials and it is determined mainly by the structural parameters such as lattice constants, Wyckoff positions and related morphology, the strain experienced by the structure and the composition of the termination elements in MXenes. In this way, the electronic structure and properties determine the dielectric properties and other optical properties of MXenes. Several methods can be utilized to enhance optical properties of structure such as altering the elemental composition, surface termination, intercalation of ions, doping and defect density, post treatment of the structure. MXenes in this way may be converted from the completely metallic to half metallic to semiconducting state by proper tuning of the band gap and inducing quantum confinement effects. Subsequently the electromagnetic wavelength response of the material can also be altered on a large scale [128, 129].

The unterminated MXenes are found to be metallic with conduction and valence band state overlapping with one another. Surface termination functionalization of MXenes results in formation of MXenes that are highly electronegative and hydrophilic in nature due to the presence of surface functional groups. This property results in the multiple modulation of electronic structure of MXenes and thereby alters the electromagnetic response and other linear and non-linear optical properties of the material. The external M layers to which the surface terminal groups are attached, contribute more to the electronic distributions of MXenes rather than the inner M-layers. Thus, the band structure of the MXene Ti_3C_2 can be changed to the semiconducting state from its initial metallic state by means of surface termination functionalization. The surface terminal functionals -OH and -F exhibit different electronegativity and depending on its value, the band gap of the corresponding MXene also change from 0.05 eV to 1 eV [130]. These differences in the electronic distribution and band structure leads to the diverse dielectric and optical properties of MXenes. The functional terminals pay a significant role in controlling the absorption and reflectivity of the MXene. For example, in the UV region, they enhance the absorption and reflectivity properties of the material whereas in the visible region, the absorption coefficients of -F and -OH terminated materials were smaller than that of their pure form. This implies that the optical response of the MXene in the UV-vis-IR region can be tuned efficiently by altering their surface functionalization [131]. The removal of the surface terminal functional groups from MXene surface by suitable methods enhances the conductivity of the material. In a similar way, a research study shows that the optical non-linearity of MXenes can be enhanced by decorating it with Ag nanoparticles. This enhances the saturated absorption (SA) and the reverse saturated absorption (RSA), thus enhancing the non-linear optical behavior of the composite as demonstrated in Figure 4.22 [132].

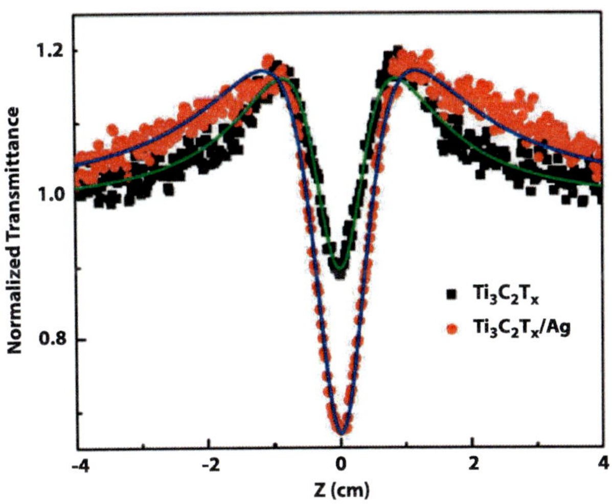

Figure 4.22 Z. scan results of $Ti_3C_2T_X$ and $Ti_3C_2T_X$/Ag composite material at 532 nm. The plot indicates enhanced transmittance when Ag nanoparticle is added to $Ti_3C_2T_X$ MXene [132]. Copyright 2021. Reproduced with permission from Optica Publishing Group.

4.7 Conclusion

This chapter discusses in detail the electronic, electrical and optical properties of the 2-D MXenes, which represent a recently invented but popular category of materials formed from the nitride, carbide or carbonitride of transition metal, exhibiting prominant applications in diverse fields including supercapacitors, sensors, photonics, photovoltaics, functional metamaterials-based device fabrication, ultrafast laser technology, biomedical and materials engineering. The research works conducted so far to assess the structural and chemical stability of MXenes are also included in detail by considering a number of specific examples. Tailoring of various properties of MXenes, viz. optical, electronic and electrical properties, are important to make them suitable for specific applications. Hence this chapter provides a brief overview of the methods that can be employed to optimize these properties of MXenes so as to make it suitable for specific applications. The chapter also discusses the non-linear optical properties of MXenes, which are controlled effectively via composition change or surface functionalization which in turn, make them apt for ultrafast laser technology applications. A brief outline of the presence of a strong non-linear behavior arising from the phenomenon of plasmonic resonances together with exceptional optical transparency, large transmission coefficient and other surface dependent optical interactions for MXenes are also presented. Most of the studies conducted so far used DFT methods as they provide very accurate estimation of electronic and optical properties of MXenes. The electrical properties also have high dependence on structure and composition of MXene and its surface functionalization. The efficient methods to control electrical properties of MXenes such as control of the synthetic route to minimize defects are also included in this review.

References

1. Sarker, P., Harrington, T., Toher, C., Oses, C., Samiee, M., Maria, J.P., Brenner, D.W., Vecchio, K.S., Curtarolo, S., High-entropy high-hardness metal carbides discovered by entropy descriptors. *Nat. Commun.*, 9, 1, 2018.
2. Gusev, A.I. and Rempel, A.A., Atomic ordering and phase equilibria in strongly nonstoichiometric carbides and nitrides, in: *Materials Science of Carbides, Nitrides and Borides*, pp. 47–64, Springer, Dordrecht, 1999.
3. Lei, Z., Liu, X., Wu, Y., Wang, H., Jiang, S., Wang, S., Hui, X., Wu, Y., Gault, B., Kontis, P., Raabe, D., Enhanced strength and ductility in a high-entropy alloy via ordered oxygen complexes. *Nature*, 563, 546, 2018.
4. Ye, R., del Angel-Vicente, P., Liu, Y., Arellano-Jimenez, M.J., Peng, Z., Wang, T., Li, Y., Yakobson, B.I., Wei, S.H., Yacaman, M.J., Tour, J.M., High-performance hydrogen evolution from $MoS_{2(1-x)}P_x$ solid solution. *Adv. Mater.*, 28, 1427, 2016.
5. Uematsu, T., Doi, T., Torimoto, T., Kuwabata, S., Preparation of luminescent $AgInS_2$–$AgGaS_2$ solid solution nanoparticles and their optical properties. *J. Phys. Chem. Lett.*, 1, 3283, 2010.
6. Li, H., Duan, X., Wu, X., Zhuang, X., Zhou, H., Zhang, Q., Zhu, X., Hu, W., Ren, P., Guo, P., Ma, L., Growth of alloy $MoS_{2x}Se_{2(1-x)}$ nanosheets with fully tunable chemical compositions and optical properties. *J. Am. Chem. Soc.*, 136, 3756, 2014.
7. Bandyopadhyay, D., Sharma, R., Chakraborti, N., The C-Nb-Ti system (carbon-niobium-titanium). *J. Phase Equilibria Diffus.*, 21, 102, 2000.

8. Zhang, W., Peng, Y., Du, Y., Chen, L., Li, Y., Wang, S., Wen, G., Xie, W., A thermodynamic description of the C–Ti–V system over the whole composition and temperature ranges. *Int. J. Refract. Met. Hard Mater.*, 48, 346, 2015.
9. Novoselov, K.S., Geim, A.K., Morozov, S.V., Jiang, D., Katsnelson, M.I., Grigorieva, I., Dubonos, S., Firsov, A., Two-dimensional gas of massless Dirac fermions in graphene. *Nature*, 438, 197, 2005.
10. Zhang, K., Feng, Y., Wang, F., Yang, Z., Wang, J., Two-dimensional hexagonal boron nitride (2D-hBN): Synthesis, properties and applications. *J. Mater. Chem. C*, 5, 11992, 2017.
11. Sun, J., Li, X., Guo, W., Zhao, M., Fan, X., Dong, Y., Xu, C., Deng, J., Fu, Y., Synthesis methods of two-dimensional MoS_2: A brief review. *Crystals*, 7, 198, 2017.
12. Cong, C., Shang, J., Wu, X., Cao, B., Peimyoo, N., Qiu, C., Sun, L., Yu, T., Synthesis and optical properties of large-area single-crystalline 2D semiconductor WS_2 monolayer from chemical vapor deposition. *Adv. Opt. Mater.*, 2, 131, 2014.
13. Asadi, Y. and Nourbakhsh, Z., Structural, electronic, mechanical, thermodynamic, and linear and nonlinear optical properties of MoS_2, $MoSe_2$, and their $MoS_{2x}Se_{2(1-x)}$ Alloys: *Ab initio* calculations. *J. Electron. Mater.*, 48, 7977, 2019.
14. Horlait, D., Middleburgh, S.C., Chroneos, A., Lee, W.E., Synthesis and DFT investigation of new bismuth-containing MAX phases. *Sci. Rep.*, 6, 1, 2016.
15. Khazaei, M., Ranjbar, A., Esfarjani, K., Bogdanovski, D., Dronskowski, R., Yunoki, S., Insights into exfoliation possibility of MAX phases to MXenes. *Phys. Chem. Chem. Phys.*, 20, 8579, 2018.
16. Kim, H. and Alshareef, H.N., MXetronics: MXene-enabled electronic and photonic devices. *ACS Mater. Lett.*, *2*, 55, 2019.
17. Mostafaei, A., Faizabadi, E., Semiromi, E.H., Theoretical studies and tuning the electronic and optical properties of Zr_2CO_2 monolayer using biaxial strain effect: Modified Becke–Johnson calculation. *Phys. E Low Dimens. Syst. Nanostruct.*, 114, 113559, 2019.
18. Blaha, P., Schwarz, K., Tran, F., Laskowski, R., Madsen, G.K., Marks, L.D., WIEN2k: An APW+ lo program for calculating the properties of solids. *J. Chem. Phys.*, 152, 074101, 2020.
19. Kresse, G. and Hafner, J., Ab initio molecular dynamics for liquid metals. *Phys. Rev. B*, 47, 558, 1993.
20. Giannozzi, P., Baroni, S., Bonini, N., Calandra, M., Car, R., Cavazzoni, C., Ceresoli, D., Chiarotti, G.L., Cococcioni, M., Dabo, I., Dal Corso, A., Quantum ESPRESSO: A modular and open-source software project for quantum simulations of materials. *J. Phys. Condens. Matter*, 21, 395502, 2009.
21. Gonze, X., Amadon, B., Anglade, P.M., Beuken, J.M., Bottin, F., Boulanger, P., Bruneval, F., Caliste, D., Caracas, R., Côté, M., Deutsch, T., ABINIT: First-principles approach to material and nanosystem properties. *Comput. Phys. Commun.*, 180, 2582, 2009.
22. Rahman, U.U., Humayun, M., Ghani, U., Usman, M., Ullah, H., Khan, A., El-Metwaly, N.M., Khan, A., MXenes as emerging materials: Synthesis, properties, and applications. *Molecules*, 27, 4909, 2022.
23. Sokol, M., Natu, V., Kota, S., Barsoum, M.W., On the chemical diversity of the MAX phases. *Trends Chem.*, 1, 210, 2019.
24. Anasori, B., Lukatskaya, M.R., Gogotsi, Y., 2D metal carbides and nitrides (MXenes) for energy storage. *Nat. Rev. Mater.*, 2, 1, 2017.
25. Zamhuri, A., Lim, G.P., Ma, N.L., Tee, K.S., Soon, C.F., MXene in the lens of biomedical engineering: Synthesis, applications and future outlook. *Biomed. Eng. Online*, 20, 1–24, 2021.
26. Lin, H., Wang, X., Yu, L., Chen, Y., Shi, J., Two-dimensional ultrathin MXene ceramic nanosheets for photothermal conversion. *Nano Lett.*, 17, 384, 2017.

27. Nayak, P., Jiang, Q., Mohanraman, R., Anjum, D., Hedhili, M.N., Alshareef, H.N., Inherent electrochemistry and charge transfer properties of few-layered two-dimensional $Ti_3C_2T_x$ MXene. *Nanoscale*, 10, 17030, 2018.
28. Urbankowski, P., Anasori, B., Makaryan, T., Er, D., Kota, S., Walsh, P.L., Zhao, M., Shenoy, V.B., Barsoum, M.W., Gogotsi, Y., Synthesis of two-dimensional titanium nitride Ti_4N_3 (MXene). *Nanoscale*, 8, 11385, 2016.
29. Hoffman, E.N., Yushin, G., El-Raghy, T., Gogotsi, Y., Barsoum, M.W., Micro and mesoporosity of carbon derived from ternary and binary metal carbides. *Microporous Mesoporous Mater.*, 112, 526, 2008.
30. Barsoum, M.W., El-Raghy, T., Farber, L., Amer, M., Christini, R., Adams, A., The topotactic transformation of Ti_3SiC_2 into a partially ordered cubic Ti ($C_{0.67}Si_{0.06}$) phase by the diffusion of Si into molten cryolite. *J. Electrochem. Soc.*, 146, 3919, 1999.
31. Xu, C., Wang, L., Liu, Z., Chen, L., Guo, J., Kang, N., Ma, X.L., Cheng, H.M., Ren, W., Large-area high-quality 2D ultrathin Mo_2C superconducting crystals. *Nat. Mater.*, 14, 1135, 2015.
32. Zhao, X., Sun, W., Geng, D., Fu, W., Dan, J., Xie, Y., Kent, P.R., Zhou, W., Pennycook, S.J., Loh, K.P., Edge segregated polymorphism in 2D molybdenum carbide. *Adv. Mater.*, 31, 1808343, 2019.
33. Fu, B., Sun, J., Wang, C., Shang, C., Xu, L., Li, J., Zhang, H., MXenes: Synthesis, optical properties, and applications in ultrafast photonics. *Small*, 17, 2006054, 2021.
34. Meshkian, R., Näslund, L.Å., Halim, J., Lu, J., Barsoum, M.W., Rosen, J., Synthesis of two-dimensional molybdenum carbide, Mo_2C, from the gallium based atomic laminate Mo_2Ga_2C. *Scr. Mater.*, 108, 147, 2015.
35. Rössler, U. (Ed.), *Solid State Theory: An Introduction*, Springer Science & Business Media, USA, 2009.
36. Grosso, G. and Parravicini, G.P., *Solid state physics*, Academic Press, USA, 2013.
37. Khazaei, M., Ranjbar, A., Arai, M., Sasaki, T., Yunoki, S., Electronic properties and applications of MXenes: A theoretical review. *J. Mater. Chem. C*, 5, 2488, 2017.
38. Bae, S., Kang, Y.G., Khazaei, M., Ohno, K., Kim, Y.H., Han, M.J., Chang, K.J., Raebiger, H., Electronic and magnetic properties of carbide MXenes—The role of electron correlations. *Mater. Today Adv.*, 9, 100118, 2021.
39. Dudarev, S.L., Botton, G.A., Savrasov, S.Y., Humphreys, C.J., Sutton, A.P., Electron-energy-loss spectra and the structural stability of nickel oxide: An LSDA+ U study. *Phys. Rev. B*, 57, 1505, 1998.
40. Buda, I.G., Lane, C., Barbiellini, B., Ruzsinszky, A., Sun, J., Bansil, A., Characterization of thin film materials using SCAN meta-GGA, an accurate nonempirical density functional. *Sci. Rep.*, 7, 1–8, 2017.
41. Chakraborty, A., Dixit, M., Aurbach, D., Major, D.T., Predicting accurate cathode properties of layered oxide materials using the SCAN meta-GGA density functional. *NPJ Comput. Mater.*, 4, 1, 2018.
42. Heyd, J., Scuseria, G.E., Ernzerhof, M., Hybrid functionals based on a screened Coulomb potential. *J. Chem. Phys.*, 118, 8207, 2013.
43. Meng, Q., Ma, J., Zhang, Y., Li, Z., Zhi, C., Hu, A., Fan, J., The S-functionalized Ti_3C_2 Mxene as a high-capacity electrode material for Na-ion batteries: A DFT study. *Nanoscale*, 10, 3385, 2018.
44. Zhao, Y. and Zhao, J., Functional group-dependent anchoring effect of titanium carbide-based MXenes for lithium-sulfur batteries: A computational study. *Appl. Surf. Sci.*, 412, 591, 2017.
45. Marini, A., Poncé, S., Gonze, X., Many-body perturbation theory approach to the electron-phonon interaction with density-functional theory as a starting point. *Phys. Rev. B*, 91, 224310, 2015.
46. Aulbur, W.G., Jönsson, L., Wilkins, J.W., Quasiparticle calculations in solids. *Solid State Phys.*, 54, 1, 2000.

47. Laasner, R., G_0W_0 band structure of $CdWO_4$. *J. Phys. Condens. Matter.*, 26, 125503, 2014.
48. Zhang, Y., Xia, W., Wu, Y., Zhang, P., Prediction of MXene based 2D tunable band gap semiconductors: GW quasiparticle calculations. *Nanoscale*, 11, 3993, 2019.
49. Sun, W., Xie, Y., Kent, P.R., Double transition metal MXenes with wide band gaps and novel magnetic properties. *Nanoscale*, 10, 11962, 2018.
50. Ding, Y.M., Nie, X., Dong, H., Rujisamphan, N., Li, Y., Many-body effects in an MXene Ti_2CO_2 monolayer modified by tensile strain: GW-BSE calculations. *Nanoscale Adv.*, 2, 2471, 2020.
51. Yang, K., Ren, S., Huang, H., Wu, B., Shen, G., Zhou, T., Liu, X., First-principles study on the half-metallicity of new MXene materials Nd_2NT_2 (T= OH, O, S, F, Cl, and Br). *Front. Chem.*, 9, 1273, 2021.
52. Rajan, A.C., Mishra, A., Satsangi, S., Vaish, R., Mizuseki, H., Lee, K.R., Singh, A.K., Machine-learning-assisted accurate band gap predictions of functionalized MXene. *Chem. Mater.*, 30, 4031, 2018.
53. Xie, Y. and Kent, P.R.C., Hybrid density functional study of structural and electronic properties of functionalized $Ti_{n+1}X_n$ (X= C, N) monolayers. *Phys. Rev. B*, 87, 235441, 2013.
54. Berdiyorov, G.R., Effect of lithium and sodium ion adsorption on the electronic transport properties of Ti_3C_2 MXene. *Appl. Surf. Sci.*, 359, 153, 2015.
55. Zha, X.H., Zhou, J., Zhou, Y., Huang, Q., He, J., Francisco, J.S., Luo, K., Du, S., Promising electron mobility and high thermal conductivity in Sc_2CT_2 (T= F, OH) MXenes. *Nanoscale*, 8, 6110, 2016.
56. Gan, L.Y., Zhao, Y.J., Huang, D., Schwingenschlögl, U., First-principles analysis of MoS_2/Ti_2C and MoS_2/Ti_2CY_2 (Y= F and OH) all-2D semiconductor/metal contacts. *Phys. Rev. B*, 87, 245307, 2013.
57. Miranda, A., Halim, J., Barsoum, M.W., Lorke, A., Electronic properties of freestanding Ti3C2Tx MXene monolayers. *Appl. Phys. Lett.*, 108, 033102, 2016.
58. Lee, Y., Cho, S.B., Chung, Y.C., Tunable indirect to direct band gap transition of monolayer Sc_2CO_2 by the strain effect. *ACS Appl. Mater. Interfaces*, 6, 14724, 2014.
59. Bafekry, A., Akgenç, B., Ghergherehchi, M., Peeters, F.M., Strain and electric field tuning of semi-metallic character $WCrCO_2$ MXenes with dual narrow band gap. *J. Phys. Condens. Matter*, 32, 355504, 2020.
60. Li, X.H., Zhang, R.Z., Cui, H.L., Strain-induced band modulation, work function, and QTAIM analysis of surface O-functionalized Ti_2C MXene. *ACS Omega*, 5, 18403, 2020.
61. Li, X.H., Shan-Shan, L., Cui, H.L., Zhang, R.Z., Pressure-induced modulation of electronic and optical properties of surface O-functionalized Ti_2C MXene. *ACS Omega*, 5, 22248, 2020.
62. Kumar, P., Shahzad, F., Yu, S., Hong, S.M., Kim, Y.H., Koo, C.M., Large-area reduced graphene oxide thin film with excellent thermal conductivity and electromagnetic interference shielding effectiveness. *Carbon*, 94, 494, 2015.
63. Naguib, M., Kurtoglu, M., Presser, V., Lu, J., Niu, J., Heon, M., Hultman, L., Gogotsi, Y., Barsoum, M.W., Two-dimensional nanocrystals produced by exfoliation of Ti_3AlC_2. *Adv. Mater.*, 23, 4248, 2011.
64. Mashtalir, O., Naguib, M., Mochalin, V.N., Dall'Agnese, Y., Heon, M., Barsoum, M.W., Gogotsi, Y., Intercalation and delamination of layered carbides and carbonitrides. *Nat. Commun.*, 4, 1, 2013.
65. Karlsson, L.H., Birch, J., Halim, J., Barsoum, M.W., Persson, P.O., Atomically resolved structural and chemical investigation of single MXene sheets. *Nano Lett.*, 15, 4955, 2015.
66. Hope, M.A., Forse, A.C., Griffith, K.J., Lukatskaya, M.R., Ghidiu, M., Gogotsi, Y., Grey, C.P., NMR reveals the surface functionalization of Ti_3C_2 MXene. *Phys. Chem. Chem. Phys.*, 18, 5099, 2016.

67. Zhang, J., Kong, N., Uzun, S., Levitt, A., Seyedin, S., Lynch, P.A., Qin, S., Han, M., Yang, W., Liu, J., Wang, X., Scalable manufacturing of free-standing, strong $Ti_3C_2T_x$ MXene films with outstanding conductivity. *Adv. Mater.*, 32, 1, 2020.
68. Shahin, N., Kazemi, S., Heidarpour, A., Mechanochemical synthesis mechanism of Ti_3AlC_2 MAX phase from elemental powders of Ti, Al and C. *Adv. Powder Technol.*, 27, 1775, 2016.
69. Lipatov, A., Alhabeb, M., Lukatskaya, M.R., Boson, A., Gogotsi, Y., Sinitskii, A., Effect of synthesis on quality, electronic properties and environmental stability of individual monolayer Ti_3C_2 MXene flakes. *Adv. Electron. Mater.*, 2, 1, 2016.
70. Jimmy, J. and Kandasubramanian, B., Mxene functionalized polymer composites: Synthesis and applications. *Eur. Polym. J.*, 122, 109367, 2020.
71. Malaki, M., Maleki, A., Varma, R.S., MXenes and ultrasonication. *JMCA*, 7, 10843, 2019.
72. Chae, Y., Kim, S.J., Cho, S.Y., Choi, J., Maleski, K., Lee, B.J., Jung, H.T., Gogotsi, Y., Lee, Y., Ahn, C.W., An investigation into the factors governing the oxidation of two-dimensional Ti_3C_2 MXene. *Nanoscale*, 11, 8387, 2019.
73. Alhabeb, M., Maleski, K., Anasori, B., Lelyukh, P., Clark, L., Sin, S., Gogotsi, Y., Guidelines for synthesis and processing of two-dimensional titanium carbide ($Ti_3C_2T_x$ MXene). *Chem. Mater.*, 29, 7633, 2017.
74. Liao, L., Jiang, D., Zheng, K., Zhang, M., Liu, J., Industry-scale and environmentally stable $Ti_3C_2T_x$ MXene based film for flexible energy storage devices. *Adv. Funct. Mater.*, 31, 2103960, 2021.
75. Shuck, C.E., Han, M., Maleski, K., Hantanasirisakul, K., Kim, S.J., Choi, J., Reil, W.E., Gogotsi, Y., Effect of Ti_3AlC_2 MAX phase on structure and properties of resultant $Ti_3C_2T_x$ MXene. *ACS Appl. Nano Mater.*, 2, 3368, 2019.
76. Wei, Q., Superhigh electromagnetic interference shielding of ultrathin aligned pristine graphene nanosheets film. *Adv. Mater.*, 32, 1, 2020.
77. Ling, Z., Ren, C.E., Zhao, M.Q., Yang, J., Giammarco, J.M., Qiu, J., Barsoum, M.W., Gogotsi, Y., Flexible and conductive MXene films and nanocomposites with high capacitance. *PNAS*, 111, 16676, 2014.
78. Han, M., Shuck, C.E., Rakhmanov, R., Parchment, D., Anasori, B., Koo, C.M., Friedman, G., Gogotsi, Y., Beyond $Ti_3C_2T_x$: MXenes for electromagnetic interference shielding. *ACS Nano*, 14, 5008, 2020.
79. Mariano, M., Mashtalir, O., Antonio, F.Q., Ryu, W.H., Deng, B., Xia, F., Gogotsi, Y., Taylor, A.D., Solution-processed titanium carbide MXene films examined as highly transparent conductors. *Nanoscale*, 8, 16371, 2016.
80. Song, D., Jiang, X., Li, Y., Lu, X., Luan, S., Wang, Y., Li, Y., Gao, F., Metal- organic frameworks-derived MnO_2/Mn_3O_4 microcuboids with hierarchically ordered nanosheets and Ti_3C_2 MXene/Au NPs composites for electrochemical pesticide detection. *J. Hazard. Mater.*, 373, 367, 2019.
81. Barsoum, M.W., The $M_{N+1}AX_N$ phases: A new class of solids: Thermodynamically stable nanolaminates. *Prog. Solid State Chem.*, 28, 201, 2000.
82. Pazniak, A., Bazhin, P., Shplis, N., Kolesnikov, E., Shchetinin, I., Komissarov, A., Polcak, J., Stolin, A., Kuznetsov, D., $Ti_3C_2T_x$ MXene characterization produced from SHS-ground Ti_3AlC_2. *Mater. Des.*, 183, 108143, 2019.
83. Sarycheva, A., Polemi, A., Liu, Y., Dandekar, K., Anasori, B., Gogotsi, Y., 2D titanium carbide (MXene) for wireless communication. *Sci. Adv.*, 4, 1, 2018.
84. Shahzad, F., Zaidi, S.A., Naqvi, R.A., 2D transition metal carbides (MXene) for electrochemical sensing: A review. *Crit. Rev. Anal. Chem.*, 52, 848, 2022.
85. Choi, Y.S., Yoo, Y.H., Kim, J.G., Kim, S.H., A comparison of the corrosion resistance of Cu–Ni–stainless steel multilayers used for EMI shielding. *Surf. Coat. Technol.*, 201, 3775, 2006.

86. Chen, Z., Xu, C., Ma, C., Ren, W., Cheng, H.M., Lightweight and flexible graphene foam composites for high-performance electromagnetic interference shielding. *Adv. Mater.*, 25, 1296, 2013.
87. Jiang, D., Murugadoss, V., Wang, Y., Lin, J., Ding, T., Wang, Z., Shao, Q., Wang, C., Liu, H., Lu, N., Wei, R., Electromagnetic interference shielding polymers and nanocomposites-A review. *Polym. Rev.*, 59, 280, 2019.
88. Weng, G.M., Li, J., Alhabeb, M., Karpovich, C., Wang, H., Lipton, J., Maleski, K., Kong, J., Shaulsky, E., Elimelech, M., Gogotsi, Y., Layer-by-layer assembly of cross-functional semi-transparent MXene-carbon nanotubes composite films for next-generation electromagnetic interference shielding. *Adv. Funct. Mater.*, 28, 1, 2018.
89. Shahzad, F., Alhabeb, M., Hatter, C.B., Anasori, B., Man Hong, S., Koo, C.M., Gogotsi, Y., Electromagnetic interference shielding with 2D transition metal carbides (MXenes). *Science*, 353, 1137, 2016.
90. Hu, T., Hu, M., Li, Z., Zhang, H., Zhang, C., Wang, J., Wang, X., Interlayer coupling in two-dimensional titanium carbide MXenes. *Phys. Chem. Chem. Phys.*, 18, 20256, 2016.
91. Khazaei, M., Arai, M., Sasaki, T., Chung, C.Y., Venkataramanan, N.S., Estili, M., Sakka, Y., Kawazoe, Y., Novel electronic and magnetic properties of two-dimensional transition metal carbides and nitrides. *Adv. Funct. Mater.*, 23, 2185, 2013.
92. Caffrey, N.M., Effect of mixed surface terminations on the structural and electrochemical properties of two-dimensional $Ti_3C_2T_2$ and V_2CT_2 MXenes multilayers. *Nanoscale*, 10, 13520, 2018.
93. Anasori, B., Shi, C., Moon, E.J., Xie, Y., Voigt, C.A., Kent, P.R., May, S.J., Billinge, S.J., Barsoum, M.W., Gogotsi, Y., Control of electronic properties of 2D carbides (MXenes) by manipulating their transition metal layers. *Nanoscale Horiz.*, 1, 227, 2016.
94. Hart, J.L., Hantanasirisakul, K., Lang, A.C., Anasori, B., Pinto, D., Pivak, Y., van Omme, J.T., May, S.J., Gogotsi, Y., Taheri, M.L., Control of MXenes' electronic properties through termination and intercalation. *Nat. Commun.*, 10, 1, 2019.
95. Miranda, A., Halim, J., Barsoum, M.W., Lorke, A., Electronic properties of freestanding $Ti_3C_2T_x$ MXene monolayers. *Appl. Phys. Lett.*, 108, 1, 2016.
96. Dillon, A.D., Ghidiu, M.J., Krick, A.L., Griggs, J., May, S.J., Gogotsi, Y., Barsoum, M.W., Fafarman, A.T., Highly conductive optical quality solution-processed films of 2D titanium carbide. *Adv. Funct. Mater.*, 26, 4162, 2016.
97. Halim, J., Moon, E.J., Eklund, P., Rosen, J., Barsoum, M.W., Ouisse, T., Variable range hopping and thermally activated transport in molybdenum-based MXenes. *Phys. Rev. B*, 98, 104202, 2018.
98. Ghidiu, M., Halim, J., Kota, S., Bish, D., Gogotsi, Y., Barsoum, M.W., Ion-exchange and cation solvation reactions in Ti_3C_2 MXene. *Chem. Mater.*, 28, 3507, 2016.
99. Lipatov, A., Alhabeb, M., Lu, H., Zhao, S., Loes, M.J., Vorobeva, N.S., Dall'Agnese, Y., Gao, Y., Gruverman, A., Gogotsi, Y., Sinitskii, A., Electrical and elastic properties of individual single-layer $Nb_4C_3T_x$ MXene flakes. *Adv. Electron. Mater.*, 6, 1, 2020.
100. Chen, X., Jiang, J., Yang, G., Li, C., Li, Y., Bioinspired wood-like coaxial fibers based on MXene@ graphene oxide with superior mechanical and electrical properties. *Nanoscale*, 12, 21325, 2020.
101. Lian, X., Shen, X., Fu, J., Gao, Z., Wan, X., Liu, X., Hu, E., Xu, J., Tong, Y., Electrical properties and biological synaptic simulation of Ag/MXene/SiO_2/Pt RRAM devices. *Electronics*, 9, 2098, 2020.
102. Wang, H., Wu, Y., Zhang, J., Li, G., Huang, H., Zhang, X., Jiang, Q., Enhancement of the electrical properties of MXene Ti_3C_2 nanosheets by post-treatments of alkalization and calcination. *Mater. Lett.*, 160, 537, 2015.

103. Sobolčiak, P., Ali, A., Hassan, M.K., Helal, M.I., Tanvir, A., Popelka, A., Al-Maadeed, M.A., Krupa, I., Mahmoud, K.A., 2D $Ti_3C_2T_x$ (MXene)-reinforced polyvinyl alcohol (PVA) nanofibers with enhanced mechanical and electrical properties. *PLoS One*, 12, 1, 2017.
104. Naguib, M., Mochalin, V.N., Barsoum, M.W., Gogotsi, Y., 25th Anniversary article: MXenes: A new family of two-dimensional materials. *Adv. Mater.*, 26, 992, 2014.
105. Yang, C., Tang, Y., Tian, Y., Luo, Y., Faraz Ud Din, M., Yin, X., Que, W., Flexible nitrogen-doped 2D titanium carbides (MXene) films constructed by an ex-situ solvothermal method with extraordinary volumetric capacitance. *Adv. Energy Mater.*, 8, 1802087, 2018.
106. Iqbal, A., Shahzad, F., Hantanasirisakul, K., Kim, M.K., Kwon, J., Hong, J., Kim, H., Kim, D., Gogotsi, Y., Koo, C.M., Anomalous absorption of electromagnetic waves by 2D transition metal carbonitride Ti_3CNT_x (MXene). *Science*, 369, 446, 2020.
107. Chaudhuri, K., Wang, Z., Alhabeb, M., Maleski, K., Gogotsi, Y., Shalaev, V., Boltasseva, A., Optical properties of MXenes, in: *2D Metal Carbides and Nitrides (MXenes)*, B. Anasori, and Y. Gogotsi, (Eds.), pp. 327–346, Springer, Cham, 2019.
108. Xuan, J., Wang, Z., Chen, Y., Liang, D., Cheng, L., Yang, X., Liu, Z., Ma, R., Sasaki, T., Geng, F., Organic-base-driven intercalation and delamination for the production of functionalized titanium carbide nanosheets with superior photothermal therapeutic performance. *Angew. Chem. Int. Ed.*, 128, 14789, 2016.
109. Wang, Z., Xuan, J., Zhao, Z., Li, Q., Geng, F., Versatile cutting method for producing fluorescent ultrasmall MXene sheets. *ACS Nano*, 11, 11559, 2017.
110. Chen, X., Sun, X., Xu, W., Pan, G., Zhou, D., Zhu, J., Wang, H., Bai, X., Dong, B., Song, H., Ratiometric photoluminescence sensing based on Ti_3C_2 MXene quantum dots as an intracellular pH sensor. *Nanoscale*, 10, 1111, 2018.
111. Xu, Q., Ding, L., Wen, Y., Yang, W., Zhou, H., Chen, X., Street, J., Zhou, A., Ong, W.J., Li, N., High photoluminescence quantum yield of 18.7% by using nitrogen-doped Ti_3C_2 MXene quantum dots. *J. Mater. Chem. C*, 6, 6360, 2018.
112. Iqbal, A. and Hamdan, N.M., Investigation and optimization of Mxene functionalized mesoporous titania films as efficient photoelectrodes. *Materials*, 14, 6292, 2021.
113. Miranda, A., Halim, J., Barsoum, M.W., Lorke, A., Electronic properties of freestanding $Ti_3C_2T_x$ MXene monolayers. *Appl. Phys. Lett.*, 108, 033102, 2016.
114. Han, M., Maleski, K., Shuck, C.E., Yang, Y., Glazar, J.T., Foucher, A.C., Hantanasirisakul, K., Sarycheva, A., Frey, N.C., May, S.J., Shenoy, V.B., Tailoring electronic and optical properties of MXenes through forming solid solutions. *J. Am. Chem. Soc.*, 142, 19110, 2020.
115. Chen, R.C., Zhang, Q.P., Ke, K., Sun, N., Xu, W.D., Liu, D.L., Yang, W., Li, Y.T., Zhou, Y.L., Yang, M.B., Yuan, J., Chemically bonding $BaTiO_3$ nanoparticles in highly filled polymer nanocomposites for greatly enhanced dielectric properties. *J. Mater. Chem. C*, 8, 8786, 2020.
116. Tu, S., Jiang, Q., Zhang, X., Alshareef, H.N., Large dielectric constant enhancement in MXene percolative polymer composites. *ACS Nano*, 12, 3369, 2018.
117. Feng, Y., Chen, P., Mao, B., Bo, M., Deng, Q., Well-coordinated dielectric properties in polymer composites bearing hybrid ceramic via interfacial effect between Ti_2C MXene particles and large-aspect-ratio ZrO_2 fibers. *Colloids Surf. A Physicochem. Eng. Asp.*, 629, 127505, 2021.
118. Tutt, L.W. and McCahon, S.W., Reverse saturable absorption in metal cluster compounds. *Opt. Lett.*, 15, 700, 1990.
119. Wang, G., Bennett, D., Zhang, C., Ó Coileáin, C., Liang, M., McEvoy, N., Wang, J.J., Wang, J., Wang, K., Nicolosi, V., Blau, W.J., Two-photon absorption in monolayer MXenes. *Adv. Opt. Mater.*, 8, 1902021, 2020.
120. Zhang, T., Chu, H., Li, Y., Zhao, S., Li, D., Third-order nonlinear properties and reverse absorption behavior in layered Ti_3C_2 MXene in the near infrared region. *Opt. Mater. Express*, 11, 4051, 2021.

121. Jhon, Y.I., Jhon, Y.M., Lee, J.H., Nonlinear optics of MXene in laser technologies. *J. Phys. Mater.*, 2, 032004, 2020.
122. Wang, Z., Li, H., Luo, M., Chen, T., Xia, X., Chen, H., Ma, C., Guo, J., He, Z., Song, Y., Liu, J., MXene photonic devices for near-infrared to mid-infrared ultrashort pulse generation. *ACS Appl. Nano Mater.*, 3, 3513, 2020.
123. Li, H., Chen, S., Boukhvalov, D.W., Yu, Z., Humphrey, M.G., Huang, Z., Zhang, C., Switching the nonlinear optical absorption of titanium carbide mxene by modulation of the surface terminations. *ACS Nano*, 16, 394, 2022.
124. Mokkath, J.H., Localized surface plasmon resonances and electric field confinement in titanium carbide (Ti_3C_2) MXene nanoclusters. *PCCP*, 23, 25807, 2021.
125. Li, X.H., Shan-Shan, L., Cui, H.L., Zhang, R.Z., Pressure-induced modulation of electronic and optical properties of surface O-functionalized Ti_2C MXene. *ACS Omega*, 5, 22248, 2020.
126. Chaudhuri, K., Alhabeb, M., Wang, Z., Shalaev, V.M., Gogotsi, Y., Boltasseva, A., Highly broadband absorber using plasmonic titanium carbide (MXene). *ACS Photonics*, 5, 1115, 2018.
127. Pandey, P.S., Singh, Y., Raghuwanshi, S.K., Theoretical analysis of the LRSPR sensor with enhance FOM for low refractive index detection using MXene and fluorinated graphene. *IEEE Sens. J.*, 21, 23979, 2021.
128. Dong, L., Kumar, H., Anasori, B., Gogotsi, Y., Shenoy, V.B., Rational design of two-dimensional metallic and semiconducting spintronic materials based on ordered double-transition-metal MXenes. *J. Phys. Chem. Lett.*, 8, 422, 2017.
129. Si, C., Jin, K.H., Zhou, J., Sun, Z., Liu, F., Large-gap quantum spin hall state in MXenes: D-band topological order in a triangular lattice. *Nano Lett.*, 16, 6584, 2016.
130. Enyashin, A.N. and Ivanovskii, A.L., Two-dimensional titanium carbonitrides and their hydroxylated derivatives: Structural, electronic properties and stability of MXenes $Ti_3C_{2-x}N_{x-}(OH)_2$ from DFTB calculations. *J. Solid State Chem.*, 207, 42, 2013.
131. Ibragimova, R., Erhart, P., Rinke, P., Komsa, H.P., Surface functionalization of 2D MXenes: Trends in distribution, composition, and electronic properties. *J. Phys. Chem. Lett.*, 12, 2377, 2021.
132. Yan, L., Xu, Y., Si, J., Li, Y., Hou, X., Enhanced optical nonlinearity of Mxene $Ti_3C_2T_x$ nanosheets decorated with silver nanoparticles. *Opt. Mater. Express*, 11, 1401, 2021.

5

Magnetic, Mechanical and Thermal Properties of MXenes

R. Ghamsarizade[1], B. Ramezanzadeh[2*], H. Eivaz Mohammadloo[3] and N. Mehranshad[1]

[1]*Department of Polymer Engineering and Color Technology, Amirkabir University of Technology, Tehran, Iran*

[2]*Surface Coating and Corrosion Department, Institute for Color Science and Technology, Tehran, Iran*

[3]*Colour, Resin & Surface Coatings Dept., Iran Polymer and Petrochemical Institute, Tehran, Iran*

Abstract

Since its first publication in 2011, attention to MXenes – two-dimensional (2D) transition-metal carbides, nitrides, and carbonitrides – has grown rapidly in recent years. Thorough studies have shed light on their expanded capabilities for utilization in various fields, including catalysis, optoelectronic devices, and photonics, amongst others. MXenes are set to offer new 2D nanoplatforms for advanced materials due to their fascinating magnetic, mechanical, and thermal features, as well as the richness of elemental composition and chemical patterning. This in-depth study, written for audiences from a wide range of academic fields, focuses on the most recent developments in the magnetic, mechanical, and thermal characteristics of MXene and the uses for these capabilities.

***Keywords*:** MXene, 2D materials, magnetic behavior, thermal conductivity, mechanical performance, transition metals

5.1 Introduction

Recent thorough explorations have shown that low dimensional compounds comprising TM[1]s having open d-orbital shells might display various fascinating features due to their distinct oxidation, spinning conditions, and fairly substantial spin-orbit connectivity. Therefore, the low dimensional compounds based on TMs offer a great suitable foundation to study and utilize the electrons 'internal degrees of freedom, including charges, valence, and spinning, along with their interrelationship both for fundamental studies and practical commercialization. Several TM-based low-dimensional compounds, such as dichalcogenide categories, have been or might be explored in the investigations [1–5].

**Corresponding author*: ramezanzadeh-bh@icrc.ac.ir

[1]Transition metal

Some of these compounds have been discussed by several researchers. MXenes [6, 7] are currently at the very forefront of efforts on substances and hope to open up new areas of study and scientific advances. MXenes are a novel family of 2D TM carbides and nitrides mostly having a general structure of $M_{n+1}X_n$ (M= Hafnium, Zirconium, Titanium, Scandium, Vanadium, Niobium, Molybdenum, Chrome, Tantalum; X= Nitrogen, Carbon; n= from 1 to 3). Etching MAX layers has recently led to the production of these compounds [8–13]. Since these 2D materials are so comparable to graphene, they have been dubbed "MXenes," and they emerge from the MAX layers after the "A" atoms are removed from the compound. The chemical vapor deposition method has successfully achieved significant advancements in creating crystals of MXenes [14, 15]. In addition, the group of MXenes has recently been enlarged to include double TM carbides [16]. It is anticipated that a substantial percentage of 2D MXenes with outstanding characteristics that have not been reported before will be generated due to the wide range of conceptual possibilities offered by the MAX layers.

5.1.1 Applications of MXenes

In the research center, MXenes have indeed proven useful as transparent conductors [17, 18], transistors [19], super-capacitors [20–22], battery industry [23, 24], shields for electro-magnetism [25], polymeric composites 'fillers [26], nanocomposites [27], purifying industry [28], surfaces with dual-responsive feature [29], catalysts [30], suitable substrates for dyes [31], promising materials for methane storage [32]. MXenes have been suggested for use in a variety of uses in the fields of electronics [33], magnetism [34], optics [35], thermo-electricity [36], and sensors [37], in addition to becoming future prospective components for catalysts and photocatalyst reactions [38, 39], hydrogen storage [40], and nano-scale superconductors [41]. It is believed that some MXenes would behave as insulators because they will have a wide band gap and involve just d-orbitals [42]. It is predicted that MXenes will be utilized as substances with a low work function and as Schottky barrier joints [43]. Due to the following points MXenes are attractive.

- MXenes are resistant to mechanical and chemical breakdown due to the ceramic characteristic of their structure.
- MXenes might exist in various shapes, including single-layer to multi-layers configurations.
- MXenes may be manufactured as complex substances by combining lighter and heavier TMs, which enables the number of valence electrons and the relativity spin-orbit correlation to be tuned. This improves the operation of its electronic components as well as its structural strength.
- The width of an MXene monolayer may be adjusted to the desired level. Over this, we are able to investigate the events associated with quantum confinement.
- The interfaces of MXenes may be chemically modified by a wide range of chemical groups, enabling surface engineering.
- Several MXenes, like graphene, show evidence of their band structures' mass-less Dirac dispersions at the Fermi level. This paves the way for several new opportunities in Dirac-based mechanics and commercialization.

5.1.2 Structure of MXenes

Even though most of their features still have not been directly seen, they are what give MXenes their distinctive identity from the other recognized 2D compounds. The MAX sheets are an extremely extensive family of ceramic compounds with hexagonal layers. Their chemical symbol is $M_{n+1}AX_n$. In this formula, "M," "A," and "X" each stand for an initial TM, an element from one of the 13 to 16 groups of periodic tables, and carbon and/or nitrogen, respectively. Over sixty distinct MAX layers chemicals have previously been successfully produced through experimentation [44].

In general, MAX layers have beneficial qualities that are usual in ceramics, including strong structural strength, as well as beneficial features that are expected from metals, including superior conductivity. MAX layers are capable of substituting for a spectrum of uses, including prevention of corrosion attacks, heat pumps, lubrication, and injectors. The elastic properties of over 240 distinct M_2AC and M_2AN substances with a variety of M and A atoms have been hypothetically explored up to this point [8], and the computations used to determine these values come from the basic principles. Once MXenes are chemically broken up by HF acidic solution, in order to get rid of the "A" atom, it has been practically shown that the outer layers frequently get loaded with F, O, and/or OH groups [45–47]. In experiments, numerous MXene structures have been explored. Some of them are outlined in Table 5.1.

The "A" atom can be extracted from the MAX layers in order to facilitate the construction of the structural parameters for the 2D MXene frameworks. Due to the fact that the MAX layers have a hexagonal structure, the resulting MXene compounds also form hexagonal lattices with similar geometry. The chemical termination of the M_2X combination might take place in one of four distinct patterns. These patterns are determined by the relative locations of the termination units that are connected to the TM elements.

- The first framework suggests that two functional units sit atop the two TMs.
- The second framework has two functional groups positioned on the upper side of empty spots A.
- Framework 3: One functional unit is placed on the upper side of a hollowed spot A, while the other functional unit is placed on the upper side of a hollowed site B. Both of these locations are on the upper side of the same empty spot.
- Framework 4 proposes that the two functional units should be situated above the hollowed spots B.

In order to decide which of these four configurations will produce very high stability, it is necessary to take into account each type of functional group›s completely relaxed atomic locations. Framework 1, which represents a typical tendency in functionalized MXenes, is chemically less robust than the other three patterns. Throughout the process of structural refinement, it is frequently changed into one of the frameworks 2, 3, or 4. It was also shown that the likely ionic state of the TMs on the surfaces is the primary factor in determining the relative morphological stability and durability of frameworks 2, 3, and 4 [55]. If the TMs can provide adequate electrons to X and the linked functional units, then Framework 2 is the most stable option.

Table 5.1 Some of the MXene structures have been studied so far.

MXene structure	Reference
Ti_2C	[7]
V_2C	[23]
Nb_2C	[23]
Mo_2C	[48]
Ti_3C_2	[49]
Zr_3C_2	[50]
Nb_4C_3	[51]
Ta_4C_3	[7]
Ti_4N_3	[52]
TiNbC	[53]
$(Ti_{0.5}Nb_{0.5})_2C$	[7]
$(V_{0.5}Cr_{0.5})_3C_2$	[7]
Ti_3CN	[7]
Mo_2TiC_2	[16]
Mo_2ScC_2	[54]
Cr_2TiC_2	[16]
$Mo_2Ti_2C_3$	[16]
$(Nb_{0.8}Ti_{0.2})_4C_3$	[51]
$(Nb_{0.8}Zr_{0.2})_4C_3$	[51]

5.2 Magnetic Characteristics of MXenes

The assessment of the magnetic characteristics displayed by MXene materials relies heavily on the surface stretching that is present in the substance. The existence of solid covalent connections between M and X, in addition to the surface characteristics, is the primary factor that causes a significant portion of the MXene components to lack magnetic properties. In addition, MXene materials have a magnetic feature because of a rise in the electrons 'density close to the FZ[2], that is attributed to d orbitals that are determined to transfer elements, or else because of an external in-plane stretch that is fine to relate potential covalent connections, including M-A and M-M links. Both of these factors contribute to the magnetic

[2]Fermi zone

character of MXene components. It has also been said that 2D chromic carbide and chromic nitride display ferromagnetic behavior, but the 2D titanium nitride and titanium carbide are known to be anti-ferromagnetic. The formation of surface stretching as a result of surface modification is the cause of the reduction of charge distribution in the vicinity of the FZ. These surface stretching could ultimately result in the disappearance of magnetization.

2D layers that belong to the MXene category have a diverse collection of magnetizations. When it comes to MXenes, the Cr_2C, Cr_2N, Ta_3C_2, and Cr_3C_2 are expected to be ferromagnets [56], and they might be exfoliated from their MAX layer [57]. On the other hand, Ti_3C_2 and Ti_3N_2 are anti-ferromagnets. It has been found that Ti_2C and Ti_2N showed almost half-metallic FM[3] [58]. The Mn_2C monolayer is shown to be an anti-ferromagnet besides a significant Neel temperature of 720 K [59], as determined by simulation analysis. Nevertheless, if the surface is functionalized correctly with F, Cl, and OH, it may be converted to an FM state that has a considerable T_C[4] of 520 K [60]. Experiments have proved that MXenes are often ended with Fluor, Oxygen, hydroxyl, or other elements [61]. Despite the fact that certain pure MXenes are expected to be FM, this is not the case. Thus, the synthesis of pure MXenes such as Cr_2C is a process that is fraught with difficulty. The latest evidence, nevertheless, demonstrates that it is possible to achieve a partly erased surface ending, which results in an increase of the conductivities of MXene [62]. It has been discovered that Cr_2C and Cr_2N MXenes, which are functionalized, are magnetic, but it has also been shown that certain of them display FM to AFM[5] changes [63, 64]. These MXenes get their magnetism qualities from the d-orbitals of the Cr elements that make up their structure. Because of the significant energy gap that exists between the FM and AFM structures of Chromium MXenes, it is possible that these structures can keep their magnetism almost all the way up to ambient temperature.

Non-terminated Cr_2C can be ferromagnetic, according to Si *et al.* [34], who found that the half-metallic gap in this compound was 2.85 eV. Comparably, they believed that peripatetic Cr d-orbital electrons caused ferromagnetism. On the other hand, the formation of the d electrons by surface endpoints (Fluor, hydroxyl, Hydrogen, or Chlore) might stimulate FM–AFM transitions, which are preceded by metal-insulator phase changes. FM half-metals that have full spin polarization of electrons near the FZ are exciting possibilities for use in spintronics. It has already been hypothesized that Cr_2N MXenes that have been functionalized with F and OH have AFM properties. Still, the oxygen-passivated Cr_2NO_2 represents an FM state and a significant 2.79 eV half-metal gap [65]. A pure single-layer of Cr_3C_2 has overall magnetic properties of 3.9 μB per formula unit. This indicates that the material is metallic and very robustly FM. Cr_3C_2 exhibits the characteristic of a half-metallic FM having a half-metallic gap 1.2 eV when subjected to stresses of a maximum of 3% strain, and this behavior may be continued even when subjected to loads of greater magnitude [66] (Figure 5.1).

Yang *et al.* explored a magnetic characteristics of different Cr-based carbides ($Cr_2M_2C_3T_2$; M = Ti, V, and Nb; T = OH, and F). The findings of calculations indicate that the FM configuration is attractive for $Cr_2Ti_2C_3O_2$ and $Cr_2V_2C_3O_2$. Still, other structures are energetically favorable AFM in situations in which the magnetism of Cr atoms is more than 1.9 μB [67] (Figure 5.2).

[3]Ferromagnetic
[4]Curie temperature
[5]Antiferromagnetic

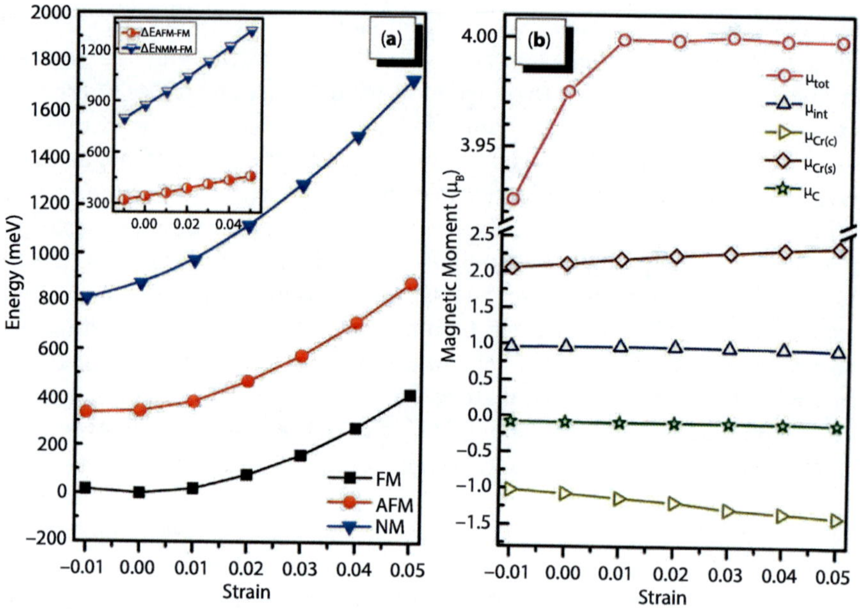

Figure 5.1 (a) The comparative energy of Cr_3C_2 in the FM, AFM, and NM forms varies with strain, and (b) Both the overall magnetism and the atomic magnetization fluctuate with the strain that is imposed [66]. Copyright 2017. Reproduced with permission from Elsevier.

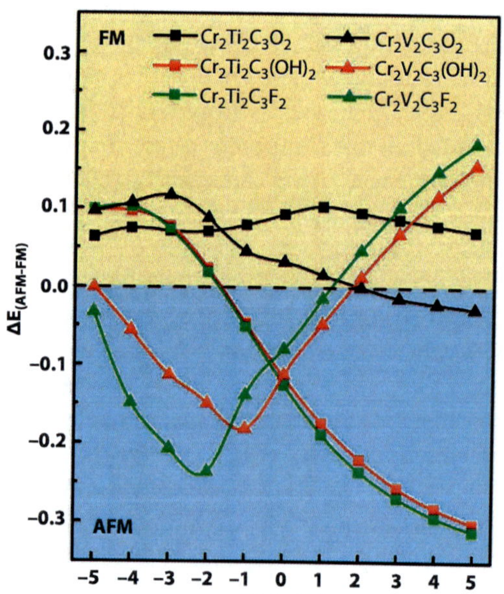

Figure 5.2 The computed value of the elastic modulus for $Cr_2M_2C_3T_2$ when subjected to extensive tension [67]. Copyright 2017. Reproduced with permission from Elsevier.

In addition, they discovered that there was a ferromagnetic-to-antiferromagnetic transition phase in the $Cr_2Ti_2C_3T_2$ and $Cr_2V_2C_3T_2$ structures when the tensile strain was changed from -5% to 5%, except for $Cr_2Ti_2C_3O_2$, for which ferromagnetic is always advantageous. The ferromagnetic ground state is produced as a consequence of a double exchange interaction, which is particularly achieved in the framework of oxygen-terminated MXenes as a consequence of electron transfer. In the meantime, the F ending causes electron confinement, which is advantageous for super-exchange interaction and primarily causes an AFM ground state to be produced. During the process of converting $Sc_2C(OH)_2$ to Sc_2CO_2, magnetism having a 1.00 μB magnitude was discovered [68].

The necessary spin uniformity may be attained by altering the chemical components of MXene nitrides, as Frey *et al.* showed in their investigation of the magnetic features in nitrides of M_2NT_x (M= Titanium, Chrome, and Manganese) [69]. It was discovered that the spin rotations might be adjusted by using the spin-orbit coupling process. This makes it possible to produce various magnetic interactions.

The magnetic properties of a large number of mixed MXenes composed of $Cr_2MC_2T_2$ (where M = Titanium and Vanadium and T = Fluor, hydroxyl, and Oxygen) have been examined. According to the ending, the materials may either be SC[6], such as $Cr_2TiC_2O_2$, AFM, such as $Cr_2TiC_2F_2$ and $Cr_2TiC_2(OH)_2$, or FM, such as $Cr_2VC_2(OH)$, $Cr_2VC_2F_2$, and $Cr_2VC_2O_2$ [70]. NM[7] compounds include $Cr_2TiC_2O_2$, whereas AFM components include Cr_2TiC_2F. The application of small electrical fields allows for the magnetism in $M_2XO_xF_{2-x}$ (M is early TM, X is C or N, and x = 0.5 to 1.5) having NS[8] functionalization of the surface to be managed in a flexible manner [71]. New studies on oxygen, hydroxyl, and Fluor-ended $Ti_2MnC_2T_x$ single-layers indicate strong FM as well as high T_C[9] (495–1133 K), which are higher than those of current 2D FM compounds [72]. Both $Hf_2MnC_2O_2$ and $Hf_2VC_2O_2$ are reported to exhibit ferromagnetism in the same research study.

Lately, it has been proven that $Hf_2VC_2F_2$ possesses an in-plane non-collinear 120° Y-type AFM ordering, which causes FM polarization. In fact, it is a type II multiferroic material having inherently high magnetic and electric connections [73]. When it comes to these types of compounds, magnetism leads to ferroelectricity, and the two properties are tightly related to one another. Because the Cr element in NS functionalized Cr_2TiC_2FCl are located in separate chemical conditions, the conduction and the valence band near FZ display opposing spinning orientations [74]. This is the case because of the NS of the functionalization. The FM half-metal, doubly mixed $TiCrCF_2$ has a significant energy gap of 3.14 eV [75] (Figure 5.3).

According to the results of simulations, the half-metallic FM nitride Ni_2NT_2 (T = Fluor, hydroxyl, and Oxygen) is a partially FM that is resistive to surface ending changes [76]. Similar to magnetic TM dichalcogenides [77, 78], MXenes allow the flexibility to manipulate magnetism by displacement, which enhances the magnetic moments of MXenes, as well as designing magnetic transitions [58, 72, 79, 80]. When just a little level of load is applied, the NM state transforms into a magnetic one, and the ferromagnetic state transforms into the antiferromagnetic one. During repeatedly imposed biaxial tension, the approximately half

[6]Semiconductor
[7]Non-magnetic
[8]Non-symmetric
[9]Curie temperatures

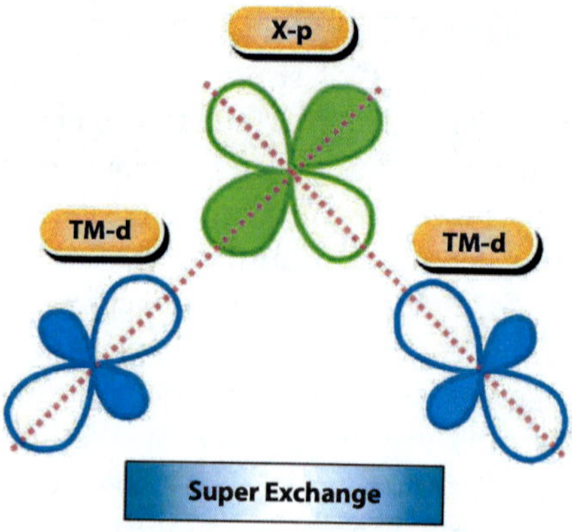

Figure 5.3 The illustration of super-exchange coupling that takes place among the TM elements of Ti_2C and the ordered double-M atoms MXenes TiMC, where M may be any of Zr, Hf, or Cr [75]. Copyright 2019. Reproduced with permission from Elsevier.

FM Ti_2C converts into an incredible half-metal, an SC gapless spin, and eventually metal. Nevertheless, a similar biaxial force approximately does not impact the Ti_2N, which maintains all of its half-metallic characteristics closely. Biaxial tensile and compressive stresses in V_2C and V_2N might produce a considerable surge in the magnetism of the material [58].

According to the phonon spectra, these structures having an O ending are steady underneath a diverse set of stresses [81]. As a consequence of suppressing crystal field splitting, functionalized Ti_2NO_2 changes from being an FM half-metal to an AFM SC [82]. Both $Cr_2Ti_2C_3T_2$ and $Cr_2V_2C_3T_2$ undergo phase transitions in a manner comparable to one another [67]. SC to TM, and indirect to direct band-gap changes, are detected in previously expected $Hf_2MnC_2O_2$ and $Hf_2VC_2O_2$ structures, having a great dependency of T_C on the induced load [83, 84]. These transitions have been reported in double MXenes, including both $Hf_2MnC_2O_2$ and $Hf_2VC_2O_2$. It was discovered that the Mo-based MXene ($Mo_3N_2F_2$) has a T_C of 237 °C and exhibits strong half-metallic FM with an EG[10] value of 0.47 eV. FM connection comes primarily from interactions of itinerant d electrons among various layers of Mo [85]. Layers in two dimensions that are based on nitrides and carbides of TM might contain of just 1 atomic plane yet possess a wide range of different characteristics to choose from. As specified, it has been expected that the low-dimensional TM nitride may present with exceptional spin-dependent features, exhibiting a huge level of spin polarization in addition to a conductivity of half-metallic type [86–88]. This would be the case if the TM nitride had a half-metallic kind of conductivity. In certain substances, the development of spin-ordered electronic states is caused by unpaired d-electrons that are positioned on the monolayer surfaces. Additionally, it is hypothesized that the lattices of this kind would have tetragonal and honeycomb polymorphs; both are magnetic. Within the class of 2D TM dichalcogenides, one may find a proper surface that maintains the single-layers half-metallicity. It is anticipated that further 2D TM mono-nitrides will exhibit auxetic [89], piezo-electric [90], or ferroelastic properties [91] and that they will be attractive features for photocatalytic applications.

[10]Energy gap

One-atom-thick TM carbides also indicate remarkable characteristics due to their specific physical structure. These features include the Hall quantum spin impact in MC (M = Titanium, Zirconium, Hafnium) [92, 93] as well as powerful FM, together with numerous h-InC nodal lines. Both of these effects can be found in the TM carbides. Therefore, various MXenes and TM nitrides might be regarded as a novel basis in order to the development of nanoscale spintronic applications in the energy sector. It is expected that the 2D material known as Fe_2C MXene would be intrinsically metallic and FM and will have an overall magnetism of 3.95 μB per unit cell. The substantial negative MAE[11] which surrounds the borders of the hexagonal Brillouin zone is responsible for the internal magnetism of Fe_2C [94].

The Heisenberg model (Equation 5.1) was used so that researchers could get an idea of how strong the magnetic coupling was in the Fe_2C MXene:

$$\widehat{H} = -\sum_{i,j} J_1 m_i m_j - \sum_{k,l} J_2 m_k m_l \tag{5.1}$$

where J_1 is the magnetic exchange interaction factor for the NN[12] and J_2 is the magnetic exchange interaction factor for the NNN[13] site pairings, m_i, m_j, m_k and m_l are the magnetism at site i, j, k, and l. The i, j, and k, l is the NN and NNN spot pairings, correspondingly. Each iron element that makes up the Fe_2C monolayer is surrounded by 3 NN and six NNN TM elements. Each iron atom in the structure of the first AFM form includes 3 NN spin-antiparallel and 6 NNN spin-parallel Fe elements associated with it. Regarding the structure of the second AFM form, each TM element is composed of one NN spin-antiparallel Fe element as well as two NN spin-parallel Fe elements. For the second AFM form, 4 of the NNN elements have spins antiparallel to one another, while the other 2 have spins parallel to one another. Consequently, the energy contained inside these magnetic nanostructures may be stated as shown in Equations 5.2–5.4, where E_0 represents the overall energy without taking into account the magnetic coupling, which is insensitive to the various magnetic phases. J_1 and J_2 are calculated by Equations 5.5, 5.6.

$$E_{FM} = E_0 - 6J_1|m|^2 - 12J_2|m|^2 \tag{5.2}$$

$$E_{first\ AFM} = E_0 + 6J_1|m|^2 - 12J_2|m|^2 \tag{5.3}$$

$$E_{second\ AFM} = E_0 - 2J_1|m|^2 + 4J_2|m|^2 \tag{5.4}$$

$$J_1 = \frac{1}{12|m|^2}(E_{first\ AFM} - E_{FM}) \tag{5.5}$$

$$J_2 = \frac{E_{second\ AFM} - E_{FM}}{16|m|^2} - \frac{1}{4}J_1 \tag{5.6}$$

---

[11]Magnetic anisotropy energy
[12]Nearest-neighbor
[13]Next-nearest-neighbor

When selecting a Hamiltonian, there are two distinct scenarios: J less than zero and J higher than zero. Both of these scenarios need careful consideration. In the scenario when J is greater than zero, the interaction energy of two magnetism prefers their being parallel; this is the case for FM. When J is less than zero, antiparallel alignments are preferred; this is the situation when the material is AFM. The factors of the exchange interaction that were estimated are as follows: J_1 = 6.17 meV and J_2 = 5.70 meV. Because J_1 and J_2 each have positive values, we may deduce that the FM arrangement is the one that the structure preferred.

In both theory and practice, the MAE is an important concept to consider when talking about 2D magnetic structures. There is a possibility that magnetic storage might take advantage of structures that have a high MAE. Most crucially, the Mermin Wagner hypothesis has claimed that the isotropic Heisenberg structure cannot provide long-distance magnetic ordering in 2D materials when the temperature is well above zero [95]. In contrast, magnetic anisotropy has the ability to produce an EG for long-wavelength spin waves, which may be used to circumvent the restrictions imposed by the Mermin Wagner theorem [96].

The XY and Ising concepts are well-known concepts that may demonstrate the magnetic ordering in two-dimensional materials [97]. Both the XY concept and the Ising concept relate to an easy plane, but the XY model also relates to an easy axis. Therefore, magnetic anisotropy is both a precondition and a basis for determining if or not a two-dimensional material possesses a magnetically ordered phase at temperatures above absolute zero. The MAE was determined by applying the following (Equation 5.7) to the data:

$$MAE = E_1 - E_2 \tag{5.7}$$

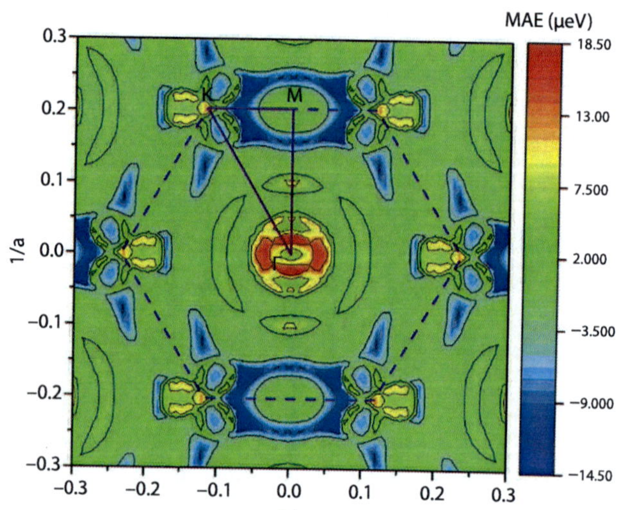

Figure 5.4 The several patterns of MAEs that may be found in the reciprocal space [97]. Copyright 1973. Reproduced with permission from Elsevier.

In which E_1 and E_2 speak for the overall energies with the involvement of the SOC[14] interactions for the in and out of plane magnetization orientation with regards to the Fe_2C MXene surface, accordingly. SOC refers to the relationship between the orbital of the electron and the electron's spin. The MAE was calculated based on the overall difference in energy, and it was found to be -22.8 eV per unit cell. Additionally, magnetism may be accomplished on an easy plane (Figure 5.4).

Due to the fact that Fe_2C is capable of displaying FM characteristics, it is essential to determine the T_C, which is an essential parameter for practical uses. The mean-field estimation is used to arrive at an approximate for the T_C, as it is provided in Equation 5.8.

$$T_C = \frac{2\Delta}{3K_B N} \tag{5.8}$$

In which N represents the amounts of magnetic elements that are included inside the cell, K_B stands for Boltzmann number and is the overall energy differences between AFM and FM alignments of Fe_2C [98]. The overall energy differences among the FM and the second AFM forms were used to compute the proportion of Δ (445.45 meV), and the proportion of N was 4. The predicted T_C was 861 K, significantly higher than a standard room's temperature.

Chrome and Manganese-doped Sc_2CT_2 MXenes (T = Oxygen, hydroxyl, or Fluor) were shown to have magnetism features [99], and these elements are considered TM magnetic elements. In their study, Lei *et al.* studied how the absence of an element affected the magnetic characteristics of monolayer Sc_2CF_2 MXene [100]. It is possible for an Sc element gap to drastically alter the crystalline phase through repulsion of the F elements that are immediately next to it and induction of an MGS[15]. The magnetism that is produced by the Sc element gap is responsive to longitudinal tensions; however, it does not change when the material is compressed. During the process of dehydrogenation from $Sc_2C(OH)_2$ to Sc_2CO_2 MXene, it is eye-opening to observe that magnetism is discovered in transitional periods with bipolar SC magnetic. It was discovered that the magnetism of the bipolar magnetic SC is roughly 1.00 μB per cell [68]. As a result of unpaired electrons in the spin-split d-orbitals, the electromagnetic features of MXenes might be substantially changed if point defects are created while this material is being synthesized [101]. This is possible because MXenes have spin-split d-orbitals. Through numerical simulations, many MXene alternatives, such as Zr_2Si, have been proposed as potential candidates for use as 2D magnets [102, 103].

The majority of the magnetic MXenes that have been anticipated are either based on magnetic TM elements or comprise magnetic TM atoms in doping topologies. To this day, there hasn't been much in the way of practical research published on the magnetic characteristics of unaltered MXenes. This is primarily because of the limitations of existing methods for manufacturing. Lately, the decreased Ti_3C_2 has been shown to provide the first scientific proof of magnetic in MXene. The research discovered that the susceptibilities of magnetic ($x = 5.752 \times 10^{-6}$) are not dependent on the temperatures above 10 K. Additionally, a Curie-like rise at temperatures lower than 10 K confirms the magnetism originations from the

[14]Spin-orbit coupling
[15]Magnetic ground state

2D layers rather than the defects of TiO_2 [104]. Because MDs[16] have such a wide range of uses in electronics, it is in everyone's best interest to progress more efficient versions of them. Three different sandwich structures made of Au–MXene–Au have been created, and it is projected that they would display outstanding MD. The magnetoresistance phenomena may be maintained during the input 0.2 V to 1.0 V voltage range. The magnetoresistance impacts have considerably relied on the thickness as well as the band-gap of the MXene [105]. The MTJ[17] composed of $Mn_2CF_2/Ti_2CO_2/Mn_2CF_2$ is also being created. This MTJ is made depending on MXene. Using voltage levels of up to 1 V, it is possible to produce a high tunneling magnetoresistance proportion [106]. The inherently strong magnetoelectric correlation that exists among these two Ferro vectors makes it possible to regulate charges via the application of a magnetic field or spin through the application of an electric field. Devices based on graphene that employ spintronics may take advantage of the half-metallic monolayer Cr_2C FM. The graphene spin polarization may account for 74% in a vertically graphene/Cr_2C heterostructure that is way higher than that of magnetic metal/graphene or heterostructures of magnetic insulator/graphene.

Siyu *et al.* showed that $MVNT_1T_2$, an asymmetrically functionalized structure, has durability intermediate between that of $MVN(T_1)_2$ and $MVN(T_2)_2$. Furthermore, it is simpler to widen the EG of MVN (M = Titanium, Vanadium, or Chrome) using the associated asymmetrical functionalization than the equivalent symmetrical functionalization. As an example, V_2NO_2 and V_2NF_2 are both metallic, but V_2NOF changes into an SC. Most significantly, the bands of MVN near the FZ are nearly distanced from others, generating distinct flat bands in the case of asymmetrical functionalization with Oxygen and Fluor groupings. This is an example of how asymmetrical functionalization may produce flattened bands [107] (Figure 5.5).

Susu *et al.* showed that when it comes to Ti_2C structures, the magnetic feature is unaffected by either the Nitrogen element amounts or the Nitrogen dispersion in the material. The magnetic ground state will never change from its antiferromagnetic arrangement. On the other hand, when there is an increase in the amount of Nitrogen, the magnetic phase of the Ti_3C_2 structures shifts from an antiferromagnetic level to a ferromagnetic level. In addition to this, Nitrogen dispersion has a significant impact on the magnetic states that are shown by the N-doped Ti_3C_2 thin films. Ti_3C_2 chooses an antiferromagnetic feature for the Nitrogen atom that is located in a separate Carbon sublayer, whilst it prefers a ferromagnetic feature for the Nitrogen atom that is located in the same Carbon sublayer. Because of an upward shift in the FZ, doping either Ti_2C or Ti_3C_2 with Nitrogen causes the material to change from a comparatively tiny SC into a metal. This occurs in both cases [108].

The magnetic characteristics of 3d TM elements adsorbed on Zr_2CO_2 Mxenes were explored in a systematic manner by Yanzong *et al.* [109]. The atoms studied range from Sc to Zn. The adsorption site that is most favorable for the Zn element is discovered to be above the Zr element, whereas the binding spot that is most ideal for the other TM elements is discovered to be higher than the Carbon element. The adsorption energies show that each of the TM elements, with the exception of the Zn element, are chemically deposited on the Zr_2CO_2. Besides, the differences in adsorption energies are proportional to the amount of 3d electrons that are present in the TM adatoms. With the exception of the Zn-deposited

[16]Magnetoresistance devices

[17]Magnetic tunnel junction

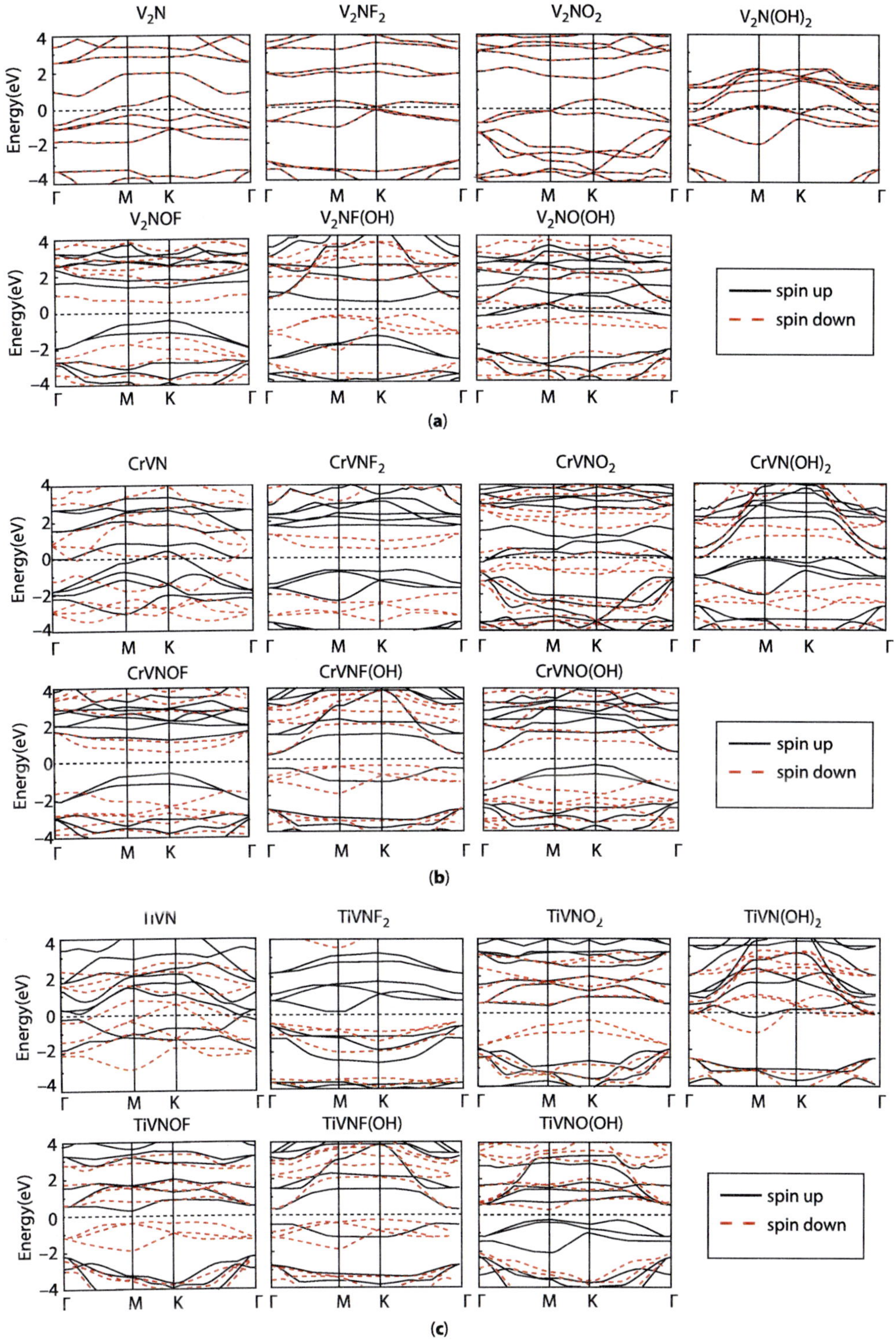

Figure 5.5 Band configurations of pure and functionalized types of (a) V_2N, (b) CrVN, and (c) TiVN, respectively [107]. Copyright 2021. Reproduced with permission from Elsevier.

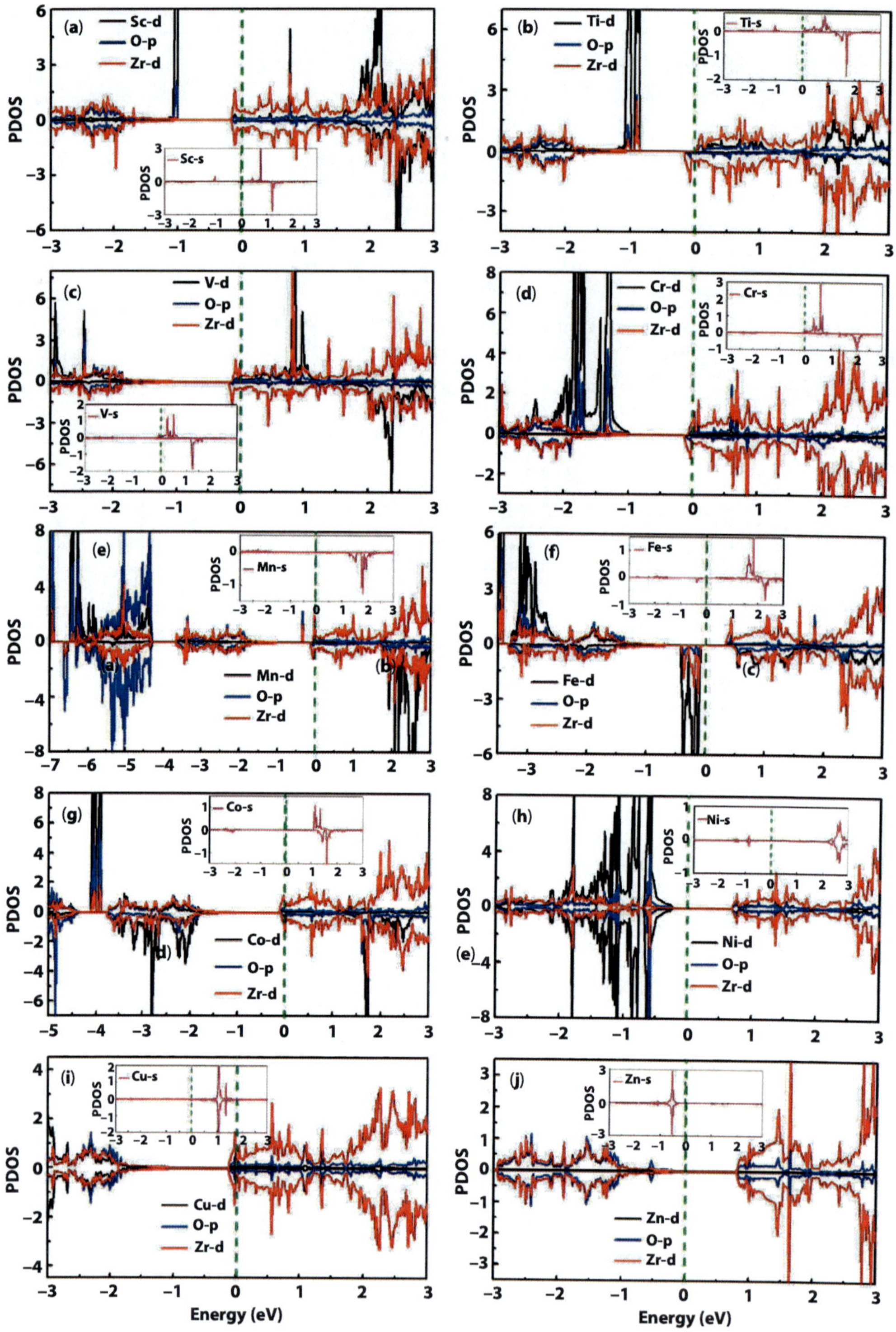

Figure 5.6 The closest Oxygen and Zirconium elements, as well as the following adatoms: (a) Scandium, (b) Titanium, (c) Vanadium, (d) Chrome, (e) Manganese, (f) Iron, (g) Cobalt, (h) Nickle, (i) Cupper, and (j) Zinc, were calculated [109]. Copyright 2021. Reproduced with permission from Elsevier.

structure, the findings indicate that there is a clear movement of electrons from the TM adatoms to the Zr_2CO_2. This is in contrast to the situation with the Zn-deposited structure, which of these possible causes might account for the majority of the TM adatom's diminished magnetism. Figure 5.6 demonstrates the electrons' movement from the 4s layer into the 3d layer for the TM adatoms. This will encourage the decline of the magnetization of the TM adatoms for the lowering of the unpaired electrons, particularly for the TM adatoms whose 3d orbitals are more than half-filled, as is the case with the TM adatoms on the boron nitride sheet or graphene.

Zhang *et al.* showed that significant magnetism is produced as a result of the replacement doping of Vanadium, Chrome, Iron, or Mangenese elements and the vacancy of an Sc element in the Sc_2CF_2 MXene. The magnetization produced by this process is 2.0, 3.0, 3.0, and 4.0 μB, accordingly [110].

The influence of straining on the magnetization of Zr_2N MXene was studied by Yue *et al.* [111]. The ground state of strain-free Zr_2N is highlighted by an inherent AFM property. The magnetic status of Zr2N, on the other hand, has a tendency to be FM whenever the induced strain is over 4%. In addition, the critical temperature for strain-free Zr_2N is around 470 K, which indicates that the AFM regime may be durable and kept at ambient temperature. After the strain has been applied, the critical temperature of phases that are AFM will start to drop. The critical temperature, on the other hand, will go up in proportion to the strain

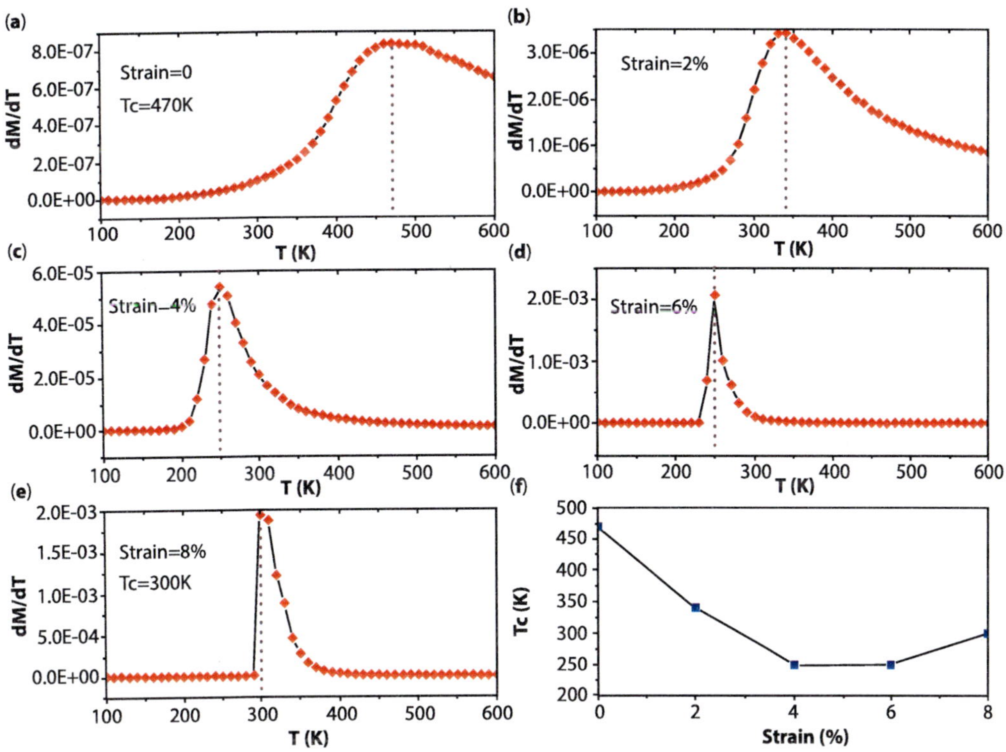

Figure 5.7 Temperature-dependent magnetic susceptibility in stresses of (a) 0%, (b) 2%, (c) 4%, (d) 6%, and (e) 8%. (f) the critical temperatures' dependency on strain [111]. Copyright 2021. Reproduced with permission from Elsevier.

that is applied as the ferromagnetic ordering becomes more advantageous. When subjected to a tensile strain of 8%, the critical temperature reaches room temperature. Furthermore, when the strain changes, both the alignment of the easy-axis as well as the MAE of Zr_2N shift and change. The MAE achieves its maximum value of 203 eV per Zr element when subjected to a strain of 2%. This is due mainly to the spin-orbit interactions that take place among filled and unfilled P_x/P_y states of Zr elements. Zr_2N MXene is suitable for use in spintronics applications because of the fact that its characteristics can be tuned (Figure 5.7).

5.3 Mechanical Characteristics of MXenes

The mechanical characteristics of MXenes are attracting a considerable attention because they have firmer (M-N) and (M-C) bonds than other materials. The previous research that was done on simulation suggested that the values for elastic factors should be twice as high as MAX structures, along with the values for 2D variables like CdS_2. Even though their elasticity characteristics are two to three times fewer than those of graphene, their bending feature has a highest of 1050 GPa that demonstrates how they may be used as composite reinforcing materials. When compared to graphene, the contact capability of MXenes with polymeric matrix for composite utility is superior because of the functionalities that are engaged [112]. Thin plates of MXene based on titanium exhibited hydrophilicity and intruded with defined contact angles in the range of 25 to 40 degrees. Additional research has shown that a surge in the proportions of layers results in a decrease in Young's modulus either for MXene nitrides or MXene carbides. Furthermore, as compared to carbide MXenes, nitride-based components have the highest possible values [113]. The occlusion of endpoints has a significant impact on the C_{11} variables, but it has also resulted in a rise in the number of critical deformations. When contrasted to the graphene sheet, which has been shown to have a great inclination for foldable electronic equipment, these numbers have the potential to be relatively high. At this time, the mechanical presence that penetrated the layers was verified along the (001) orientation.

Even though there are many different methods available for the recognition of materials that are bulk, the evaluation of 2D materials coupled with their mechanical characteristics is still a difficult task. The majority of these investigational computations are calculated using a technique called nano-indentation. This approach involves applying stress to the central point of a 2D material utilizing an AFM tip. Using this method, the practical Young's modulus of the $Ti_3C_2T_x$ monolayer was determined (333±20 GPa), which is considerably near the figure of 386 GPa for the $Ti_3C_2O_2$ compound but is still much greater than that of graphene oxide and MoS_2 [114]. Because of hardships in monitoring systems, the inadequacy of MXenes surface direct authority, the intrusion of dimensional vacancies, and the interactions between loose composite materials were the most significant considerations that need to be resolved. As a result, the mechanical prediction of MXenes was negatively impacted. Hence, novel exploratory discoveries must concentrate mainly on the progression of fabrication techniques to locate its defect nature and differentiate between numerous functional units of pure components [115]. It's possible that a comprehensive evaluation of the mechanical properties together, including the essential functional groupings rooted in concepts and experimentation, has not yet been developed.

Barsoum *et al.* reviewed the mechanical and elastic features of MAX bulk structures. They express the difference between the bulk MAX phase and the MX counterparts. Generally, the MAX zones are pretty stiff and elastically isotropic. They can withstand breakage and are easily machined due to their low hardness (2–8 GPa). Some of them weigh nearly nothing while yet being resistant to things like heat extremes, oxidation, fatigue, and creep. However, under compressive cycling loading up to 1 GPa at ambient temperature, they behave as nonlinear elastic materials, dissipating just 25% of the mechanical energy. At higher temperatures, they undergo a brittle-to-plastic transformation, and the distortion rate plays a pivotal role in determining their mechanical performance. One of the most exciting aspects of the MAX phases is their hysteretic, nonlinear, and elastic properties. As it is known, this behavior is uncommon for elastically stiff materials. When the MAX phases are subjected to cyclic stress or strain, hysteretic loops of varying sizes and shapes form as a result. Barsoum *et al.* expressed that Ti_3SiC_2 experiences primary, secondary, and tertiary creep when the tension or compression is loaded in the temperature range of 1000–1300°C. firstly a short primary-creep stage occurs in which a rapid decrease in deformation rate can be seen. Going through, a secondary-creep regime takes place during which no significant change in the creep rate over a considerable period can be seen. In this stage, the minimum creep rate, $\dot{\varepsilon}_{min}$, is characterized. This creep rate occurs during either tension or compression and can be stated within Equation 5.9:

$$\dot{\varepsilon}_{min} = \dot{\varepsilon}_0 A \left(\frac{\sigma}{\sigma_0} \right)^n \exp\left(\frac{-Q}{RT} \right) \tag{5.9}$$

Where A defines the stress-independent constant, n is the exponent of stress, and Q is the creep activation energy. R and T are the universal gas constant and the absolute temperature, and $\dot{\varepsilon}_0 = 1\ s^{-1}$ and $\sigma_0 = 1$ MPa.

Barsoum *et al.* also investigated the fatigue property of Ti_3SiC_2. They found that at ambient temperature, both fine and coarse-grained Ti_3SiC2 microstructures had a significant resistance curve characteristic, in which the tension required to expand a fracture grew with the length of the crack. The fine-grained specimens had a range of K_{1c} = 8 MPam√m to 9.5 MPam√m with a fracture length of 1.5 mm. Crack development initiation in the coarse-grained specimens occurred between 8.5 and 11 MPam√m, and peaked at 14 to 16 MPam√m after 2.7 to 4 mm of crack growth, indicating that the materials were much more resistant to failure [116].

Borysiuk *et al.* studied the mechanical features of MXenes using classical molecular dynamics. They obtained the strain–stress curves under tensile deformation of the samples and used them to calculate the Young's modulus. It can be understood that the simulated samples' structure is preserved in the process of deformation from the radial distribution function. They demonstrated that the mechanical characteristics and the development of the atomic structure of Tin+1Cn under tensile stress are affected by the number of atomic layers "n" in MXenes. This may suggest that $Ti_{n+1}C_n$ with varying "n" has distinct mechanical failure causes. Thin Ti_2C carbide has the greatest Young's modulus. They determined that once a free boundary edge is present in a sample, MXene starts to roll up upon itself. The large surface energy in a vacuum is responsible for this phenomenon. Smaller pieces generally keep their initial flat sheet-like shape in the Ti_4C_3 structure because the surface tension is inadequate to conquer the high stiffness of the MXenes explored in this study. A crack results in the formation of these pieces. After a crack, the local $Ti_{n+1}C_n$ structure is preserved in all specimens. Yield stress increased somewhat with an accelerating trend, as shown by Borysiuk *et al.* This suggests that $Ti_{n+1}C_n$ MXenes might have a defect generation and plastic deformation process similar to metals [117].

The researchers Khazaei *et al.* calculated the mechanical characteristics of the M_2AB phases and compared them to those of the equivalent M_2AC and M_2AN phases. M_2AlB phases have been the primary concern of their attention because they are still in theoretical stages. According to their analysis, only two of the M_2AlX substances under consideration—Sc_2AlB and Mo_2AlN—have forms that aren't mechanically stable. Additionally, they concurred that Mo_2AlN is mechanically unsteady [118].

The elastic constants, shear and Young's moduli, ductility, Poisson's ratio, shear and compression anisotropy, the Debye temperature, and the minimum T_C of M_2AlC, M_2AlN, and the hypothetical M_2AlB phase were all investigated in greater detail by Khazaei *et al.* Some of the M_2AlB phases were discovered to be metastable. Bulk, shear, and Young's moduli of Ta_2AlX, Cr_2AlB/C, and Mo_2AlB/C were all said to be quite significant. Based on the elasticity index, Zr_2AlB, Ta_2AlN, and Cr_2AlN are promising candidates for structural purposes. A number of materials, including Ta_2AlB, Zr_2AlC, Hf_2AlC, and Nb_2AlN, have the potential to be exfoliated into unique 2D nanosheets due to the large disparities in total binding capabilities along the ab plane and the c-axis [11].

First-principles calculations using $Ti_{n+1}C_n$ (n= 1 to 3) were used to thoroughly study the stress-strain profiles and the deformation processes in reaction to tensile stress by Guo *et al.* They showed that Ti_2C can withstand significant stresses of 9.5%, 18%, and 17% under biaxial and uniaxial pressures along x and y. By adding oxygen functional groups to the surface, the Ti_2CO_2's stability may be improved beyond that of graphene (15% biaxial) to 20%, 28%, and 26.5%. In $Ti_{n+1}C_n$, the breakdown of the material is because of the deformation of the outermost atomic layer. Surface functionalization may mitigate this breakdown, thereby increasing mechanical adaptability. Additionally, this research elaborated on Young's modulus and critical stresses of $Ti_{n+1}C_n$ and $Ti_{n+1}C_nO_2$. Its research shed light on the nanoscale deformation process of MXenes and their potential use in bendable electronics. According to them, Ti_2C mechanical flexibility is greatly enhanced by the ending of surface groups, which is necessary to fulfill the need for flexible electronics [119].

First-principles calculations are being used for examining the mechanical features of V-based nitride MXene structures, MVN (M = Titanium, Vanadium, Chrome) by Li *et al.* These MXenes function by Fluor, hydroxyl, and Oxygen groups, symmetrically and asymmetrically. It is investigated by Li *et al.* that because of the negative formation energy of all functionalized structures, they are thermodynamically stable. It is stated by Zha *et al.* that Oxygen functionalized MXenes are stronger than Fluor and hydroxyl functionalized MXenes mechanically [120].

The stability of MXene structures that are functionalized with Fluor and hydroxyl groups has been confirmed by Li *et al.*; however, the mechanical and dynamical stability of nitride MXenes functionalized with the Oxygen group has declined [121]. It is suggested that by modulating the type and degree of MXene surface ending, a wide variety of MXenes can be reached.

To confirm the mechanical stability of hexagonal MXenes, Equations 5.10–5.14 have been used.

$$\frac{\Delta E}{V_0} = (C_{11} + C_{12})\delta^2 \tag{5.10}$$

$$\frac{\Delta E}{V_0} = \frac{1}{4}(C_{11} + C_{12})\delta^2 \tag{5.11}$$

$$Y = \frac{(C_{11}^2 - C_{12}^2)}{C_{11}} \tag{5.12}$$

$$G = \frac{(C_{11} - C_{12})}{C_2} \tag{5.13}$$

$$v = \frac{C_{12}}{C_{11}} \tag{5.14}$$

Where Y is young' modulus, G represents shear modulus, v refers to Poisson's ratio, and C_{ij} is two independent elastic constants.

The greater young's modulus, the harder the material. Young's modulus refers to the resistance of the material to mechanical deformations. Y data refers to the stiffness of MXene when it is symmetrically functionalized by O groups. G data obtains information about deformation resistance. Symmetrically O-functionalized structures present higher deformation resistance. The type of chemical bonding is deduced by Poisson's ratio. It is understood that pure CrVN has a more ionic bonding nature, while V_2N and TiVN have covalent bonding nature. Moreover, there is an increase in Poisson's ratio of V_2N/CrVN/TiVN after being surface functionalized. This results in better elasticity and flexibility because of the adsorption of the terminal groups [107].

Morphology affects the mechanical features of Mxenes, which is why it has been studied by Wing *et al.* Moreover, some applications of MXenes for their mechanical properties are being shared [122]. The stress-strain curves for MXenes with different functional groups and pristine Ti_3C_2 are given in Wing *et al.* work. It is understood that the mechanical characteristics of MXenes are affected by the functional group addition. There are two modes for deformation: first, the uniaxial test in the y and x, and second, a biaxial test in which the load is induced in both x and y directions. It has been noted that all functionalized $Ti_3C_2T_x$ is stronger than bare Ti_3C_2 for all modes. It has been stated that the greatest improvement in the mechanical tests belongs to the oxygen functionalization group (Figure 5.8).

It has been stated by Gue *et al.* that 9.5% and 18% of strain under biaxial and uniaxial tension has been endured by Ti_2C [119]. By introduction of O-terminations, Ti_2CO_2 stretches up to 20% and 28% (Figure 5.9) [123].

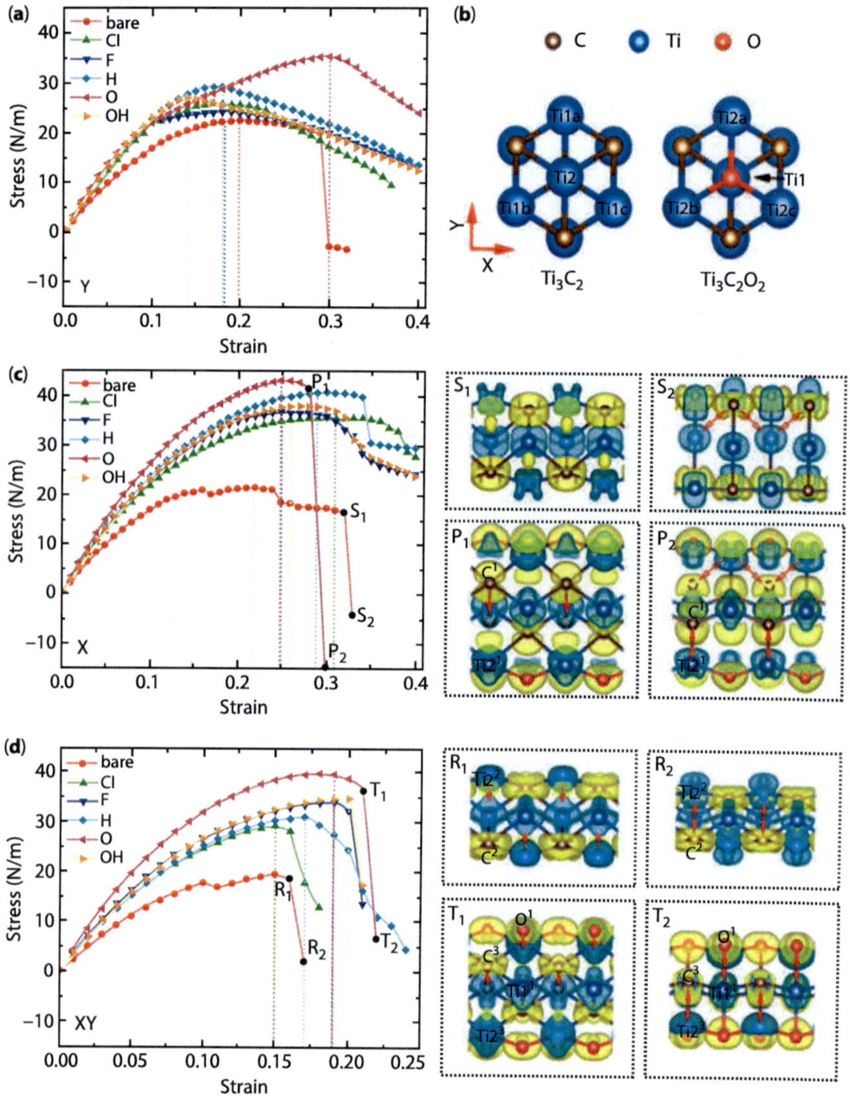

Figure 5.8 Uniaxial tension y stress-strain curves (a), Tensile orientations, and stacking orders of Ti_3C_2 and $Ti_3C_2O_4$ are shown schematically in (b), Valence charge density differences (VCDDs) between S_2 for Ti_3C_2 and P_1 and P_2 for TiC_2O_2 along the uniaxial tension x direction (c), and VCDDs R_1 and R_2 for Ti_3C_2 and T_1 and T_2 for TiC_2O_4 and stress-strain curves under biaxial tension (d) [122]. Copyright 2022. Reproduced with permission from Elsevier.

Moreover, the mechanical features of $Ti_3C_2F_x{}^nO_{2-x}$ structures are studied by Li *et al.* The stress-strain curve under uniaxial and biaxial tension has been drawn in Figure 5.10. There is a strong anisotropy of in-plane planar, which is presented by five configurations [124].

Calculation and analysis of affection of the different strains on mechanical properties of Ti_2CT_2 (T = Fluor, hydroxyl) and M_2CO_2 (M = Hafnium, Zirconium, Vanadium, Niobium) materials are done by Zhang *et al.* The calculations were based on the first-principles density

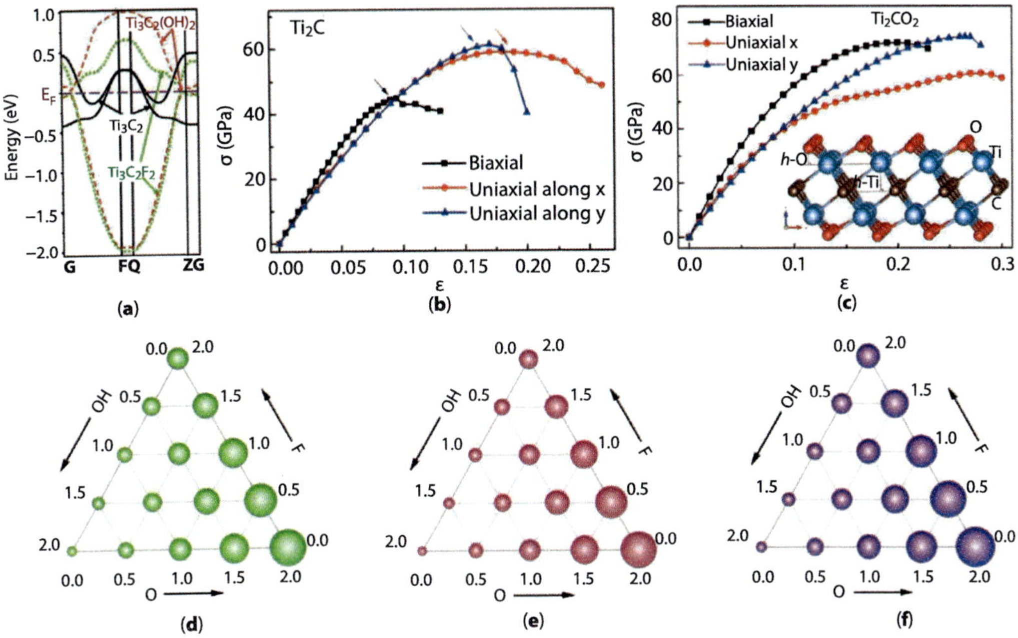

Figure 5.9 (a) The monolayer Ti_3C_2 calculated band structure with OH and F functional groups and no end, according to SC and metal properties. Stress-strain profiles for Ti_2C were computed indefinitely (b). O functional group Ti_2C stress-strain curves were calculated (c). How stable (d) Ti_3C_2 and (e) Ti_2C are in comparison. Nb_4C_3 has a wide variety of functional groups (F, O, and OH). The relative stability of MXenes improves as the bubble size increases (f) [123]. Copyright 2021. Reproduced with permission from Elsevier.

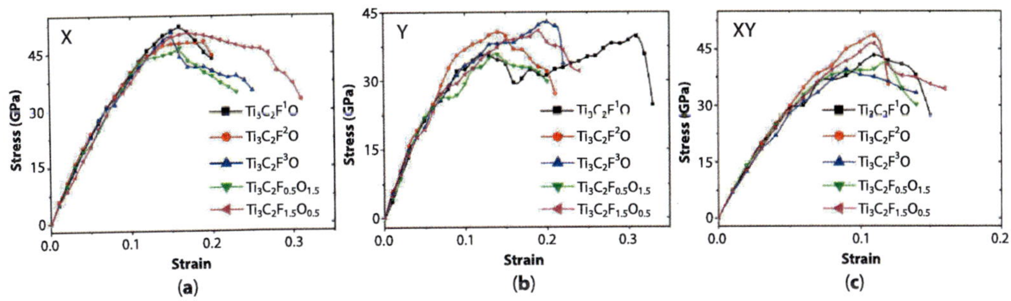

Figure 5.10 The stress-strain curve via (a) uniaxial x-direction, (b) uniaxial y direction, and (c) biaxial tension [124]. Copyright 2021. Reproduced with permission from Elsevier.

functions. It is stated that “AA stacking” has the best stability of the two-dimensional material of Ti_2CF_2 and also M_2CO_2 (M = Hf, Zr, V, Nb); the strain-stress curves are provided in Figure 5.11. It is written that there has been no significant change in the metallic nature of Ti_2CF_2 with the increase of compressive strain [125].

The affection of the binding site of the oxygen atoms on the mechanical properties was investigated by Yengejeh *et al.* The mechanical stiffness can be significantly influenced by

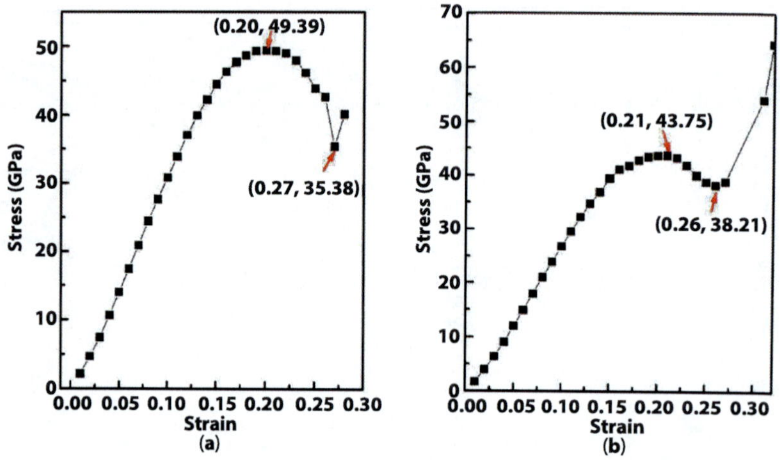

Figure 5.11 Stress-strain curve of (a) Hf_2CO_2 (b) Zr_2CO_2 multilayer material with increasing compressive strain [125]. Copyright 2020. Reproduced with permission from Elsevier.

(M–O) binding strength, suggested by COHP[18]. The in-plane mechanical properties include 2D stiffness and shear modulus and Poisson's ratio. For calculating these properties, the following Equations (5.15–5.17) can be used. The results are provided in Figure 5.12.

$$E^{2D} = \frac{C_{11}C_{22} - C_{12}C_{21}}{C_{22}} \tag{5.15}$$

$$G^{2D} = C_{66} \tag{5.16}$$

$$\nu^{2D} = \frac{C_{12}}{C_{22}} \tag{5.17}$$

According to Equation (5.18), the cohesive energy of MXenes (E_{coh}) can be calculated [126].

$$E_{coh} = \frac{(E_{tot} - \Sigma E_{atom})}{N} \tag{5.18}$$

In this equation, E_{tot} refers to the overall energy of MXene, E_{atom} shows the energy of the single atom, and N indicates the amounts of elements per unit cell.

Classical molecular dynamics simulations are used for studying the mechanical features of both pure and functionalized $Ti_{n+1}C_nO_2$ and $Ti_{n+1}N_nO_2$ (n =1 to 2) by Hatam-Lee *et al.* The stress-strain curves, strength, Young's modulus, and fraction strain curves have been drawn [127]. It is reported that the greatest Young's modulus belongs to Ti_2N (around

[18]Crystal orbital Hamilton population

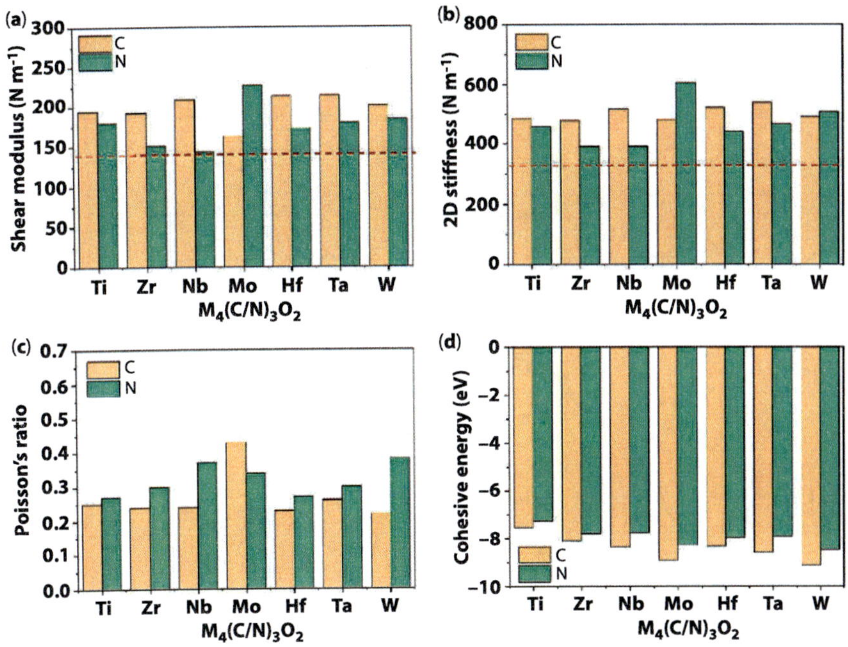

Figure 5.12 (a) Shear modulus, (b) 2D stiffness, (c) Poisson's ratio, and (d) Cohesive energy of functionalized MXenes. The red dotted lines show the corresponding values of graphene [126]. Copyright 2021. Reproduced with permission from Elsevier.

520 GPa) and the lowest belongs to Ti_3C_2 (about 130 GPa). It is also stated that oxidation that might happen during the experiments has a positive effect on the Young's modulus in Titanium carbide MXene and increases it. Still, it appears to decrease Young's modulus in titanium nitride. However, an increase in fracture stress has occurred in both groups. The stress-strain curve for Ti_2CO_2 is presented in Figures 5.13–5.15.

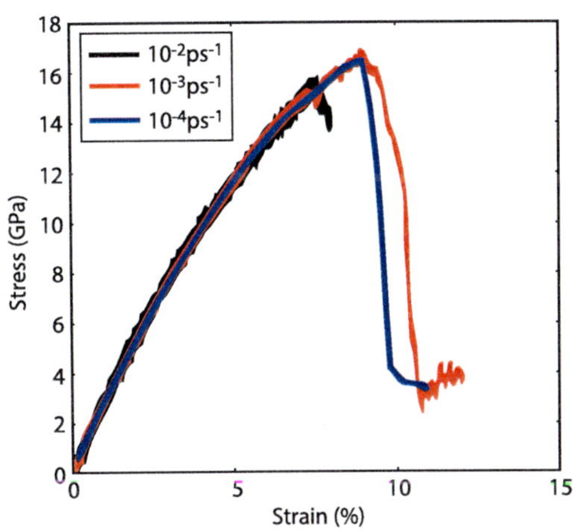

Figure 5.13 Ti_2CO_2 stress-strain curves obtained at different strain rates and in the armchair orientation [127]. Copyright 2021. Reproduced with permission from Elsevier.

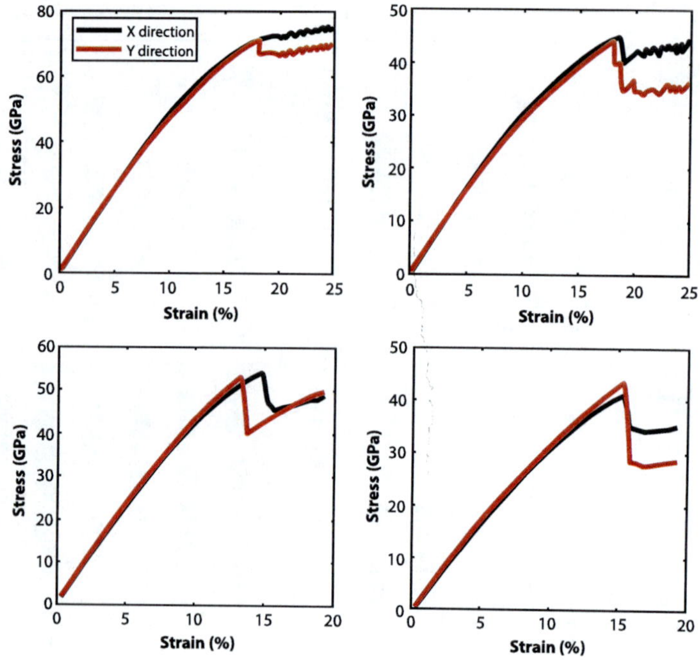

Figure 5.14 Titanium nitride MXene stress-strain diagrams in both the armchair and zigzag orientations [127]. Copyright 2021. Reproduced with permission from Elsevier.

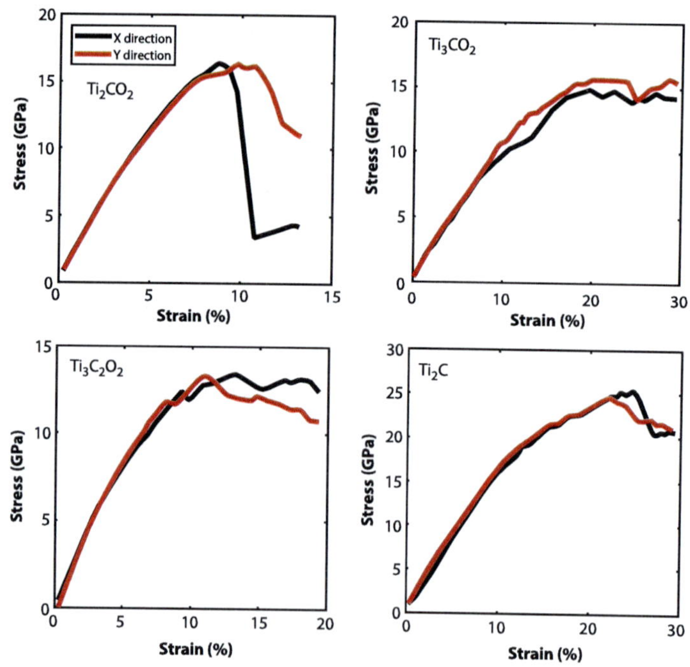

Figure 5.15 Titanium carbide MXene stress-strain diagrams in both the armchair and zigzag orientations [127]. Copyright 2021. Reproduced with permission from Elsevier.

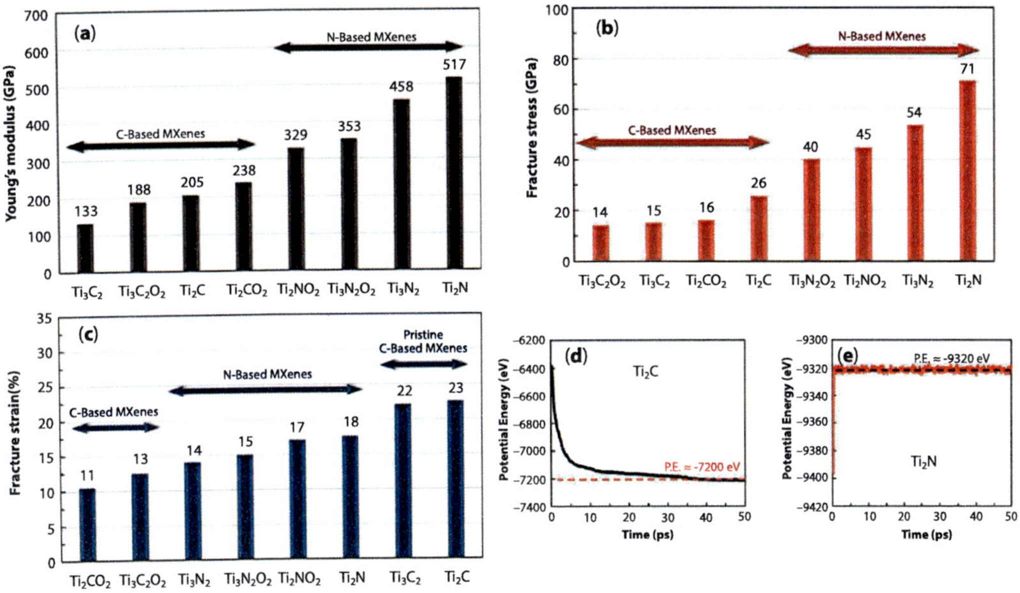

Figure 5.16 (a) MXene structures young's modulus, (b) Fracture stress in MXene structures, (c) Fracture strain in MXene structures, and (d and e) MXene structures' potential energy curves [127]. Copyright 2021. Reproduced with permission from Elsevier.

By comparing the Young's modulus (Figure 5.16a), the superiority of Nitrogen-based ones is understood. Fatigue happens at a greater stress level in nitrogen-based MXenes, as shown by a comparison of the fracture stress in Mxenes (Figure 5.16b). When evaluating MXene structures with varying Young's moduli, it is found that the levels of fracture strain are quite similar (Figure 5.16-c). The potential energy curves of MXene structures are shown in Figures 5.16d and 5.16e. The Nitrogen-based curves were found to have much greater atomic bonding, as seen by the results.

5.4 Thermal Characteristics of MXenes

When analyzing the performance of MXene, the thermal stability is among the most important parameters to take into consideration. Using thermogravimetric and mass spectrometric analyses, Seredych *et al.* assessed the thermal stability of $Ti_3C_2T_x$, Nb_2CT_x, and Mo_2CT_x MXene until 1500°C [128]. The research focused on this temperature range. The researchers came to the conclusion that the thermal characteristics are highly influenced by the content of MXene. This suggests that the kind of TM and the etching technique have a significant role in determining the surface characteristics. The rise in the proportion of water that was trapped in the $Ti_3C_2T_x$ MXene layers was observed by increasing the content of HF, and the thermal stability of the material was increased by employing a lower range of the etching chemical. When compared to Nb_2CT_x and Mo_2CT_x MXene, the thermal stability of $Ti_3C_2T_x$ MXene was significantly superior. The thermal properties of the multi-layered V_2CT_x MXene were evaluated by Thakur *et al.* [129]. The materials were exposed to a temperature of 600°C in an atmosphere, including Nitrogen gas, Hydrogen gas, and air. While warmed to the temperatures over 300°C in the presence of oxygen, it was discovered that V_2CT_x MXene underwent minor oxidation. The findings of the SEM showed that the microstructure of V_2CT_x MXene was not affected when subjected to heat in a Nitrogen gas or Hydrogen gas but the layered structure broke when subjected to high temperatures in air. The findings of an XPS showed that throughout the dehydration process, the vanadium atom was oxidized at temperatures over 300°C, and complete dehydration was accomplished at temperatures above 700°C.

The $Ti_3C_2T_x$ films were used by Lin *et al.* The delaminated nano-flakes of $Ti_3C_2T_x$ were used to make these sheets, which were then filtered with the help of suction. Both the intrinsic T_C of the $Ti_3C_2T_x$ flake and the inter-flake thermal resistance contribute to the final T_C, as realized by Lin *et al.* A mono flake of $Ti_3C_2T_x$ was utilized since its T_C is representative of the full film. $Ti_3C_2T_x$ flake's effective and intrinsic thermal conductivities are related, and this is demonstrated by Equation 5.19.

$$\frac{D}{k_{eff}} = \frac{D}{k_p} + \sum R \tag{5.19}$$

$Ti_3C_2T_x$ flakes have a typical lateral dimension of D, and the total thermal resistance between the flakes is R, where k_{eff} and k_p are the effective and inherent thermal conductivities of the flake, accordingly. It has been discovered that $k_{eff}=k_p$ when there is no separation and complete interaction among the $Ti_3C_2T_x$ flakes ($\Sigma R = 0$). The interface thermal resistance at the interconnections of individual flakes and phonon scattering between nearby flakes ensure that R constantly exceeds zero in actuality.

The particle sizes of the initial MAX materials and any reduced size that may have occurred during the delamination procedures, such as ultrasonic treatment, are the primary determinants of the lateral dimension D of the MXene flake. If all other factors are maintained constant, a bigger value for D results in a higher T_C. Contact resistance at surfaces and phonon scattering between neighboring $Ti_3C_2T_x$ nano-flakes are the primary sources of the total thermal resistance. Therefore, the reduction of ΣR by improvement of the interaction between the $Ti_3C_2T_x$ nano-flakes leads to the improvement of the T_C of MXene film (Figure 5.17) [130].

Khazaei *et al.* expressed that for T_C, $k_{min} = k_B v_a A_{min}/V_{atomic}$ in which k_{min} refers to the minimum T_C at high temperatures. Phonons have the shortest possible mean free route inside the Debye model (A_{min}). Under the assumption that A_{min} is proportional to the cube root of V_{atomic}, k_{min} may be calculated using the following Equation 5.20 [11].

$$k_{min} = k_B v_a \left(V_{atomic}\right)^{-\frac{2}{3}} \tag{5.20}$$

Grandi *et al.* demonstrated that T_C has inverse temperature dependence. They show no differences between the lattice thermal conductivities of the heterostructures and the bare MXenes [131]. Kumar *et al.* explored the thermoelectric behavior of the MXene Sc_2C functionalized by Oxygen, Fluor, and hydroxyl groups. By utilizing the density functional

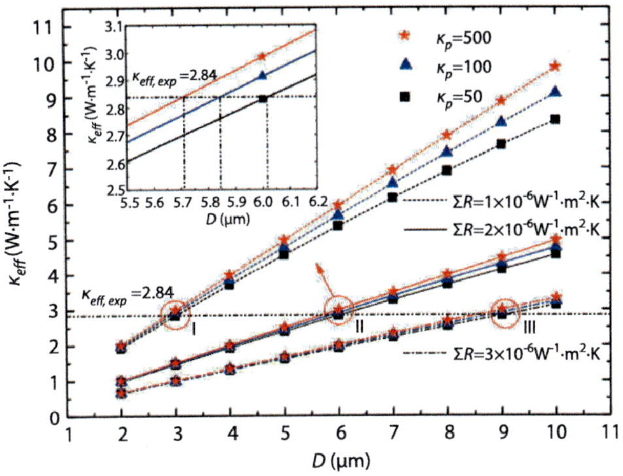

Figure 5.17 k_{eff} vs. dimension of the flakes of $Ti_3C_2T_X$ [130]. Copyright 2018. Reproduced with permission from MDPI.

theory and the Boltzmann transport formalism, they accounted for the motion of electrons and phonons. They expressed that suppressing the bipolar conduction at elevated temperature leads to the enhancement of the Seebeck coefficient for hydroxyl functionalization. The mixed grain approach can reduce the large electronic contribution to the T_C obtained for OH functionalization at small carrier concentrations [132]. Bouhemadou *et al.* expressed that minimum theoretical T_C can be calculated as shown in Equation 5.21 [133].

$$k_{min} = k_B v_m \left(\frac{M_m}{\rho}\right)^{-\frac{2}{3}} \tag{5.21}$$

Where M_m means atomic mass, and v_m is the average sound that can be calculated through Equation 5.22.

$$v_m = \left[\frac{1}{3}\left(\frac{2}{v_t^3} + \frac{1}{v_l}\right)\right]^{-\frac{1}{3}} \tag{5.22}$$

Utilizing the shear modulus G and the bulk modulus B from Navier's relation, we can calculate the longitudinal and transverse sound velocities, v_l and v_t (Equations 5.23, 5.24).

$$v_t = \left(\frac{G}{\rho}\right)^{\frac{1}{2}} \tag{5.23}$$

$$v_l = \left(\frac{3B+4G}{3\rho}\right)^{\frac{1}{2}} \tag{5.24}$$

Debye temperature (θ_D) is one of the most critical factors for thermal characteristics. Higher θ_D implies a higher TC. θ_D can be calculated as indicated in Equation 5.25.

$$\theta_D = \frac{h}{k_B}\left[\frac{3}{4\pi V_a}\right]^{\frac{1}{3}} v_m \tag{5.25}$$

Bouhemadou *et al.* studied M_2PC (M= V, Nb, Ta). They showed that these ceramics indicate high θ_D, which shows they possess a rather stiff lattice and, thus, good TC (Table 5.2).

Thermal features of semi-conductive Sc_2CT_2 (T= Fluor, hydroxyl) and O-functionalized M_2CO_2 (M=Titanium, Zirconium, Hafnium) are studied by Zha *et al.* TC of most metals, and semi-conducting low dimensional materials is higher than the predicted values of Sc_2CT_2 (T=Flour, hydroxyl) MXenes.

TC and the Sc_2CF_2 armchair direction have also been determined. The results have shown roughly 470 Wm^{-1} K^{-1} based on a 5 μm length flake at ambient temperature. This value is higher than the value of the best traditional conductor silver. It is demonstrated that $Sc_2C(OH)_2$ has lower and less anisotropic TC compared to Sc_2CF_2. It has been mentioned that the TC of Hf_2CO_2 is greater than MoS_2 and phosphorene. This value along the armchair direction is approximately 85~130 $Wm^{-1}K^{-1}$ with a flake length of 5~100 μm. The thermal expansion coefficient of Hf_2CO_2 has been calculated to be around 6×10^{-6} K^{-1} at ambient temperature. This value is higher for most metals. In addition, this material can be counted as SC, and carrier mobility has been considered high and anisotropic. The TC of Ti_2CO_2 has been noted to be lower, and carrier mobility has been noted to be higher than those of Hf_2CO_2. According to Zha *et al.*, all these materials are promising to be used in electronic devices [38, 39]. The TC of Mo_2C has also been studied by Zha *et al.* This value has been determined to be approximately 48 $Wm^{-1}K^{-1}$. The TC of Mo_2C can increase by increasing the temperature. The specific heat value has been mentioned to be 290 $JKg^{-1}K^{-1}$, and the thermal expansion coefficient value is roughly 2×10^{-6} K^{-1} at room temperature [48].

Table 5.2 Calculated parameters from the mentioned equations [133]. Copyright 2011. Reproduced with permission from Elsevier.

	ρ	v_l	v_t	v_m	k_{min}	θ_D
V_2PC	5.40	8755	5186	5744	1.5914	766
Nb_2PC	6.99	7748	4561	5055	1.2254	631
Ta_2PC	11.91	5953	3452	3831	0.9061	472

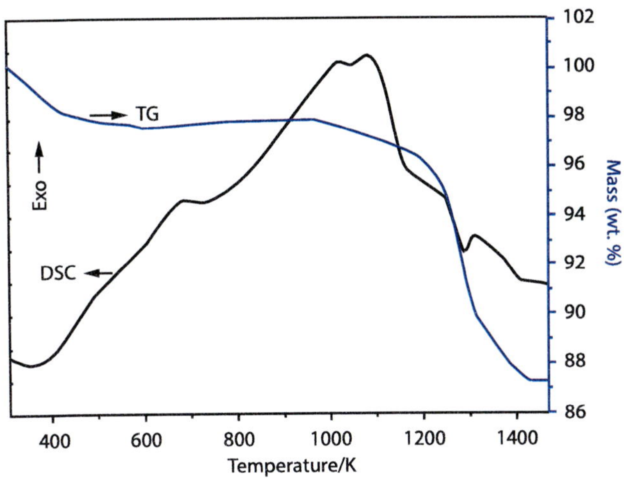

Figure 5.18 TG-DSC curve of Ti_2C nanosheets [134]. Copyright 2015. Reproduced with permission from Elsevier.

The DSC-TG[19] curve of Ti_2C nanosheets from room temperature to 1473 K was prepared by Li *et al.* (Figure 5.18). It can be understood that by increasing the temperature, a gradual decrease in the mass of Ti_2C from room temperature to around 500 K occurs [134]. It can be observed from the TG curve that there is no obvious loss of weight between 673–1173 K, while a broad exothermic peak occurs in the DSC curve. This was described by Li *et al.* as some exothermic processes such as the growth of the nanocrystals, recrystallization and etc., around 11 wt,% loss of weight appeared in the 1170–1390 K temperature range. Moreover, around 1310 K, a sharp endothermic peak was observed, which may be explained by the removal of Fluor groups.

Minimum TC (k_{min}) of TM_2C carbides was measured through Clarke's and Cahill's model by Bao *et al.* It is said that TC decreases with the increase in temperature. Clarke's model is as shown in (Equations 5.26, 5.27) [135–138].

$$k_{min} = 0.87 K_B M_a^{-\frac{2}{3}} E^{-\frac{1}{2}} \rho^{\frac{1}{6}} \tag{5.26}$$

$$M_a = \left[\frac{M}{mNl} \right] \tag{5.27}$$

Where M_a describes the mean atomic mass in the TM_2C unit cell, K_B indicates Boltzmann's constant, E is young's modulus, ρ refers to the density, M represents the molar mass, m denotes the overall number of atoms in each unit cell, and Nl indicates Avogadro's constant. Cahill's concept is as shown in Equation 5.28.

[19]Differential scanning calorimetry-thermogravimetry.

$$k_{min} = \left(\frac{K_B}{2.48}\right) n^{\frac{2}{3}} (v_1 + 2v_t) \tag{5.28}$$

In this equation, n shows the density of the amounts of elements per unit volume.

From the value of k_{min}, it can be understood that the highest Debye temperature belongs to V_2C and the lowest Debye temperature belongs to Ta_2C. It is shown that the values of k_{min}, which have been provided by Cahill's concept, are somewhat higher than the ones calculated by Clarke's concept. This is because of the accountancy of spectrum phonon in Cahill's model and ignorance of optical phonon in Clarke's concept [139].

The Mo_2CT_x thermal stability was studied by Guo *et al.* This MXene was demonstrated to be steady in the Ar atmosphere at a temperature of up to 500°C. With an increase in temperature to 700°C, a transformation in Mo_2CT_x and the coexistence of β-Mo_2C/MoC/Mo_3C_2 was seen. Mo_2CT_x MXene was stable at 200°C in the air atmosphere, and by increasing the temperature to 590°C, complete oxidation to MoO_3 occurred.

TG and DTA[20] curves in Ar and air atmosphere for Mo_2CT_x have been taken place in Guo *et al.* work (Figures 5.19a and 5.19b). Also, SEM[21] images (Figure 5.20) of Mo_2CT_x after and before heating has been shown in Guo *et al.* study [140]. As it is shown in the figures, Mo_2CT_x is thermally stable below 500°C. TG curve in Ar atmosphere had three stages. In the beginning, Mo_2CT_x has adsorbed the water physically, and this has resulted in the first stage of the TG curve, which is 3.19% weight loss from ambient temperature to 202°C. A slow and weak loosing of weight in the second stage happened from 202°C to 498°C due to slowly loosing of OH/O terminated groups. No significant endothermal and exothermal peaks can be seen in the DTA curve. From around 498°C to 700°C, the last stage of TG starts, which is approximately 3% of weight loss. This stage has occurred because of the fast loss of ending functional groups. In the DTA curve, an endothermic peak has taken place.

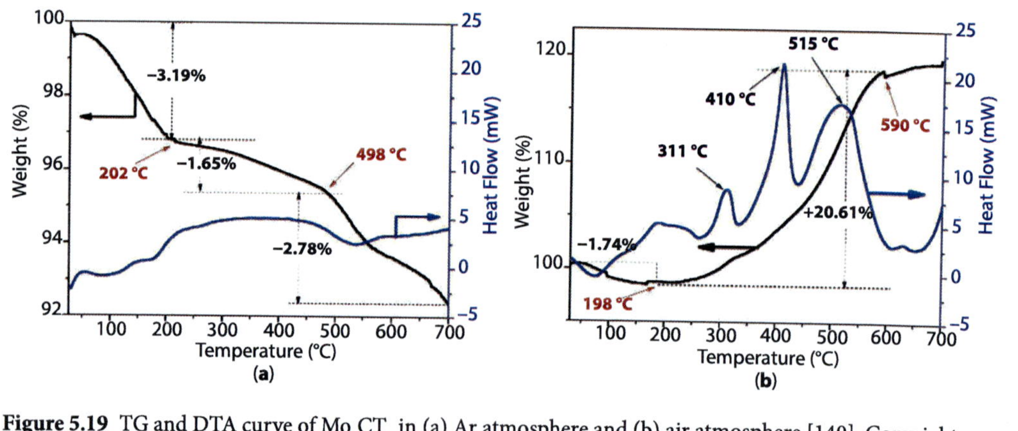

Figure 5.19 TG and DTA curve of Mo_2CT_x in (a) Ar atmosphere and (b) air atmosphere [140]. Copyright 2020. Reproduced with permission from Elsevier.

[20]Differential thermal analysis

[21]Scanning electron microscope

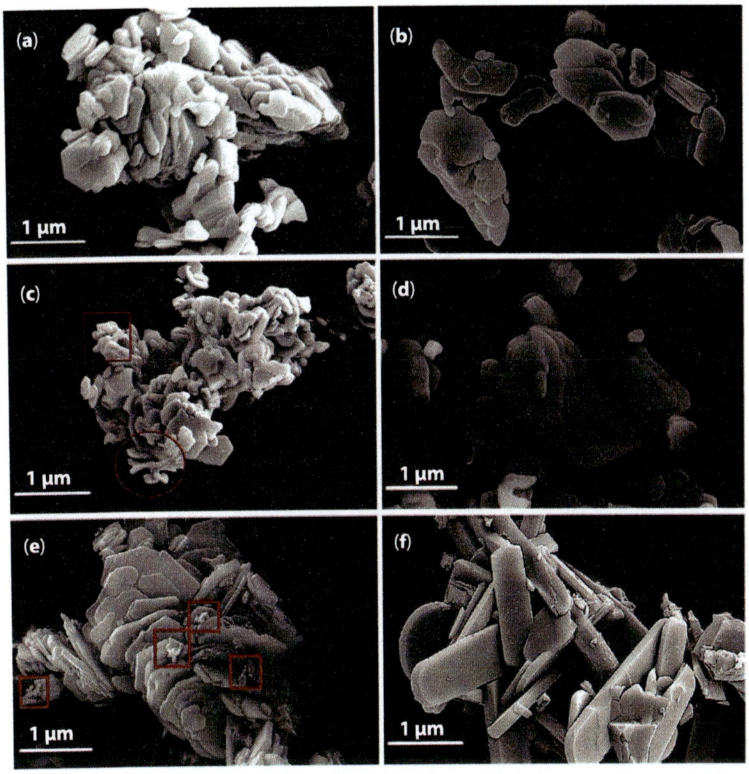

Figure 5.20 SEM images of Mo_2CT_x-N MXene after heat treatment at (a) 200°C in Ar, (b) 500°C in Ar, (c) 700°C in Ar, (d) at 200°C in air, (e) 450°C in air, and (f) 570°C in the air [140]. Copyright 2020. Reproduced with permission from Elsevier.

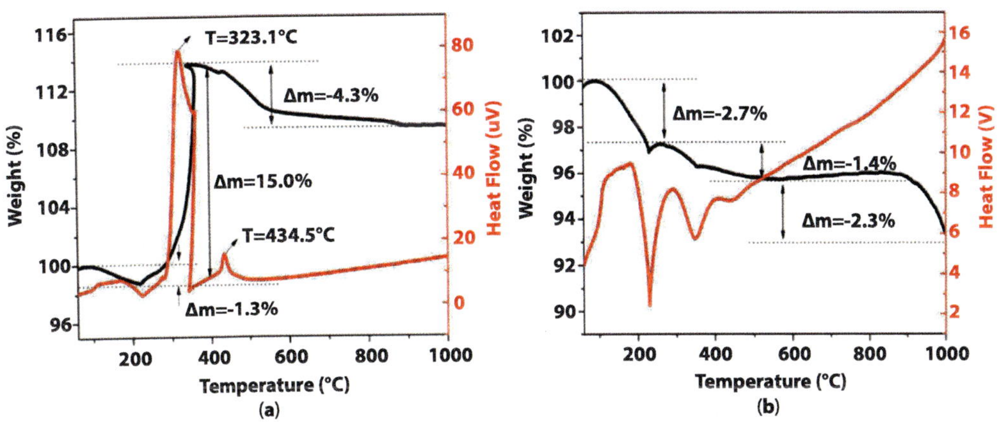

Figure 5.21 TG-DTA curve of Ti_3C_2 in (a) air (b) argon atmosphere [141]. Copyright 2017. Reproduced with permission from Elsevier.

TG and DTA curves of Ti_3C_2 in the air atmosphere were studied by Feng *et al.* Ti_3C_2 was exfoliated by NH_4HF_2. The TG curve from ambient temperature to 100°C in different atmospheres is shown in Figure 5.21a [141]. In the first stage, a 1.3% weight loss from 220°C has been demonstrated. This has happened because of the physically adsorbed water on the surface of the MXene. From 220°C to 360°C, the second stage started. In this stage, Ti_3C_2 gained 15% weight because of the oxidation of Ti_3C_2, which results in a sharp increase in weight. This can also be seen as a sharp exothermic peak at around 320°C in the DTA curve. Another weight loss happened from 360°C to 1000°C, which indicates the third stage. The TG-DTA curve of Ti_3C_2 in the Ar atmosphere (Figure 5.21b) can also be seen in Fenf *et al.* study.

5.5 Conclusion

MXenes are new 2D structures where their layered structure as well as the unique interactions of atoms have made them possible to achieve unique chemical and physical features, such as magnetism, thermal conductivity, thermal stability, and mechanical resistance. As discussed in this chapter, with slight changes in the structure of the magazines, their properties change significantly, and we can expect complex properties and applications from them. It is hoped that as new MXene structures are introduced, their use in everyday technology will become more visible in the future. In addition, other scientific fields such as the medical industry, protective coatings, optics, etc., can also make more use of the unique features of this category of new materials in the future.

References

1. Wang, Q.H. *et al.*, Electronics and optoelectronics of two-dimensional transition metal dichalcogenides. *Nat. Nanotechnol.*, 7, 11, 699–712, 2012.
2. Butler, S.Z. *et al.*, Progress, challenges, and opportunities in two-dimensional materials beyond graphene. *ACS Nano*, 7, 4, 2898–2926, 2013.
3. Fiori, G. *et al.*, Electronics based on two-dimensional materials. *Nat. Nanotechnol.*, 9, 10, 768–779, 2014.
4. Bhimanapati, G.R. *et al.*, Recent advances in two-dimensional materials beyond graphene. *ACS Nano*, 9, 12, 11509–11539, 2015.
5. Zhang, H., Ultrathin two-dimensional nanomaterials. *ACS Nano*, 9, 10, 9451–9469, 2015.
6. Naguib, M. *et al.*, Two-dimensional nanocrystals produced by exfoliation of Ti3AlC2. *Adv. Mater.*, 23, 37, 4248–4253, 2011.
7. Naguib, M. *et al.*, Two-dimensional transition metal carbides. *ACS Nano*, 6, 2, 1322–1331, 2012.
8. Cover, M.F. *et al.*, A comprehensive survey of M2AX phase elastic properties. *J. Phys.: Condens. Matter*, 21, 30, 305403, 2009.
9. Barsoum, M.W. and El-Raghy, T., The MAX phases: Unique new carbide and nitride materials: Ternary ceramics turn out to be surprisingly soft and machinable, yet also heat-tolerant, strong and lightweight. *Am. Sci.*, 89, 4, 334–343, 2001.
10. Sun, Z.M., Progress in research and development on MAX phases: A family of layered ternary compounds. *Int. Mater. Rev.*, 56, 3, 143–166, 2011.

11. Khazaei, M. *et al.*, Trends in electronic structures and structural properties of MAX phases: A first-principles study on M_2AlC (M = Sc, Ti, Cr, Zr, Nb, Mo, Hf, or Ta), M_2AlN, and hypothetical M_2AlB phases. *J. Phys.: Condens. Matter*, 26, 50, 505503, 2014.
12. Khazaei, M. *et al.*, The effect of the interlayer element on the exfoliation of layered Mo_2AC (A = Al, Si, P, Ga, Ge, As or In) MAX phases into two-dimensional Mo_2C nanosheets. *Sci. Technol. Adv. Mater.*, 15, 1, 014208, 2014.
13. Ashton, M. *et al.*, Computational discovery of stable M_2AX phases. *Phys. Rev. B*, 94, 5, 054116, 2016.
14. Xu, C. *et al.*, Large-area high-quality 2D ultrathin Mo_2C superconducting crystals. *Nat. Mater.*, 14, 11, 1135–1141, 2015.
15. Geng, D. *et al.*, Controlled growth of ultrathin Mo_2C superconducting crystals on liquid Cu surface. *2D Mater.*, 4, 1, 011012, 2016.
16. Anasori, B. *et al.*, Two-dimensional, ordered, double transition metals carbides (MXenes). *ACS Nano*, 9, 10, 9507–9516, 2015.
17. Mariano, M. *et al.*, Solution-processed titanium carbide MXene films examined as highly transparent conductors. *Nanoscale*, 8, 36, 16371–16378, 2016.
18. Yang, Y. *et al.*, Large-area highly conductive transparent two-dimensional Ti2CTx film. *J. Phys. Chem. Lett.*, 8, 4, 859–865, 2017.
19. Lai, S. *et al.*, Surface group modification and carrier transport properties of layered transition metal carbides (Ti2CTx, T: –OH, –F and –O). *Nanoscale*, 7, 46, 19390–19396, 2015.
20. Lukatskaya, M.R. *et al.*, Cation intercalation and high volumetric capacitance of two-dimensional titanium carbide. *Science*, 341, 6153, 1502–1505, 2013.
21. Ghidiu, M. *et al.*, Conductive two-dimensional titanium carbide 'clay' with high volumetric capacitance. *Nature*, 516, 7529, 78–81, 2014.
22. Rakhi, R.B. *et al.*, Effect of postetch annealing gas composition on the structural and electrochemical properties of Ti2CTx MXene electrodes for supercapacitor applications. *Chem. Mater.*, 27, 15, 5314–5323, 2015.
23. Naguib, M. *et al.*, New two-dimensional niobium and vanadium carbides as promising materials for Li-ion batteries. *J. Am. Chem. Soc.*, 135, 43, 15966–15969, 2013.
24. Xie, Y. *et al.*, Prediction and characterization of MXene nanosheet anodes for non-lithium-ion batteries. *ACS Nano*, 8, 9, 9606–9615, 2014.
25. Shahzad, F. *et al.*, Electromagnetic interference shielding with 2D transition metal carbides (MXenes). *Science*, 353, 6304, 1137–1140, 2016.
26. Zhang, X. *et al.*, Ultrathin nanosheets of MAX phases with enhanced thermal and mechanical properties in polymeric compositions: Ti3Si0.75Al0.25C2. *Angew. Chem.*, 125, 16, 4457–4461, 2013.
27. Xue, M. *et al.*, Preparation of TiO2/Ti3C2Tx hybrid nanocomposites and their tribological properties as base oil lubricant additives. *RSC Adv.*, 7, 8, 4312–4319, 2017.
28. Peng, Q. *et al.*, Unique lead adsorption behavior of activated hydroxyl group in two-dimensional titanium carbide. *J. Am. Chem. Soc.*, 136, 11, 4113–4116, 2014.
29. Chen, J. *et al.*, CO2 and temperature dual responsive "smart" MXene phases. *Chem. Commun.*, 51, 2, 314–317, 2015.
30. Fan, G. *et al.*, Magnetic, recyclable PtyCo1-y/Ti3C2X2 (X = O, F) catalyst: A facile synthesis and enhanced catalytic activity for hydrogen generation from the hydrolysis of ammonia borane. *New J. Chem.*, 41, 7, 2793–2799, 2017.
31. Mashtalir, O. *et al.*, Dye adsorption and decomposition on two-dimensional titanium carbide in aqueous media. *J. Mater. Chem. A*, 2, 35, 14334–14338, 2014.
32. Liu, F. *et al.*, Preparation and methane adsorption of two-dimensional carbide Ti2C. *Adsorption*, 22, 7, 915–922, 2016.

33. Enyashin, A.N. and Ivanovskii, A.L., Structural and electronic properties and stability of MXenes Ti2C and Ti3C2 functionalized by methoxy groups. *J. Phys. Chem. C*, 117, 26, 13637–13643, 2013.
34. Si, C., Zhou, J., Sun, Z., Half-metallic ferromagnetism and surface functionalization-induced metal–insulator transition in graphene-like two-dimensional Cr2C crystals. *ACS Appl. Mater. Interfaces*, 7, 31, 17510–17515, 2015.
35. Lashgari, H. *et al.*, Electronic and optical properties of 2D graphene-like compounds titanium carbides and nitrides: DFT calculations. *Solid State Commun.*, 195, 61–69, 2014.
36. Khazaei, M. *et al.*, Two-dimensional molybdenum carbides: Potential thermoelectric materials of the MXene family. *Phys. Chem. Chem. Phys.*, 16, 17, 7841–7849, 2014.
37. Yu, X.-f. *et al.*, Monolayer Ti2CO2: A promising candidate for NH3 sensor or capturer with high sensitivity and selectivity. *ACS Appl. Mater. Interfaces*, 7, 24, 13707–13713, 2015.
38. Zha, X.-H. *et al.*, Promising electron mobility and high thermal conductivity in Sc2CT2 (T = F, OH) MXenes. *Nanoscale*, 8, 11, 6110–6117, 2016.
39. Zha, X.-H. *et al.*, The thermal and electrical properties of the promising semiconductor MXene Hf2CO2. *Sci. Rep.*, 6, 1, 27971, 2016.
40. Hu, Q. *et al.*, MXene: A new family of promising hydrogen storage medium. *J. Phys. Chem. A*, 117, 51, 14253–14260, 2013.
41. Zhang, J.-J. and Dong, S., Superconductivity of monolayer Mo2C: The key role of functional groups. *J. Chem. Phys.*, 146, 3, 034705, 2017.
42. Si, C. *et al.*, Large-gap quantum spin hall state in MXenes: D-band topological order in a triangular lattice. *Nano Lett.*, 16, 10, 6584–6591, 2016.
43. Zhao, H. *et al.*, First-principles design of silicene/Sc2CF2 heterojunction as a promising candidate for field effect transistor. *J. Appl. Phys.*, 117, 8, 085306, 2015.
44. Barsoum, M.W., The MN+1AXN phases: A new class of solids: Thermodynamically stable nanolaminates. *Prog. Solid State Chem.*, 28, 1, 201–281, 2000.
45. Hope, M.A. *et al.*, NMR reveals the surface functionalisation of Ti3C2 MXene. *Phys. Chem. Chem. Phys.*, 18, 7, 5099–5102, 2016.
46. Wang, H.-W. *et al.*, Resolving the structure of Ti3C2Tx MXenes through multilevel structural modeling of the atomic pair distribution function. *Chem. Mater.*, 28, 1, 349–359, 2016.
47. Sharma, G. *et al.*, Calorimetric determination of thermodynamic stability of MAX and MXene phases. *J. Phys. Chem. C*, 120, 49, 28131–28137, 2016.
48. Zha, X.-H. *et al.*, Intrinsic structural, electrical, thermal, and mechanical properties of the promising conductor Mo2C MXene. *J. Phys. Chem. C*, 120, 28, 15082–15088, 2016.
49. Naguib, M. *et al.*, Two-dimensional nanocrystals: Two-dimensional nanocrystals produced by exfoliation of Ti3AlC2 (Adv. Mater. 37/2011). *Adv. Mater.*, 23, 37, 4207–4207, 2011.
50. Zhou, J. *et al.*, A two-dimensional zirconium carbide by selective etching of Al3C3 from nanolaminated Zr3Al3C5. *Angew. Chem. Int. Ed.*, 55, 16, 5008–5013, 2016.
51. Yang, J. *et al.*, Two-dimensional nb-based M4C3 solid solutions (MXenes). *J. Am. Ceram. Soc.*, 99, 2, 660–666, 2016.
52. Urbankowski, P. *et al.*, Synthesis of two-dimensional titanium nitride Ti4N3 (MXene). *Nanoscale*, 8, 22, 11385–11391, 2016.
53. Li, Y.-M. *et al.*, Theoretical investigations of TiNbC MXenes as anode materials for Li-ion batteries. *J. Alloys Compd.*, 778, 53–60, 2019.
54. Meshkian, R. *et al.*, Theoretical stability and materials synthesis of a chemically ordered MAX phase, Mo2ScAlC2, and its two-dimensional derivate Mo2ScC2 MXene. *Acta Mater.*, 125, 476–480, 2017.
55. Khazaei, M. *et al.*, Novel electronic and magnetic properties of two-dimensional transition metal carbides and nitrides. *Adv. Funct. Mater.*, 23, 17, 2185–2192, 2013.

56. Zhang, Z. *et al.*, Tunable electronic and magnetic properties of two-dimensional materials and their one-dimensional derivatives. *WIREs Comput. Mol. Sci.*, 6, 4, 324–350, 2016.
57. Ingason, A.S., Dahlqvist, M., Rosen, J., Magnetic MAX phases from theory and experiments; a review. *J. Phys.: Condens. Matter*, 28, 43, 433003, 2016.
58. Gao, G. *et al.*, Monolayer MXenes: Promising half-metals and spin gapless semiconductors. *Nanoscale*, 8, 16, 8986–8994, 2016.
59. Hu, L., Wu, X., Yang, J., Mn2C monolayer: A 2D antiferromagnetic metal with high Néel temperature and large spin–orbit coupling. *Nanoscale*, 8, 26, 12939–12945, 2016.
60. He, J., Lyu, P., Nachtigall, P., New two-dimensional Mn-based MXenes with room-temperature ferromagnetism and half-metallicity. *J. Mater. Chem. C*, 4, 47, 11143–11149, 2016.
61. Alhabeb, M. *et al.*, Guidelines for synthesis and processing of two-dimensional titanium carbide (Ti3C2Tx MXene). *Chem. Mater.*, 29, 18, 7633–7644, 2017.
62. Hart, J.L. *et al.*, Control of MXenes' electronic properties through termination and intercalation. *Nat. Commun.*, 10, 1, 522, 2019.
63. Je, M., Lee, Y., Chung, Y.-C., Structural stability and electronic properties of multi-functionalized two-dimensional chromium carbides. *Thin Solid Films*, 619, 131–136, 2016.
64. He, J. *et al.*, High temperature spin-polarized semiconductivity with zero magnetization in two-dimensional Janus MXenes. *J. Mater. Chem. C*, 4, 27, 6500–6509, 2016.
65. Wang, G., Theoretical prediction of the intrinsic half-metallicity in surface-oxygen-passivated Cr2N MXene. *J. Phys. Chem. C*, 120, 33, 18850–18857, 2016.
66. Zhang, Y. and Li, F., Robust half-metallic ferromagnetism in Cr3C2 MXene. *J. Magn. Magn. Mater.*, 433, 222–226, 2017.
67. Yang, J. *et al.*, Tuning magnetic properties of Cr2M2C3T2 (M=Ti and V) using extensile strain. *Comput. Mater. Sci.*, 139, 313–319, 2017.
68. Zha, X.-H. *et al.*, Bipolar magnetic semiconductors among intermediate states during the conversion from Sc2C(OH)2 to Sc2CO2 MXene. *Nanoscale*, 10, 18, 8763–8771, 2018.
69. Frey, N.C. *et al.*, Tuning noncollinear spin structure and anisotropy in ferromagnetic nitride MXenes. *ACS Nano*, 12, 6, 6319–6325, 2018.
70. Yang, J. *et al.*, Tunable electronic and magnetic properties of Cr2M′ C2T2 (M′= Ti or V; T= O, OH or F). *Appl. Phys. Lett.*, 109, 20, 203109, 2016.
71. Frey, N.C. *et al.*, Surface-engineered MXenes: Electric field control of magnetism and enhanced magnetic anisotropy. *ACS Nano*, 13, 3, 2831–2839, 2019.
72. Dong, L. *et al.*, Rational design of two-dimensional metallic and semiconducting spintronic materials based on ordered double-transition-metal MXenes. *J. Phys. Chem. Lett.*, 8, 2, 422–428, 2017.
73. Zhang, J.-J. *et al.*, Type-II multiferroic Hf2VC2F2 MXene monolayer with high transition temperature. *J. Am. Chem. Soc.*, 140, 30, 9768–9773, 2018.
74. He, J. *et al.*, Cr2TiC2-based double MXenes: Novel 2D bipolar antiferromagnetic semiconductor with gate-controllable spin orientation toward antiferromagnetic spintronics. *Nanoscale*, 11, 1, 356–364, 2019.
75. Hu, Y. *et al.*, Ordered double-M elements MXenes TiMC: Large in-plane stiffness and ferromagnetism. *J. Magn. Magn. Mater.*, 486, 165280, 2019.
76. Wang, G. and Liao, Y., Theoretical prediction of robust and intrinsic half-metallicity in Ni2N MXene with different types of surface terminations. *Appl. Surf. Sci.*, 426, 804–811, 2017.
77. Ma, Y. *et al.*, Evidence of the existence of magnetism in pristine VX2 monolayers (X = S, Se) and their strain-induced tunable magnetic properties. *ACS Nano*, 6, 2, 1695–1701, 2012.
78. Zhou, Y. *et al.*, Tensile strain switched ferromagnetism in layered NbS2 and NbSe2. *ACS Nano*, 6, 11, 9727–9736, 2012.

79. Khazaei, M. *et al.*, Electronic properties and applications of MXenes: A theoretical review. *J. Mater. Chem. C*, 5, 10, 2488–2503, 2017.
80. Zhao, S., Kang, W., Xue, J., Manipulation of electronic and magnetic properties of M2C (M= Hf, Nb, Sc, Ta, Ti, V, Zr) monolayer by applying mechanical strains. *Appl. Phys. Lett.*, 104, 13, 133106, 2014.
81. Khan, S.A. *et al.*, Strain engineering of electronic structures and photocatalytic responses of MXenes functionalized by oxygen. *Phys. Chem. Chem. Phys.*, 19, 22, 14738–14744, 2017.
82. Chen, W. *et al.*, Tension-tailored electronic and magnetic switching of 2D Ti2NO2. *J. Phys. Chem. C*, 121, 46, 25729–25735, 2017.
83. Siriwardane, E.M. and Çakir, D., Strain engineering of electronic and magnetic properties of double-transition metal ferromagnetic semiconductor MXenes. *J. Appl. Phys.*, 125, 8, 082527, 2019.
84. Siriwardane, E.M.D. *et al.*, Strain–spintronics: Modulating electronic and magnetic properties of Hf2MnC2O2 MXene by uniaxial strain. *J. Phys. Chem. C*, 123, 19, 12451–12459, 2019.
85. Li, S.-s. *et al.*, Emergence of ferrimagnetic half-metallicity in two-dimensional MXene Mo3N2F2. *Appl. Phys. Lett.*, 111, 20, 202405, 2017.
86. Zhang, S. *et al.*, Robust ferromagnetism in monolayer chromium nitride. *Sci. Rep.*, 4, 1, 5241, 2014.
87. Kuklin, A.V. *et al.*, Two-dimensional hexagonal CrN with promising magnetic and optical properties: A theoretical prediction. *Nanoscale*, 9, 2, 621–630, 2017.
88. Kuklin, A.V., Shostak, S.A., Kuzubov, A.A., Two-dimensional lattices of VN: Emergence of ferromagnetism and half-metallicity on nanoscale. *J. Phys. Chem. Lett.*, 9, 6, 1422–1428, 2018.
89. Zhou, L. *et al.*, Computational dissection of two-dimensional rectangular titanium mononitride TiN: Auxetics and promises for photocatalysis. *Nano Lett.*, 17, 7, 4466–4472, 2017.
90. Anand, S., Thekkepat, K., Waghmare, U.V., Two-dimensional rectangular and honeycomb lattices of NbN: Emergence of piezoelectric and photocatalytic properties at nanoscale. *Nano Lett.*, 16, 1, 126–131, 2016.
91. Xu, B. *et al.*, A two-dimensional tetragonal yttrium nitride monolayer: A ferroelastic semiconductor with switchable anisotropic properties. *Nanoscale*, 10, 1, 215–221, 2018.
92. Zhang, Z. *et al.*, Two-dimensional tetragonal TiC monolayer sheet and nanoribbons. *J. Am. Chem. Soc.*, 134, 47, 19326–19329, 2012.
93. Zhou, L. *et al.*, Prediction of the quantum spin hall effect in monolayers of transition-metal carbides MC (M = Ti, Zr, Hf). *2D Mater.*, 3, 3, 035022, 2016.
94. Yue, Y., Fe2C monolayer: An intrinsic ferromagnetic MXene. *J. Magn. Magn. Mater.*, 434, 164–168, 2017.
95. Mermin, N.D. and Wagner, H., Absence of ferromagnetism or antiferromagnetism in one- or two-dimensional isotropic heisenberg models. *Phys. Rev. Lett.*, 17, 22, 1133–1136, 1966.
96. Yue, Y. *et al.*, Pure-phase zeolite beta synthesized from natural aluminosilicate minerals and its catalytic application for esterification. *Appl. Clay Sci.*, 126, 1–6, 2016.
97. Kosterlitz, J.M. and Thouless, D.J., Ordering, metastability and phase transitions in two-dimensional systems. *J. Phys. C: Solid State Phys.*, 6, 7, 1181–1203, 1973.
98. Hynninen, T. *et al.*, High Curie temperatures in (Ga, Mn) N from Mn clustering. *Appl. Phys. Lett.*, 88, 12, 122501, 2006.
99. Yang, J. *et al.*, Investigation of magnetic and electronic properties of transition metal doped Sc2CT2 (T = O, OH or F) using a first principles study. *Phys. Chem. Chem. Phys.*, 18, 18, 12914–12919, 2016.
100. Ao, L. *et al.*, Engineering the electronic and magnetic properties of Sc2CF2 MXene material through vacancy doping and lattice straining. *Mater. Sci. Forum*, 900, 61–64, 2017.

101. Bandyopadhyay, A., Ghosh, D., Pati, S.K., Effects of point defects on the magnetoelectronic structures of MXenes from first principles. *Phys. Chem. Chem. Phys.*, 20, 6, 4012–4019, 2018.
102. Shao, X. *et al.*, Zr2Si: An antiferromagnetic Dirac MXene. *Phys. Chem. Chem. Phys.*, 20, 6, 3946–3952, 2018.
103. Jiang, Z. *et al.*, MBene (MnB): A new type of 2D metallic ferromagnet with high Curie temperature. *Nanoscale Horiz.*, 3, 3, 335–341, 2018.
104. Yoon, Y. *et al.*, Low temperature solution synthesis of reduced two dimensional Ti3C2 MXenes with paramagnetic behaviour. *Nanoscale*, 10, 47, 22429–22438, 2018.
105. Cui, C. *et al.*, Intercorrelated in-plane and out-of-plane ferroelectricity in ultrathin two-dimensional layered semiconductor In2Se3. *Nano Lett.*, 18, 2, 1253–1258, 2018.
106. Balcı, E., Akkuş, Ü.Ö., Berber, S., High TMR in MXene-based Mn2CF2/Ti2CO2/Mn2CF2 magnetic tunneling junction. *ACS Appl. Mater. Interfaces*, 11, 3, 3609–3616, 2019.
107. Li, S. *et al.*, Modulating electronic, magnetic and mechanical properties of MVN (M = Ti, V, Cr) MXenes by surface functionalization. *Solid State Commun.*, 336, 114411, 2021.
108. Liao, S. *et al.*, Electronic and magnetic properties in nitrogen-doped 2D titanium carbides (MXenes): Insight from first-principles calculations. *Solid State Commun.*, 340, 114549, 2021.
109. Wang, Y. *et al.*, Electronic and magnetic properties of 3d transition metal atom adsorbed Zr2CO2 Mxene: First-principles study. *Solid State Commun.*, 325, 114140, 2021.
110. Zhang, R.-Z. *et al.*, First-principles study of structural, electronic and magnetic properties of transition metal doped Sc2CF2 MXene. *Appl. Surf. Sci.*, 581, 152360, 2022.
111. Yue, Y. *et al.*, Tuning the magnetic properties of Zr2N MXene by biaxial strain. *Ceram. Int.*, 47, 2, 2367–2373, 2021.
112. Khazaei, M. *et al.*, Topological insulators in the ordered double transition metals M2MC2 MXenes. *Phys. Rev. B*, 94, 12, 125152, 2016.
113. Pargoletti, E. *et al.*, Engineering of SnO2–graphene oxide nanoheterojunctions for selective room-temperature chemical sensing and optoelectronic devices. *ACS Appl. Mater. Interfaces*, 12, 35, 39549–39560, 2020.
114. Zhang, P. *et al.*, Aryl diazonium-assisted amidoximation of MXene for boosting water stability and uranyl sequestration via electrochemical sorption. *ACS Appl. Mater. Interfaces*, 12, 13, 15579–15587, 2020.
115. Xiu, L. *et al.*, Aggregation-resistant 3D MXene-based architecture as efficient bifunctional electrocatalyst for overall water splitting. *ACS Nano*, 12, 8, 8017–8028, 2018.
116. Barsoum, M.W. and Radovic, M., Elastic and mechanical properties of the MAX phases. *Annu. Rev. Mater. Res.*, 41, 195–227, 2011.
117. Borysiuk, V.N., Mochalin, V.N., Gogotsi, Y., Molecular dynamic study of the mechanical properties of two-dimensional titanium carbides $Ti_{n+1}C_n$ (MXenes). *Nanotechnology*, 26, 26, 265705, 2015.
118. Hettinger, J.D. *et al.*, Electrical transport, thermal transport, and elastic properties of M2AlC (M=Ti, Cr, Nb, and V). *Phys. Rev. B*, 72, 11, 115120, 2005.
119. Guo, Z. *et al.*, Flexible two-dimensional Tin+1Cn (n = 1, 2 and 3) and their functionalized MXenes predicted by density functional theories. *Phys. Chem. Chem. Phys.*, 17, 23, 15348–15354, 2015.
120. Zha, X.-H. *et al.*, Role of the surface effect on the structural, electronic and mechanical properties of the carbide MXenes. *EPL (Europhys. Lett.)*, 111, 2, 26007, 2015.
121. Yorulmaz, U. *et al.*, Vibrational and mechanical properties of single layer MXene structures: A first-principles investigation. *Nanotechnology*, 27, 33, 335702, 2016.
122. Wang, D. *et al.*, Chapter 4 - Optical and mechanical properties of MXenes, in: *Mxenes and their Composites*, K.K. Sadasivuni, *et al.* (Eds.), pp. 131–169, Elsevier, Amsterdam, Netherlands, 2022.

123. Tang, H. *et al.*, Post-processing strategies for improving the electrical and mechanical properties of MXenes. *Chem. Eng. J.*, 425, 131472, 2021.
124. Li, C. *et al.*, Two-dimensional Janus Ti3C2FxO2-x MXene with tunable electronic and mechanical properties. *Scr. Mater.*, 194, 113688, 2021.
125. Zhang, Y. *et al.*, Theoretical study on the electrical and mechanical properties of MXene multilayer structures through strain regulation. *Chem. Phys. Lett.*, 760, 137997, 2020.
126. Imani Yengejeh, S. *et al.*, Oxygen-terminated M4X3 MXenes with superior mechanical strength. *Mech. Mater.*, 160, 103957, 2021.
127. Hatam-Lee, S.M., Esfandiar, A., Rajabpour, A., Mechanical behaviors of titanium nitride and carbide MXenes: A molecular dynamics study. *Appl. Surf. Sci.*, 566, 150633, 2021.
128. Seredych, M. *et al.*, High-temperature behavior and surface chemistry of carbide MXenes studied by thermal analysis. *Chem. Mater.*, 31, 9, 3324–3332, 2019.
129. Thakur, R. *et al.*, Insights into the thermal and chemical stability of multilayered V2CTx MXene. *Nanoscale*, 11, 22, 10716–10726, 2019.
130. Chen, L. *et al.*, Measurement and analysis of thermal conductivity of Ti3C2Tx MXene films. *Materials*, 11, 9, 1701, 2018.
131. Gandi, A.N., Alshareef, H.N., Schwingenschlögl, U., Thermal response in van der Waals heterostructures. *J. Phys.: Condens. Matter*, 29, 3, 035504, 2016.
132. Kumar, S. and Schwingenschlögl, U., Thermoelectric performance of functionalized Sc2C MXenes. *Phys. Rev. B*, 94, 3, 035405, 2016.
133. Bouhemadou, A., Khenata, R., Binomran, S., Structural parameters, electronic structures, elastic stiffness and thermal properties of M2PC (M=V, Nb, Ta). *Physica B: Condens. Matter*, 406, 14, 2851–2857, 2011.
134. Li, J. *et al.*, Thermal stability of two-dimensional Ti2C nanosheets. *Ceram. Int.*, 41, 2, Part A, 2631–2635, 2015.
135. Shen, Y., Clarke, D.R., Fuierer, P.A., Anisotropic thermal conductivity of the aurivillus phase, bismuth titanate (Bi4Ti3O12): A natural nanostructured superlattice. *Appl. Phys. Lett.*, 93, 10, 102907, 2008.
136. Clarke, D.R., Materials selection guidelines for low thermal conductivity thermal barrier coatings. *Surf. Coat. Technol.*, 163-164, 67–74, 2003.
137. Clarke, D. and Levi, C., Materials design for the next generation thermal barrier coatings. *Annu. Rev. Mater. Res.*, 33, 1, 383–417, 2003.
138. Cahill, D.G., Watson, S.K., Pohl, R.O., Lower limit to the thermal conductivity of disordered crystals. *Phys. Rev. B*, 46, 10, 6131–6140, 1992.
139. Bao, L. *et al.*, Anisotropies in elastic properties and thermal conductivities of trigonal TM2C (TM = V, Nb, Ta) carbides. *Solid State Sci.*, 98, 106027, 2019.
140. Guo, Y. *et al.*, Synthesis of two-dimensional carbide Mo2CTx MXene by hydrothermal etching with fluorides and its thermal stability. *Ceram. Int.*, 46, 11, Part B, 19550–19556, 2020.
141. Feng, A. *et al.*, Fabrication and thermal stability of NH4HF2-etched Ti3C2 MXene. *Ceram. Int.*, 43, 8, 6322–6328, 2017.

6

MXene-Reinforced Polymer Composites: Fabrication Methods, Processing, Properties and Applications

Zhenting Yin, Pengfei Jia and Bibo Wang*

State Key Laboratory of Fire Science, University of Science and Technology of China, Hefei, Anhui, P. R. China

Abstract

As a rising 2D material comparable to graphene, MXene has mushroomed in the past 10 years. The intrinsic composition composed of more than 30 types of transition metal elements and abundant surface groups make MXene have the characteristics of changeable properties and become the key to its wide use. The magic properties of MXenes encourage their application in sensing, energy applications, electromagnetic interference, and catalytical conversion fields. Researchers use different methods, such as physical mixing, surface modification and *in situ* polymerization, to combine MXene with polymers, investigating the influence of MXene on the mechanical, thermal and conductive properties of polymer (nano) composites. This chapter is divided into five perspectives: *Introduction, Fabrication Methods, Processing, Properties, Applications and conclusion*, which demonstrate the improvement for MXene applied in reinforced polymer (nano) composites through a summary of previous reports. Finally, based on a review of the MXene literature, the prospects of large-scale industrial production of MXene-reinforced polymers are discussed.

Keywords: 2D material, MXene, polymer (nano) composites, reinforced polymer composites, fabrication methods, processing, properties, applications

6.1 Introduction

Research on polymers has been ongoing for a long time. Before the discovery of MXenes, many 2D materials have been introduced into polymers to enhance their properties [1, 2], for example, graphene [3], carbon nanotubes [4], nanospheres [5], metals [6], metal sulfides [7] and polymers [8–10]. Long polymer chains contain many carbon molecules that can bind to a variety of molecules to regulate the chemical and mechanical performances of polymers. These polymer composites demonstrate attractive performance with high stiffness, strength, fatigue resistance, wear resistance, fracture resistance, impact resistance, and corrosion resistance. Owing to the interfacial interaction between the continuous and

**Corresponding author*: wbibo@ustc.edu.cn

dispersed phases in polymer composites, polymer composites have achieved remarkable achievements in engineering applications [11].

To date, many kinds of MXene-reinforced polymer composites have been successfully fabricated by combining MXenes and polymers. Many papers about "MXene Reinforced Polymer Composites" have been retrieved by the Web of Science in the past 6 years. The retrieval results show that the number of related publications has increased significantly since 2018 (Figure 6.1). These research results prove that MXene-reinforced polymer composites have attracted increasing interest from researchers and have very broad application prospects. Owing to the magic properties of MXene, the MXene reinforced polymer composites are gifted with outstanding performances, such as electrical conductivity, thermal and mechanical performances, which could be expressly regulated by changing the type of polymer, MXene and their composition. The review summarizes the manufacture, properties and application of the MXene reinforced polymer composites (Figure 6.2). The fabrication method of MXene reinforced polymer composites was reviewed in the second part, such as physical mixing (PM) [12], surface modification (SM) [13] and *in situ* polymerization (IP) [14]. The interesting properties of MXene-reinforced polymer composites are reviewed in the third part, including electrical properties, thermal properties, mechanical properties, photothermal properties, and flame retardant properties. The application of MXene-reinforced polymer composites is thoroughly summarized in the fourth section, including sensing, energy applications, electromagnetic interference, catalytical conversion and oil/water separation. Finally, based on a review of the MXene literature, the conclusion and prospects of the large-scale industrial production of MXene-reinforced polymers are discussed [15].

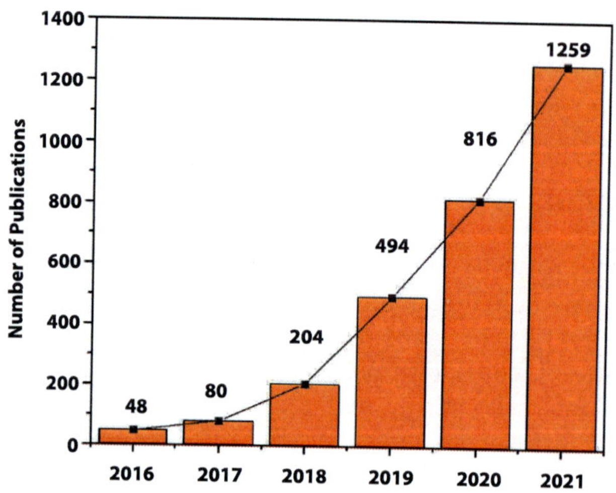

Figure 6.1 Number change of publications by "polymer (topic) and MXene (topic)" or "composite (topic) and MXene (topic)" on Web of Science.

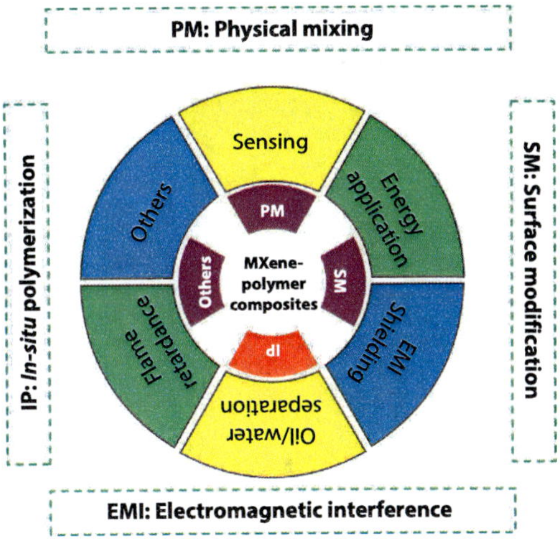

Figure 6.2 The fabrication methods and usages of MXene-reinforced polymer composites.

6.2 Fabrication Methods and Processing

6.2.1 Direct Physical Mixing

The most widely applied strategy for preparing MXene-reinforced polymer composites is direct physical mixing of MXene with polymers. One direct physical mixing method is melt blending. By using the melt blending method, MXenes can combine with some linear thermoplastic polymer materials for preparing MXene-reinforced polymer composites, such as thermoplastic polyurethane (TPU), polylactic acid (PLA) and polypropylene (PP) [16–18]. The linear thermoplastic polymer is heated to a molten state at a specific temperature and then uniformly mixed with the inorganic MXene in a machine by constant stirring. The MXene and TPU particles were blended in an internal mixer according to the desired stoichiometry and stirred at 160 °C and 40 rpm for 5 min to obtain the mixture. After that, the mixture is compressed to obtain these samples [19]. In the same way, Shen [20] combined two-dimensional MXene nanosheets with short aramid fibers (SAF) to form $Ti_3C_2T_x$ nanosheet-modified SAF ($Ti_3C_2T_x$/SAF). $Ti_3C_2T_x$/SAF-reinforced PP ($Ti_3C_2T_x$/SAF/PP) composites were prepared by melt blending (Figure 6.3a).

Solution casting [21–23] is a common method that blends the polymer matrix and the filler in the solvent to obtain the composite material by mold casting and solvent evaporation. To thoroughly mix the polymer and the filler, the mixture is supposed to be completely mixed in the solution casting process (Figure 6.3b). Polyvinyl alcohol (PVA) and MXene were well mixed by sonication and stirring, while the mixture was poured into unprepared molds. After that, the mixture was placed to dry the solvent, yielding the well-dispersed composite [24, 25]. In the same approach, Pan [26] obtained three MXene-reinforced polymer composites, which were prepared by dip coating and solution blending. They explored the potential use of MXene to enhance the anti-dripping properties of polymers. These advantages of the solution casting method are simplicity and ease of control; thus, it can also be used to prepare polystyrene (PS), poly(3,4-ethylenedioxythiophene) (PEDOT) [27, 28], and polyvinylchloride (PVC) composites [29, 30].

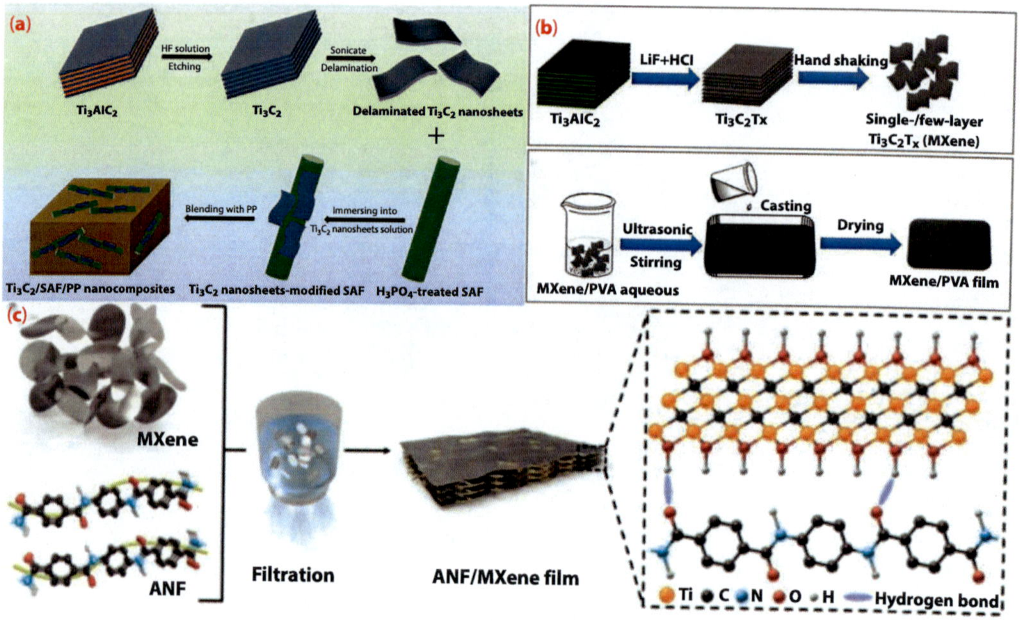

Figure 6.3 By means of physical mixing from MXene and polymer composites acquires MXene reinforced polymer composites. (a) Preparation of $Ti_3C_2T_x$/SAF/PP composites by the melt-mixing method. Source: Tang *et al.* [20]. Reproduced with permission from Wiley-VCH. (b) Preparation of PVA/MXene composite films by the casting method. Source: Li *et al.* [25]. Reproduced with permission from Wiley-VCH. (c) The fabrication of the ANF/MXene film through filtration. Source: Wen *et al.* [35]. Reproduced with permission from Elsevier.

The solution suction filtration method could be used to prepare MXene-reinforced polymer membrane materials [31–34]. To solution, a certain amount of polymer is first dissolved in a polar solvent. Then, the MXene and polymer are blended in solution under vigorous stirring as well as sonication. Finally, the solution was filtered to prepare a composite membrane material. Because of its excellent hydrophilic nature, MXene was combined with aramid nanofibers (ANFs) to develop a film composite with a layered structure through a simple vacuum-assisted filtration method (Figure 6.3c) [35].

6.2.2 Surface Modification

Although the physical mixing of MXene with polymers could simply prepare MXene-reinforced polymer composites, many issues remain to be addressed (Figure 6.4). For example, the presence of hydrophilic groups from MXene [36–38] makes it incompatible with hydrophobic polymers, such as polydimethylsilane (PDMS) [39–42], epoxy resin (EP) and polyvinylidene fluoride (PVDF) [43–46]. The hydrophilicity/hydrophobicity of the MXene surface could be changed by using surface modification methods so that the MXene can be scattered among the hydrophilic or hydrophobic polymer network. In addition, the treatment of the MXene surface can also provide chemical cross-linking sites on the surface of MXene to enhance the stability and mechanical property from the MXene reinforced polymer composites [47–49].

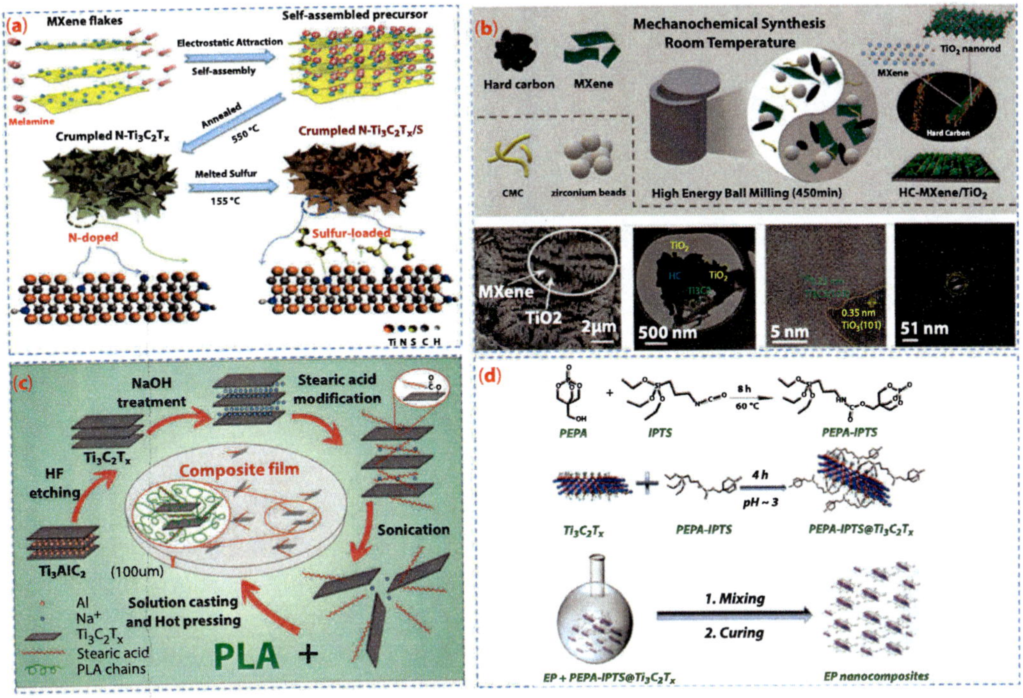

Figure 6.4 Surface modification of MXene produces MXene-reinforced polymer composites. (a) The preparation process of N-MXene/S composites. Source: Bao *et al.* [55]. Reproduced with permission from Wiley-VCH. (b) The mechanochemical fabrication of HC-MXene/TiO2 electrodes with $Ti_3C_2T_x$ as a multifunctional conductive agent. Source: Gao *et al.* [58]. Reproduced with permission of the Royal Society of Chemistry. (c) The preparation process of PLA composites. Source: Yi *et al.* [63]. Reproduced with permission from Wiley-VCH. (d) Synthesis of PEPA-IPTS and PEPA-IPTS@$Ti_3C_2T_x$ and preparation of EP. Source: Yin *et al.* [68]. Reproduced with permission from Elsevier.

This is a valid way to introduce functional groups by using nitrogen doping (e.g., $-NH_2$ [50–52], $-NO_2$ [53] and $-SO_3$ [54]) on the MXene surface. By regulating the oxidation state of the Ti elements in the MXene, it is beneficial to manipulate the conductivity and capacitance of MXene-reinforced polymer composites. Nitrogen-doped MXene nanosheets with strong physicochemical coadsorption were synthesized by a novel one-step method and used as a new sulfur host for lithium-sulfur batteries [55, 56]. The nitrogen doping strategy can introduce heteroatoms into MXene nanosheets and induce a well-defined porous structure with a high surface area and large pore volume at the same time (Figure 6.4a). A covalent bond can also exist between the amine and manganese, as determined by the aryldiazide reaction [57]. For example, the phenyl sulfonate group can exist on the surface of MXene by diazosalt [58] in the presence of sodium nitrite and hydrochloric acid in reaction with Na^+ insertion MXene, producing diazine salt and releasing N_2 [58]. MXene and phenylsulfonic acid are linked by the Ti-O-C bond (Figure 6.4b).

There is a mass of hydroxyl on the surface of MXene. It could bring into some other functional groups through chemical reactions, such as carboxyl, amino and long-chain alkanes [59–62]. Through this strategy, the chemical properties on the surface of MXene can adjust the interaction between MXene and polymer. For example, for improving the hydrophobicity of the MXene surface, stearic acid was used to modify the MXene [63]. After modification, the hydrophobicity of the MXene surface was enhanced. Finally, stearic acid-modified MXene was added to the polylactic acid (PLA) matrix with the aim of enhancing compatibility and physical properties (Figure 6.4c).

Functional alkoxysilanes are commonly used to connect inorganic and organic materials by forming covalent bonds. Meanwhile, the silane precursor has different kinds of functional groups, including ammonia, methacrylate, chloride, phenyl, vinyl and epoxy, which enhance the interaction of MXene with hydrophilic or hydrophobic polymers [64–66]. Modification of the MXene surface created a Ti-O-Si bond between the silane conjugates and the MXene [67]. In the presence of water, silane conjugates contain three alkyl groups that can be catalyzed by base or acid into silanol groups. The hydroxyl group on the surface of the MXene condenses the resulting silanol group of the silane-conjugate for forming Ti-O-Si bond. For example, the $Ti_3C_2T_x$ grafts PEPA (Pentaerythritol tetrakis (dihydrogen phosphate)-IPTS (3-Isocyanatopropyltrimethoxysilane) for forming a nanohybrid (PEPA-IPTS@$Ti_3C_2T_x$) [68] (Figure 6.4d).

MXene (-OH, =O, -F) is rich in surface groups and can fabricate hydrogen bonds via the hydrophilic functional groups of the polymer, such as amine, hydroxyl, carboxyl, etc.) [69–73]. Hydrogen bonding could make MXene disperse stably and uniformly in a dimer complex, which prepares MXene-reinforced polymer composites with a 3D structure. Yellow collagen gum is a hydrophilic polysaccharide rich in aerobic functional groups that can be used to fabricate film composites by hydrogen bonds on the surface of MXene [74]. Because of the strong hydrogen bond between yellow collagen glue and MXene, which make MXene combine with yellow collagen (Figure 6.5a).

The -O and -OH on MXene surface negatively charge MXene surface and are tightly bound with positively charged polymers [75–77]. A self-supporting MXene hydrogel is prepared by electrostatic interactions between MXene nanosheets. The prepared hydrogel structure is stable and can be used to prepare a highly oriented fiber. This can be attributed to the three-dimensional network of MXene hydrogels enduring structural deformation, with very excellent mechanical properties, so there is a very wide range of application prospects in the biomedical field (Figure 6.5b).

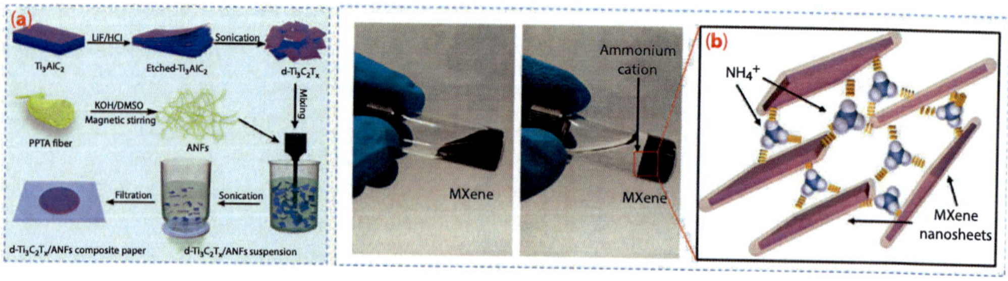

Figure 6.5 Surface modification of MXene produces MXene-reinforced polymer composites. (a) The preparation process of d-Ti3C2Tx/ANF paper. Source: Xie *et al.* [74]. Reproduced with permission of the Royal Society of Chemistry. (b) Digital image of MXene and gelled MXene, as well as the process of the electrostatic interaction among MXene nanosheets. Source: Shin *et al.* [75]. Reproduced with permission of the American Chemical Society.

6.2.3 *In Situ* Polymerization

In situ polymerization can also be carried out on the surface of MXene without any initiator (*in situ* self-polymerization). The charge transfer [78–80] between the polymer and MXene could illustrate the process. Two major electron transfer processes of 3,4-ethyl doxythiophene (EDOT) charge transfer occur during polymerization, where electrons can transfer from energy tracks (HOMOs) of MXene to low energy tracks (LUMOs), or these electrons can transfer from HOMOs to LUMOs of MXene [81]. Furthermore, to accomplish the electron shift, monomers must approach the MXene surface in the process of polymerization. Hydroxyl and fluorinated ends make 2-(ethyl) methacrylate (DMAEMA) aggregate on the surface of V_2C [82]. The resulting grafted V_2C is sensitive to carbon and temperature. Similar to EDOT, aniline and pyrrole can also polymerize on the surface of MXene. After that, these conductive polymers could expand the MXene layer spacing, connect the MXene sheet and form more conductive pathways. The electrical conductivity and capacitance of MXene reinforced polymer composites are remarkably enhanced by the way of further promoting electron-ion transport (Figure 6.6a).

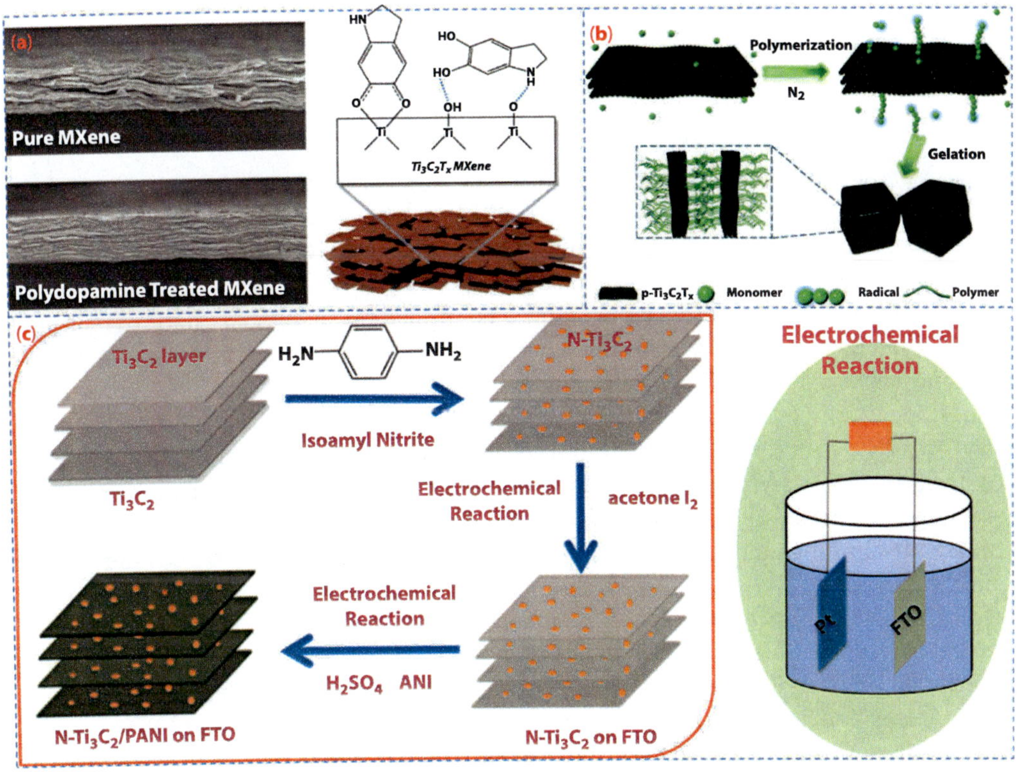

Figure 6.6 By *in situ* polymerization, MXene-reinforced polymer composites can be obtained. (a) The preparation process and SEM cross-sectional images from different types of MXene films (scale bar: 20 μm). Source: Lee *et al.* [83]. Reproduced with permission of the American Chemical Society. (b) Schematic illustration of p-$Ti_3C_2T_x$-initiated polymerization and subsequent gelation. Source: Tao *et al.* [78]. Reproduced with permission of the Royal Society of Chemistry. (c) The fabrication process from N-$Ti_3C_2T_x$/PANI. Source: Wu *et al.* [89]. Reproduced with permission from Elsevier.

In situ polymerization refers to the mixing of the polymer precursor, curing agent (or initiator) and filler in solution and then polymerizing the polymer under certain conditions to form a macromolecule. The quantified MXene powder was mixed with the unsaturated polyester resin (UPR) prepolymer solution, the accelerator and curing agent were added, and the uniform dispersion solution was poured into the preprepared mold to obtain the sample [84]. This method has been diffusely applied in the preparation of MXene reinforced polymer composites, like PU [85], polyacrylamide (PAM) [86], EP and poly methyl methacrylate (PMMA) [87]. When MXene is mixed with EP, the EP molecule chain is inserted into MXene. The above method will increase the layer spacing and make the molecular structure more stable after curing [78] (Figure 6.6b).

In situ electrochemical polymerization has been shown to be another convenient method for preparing MXene-reinforced polymer membrane composites, mainly conductive polymers [88]. A two-step method for preparing $Ti_3C_2T_x$/PPy hybrid membranes was developed. First, uniform $Ti_3C_2T_x$ particle membranes were formed by electrophoresis deposition, followed by *in situ* polymerization of the inserted pyrrole [89]. First, an independent $Ti_3C_2T_x$/PPy membrane was removed from the electrode. After that, ultrathin all-solid-state supercapacitors were prepared, showing excellent cycle stability and capacitance (Figure 6.6c). The electrochemical polymerization method shows a new strategy for designing and fabricating detached MXene/conductive polymer membranes with outstanding electrochemical performances, with great prospects in the field of energy storage.

6.2.4 Others

Besides the widely used direct physical mixing, surface modification and *in situ* polymerization methods, some other methods [90, 91] have been applied on preparing MXene reinforced polymer composites with outstanding performances, like layer-by-layer (LbL) assembly and electrospinning (Figure 6.7). LbL assembly strategies are often used to fabricate functionalized films, whose acting force is controlled through electrostatic or hydrogen bonding interactions (Figure 6.7a). In general, by the LbL method, the electrostatic phase by oppositely charged polyelectrolyte interactions alternately deposits on substrates to form multilayer heterostructure ultrathin films. Lu realized [92] a facile and environmentally friendly method for the preparation of flame-retardant polyurethane (PU) sponges by using the LbL assembly method containing octaamino-polyhedral oligomeric silsesquioxane (OAPOSS) (positively charged) and MXene nanosheets (negatively charged). The electrospinning method is a good way to fabricate nanofibers and fibrous membranes (Figure 6.7b). Jiang [93] used polyacrylonitrile (PAN) electrostatic spinning MXene flakes and carbonized fiber networks, and an independent MXene/carbon nanofiber electrode was prepared. The LbL and electrospinning strategies provide some typical approaches to design and prepare independent MXene-reinforced polymer composites with special performance, extending their applications in various fields.

To date, owing to the special properties of MXene-reinforced polymer composites, researchers are still working hard to develop new strategies for making this polymer material. In addition to physical mixing, surface modification, *in situ* polymerization, LbL and electrospinning, other multimolecular interactions have been introduced into the preparation of MXene-reinforced polymer composites, such as supramolecular interactions, complexation interactions, biospecific interactions, and coordination interactions [94].

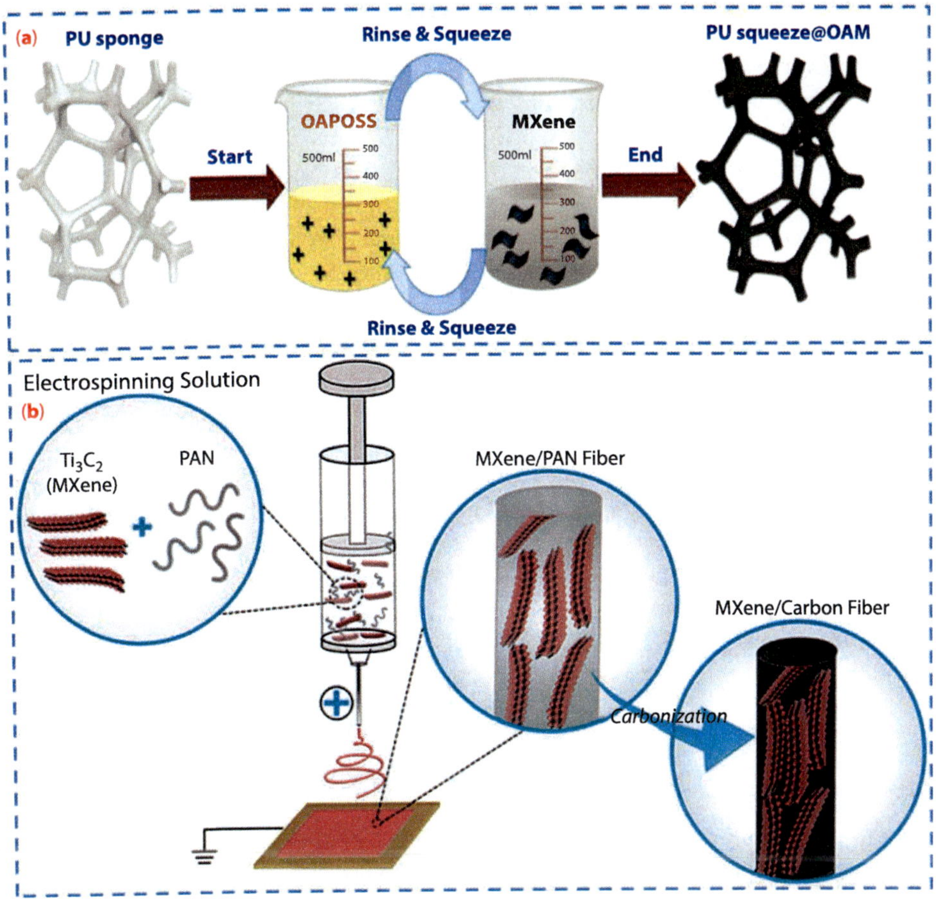

Figure 6.7 MXene reinforced polymer composites could be acquired by the technology of LbL assembly and electrospinning. (a) The fabrication process of the FPUF composite. Source: Lu *et al.* [92]. Reproduced with permission from Elsevier. (b) The electrospinning technology acquires $Ti_3C_2T_x$ MXene/carbon nanofibers. Source: Levitt *et al.* [93]. Reproduced with permission of the Royal Society of Chemistry.

6.3 Properties

6.3.1 Electrical Properties

Currently, polymer composites generally encounter the problem of poor electrical conductivity; fortunately, they can be modified through nanofillers, coatings, etc., to meet their application in the fields of electronics and electrical appliances. Polymer intercalates MXene, which not only expands the layer spacing of MXene but also MXene can achieve molecular-level coupling with the polymers, which contributes to the conductivity, dielectric, and piezoelectricity promotion of polymers [95].

The first step in exploring the potential utilization of MXene in the polymer was made by Yury's group in 2014. $Ti_3C_2T_x$ was mixed with electrically charged neutral polyvinyl alcohol (PVA). The as-designed composites showed a high conductivity up to 2.2×10^4 $S\ m^{-1}$ and were flexible [8]. The surface terminations of MXene play a large role in the electrical conductivity. To achieve a high conductivity of MXene, Yury's group reported that high-quality MXene displayed electrical conductivities (11,000 S cm^{-1}) and field-effect electron morbidities (6 cm^2 V^{-1} S^{-1}) [95]. Yury (Figure 6.8) also modified the synthesis of Ti_3AlC_2 to make a much more stoichiometric MAX, where the Ti:C ratio was close to 3:2. In comparison, the conductivity of the MXene stored for six months is dramatically high (6000 S cm^{-1}), indicating that the MXene prepared by this method has excellent oxidation resistance in an air atmosphere [96]. For $M_{n+1}X_n$-type MXene, the surface d-orbital electrons have a great influence on the Fermi energy level. In addition, the transition metal elements (M' in $M'_2M''CT_x$ and $M'_2M''_2C_3T_x$) in the outer layer exert a greater influence on the electronic performance of MXene than the transition metal elements in the inner layer. It can be inferred that the electronic structure of the surface metal element determines the conductivity of MXene. Meanwhile, the electronic state of the metal element on the surface can be changed by chemical modification. Therefore, the surface group of MXene has a significant influence on the conductivity of MXene [97].

A facile hydrothermal strategy was utilized to acquire highly conductive MXene films, and the surface termination was modified by different temperatures, concentrations and reaction solvents. Cation intercalation was the key for MXene to maintain a high electrical conductivity of up to 1652 S cm^{-1}. Meanwhile, the heat-related Seebeck coefficient of MXene was enhanced threefold by up to 44.98 μW m^{-1} k^{-1}, implying its potential application in thermoelectric and semiconductor fields [98]. PVDF/Co/MXene films were fabricated through the CO_2-assisted foaming method (Figures 6.8i, j). With the loading of 8 wt% MXene, the

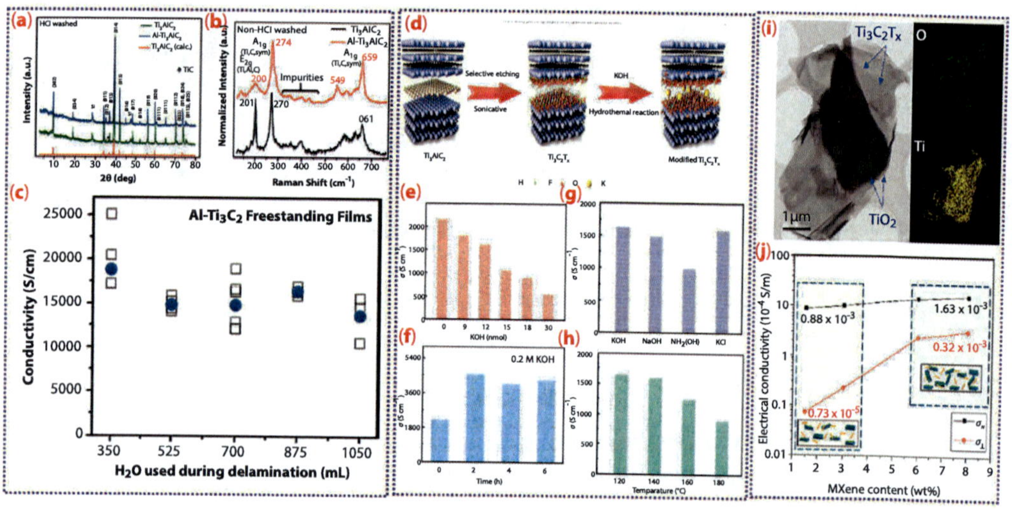

Figure 6.8 (a) XRD of Al-Ti_3AlC_2 and Ti_3AlC_2. (b) Polarized Raman spectra of Al-Ti_3AlC_2 and Ti_3AlC_2. (c) $Ti_3C_2T_x$ surface termination. Source: Mathis *et al.* [96]. Reproduced with permission of the American Chemical Society. (d) Illustration of the preparation of modified $Ti_3C_2T_x$. (e-h) Electrical conductivity of K-$Ti_3C_2T_x$ by different concentrations of KOH. Source: Peng *et al.* [98]. Reproduced with permission of Elsevier. (f) Electrical conductivity of K-$Ti_3C_2T_x$ in KOH at different times. (g) Electrical conductivity of K-$Ti_3C_2T_x$ by different alkalis. (h) Electrical conductivity of K-$Ti_3C_2T_x$ at different temperatures. (i) TEM image of annealed MXene. (j) Values of $S_\parallel$ and $S_\perp$ of MCS. Source: Li *et al.* [99]. Reproduced with permission of Elsevier.

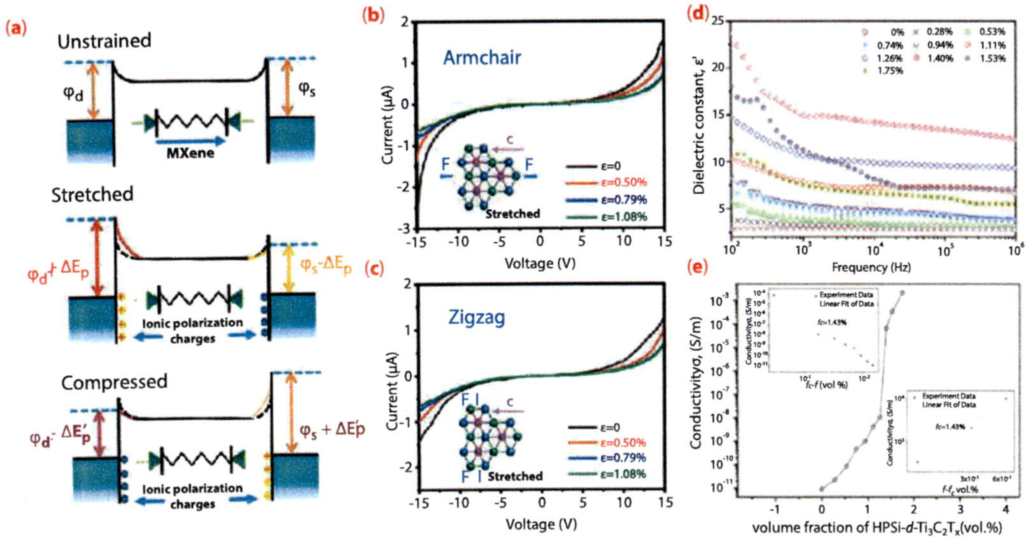

Figure 6.9 (a) Piezoelectric effect demonstration for monolayer $Ti_3C_2T_x$. (b) Characterization of the piezoelectric effect in $Ti_3C_2T_x$ with armchair directional strain as asymmetric modulation of carrier transport. (c) Characterization of the piezoelectric effect under sawtooth directional strain with asymmetric modulation of carrier transport. Source: Tan *et al.* [100]. Reproduced with permission from Elsevier. (d) Dielectric constant of d-$Ti_3C_2T_x$/PDMS. (e) Dielectric constant of two types of PDMS-based composites. Source: Wei *et al.* [101]. Reproduced with permission from the American Chemical Society.

conductivity of PVDF composites exhibited an increase of 4.4 times to 0.32×10^{-3} S m^{-1}, achieving a high EMW absorption with a minimum reflection loss of -45.6 dB [99].

In the armchair direction, the intrinsic current output of $Ti_3C_2T_x$ MXene was 0.3 nA under 1.08% tensile strain, and its power density and conversion efficiency corresponded to 6.5 mW/m^2 and 11.15%, respectively [100]. Achieving superior dielectric and loss was vital for electromechanical transducers. Hyperbranched polysiloxane (HPSi)-modified $Ti_3C_2T_x$ was incorporated into PDMS. After loading to 1.43 vol%, the dielectric constant was 23.7 at 100 Hz, which was 8.5 times higher than that of pure PDMS (Figure 6.9) [101].

6.3.2 Thermal Properties

The thermal properties of polymer composites are of great significance to the practical processing and applications of common thermal parameters, e.g., thermal conductivity, thermal stability, and thermal expansion coefficient (TEC). MXene-reinforced polymer composites are formed by one or more structural units through covalent bonds, and most polymers do not contain free electrons inside. Herein, the transmission of heat through the movement of phonons is dominant, but phonons need to pass through different crystal regions or a large number of anharmonic coupling motions. The results are that the scattering of phonons is serious, causing the low mean free path [102]. As a two-dimensional material, according to DFT calculations, the thermal conductivity of MXenes surpasses that of most metal materials and is comparable to that of graphene, which shows application prospects for polymers with improved thermal properties [103].

Wu *et al.* [104] investigated the thermal conductivity of CNF/MXenes paper. The thermal conductivity of CNF was almost zero, but that value of CNF/MXene paper was much higher than that of neat CNF. 14.93 W m^{-1} K^{-1} of in-plane thermal conductivity for laminated paper was achieved when the loading amount of MXene was 60 wt%. In addition, morphologies and the heat conductive network between fillers can affect the thermal conductivity of polymers. Experimental results showed that compared with CNF/MXene (9.18 W mk^{-1}) as zero-dimensional filling, the in-plane value of CM-60 was increased by 62.3%. Because zero-dimensional filled MXene showed an obvious sea-island structure, the filler was randomly dispersed in the polymer, inhibiting the continuous heat conduction path. However, two-dimensional filled MXene easily formed a 3D conductive network and increased the bonding length forming hydrogen bonds, which was crucial for the enhancement of thermal conductivity.

Yu *et al.* investigated the thermal decomposition properties of PS composites. Pure PS experienced a stage of thermal weight loss at 20 - 700 °C, and there was little residual carbon remaining (0.82 wt%). The addition of POSS-MXene made the T_i of the PS composite, and T_{max} gradually increased. The addition of 2 wt% POSS-MXene increased T_i and T_{max} to 382 °C and 421 °C (PS) to 404 °C and 432 °C, respectively. The content of residual char also increased to 2.33 wt%, which can be explained by 2D POSS-MXene acting as a barrier to effectively inhibit the diffusion of flammable volatile gases to the cracking zone [105]. $Ti_3C_2T_x$ was incorporated into EP, and the TCE value decreased from 46.3 for pure EP to 40.7 ppm/K for EP-1.0M at 40 °C. From the results (Figure 6.10), it could be concluded that MXene can restrict the movement of EP to enhance the thermal resistance of EP composites [106].

6.3.3 Mechanical Properties

Mechanical properties are a basic property of polymer-based composites, which is the key to their wide application. Adding different nanofillers is a common strategy to improve the mechanical performance of polymers. The unique properties and structure of nanoparticles make it possible to be applied to polymers. S. Mishra added 6 wt% nano-$Mg(OH)_2$ to PP, and compared to pure PP, the tensile strength and Young's modulus showed 91% and 266% increases, respectively. However, when $Mg(OH)_2$ increased to 12 wt%, in contrast to the tensile strength of PP-6, the tensile strength of PP-12 showed a 21.4% reduction. The experiments showed that excessive addition of $Mg(OH)_2$ significantly reduced the mechanical performance of PP composites [107]. Therefore, finding low-addition fillers has become a research hotspot. The theoretical calculations provide valuable insights and guidance for MXene-reinforced polymers, and theoretical simulations comprising molecular dynamics (MD), DFT and finite element (FE) have been applied. The studies showed that the predicted elastic modulus of MXene was above 500 GPa, which was much higher than that of the polymers. Atomistic simulations acquired simulated results showing Young's moduli for $Ti_3C_2O_2$ (466 GPa) and Ti_2CO_2 (983 GPa), which were higher than the Young's moduli of pure GO. In addition, this work shows that a clear fracture mechanism displays defects that fail to propagate along the edges [108]. Meanwhile, MD nanoindentation simulation results exhibited that a Young's modulus of 386 GPa was predicted for a monolayer $Ti_3C_2T_x$ mixture with titanium (1%) and carbon (10%) atom defects. The formation of wrinkles, flakes, layered structures and atomic imperfections prompts decreased mechanical properties (Figures 6.11a–f) [109].

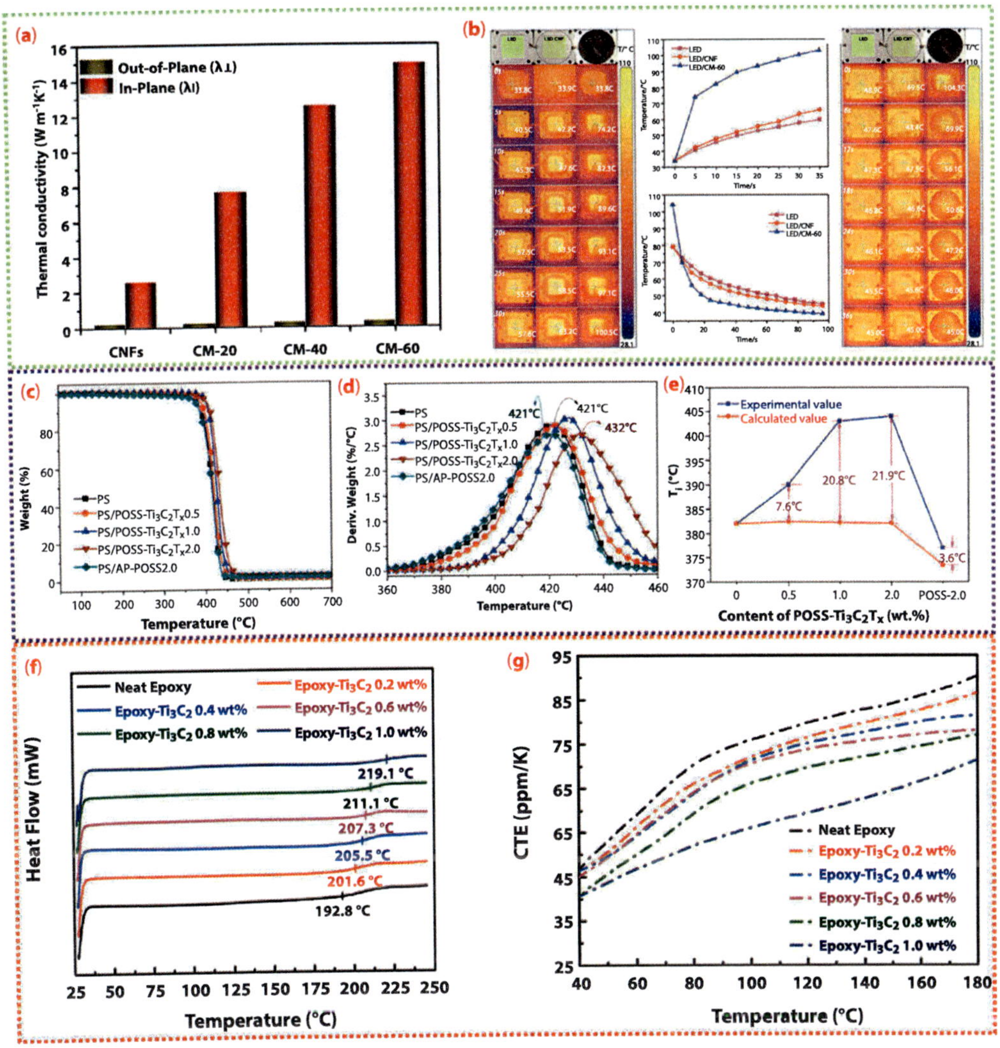

Figure 6.10 (a) Two types of thermal conductivities (in-plane and out-of-plane) for papers. (b) Infrared images of different papers. Source: Jiao *et al.* [104]. Reproduced with permission of Elsevier. (c) TGA and (d) DTG curves of PS composites. (e) Study of experimental values and calculated values of char residues at 800 °C. Source: Yu *et al.* [105]. Reproduced with permission from Elsevier. (f) DSC curves, (g) CTE curves of pure EP and Ti_3C_2/EP composites. Source: Reproduced with permission of Zhang *et al.* 2019 [106]. Copyright 2014, Springer Nature.

An emulsion method was utilized to prepare polyurethane nanocomposites with the incorporation of MXene. After loading MXene (0.5 wt%), improvements in yield stress (70%), tensile strength (20%) and hardness (13%) were achieved. This was ascribed to ball milling ensuring the stable existence of PU emulsions, which was the key factor in prompting mechanical enhancement of PU composites [110]. It was acknowledged that homogeneous dispersion was a key challenge to fabricating high-performance polymers. To achieve uniform MXene dispersion in TPU, MXene was first surface-treated with polyethylene glycol (PEG) and then blended into TPU. After loading MXene (0.5 wt%), improvements

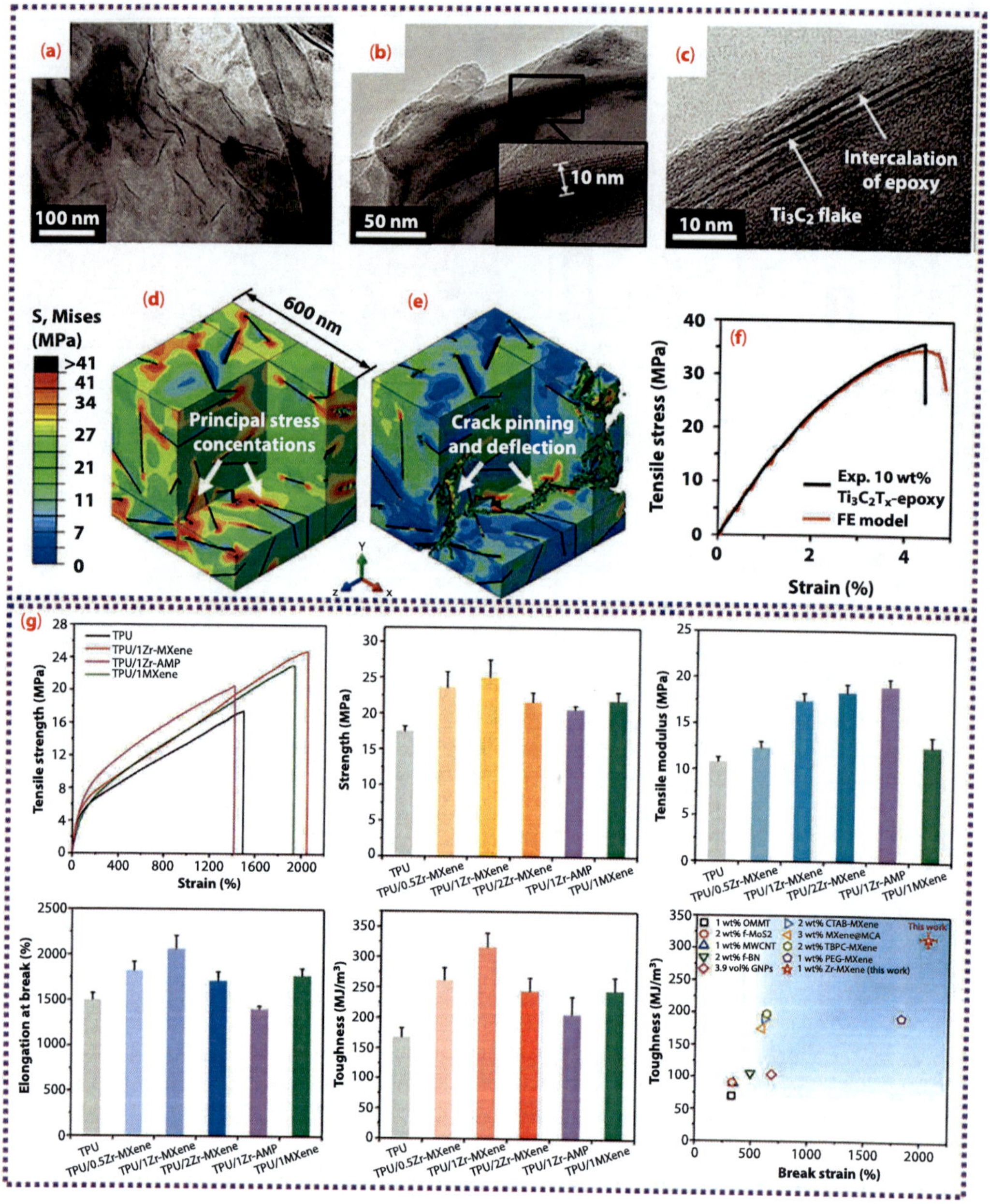

Figure 6.11 TEM characterization results (a) for EP/MXenes with randomly filled MXene, (b) Multilayered MXene nanosheets, (c) EP, (d) Stress distribution simulation for EP/MXene (3.6 vol% filler) and crack formation demonstration. (e) Crack propagation within EP/MXene (3.6 vol% filler) and stress drops after a total failure. (f) Stress-strain results of the 3D RVE model with EP/MXene (3.6 vol% filler) and comparison with the tensile strength experiment results. Source: Monastyreckis *et al.* [109]. Reproduced with permission of Elsevier. (g) Different types of mechanical performance of TPU nanocomposites and comparison results of TPU nanocomposites with previous reports. Source: Liu *et al.* [114]. Reproduced with permission of Elsevier.

in the tensile strength (47.1%) and storage modulus (39.8%) of TPU composites were achieved. PEG successfully interacted with MXene to avoid agglomeration. Meanwhile, the load transfer and strong interaction effect between MXene and the TPU matrix further enhanced the mechanical performance of TPU composites [111].

Jeon Hyeon Woo reported a novel strategy to prepare PVA films using sulfanilic acid-modified MXene. Sulfanilic acid was employed as a crosslinking agent, and MXene functioned as a reinforcing agent. With a loading of 0.5 wt% (MXene and boric acid (BA)), the tensile strength and modulus of PVA films showed 67% and 49% increases, respectively. The addition of BA was helpful for covalent bonding, leading to the improvement of interfacial force, and the formed covalent bond also acted as a bridge for force transfer and minimized the void between MXene and PVA [112]. Through electrostatic repulsion forces, a highly flexible and conductive MXene/NR film was fabricated between the negative MXene sheet and NR. With the addition of 6.71 vol% MXene, in contrast to pure NR, 7- and 150-fold improvements in the tensile strength and modulus of NR composites were achieved, respectively. This facile and effective strategy has potential for the large-production of flexible and foldable polymer composites [113]. (Zr-MXene) was added to TPU. Compared to neat TPU, the tensile strength of TPU/1Zr-MXene was increased by 43.4% [114]. The formed H-bonds accelerated the ring-opening polymerization and the interfacial interactions. Thus, an enhancement of 50% in impact strength was achieved after the loading of 2 wt% $Ti_3C_2T_x$ MXene (Figure 6.11g) [115]. The mechanism of mechanical reinforcement has been gradually studied.

6.3.4 Photo Thermal Properties

Light is a kind of sustainable clean energy with abundant resources and widespread distribution. Therefore, it can effectively use light energy and apply it to photothermal conversion, photothermal treatment and other fields. MXene has superior electrical conductivity, a relatively large specific surface area, low toxicity, excellent biocompatibility and high thermal conductivity, making it an excellent photothermal conversion material. It has been used to prepare actuators driven by solar evaporation, biomedicine and light. The hydrophilic MXene interface can effectively form a channel for small molecular water molecules, and the polymer segment acts as an intercalating agent to further increase the layer spacing of MXene. Wang *et al.* reported a general procedure to develop a light heating system. For various nanocomposites (Figure 6.12), the results showed that the light to water evaporation efficiency (84%) for the self-floating MXene film was achieved when treated with only one sun irradiation, which was the highest among various photothermal evaporation systems [116]. MXene enhances the heat storage capacity of polymers, which creates the possibility for its application in the field of photothermal therapy.

A series of studies have shown that MXene can absorb the energy of photons and specifically heat tumor cells without causing damage to surrounding healthy cells. However, according to current research, MXene is easily oxidized to titanium dioxide and loses its biological activity in water and aerobic environments. Fortunately, polymer-modified MXene can effectively improve the biocompatibility of MXene without reducing the photothermal efficiency. Chen *et al.* reported that molybdenum carbide (Mo_2C) engineered with

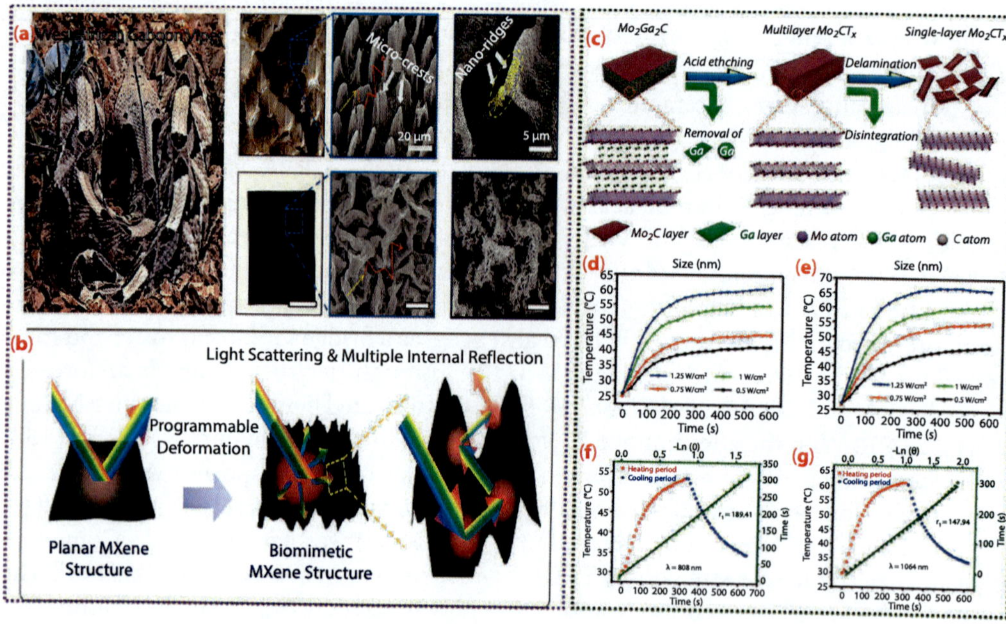

Figure 6.12 (a) Hierarchical MXene structures with light absorption and enhanced light-to-heat performance, (b) Demonstration of the MXene with light absorption and enhanced light-to-heat performance. Source: Li *et al.* [116]. Reproduced with permission from Wiley-VCH. (c) Demonstration of preparation and delamination of Mo2C MXenes. (c) Mo_2C-PVA dispersions treated with two types of lasers (808 and 1064 nm). (d–g) Light-induced heat efficiency under two types of lasers (808 and 1064 nm). Source: Feng *et al.* [117]. Reproduced with permission from Wiley-VCH.

PVA was applied as an effective photonic hyperthermia. Through the analysis (Figure 6.12), it was found that Mo_2C-PVA showed superior biocompatibility and quick degradability. Furthermore, the desirable photothermal conversion efficiencies for near infrared (NIR)-I and NIR-II were 24.5% and 43.3%, respectively [117].

6.3.5 Flame Retardant Properties

Two-dimensional lamellar nanoparticles have a large specific surface area and changeable composition characteristics and have been widely used in flame retardants. Metal elements contained in two-dimensional materials can exert excellent metal catalytical effect and suppress the release of toxicity. Furthermore, the lamellar structure can play a barrier effect, reducing the mass and heat transfer process for between burned area and underlying polymers (Figure 6.13).

The different flame retardancy performance between MAX and MXenes was studied by Jiang *et al.* 2 wt% MAX decreased the pHRR, TSP and COP of UPR composites by 11.04%, 19.08% and 15.78%, respectively. In contrast, after the incorporation of 2 wt% MXene, the pHRR, TSP and COP of UPR composites were reduced by 29.56%, 25.6% and 31.58%,

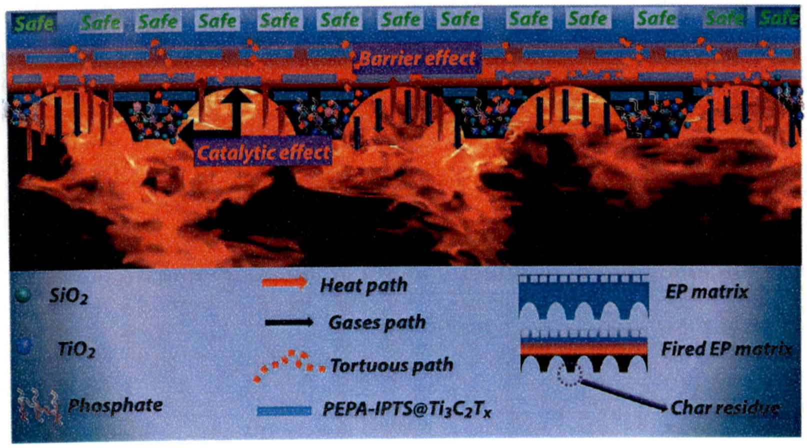

Figure 6.13 Demonstration of flame-retardant mechanism for EP composites. Source: Yin *et al.* [68]. Reproduced with permission of Elsevier.

respectively. The difference in the flame retardancy of UPR composites was due to the large layer spacing and specific surface area between MAX and MXene [84]. Wet ball milling was provided to produce MXene on a large scale through mixing by Hu *et al.* With 3 wt% PDDA-MXene into TPU, the pHRR and TSP were reduced by 50% and 47%, respectively [118]. A copper-organophosphate modified MXene (Cup-MXene) was designed by rational design (Figure 6.14). The results showed that UL-94 (V-0 rating) and pHRR (64.4% reduction compared with pure EP) were achieved after loading 5 wt% CuP-MXene. The designed nanohybrid combined with 1D CuP and 2D MXene clearly showed a synergistic effect on flame retardant performance [119].

Unmodified nanoparticles tend to aggregate in polymers, and the organic treatment method was effective in inhibiting aggregation. Yang *et al.* reported BODIPY-treated MXene (BODIPY-MXene) incorporated into ABS resin, and the LOI from 19.5% (pure ABS) shifted to 23.5% (ABS/BODIPY-MXene2.0). The cone calorimeter showed that the pHRR and PSPR of ABS/BODIPY-MXene2.0 were reduced by 24.5% and 18.4%, respectively [120]. Jiang *et al.* studied a UV-curable MXene composite coating and dipped the coating onto rigid polyurethane foam (RPUF) by spraying. The results showed that it was self-extinguished and that heat and smoke were significantly reduced. In summary, MXene and its hybrids can be applied to reduce the fire hazards of polymers.

6.3.6 Others

With the in-depth study of MXenes, other properties, such as optical properties and tribological properties, are gradually being explored. It has a layered structure similar to graphene and excellent mechanical properties. These characteristics have potential for the application of MXenes in friction fields. Philipp found that the laminate structure caused weak

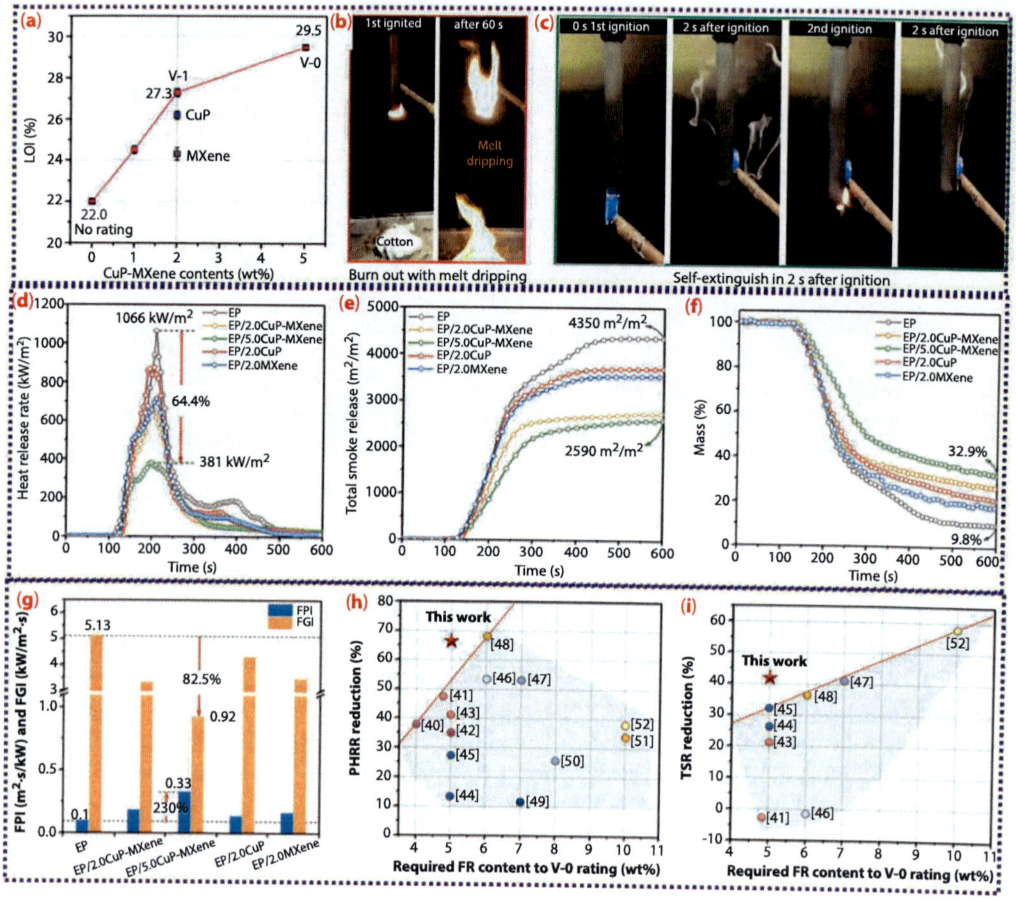

Figure 6.14 (a) LOI, digital photos for VBT of (b) pure EP and (c) EP-5.0CuP-MXene. (d) HRR, (e) TSR, (f) ML, (g) FPI and FGI of EP composites. Comparisons of (h) pHRR reduction and (i) TSR reduction of EP composites treated with various flame retardants. Source: Liu *et al.* [119]. Reproduced with permission of Elsevier.

interlayer interactions, which were relatively difficult to shear, promising its application as solid lubrication. Through the results (Figure 6.15), MXene-based coatings demonstrated a low wear rate (4×10^{-9} mm^3 N^{-1} m^{-1}) after 100,000 sliding cycles, revealing the excellent tribological properties of MXenes [121]. However, the in-depth relationship among the structure, composition and physicochemical properties of MXene-reinforced polymers is not clear, and industrial production is unforeseeable. Therefore, more focus should be concentrated on how to elucidate the fundamental properties of MXene-reinforced polymers, directing the fabrication of MXene-reinforced polymers.

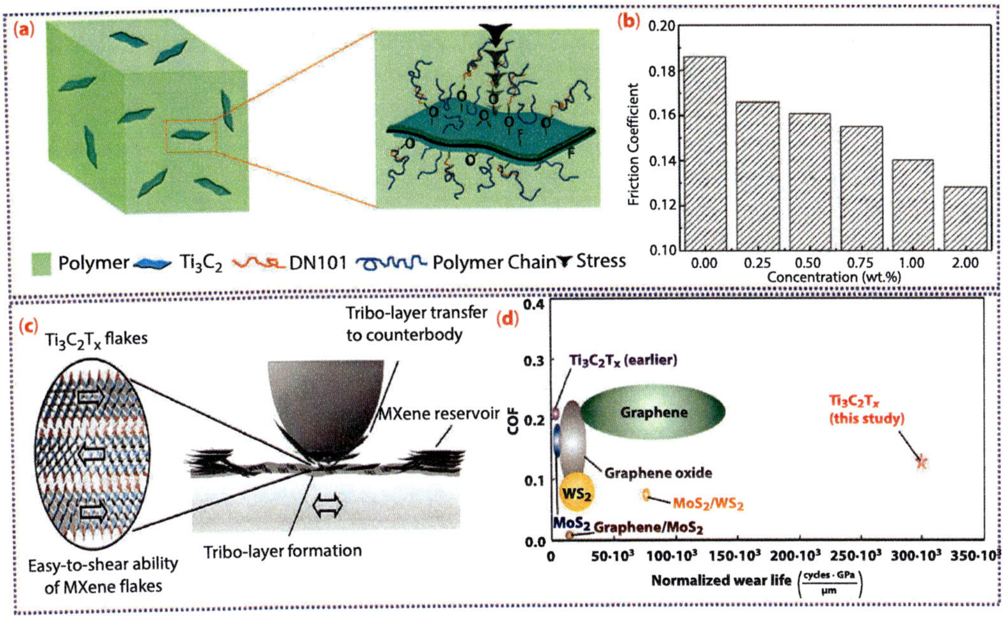

Figure 6.15 (a) Schematic demonstration of the improvement mechanism by Ti_3C_2. (b) Friction coefficient of nanocomposites. Source: Zhang *et al.* [122]. Reproduced with permission of Elsevier. (c) Illustration of the underlying mechanism. (d) Comparisons between normalized wear life, measured COF, and previous reports. Source: Grützmacher *et al.* [121]. Reproduced with permission of the American Chemical Society.

6.4 Applications

6.4.1 Sensors

According to current research on MXenes applied for electric sensors, pressure sensors, humidity sensors, temperature sensors and strain sensors, MXene-reinforced polymers have a stable response and outstanding performance, and the development of MXene-reinforced polymer sensors is similar to that of bamboo shoots after rain.

A piezoelectric film was acquired by electrospun polyvinylidene-fluoride-trifluoroethylene (PVDF-TrFE) and MXene. After loading 20 wt% MXene, the film produced a continuous output power density of 3.64 mW m^{-2} under 2 N at a frequency of 1 Hz, which was capable of sensing the motion of the human body [123]. A wearable and washing fabric was made through MXene absorbed onto the naturally rough surface of cotton fabric; it has an ultrafast response and relaxation time of less than 35 ms and superior cyclical stability higher than 2000 times [124]. A flexible 3D MXene framework was fabricated by the electrospinning and self-assembly technique. It can trace the ppb range of VOCs [125].

It was convenient to construct an early fire warning sensor with high flame retardancy when exposed to fire hazards. Herein, a flame-retardant paper comprised of thermoplastic polyurethane (SMPU) and MXene was fabricated through electrospinning and vacuum filtering. The flexible shape responding to temperature was outstanding to send out the timely warning when connected to the cutoff circuit. Shape memory was also triggered by electricity to realize self-cutting in the circuit [126]. Polyvinyl pyrrolidone-modified $Ti_3C_2T_x$ MXene was obtained, exhibiting superior performance in cyclic warning and weather-resistant performance (Figure 6.16). Briefly, the as-designed film displayed an ultrafast fire warning (1.8 s) and recovery time (1.0 s). Through structural analysis, it was found that thermal pyrolysis of the polymer chain prompted MXene oxidation, forming a compact C/N-doped titanium network, and activated electron transfer produced a fast cyclical warning signal [127].

MXene-based polymers can respond to outer fire at a high speed and produce a stable and strong signal. More importantly, the composites exhibited extraordinary cyclical stability, which has potential for the fabrication of next-generation intelligent sensors.

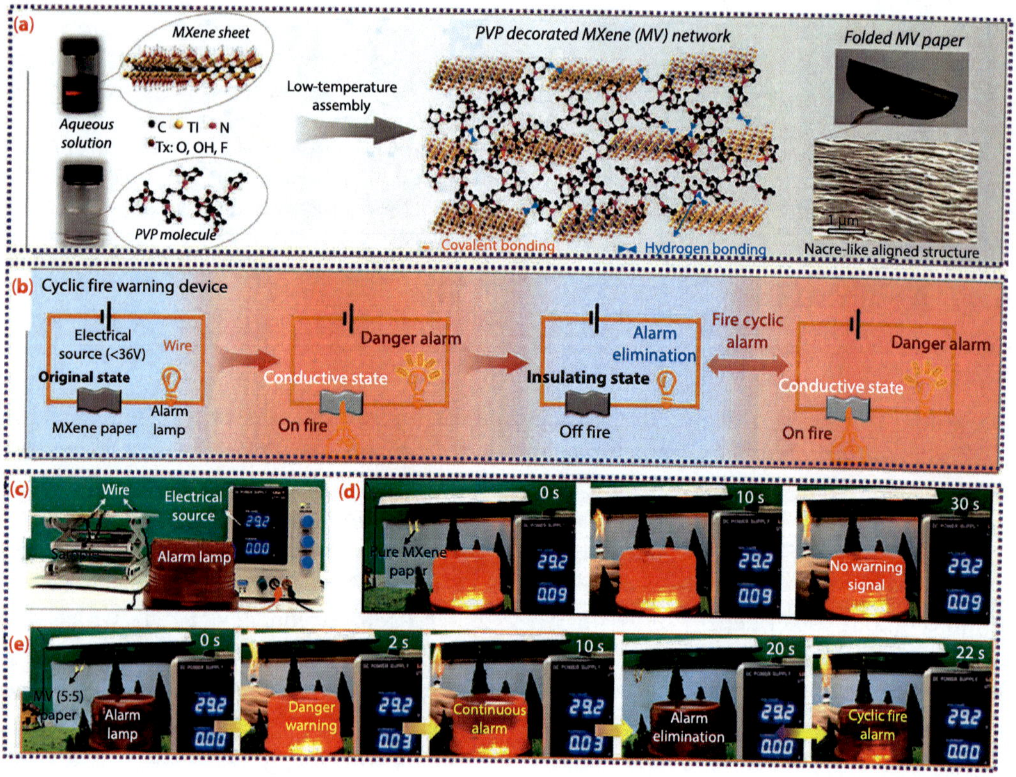

Figure 6.16 (a) Fabrication demonstration of MXene paper. (b) Model demonstration and (c) Digital photographs of the fire sensor. Fire sensor running process of (d) Pure MXene paper and (e) Conductivity parameters of PVP-modified MXenes. Source: Mao *et al.* [127]. Reproduced with permission of Elsevier.

6.4.2 Energy Applications

The good electrical properties, stability and rich surface groups of MXenes have attracted attention in the energy storage field. The excellent electrical performance of MXene can be used to prepare lithium batteries and supercapacitors (Figure 6.17a).

Due to the outstanding property of MXene, it was applied to prepare MXene/polymer electrochemical capacitors. At 120 Hz, a scalable processing method was provided for MXene/conductive polymers with high areal (0.56 mF cm^{-2}) and volumetric capacitances (24.2 F cm^{-3}). In addition, the ECS displayed tandem operation at frequencies of 60 to 10,000 Hz and exhibited long-term durability over 3,000 cycles during the bending test. The changes in thickness and design ratio between $Ti_3C_2T_x$ and PEDOT:PSS also exerted a significant influence on the charge storage kinetics performance [11]. All-solid-state micro-supercapacitors were prepared by painting MXene ink. After the increased deposited layer and enhanced electrode height, at 2 mV S^{-1}, it showed excellent areal capacitances of 1035 mF cm^{-2} [128]. 3D FeNi-LDH was fabricated onto 2D $Ti_3C_2T_x$ MXene through ionic hetero-assembly. The obtained MXene-based nanosheets exhibited superior electrical conductivity and electrolyte accessibility. The gravimetric capacitance was 922.6 F g^{-1} and retained 88% after 10,000 cycles [129]. After the incorporation of 15 wt% MXene into polyaniline (PANI), a superior capacitance retention was still maintained at 96% after 10,000 cycles at 50 mV S^{-1},

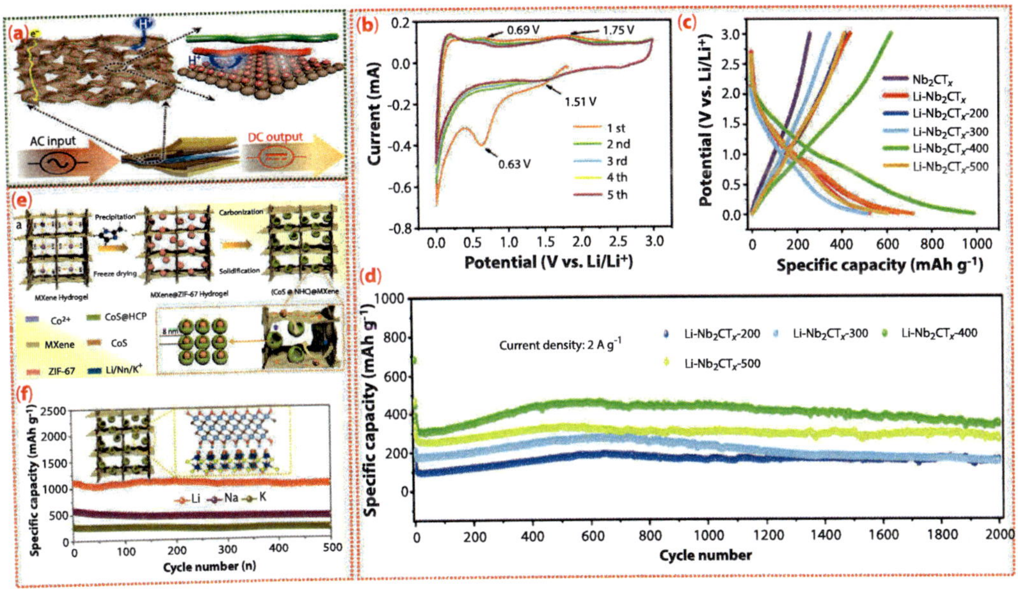

Figure 6.17 (a) MXene-based ECs. Source: Gund *et al.* [11]. Reproduced with permission of Elsevier. (b) CV of the Li-Nb_2CT_x-400 electrode (scanning rate: 0.5 mV s^{-1}). (c) GDC results of the first circle of various Nb_2CT_x electrodes at 0.05 A g^{-1} and (d) specific capacity cycling properties of samples annealed at 2.0 A g^{-1}. Source: Zhao *et al.* [132]. Reproduced with permission of Elsevier. (e) Manufacturing process of the (CoS@NP@NHC)@MXene aerogel. (f) Capacity cycle performance of the aerogel. Source: Yao *et al.* [133]. Reproduced with permission of the American Chemical Society.

and the capacitance was 434 F g^{-1}. In addition, through DFT calculations, it was found that the bonding between PANI and the surface group of MXene effectively enhanced the binding energy to -2.11 eV [130]. To overcome the aggregation and restacking of Ti_3C_2 MXene, a 3D porous MXene foam was applied as an electrode for lithium-ion batteries, and the results showed that the developed architecture significantly enhanced the capacity of up to 455.5 mAh g^{-1} at 50 mA g^{-1} and had an outstanding rate property (101 mAh g^{-1} at 18 A g^{-1}) [131]. Li^+-intercalated Nb_2CT_x annealed (Figures 6.17b–d) at 400 °C was applied as an electrode for lithium-ion batteries. The 448 mAh g^{-1} at 0.05 Ah g^{-1} of capacity was achieved, and the capacitance retention was 75% after 2000 cycles at 2 A g^{-1} [132]. An aerogel structure comprised of MOF@MXene was prepared and filled with hollow CoS, exhibiting outstanding electrochemical performance in batteries. After 800 cycles at 1 A g^{-1} and 1000 cycles at 5 A g^{-1} for LIBs, the capacities were 1145 mAh g^{-1} and 574.1 mAh g^{-1}, respectively. The theoretical calculation demonstrated charge transfer, especially at the interface (Figures 6.17e, f) [133].

From the development of MXene-based energy storage devices, it can be seen that MXene has potential for the large-scale production of MXene-reinforced polymers. However, there are still challenges to be solved, such as the aggregation, short storage time, and controllable quality of MXene.

6.4.3 Electromagnetic Interference Shielding

In response to the advent of the 5G era, the development of miniaturized and lightweight electronic equipment has become a research hotspot, but EM waves can emit radiation during transmission. Despite the light weight and low cost of most polymers, they usually cannot resist EM waves. Fortunately, MXene-reinforced polymers pave the way for the preparation of conductive devices with tunable EMI performance [134–140].

In 2016, Professor Koo *et al.* reported the potential EMI application of MXene for polymers. The results showed that a 45 μm $Ti_3C_2T_x$ film displayed EMI SE up to 92 dB, which was the highest before 2016 [141].

To meet the multifunctional scenario, an ultrathin, lightweight and flexible EMI foam was prepared by hydrazine-induced foaming. It exhibited an EMI SE of up to 70 dB, much higher than that of unfoamed MXene, which may have potential applications in aerospace and smart electronics [142]. A few layers of $Ti_3C_2T_x$ (f-$Ti_3C_2T_x$) were incorporated into PVA to fabricate a porous and absorption-dominated composite with a loading of 0.15 vol% MXene. The sample exhibited an excellent SSE/t (5136 dB cm^{-2} g^{-1}) and an SE_R of 2 dB [143]. MXene was decorated on the surface of PS beads and encapsulated by PDMS. After loading 0.0225-0.04449 vol% MXene, the composites showed an excellent EMI SE of 23.5-39.8 dB in the X band. In addition, these composites were highly compressional [144].

Currently, the consensus on preparing green and degradable EMI composites has been reached. The delignified wood and d-MXene formed a "mortar-brick" structure. After immersion in the MXenes for 8 h, the electrical conductivity and EMI SE of the wood composites were 6333 s m^{-1} and 39.3 dB, respectively, at a frequency of 12.4 GHz, respectively [145]. A solution casting method was applied to farbricate annealed MXene/EP composites. After loading 15 wt% annealed MXene, the EMI SE achieved 41 dB, which was 37% higher than that of pure $Ti_3C_2T_x$ [146]. An electrical insulating sandwich-structure MXene-based film was fabricated. In contrast to the pure $Ti_3C_2T_x$ MXene layer, this novel structure could retain an EMI SE of 50.01 dB after the loading of Ca^{2+}-crosslinked sodium alginate and montmorillonite (MMT) [147]. Waterborne polyurethane (ADWPU) blended with $Ti_3C_2T_x$ MXene revealed a tunable analogical structure and displayed a superior EMI performance of 51.37 dB at 12.4 GHz [12]. A stretchable MXene was coated on the surface of TPU fabric through the spraying method. The obtained wrinkled TPU film showed a value of 30 dB after the loading of 0.417 mg cm^{-2} of MXene under 50% strain pressure [148].

In summary, there have been many applications on different polymers through various production methods, which demonstrates the superiority of MXene. However, inconsistencies, such as the size of the lamella and the shortcomings of dispensability and oxidation, still hinder the development of MXene-reinforced polymers.

6.4.4 Catalytically Conversion

Catalysts have always been an important contributor to the world economy. In the process of industrialization over the past two centuries, practice has shown that more than 90% of chemical reactions require catalysts. Catalysts can greatly reduce the harsh conditions of chemical reactions and are indispensable in the modern society of the energy crisis and oil crisis. MXene demonstrates high reactivity in various catalytic reactions due to the abundant hydrophilic groups on the surface, such as =O, -F and -OH. By controlling the size and surface groups of the lamellae, MXene can have different selectivity to meet the requirements of MXene applications. Moreover, MXene has metal-like conductivity, which can effectively promote the transfer of electrons and protons. In recent years, MXene has been applied in the fields of flue gas toxicity conversion, polluted wastewater treatment and carbon dioxide conversion, photothermal catalysis and other fields, which has brought MXene to applications with broad prospects.

6.4.5 Oil/Water Separation

Crude oil leakage and oil pollution have caused great pollution to the environment. Compared with previous treatment methods, this relatively environmentally friendly and energy-saving treatment method has developed vigorously in recent years.

In 2019, Yang reported that a polyimide aerogel comprised of polyimide and MXene through freeze-drying treatment showed excellent separation of water and oil. The aerogel could absorb 18-58 times its weight to various organic solvents, hoping to be applied for oil/water separation [149]. Halloysite nanotubes and polydopamine (PDA) were deposited on the surface of MXene. The results showed that the pure water flux was 5036.2 L m^{-2} bar^{-1}, and the efficiency for rejections of petroleum ether and lubricating oil was 99.8%, implying that it was proper for oil/water separation [150]. A polyacrylonitrile (PAN) membrane was modified by MXene and O-MWCNTs, exhibiting excellent oil/water separation performance and having a high water permeability [151]. Yang reported a superwetting membrane via MXene self-assembly on the surface of porous polyvinylidene fluoride (PVDF) and then β-FeOOH. The results showed a high permeation flux (500.38-1022.7 $m^{-2}.h^{-1}$) and separation efficiency (99.32-99.72%) for various oil/water emulsions [152]. A conductive PI/MXene aerogel was prepared by freeze-drying method (Figures 6.18c, d). The aerogel exhibited excellent oil/water separation performance with a high absorption capacity (55.85-135.29 g. g^{-1}) [153].

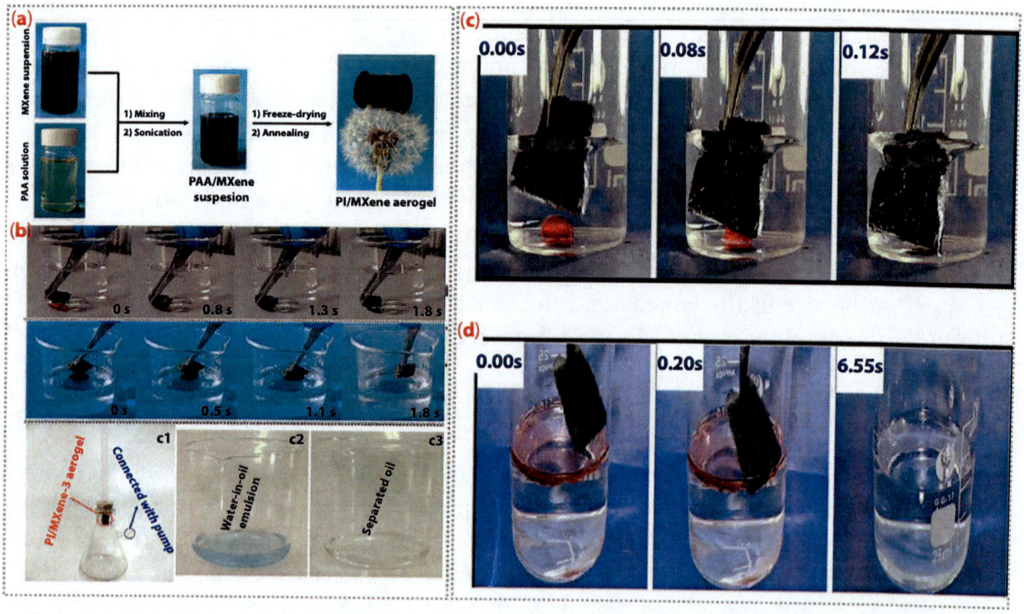

Figure 6.18 (a) Preparation of PI/MXene aerogels. (b) Water-oil system (PI/MXene-3 aerogel) to absorb chloroform and soybean oil. (c) Oil/water mixture separation result. Source: Wang *et al.* [149]. Reproduced with permission of the American Chemical Society. and (d) Soybean oil absorption test using the PINF/MXene aerogel. Source: Liu *et al.* [153]. Reproduced with permission from Wiley-VCH.

In summary, MXene-based oil/water separation polymers have attracted the attention of researchers around the world. Its unique structure and properties endow it with potential applications in industry.

6.4.6 Others

Ascribed to the merits of MXenes, they have also been employed in the fields, e.g., optical applications, wear-resistant applications, water purification, biomedicine (antibacterial, drug transportation, anti-aging) and gas separation. For example, Koo *et al.* investigated MXene-reinforced polymer light-emitting dioides under alternating current [154]. Later, Zhang *et al.* reported that MXene with a superior nonlinear optical response was potentially employed in ultrafast photonics [155]. Chen and his colleagues also explored $Ti_3C_2T_x$ MXene modified by polydopamine, and amino silane was utilized to enhance the wear-resistant property of nitrile butadiene rubber (NBR) [156]. Peng *et al.* studied PdNP modified MXene/polymer composites, which was applied to degrade 4-nitrophenol (4-NP) and 2-nitrophenol (2-NA) in wastewater [157]. Wang *et al.* fabricated a 3D MXene-decorated honeycomb-like fabric to achieve a solar evaporator [158]. Mahmoud *et al.* studied the antibacterial mechanism between MXene nanosheets and a model bacterial membrane through molecular dynamics simulations [159]. Yury and Wang's group reported a lamellar stacked MXene membrane for gas separation. The molecular dynamics simulations confirmed that the interlayer space among MXene was responsible for the enhancement of gas separation [160].

6.5 Conclusion and Outlook

The past ten years have seen the tremendous development of MXene. Since 2011, countless MAX and MXene have been synthesized and investigated, indicating that MXene-reinforced polymers are endowed with unparalleled properties, such as superior mechanical, thermal and conductive performance. We thoroughly reviewed the fabrication and practical applications of MXene-reinforced polymers. In addition, we summarize the MXene polymers utilized in the field, such as sensors, energy applications, EMI shielding, oil/water separation and others. The current research shows that MXene is superior for preparing MXene-based polymers. However, the industrial application of MXene polymers is still limited. The remaining questions are mainly divided into three perspectives: (1) It is still difficult to obtain controllable MXenes with extremely similar sizes and flake distributions. (2) The storage of MXene cannot reach a satisfactory time date due to the trace amount of water, oxygen and uncontrolled temperature. (3) The stacking of MXene is serious when added into the polymer matrix, severely deteriorating the overall performance of polymer composites. As a typical 2D material, we believe that the research on MXene-reinforced polymers will be step by step, eventually satisfying practical large-scale production, such as 2D graphene, benefiting the development of polymer (nano) composites.

Acknowledgment

This work was supported by the National Natural Science Foundation of China (No. U1833113, 51973203, 51303167, and 51403004) and the Fundamental Research Funds for the Central Universities (No. WK2320000054 and WK2320000044).

References

1. Zha, X.-H., Huang, Q., He, J., He, H., Zhai, J., Francisco, J.S., Du., S., The thermal and electrical properties of the promising semiconductor MXene Hf_2CO_2. *Sci. Rep.*, 6, 1, 1–10, 2016.
2. Jia, P., Zhu, Y., Lu, J., Wang, B., Song, L., Wang, B., Hu, Y., Multifunctional fireproof electromagnetic shielding polyurethane films with thermal management performance. *Chem. Eng. J.*, 439, 135673, 2022.
3. Xu, C., Song, S., Liu, Z., Chen, L., Wang, L., Fan, D., Kang, N., Ma, X., Cheng, H.-M., Strongly coupled high-quality graphene/2D superconducting Mo_2C vertical heterostructures with aligned orientation. *ACS Nano*, 11, 6, 5906–5914, 2017.
4. Liu, Y., Wang, W., Ying, Y., Wang, Y., Peng, X., Binder-free layered Ti_3C_2/CNTs nanocomposite anodes with enhanced capacity and long-cycle life for lithium-ion batteries. *Dalton Trans.*, 44, 16, 7123–7126, 2015.
5. Nan, J., Guo, X., Xiao, J., Li, X., Chen, W., Wu, W., Liu, H., Wang, Y., Wu, M., Wang, G., Nanoengineering of 2D MXene-based materials for energy storage applications. *Small*, 17, 9, 1902085, 2021.
6. Pandey, R.P., Rasool, K., Madhavan, V.E., Aïssa, B., Gogotsi, Y., Mahmoud, K.A., Ultrahigh-flux and fouling-resistant membranes based on layered silver/MXene ($Ti_3C_2T_x$) nanosheets. *J. Mater. Chem. A*, 6, 8, 3522–3533, 2018.

7. Chandran, M., Thomas, A., Raveendran, A., Vinoba, M., Bhagiyalakshmi, M., MoS_2 confined MXene heterostructures as electrode material for energy storage application. *J. Energy Storage*, 30, 101446, 2020.
8. Ling, Z., Ren, C.E., Zhao, M.-Q., Yang, J., Giammarco, J.M., Qiu, J., Barsoum, M.W., Gogotsi, Y., Flexible and conductive MXene films and nanocomposites with high capacitance. *Proc. Natl. Acad. Sci.*, 111, 47, 16676–16681, 2014.
9. Yue, Y., Liu, N., Ma, Y., Wang, S., Liu, W., Luo, C., Zhang, H., Cheng, F., Rao, J., Hu, X., Su, J., Gao, Y., Highly self-healable 3D microsupercapacitor with MXene-graphene composite aerogel. *ACS Nano*, 12, 5, 4224–4232, 2018.
10. Wang, S., Shao, H.-Q., Liu, Y., Tang, C.-Y., Zhao, X., Ke, K., Bao, R.-Y., Yang, M.-B., Yang, W., Boosting piezoelectric response of PVDF-TrFE via MXene for self-powered linear pressure sensor. *Compos. Sci. Technol.*, 202, 108600, 2021.
11. Gund, G.S., Park, J.H., Harpalsinh, R., Kota, M., Shin, J.H., Kim, T., Gogotsi, Y., Park, H.S., MXene/polymer hybrid materials for flexible AC-filtering electrochemical capacitors. *Joule*, 3, 1, 164–176, 2019.
12. Lu, J., Zhang, Y., Tao, Y., Wang, B., Cheng, W., Jie, G., Song, L., Hu, Y., Self-healable castor oil-based waterborne polyurethane/MXene film with outstanding electromagnetic interference shielding effectiveness and excellent shape memory performance. *J. Colloid Interface Sci.*, 588, 164–174, 2021.
13. Zeng, G., Wei, K., Zhang, H., Zhang, J., Lin, Q., Cheng, X., Sengupta, A., Chiao, Y.-H., Ultra-high oil-water separation membrane based on two-dimensional MXene ($Ti_3C_2T_x$) by co-incorporation of halloysite nanotubes and polydopamine. *Appl. Clay Sci.*, 211, 106177, 2021.
14. Lu, J., Wang, B., Jia, P., Cheng, W., Liao, C., Xu, Z., Cheng, L., Hu, Y., Designing advanced 0D-2D hierarchical structure for Epoxy resin to accomplish exceeding thermal management and safety. *Chem. Eng. J.*, 427, 132046, 2022.
15. Yin, J., Zhan, F., Jiao, T., Deng, H., Zou, G., Bai, Z., Zhang, Q., Peng, Q., Highly efficient catalytic performances of nitro compounds via hierarchical PdNPs-loaded MXene/polymer nanocomposites synthesized through electrospinning strategy for wastewater treatment. *Chin. Chem. Lett.*, 31, 4, 992–995, 2020.
16. Tang, T., Wang, S., Jiang, Y., Xu, Z., Chen, Y., Peng, T., Khan, F., Feng, J., Song, P., Zhao, Y., Flexible and flame-retarding phosphorylated MXene/polypropylene composites for efficient electromagnetic interference shielding. *J. Mater. Sci. Technol.*, 111, 66–75, 2022.
17. Chen, Q., Gao, Q., Wang, X., Schubert, D.W., Liu, X., Flexible, conductive, and anisotropic thermoplastic polyurethane/polydopamine/MXene foam for piezoresistive sensors and motion monitoring. *Compos. Part A-Appl. Sci.*, 106838, 2022.
18. Zhou, Y., Lin, Y., Tawiah, B., Sun, J., Yuen, R.K.K., Fei, B., DOPO-decorated two-dimensional MXene nanosheets for flame-retardant, ultraviolet-protective, and reinforced polylactide composites. *ACS Appl. Mater. Interfaces*, 13, 18, 21876–21887, 2021.
19. Gao, Q., Feng, M., Li, E., Liu, C., Shen, C., Liu, X., Mechanical, thermal, and rheological properties of $Ti_3C_2T_x$ MXene/thermoplastic polyurethane nanocomposites. *Macromol. Mater. Eng.*, 305, 10, 2000343, 2020.
20. Tang, B., Yang, Y., Shi, Y., Nie, H., Xia, H., Shen, X., Improved mechanical performances of short aramid fiber-reinforced polypropylene composites by Ti_3C_2 Mxene nanosheets. *Polym. Compos.*, 42, 4, 2010–2018, 2021.
21. Mirkhani, S.A., Shayesteh Zeraati, A., Aliabadian, E., Naguib, M., Sundararaj, U., High dielectric constant and low dielectric loss via poly (vinyl alcohol)/$Ti_3C_2T_x$ MXene nanocomposites. *ACS Appl. Mater. Interfaces*, 11, 20, 18599–18608, 2019.

22. Ma, L., Hamidinejad, M., Liang, C., Zhao, B., Habibpour, S., Yu, A., Filleter, T., Park, C.B., Enhanced electromagnetic wave absorption performance of polymer/SiC-nanowire/MXene ($Ti_3C_2T_x$) composites. *Carbon*, 179, 408–416, 2021.
23. Shao, J., Wang, J.-W., Liu, D.-N., Wei, L., Wu, S.-Q., Ren, H., A novel high permittivity percolative composite with modified MXene. *Polymer*, 174, 86–95, 2019.
24. Deng, Q., Zhou, F., Qin, B., Feng, Y., Xu, Z., Eco-friendly poly (vinyl alcohol)/delaminated V_2C MXene high-k nanocomposites with low dielectric loss enabled by moderate polarization and charge density at the interface. *Ceram. Int.*, 46, 17, 27326–27335, 2020.
25. Li, L., Liu, X., Wang, J., Yang, Y., Cao, Y., Wang, W., New application of MXene in polymer composites toward remarkable anti-dripping performance for flame retardancy. *Compos. Part A-Appl. Sci.*, 127, 105649, 20192019.
26. Pan, Y., Fu, L., Zhou, Q., Wen, Z., Lin, C.-T., Yu, J., Wang, W., Zhao, H., Flammability, thermal stability and mechanical properties of polyvinyl alcohol nanocomposites reinforced with delaminated $Ti_3C_2T_x$ (MXene). *Polym. Compos.*, 41, 1, 210–218, 2020.
27. Li, L., Cheng, Y., Cao, H., Liang, Z., Liu, Z., Yan, S., Li, L., Jia, S., Wang, J., Gao, Y., MXene/rGO/PS spheres multiple physical networks as high-performance pressure sensor. *Nano Energy*, 95, 106986, 2022.
28. Wang, J., Yang, K., Wang, H., Li, H., A new strategy for high-performance electromagnetic interference shielding by designing a layered double-percolated structure in PS/PVDF/MXene composites. *Eur. Polym. J.*, 151, 110450, 2021.
29. Mazhar, S., Qarni, A.A., Haq, Y.U., Ul. Haq, Z., Murtaza, I., Promising PVC/MXene based flexible thin film nanocomposites with excellent dielectric, thermal and mechanical properties. *Ceram. Int.*, 46, 8, 12593–12605, 2020.
30. Feng, Y., Qiu, H., Mao, B., Bo, M., Deng, Q., Preparation of hybrid ceramic/PVC composites showing both high dielectric constant and breakdown strength ascribed to interfacial effect between V_2C MXene and Cu_2O. *Colloids Surf. A*, 630, 127650, 2021.
31. Kim, H., Anasori, B., Gogotsi, Y., Alshareef, H.N., Thermoelectric properties of two-dimensional molybdenum-based MXenes. *Chem. Mater.*, 29, 15, 6472–6479, 2017.
32. Xue, J., Shi, Y., Wang, W., Yu, Y., Tang, C., Self-assembled Ti_3C_2Tx-MXene/PTh composite electrodes for electrochemical capacitors. *J. Mater. Sci.-Mater. Electron.*, 33, 6636–6645, 2022.
33. Luo, W., Niu, Z., Mu, P., Li., J., MXene/poly (ethylene glycol) mixed matrix membranes with excellent permeance for highly efficient separation of CO_2/N_2 and CO_2/CH_4. *Colloids Surf. A: Physicochem. Eng. Asp.*, 640, 128481, 2022.
34. Kotov, N.A., Layered biomimetic composites from MXenes with sequential bridging. *Angew. Chem. Int. Ed.*, 2021, 14140, 2022.
35. Weng, C., Xing, T., Jin, H., Wang, G., Dai, Z., Pei, Y., Liu, L., Zhang, Z., Mechanically robust ANF/MXene composite films with tunable electromagnetic interference shielding performance. *Compos. Part A-Appl. Sci.*, 135, 105927, 2020.
36. Lee, J.T., Wyatt, B.C., Davis Jr., G.A., Masterson, A.N., Pagan, A.L., Shah, A., Anasori, B., Sardar, R., Covalent surface modification of $Ti_3C_2T_x$ MXene with chemically active polymeric ligands producing highly conductive and ordered microstructure films. *ACS Nano*, 15, 12, 19600–19612, 2021.
37. VahidMohammadi, A., Moncada, J., Chen, H., Kayali, E., Orangi, J., Carrero, C.A., Beidaghi, M., Thick and freestanding MXene/PANI pseudocapacitive electrodes with ultrahigh specific capacitance. *J. Mater. Chem. A*, 6, 44, 22123–22133, 2018.
38. Wu, W., Wang, C., Zhao, C., Wei, D., Zhu, J., Xu, Y., Facile strategy of hollow polyaniline nanotubes supported on Ti_3C_2-MXene nanosheets for high-performance symmetric supercapacitors. *J. Colloid Interface Sci.*, 580, 601–613, 2020.

39. Xu, X., Chen, Y., He, P., Wang, S., Ling, K., Liu, L., Lei, P., Huang, X., Zhao, H., Cao, J., Yang, J., Wearable CNT/$Ti_3C_2T_x$ MXene/PDMS composite strain sensor with enhanced stability for real-time human healthcare monitoring. *Nano Res.*, 14, 8, 2875–2883, 2021.
40. Song, T.E., Yun, H., Kim, Y.J., Jeon, H.S., Ha, K., Jung, H.-T., Han, H., Gogotsi, Y., Ahn, C.W., Lee, Y., Vertically aligned nanopatterns of amine-functionalized Ti_3C_2 MXene via soft lithography. *Adv. Mater. Interfaces*, 7, 18, 2000424, 2020.
41. Wu, X., Han, B., Zhang, H.-B., Xie, X., Tu, T., Zhang, Y., Dai, Y., Yang, R., Yu, Z.-Z., Compressible, durable and conductive polydimethylsiloxane-coated MXene foams for high-performance electromagnetic interference shielding. *Chem. Eng. J.*, 381, 122622, 2020.
42. Zhang, K., Sun, J., Song, J., Gao, C., Wang, Z., Song, C., Wu, Y., Liu, Y., Self-healing Ti_3C_2 MXene/PDMS supramolecular elastomers based on small biomolecules modification for wearable sensors. *ACS Appl. Mater. Interfaces*, 12, 40, 45306–45314, 2020.
43. Sang, M., Liu, S., Li, W., Wang, S., Li, J., Li, J., Xuan, S., Gong, X., Flexible polyvinylidene fluoride (PVDF)/MXene ($Ti_3C_2T_x$)/Polyimide (PI) wearable electronic for body Monitoring, thermotherapy and electromagnetic interference shielding. *Compos. Part A-Appl. Sci.*, 153, 106727, 2022.
44. Zhao, Q., Yang, L., Ma, Y., Huang, H., He, H., Ji, H., Wang, Z., Qiu, J., Highly sensitive, reliable and flexible pressure sensor based on piezoelectric PVDF hybrid film using MXene nanosheet reinforcement. *J. Alloys Compd.*, 886, 161069, 2021.
45. Li, W., Song, Z., Zhong, J., Qian, J., Tan, Z., Wu, X., Chu, H., Nie, W., Ran, X., Multilayer-structured transparent MXene/PVDF film with excellent dielectric and energy storage performance. *J. Mater. Chem. C*, 7, 33, 10371–10378, 2019.
46. Cao, Y., Deng, Q., Liu, Z., Shen, D., Wang, T., Huang, Q., Du, S., Jiang, N., Lin, C.-T., Yu, J., Enhanced thermal properties of poly (vinylidene fluoride) composites with ultrathin nanosheets of MXene. *RSC Adv.*, 7, 33, 20494–20501, 2017.
47. Huang, H., Jiang, R., Feng, Y., Ouyang, H., Zhou, N., Zhang, X., Wei, Y., Recent development and prospects of surface modification and biomedical applications of MXenes. *Nanoscale*, 12, 3, 1325–1338, 2020.
48. Salauddin, M., Rana, S.S., Rahman, M.T., Sharifuzzaman, Md., Maharjan, P., Bhatta, T., Cho, H., Lee, S.H., Park, C., Shrestha, K., Sharma, S., Park, J.Y., Fabric-assisted MXene/silicone nanocomposite-based triboelectric nanogenerators for self-powered sensors and wearable electronics. *Adv. Funct. Mater.*, 32, 2107143, 2021.
49. Liu, L., Ying, G., Sun, C., Min, H., Zhang, J., Zhao, Y., Wen, D., Ji, Z., Liu, X., Zhang, C., Wang, C., MXene ($Ti_3C_2T_x$) functionalized short carbon fibers as a cross-scale mechanical reinforcement for epoxy composites. *Polymers*, 13, 11, 1825, 2021.
50. Wang, X., Wang, J., Qin, J., Xie, X., Yang, R., Cao, M., Surface charge engineering for covalently assembling three-dimensional MXene network for all-climate sodium ion batteries. *ACS Appl. Mater. Interfaces*, 12, 35, 39181–39194, 2020.
51. Shang, T., Lin, Z., Qi, C., Liu, X., Li, P., Tao, Y., Wu, Z., Li, D., Simon, P., Yang, Q.-H., 3D macroscopic architectures from self-assembled MXene hydrogels. *Adv. Funct. Mater.*, 29, 33, 1903960, 2019.
52. Qin, L., Tao, Q., Liu, X., Fahlman, M., Halim, J., Persson, P.O.Å., Rosen, J., Zhang, F., Polymer-MXene composite films formed by MXene-facilitated electrochemical polymerization for flexible solid-state microsupercapacitors. *Nano Energy*, 60, 734–742, 2019.
53. Tian, Y., Que, W., Luo, Y., Yang, C., Yin, X., Kong, L.B., Surface nitrogen-modified 2D titanium carbide (MXene) with high energy density for aqueous supercapacitor applications. *J. Mater. Chem. A*, 7, 10, 5416–5425, 2019.

54. Wang, H., Zhang, J., Wu, Y., Huang, H., Jiang, Q., Chemically functionalized two-dimensional titanium carbide MXene by *in situ* grafting-intercalating with diazonium ions to enhance supercapacitive performance. *J. Phys. Chem. Solids*, 115, 172–179, 2018.
55. Bao, W., Liu, L., Wang, C., Choi, S., Wang, D., Wang, G., Facile synthesis of crumpled nitrogen-doped mxene nanosheets as a new sulfur host for lithium-sulfur batteries. *Adv. Energy Mater.*, 8, 13, 1702485, 2018.
56. Lu, C., Yang, L., Yan, B., Sun, L., Zhang, P., Zhang, W., Sun., Z., Nitrogen-doped Ti_3C_2 MXene: Mechanism investigation and electrochemical analysis. *Adv. Funct. Mater.*, 30, 47, 2000852, 2020.
57. Chen, C., Boota, M., Urbankowski, P., Anasori, B., Miao, L., Jiang, J., Gogotsi, Y., Effect of glycine functionalization of 2D titanium carbide (MXene) on charge storage. *J. Mater. Chem. A*, 6, 11, 4617–4622, 2018.
58. Gao, P., Shi, H., Ma, T., Liang, S., Xia, Y., Xu, Z., Wang, S., Min, C., Liu, L., MXene/TiO_2 heterostructure-decorated hard carbon with stable Ti-O-C bonding for enhanced sodium-ion storage. *ACS Appl. Mater. Interfaces*, 13, 43, 51028–51038, 2021.
59. Liu, M., Zhou, J., He, Y., Cai, Z., Ge, Y., Zhou, J., Song, G., ε-Poly-L-lysine-protected Ti_3C_2 MXene quantum dots with high quantum yield for fluorometric determination of cytochrome c and trypsin. *Microchim. Acta*, 186, 12, 1–9, 2019.
60. Liu, H., Fu, R., Su, X., Wu, B., Wang, H., Xu, Y., Liu, X., Electrical insulating MXene/PDMS/BN composite with enhanced thermal conductivity for electromagnetic shielding application. *Compos. Commun.*, 23, 100593, 2021.
61. Guo, Q., Zhang, X., Zhao, F., Song, Q., Su, G., Tan, Y., Tao, Q., Zhou, T., Yu, Y., Zhou, Z., Lu, C., Protein-inspired self-healable Ti_3C_2 MXenes/rubber-based supramolecular elastomer for intelligent sensing. *ACS Nano*, 14, 3, 2788–2797, 2020.
62. Guan, W., Yang, X., Dong, C., Yan, X., Zheng, W., Xi, Y., Ruan, X., Dai, Y., He, G., Prestructured MXene fillers with uniform channels to enhance CO_2 selective permeation in mixed matrix membranes. *J. Appl. Polym. Sci.*, 138, 8, 49895, 2021.
63. Yi, Z., Yang, J., Liu, X., Mao, L., Cui, L., Liu, Y., Enhanced mechanical properties of poly(lactic acid) composites with ultrathin nanosheets of MXene modified by stearic acid. *J. Appl. Polym. Sci.*, 137, 48621, 2019.
64. Wang, H., Zhang, J., Wu, Y., Huang, H., Li, G., Zhang, X., Wang, Z., Surface modified MXene Ti_3C_2 multilayers by aryl diazonium salts leading to large-scale delamination. *Appl. Surf. Sci.*, 384, 287–293, 2016.
65. Hao, L., Zhang, H., Wu, X., Zhang, J., Wang, J., Li, Y., Novel thin-film nanocomposite membranes filled with multi-functional $Ti_3C_2T_x$ nanosheets for task-specific solvent transport. *Compos. Part A-Appl. Sci.*, 100, 139–149, 2017.
66. Wu, Q., Li, N., Wang, Y., Xu, Y., Wu, J., Jia, G., Ji, F., Fang, X., Chen, F., Cui, X., Ultrasensitive and selective determination of carcinoembryonic antigen using multifunctional ultrathin amino-functionalized Ti_3C_2-MXene nanosheets. *Anal. Chem.*, 92, 4, 3354–3360, 2020.
67. Ding, C., Liang, J., Zhou, Z., Li, Y., Peng, W., Zhang, G., Zhang, F., Fan, X., Photothermal enhanced enzymatic activity of lipase covalently immobilized on functionalized $Ti_3C_2T_x$ nanosheets. *Chem. Eng. J.*, 378, 122205, 2019.
68. Yin, Z., Wang, B., Tang, Q., Lu, J., Liao, C., Jia, P., Cheng, L., Song, L., Inspired by placoid scale to fabricate MXene derivative biomimetic structure on the improvement of interfacial compatibility, mechanical property, and fire safety of epoxy nanocomposites. *Chem. Eng. J.*, 431, 133489, 2022.
69. Li, X., Ma, Y., Shen, P., Zhang, C., Yan, J., Xia, Y., Luo, S., Gao, Y., Self-healing microsupercapacitors with size-dependent 2D MXene. *ChemElectroChem*, 7, 3, 821–829, 2020.

70. Shin, H., Eom, W., Lee, K.H., Jeong, W., Kang, D., Han, T., Highly electroconductive and mechanically strong MXene fibers using a deformable MXene gel. *ACS Nano*, 15, 2, 3320–3329, 2021.
71. Chen, K., Chen, Y., Deng, Q., Jeong, S.-H., Jang, T.-S., Du, S., Kim, H.-E., Huang, Q., Han, C.-M., Strong and biocompatible poly(lactic acid) membrane enhanced by $Ti_3C_2T_z$ (MXene) nanosheets for guided bone regeneration. *Mater. Lett.*, 229, 114–117, 2018.
72. Ronchi, R.M., Marchiori, C.F., Araujo, C.M., Arantes, J.T., Santos, S.F., Thermoplastic polyurethane-$Ti_3C_2T_x$ MXene nanocomposite: The influence of functional groups upon the matrix-reinforcement interaction. *Appl. Surf. Sci.*, 528, 146526, 2020.
73. Chen, S., Xu, J., Shi, M., Yu, Y., Xu, Q., Duan, X., Gao, Y., Lu, L., Polydopamine bridged MXene and NH_2-MWCNTs nanohybrid for high-performance electrochemical sensing of acetaminophen. *Appl. Surf. Sci.*, 570, 151149, 2021.
74. Xie, F., Jia, F., Zhuo, L., Lu, Z., Si, L., Huang, J., Zhang, M., Ma, Q., Ultrathin MXene/aramid nanofiber composite paper with excellent mechanical properties for efficient electromagnetic interference shielding. *Nanoscale*, 11, 48, 23382–23391, 2019.
75. Shin, H., Eom, W., Lee, K.H., Jeong, W., Kang, D.J., Han, T.H., Highly electroconductive and mechanically strong $Ti_3C_2T_x$ MXene fibers using a deformable MXene gel. *ACS Nano*, 15, 3320–3329, 2021.
76. Zhao, X., Zha, X.-J., Tang, L.-S., Pu, J.-H., Ke, K., Bao, R.-Y., Liu, Z., Yang, M.-B., Yang, W., Self-assembled core-shell polydopamine@MXene wit synergistic solar absorption capability for highly efficient solar-to-vapor generationm. *Nano Res.*, 13, 1, 255–264, 2020.
77. Zukiene, K., Monastyreckis, G., Kilikevicius, S., Procházka, M., Micusik, M., Omastová, M., Aniskevich, A., Zeleniakiene, D., Wettability of MXene and its interfacial adhesion with epoxy resin. *Mater. Chem. Phys.*, 257, 123820, 2021.
78. Tao, N., Zhang, D., Li, X., Lou, D., Sun, X., Wei, C., Li, J., Yang, J., Liu, Y.-N., Near-infrared light-responsive hydrogels via peroxide-decorated MXene-initiated polymerization. *Chem. Sci.*, 10, 46, 10765–10771, 2019.
79. Wang, H., Zheng, R., Shu, C., Long, J., Promoting the electrocatalytic activity of $Ti_3C_2T_x$ MXene by modulating CO_2 adsorption through oxygen vacancies for high-performance lithium-carbon dioxide batteries. *ChemElectroChem*, 7, 24, 4922–4930, 2020.
80. Cai, M., Yan, H., Li, Y., Li, W., Li, H., Fan, X., Zhu, M., $Ti_3C_2T_x$/PANI composites with tunable conductivity towards anticorrosion application. *Chem. Eng. J.*, 410, 128310, 2021.
81. Li, J., Wang, X., Sun, W., Maleski, K., Shuck, C.E., Li, K., Urbankowski, P., Hantanasirisakul, K., Wang, X., Kent, P., Wang, H., Gogotsi, Y., Intercalation-induced reversible electrochromic behavior of two-dimensional $Ti_3C_2T_x$ MXene in organic electrolytes. *ChemElectroChem*, 8, 1, 151–156, 2021.
82. Hu, Y., Cui, Y., Que, X., Zhang, Z., Peng, J., Li, J., Zhai, M., Super adhesive MXene-based nanocomposite hydrogel with self-healable and conductivity properties via radiation synthesis. *Adv. Eng. Mater.*, 24, 9, 2101692, 2021.
83. Lee, G.S., Yun, T., Kim, H., Kim, I.H., Choi, J., Lee, S.H., Lee, H.J., Hwang, H.S., Kim, J.G., Kim, D.-W., Lee, H.M., Koo, C.M., Kim, S.O., Mussel inspired highly aligned $Ti_3C_2T_x$ MXene film with synergistic enhancement of mechanical strength and ambient stability. *ACS Nano*, 14, 9, 11722–11732, 2020.
84. Hai, Y., Jiang, S., Zhou, C., Sun, P., Huang, Y., Niu, S., Fire-safe unsaturated polyester resin nanocomposites based on MAX and MXene: A comparative investigation of their properties and mechanism of fire retardancy. *Dalton Trans.*, 49, 18, 5803–5814, 2020.
85. Sun, Y., Wang, S., Du, X., Du, Z., Wang, H., Cheng, X., Skin-conformal MXene-doped wearable sensors with self-adhesive, dual-mode sensing, and high sensitivity for human motions and wireless monitoring. *J. Mater. Chem. B*, 9, 41, 8667–8675, 2021.

86. Xu, G., Wang, B., Song, S., Ren, Z., Li, J., Xu, X., Zhou, W., Li, H., Yang, H., Zhang, L., Li, Y., High-performance and robust dual-function electrochromic device for dynamic thermal regulation and electromagnetic interference shielding. *Chem. Eng. J.*, 422, 130064, 2021.
87. Seyedin, S., Uzun, S., Levitt, A., Anasori, B., Dion, G., Gogotsi, Y., Razal, J.M., MXene composite and coaxial fibers with high stretchability and conductivity for wearable strain sensing textiles. *Adv. Funct. Mater.*, 30, 12, 1910504, 2020.
88. Chen, Z., Han, Y., Li, T., Zhang, X., Wang, T., Zhang, Z., Preparation and electrochemical performances of doped MXene/poly (3, 4-ethylenedioxythiophene) composites. *Mater. Lett.*, 220, 305–308, 2018.
89. Wu, W., Niu, D., Zhu, J., Gao, Y., Wei, D., Liu, X., Wang, F., Wang, L., Yang, L., Organ-like Ti_3C_2 Mxenes/polyaniline composites by chemical grafting as high-performance supercapacitors. *J. Electroanal. Chem.*, 847, 113203, 2019.
90. An, H., Habib, T., Shah, S., Gao, H., Patel, A., Echols, I., Zhao, X., Radovic, M., Green, M.J., Lutkenhaus, J.L., Water sorption in MXene/polyelectrolyte multilayers for ultrafast humidity sensing. *ACS Appl. Nano Mater.*, 2, 2, 948–955, 2019.
91. An, H., Habib, T., Shah, S., Gao, H., Radovic, M., Green, M.J., Lutkenhaus, J.L., Surface-agnostic highly stretchable and bendable conductive MXene multilayers. *Sci. Adv.*, 4, 3, 0118, 2018.
92. Lu, J., Liao, C., Cheng, L., Jia, P., Yin, Z., Song, L., Wang, B., Hu, Y., Cleaner production to a multifunctional polyurethane sponge with high fire safety and low toxicity release. *J. Cleaner Prod.*, 333, 130172, 2022.
93. Levitt, A.S., Alhabeb, M., Hatter, C.B., Sarycheva, A., Dion, G., Gogotsi, Y., Electrospun MXene/carbon nanofibers as supercapacitor electrodes. *J. Mater. Chem. A*, 7, 1, 269–277, 2019.
94. Borges, J. and Mano, J.F., Molecular interactions driving the layer-by-layer assembly of multilayers. *Chem. Rev.*, 114, 18, 8883–8942, 2014.
95. Han, G., Zhao, X., Feng, Y., Ma, J., Zhou, K., Shi, Y., Liu, C., Xie, X., Highly flame-retardant epoxy-based thermal conductive composites with functionalized boron nitride nanosheets exfoliated by one-step ball milling. *Chem. Eng. J.*, 407, 127099, 2021.
96. Mathis, T.S., Maleski, K., Goad, A., Sarycheva, A., Anayee, M., Foucher, A.C., Hantanasirisakul, K., Shuck, C.E., Stach, E.A., Gogotsi, Y., Modified MAX phase synthesis for environmentally stable and highly conductive Ti_3C_2 MXene. *ACS Nano*, 15, 4, 6420–29, 2021.
97. Hantanasirisakul, K. and Gogotsi, Y., Electronic and optical properties of 2D transition metal carbides and nitrides (MXenes). *Adv. Mater.*, 30, 52, 1804779, 2018.
98. Liu, P., Ding, W., Liu, J., Shen, L., Jiang, F., Liu, P., Zhu, Z., Zhang, G., Liu, C., Xu, J., Surface termination modification on high-conductivity MXene film for energy conversion. *J. Alloys Compd.*, 829, 154634, 2020.
99. Li, R., Gao, Q., Xing, H., Su, Y., Zhang, H., Zeng, D., Fan, B., Zhao, B., Lightweight, multifunctional MXene/polymer composites with enhanced electromagnetic wave absorption and high-performance thermal conductivity. *Carbon*, 183, 301–12, 2021.
100. Tan, D., Jiang, C., Sun, N., Huang, J., Zhang, Z., Zhang, Q., Bu, J., Bi, S., Guo, Q., Song, J., Piezoelectricity in monolayer MXene for nanogenerators and piezotronics. *Nano Energy*, 90, 106528, 2021.
101. Wei, L., Wang, J.-W., Gao, X.-H., Wang, H.-Q., Wang, X.-Z., Ren, H., Enhanced dielectric properties of a poly (dimethyl siloxane) bimodal network percolative composite with MXene. *ACS Appl. Mater. Interfaces*, 12, 14, 16805–14, 2020.
102. Wang, Y., Zhang, F., Chen, X., Jin, Y., Zhang, J., Burning and dripping behaviors of polymers under the UL94 vertical burning test conditions. *Fire Mater.*, 34, 4, 203–15, 2010.
103. Gao, L., Li, C., Huang, W., Mei, S., Lin, H., Ou, Q., Zhang, Y., Guo, J., Zhang, F., Xu, S., MXene/polymer membranes: Synthesis, properties, and emerging applications. *Chem. Mater.*, 32, 5, 1703–47, 2020.

104. Jiao, E., Wu, K., Liu, Y., Lu, M., Hu, Z., Chen, B., Shi, J., Lu, M., Ultrarobust MXene-based laminated paper with excellent thermal conductivity and flame retardancy. *Compos. Part A Appl. Sci. Manuf.*, 146, 106417, 2021.
105. Yu, B., Yuen, A.C.Y., Xu, X., Zhang, Z.-C., Yang, W., Lu, H., Fei, B., Yeoh, G.H., Song, P., Wang, H., Engineering MXene surface with POSS for reducing fire hazards of polystyrene with enhanced thermal stability. *J. Hazard. Mater.*, 401, 123342, 2021.
106. Kang, R., Zhang, Z., Guo, L., Cui, J., Chen, Y., Hou, X., Wang, B., Lin, C.-T., Jiang, N., Yu, J., Enhanced thermal conductivity of epoxy composites filled with 2D transition metal carbides (MXenes) with ultralow loading. *Sci. Rep.*, 9, 1, 1–14, 2019.
107. Mishra, S., Sonawane, S., Singh, R., Bendale, A., Patil, K., Effect of nano-$Mg(OH)_2$ on the mechanical and flame-retarding properties of polypropylene composites. *J. Appl. Polym. Sci.*, 94, 1, 116–22, 2004.
108. Plummer, G., Anasori, B., Gogotsi, Y., Tucker, G.J., Nanoindentation of monolayer $Ti_{n+1}C_nT_x$ MXenes via atomistic simulations: The role of composition and defects on strength. *Comput. Mater. Sci.*, 157, 168–74, 2019.
109. Monastyreckis, G., Mishnaevsky Jr., L., Hatter, C., Aniskevich, A., Gogotsi, Y., Zeleniakiene, D., Micromechanical modeling of MXene-polymer composites. *Carbon*, 162, 402–09, 2020.
110. Zhi, W., Xiang, S., Bian, R., Lin, R., Wu, K., Wang, T., Cai, D., Study of MXene-filled polyurethane nanocomposites prepared via an emulsion method. *Compos. Sci. Technol.*, 168, 404–11, 2018.
111. Sheng, X., Zhao, Y., Zhang, L., Lu, X., Properties of two-dimensional Ti_3C_2 MXene/thermoplastic polyurethane nanocomposites with effective reinforcement via melt blending. *Compos. Sci. Technol.*, 181, 107710, 2019.
112. Woo, J.H., Kim, N.H., Kim, S.I., Park, O.-K., Lee, J.H., Effects of the addition of boric acid on the physical properties of MXene/polyvinyl alcohol (PVA) nanocomposite. *Compos. B. Eng.*, 199, 108205, 2020.
113. Luo, J.-Q., Zhao, S., Zhang, H.-B., Deng, Z., Li, L., Yu, Z.-Z., Flexible, stretchable and electrically conductive MXene/natural rubber nanocomposite films for efficient electromagnetic interference shielding. *Compos. Sci. Technol.*, 182, 107754, 2019.
114. Liu, L., Zhu, M., Shi, Y., Xu, X., Ma, Z., Yu, B., Fu, S., Huang, G., Wang, H., Song, P., Functionalizing MXene towards highly stretchable, ultratough, fatigue-and fire-resistant polymer nanocomposites. *Chem. Eng. J.*, 424, 130338, 2021.
115. Liu, J., Jia, X., Liu, Y., Wuliu, Y., Dai, J., Zhu, X., Liu, X., A synergetic strategy of well dispersing hydrophilic $Ti_3C_2T_x$ MXene into hydrophobic polybenzoxazine composites for improved comprehensive performances. *Compos. Sci. Technol.*, 219, 109248, 2022.
116. Li, K., Chang, T.H., Li, Z., Yang, H., Fu, F., Li, T., Ho, J.S., Chen, P.Y., Biomimetic MXene textures with enhanced light-to-heat conversion for solar steam generation and wearable thermal management. *Adv. Energy Mater.*, 9, 34, 1901687, 2019.
117. Feng, W., Wang, R., Zhou, Y., Ding, L., Gao, X., Zhou, B., Hu, P., Chen, Y., Ultrathin molybdenum carbide MXene with fast biodegradability for highly efficient theory-oriented photonic tumor hyperthermia. *Adv. Funct. Mater.*, 29, 22, 1901942, 2019.
118. He, L., Wang, J., Wang, B., Wang, X., Zhou, X., Cai, W., Mu, X., Hou, Y., Hu, Y., Song, L., Large-scale production of simultaneously exfoliated and functionalized Mxenes as promising flame retardant for polyurethane. *Compos. B. Eng.*, 179, 107486, 2019.
119. Liu, L., Zhu, M., Ma, Z., Xu, X., Seraji, S.M., Yu, B., Sun, Z., Wang, H., Song, P., A reactive copper-organophosphate-MXene heterostructure enabled antibacterial, self-extinguishing and mechanically robust polymer nanocomposites. *Chem. Eng. J.*, 430, 132712, 2022.

120. Zhu, S.-E., Wang, F.-D., Liu, J.-J., Wang, L.-L., Wang, C., Yuen, A.C.Y., Chen, T.B.Y., Kabir, I.I., Yeoh, G.H., Lu, H.-D., BODIPY coated on MXene nanosheets for improving mechanical and fire safety properties of ABS resin. *Compos. B. Eng.*, 223, 109130, 2021.
121. Grutzmacher, P.G., Suarez, S., Tolosa, A., Gachot, C., Song, G., Wang, B., Presser, V., Mücklich, F., Anasori, B., Rosenkranz, A., Superior wear-resistance of $Ti_3C_2T_x$ multilayer coatings. *ACS Nano*, 15, 5, 8216–24, 2021.
122. Zhang, H., Wang, L., Chen, Q., Li, P., Zhou, A., Cao, X., Hu, Q., Preparation, mechanical and anti-friction performance of MXene/polymer composites. *Mater. Des.*, 92, 682–89, 2016.
123. Wang, S., Shao, H.-Q., Liu, Y., Tang, C.-Y., Zhao, X., Ke, K., Bao, R.-Y., Yang, M.-B., Yang, W., Boosting piezoelectric response of PVDF-TrFE via MXene for self-powered linear pressure sensor. *Compos. Sci. Technol.*, 202, 108600, 2021.
124. Liu, R., Li, J., Li, M., Zhang, Q., Shi, G., Li, Y., Hou, C., Wang, H., MXene-coated air-permeable pressure-sensing fabric for smart wear. *ACS Appl. Mater. Interfaces*, 12, 41, 46446–54, 2020.
125. Yuan, W., Yang, K., Peng, H., Li, F., Yin, F., A flexible VOCs sensor based on a 3D Mxene framework with a high sensing performance. *J. Mater. Chem. A*, 6, 37, 18116–24, 2018.
126. Zhang, L., Huang, Y., Dong, H., Xu, R., Jiang, S., Flame-retardant shape memory polyurethane/MXene paper and the application for early fire alarm sensor. *Compos. B. Eng.*, 223, 109149, 2021.
127. Mao, M., Yu, K.-X., Cao, C.-F., Gong, L.-X., Zhang, G.-D., Zhao, L., Song, P., Gao, J.-F., Tang, L.-C., Facile and green fabrication of flame-retardant $Ti_3C_2T_x$ MXene networks for ultrafast, reusable and weather-resistant fire warning. *Chem. Eng. J.*, 427, 131615, 2022.
128. Orangi, J., Hamade, F., Davis, V.A., Beidaghi, M., 3D printing of additive-free 2D $Ti_3C_2T_x$ (MXene) ink for fabrication of micro-supercapacitors with ultra-high energy densities. *ACS Nano*, 14, 1, 640–50, 2019.
129. Zhang, R., Dong, J., Zhang, W., Ma, L., Jiang, Z., Wang, J., Huang, Y., Synergistically coupling of 3D FeNi-LDH arrays with $Ti_3C_2T_x$-MXene nanosheets toward superior symmetric supercapacitor. *Nano Energy*, 91, 106633, 2022.
130. Boota, M., Jung, E., Ahuja, R., Hussain, T., MXene binder stabilizes pseudocapacitance of conducting polymers. *J. Mater. Chem. A*, 9, 36, 20356–61, 2021.
131. Zhao, Q., Zhu, Q., Miao, J., Zhang, P., Wan, P., He, L., Xu, B., Flexible 3D porous MXene foam for high-performance lithium-ion batteries. *Small*, 15, 51, 1904293, 2019.
132. Zhao, J., Wen, J., Xiao, J., Ma, X., Gao, J., Bai, L., Gao, H., Zhang, X., Zhang, Z., Nb_2CT_x MXene: High capacity and ultra-long cycle capability for lithium-ion battery by regulation of functional groups. *J. Energy Chem.*, 53, 387–95, 2021.
133. Yao, L., Gu, Q., Yu, X., Three-dimensional MOFs@MXene aerogel composite derived MXene threaded hollow carbon confined CoS nanoparticles toward advanced alkali-ion batteries. *ACS Nano*, 15, 2, 3228–40, 2021.
134. Zhang, Y., Tian, W., Liu, L., Cheng, W., Wang, W., Liew, K.M., Wang, B., Hu, Y., Eco-friendly flame retardant and electromagnetic interference shielding cotton fabrics with multi-layered coatings. *Chem. Eng. J.*, 372, 1077–90, 2019.
135. Zhang, Y., Cheng, W., Tian, W., Lu, J., Song, L., Liew, K.M., Wang, B., Hu, Y., Nacre-inspired tunable electromagnetic interference shielding sandwich films with superior mechanical and fire-resistant protective performance. *ACS Appl. Mater. Interfaces*, 12, 5, 6371–82, 2020.
136. Cheng, W., Zhang, Y., Tian, W., Liu, J., Lu, J., Wang, B., Xing, W., Hu, Y., Highly efficient MXene-coated flame retardant cotton fabric for electromagnetic interference shielding. *Ind. Eng. Chem. Res.*, 59, 31, 14025–36, 2020.
137. Lu, J., Cheng, W., Shi, Y., Jia, P., Liao, C., Zhang, K., Song, L., Wang, B., Hu, Y., A facile strategy for lightweight, anti-dripping, flexible polyurethane foam with low smoke emission tendency and superior electromagnetic wave blocking. *J. Colloid Interface Sci.*, 603, 25–36, 2021.

138. Cheng, W., Zhang, Y., Tao, Y., Lu, J., Liu, J., Wang, B., Song, L., Jie, G., Hu, Y., Durable electromagnetic interference (EMI) shielding ramie fabric with excellent flame retardancy and self-healing performance. *J. Colloid Interface Sci.*, 602, 810–21, 2021.
139. Liu, J., Zhang, Y., Cheng, W., Lei, S., Song, L., Wang, B., Hu, Y., Anti-fogging, frost-resistant transparent and flexible silver nanowire-$Ti_3C_2T_x$ MXene based composite films for excellent electromagnetic interference shielding ability. *J. Colloid Interface Sci.*, 608, 2493–504, 2022.
140. Lu, J., Cheng, L., Liao, C., Jia, P., Song, L., Wang, B., Hu, Y., Ultrathin and mechanically robust mussel byssus-inspired MXene@aramid nanofibers materials with superior endurance in harsh environments for tunable EMI shielding performance. *Adv. Mater. Interfaces*, 9, 5, 2101359, 2022.
141. Shahzad, F., Alhabeb, M., Hatter, C.B., Anasori, B., Man Hong, S., Koo, C.M., Gogotsi, Y., Electromagnetic interference shielding with 2D transition metal carbides (MXenes). *Science*, 353, 6304, 1137–40, 2016.
142. Liu, J., Zhang, H.B., Sun, R., Liu, Y., Liu, Z., Zhou, A., Yu, Z.Z., Hydrophobic, flexible, and lightweight MXene foams for high-performance electromagnetic-interference shielding. *Adv. Mater.*, 29, 38, 1702367, 2017.
143. Xu, H., Yin, X., Li, X., Li, M., Liang, S., Zhang, L., Cheng, L., Lightweight $Ti_3C_2T_x$ MXene/poly (vinyl alcohol) composite foams for electromagnetic wave shielding with absorption-dominated feature. *ACS Appl. Mater. Interfaces*, 11, 10, 10198–207, 2019.
144. Jia, X., Shen, B., Zhang, L., Zheng, W., Construction of compressible polymer/MXene composite foams for high-performance absorption-dominated electromagnetic shielding with ultra-low reflectivity. *Carbon*, 173, 932–40, 2021.
145. Wang, Z., Han, X., Han, X., Chen, Z., Wang, S., Pu, J., MXene/wood-derived hierarchical cellulose scaffold composite with superior electromagnetic shielding. *Carbohydr. Polym.*, 254, 117033, 2021.
146. Wang, L., Chen, L., Song, P., Liang, C., Lu, Y., Qiu, H., Zhang, Y., Kong, J., Gu, J., Fabrication on the annealed $Ti_3C_2T_x$ MXene/epoxy nanocomposites for electromagnetic interference shielding application. *Compos. B. Eng.*, 171, 111–18, 2019.
147. Wang, B., Yin, Z., Zhang, Y., Jia, P., He, R., Yu, F., Hu, Y., Robust flame-retardant, super mechanical laminate epoxy composites with tunable electromagnetic interference shielding. *Mater. Today Phys.*, 26, 100724, 2022.
148. Dong, J., Luo, S., Ning, S., Yang, G., Pan, D., Ji, Y., Feng, Y., Su, F., Liu, C., MXene-coated wrinkled fabrics for stretchable and multifunctional electromagnetic interference shielding and electro/photo-thermal conversion applications. *ACS Appl. Mater. Interfaces*, 13, 50, 60478–88, 2021.
149. Wang, N.-N., Wang, H., Wang, Y.-Y., Wei, Y.-H., Si, J.-Y., Yuen, A.C.Y., Xie, J.-S., Yu, B., Zhu, S.-E., Lu, H.-D., Robust, lightweight, hydrophobic, and fire-retarded polyimide/MXene aerogels for effective oil/water separation. *ACS Appl. Mater. Interfaces*, 11, 43, 40512–23, 2019.
150. Zeng, G., Wei, K., Zhang, H., Zhang, J., Lin, Q., Cheng, X., Sengupta, A., Chiao, Y.-H., Ultra-high oil-water separation membrane based on two-dimensional MXene ($Ti_3C_2T_x$) by co-incorporation of halloysite nanotubes and polydopamine. *Appl. Clay Sci.*, 211, 106177, 2021.
151. Feng, X., Yu, Z., Long, R., Sun, Y., Wang, M., Li, X., Zeng, G., Polydopamine intimate contacted two-dimensional/two-dimensional ultrathin nylon basement membrane supported RGO/PDA/MXene composite material for oil-water separation and dye removal. *Sep. Purif. Technol.*, 247, 116945, 2020.
152. Hu, J., Zhan, Y., Zhang, G., Feng, Q., Yang, W., Chiao, Y.-H., Zhang, S., Sun, A., Durable and super-hydrophilic/underwater super-oleophobic two-dimensional MXene composite lamellar membrane with photocatalytic self-cleaning property for efficient oil/water separation in harsh environments. *J. Membr. Sci.*, 637, 119627, 2021.

153. Liu, H., Chen, X., Zheng, Y., Zhang, D., Zhao, Y., Wang, C., Pan, C., Liu, C., Shen, C., Lightweight, superelastic, and hydrophobic polyimide nanofiber/MXene composite aerogel for wearable piezoresistive sensor and oil/water separation applications. *Adv. Funct. Mater.*, 31, 13, 2008006, 2021.

154. Lee, S., Kim, E.H., Yu, S., Kim, H., Park, C., Park, T.H., Han, H., Lee, S.W., Baek, S., Jin, W., Alternating-current MXene polymer light-emitting diodes. *Adv. Funct. Mater.*, 30, 32, 2001224, 2020.

155. Fu, B., Sun, J., Wang, C., Shang, C., Xu, L., Li, J., Zhang, H., MXenes: Synthesis, optical properties, and applications in ultrafast photonics. *Small*, 17, 11, 2006054, 2021.

156. Qu, C., Li, S., Zhang, Y., Wang, T., Wang, Q., Chen, S., Surface modification of $T_{i3}C_2$-MXene with polydopamine and amino silane for high performance nitrile butadiene rubber composites. *Tribol. Int.*, 163, 107150, 2021.

157. Yin, J., Zhan, F., Jiao, T., Deng, H., Zou, G., Bai, Z., Zhang, Q., Peng, Q., Highly efficient catalytic performances of nitro compounds via hierarchical PdNPs-loaded MXene/polymer nanocomposites synthesized through electrospinning strategy for wastewater treatment. *Chin. Chem. Lett.*, 31, 4, 992–95, 2020.

158. Lei, Z., Sun, X., Zhu, S., Dong, K., Liu, X., Wang, L., Zhang, X., Qu, L., Zhang, X., Nature inspired MXene-decorated 3D honeycomb-fabric architectures toward efficient water desalination and salt harvesting. *Nanomicro Lett.*, 14, 1, 1–16, 2022.

159. Lee, O.-S., Madjet, M.E., Mahmoud, K.A., Antibacterial mechanism of multifunctional MXene nanosheets: Domain formation and phase transition in lipid bilayer. *Nano Lett.*, 21, 19, 8510–17, 2021.

160. Ding, L., Wei, Y., Li, L., Zhang, T., Wang, H., Xue, J., Ding, L.-X., Wang, S., Caro, J., Gogotsi, Y., MXene molecular sieving membranes for highly efficient gas separation. *Nat. Commun.*, 9, 1, 1–7, 2018.

7

Structural, Morphological and Tribological Properties of Polymer/MXene Composites

Humira Assad[1], Ishrat Fatma[1], Praveen Kumar Sharma[1] and Ashish Kumar[2*]

[1]*Department of Chemistry, School of Chemical Engineering and Physical Sciences, Lovely Professional University, Phagwara, Punjab, India*
[2]*Department of Chemistry, Nalanda College of Engineering, Bihar Engineering University, Science, Technology and Technical Education Department, Government of Bihar, India*

Abstract
MXenes, a current recruit of the two-dimensional (2-D) material family, is a remarkably fresh and amazing category of substances with diverse compositions and interesting features for instance dissolution rate as well as metallic conductivity. These excellent attributes have piqued the interest of many researchers, and substantial investigation on MXene has been initiated, with emerging concepts identified in the disciplines of power storage, detection, electromagnetic shielding, and so on. Because of these characteristics, they are promising alternatives to be used as reinforcement in polymer nanocomposites. However, because MXenes contain an abundance of surface functional groups, their physical and chemical characteristics can be optimized by the establishment of MXene-polymer composite materials. Polymers can be used to increase interlamellar separation distance, decrease the separation of ion/electron mobility, increase the surface hydrophilicity, and therefore mentor the gathering of MXene-polymer constructions. Furthermore, numerous MXene/polymer nanocomposites have been established to date, with many astounding results reported. Nevertheless, synopses of the current state and advancement in this profession are uncommon. The main motive of this chapter is to review the various synthesis method as well as the scientific pathway of MXene/polymer nanocomposites. Furthermore, the numerous different Structural, Morphological, and Tribological Characteristics of Polymer/MXene Composites are summarized, accompanied by novel applications and noteworthy consequences of the composites in biomedical, perceiving, electromagnetic safeguards, and other pastures. Ultimately, the Polymer/MXene Composite's advancement opportunities and strategic issues are elucidated.

***Keywords*:** MXenes, polymer, polymer nanocomposites, fabrication, tribological, applications

**Corresponding author*: drashishchemlpu@gmail.com

Abbreviations

2-D	Two-Dimensional
DMF	N, N-Dimethylformamide
DMS	Dimethyl Sulfoxide
FEA	Finite Element Analysis
GBR	Guided Bone Regeneration
GO_x	Glucose Oxidase
HF	Hydrofluoric Acid
LBL	Layer-By-Layer
MD	Molecular Dynamics
MWCNT	Multi-Walled Carbon Nanotube
OTES	N-Octyltriethoxysilane
PAA	Poly Acrylic Acid
PAA	Polyacrylic Acid
PAM	Polyacrylamide
PAN	Polyacrylonitrile
PDDA	Poly Di-Allyl Dimethyl Ammonium Chloride
PEBA	Poly Ether-Block-Amide
PEG	Polyethylene Glycol
PEI	Poly ethylene imine
PS	Polystyrene
PTT	Photothermal Treatment
PU	Polyurethane
PVA	Polyvinyl Alcohol
PVDF	Polyvinylidene Fluoride
PVP	Poly Vinylpyrrolidone
Tg	Glass Transition Temperature
TPU	Thermoplastic Poly Urethane
UHMWPE	Ultrahigh Molecular Weight Polyethylene
UTS	Ultimate Tensile Strength
VAF	Vacuum-Assisted Filtration
YS	Yield Stress

7.1 Introduction

Two-dimensional (2D) nanoparticles, with dazzling qualities than that of their bulk equivalents, have flourished in multiple investigation areas from the time when graphene was emerged in 2004 [1–4]. Due to the decrease in dimension as well as size, 2-D particles have unveiled several fascinating characteristics, holding remarkable potential for a host of solicitations right from electronic as well as optoelectronic appliances to electrochemical catalysis [5–7]. In current years, with unlimited improvements in artificial practices, several 2D materials other than graphene have been efficiently manufactured [8]. Out of which, a recently detected huge family of 2D transition metal carbides, as well as nitrides called MXenes, is a speedy highflier [9–12]. MXenes are usually created through etching of A layers from several MAX phases, [13–15] that are ternary carbides/nitrides mainly indicated by $M_{n+1}AX_n$, in which M signifies a transition metal, A signifies elements of 3rd A or 4th A group, X may be C/N, while n is equal to 1, 2, 3 and so on [16]. MAX phases bear layered hexagonal assemblies where the $M_{n+1}X_n$ components, as well as the A layers, are present one after another [17]. Since the M–X bonds are considerably tougher than that of the M-A bonds, the A layers may be etched without abolishing the M–X bonds, that form insipidly bonded $M_{n+1}Xn$ stratums which may be freely unglued through sonication. The resultant 2D particles were known as MXenes whose nature is similar to graphene. It must be pointed out that during etching, the outer layers of the $M_{n+1}X_n$ units are eternally concealed with functional moieties like -O, -OH, and -F [18, 19]. Accordingly, the MXenes are represented by $M_{n+1}X_nT_x$, in which T_x signifies the external functional assemblies. Analytically, the amounts of dissimilar functional moieties on the external surfaces of MXene are indefinite which modify with the etching situation. The adaptable chemistries of MXenes endow them to acquire many mechanical, electronic, magnetic as well as electrochemical characteristics. Specifically, the decent elasticity of MXenes, conjoined with their 2D appearance as well as incrusted configurations, forms it simple for MXenes to make composites with many other moieties, that offer a chance of assimilating the exceptional characteristics of dissimilar elements in a coordinating mode. The bandgap, as well as conductivity characteristics, are tunable in these assemblies. MXene containing metals also reinforces the electronic properties. The existence of some defects on the exterior of MXene could further stimulate the movement of the charge. As MXene nano-sheets may be incrusted to create the multi stratum assembly. Weak forces occur among the nano-sheets. Accordingly, because of poor forces, guest particles may be interpolated utilizing solvent dispersion techniques. Interpolation declines the forces among the adjoining sheets as well as increases the interlayer distances. MXene possesses an ionic surface that can invite the adjoining electrons/charges. Moreover, the functional groups present in the MXenes turn them hydrophilic. Stable formulations in H_2O as well as in numerous polar solvents may be created, often via probing or bath sonication of an MXene-comprising solution at neutral pH. These groups also enable certain reactions as well as interactions with hydrophilic polymers like polyvinyl alcohol (PVA), poly di-allyl dimethyl ammonium chloride, polyacrylic acid, and a variety of biological polymers. Polymer addition causes MXene to exfoliate because of an increase in the interlayer distance, as displayed in Figure 7.1.

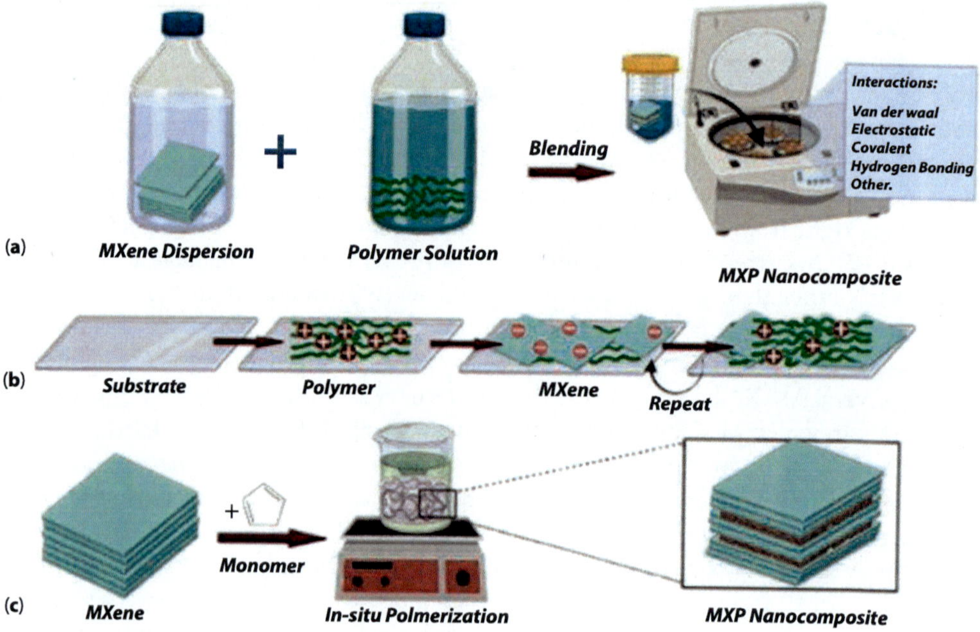

Figure 7.1 Schematic representation of MXene/Polymer nanocomposite formation [20]. Copyright 2021. Reproduced with permission from MDPI.

It prevents MXene layers from mounting and causes hierarchal lamellar formations to develop. Interfacial multi-interactions (covalent, electrostatic, etc.) lead to the development of MXP-NCs with MXene sheets layered through polymer chains, relying on the type of polymer used. The frequent deployment of fillers in polymers to generate greater functioning composite substances with enhanced characteristics emerged in the 1960s. Nanomaterials were not presented as extracts to generate a new composite, known now as NCs, until the late 1980s and early 1990s, with Toyota's research on nylon/clay hybrids. Nanoparticles, especially clays in addition to other 2D materials, have significant benefits than that of other 3D equivalents, primarily because of their extraordinarily surface area-volume ratios as well as high particle aspect ratios. With modest nanofiller loadings of 2% vol. percent, significant gains in mechanical, melting, viscoelastic, as well as gas barrier characteristics can be realized if these nanosheets can be molted as well as homogeneously distributed into a polymer medium. Because of MXenes' distinctive qualities and different conformations, analysis into their usage as nanofillers has garnered momentum as their detection, particularly as the chemical link between MXenes and nanofillers is strengthened. $Ti_3C_2T_z$, the most frequently investigated MXene till now, was combined with PDDA as well as PVA, both hydrophilic polymers, in the first information of an MXene polymer NC in 2014 [21]. Since the publication of the first paper, there has been a surge in enthusiasm in MXene polymer NC research; nonetheless, the scope is significantly narrower than that of MXene research. Till now, the main analysis has concentrated on the usage of $Ti_3C_2T_z$ since it is well-studied, simple to synthesize, and relatively stable in the face of oxidation when compared to other MXenes. Consequently, not only MXenes however MXene-based nanocomposites have involved extensive research concerns, owning unlimited potential for many solicitations. Because of their extraordinary conductivity and outstanding electrochemical doings, MXenes in addition to MXene-based nanocomposites, primarily discover solicitations in various energy storage appliances like Li-ion batteries as well as in supercapacitors [22–24]. Incredibly, in recent times they have further grown to protuberance in the environment-allied areas. Particularly, (i) they can be utilized as competent catalysts or co-catalysts for various photo-catalytic phenomena [25–27]. (ii) They can eliminate impurities in H_2O like large metal ions, biological dyes, eutrophic elements, as well as nuclear trash; [28, 29]. (iii) They can also be utilized in various sensors including biosensors as well as gas sensors, showing outstanding performances [30, 31]. As a result, the main objective of this chapter is to offer an ephemeral review of MXene/polymer nanocomposites, as well as their synthesis, properties, and innovative solicitations.

7.2 Overview of MXene

MXenes are the names for 2D transition metal carbides as well as nitrides [32]. Mn+1XnTz (n = 1, 2, or 3) is the universal formula for MXene. Here M stands for a transition metal, X for carbon or nitrogen, and Tz for terminal units like -OH, -F, etc. Mono-transition metal complexes, like M_2C, M_3C_2, and M_4C_3, are discovered here with one metal in the effete. The double-transition metal complexes, on the other hand, can be denoted by the symbols $M'_2M''C_2$, $M'_2M''C_3$, or $M'_2M''C_4$, where M' and M" indicate separate transition metals. MXene nanosheets can range in size from 1 to 100 nm. A solo MXene nanosheet could bear an atom or two of thickness.

Several techniques, including liquid-phase exfoliation, etching/delamination, ultrasonication, intercalation, fluorine-free approach, as well as synthetic procedures, have been employed to create MXene. Both Covalent as well as metallic bonds are present in the structure of MXene. In the beginning, Ti_3C_2 was produced in 2011 using hydrofluoric acid (HF) [33]. Above thirty MXenes have been created so far utilizing various methods as well as materials since the discovery of MXene [34]. These structures allow for the tuning of the band gap and conductivity characteristics. The choice of metal in MXene affects the electrochemical properties as well. The charge mobility may be further enhanced by the existence of imperfections on the MXene surface. Due to the MXene nanosheet's ability to be stacked to create multilayered structures. The nanosheets are held together by weak forces.

Therefore, employing solvent dispersion techniques, guest molecules can be intercalated due to weak forces. The forces between adjacent sheets are weakened by intercalation, which increases interdistances. MXene has a charged surface that might draw in neighboring charges or electrons. In contrast to nanocarbons like graphene, the addition of a hydroxyl group makes the structure more hydrophilic. Solicitations of MXene have been found in energy storage, thermal constancy, electronics, and other promising fields [35].

7.3 MXene/Polymer Nanocomposites

When creating nanocomposites using polymers, MXenes with exceptional mechanical characteristics, hydrophilic surfaces, as well as metallic conductivity can develop the mechanical in addition to thermal characteristics of polymers. As compared to multi stratum MXenes, single stratum MXenes have greater approachable surface hydrophilicity as well as improved congeniality with polymer materials. Therefore, MXenes are typically exfoliated before conjoining with polymer materials. Ling *et al.* explored the production of a unilayered $Ti_3C_2T_x$–polyvinyl alcohol composite via mingling the colloidal solution of Ti_3C_2Tx layers using a PVA solution in water [36]. The $Ti_3C_2T_x$-PVA nanocomposite is an incrusted assembly in which PVA is dispersed among $Ti_3C_2T_x$ layers. It owns extraordinary thermo-stability since extremely hydrophilic PVA may develop tough hydrogen bonds beside $Ti_3C_2T_x$. Related to single $Ti_3C_2T_x$ as well as PVA, it is the $Ti_3C_2T_x$-PVA that bears extra elasticity as well as ductility. Specifically, the ductility of the $Ti_3C_2T_x$-PVA nanocomposite having 40 wt. % $Ti_3C_2T_x$ extents 91±10 MPa, that is approximately four-fold higher as compared to Ti_3C_2Tx film. Naguib and his coworkers create a $Ti_3C_2T_x$-PAM nanocomposite utilizing a two-step scheme. in the first step, the Dimethylsulfoxide (DMSO) is inserted in between the stratums of $Ti_3C_2T_x$ to upsurge the stratum gaps of $Ti_3C_2T_x$ as well as to obtain complete exfoliation of specific $Ti_3C_2T_x$ stratums. After that, the $Ti_3C_2T_x$, as well as PAM solutions, are consistently stirred followed by dehydration at normal temperature for about 4 to 5 consecutive days. Several $Ti_3C_2T_x$-PAM composites are manufactured via changing the mass ratio of $Ti_3C_2T_x$ towards PAM. As soon as $Ti_3C_2T_x$ holds for 6 wt. %, the resultant composite bear finest mechanical characteristics in addition to the maximum electrical conductivity [37].

UHMWPE (ultrahigh molecular weight polyethylene) is also described that efficiently make nanocomposites by using $Ti_3C_2T_x$. Just earlier than fabrication of the Ti_3C2T_x-UHMWPE nanocomposite, the surface improved $Ti_3C_2T_x$ fine particles were prepared, and the altered surface can increase the congeniality as well as distribution stability of $Ti_3C_2T_x$ in UHMWPE [38]. The combinations of surface improved $Ti_3C_2T_x$ plus UHMWPE are shaped on a press vulcanizer, followed by heating and then kept under pressure of about 10 MPa. The molded $Ti_3C_2T_x$-UHMWPE composites using several weights of $Ti_3C_2T_x$ bear a greater strength as compared to UHMWPE. Out of them, the 10 wt. % $Ti_3C_2T_x$-UHMWPE composite bears maximum yield strength, final tensile strength as well as elongation [38]. MXene/polymer composites that have been reported thus far typically consist of filled composites and complexes. In this context, the term "filled composite" signifies a bulk material wherein MXene is utilized as a filler or component and is scattered randomly with the polymer. In a complex, MXene and polymer are created as a particular material product, with a particular material structure, or both. MXene/polymer complex is specifically divided into layered composite (2-2) and structure, MXene coated polymeric fiber or fabric composite, MXene/polymer composite fiber or fabric, MXene/polymer composite aerogel, foam, or sponge, and MXene/polymer composite fiber or fabric composite [39]. The most extensively researched filled composites involve a thermoplastic and a thermosetting polymer matrix. Complexes haven't been discussed as much, however Zhou and co-workers. did discuss a encrusted composite (structure) they created via spin-coating an MXene dispersion onto electrospun polycaprolactone (PCL) fiber membranes [40]. Besides the above-discussed polymers, poly (acrylic acid), and alginate/PEO can also make composites with MXenes. Usually, the MXene - polymer composites show excellent mechanical functions as compared to single MXenes as well as polymers. Furthermore, numerous of them exhibit wonderful electric conductance and consequently are promising candidates for solicitations in various electronic appliances.

7.4 MXene/Polymer Nanocomposite Fabrication Methods

Fabricating composites has become a popular approach for developing robust and adaptable substances in recent times. MXenes are regarded as ideal prospects for the production of multifunctional composites owing to their 2D geometry, multilayer architectures, and remarkable elasticity, which has sparked a rush in research into MXene-based composites. Unilayered MXenes possess greater affable outer-layer hydrophilicity as well as superior polymer compatibility than multi-layered MXenes. As a result, MXenes are frequently exfoliated before being combined with several polymers. Many new composites have been created so far by mixing MXenes with a diversity of materials including polymers, metal oxides, as well as carbon nanotubes [41]. Naguib *et al.* use a two-step approach to make a $Ti_3C_2T_x$ -polyacrylamide (PAM) hybrid [37]. To begin, DMSO is administered into the $Ti_3C_2T_x$ interlayers to enhance layer separation and gain complete delamination of particular $Ti_3C_2T_x$ sheets. The $Ti_3C_2T_x$ and PAM solutions are then mixed thoroughly and cured for 4-5 days at room temperature. By changing the mass ratio of $Ti_3C_2T_x$ to PAM, several $Ti_3C_2T_x$-PAM composites can be made. The composite exhibits excellent mechanical qualities and the maximum electrical conductivity of 3.3×10^{-2} Sm^{-1} when $Ti_3C_2T_x$ accounts for 6 wt.%. The results showed a significant quantity of indentations, suggesting that the rough surface is a conventional ductile fracture and that this specimen is sturdy. In addition, MXenes can also be combined with poly (acrylic acid), polyvinylpyrrolidone (PVP), poly (ethylene oxide) (PEO, and alginate/PEO to produce composites [40, 42]. In general, MXene–polymer composites outperform isolated MXenes and polymers in terms of mechanical endurance. Furthermore, many of them have strong electric conductivity, making them suitable for use in portable electronics. Nonetheless, MXene-reinforced composites are made using a variety of procedures, spanning from solution mixing to powder metallurgy-based tactics. A brief description of the existing production techniques for MXene-based composites (Figure 7.2) is presented in this section.

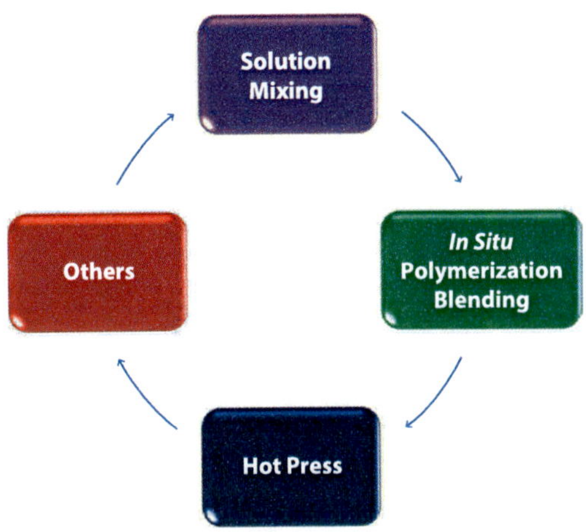

Figure 7.2 Various production techniques for MXene-based composites.

7.4.1 Solution Mixing

Because of the hydrophilic nature of MXene nanosheets, equipped with functional groups, the majority of MXene-reinforced polymer nanocomposites have been generated through solution mixing procedures too far. MXene nanoparticles are frequently disseminated in polar media such as H_2O, dimethyl sulfoxide (DMSO), N, N-dimethylformamide (DMF), and others in this technique; polymer substance could also be dissolved in the very same dispersion or another with reciprocal solubilities. The polymer, as well as MXene mixtures, are then combined and merged to form a homogeneous mixture of MXene composites. It should be emphasized that the solubility of MXene in the presence of nonpolar polymers or mildly polar moieties remains a challenge, necessitating the use of an appropriate surface pre-modification to improve dissolvability. Various MXene nanoparticles have been found in a variety of polymers, including polyurethane, polyfluorenes, polylactic acid, polyethyleneimine, [43–47] Nafion, polyacrylamide, cellulose, polyethylene oxide, polyvinyl alcohol, chitosan, polybenzimidazole [48–54]. Even though solution blending is a simple procedure that takes benefit of MXene nanomaterials' hydrophilic nature, it has a couple of significant disadvantages, including the creation of excessive amounts of waste, poor mechanical properties affiliated with the resulting composites, and painstaking solvent evaporation.

7.4.2 *In Situ* Polymerization Blending

Under a wet processing situation, polymer monomers, activators, and sufficient curing reagents are properly integrated with MXene nano-flakes, and then these subunits are *in situ* synthesized [55, 56]. This approach could lead to well-dispersed MXene nanosheets in a polymeric matrix with a high interfacial stiffness at the particle-matrix contact, resulting in improved mechanical, Tribological, thermal, as well as electrical characteristics. The approach works best with thermosets, polymers with linear molecules, and polymers with heterocyclic compounds [57]. MXene plates have infused polythiophene, polyaniline, polydopamine, and polypyrrole via *in situ* synthesis [58–62].

7.4.3 Hot Press

When contrasted to polymers, MXenes have exceptional thermal stability and a much higher breakdown temperature [63]. MXene is mixed with a melted polymer matrix in a hot-press production process, where it is disseminated by a proper intensity stirrer at a regulated temperature, and then the melt is exposed to a hot-pressing procedure to generate the ultimate nanocomposites. Because the technology is solvent-free, versatile, cost-effective, and ecologically benign, it might be implemented in large-scale manufacturing techniques. MXenes have filled many types of polymers using hot-pressing processes, including UHMWPE, linear low-density polyethylene (LLDPE), polyurethane (PU), polystyrene (PS), polyaniline (PANI), as well as polyvinylidene fluoride (PVDF), [64–67]. When contrasted to solution-based procedures where the dispersibility in addition to the solubility of MXene flakes are insufficient, hot pressing is a more suitable production method for hydrophobic polymers.

7.4.4 Other Methods

Various MXene-reinforced composites have indeed been created to date, with the most popular ways succinctly listed above. Other approaches include powder metallurgy along with cold/hot pressing, layer-by-layer (LbL) assembly, vacuum-assisted filtration (VAF), emulsification, [68] electrospinning, as well as electrochemical deposition processes, to name a few [69, 70]. In addition, substantial investigation is being conducted to optimize the affinity of MXene sheets with polymeric matrix, resulting in a composite with improved targeted qualities. For MXene-based composites with a metal or ceramic host matrix, interfacial bonding is critical.

Sheng *et al.* [71] along with Cao *et al.* [72] utilized the melt blending equipment to make MXene/polymer nanocomposites in which the extruder as well as the internal mixer were employed as dispersion apparatuses. Certain polymers are created in emulsion systems, that are more commonly present in polymers employed in several coatings industries [73]. The emulsions are also utilized along with MXene in order to create MXene/polymer composites. The group in Nanjing Tech. University presented the method of producing various composites in the presence of PU + MXene. It was established that the PU emulsion as well as MXene were extremely dispersed that becomes extra stable after doing ball milling. Based on emulsion blending, Li *et al.* mix the two-roll mill procedure to make the very first MXene nanocomposites [74]. High-speed mechanical method was also employed for obtaining MXene/polymer nanocomposites, like Ti_3C_2 [38].

Moreover, the formation of MXene/Polymer Composites also comprises (i) layering composites which includes filtered deposition, spin, as well as spraying coating; (ii) composite products for example foams produced from freeze-drying while fabrics made through electrospinning method [75–77]. Layered composites are mainly utilized in sensing, purification of H_2Omaswell as in numerous electromagnetic protection solicitations. Zhou along his team explored that spin coating of anionic Ti_3C_2Tx as well as cationic MWCNT results in the formation of modified PCL film which inter is used for the production of Ti_3C_2Tx/PCL as well as MWCNT/PCL laminate after which these two are slacked in to a layered composite films utilized in energy storage electrode [40, 78]. Yang *et al.* produced MXene/PVB//Co_2Z/PVB slaked composites used in electromagnetic sheltering [79]. In addition to this, layered composites also bears a polymer-fiber filter layer as substrate by keeping Mxene suspension on its surface via filtering and results in the production of some layered composite substances. For instance, Rasool along his co-workers described the usage of a PVDF fiber filter as a substrate used in the production of numerous MXene layered composites [80]. More over some MXene fiber composites were also reported. Jiang and his team circulated flexible nanogenerators which exhibited excellent power generation working [81]. Cao *et al.* presented various types of MXene coated fiber composites including a fiber composite known as cellulose nanofiber [82]. Mayerberger *et al.* [83] reported various types of MXene/polymer nanocomposite fibers and combine them with MXene/poly(acrylic acid)solutions via electrospinning method. A MXene/CS composite fibers were also produced as described in previous studies [84].

7.5 Characteristics of Polymer/MXene Composites

There are numerous procedures of creating MXenes, and merging them into polymers to alter their features. MXene/polymer nanocomposites profit from the desirable characteristics of MXenes as well as the adaptability and simple processing of polymers. In comparison to other nanocomposites, these composites have demonstrated superior features like high light-to-heat conversion, efficient electromagnetic interference shielding, as well as high charge storage. This section summarizes a few properties of the Polymer/MXene Composites, such as their mechanical, thermal, tri-biological properties, etc. as shown in Figure 7.3.

7.5.1 Structural Properties

Despite the fact that MXene may give polymer matrices exceptional qualities, it is still necessary to modify MXene in order to increase its excellent performances to fulfill the growing need for variety as well as diversity in functional polymers. The 2D architecture of MXenes and their tunable functional groups and layer spacing make it simple to incorporate the exceptional features of various modifiers to further enhance MXene/polymer composites. MXenes can be difficult to apply since they are simple to rearrange and oxidize in the air, leading to the production of TiO_2 [85]. The foregoing problems can be effectively solved by the functional transformation of MXene, which can also provide MXene outstanding dispersibility and compatibility in polymer matrices to produce polymer composites with extraordinary performance. According to the literature, MXene-based nano-fillers have enormous potential for producing polymer composites that function superbly. Tanvir *et al.* investigated the polymeric nanocomposites modified by 2D $Ti_3C_2T_x$ (MXene) characteristics and revealed that with increase in the MXene content the storage modulus of the composites increase rendering these materials promising for various applications. Moreover the structural pattern of the neat coPA films and composites of coPA and MXene were analyzed by XRD as shown in Figure 7.4, which mainly approves the amorphous nature of substance, with small quantities of the crystalline phase [86].

Moreover, MXenes give composite materials based on polymers better characteristics [87]. MXenes offer excellent potential as composite components with a range of polymers due to their diverse chemistry. The innovative Mxene/alginate composites were created by Dong *et al.* [88] who also evaluated their potential for removing lead and copper ions from wastewater. The functional moieties of sodium alginate and the Mxene/alginate composites

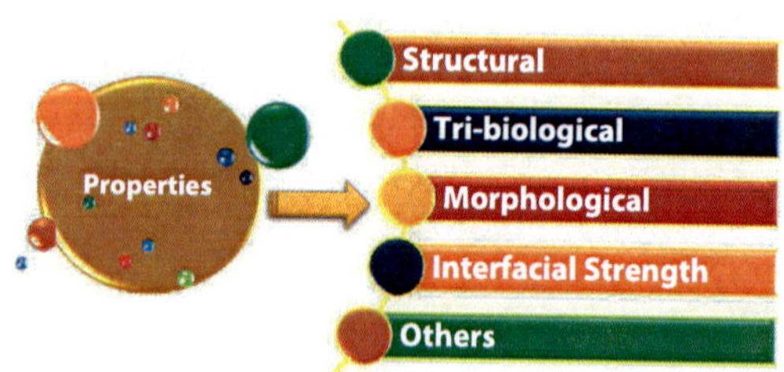

Figure 7.3 Numerous characteristic properties of polymer/MXene composites.

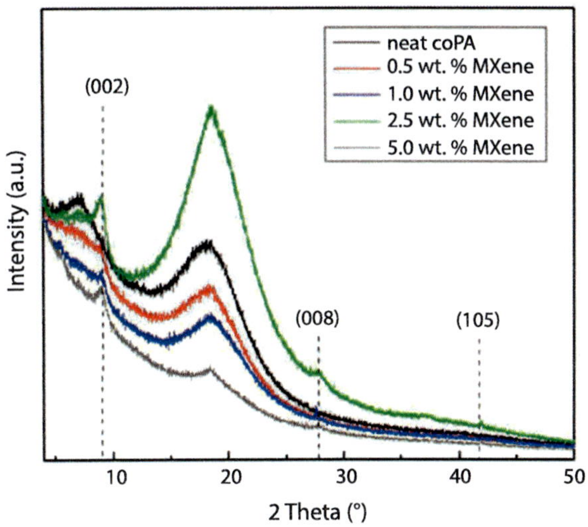

Figure 7.4 XRD spectra of the coPA/MXene films [86]. Copyright 2019. Reproduced with permission from MDPI.

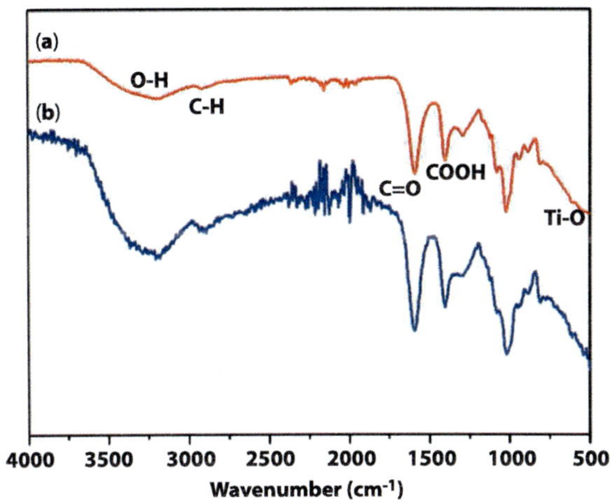

Figure 7.5 FT-IR spectra of (a) Mxene/alginate composites and (b) sodium alginate [88]. Copyright 2019. Reproduced with permission from RSC.

are investigated using FT-IR analysis as shown in Figure 7.5. The spectra of the Mxene/alginate composites show a novel adsorption peak at 596 cm^{-1} that contrasts with the infrared spectrum of sodium alginate, indicating that the sodium alginate was effectively absorbed into the Mxene. The lead and copper ions' ability to chelate is improved by the Mxene/alginate composites created in this work, which also improve ion transport effectiveness.

Also, Spherulites' development and the behavior of polymeric materials during crystallization could both be dramatically impacted by MXenes [89]. Liu *et al.* [90] explored that the insertion of $Ti_3C_2T_x$ significantly altered the glass transition temperature (Tg); additionally, the mechanical aspect increased by 23% from 104.6 MPa for the neat Nafion to

128.4 MPa for the composite specimen, most likely as a result of the elevated aspect ratio and -OH termination units of MXene layers that provided H-bonding interconnections. Compared to graphene, MXene's functional moieties like $-O_2$, -OH, and -F, have a much higher affinity for polymeric components. It should be mentioned that the usage of pristine graphene is typically restricted in composite fabrication caused by the lack of surface terminations. Because of its hydrophilic nature, MXene sheets have a greater ability to wettably adhere to a wide range of materials [91]. Zhang *et al.* [92] examined a range of characteristics, primarily the structural characteristics of 1.0 wt.% Epoxy composites reinforced with Ti_2CTx MXene report fracture toughness, flexural strength, and impact strength of 17.8 kJ m^{-2}, 98 MPa, and17.8 KJ m^{-2} with improvements of 76%, 66%, and 76%, respectively. The main causes of the enhancements in characteristics were believed to be good interfacial compatibility among the host epoxy matrix and the well-dispersed reinforcing particles. Additionally, it was discovered that when the epoxy substance was strengthened by the nano-sized 2D MXene sheets, the wear rate was dramatically decreased and the antifriction qualities extensively increased. Despite the fact that MXenes are hydrophilic substances, Raagulan *et al.* [93] were able to create a nanocomposite coating by using graphene oxide as well as MXene as reinforcing agents in the PVDF matrix. The resulting coatings showed extremely high adaptability, lightweight, wettability (126° contact angle), and (41 dB and 99.99 % efficiency).

Compared to a Ti_3C_2 reinforced UHMWPE nanocomposite, the aforementioned composite, demonstrated superior toughness, creep resistance, and abrasive qualities [38]. The best overall thermal characteristics were found in 0.5 wt. % d-Ti_3C_2 strengthened polyvinyl alcohol (PVA) composites made using a solution casting technique, according to Liu *et al.* [94]. The findings suggest that the use of MXene reinforcing nano-sheets improved dramatically the tensile strength at 0.5 weight % reinforcement, the expansion at break, as well as the strain-stress performance. The aforesaid mechanical improvements were achieved through hydrogen bonding at the reinforcement-matrix interface and uniform d- Ti_3C_2 distribution. Polyurethane nanocomposites packed with various amounts of d- Ti_3C_2 MXene nano-flakes were examined by Wang *et al.* [95].

When 1.0 wt. % of the d- Ti_3C_2 loading was used, the specific strength, elongation, as well as dissipated energy statistics increased by 29 %, 53 %, and 3.4 %, respectively. The researchers thought the characteristics of the created composites could still be enhanced if a suitable binding moiety were used among the d- Ti_3C_2 MXene and the PU matrix. MXenes have already been thought of as a functionalizing agent for nano reinforcing particles and fibers due to their rich physicochemical characteristics. In attempt to improve the water solubility between the carbon fiber and epoxy matrix material, Liu *et al.* [96] used Ti_3C_2 MXene nanolayers. This enhanced wettability led to greater engineering properties; MXene flakes could efficaciously improve the interface behavior and topography of the reinforcing carbon fibers and enhanced the interfacial shear capacity of the functionalized specimen by about 186 % in comparison to that of non-functionalized composite. The hardness, yield stress (YS), as well as TS of the MXene-filled polyurethane (PU) nanocomposites studied by Zhi *et al.* [97] were increased by 10%, 70%, and 20%, respectively, as a result of a robust interfacial communication between the MXene and the PU, which was confirmed by FTIR analysis. The relevance of homogeneous scattering and the appropriate ball milling procedure of MXene substances was underlined in this study because they have a substantial impact not only on the mechanical properties but also on other characteristics like the reduction of water absorption.

The homogeneous dispersion and appropriate ball milling of MXene materials were addressed in this study because they have a major impact on not only the materials' characteristics but also on other aspects like the reduction of water absorption.

Zhang and team [98] engineered TiO_2/Ti_3C_2 composite particles. Using an XRD pattern (Figure 7.6), the phase composition of the prepared sample was examined. From the pattern of Ti_3AlC_2, it is clear that three significant diffraction peaks are present. Peaks at 9.6, 39.1, and 41.9 degrees are each strongly indexed to the (002), (104), and (105) lattice planes of hexagonal Ti_3AlC_2, respectively. The distinctive Ti_3AlC_2 (104) diffraction peak has completely vanished after HF etching. Additionally, the (002) diffraction peak broadens and shifts to lower angles in comparison to where they were originally, which shows that Ti_3AlC_2 has transformed into Ti_3C_2. Following hydrothermal processing, the TiO_2/Ti_3C_2 composite particles show three additional diffraction peaks, with peaks at 25.3°, 37.9°, and 48.0° corresponding to the (101), (004), and (200) lattice planes of anatase, respectively. A further significant finding is that the molar mass of TBOT has a significant impact on the intensity of the (101) diffraction peak at 25.3°, indicating that the density of TiO_2 nano-dots can be controlled by the TiO_2 precursor solution. Four distinct adsorption bands may be seen in Ti_3C_2 and TiO_2/Ti_3C_2 according to FTIR spectra. The findings show that the spectra of TiO_2/Ti_3C_2 exhibit a prominent adsorption band at 542 cm^{-1}, which is attributed to the stretching vibration of TiO_2's Ti-O bond.

Through the use of a melt blending approach, Sheng *et al.* [71] entangled different concentrations of Ti_3C_2 MXene nano-flakes to the thermoplastic polyurethane component demonstrated a significant degree of improvements in the characteristics (thermal). Before being added to the TPU matrix, the Ti_3C_2 MXene constituents were magnetically combined with polyethylene glycol (PEG) to widen the interlayer spacing's of MXene layers. Surprisingly, the findings show that Ti_3C_2 increases the PU matrix's fracture toughness in all weight fractions. It also became clear that tiny weight fraction additions might significantly improve the hardenability of Ti_3C_2 filled polyurethane composites. It was determined that 2D MXene flakes have a good prospect to be utilized in new composite production methods due to the expanded MXene interlayer lamina offered by PEG pretreatment, well exfoliation and homogenously dispersion, and good hydrophilicity of MXene flakes via the contacts amid $-O_2$ and -OH units of the MXene layers and that of those charged functional parts of the polyurethane polymeric chains [71, 99]. Sliozberg *et al.* [100] established an MXene reinforced thermosetting polymer matrix (diglycidyl ether of bisphenol A

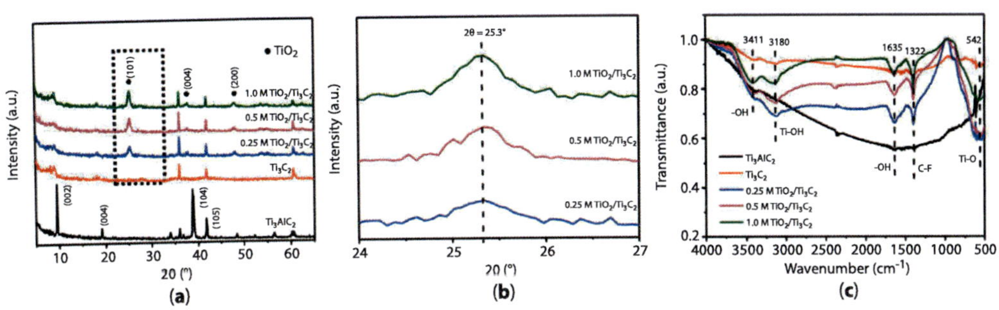

Figure 7.6 XRD patterns (a,b) and FTIR spectra (c) of Ti_3AlC_2, Ti_3C_2, and TiO_2/Ti_3C_2 [98]. Copyright 2021. Reproduced with permission from MDPI.

resin) with the goal of increasing the flexural and cracking tenacity of thermosetting epoxy polymer. They did this by using a DFT strategy and coarse-grained molecular dynamics (MD) simulations to comprehend the fundamental fracture processes and anticipate the interfacial adhesion strength in the MXene-reinforced. They discovered that the interfacial interaction is largely unaltered by the strengthening MXene groups depending on their computational and experimental work ($Ti_3C_2T_x$ or Ti_2CTx). Efficient interfacial potency was obtained in the event of decreased H- coverage of the $Ti_3C_2T_x$ outer layer due to the higher electrostatic interactions.

The ABAQUS finite element analysis (FEA) showed that the insertion of 30 vol. % associated $Ti_3C_2T_x$ MXene nanolayers to a PVA matrix tends to result in a markedly better TS of 91.4 % in addition to the elasticity modulus of 743 %. Monastyreckis and coworkers [101] devised a micromechanical modeling of multilayered MXene-reinforced PVA matrix NC. Investigations were done into how size of the particles, granular density divisions, and alignment levels affected mechanical characteristics. It has been shown that epoxy composites reinforced with sheets of -OH terminated Ti_3C_2 MXene as well as graphene (MG-EP) have the potential to significantly reduce wear rate. This wear lessening is not only significantly greater than that of neat epoxy (pure-EP), but also significantly greater than that of specimens loaded with either graphene (G-EP) or MXene (M-EP) reinforcing units [102]. The improvement in wear behavior is likely brought on by the 2D MXenes' superior thermal properties, high load-bearing capacity, wrapping structure, and the tribo-films that formed amid the mating surfaces at the time of wear tests. Although the clean polymer was primarily brittle when it cracked, the inclusion of reinforcing chemicals changed how the fractured surfaces looked. The study clearly shows that the pure epoxy has a nine-fold higher wear rate than the MG-EP covering nanocomposite. The results also demonstrates the presence of visible furrows as well as lengthy crack tracks which suggest that the unreinforced sample has more severe ploughing damage, significant plastic buckles, and shallow abrasions than the reinforced varieties. When 0.5 wt. % $Ti_3C_2T_x$ was utilized, Chen and associates [103] created a $Ti_3C_2T_x$ MXene reinforced PLA matrix bio-composite with the ability to regenerate bone and a superior TS of 72 MPa, which is an improvement of 33 % over plain PLA. Moreover, Chitosan was used by Hu *et al.* [104] to strengthen Ti_3C_2 films, resulting in a 430 % rise in strength properties from 8.20 to 43.52 MPa. Nevertheless, resistance value has risen substantially with chitosan concentration levels, rising from 0.39 mΩ cm for unudeltered MXene to 54.91 mΩ cm for the MXene-reinforced composite having a 14 wt. % chitosan additive.

Surprisingly, because nanotechnology is a rapidly expanding subject, MXene/polymer composites have very promising futures and can be further investigated in all theoretical, computational, and experimental studies from atomic to macroscopic scales. The structural characteristics of MXene-based nanocomposites still need to be thoroughly characterized and optimized before good targeted materials can be produced.

7.5.2 Tri-Biological Properties

Due to attributes including simplicity of introducing reinforcements, particularly nanoparticles, to their matrices, thermosetting polymers are extensively employed as the matrix of composite materials in numerous solicitations. Some research has been done so far on the inclusion of MXenes into thermosetting polymers [105]. MXenes are the best alternative for increasing the mechanical and tribological capabilities of various polymers because of their layered architecture and the availability of functional moietes including OH (hydroxyl), O_2 (oxygen), and F (fluorine) units. Epoxy resins are one of the most widely employed thermosetting polymers with a wide range of uses. Epoxy resins are widely used in tribological solicitations like polymeric gears, bearings, and coatings, therefore improvements in their mechanical, as well as tribological performance, are critical. Some studies have found significant improvements in the mechanical as well as tribological operation of epoxy composites by including MXenes in their matrixes. Zhang and colleagues [98] created TiO_2/Ti_3C_2 composite particles, as elaborated in section 7.5.1, to enhance the thermo-mechanical and tribological characteristics of epoxy resin. According to the results (Figure 7.7X) of a tribological test, adding TiO_2/Ti_3C_2 can significantly reduce the rate at which epoxy resin wears out. The medium density TiO_2/Ti_3C_2/epoxy nanocomposites among them saw the lowest

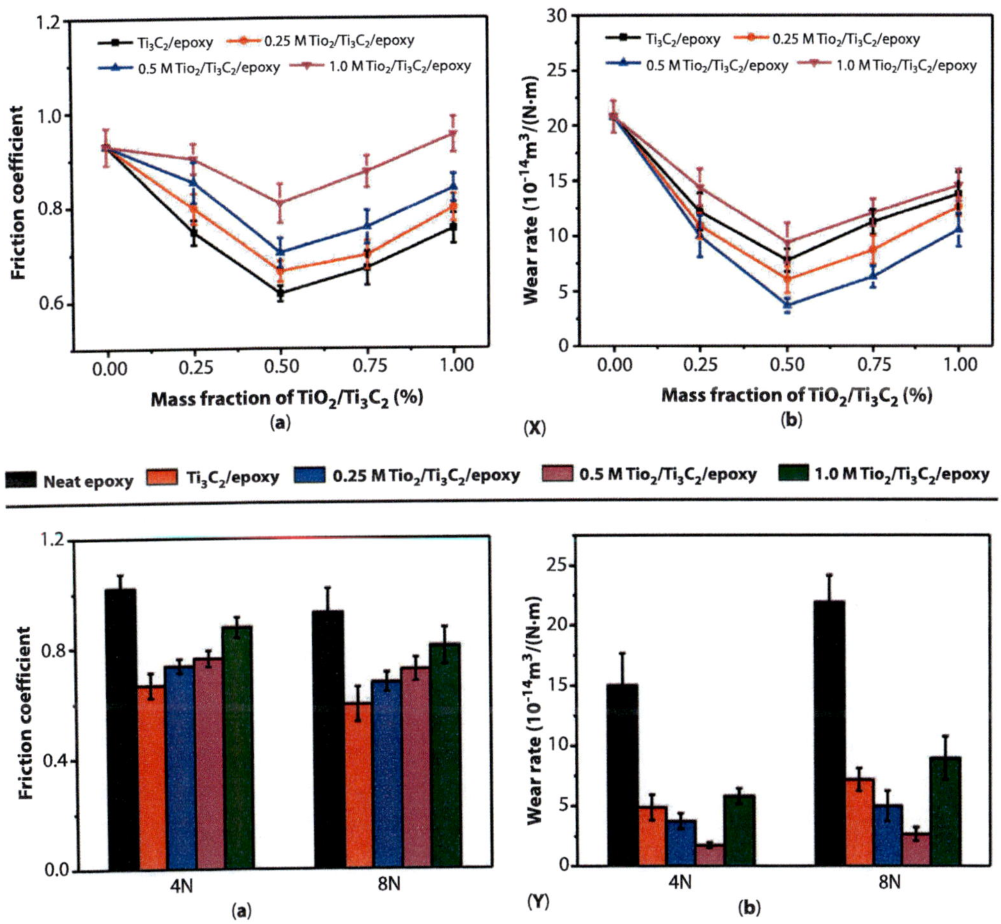

Figure 7.7 (X) Schematic representation of (a)Friction coefficient and (b) wear rate of TiO_2/Ti_3C_2/epoxy nanocomposites with different mass friction under a normal load of 8 N, and (Y) Friction coefficient (a) and wear rate (b) of neat epoxy, Ti_3C_2/epoxy, 0.25 M TiO_2/Ti_3C_2/epoxy, 0.5 M TiO_2/Ti_3C_2/epoxy, and 1.0 M TiO_2/Ti_3C_2/epoxy nanocomposites at 0.5% mass fraction under the normal load of 4 N and 8 N [98]. Copyright 2021. Reproduced with permission from MDPI.

wear rate. This may be explained by the fact that the medium-density TiO_2/Ti_3C_2 and the epoxy matrix, which can withstand a greater than usual shear stress during sliding friction, were mechanically interlocked by the moderate TiO_2 nanodot protuberances on the Ti_3C_2 surface. The wear form was gradually changed from fatigue wear in neat epoxy to abrasive wear in TiO_2/Ti_3C_2/epoxy nanocomposites, according to the morphologies of worn surfaces and wear debris.

Furthermore, the study found that, regardless of the normal load of 4 N or 8 N, 0.5 M TiO_2/Ti_3C_2/epoxy nanocomposite experiences the lowest wear rate (Figure 7.7Y). This finding may be explained by the fact that the medium density TiO_2 nanodots grown on Ti_3C_2 offer a plentiful supply of fasteners for the robust communication between the TiO_2/Ti_3C_2 composite particles and epoxy matrix. A increased shear stress induced by the enlarged normal load caused more epoxy granules or additives to come loose from the exterior of epoxy nanocomposites, which is why the wear rate of all the specimens under normal load of 8 N is significantly greater than that under the normal load of 4 N.

Zhang *et al.* [92], for instance, investigated the mechanical as well as tribological characteristics of Ti_2CT_x MXene/epoxy nanocomposites. They employed various weight percentages of Ti_2C MXene ranging from 0.5 to 2.0. The findings demonstrate that the Al element was selectively etched and the layered 2D Ti_2CT_x MXene was successfully produced. Furthermore, proper Ti_2CT_x dispersion in the epoxy resin, as well as a firm contact, have been identified. Because epoxy molecular chains were intercalated into Ti_2CT_x layers, the fissure surface of Ti_2CT_x/epoxy nanocomposites was coarser as compared to pure epoxy, according to their fracture surface studies. The introduction of MXenes to the epoxy increases its toughness, which can be taken as evidence of improved micro-hardness. They claimed that after curing, the molecular manacles of the epoxy resin were grafted with the –OH units of the Ti_2CT_x MXene compounds, forming a 3D structure with a robust Ti_2CT_x/ epoxy interface. Furthermore, the existence of functional moieties on the architecture of the MXenes was linked to the reported excellent dispersion of Ti_2CT_x particulates in the epoxy matrix. However, variations in the mechanical characteristics of the composites, like fracture toughness, tensile properties, as well as flexural modulus, as well as the composites' friction coefficient levels, were detected as a function of Ti_2CT_x. Owing to the mobility of molecular manacles on the interface of the epoxy polymeric material, the micro-hardness of samples decreased from 23.8 Hv to 18.2 Hv when Ti_2CT_x concentration was 1.0 wt. %. However, increasing the concentration of Ti_2CTx improved the fracture toughness of Ti_2CT_x/epoxy nanocomposite due to the capacity to sustain a load and withstand the distortion of excessive pressure by a set quantity of Ti_2CT_x. The inclusion of Ti_2CT_x had a favorable effect on the impact and flexural modulus results. Because the mass content of Ti_2CT_x was 1.0 wt. %, the tensile properties, and flexural modulus were enhanced by 76 and 66 %, respectively. The excellent bonding among the Ti_2CT_x nanosheets as well as epoxy resin attributed to the prevalence of functional moieties is the fundamental explanation for the indicated enhancements in the compressive and flexural strength of the composite. In other terms, because of the robust Ti_2CT_x/epoxy junction and the Ti_2CT_x 's specific strength, interface debonding and fracture propagation demand a huge amount of energy, which improves the composites' strength and durability. It was reported that the Ti_2CT_x had the greatest impact on the composites' tribological behavior. With the 2.0 % mass fraction incorporation of Ti_2CT_x, the friction coefficient of the epoxy nanocomposite dropped by more than double as compared to pure epoxy with a sliding velocity of 3×10^{-1} m/s.

The functionalization of nanoparticles is an effective method for improving dispersion in polymeric matrixes. A novel epoxy-Ti_3C_2 with three-dimensional nanosheets (3DNS) composite was created by MENG *et al.* using the freeze-drying process [106]. The findings indicated that when compared to epoxy resin, composites have improved CTE, TC, and fracture toughness. Additionally, epoxy-Ti 3C2 3DNS with 3.0 g nanosheets had the best thermal (TC and CTE) and frictional (friction coefficient) capabilities, which were crucial for pragmatic application in high-performance devices. Figure 7.8 displays the histogram of the friction coefficient for epoxy resin and epoxy- Ti_3C_2 3DNS composites with various contents of few-layer Ti_3C_2 MXene. Figure 7.9, on the other hand, depicts the wear rate and wear volumes of epoxy resin and epoxy- Ti_3C_2 3DNS composites with various contents of few-layer Ti_3C_2 MXene. Both the results demonstrate that the three-dimensional structure of few-layer Ti_3C_2 MXene can significantly enhance the wear and friction capabilities of the epoxy resin matrix. Epoxy- Ti_3C_2 3DNS 1.0 composites have the maximum friction coefficient and the minimum wear rate, both of which are 76.3% lower than those of epoxy resin. Epoxy- Ti_3C_2 3DNS 3.0 composites have the lowest wear rate and are 67.3% lower than those of epoxy resin.

Furthermore, Yan and his associates [107] investigated the anti-wear capabilities of epoxy coverings covered with single-layer $Ti_3C_2T_x$ sheets and amino-functionalized $Ti_3C_2T_x$ nanosheets. By adding $Ti_3C_2T_x$ sheets to epoxy coatings, the tribological performance of the coatings shows a declining trend, which is observable for friction coefficient and wear rate values. Higher $Ti_3C_2T_x$ content and functionalization, on the other hand, had a greater impact on coating wear resistance. Adhesive wear has been proposed as the principal wear procedure of unadulterated epoxy coatings, according to Atherir's studies. Adhesive bonding is caused by substantial plastic distortion and a larger interaction between the epoxy covering as well as the counter-body. Nevertheless, by improving the hardness and strength of the specimens, the introduction of $Ti_3C_2T_x$ lowers plastic deformation and, as a result, the attrition rate of the specimens. When contrasted to as-prepared $Ti_3C_2T_x$, amino-functionalized

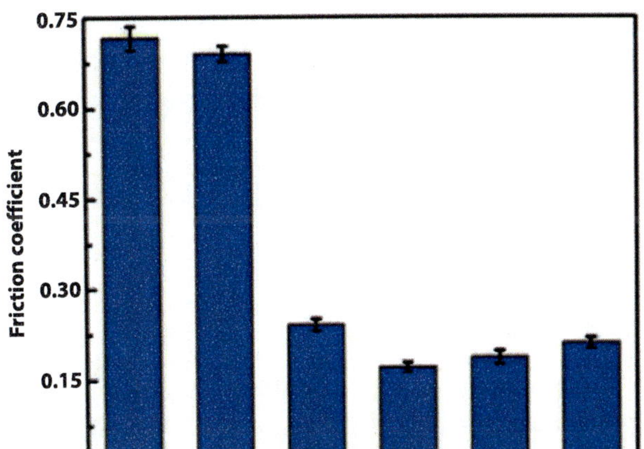

Figure 7.8 Friction coefficient for epoxy resin and epoxy-Ti_3C_2 3DNS composites [106]. Copyright 2021. Reproduced with permission from Springer.

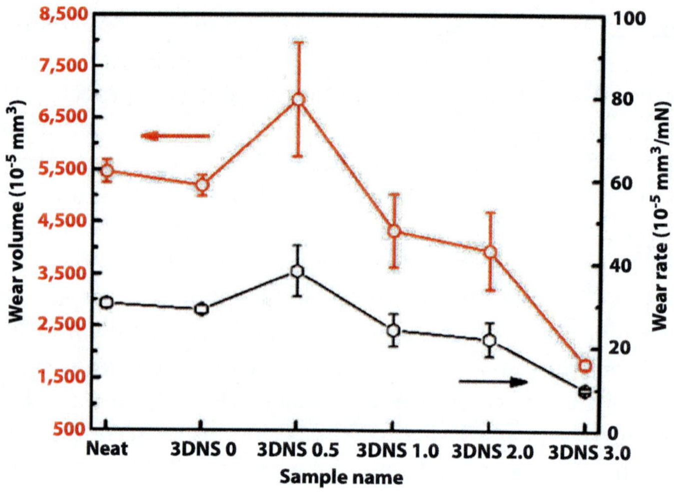

Figure 7.9 Wear rate and wear volumes of epoxy resin and epoxy-Ti_3C_2 3DNS composites [106]. Copyright 2021. Reproduced with permission from Springer.

$Ti_3C_2T_x$ has a better influence on improving the tribological features of the nanocomposite due to its superior dispersion and contact with the epoxy. MXene's synergistic effects with other nanoparticles have rarely been studied. Yan *et al.* [102], for example, investigated the synergistic benefits of $Ti_3C_2T_x$ MXene as well as graphene as epoxy coating supports in increasing tribological performance. The tribological experiments were conducted under a variety of situations, including friction counterparts (Al_2O_3 and Si_3N_4 balls) as well as test conditions (dry sliding plus 3.5 wt. % NaCl solution). In general, their findings revealed that the inclusion of both $Ti_3C_2T_x$, as well as graphene, had a greater impact on the epoxy coating's wear resistance in all scenarios. Furthermore, in dry conditions, the wear rates of $Ti_3C_2T_x$/epoxy, graphene/epoxy, and $Ti_3C_2T_x$/graphene/epoxy coverings were 271 × 10^{-2}, 29 × 10^{-1}, and 9 times lesser than that of pure epoxy veneers. Furthermore, at 3.5 wt. % NaCl, the friction coefficients of pure epoxy coating, $Ti_3C_2T_x$/epoxy, graphene/epoxy, and $Ti_3C_2T_x$/graphene/epoxy coating were 14 × 10^{-2}, 10 × 10^{-2}, 9 × 10^{-2}, and 8 × 10^{-2}, respectively. The wear rate of the $Ti_3C_2T_x$/graphene/epoxy coating was nearly 26 times lower than that of the corresponding epoxy coating when uncovered to 3.5 wt. % NaCl. The improved wear resistance of $Ti_3C_2T_x$/graphene/epoxy can be attributed to their synergistic impacts in the development of tribo-layers on the interface of epoxy coverings, which improves the powered as well as tribological effectiveness [102].

7.5.3 Morphological Properties

A composite is a material made up of two or more physical components that have much greater qualities than the individual components, such as the matrix and filler. Composites are typically made of several matrices, like polymers, metals, ceramics, etc. Chemical resistance, low density, good wettability, and ease of moulding into desired and engineered shapes are among the benefits of polymers [108]. As a result, polymers are utilized as matrices more frequently than metals and ceramics. Polymers, on the other hand, have lower strengths and moduli. The insertion of additives and natural fillers (of plant, animal, and mineral origin) is the most common strategy for improving mechanical qualities to solve this difficulty. However, nanomaterials being a new type of filler with unique characteristics and nanostructures, are perfectly proficient for usage in polymer matrix composite forms. As a result, investigational and hypothetical research on MXene/polymer has substantial theoretical and real-world implications. MXene (Ti_3C_2) was investigated by Kang *et al.* as a reinforcing addition to enhance the thermal characteristics of polymers. Using a FE-SEM, it was possible to examine the microcosmic morphologies of the samples and determine how Ti_3C_2 MXene was dispersed in the epoxy matrix and how it adhered to the epoxy matrix on the interfacial surface. The findings show that the stripes that form on the fracture surface are shaped like rivers and that the spaces between the stripes are exceptionally smooth Figure 7.10(a). These stripes are a sign of brittle thermosetting polymer, which is normal. There are also some stripes on the fracture surfaces, as seen in Figures 7.10(b-f), but Ti_3C_2/epoxy composites have coarser fracture surfaces than plain epoxy [108].

Nanomaterials including ZnO, Al_2O_3, carbon nanotubes, and graphene have all been researched comprehensively as fillers for UHMWPE. Because of nanoscale effects, the results demonstrate that adding nanoparticles can significantly improve thermal and morphological qualities. However, 2-D nanomaterials, like MXene, have sparked a great flurry of basic and technological research and have emerged as interesting alternatives for usage as reinforcement in polymer nanocomposites (PNCs). Si and coworkers also explored several properties of polystyrene/MXene nanosheets. The results demonstrate that the qualities of amended Ti_3C_2/PS nanocomposites were significantly enhanced in terms of thermal stability and flame-retardant capabilities. By using SEM, the dispersion of Ti_3C_2 modified with DTAB, OTAB, and DDAB in PS was investigated. The outcomes are displayed in Figure 7.11. The improvements in PS nanocomposites' properties are attributable to the better catalytic, barrier, and dispersion properties of Ti_3C_2 nanosheets [109].

Sobolčiak *et al.* used an electrospinning process to effectively generate Novel 2D Ti_3C_2Tx-strengthened PVA nanofibers and tested them for various qualities [110]. The electrospun nanofibers were made with a mixed mixture of PVA comprising varying wt. the percent of MXene as well as CNC. The physicochemical characteristics of $Ti_3C_2T_x$CNC/PVA aqueous blends were blended to be excellent. The mixed ratio nanofibers were electrospun directly from the well-dispersed aqueous solutions. Demonstrative SEM micrographs of the four nanofiber composites show that the nanofibers are homogeneous and smooth, with thicknesses ranging from 174-194 nm, with no beads, in addition to the uniform dispersion of fiber thickness. The nanofiber diameter reduced as the $Ti_3C_2T_x$ ratio in the composite increased, as perceived for blends C1M1, C0M2 than that of those encompassing only PVA (C_0M_0). The normal widths of the nanofibers evaluated by SEM for C_0M_0, C_2M_0, C_0M_2, and C_1M_1 were 182 (±53), 195 (±33), 174 (±53), and 179 (±36) nm, respectively. Identifying the Ti unit in the EDS spectra reported the existence of Ti_3C_2Tx within C_0M_2 as well as C_1M_1 electrospun nanofibers. The distribution of Ti_3C_2Tx nanoparticles within the PVA matrix was confirmed by EDS mapping of C_0M_2 possessing Ti_3C_2Tx (green hue). The outer shape of the fibers was also investigated using AFM. The samples outperformed axially aligned fibers with relatively smooth surfaces, according to the data (2D AFM-phase as well as 3D AFM-height images). In the occurrence of CNC/Ti_3C_2Tx that were appropriately positioned within the fibers, no apparent aggregation was seen. Furthermore, no significant changes in roughness were identified in longitudinal or lateral directions, indicating that they were

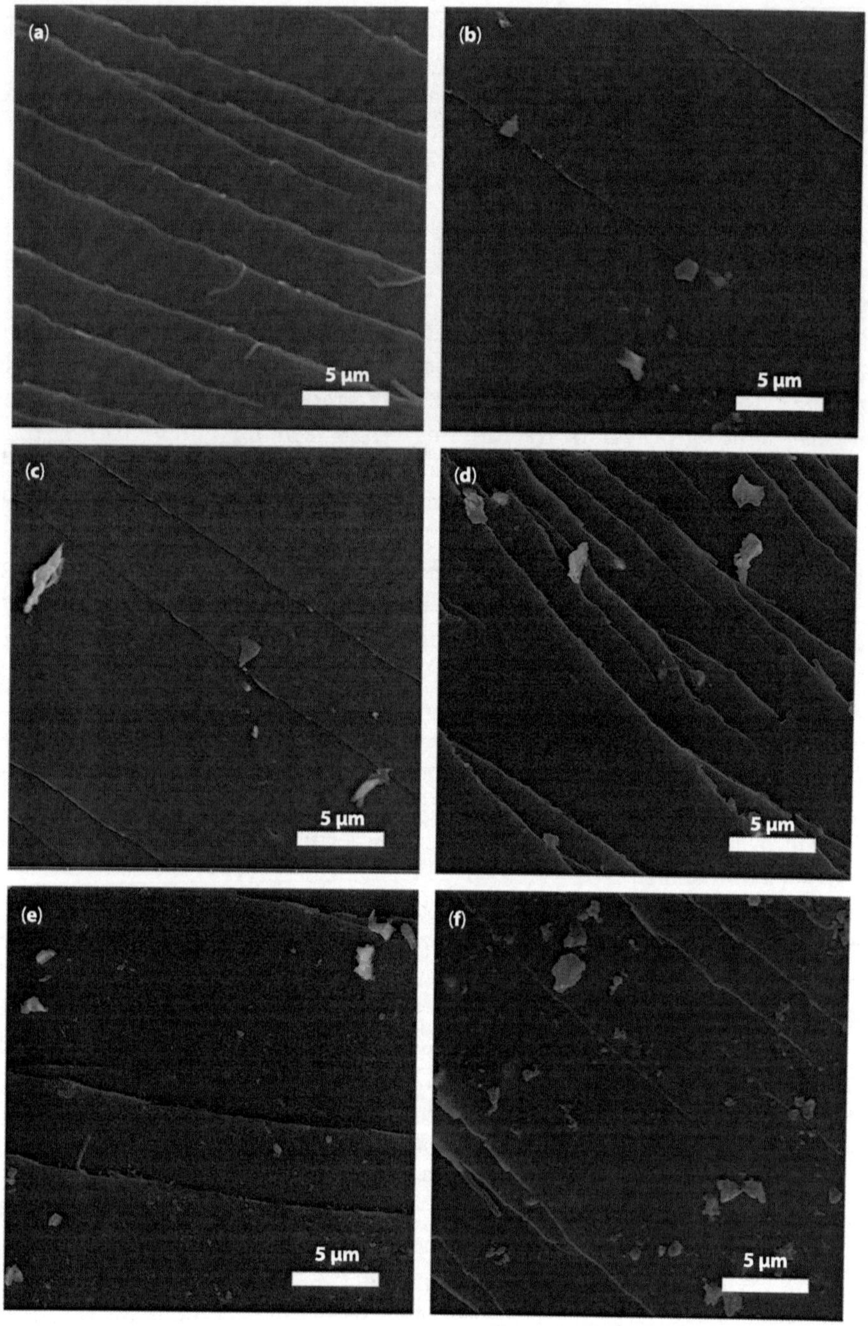

Figure 7.10 FE-SEM images of the fractured surfaces of the Ti_3C_2/epoxy composites with different amounts (a) 0 wt.%, (b) 0.2 wt.%, (c) 0.4 wt.%, (d) 0.6 wt.%, (e) 0.8 wt.%, and (f) 1.0 wt.%, respectively of Ti_3C_2 MXene. Copyright 2019. Reproduced with permission from Scientific Reports.

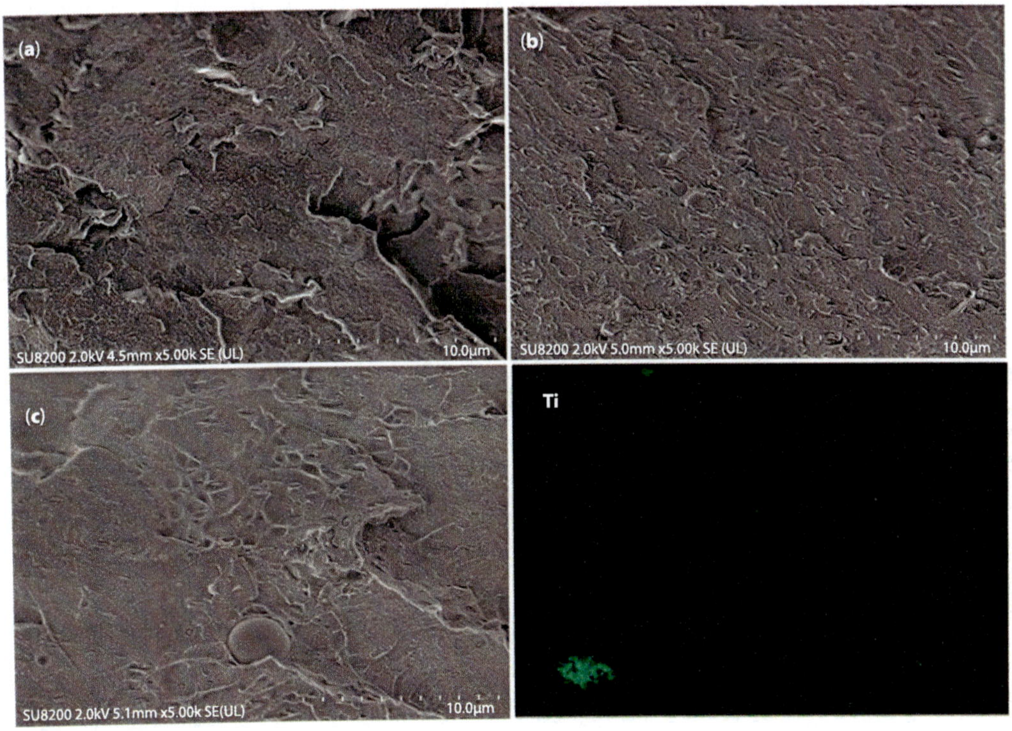

Figure 7.11 SEM images (a) polystyrene (PS)/DTAB-Ti_3C_2, (b) PS/OTAB-Ti_3C_2, (c) PS/DDAB-Ti_3C_2 and (d) the Ti element for PS/DDAB-Ti_3C_2 surface [109]. Copyright 2019. Reproduced with permission from MDPI.

well dispersed among the PVA fibers. However, some of the fibers have been observed to be bundled together, resulting in multi-fiber structures. In addition to structural characteristics mentioned above, the composites' morphology was also evaluated by the researchers [88]. The morphology of the original Ti_3AlC_2 (MAX) is erratic (Figure 7.12a). Ti_3C_2Tx (Mxene) exhibits a characteristic two-dimensional organ-like shape following hydrofluoric acid etching (Figure 7.12b). When alginate is introduced to Ti_3C_2Tx, it has a tendency to first occupy the interlayer before partially covering the surface (Figure 7.12c). The Mxene/alginate composites have a coarser interface than pure Ti_3C_2Tx , which creates a better setting for Pb^{2+} and Cu^{2+} adsorption.

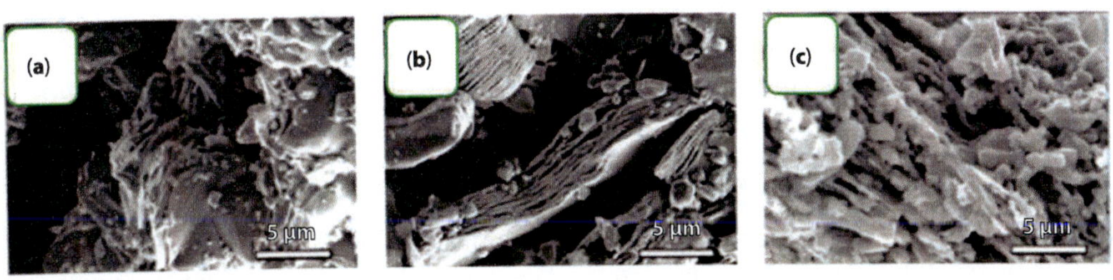

Figure 7.12 Surface morphology of (a) Ti_3AlC_2; (b) Ti_3C_2Tx; (c) Mxene/alginate composites [88]. Copyright 2019. Reproduced with permission from RSC.

The morphological features of MXene/PNCs is typically documented as "brick-and-mortar", which is advantageous for enhancing the mechanical characteristics of hybrid assemblies. According to "brick-and-mortar", a solo stratum of MXene behave as the brick while the polymer acts as mortar that expedites stress transfer to MXene fakes. Besides, the polymer acts as an adhesive to mend the mechanical characteristics of the MXene nanocomposites. But, Zhou along his team [111] presented that the multi-stratum configuration of an MXene/polymer system is functions more efficiently as compared to "brick-and-mortar" architecture for refining the mechanical features of MXene nanocomposites. They also revealed that the multi-stratum Ti_3C_2/CNF nanocomposites possess greater tensile strength, toughness as well strain that that of single stratum Ti_3C_2/CNF nanocomposites. In MXene based nanocomposites possessing "brick-and-mortar" morphology, various interfaces are present amid MXene flakes for wave dispersion as well as repetitive reflections. Accordingly, the hybrid nanocomposites works as an absorption-prevailing EMI shielding substance. Zhou and team [111] manufactured the single-stratum Ti_3C_2/CNF PNCs by employing another vacuum filtration method and obtain the EMI shielding ability of the nanocomposites rely on the sum of Ti_3C_2 as well as CNF stratums. Extreme protecting was achieved in the presence of four stratums of Ti_3C_2 as well as five stratums of CNFs. Above 90% of the incident EM waves were reflected immediately after accessing the arrangement because there occurs a misalliance in the conductivity as well as impedance of nanocomposites. Also, when the impedance mismatch at the edge amid the air and the nanocomposites upsurges, the possibility of reflection of the electromagnetic waves also rises. They explored that the multi-stratum Ti_3C_2/polymer systems are more operational than that of single-layer mixed Ti_3C_2/polymer systems because of greater impedance mismatch. Pure Ti_3C_2 foams having a hydrophobic interface are considered as exceptional EMI shielding substances for functioning in wet environments. H_2O soluble single-stratum Ti3C2 nanocomposites are also employed as EMI shields [112].

MXenes bearing encrusted morphology having a wideness of approximately 1 nm hold an extremely active surface comprising termination moieties like –COOH, –O, –OH, –F, allowing effective as well as positive interactions inside the assimilating matrix. Because of its lamellar assembly, the channels produced in the hybrid matrix offer choosy movement of ions as well as constituents [113]. Moreover, extraordinary solubility in H_2O laterally with specifically adjustable interlayer distance of MXene sheets may support the restriction of a particular particle but permitting other moieties to pass effortlessly over the porous inter-planar passages.

The stacking actions of MXenes, changeable interlayer distance as well as surface characteristics have produced countless possibilities to examine MXene-based composites for various membrane solicitations. Small membrane resistance is advantageous for individual separations. Besides, active surface of MXene may be chemically operationalized to alter specific interaction among various which depends upon the nature of the infiltrating moieties. For instance, membranes along with reversible carrier action are proposed that are proficient of interacting particularly with one gas constituent but permitting free penetration of other constituents in the mixture. Pristine Ti_3C_2 films showed diffusion control approaches initiating molecular sieving impacts which are perfect for the separation membranes. The single-stratum Ti_3C_2 MXene films showed a H_2 permeability, more than 2000 barrier surpassing the Robeson upper bound [114]. In separation of membrane gas, there is constantly a trade-off amid choosiness and penetrability. The function of a membrane during the separation of a gas pair is assessed, comparative to the new Robeson bound for the gas pair. A membrane whose isolating potency is more than Robeson bound is superior to that one whose separation capacity is less than that of bound. A self-supporting Ti_3C_2 MXene lamellar membrane was established to yield osmotic power produced via salinity gradient. These ion specific membranes displayed an osmotic energy conversion efficiency of 40% at 298.15 K. Ti_3C_2 MXene-based flexible polymer hybrids were established for CO_2 absorption by means of PEBAX-1657 through high CO_2/N_2 permeability more than 2008 CO_2/N_2 upper bound [115]. Liu and co-workers established a stable Ti_3C_2-based poly(ether-block-amide) hybrid membrane for CO_2 absorption through a Ti_3C_2 loading less than 0.15% and developed permeation rates more than ~22 GPU. In addition to this, single stratum Ti_3C_2 draped PAN fibers were manufactured for specific trapping of atmospheric matter facilitating single-step air decontamination in cycle showing anti-bacterial functionality [116]. The membranes showed comprehensive potency lifecycle combined having a constant architectural morphology. From other investigation, borate as well as polyethylenimine moieties intertwined amid Ti_3C_2 MXene sheets demonstrated H_2/CO_2 selective passage as well as separation [117]. MXene's extra performance as an active bacteriostatic entity may be extra gentle to assimilate and create multi-role sanitization as well as anti-biofouling capacities for membranes in industrialized scale air/H_2O management services.

7.5.4 Interfacial Strength

The enhancement in interfacial contact between the particles and the adjacent polymer matrix is often attributed to the reinforcing effect on MXene NCs. However, this relationship is rarely defined, and material qualities are often considered as an clue of enhanced interfacial strength. Taloub *et al.* [118] explored the impact of $Ti_3C_2T_z$ grafting on PIPD fibers by silane coupling, as mentioned earlier. When MXene was applied to the interface of these fibers, the IFSS improved by 61.5 percent, from approximately (25 - 40.6) MPa, an upsurge of around 15.6 MPa, as determined via the pullout of these fibers from an epoxy matrix. While this isn't a conventional NC because the interface is among the PIPD fiber and the epoxy relatively than the resilience of the contact surface amid the MXene outer-layer and the polymer matrix, the results indicate that surface functional moietes, as well as coupling agencies, can enhance contact as well as dispersion by restricting particle aggregation. NC yield stress rises when the filler load of 2D nanomaterials like graphene, MXene, in addition to layered silicates, among others, develops. This output stress is a result of the filler's and host's interfacial contact. For short fiber-encompassing PNCs, the dependency of TS on interfacial communication, fiber diameter, as well as fraction can be described using the following formula:

$$\sigma_c = \sigma_p + \left(\frac{Lf\tau}{2R} - \sigma p\right)\varphi_f \tag{7.1}$$

where τ denotes the interfacial stress transfer parameter, σ_c, and σ_p denotes the composite and polymer tensile yield stresses, L_f denotes the fiber length, and R denotes the fiber radius. For 2D sheets like MXene, this relationship can be simplified by replacing $\frac{Lf}{2R}$ with the aspect ratio (A_f). This statement can be stated more succinctly in accordance of relative tensile YS, i.e., the ratio of composite YS to unfilled polymer yield stress, which will be demarcated as σ_p, yielding the below equation.

$$\sigma_{R-1} = \left(\frac{(Af\tau)}{\sigma p} - \sigma p \right) \varphi_f \tag{7.2}$$

Values of τ should be modest with weak contacts, typically about < 10 MPa; nevertheless, fillers that are covalently linked to polymer manacles should have greater rates. Simulations of cross-linked PE/CNT NCs, for instance, indicate τ values ranging from 30 to 110 MPa. The quantifying and publishing of something like the interfacial stress transfer parameter would be a wonderful contribution to the area of MXene polymer NC survey, especially since this statistics can be collected from a standard tensile test providing the volume fraction is adequately altered.

Moreover, MXene-filled polyurethane (PU) NCs were fabricated thru an emulsion procedure by Zhi *et al.* [98], and a 5×0^{-1} wt. percent MXene additive-enhanced the hardness, yield stress (YS), as well as TS by 10%, 70%, and 20%, respectively, due to a strong interfacial interaction of MXene-PU, which was authenticated via FTIR analysis. The relevance of homogeneous dispersion and correct ball milling of MXene materials was highlighted in this study, as it has a major impact not only on mechanical characteristics but also on other features like H_2O absorption decrease [119].

7.5.5 Other Properties

In addition to the above-mentioned characteristics, MXene/PNCs are also showed excellent biocompatible, antibacterial characteristics, and filtration and separation uses, electromagnetic wave absorption and protecting characteristics, and flame retardancy. For instance Yu and his co-workers [120] informed that MXene may enhance the flame retardant characteristics of poly urethane. Addition of 2 wt% Ti_3C_2 improved by tetrabutyl phosphine chloride considerably declined the peak heat release rate (52.2%, from 1422 of TPU to 680 kW/m^2), peak smoke yield (57.4%, from 0.406 of TPU to 0.173 m^2/s), CO (41.6%, from 0.0226 of thermoplastic poly urethane (TPU) to 0.0132 g/s), and CO_2 (51.7%, from 1.353 of TPU to 0.653 g/s) peak emission of thermoplastic polyurethane (TPU), attaining a remarkable flame resistance. Gao's *et al.* manufactured a $Ti_3C_2T_x$/polyacrylonitrile nanofiber composite membrane to attain proficient elimination of PM2.5 in air (99.7% removal rate). These estimations confirmed the extraordinary inclusive functioning of the MXene/PNCs.

7.6 Novel Applications of Polymer/MXene Composites

The electrical, photonic, electrical, thermodynamic, tensile, and other physicochemical characteristics of polymer/MXene nanocomposites are fascinating and offer advantages in many sectors as shown in Figure 7.13 [121].

Power storage devices, electrodes, transistors, absorbers, photonics, sensor, and biomedical equipment have all used MXene-based nanoparticles [122]. Because of its electrochemical

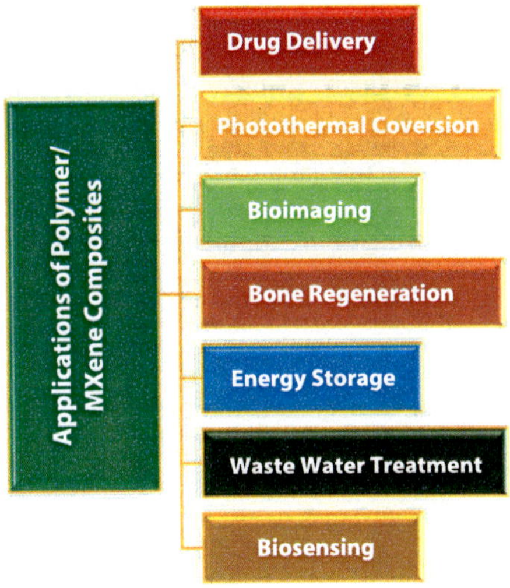

Figure 7.13 Applications of Polymer/MXene composites.

energy storage, MXene/PNCs have been employed in super-capacitor applications. The polymer/MXene NCss may have characteristics such as fluorine, OH, and others, as well as good ion intercalation behavior. Polymer/MXene has excellent mechanical qualities for electrode fabrication. Metal nitrides, carbides, and other developing two-dimensional inorganic compounds have extremely excellent electrical conductivity (8,000 Sm^{-1}) as well as specific capacitance [123]. MXene-based nanomaterials exhibit better capacitive performance than graphene, carbon nanotubes, carbon nanofibers, and other nanomaterials in certain circumstances. Producing malleable electrodes with elevated energy and power densities has proven difficult. Electrospinning was used to create the electrodes for the super-capacitor. The MXene-based nanofiber electrode has an enormous specific surface area as well as a high specific capacitance. The electrode containing 1% MXene nanofillers has a good energy stowage capacity. In addition, different fabrication methods for energy storage devices have been applied, such as coating, moulding, and so on. The super-capacitor electrodes were made with a polymer/MXene nanocomposite that was electrochemically polymerized *in situ*. Zhou *et al.* [124] used spray coverings LBL assembly to make $Ti_3C_2T_x$ MXene. Polycaprolactone/carbon nanotube was used to nanocomposite $Ti_3C_2T_x$ MXene. The electrode that resulted has a capacitance of 30–50 mF cm^{-2}. Using a $Ti_3C_2T_x$ MXene-based electrode improved the super-capacitors capacitance stability by 14–16 percent. Qin *et al.* [125] used electrochemical polymerization to create a polypyrrole/MXene nanocomposite. The research explores the electrochemical polymerization of polypyrrole and MXene to generate a nanocomposite. The pyrrole monomer dispersion in MXene galleries can be detected. Following that, the pyrrole monomer was cross-linked, resulting in polypyrrole chains bearing a MXene nanofiller. The capacitance and energy density were measured in the regularly constructed super-capacitor. Relying on polyaniline/polypyrrole as well as MXene nanofiller, Wu *et al.* [126] created bendable super-capacitor electrodes. Using the *in situ* solution approach, the unsupported nanocomposite film electrode was made. The $Ti_3C_2T_x$ MXene has a capacitance of 284 $mFcm^{-2}$ when combined with conducting macromolecules that depicts the polymer/$Ti_3C_2T_x$ MXene nanocomposite's capacitance preservation. After 10^3 charge-discharge cycles, the nanocomposite's capacitance remained at 100%. Capacitance retention in the neat polymer was found to be 80%. The solution approach was found to be successful in improving polymer- $Ti_3C_2T_x$ MXene friendliness and blending, resulting in higher conductivity and specific capacitance. Because of the MXene nanofillers, the polymer/MXene nanocomposites have significantly improved capacitive performance, capacitance retention, energy density, and charge-discharge. Moreover, to make the enhanced screens, nanoparticles were also added to the polymer/MXene. The polymer/MXene was filled with silver and iron oxide nanoparticles. Electrocatalytic activity has been demonstrated for this kind of membrane. The catalytic performance of the polypyrrole and $Ti_3C_2T_x$ nanocomposite membrane has also been investigated. PVA, poly (acrylic acid) (PAA), MXene, and Pd nanoparticle-centered NCs (PVA/PAA/MXene @ Pd-NP) was created by Yin *et al* [127]. The PVA and PAA were dissolved in water. MXene was distributed in the solvent as well. To make the nanocomposite nanofiber, the polymer and MXene dispersions were mixed as well as electrospun. The PVA/PAA/MXene/Pd nanofiber membrane was then formed by treating the nanofiber barrier with a palladium (II) chloride ($PdCl_2$) mixture. With an effective surface area of 13.48 m^2g^{-1}, the membrane has a large specific surface area. For 2-nitrophenol (2-NP) or 4-nitrophenol (4-NP), the NC nanofiber membranes demonstrated excellent catalytic activity. The homogenous dispersion of MXene in the polymers was responsible for the nanocomposite membrane's functionality. However, more research is required to find the contaminants that will be removed by the membrane. Because of its superior biocompatibility, conductivity, as well as hydrophilicity, the polymer/MXene nanocomposite has also been used in the biomedical sector. Because of its wide aspect ratio, MXene's functions may interact better with polymers and bio-systems. The outstanding qualities demonstrated by polymer functionalized MXenes and their 2D planar structure are therefore responsible for their applications in the medical field, such as antibacterial, photothermal treatment (PTT), drug administration, diagnostic imaging, biosensors, as well as bone regeneration. The Ti_3C_2 MXene has been investigated and utilized by Han *et al.* to treat malignant cells [128]. In this sense, the Ti_3C_2 MXene demonstrated a 212 % drug loading capacity. The nanomaterial exhibited an 82 % medication release rate. Rakhi *et al.* [129] created a biosensor out of glucose oxidase (GO_x). Nafion and MXene were used to make the GO_x. The Ti_3C_2 MXene was proved to be biocompatible and cytotoxic in the least amount of time. In the nano DD (drug-delivery) system, the polymer/MXene nanocomposite also behaved well. A vast surface area was used in this biosensor to increase charge transfer and excellent sensitivity. The enzyme GO_x was utilized to convert $C_6H_{12}O_6$ to gluconolactone, and hydrogen peroxidase was used to convert the gluconolactone to gluconolactone (H_2O_2). As a result, the MXene-based GO_x has been effectively employed to sense glucose and deliver drugs. Moreover, the inference of the tumor site and the PTT monitoring are made possible by diagnostic imaging. Numerous diagnostic imaging techniques exist. MRI is a powerful imaging technology used to examine the anatomical structure of soft tissues by nonionizing electromagnetic waves that are generated from the tissues. Computed tomography is regarded as the best imaging modality because to its high spatial resolution and pertinent anatomical tomography. As a result of the high tissue penetration in addition to optical imaging penetration limit-exceeding photoacoustic signals produced by NIR irradiation, PA imaging is used to detect stress waves in irradiated tissues in real-time [130]. Water accelerates the formation of new bones whereas the Ti-based units discharged from the MXene's polymer membranes break down in the existence of O_2. Although polymer/MXene composites can be used to regenerate bones, guided bone regeneration (GBR), which is increasingly used for orthodontic implantation and periodontal regeneration, works on the premise of using a GBR membrane to guard the bone from soft tissue interference. In order to create a biocompatible and durable membrane for GBR, Chen *et al.* combined PLA matrix with Ti_3C_2Tz nanosheets and used n-octyltriethoxysilane (OTES) as the interfacial facilitating ingredient. The ultimate TS (UTS) of the solvent cast OTES-Ti_3C_2Tz/PLA nanocomposite membrane, which was 72 MPa and 33 % more than that of the clean PLA membrane, demonstrated exceptional mechanical characteristics. The improvement of mechanical characteristics with efficient load transfer is what causes the good compatibility between PLA and OTES-Ti_3C_2Tz. Compared to GO on the membrane, the Ti_3C_2Tz nanosheets' enhanced biocompatibility boosted preosteoblast attachment, growth, and differentiation potential. Nevertheless, MXene's usage in biological applications is limited due to its poor stability. There is also a lack of controlled and sustained medication distribution and endurance. Moreover, current issues with the nanomaterials, such as mass fabrication, stowage, arrangement accuracy, *in vivo* retention, and biosafety over an protracted length of time, among others, are barriers to their broad usage in nanomedicine and must be tackled. Deep tissue penetration was made possible by NIR II radiation, although very few photothermal agents have been observed in this biological window. By investigating its stage of genotoxicity, immunotoxicity, and reproductive toxicity, MXenes can be optimized in provisional implants, fertility cure, tissue regeneration, and genetics. Although NIR II radiation allowed for deep tissue penetration, there are very few reports of photothermal agents in this biological window. To measure and utilize conducting polymers such as polyaniline as well as polyvinylpyrrolidone on MXene substances for detecting numerous analytes and for sarcoma diagnosis through photothermal therapy, the market needs to be stimulated. It is crucial to do research to find out whether these nanocomposites may be employed to remove deposits from artery walls and smash kidney stones. MXene-polymer composites have only recently been used in the medical field, hence more research into polymer-based MXenes for bio applications is unquestionably important.

7.7 Conclusion and Outlook

MXene is a flexible 2D NM that improves the mechanical, electrical, and thermal characteristics of polymers dramatically. MXene has considerable hydrophilic characteristics as a result of the use of strong acid for etching formation, making it ideal for solution mixing to make nanocomposites. MXene/polymer NMs have a variety of outstanding functional capabilities when being mixed with a polymer, including electric, force, and heat attributes. A modest quantity of MXene, for instance, could dramatically boost the conductivity of polymeric composites. The synergistic consequences of increased conductivity, wide specific area nano-sheet architecture, and diversified surface composition are responsible for the remarkable performance advantages of MXene/polymer nanocomposites, according to researchers. MXene/polymer nanocomposites' adaptability has piqued interest, with applications including tumor eradication, flexible monitoring, energy stowage and transformation, and electromagnetic protection. However, the mixing impact of MXene in a polymer solution, nevertheless, necessities to be tweaked because of the macromolecular structure of the polymer. Furthermore, MXene polymer NCs show great promise for an extensive range of solicitations and sectors, and the area is still in its early stages, with a lot of studies ahead of it. Although several of the reported outcomes are truly astounding, there are numerous problems related to the manufacture of these NCs on a large scale. The susceptibility of MXenes to oxidation is the most important factor when deciding. While there is a lot of research going on in this domain, and oxidation paces are slower in solid settings (like PVA), the polymer doesn't fully inhibit oxidation. Additionally, the extent to which MXene is oxidized in individual polymers is uncertain, as it will most likely be determined by the polymer's chemistry and/or structure. This needs to be addressed since oxidation can be harmful to the usage of MXene as a filler in solicitations where oxidation affects functioning. Besides, the following significant technological obstacles still need to be resolved before MXene/polymer nanocomposites may be used in industrial settings and produced on a wide scale. Since, MXene is safe, effective, and requires little expense to separate multilayer MXene into individual flakes of MXene. Nevertheless, because of the non-Van der Waals interstrata communication amid the MAX phase stratums, the method of stripping to yield MXene is highly reliant on aggressive and hazardous chemical etching pathways, which significantly impedes the production of MXene/polymer nanocomposites on a large scale at low cost. Furthermore, MXene must be dispersed into the polymer matrix. MXene is an inorganic functional hydrophilic moiety, while most polymer materials use hydrophobic polymers. The main obstacle to the development and use of MXene/PNCs will be in which way to disband MXene with outstanding amalgamation presentation in a hydrophobic polymer. MXene/PNCs have demonstrated their considerable capacity and usage value in spite of the aforementioned difficulties. In order to create MXene/polymer nanocomposites with good and complete performance, it will be advantageous for the researchers to research on the MXene stripping including dispersion methodologies in the forthcoming period.

References

1. Novoselov, K.S., Geim, A.K., Morozov, S.V., Jiang, D.E., Zhang, Y., Dubonos, S.V., Grigorieva, I.V., Firsov, A.A., Electric field effect in atomically thin carbon films. *Science*, 306, 5696, 666–9, 2004.
2. Novoselov, K.S., Colombo, L., Gellert, P.R., Schwab, M.G., Kim, K., A roadmap for graphene. *Nature*, 490, 7419, 192–200, 2012.
3. Mattevi, C., Kim, H., Chhowalla, M., A review of chemical vapour deposition of graphene on copper. *J. Mater. Chem.*, 21, 10, 3324–34, 2011.
4. Wang, H., Maiyalagan, T., Wang, X., Review on recent progress in nitrogen-doped graphene: Synthesis, characterization, and its potential applications. *ACS Catal.*, 2, 5, 781–94, 2012.
5. Li, L., Yu, Y., Ye, G.J., Ge, Q., Ou, X., Wu, H., Feng, D., Chen, X.H., Zhang, Y., Black phosphorus field-effect transistors. *Nat. Nanotechnol.*, 9, 5, 372–7, 2014.
6. Dávila, M.E., Xian, L., Cahangirov, S., Rubio, A., Le Lay, G., Germanene: A novel two-dimensional germanium allotrope akin to graphene and silicene. *New J. Phys.*, 16, 9, 095002, 2014.
7. Liu, H., Neal, A.T., Zhu, Z., Luo, Z., Xu, X., Tománek, D., Ye, P.D., Phosphorene: An unexplored 2D semiconductor with a high hole mobility. *ACS Nano*, 8, 4, 4033–41, 2014.

8. Bhimanapati, G.R., Lin, Z., Meunier, V., Jung, Y., Cha, J., Das, S., Xiao, D., Son, Y., Strano, M.S., Cooper, V.R., Liang, L., Recent advances in two-dimensional materials beyond graphene. *ACS Nano*, 9, 12, 11509–39, 2015.
9. Tang, X., Guo, X., Wu, W., Wang, G., 2D metal carbides and nitrides (MXenes) as high-performance electrode materials for lithium-based batteries. *Adv. Energy Mater.*, 8, 33, 1801897, 2018.
10. Rasool, K., Helal, M., Ali, A., Ren, C.E., Gogotsi, Y., Mahmoud, K.A., Antibacterial activity of Ti3C2T x MXene. *ACS Nano*, 10, 3, 3674–84, 2016.
11. Hantanasirisakul, K. and Gogotsi, Y., Electronic and optical properties of 2D transition metal carbides and nitrides (MXenes). *Adv. Mater.*, 30, 52, 1804779, 2018.
12. Khazaei, M., Ranjbar, A., Arai, M., Sasaki, T., Yunoki, S., Electronic properties and applications of MXenes: A theoretical review. *J. Mater. Chem. C*, 5, 10, 2488–503, 2017.
13. Barsoum, M.W. and El-Raghy, T., The MAX phases: Unique new carbide and nitride materials: Ternary ceramics turn out to be surprisingly soft and machinable, yet also heat-tolerant, strong and lightweight. *Am. Sci.*, 89, 4, 334–43, 2001.
14. Barsoum, M.W., The MN+ 1AXN phases: A new class of solids: Thermodynamically stable nanolaminates. *Prog. Solid State Chem.*, 28, 1-4, 201–81, 2000.
15. Sun, Z.M., Progress in research and development on MAX phases: A family of layered ternary compounds. *Int. Mater. Rev.*, 56, 3, 143–66, 2011.
16. Högberg, H., Hultman, L., Emmerlich, J., Joelsson, T., Eklund, P., Molina-Aldareguia, J.M., Palmquist, J.P., Wilhelmsson, O., Jansson, U., Growth and characterization of MAX-phase thin films. *Surf. Coat. Technol.*, 193, 1-3, 6–10, 2005.
17. Zhang, X., Xu, J., Wang, H., Zhang, J., Yan, H., Pan, B., Zhou, J., Xie, Y., Ultrathin nanosheets of MAX phases with enhanced thermal and mechanical properties in polymeric compositions: Ti3Si075Al0.25C2. 25C2. *Angew. Chem.*, 125, 16, 4457–61, 2013.
18. Berdiyorov, G.R. and Mahmoud, K.A., Effect of surface termination on ion intercalation selectivity of bilayer $Ti_3C_2T_2$ (T= F, O and OH) MXene. *Appl. Surf. Sci.*, 416, 725–30, 2017.
19. Li, S., Tuo, P., Xie, J., Zhang, X., Xu, J., Bao, J., Pan, B., Xie, Y., Ultrathin MXene nanosheets with rich fluorine termination groups realizing efficient electrocatalytic hydrogen evolution. *Nano Energy*, 47, 512–8, 2018.
20. Chaudhary, V., Gautam, A., Mishra, Y.K., Kaushik, A., Emerging MXene–polymer hybrid nanocomposites for high-performance ammonia sensing and monitoring. *Nanomaterials*, 11, 10, 2496, 2021.
21. Carey, M. and Barsoum, M.W., MXene polymer nanocomposites: A review. *Mater. Today Adv.*, 9, 100120, 2021.
22. Liang, X., Rangom, Y., Kwok, C.Y., Pang, Q., Nazar, L.F., Interwoven MXene nanosheet/carbon-nanotube composites as Li–S cathode hosts. *Adv. Mater.*, 29, 3, 1603040, 2017.
23. Zhao, M.Q., Ren, C.E., Ling, Z., Lukatskaya, M.R., Zhang, C., Van Aken, K.L., Barsoum, M.W., Gogotsi, Y., Flexible MXene/carbon nanotube composite paper with high volumetric capacitance. *Adv. Mater.*, 27, 2, 339–45, 2015.
24. Xie, Y., Naguib, M., Mochalin, V.N., Barsoum, M.W., Gogotsi, Y., Yu, X., Nam, K.W., Yang, X.Q., Kolesnikov, A.I., Kent, P.R., Role of surface structure on Li-ion energy storage capacity of two-dimensional transition-metal carbides. *J. Am. Chem. Soc.*, 136, 17, 6385–94, 2014.
25. Cheng, Y., Zhang, Y., Li, Y., Dai, J., Song, Y., Hierarchical Ni 2 P/Cr 2 CT x (MXene) composites with oxidized surface groups as efficient bifunctional electrocatalysts for overall water splitting. *J. Mater. Chem. A*, 7, 15, 9324–34, 2019.
26. Guo, Z., Zhou, J., Zhu, L., Sun, Z., MXene: A promising photocatalyst for water splitting. *J. Mater. Chem. A*, 4, 29, 11446–52, 2016.

27. Ye, M., Wang, X., Liu, E., Ye, J., Wang, D., Boosting the photocatalytic activity of P25 for carbon dioxide reduction by using a surface-alkalinized titanium carbide MXene as cocatalyst. *Chem Sus Chem*, 11, 10, 1606–11, 2018.
28. Peng, Q., Guo, J., Zhang, Q., Xiang, J., Liu, B., Zhou, A., Liu, R., Tian, Y., Unique lead adsorption behavior of activated hydroxyl group in two-dimensional titanium carbide. *J. Am. Chem. Soc.*, 136, 11, 4113–6, 2014.
29. Zhang, Q., Teng, J., Zou, G., Peng, Q., Du, Q., Jiao, T., Xiang, J., Efficient phosphate sequestration for water purification by unique sandwich-like MXene/magnetic iron oxide nanocomposites. *Nanoscale.*, 8, 13, 7085–93, 2016.
30. Wang, F., Yang, C., Duan, M., Tang, Y., Zhu, J., TiO2 nanoparticle modified organ-like Ti_3C_2 MXene nanocomposite encapsulating hemoglobin for a mediator-free biosensor with excellent performances. *Biosens. Bioelectron.*, 74, 1022–8, 2015.
31. Yu, X.F., Li, Y.C., Cheng, J.B., Liu, Z.B., Li, Q.Z., Li, W.Z., Yang, X., Xiao, B., Monolayer Ti_2CO_2: A promising candidate for NH3 sensor or capturer with high sensitivity and selectivity. *ACS Appl. Mater. Interfaces*, 7, 24, 13707–13, 2015.
32. Verger, L., Natu, V., Carey, M., Barsoum, M.W., MXenes: An introduction of their synthesis, select properties, and applications. *Trends Chem.*, 1, 7, 656–69, 2019.
33. Naguib, M., Mashtalir, O., Carle, J., Presser, V., Lu, J., Hultman, L., Gogotsi, Y., Barsoum, M.W., Carbides, T.D., Two-dimensional transition metal carbides. *ACS Nano*, 6, 1322, 2012, Crossref, ISI.
34. Verger, L., Xu, C., Natu, V., Cheng, H.M., Ren, W., Barsoum, M.W., Overview of the synthesis of MXenes and other ultrathin 2D transition metal carbides and nitrides. *Curr. Opin. Solid State Mater. Sci.*, 23, 3, 149–63, 2019.
35. Assad, H., Fatma, I., Kumar, A., Kaya, S., Vo, D.V., Al-Gheethi, A., Sharma, A., An overview of MXene-based nanomaterials and their potential applications towards hazardous pollutant adsorption. *Chemosphere*, 298, 134221, 2022.
36. Ling, Z., Ren, C.E., Zhao, M.Q., Yang, J., Giammarco, J.M., Qiu, J., Barsoum, M.W., Gogotsi, Y., Flexible and conductive MXene films and nanocomposites with high capacitance. *Proc. Natl. Acad. Sci.*, 111, 47, 16676–81, 2014.
37. Naguib, M., Saito, T., Lai, S., Rager, M.S., Aytug, T., Paranthaman, M.P., Zhao, M.Q., Gogotsi, Y., $Ti_3C_2T_x$ (MXene)–polyacrylamide nanocomposite films. *RSC Adv.*, 6, 76, 72069–73, 2016.
38. Zhang, H., Wang, L., Chen, Q., Li, P., Zhou, A., Cao, X., Hu, Q., Preparation, mechanical and anti-friction performance of MXene/polymer composites. *Mater. Des.*, 92, 682–9, 2016.
39. Aghamohammadi, H., Amousa, N., Eslami-Farsani, R., Recent advances in developing the MXene/polymer nanocomposites with multiple properties: A review study. *Synth. Met.*, 273, 116695, 2021.
40. Zhou, Z., Panatdasirisuk, W., Mathis, T.S., Anasori, B., Lu, C., Zhang, X., Liao, Z., Gogotsi, Y., Yang, S., Layer-by-layer assembly of MXene and carbon nanotubes on electrospun polymer films for flexible energy storage. *Nanoscale*, 10, 13, 6005–13, 2018.
41. Zhan, X., Si, C., Zhou, J., Sun, Z., MXene and MXene-based composites: Synthesis, properties and environment-related applications. *Nanoscale Horiz.*, 5, 2, 235–58, 2020.
42. Luo, J., Tao, X., Zhang, J., Xia, Y., Huang, H., Zhang, L., Gan, Y., Liang, C., Zhang, W., Sn4+ ion decorated highly conductive Ti3C2 MXene: Promising lithium-ion anodes with enhanced volumetric capacity and cyclic performance. *ACS Nano*, 10, 2, 2491–9, 2016.
43. Zheng, X., Shen, J., Hu, Q., Nie, W., Wang, Z., Zou, L., Li, C., Vapor phase polymerized conducting polymer/MXene textiles for wearable electronics. *Nanoscale*, 13, 3, 1832–41, 2021.
44. Jia, X., Shen, B., Zhang, L., Zheng, W., Construction of compressible polymer/MXene composite foams for high-performance absorption-dominated electromagnetic shielding with ultra-low reflectivity. *Carbon*, 173, 932–40, 2021.

45. Boota, M., Pasini, M., Galeotti, F., Porzio, W., Zhao, M.Q., Halim, J., Gogotsi, Y., Interaction of polar and nonpolar polyfluorenes with layers of two-dimensional titanium carbide (MXene): Intercalation and pseudocapacitance. *Chem. Mater.*, 29, 7, 2731–8, 2017.
46. Seroka, N.S. and Mamo, M.A., Application of functionalised MXene-carbon nanoparticle-polymer composites in resistive hydrostatic pressure sensors. *SN Appl. Sci.*, 2, 3, 1–1, 2020.
47. Wu, X., Hao, L., Zhang, J., Zhang, X., Wang, J., Liu, J., Polymer-$Ti_3C_2T_x$ composite membranes to overcome the trade-off in solvent resistant nanofiltration for alcohol-based system. *J. Membr. Sci.*, *515*, 175–188, 2016.
48. Sajid, M., Kim, H.B., Siddiqui, G.U., Na, K.H., Choi, K.H., Linear bi-layer humidity sensor with tunable response using combinations of molybdenum carbide with polymers. *Sens. Actuators A: Phys.*, 262, 68–77, 2017.
49. Jiao, S., Zhou, A., Wu, M., Hu, H., Kirigami patterning of MXene/bacterial cellulose composite paper for all-solid-state stretchable micro-supercapacitor arrays. *Adv. Sci.*, 6, 12, 1900529, 2019.
50. Tu, S., Jiang, Q., Zhang, X., Alshareef, H.N., Large dielectric constant enhancement in MXene percolative polymer composites. *ACS Nano*, 12, 4, 3369–77, 2018.
51. Liu, R. and Li, W., High-thermal-stability and high-thermal-conductivity $Ti_3C_2T_x$ MXene/poly (vinyl alcohol) (PVA) composites. *ACS Omega*, 3, 3, 2609–17, 2018.
52. Mirkhani, S.A., Shayesteh Zeraati, A., Aliabadian, E., Naguib, M., Sundararaj, U., High dielectric constant and low dielectric loss via poly (vinyl alcohol)/Ti3C2Tx MXene nanocomposites. *ACS Appl. Mater. Interfaces*, 11, 20, 18599–608, 2019.
53. Faruk, M.O., Ahmed, A., Adak, B., Marzana, M., Hossain, M.M., Mukhopadhyay, S., High performance 2D MXene based conducting polymer hybrids: Synthesis to emerging applications. *J. Mater. Chem. C*, 9, 32, 10193–215, 2021.
54. Fei, M., Lin, R., Deng, Y., Xian, H., Bian, R., Zhang, X., Cheng, J., Xu, C., Cai, D., Polybenzimidazole/Mxene composite membranes for intermediate temperature polymer electrolyte membrane fuel cells. *Nanotechnology*, 29, 3, 035403, 2017.
55. Chen, Z., Han, Y., Li, T., Zhang, X., Wang, T., Zhang, Z., Preparation and electrochemical performances of doped MXene/poly (3, 4-ethylenedioxythiophene) composites. *Mater. Lett.*, 220, 305–8, 2018.
56. Liu, R., Miao, M., Li, Y., Zhang, J., Cao, S., Feng, X., Ultrathin biomimetic polymeric $Ti_3C_2T_x$ MXene composite films for electromagnetic interference shielding. *ACS Appl. Mater. Interfaces*, 10, 51, 44787–95, 2018.
57. Carey, M., Hinton, Z., Sokol, M., Alvarez, N.J., Barsoum, M.W., Nylon-6/$Ti_3C_2T_z$ MXene nanocomposites synthesized by *in situ* ring opening polymerization of ε-caprolactam and their water transport properties. *ACS Appl. Mater. Interfaces*, 11, 22, 20425–36, 2019.
58. Ren, Y., Zhu, J., Wang, L., Liu, H., Liu, Y., Wu, W., Wang, F., Synthesis of polyaniline nanoparticles deposited on two-dimensional titanium carbide for high-performance supercapacitors. *Mater. Lett.*, 214, 84–7, 2018.
59. Wei, H., Dong, J., Fang, X., Zheng, W., Sun, Y., Qian, Y., Jiang, Z., Huang, Y., Ti3C2Tx MXene/polyaniline (PANI) sandwich intercalation structure composites constructed for microwave absorption. *Compos. Sci. Technol.*, 169, 52–9, 2019.
60. Wang, H., Li, L., Zhu, C., Lin, S., Wen, J., Jin, Q., Zhang, X., *In situ* polymerized $Ti_3C_2T_x$/PDA electrode with superior areal capacitance for supercapacitors. *J. Alloys Compd.*, 778, 858–65, 2019.
61. Tang, T., Wang, S., Jiang, Y., Xu, Z., Chen, Y., Peng, T., Khan, F., Feng, J., Song, P., Zhao, Y., Flexible and flame-retarding phosphorylated MXene/polypropylene composites for efficient electromagnetic interference shielding. *J. Mater. Sci. Technol.*, 111, 66–75, 2022.

62. Yue, Q., Kang, J., Shao, Z., Zhang, X., Chang, S., Wang, G., Qin, S., Li, J., Mechanical and electronic properties of monolayer MoS2 under elastic strain. *Phys. Lett. A*, 376, 12-13, 1166–70, 2012.
63. Lipatov, A., Lu, H., Alhabeb, M., Anasori, B., Gruverman, A., Gogotsi, Y., Sinitskii, A., Elastic properties of 2D $Ti_3C_2T_x$ MXene monolayers and bilayers. *Sci. Adv.*, 4, 6, eaat0491, 2018.
64. Aakyiir, M., Kingu, M.A., Araby, S., Meng, Q., Shao, J., Amer, Y., Ma, J., Stretchable, mechanically resilient, and high electromagnetic shielding polymer/MXene nanocomposites. *J. Appl. Polym. Sci.*, 138, 22, 50509, 2021.
65. Zhong, Q., Li, Y., Zhang, G., Two-dimensional MXene-based and MXene-derived photocatalysts: Recent developments and perspectives. *Chem. Eng. J.*, 409, 128099, 2021.
66. Wang, W., Yuen, A.C., Long, H., Yang, W., Li, A., Song, L., Hu, Y., Yeoh, G.H., Random nano-structuring of PVA/MXene membranes for outstanding flammability resistance and electromagnetic interference shielding performances. *Compos. Part B: Eng.*, 224, 109174, 2021.
67. Sun, R., Zhang, H.B., Liu, J., Xie, X., Yang, R., Li, Y., Hong, S., Yu, Z.Z., Highly conductive transition metal carbide/carbonitride (MXene)@ polystyrene nanocomposites fabricated by electrostatic assembly for highly efficient electromagnetic interference shielding. *Adv. Funct. Mater.*, 27, 45, 1702807, 2017.
68. Chen, X., Zhao, Y., Li, L., Wang, Y., Wang, J., Xiong, J., Du, S., Zhang, P., Shi, X., Yu, J., MXene/polymer nanocomposites: Preparation, properties, and applications. *Polym. Rev.*, 61, 1, 80–115, 2021.
69. Gao, L., Li, C., Huang, W., Mei, S., Lin, H., Ou, Q., Zhang, Y., Guo, J., Zhang, F., Xu, S., Zhang, H., MXene/polymer membranes: Synthesis, properties, and emerging applications. *Chem. Mater.*, 32, 5, 1703–47, 2020.
70. Prakash, N.J. and Kandasubramanian, B., Nanocomposites of MXene for industrial applications. *J. Alloys Compd.*, 862, 158547, 2021.
71. Sheng, X., Zhao, Y., Zhang, L., Lu, X., Properties of two-dimensional Ti_3C_2 MXene/thermoplastic polyurethane nanocomposites with effective reinforcement via melt blending. *Compos. Sci. Technol.*, 181, 107710, 2019.
72. Cao, X., Wu, M., Zhou, A., Wang, Y., He, X., Wang, L., Non-isothermal crystallization and thermal degradation kinetics of MXene/linear low-density polyethylene nanocomposites. *E-Polym.*, 17, 373–381, 2017, Search in.
73. Kausar, A., Polymer/MXene nanocomposite–A new age for advanced materials. *Polym.-Plast. Technol. Mater.*, 60, 13, 1377–92, 2021.
74. Li, Q., Zhong, B., Zhang, W., Jia, Z., Jia, D., Qin, S., Wang, J., Razal, J.M., Wang, X., Ti_3C_2 MXene as a new nanofiller for robust and conductive elastomer composites. *Nanoscale*, 11, 31, 14712–9, 2019.
75. Liu, C., Tian, Y., An, Y., Yang, Q., Xiong, S., Feng, J., Qian, Y., Robust and flexible polymer/MXene-derived two dimensional TiO_2 hybrid gel electrolyte for dendrite-free solid-state zinc-ion batteries. *Chem. Eng. J.*, 430, 132748, 2022.
76. Levitt, A.S., Alhabeb, M., Hatter, C.B., Sarycheva, A., Dion, G., Gogotsi, Y., Electrospun MXene/carbon nanofibers as supercapacitor electrodes. *J. Mater. Chem. A*, 7, 1, 269–77, 2019.
77. Huang, X., Wang, R., Jiao, T., Zou, G., Zhan, F., Yin, J., Zhang, L., Zhou, J., Peng, Q., Facile preparation of hierarchical AgNP-loaded MXene/Fe3O4/polymer nanocomposites by electrospinning with enhanced catalytic performance for wastewater treatment. *ACS Omega*, 4, 1, 1897–906, 2019.
78. Feng, Y., Zhou, F., Bo, M., Huang, Y., Deng, Q., Peng, C., Enabling high dielectric response in PVDF/V2C MXene–TiO_2 composites based on nontypical V–F–Ti bonding and fermi-level overlapping mechanisms. *J. Phys. Chem. C*, 124, 50, 27780–9, 2020.

79. Yang, H., Dai, J., Liu, X., Lin, Y., Wang, J., Wang, L., Wang, F., Layered PVB/Ba3Co2Fe24O41/ Ti_3C_2 Mxene composite: Enhanced electromagnetic wave absorption properties with high impedance match in a wide frequency range. *Mater. Chem. Phys.*, 200, 179–86, 2017.
80. Rasool, K., Mahmoud, K.A., Johnson, D.J., Helal, M., Berdiyorov, G.R., Gogotsi, Y., Efficient antibacterial membrane based on two-dimensional $Ti_3C_2T_x$ (MXene) nanosheets. *Sci. Rep.*, 7, 1, 1–1, 2017.
81. Jiang, C., Wu, C., Li, X., Yao, Y., Lan, L., Zhao, F., Ye, Z., Ying, Y., Ping, J., All-electrospun flexible triboelectric nanogenerator based on metallic MXene nanosheets. *Nano Energy*, 59, 268–76, 2019.
82. Cao, W.T., Chen, F.F., Zhu, Y.J., Zhang, Y.G., Jiang, Y.Y., Ma, M.G., Chen, F., Binary strengthening and toughening of MXene/cellulose nanofiber composite paper with nacre-inspired structure and superior electromagnetic interference shielding properties. *ACS Nano*, 12, 5, 4583–93, 2018.
83. Mayerberger, E.A., Urbanek, O., McDaniel, R.M., Street, R.M., Barsoum, M.W., Schauer, C.L., Preparation and characterization of polymer-$Ti_3C_2T_x$ (MXene) composite nanofibers produced via electrospinning. *J. Appl. Polym. Sci.*, 134, 37, 45295, 2017.
84. Mayerberger, E.A., Street, R.M., McDaniel, R.M., Barsoum, M.W., Schauer, C.L., Antibacterial properties of electrospun $Ti_3C_2T_z$ (MXene)/chitosan nanofibers. *RSC Adv.*, 8, 62, 35386–94, 2018.
85. Gong, K., Zhou, K., Qian, X., Shi, C., Yu, B., MXene as emerging nanofillers for high-performance polymer composites: A review. *Composites Part B: Eng.*, 217, 108867, 2021.
86. Tanvir, A., Sobolčiak, P., Popelka, A., Mrlik, M., Spitalsky, Z., Micusik, M., Prokes, J., Krupa, I., Electrically conductive, transparent polymeric nanocomposites modified by 2D $Ti_3C_2T_x$ (MXene). *Polymers*, 11, 8, 1272, 2019.
87. Jimmy, J. and Kandasubramanian, B., Mxene functionalized polymer composites: Synthesis and applications. *Eur. Polym. J.*, 122, 109367, 2020.
88. Dong, Y., Sang, D., He, C., Sheng, X., Lei, L., Mxene/alginate composites for lead and copper ion removal from aqueous solutions. *RSC Adv.*, 9, 50, 29015–22, 2019.
89. Huang, Z., Wang, S., Kota, S., Pan, Q., Barsoum, M.W., Li, C.Y., Structure and crystallization behavior of poly (ethylene oxide)/$Ti_3C_2T_x$ MXene nanocomposites. *Polymer*, 102, 119–26, 2016.
90. Liu, Y., Zhang, J., Zhang, X., Li, Y., Wang, J., $Ti_3C_2T_x$ filler effect on the proton conduction property of polymer electrolyte membrane. *ACS Appl. Mater. Interfaces*, 8, 31, 20352–63, 2016.
91. Ronchi, R.M., Arantes, J.T., Santos, S.F., Synthesis, structure, properties and applications of MXenes: Current status and perspectives. *Ceram. Int.*, 45, 15, 18167–88, 2019.
92. Zhang, H., Wang, L., Zhou, A., Shen, C., Dai, Y., Liu, F., Chen, J., Li, P., Hu, Q., Effects of 2-D transition metal carbide Ti_2CT_x on properties of epoxy composites. *RSC Adv.*, 6, 90, 87341–52, 2016.
93. Raagulan, K., Braveenth, R., Jang, H.J., Seon Lee, Y., Yang, C.M., Mi Kim, B., Moon, J.J., Chai, K.Y., Electromagnetic shielding by MXene-graphene-PVDF composite with hydrophobic, lightweight and flexible graphene coated fabric. *Materials*, 11, 10, 1803, 2018.
94. Liu, S., Wang, L., Wang, X., Liu, L., Zhou, A., Cao, X., Preparation, mechanical and thermal characteristics of d-Ti_3C_2/PVA film. *Mater. Today Commun.*, 22, 100799, 2020.
95. Wang, X., Wang, L., He, Y., Wu, M., Zhou, A., The effect of two-dimensional d-Ti_3C_2 on the mechanical and thermal conductivity properties of thermoplastic polyurethane composites. *Polym. Compos.*, 41, 1, 350–9, 2020.
96. Liu, L., Ying, G., Hu, C., Zhang, K., Ma, F., Su, L., Zhang, C., Wang, C., Functionalization with MXene (Ti_3C_2) enhances the wettability and shear strength of carbon fiber-epoxy composites. *ACS Appl. Nano Mater.*, 2, 9, 5553–62, 2019.

97. Zhi, W., Xiang, S., Bian, R., Lin, R., Wu, K., Wang, T., Cai, D., Study of MXene-filled polyurethane nanocomposites prepared via an emulsion method. *Compos. Sci. Technol.*, 168, 404–11, 2018.
98. Zhang, Y., He, X., Cao, M., Shen, X., Yang, Y., Yi, J., Guan, J., Shen, J., Xi, M., Zhang, Y., Tang, B., Tribological and thermo-mechanical properties of TiO_2 nanodot-decorated Ti_3C_2/epoxy nanocomposites. *Materials*, 14, 10, 2509, 2021.
99. Jiao, C., Deng, Z., Min, P., Lai, J., Gou, Q., Gao, R., Yu, Z.Z., Zhang, H.B., Photothermal healable, stretchable, and conductive MXene composite films for efficient electromagnetic interference shielding. *Carbon*, 198, 179–187, 2022.
100. Sliozberg, Y., Andzelm, J., Hatter, C.B., Anasori, B., Gogotsi, Y., Hall, A., Interface binding and mechanical properties of MXene-epoxy nanocomposites. *Compos. Sci. Technol.*, 192, 108124, 2020.
101. Monastyreckis, G., Mishnaevsky, Jr. L., Hatter, C.B., Aniskevich, A., Gogotsi, Y., Zeleniakiene, D., Micromechanical modeling of MXene-polymer composites. *Carbon*, 162, 402–9, 2020.
102. Yan, H., Zhang, L., Li, H., Fan, X., Zhu, M., Towards high-performance additive of Ti_3C_2/graphene hybrid with a novel wrapping structure in epoxy coating. *Carbon*, 157, 217–33, 2020.
103. Chen, K., Chen, Y., Deng, Q., Jeong, S.H., Jang, T.S., Du, S., Kim, H.E., Huang, Q., Han, C.M., Strong and biocompatible poly (lactic acid) membrane enhanced by $Ti_3C_2T_z$ (MXene) nanosheets for Guided bone regeneration. *Mater. Lett.*, 229, 114–7, 2018.
104. Hu, C., Shen, F., Zhu, D., Zhang, H., Xue, J., Han, X., Characteristics of Ti_3C_2X–chitosan films with enhanced mechanical properties. *Front. Energy Res.*, 4, 41, 2017.
105. Saba, N., Mohammad, F., Pervaiz, M., Jawaid, M., Alothman, O.Y., Sain, M., Mechanical, morphological and structural properties of cellulose nanofibers reinforced epoxy composites. *Int. J. Biol. Macromol.*, 97, 190–200, 2017.
106. Meng, F., Zhang, Z., Gao, P., Kang, R., Boyjoo, Y., Yu, J., Liu, T., Excellent tribological properties of epoxy—Ti_3C_2 with three-dimensional nanosheets composites. *Friction*, 9, 4, 734–46, 2021.
107. Yin, Y., Kent, D., Tan, Q., Bermingham, M., Zhang, M.X., Spheroidization behaviour of a Fe-enriched eutectic high-entropy alloy. *J. Mater. Sci. Technol.*, 51, 173–9, 2020.
108. Kang, R., Zhang, Z., Guo, L., Cui, J., Chen, Y., Hou, X., Wang, B., Lin, C.T., Jiang, N., Yu, J., Enhanced thermal conductivity of epoxy composites filled with 2D transition metal carbides (MXenes) with ultralow loading. *Sci. Rep.*, 9, 1, 1–4, 2019.
109. Si, J.Y., Tawiah, B., Sun, W.L., Lin, B., Wang, C., Yuen, A.C., Yu, B., Li, A., Yang, W., Lu, H.D., Chan, Q.N., Functionalization of MXene nanosheets for polystyrene towards high thermal stability and flame retardant properties. *Polymers*, 11, 6, 976, 2019.
110. Sobolčiak, P., Ali, A., Hassan, M.K., Helal, M.I., Tanvir, A., Popelka, A., Al-Maadeed, M.A., Krupa, I., Mahmoud, K.A., 2D $Ti_3C_2T_x$ (MXene)-reinforced polyvinyl alcohol (PVA) nanofibers with enhanced mechanical and electrical properties. *PLoS One*, 12, 8, e0183705, 2017.
111. Zhou, B., Zhang, Z., Li, Y., Han, G., Feng, Y., Wang, B., Zhang, D., Ma, J., Liu, C., Flexible, robust, and multifunctional electromagnetic interference shielding film with alternating cellulose nanofiber and MXene layers. *ACS Appl. Mater. Interfaces*, 12, 4, 4895–905, 2020.
112. Shahzad, F., Alhabeb, M., Hatter, C.B., Anasori, B., Man Hong, S., Koo, C.M., Gogotsi, Y., Electromagnetic interference shielding with 2D transition metal carbides (MXenes). *Science*, 353, 6304, 1137–40, 2016.
113. Ding, L., Wei, Y., Wang, Y., Chen, H., Caro, J., Wang, H., A two-dimensional lamellar membrane: MXene nanosheet stacks. *Angew. Chem. Int. Ed.*, 56, 7, 1825–9, 2017.
114. Ding, L., Wei, Y., Li, L., Zhang, T., Wang, H., Xue, J., Ding, L.X., Wang, S., Caro, J., Gogotsi, Y., MXene molecular sieving membranes for highly efficient gas separation. *Nat. Commun.*, 9, 1, 1–7, 2018.

115. Wu, X., Liao, H., Ma, D., Chao, M., Wang, Y., Jia, X., Wan, P., Zhang, L., A wearable, self-adhesive, long-lastingly moist and healable epidermal sensor assembled from conductive MXene nanocomposites. *J. Mater. Chem. C*, 8, 5, 1788–95, 2020.
116. Gao, X., Li, Z.K., Xue, J., Qian, Y., Zhang, L.Z., Caro, J., Wang, H., Titanium carbide $Ti_3C_2T_x$ (MXene) enhanced PAN nanofiber membrane for air purification. *J. Membr. Sci.*, 586, 1, 162–9, 2019.
117. Xing, C., Chen, S., Liang, X., Liu, Q., Qu, M., Zou, Q., Li, J., Tan, H., Liu, L., Fan, D., Zhang, H., Two-dimensional MXene (Ti3C2)-integrated cellulose hydrogels: Toward smart three-dimensional network nanoplatforms exhibiting light-induced swelling and bimodal photothermal/chemotherapy anticancer activity. *ACS Appl. Mater. Interfaces*, 10, 33, 27631–43, 2018.
118. Taloub, N., Henniche, A., Liu, L., Li, J., Rahoui, N., Hegazy, M., Huang, Y., Improving the mechanical properties, UV and hydrothermal aging resistance of PIPD fiber using MXene (Ti3C2 (OH) 2) nanosheets. *Compos. B. Eng.*, 163, 260e271, 2019.
119. Jasim, S.A., Hadi, J.M., Opulencia, M.J., Karim, Y.S., Mahdi, A.B., Kadhim, M.M., Bokov, D.O., Jalil, A.T., Mustafa, Y.F., Falih, K.T., MXene/metal and polymer nanocomposites: Preparation, properties, and applications. *J. Alloys Compd.*, 13, 165404, 2022.
120. Yu, B., Tawiah, B., Wang, L.Q., Yuen, A.C., Zhang, Z.C., Shen, L.L., Lin, B., Fei, B., Yang, W., Li, A., Zhu, S.E., Interface decoration of exfoliated MXene ultra-thin nanosheets for fire and smoke suppressions of thermoplastic polyurethane elastomer. *J. Hazard. Mater.*, 374, 110–9, 2019.
121. Song, Y., Chen, Y., Jiang, X., Ge, Y., Wang, Y., You, K., Wang, K., Zheng, J., Ji, J., Zhang, Y., Li, J., Nonlinear few-layer MXene-assisted all-optical wavelength conversion at telecommunication band. *Adv. Opt. Mater.*, 7, 18, 1801777, 2019.
122. Nan, J., Guo, X., Xiao, J., Li, X., Chen, W., Wu, W., Liu, H., Wang, Y., Wu, M., Wang, G., Nanoengineering of 2D MXene-based materials for energy storage applications. *Small*, 17, 9, 1902085, 2021.
123. Xiong, D., Li, X., Bai, Z., Lu, S., Recent advances in layered Ti3C2Tx MXene for electrochemical energy storage. *Small*, 14, 17, 1703419, 2018.
124. Zhou, Z., Panatdasirisuk, W., Mathis, T.S., Anasori, B., Lu, C., Zhang, X., Liao, Z., Gogotsi, Y., Yang, S., Layer-by-layer assembly of MXene and carbon nanotubes on electrospun polymer films for flexible energy storage. *Nanoscale*, 10, 13, 6005–13, 2018.
125. Qin, L., Tao, Q., Liu, X., Fahlman, M., Halim, J., Persson, P.O., Rosen, J., Zhang, F., Polymer-MXene composite films formed by MXene-facilitated electrochemical polymerization for flexible solid-state microsupercapacitors. *Nano Energy*, 60, 734–42, 2019.
126. Wu, X., Huang, B., Lv, R., Wang, Q., Wang, Y., Highly flexible and low capacitance loss supercapacitor electrode based on hybridizing decentralized conjugated polymer chains with MXene. *Chem. Eng. J.*, 378, 122246, 2019.
127. Yin, J., Zhan, F., Jiao, T., Deng, H., Zou, G., Bai, Z., Zhang, Q., Peng, Q., Highly efficient catalytic performances of nitro compounds via hierarchical PdNPs-loaded MXene/polymer nanocomposites synthesized through electrospinning strategy for wastewater treatment. *Chin. Chem. Lett.*, 31, 4, 992–5, 2020.
128. Han, X., Huang, J., Lin, H., Wang, Z., Li, P., Chen, Y., 2D ultrathin MXene-based drug-delivery nanoplatform for synergistic photothermal ablation and chemotherapy of cancer. *Adv. Healthcare Mater.*, 7, 9, 1701394, 2018.
129. Rakhi, R.B., Nayak, P., Xia, C., Alshareef, H.N., Novel amperometric glucose biosensor based on MXene nanocomposite. *Sci. Rep.*, 6, 1, 1–0, 2016.
130. Riazi, H., Nemani, S.K., Grady, M.C., Anasori, B., Soroush, M., Ti_3C_2 MXene–polymer nanocomposites and their applications. *J. Mater. Chem. A*, 9, 13, 8051–98, 2021.

8

MXene-Reinforced Polymer Composites for Dielectric Applications

Karuppasamy P.[1], Sennappan M.[1], Hemavathi B.[1*], Manjunath H. R.[2] and Anjanpura V. Raghu[2,3†]

[1]Department of Chemistry, Dayananda Sagar College of Engineering, Kumaraswamy Layout, Bangalore, India
[2]Department of Basic Sciences, Faculty of Engineering and Technology JAIN (Deemed-to-be University), Bengaluru, India
[3]Faculty of Science and Technology, BLDE (Deemed-to-be University), Vijayapura, India

Abstract
Two-dimensional (2D) materials resemble graphene structure and lead to the discovery of novel 2D materials called MXenes having transition metal carbides/nitrides. Since its inception in 2011, these 2D materials are investigated not only experimentally but also theoretically and receiving great attention recently owing to their exceptional insights into the structural features. However, incorporation of MXene materials with polymer is employed for fabricating hybrid composite materials that paves a way to find novel materials for various applications including electromagnetic interference shielding, sensing/biosensing, energy storage, photo-thermal, and filtration and biomedical fields. MXenes/polymer composites exhibit excellent di-electric properties due to their abundant surface terminations, multi-functionalities, specific surface area, hydrophilic nature, conductivity, and synergistic effects. Even though, these hybrid composites have exquisite properties and applications, but still challenges are there for developing novel MXene/polymer composites with respect to mass production and commercialization. However, the recent performance and development of MXene/polymer composites have been proved as potential candidates for high-performance dielectric applications.

Keywords: MXene/polymer composite, dielectric constant, dielectric loss, conductivity

8.1 Introduction

In the recent years, mission to explore new and exquisite material remain main focus of research for industry and academia. Single-layered graphene family has been an indubitable discovery as two dimensional (2D) material ideal for industrial application. However, inquisitive minds driven to find new 2D materials resembling graphene like hexagonal structure has propelled the discovery of new 2D materials. MXene, are among the new class

**Corresponding author*: hemavathi-chem@dayanandasagar.edu
†*Corresponding author*: avraghu23@gmail.com

of 2D materials discovered in the year 2011 has more viable academic and practical applications [1]. In the past decade, MXene have been synthesized using early transition metals and non-metals such as carbon and nitrogen with functional groups like O, F, OH, Cl, etc. in terminal position [1, 2]. The general formula of MXene is $M_{n+1}X_n$, with surface termination groups like depending on the environment and chemicals employed [3]. The synthesis techniques and the modification process of MXene renders them with functional groups over the surface of the 2D flakes. Hence, MXene having rich hetero-atoms possessing lone pair of electrons and their physical properties includes flexibility, morphology, layered structure and hydrophilicity, endows facility to form composites with polymers. MXene/polymer composites are thin layer of 2D material displays remarkable physical, chemical and electrochemical properties owing to outstanding mechanical properties, hydrophilic surfaces and conductivity of MXene. However, the method of the mixing the MXene and polymer and fabrication of hybrid composite endows them with morphology and properties which are widely explored and applied in various fields includes chemistry, biology, material science, nanotechnology etc.

It is evident now that the utilization of the MXene/polymer composites majorly depends on the synthesis, modification, and fabrication processes. Primarily, fully etched and exfoliated MXene which is synthesized by appropriate process is the most desired property of the MXene filler. MXene was originally synthesized from the ternary metal carbide or nitride represented as MAX (M-transition metal, A-main-group III or IV element, X-carbon or nitrogen elements) by HF etching process. Other etching process such as fluoride salt etching, fluoride free etching is also known to yield flaking 2D materials with appropriate properties. The etching process selectively removes the main group element to afford fully exfoliated MXene. This exfoliated filler need to be endowed with suitable functional group to make them suitable for dispersion in the polymer matrix with formation strong hybrid interaction between the filler and matrix. This is achieved by either by covalent or non-covalent modification of the MXene. Now the functionalized MXene is more suitable for the preparation of MXene/polymer composite. The incorporation of MXene flake in the polymer can be done by *in situ* polymerization or *ex situ* blending techniques and further, the composite is fabricated in to a useful product by drop casting, vacuum evaporation of hot press method. Now, the fully fabricated MXene/polymer with suitable physical and electrical properties are widely applicable as energy storage materials which is ascribed to its excellent dielectric properties such as good dielectric constant, high breakdown strength, low dielectric loss and high charge density [4, 5]. These properties are optimized in a dielectric capacitor by varying the composition, fabrication method and the processing technique of the MXene/polymer nanocomposite.

8.2 Synthesis of MXene

Synthesis technique for MXene-based materials is a crucial aspect to tune the property of the composites for desired performances. This will also help in the dispersion of the MXene nanofillers which influences the synergistic interactions between the MXenes and polymer matrices. Most commonly used technique for the synthesis of MXene is by etching of MAX precursors followed by mechanical sonication to exfoliate the filler and further sintering at high temperature. MAX phase generally exists in hexagonal layered structure comprising

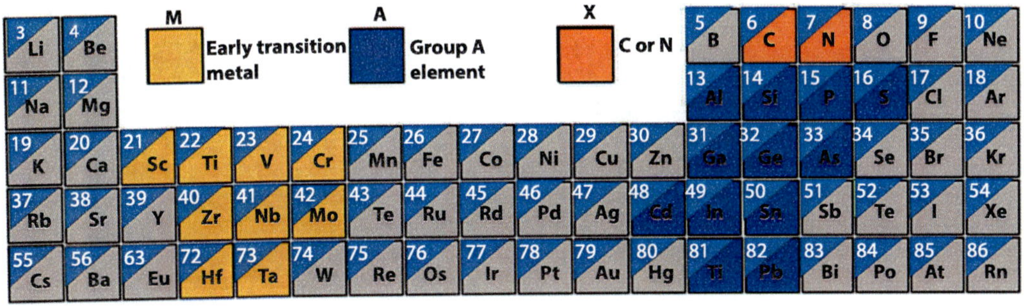

Figure 8.1 Elements required in MAX phase are depicted in periodic table with early transition metal (M) denoted with yellow color, group 13 and 14 elements (A) denoted by blue color, and carbon or nitrogen (X) denoted by orange color.

transition metal carbides and nitrides expressed as having $M_{n+1}AX_n$ formula, where, M stands for early transition metal, A will be elements from group 13 or 14 in the periodic table, and X is C or N. By computing various transition metal with C or N, theoretically various MAX based materials are studied (Figure 8.1). The synthetic techniques of MXenes is principally classified as top-down technique involving chemical etching from MAX or non-MAX precursors and bottom-up process involving chemical vapor deposition (CVD) technique.

8.2.1 Etching of MAX Phases

Unlike graphene, mechanical exfoliation of MAX phase to obtain MXene is not feasible as there exists a stronger bond in the MAX layers, however, the bonding between the M and A is chemically vulnerable compared to the M and X bond which make the chemical etching method successful in exfoliating the MAX phase to get MXene. Since, the bonding between the layers in MAX phased is stronger, the synthesis necessitates stronger etchants. The stronger etchant removes the 'A' layers (like Al) by corrosion leading to stable $M_{n+1}X_n$ layers. The MAX phase unit cell consists of M-metal and X-atom held by stronger chemical bond (ionic/metallic/covalent) but A- atom is held with pure metallic bond in the unit cell which make the M-A more susceptible and it can be easily cleaved with the help of stronger etchants. At room temperature (RT), wet chemical etching of the precursor MAX phase separates selective layers using HF [1, 6, 7], or any fluoride-containing etchant [8, 9]. However, selective etching of M-A layer is very crucial and only way to obtain the MXene flakes hence, successful conversion of MAX to MXene phase (Figure 8.2) [6]. Also it will be achieved by carefully varying the concentration of the etchant and the time duration of the etching process [10]. Further, together with the chemical etching, mechanical sonication is required for the exfoliation of the MXene layers. The time duration of sonication is also crucial aspect in the exfoliation process and it will facilitate effective delamination of the MAX phase.

Naguib *et al.* synthesized the first 2D MXene using Ti_3AlC_2 powder and treated it with Concentrated HF at RT [1]. This reaction facilitated the selective removal of interfacial Al in the MAX phase and the reactions can be summarized as below.

$$3HF + Ti_3AlC_2 \rightarrow Ti_3C_2 + 3/2\ H_2 + AlF_3 \tag{8.1}$$

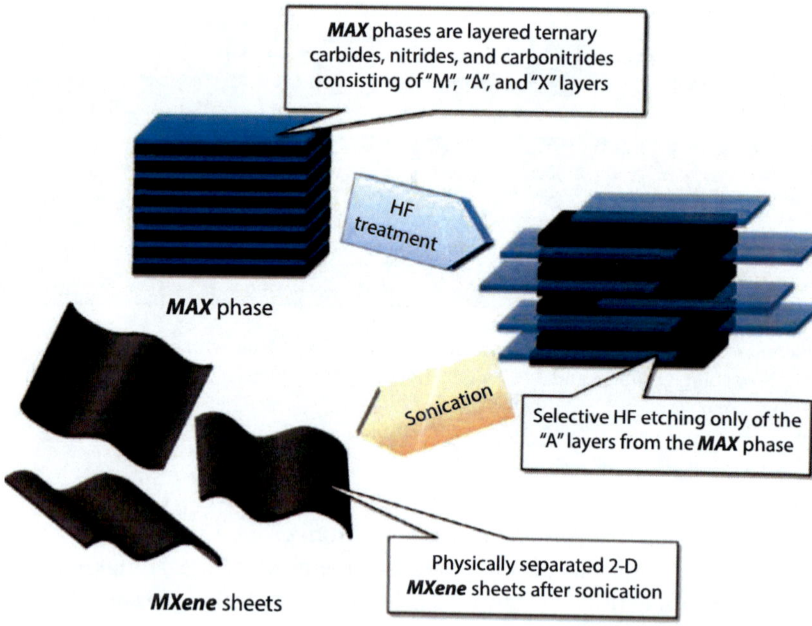

Figure 8.2 Exfoliation technique of Ti_3AlC_2 by wet HF etching to get Ti_3C_2. Reproduced with permission from Ref. [6], Copyright 2012, American Chemical Society.

$$Ti_3C_2 + 2H2O \rightarrow H_2 + Ti_3C_2(OH)_2 \tag{8.2}$$

$$2HF + Ti_3C_2 \rightarrow H_2 + Ti_3C_2F_2 \tag{8.3}$$

Further, other MAX phases were etched using HF to furnish new MXenes family consisting of Nb_2CT_x, V_2CT_x, or $Mo_2Ti_2C_3$ [11–13]. However, these families of 2D materials carry other miscellaneous terminations groups such as (OH, F, O, and H). Hence, these termination groups, T included in the general formulas of MXenes as $M_{n+1}X_nT_x$. The time duration and the temperature of the etching vary with the kind of the MXene and the bonding existing between the M-A layers. For instance, in case of Nb_2AlC, where the bonding between the M-A is stronger, then the etching process is carried out for 90h using 50% HF. However, with increase in the temperature from RT to 55°C, the time duration of etching was reduced to 40h [14]. The MAX precursors have Al layer which need to be etched and the kind of etchant and the concentration of the etchant usually depend on the nature of M-A bonding. Also, the stability of the MXene depends on the precursor and the condition of etching process. Hence, it is imperative to carefully choose the precursor MAX phase and economic feasibility of the process in getting the MXene for the production in the industrial scale.

8.2.2 Modified Acid Etching Methods of MAX Phases

Use of HF for a large scale production of MXene is not preferred as HF is harmful and also it makes this process of etching with concentrated HF non feasible. Other methods of etching are by *in situ* production of HF is used as an alternative technique by many groups.

One such method is synthesis of Ti_3AlC_2 involving use of hydrochloric acid (HCl) and lithium fluoride (LiF) [8]. *In situ* formation of HF due to presence of H^+ and F^- ions in the LiF and HCl solution, exhibited the same etching capability as HF at 35°C [8, 15]. *In situ* generation of HF can also be done by using ammonium fluoride (NH_4F) solution and this techniques was utilized to produce Ti_3C_2 MXene by soaking Ti_3AlC_2 in NH_4F solution [16]. Furthermore, use of fluoride salts (NaF, KF, LiF, and CaF_2) along with HCl or sulfuric acid (H_2SO_4), is explored as a synthetic approaches for the synthesis of 2D MXenes [12, 13].

Another attempt is to use milder etchants such as ammonium bifluoride (NH_4HF_2) as etchant. Compared to the etching using HF, etching of Ti_3AlC_2 using NH_4HF_2 provides one-step synthesis of Ti_3C_2, where simultaneous intercalation of ammonia (cations) in the MXene flakes gives better etching process far superior compared to HF etching [17]. The process is described as follows,

$$Ti_3AlC_2 + 3NH_4HF_2 \rightarrow Ti_3C_2 + (NH4)_3AlF_6 + 3/2\, H_2 \tag{8.4}$$

$$Ti_3C_2 + H_2O + aNH_4HF_2 \rightarrow (NH_3)_c(NH_4)_dTi_3C_2(OH)_xF_y \tag{8.5}$$

Reaction (8.4) indicates the etching process by NH_4HF_2 and the reaction (8.5) indicates the intercalation of NH_3 and NH_4^{+1} between $Ti_3C_2T_x$ layers. The enlarged inter-planar spacing in the 2D flakes will increase the spacing and further assist in maintaining the 2D structure. This increased inter-layer spacing not only reduces the etching time, it also forbid the process of sonication which follows etching process [18].

In case of synthesis of Ti_3N_2 from $Ti_{n+1}AlN_n$ precursor, the usual trend of formation of Ti_3C_2 is not followed, since, the formation energy of Ti–N bond is much greater than in the case of Ti–C bond, which signifies that energy is more for the extraction of Al element from $Ti_{n+1}AlN_n$ phase. Hence, nitride MXene synthesis requires higher temperature treatment. Ti_3C_2 fabricated from Ti_4AlN_3 powders reacting with fluoride salts such as LiF/NaF at 550°C results in the etching of Al by molten fluoride in the Ar atmosphere (Figure 8.3) [19].

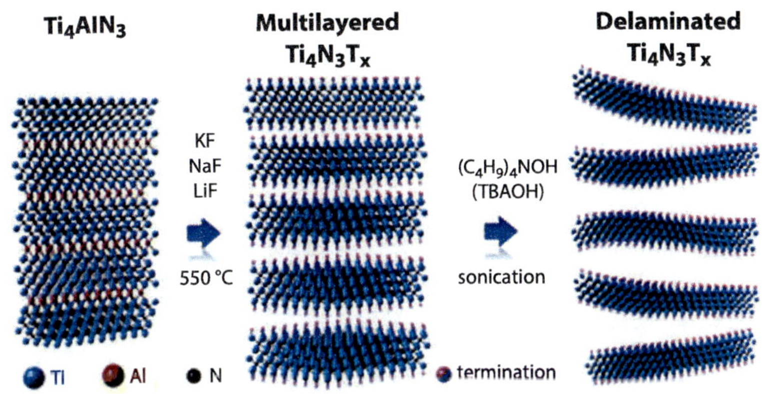

Figure 8.3 Synthesis of Ti_4N_3 from Ti_4AlN_3 using molten salts of fluorine. Reproduced with permission from Ref. [19]. Copyright 2016, The Royal Society of Chemistry.

Care should be taken as at high temperature Ti_3C_2 may oxidize to TiO_2 along with the formation of oxides of carbon in the O_2 atmosphere. Hence, synthesis of new nitride containing MXenes need to be investigated further to achieve greater understanding of the fabrication process.

8.2.3 Fluoride Salts as Etchants

To alleviate the damage of liquid and gas phase HF, a hydrothermal method of synthesis is preferred. This strategy for the synthesis of MXene is to treat the molten fluorinated salts such as NH_3F, NaF, KF, LiF, as the etchant followed by the intercalation of tetrabutylammonium hydroxide (TBAOH) to produce good results especially to obtain $Ti_4N_3T_x$ flakes [19]. Synthesis of Ti_3AlC_2 MAX in NH_4F aqueous solution in autoclave furnished black precipitates which showed better etching effect with increased duration and concentration of NH_4F, since, degree of hydrolysis is better leading to better etching process. However, the temperature of the reaction needs to be controlled to avoid the high temperature oxidation of $Ti_3C_2T_x$ to TiO_2. Generally, etching with fluorinated salts offers an alternative and harmless route to the synthesis high quality MXenes [20].

8.2.4 Fluoride-Free Synthesis Methods

Alternative pathways for the fluoride-free synthesis use other etching agents which can also produce MXenes, effectively. However, commercialization of these methods still needs to be ascertained. The fluoride free techniques used for the synthesis of MXenes are: chemical vapor deposition (CVD) method, hydrothermal method, and electrochemical synthesis. CVD is a bottom-up approaches used to synthesize bare MXenes by depositing individually atoms of Ti, Al, followed by C onto an insulating sapphire substrate with the help of direct current magnetron sputtering technique [17]. CVD technique is used to get the bare MXene material directly over a substrate. A 2D ultrathin α-$Mo_2C$was produced by CVD technique from carbon source at 1085°C, over a Cu/Mo substrate [21]. The prepared films were found to be superconducting and defect free.

Hydrothermal method involves the synthesis of MXene from an alkaline solution at high temperature. $Ti_3C_2T_x$ (T=OH,O) prepared from of Ti_3AlC_2 using alkaline solution of NaOH at 270°C which dissolves Al elements from the MAX phase [22]. Further, after the hydrothermal process, MXene is treated with aqueous tetramethylammonium hydroxide (TMAOH) to exfoliate the flakes through intercalation which leads to good photo-thermal properties. In this alkali etching, formation of $Al(OH)_4$ at the intercalation will facilitate the separation of titanium carbide nanosheets.

Electrochemical etching technique is another fluoride-free synthesis pathway which is less hazardous method where, electrochemical etching of Ti_2AlC is done by making it as an electrode in dilute HCL to get Ti_2CT_x. The MXene prepared by this method has only –Cl terminal groups, along with –O and –OH. However, etching parameters need to be controlled to avoid over-etching of the electrode [23].

8.3 Modification Strategies of MXene

Certainly, MXene have been explored in various applications particularly in combination with polymer matrix. However, the interaction of MXene filler and polymer matrix need to be improved to avoid the phase separation between the two dissimilar components in the composite, which will influence the application and its utilization. To improve the interaction between the polymer and the filler, surface modification of the MXene is proved to be a more conducive process [24]. Surface modifications endows the MXene with multi-functional properties and increases the interaction between the MXene and the polymer. It also helps in achieving homogenous dispersion of MXene filler into a polymer matrix. Modification of MXene and grafting of various functional groups can be accomplished either by covalent interaction or by non-covalent interaction.

8.3.1 Covalent Interaction

Covalent modification of the MXene renders the rich functional groups in the material which guarantees more interaction in the MXene-based composite materials. Various reports on the modification of the MXene reveal the approach of grafting of functional groups such as $-NH_2$, -COOR, $-C_6H_6$, and $-C_{12}H_{26}$on the surface of the MXene. Thus modified $Ti_3C_2T_z$ is used to prepare nano-hybrid composite with various other polymers such as polyethyleneimine (PEI)- or poly (dimethyl siloxane) (PDMS). The functionalization of powdered Ti_3C_2 to get colloidal $Ti_3C_2T_z$i s simply attained by adding (3-aminopropyl) triethoxysilane, (*g*-methacryloxypropyl)trimethoxysilane, (methyl aniline)triethoxysilane, or (dodecyl)triethoxysilane to graft $-NH_2$, -COOR, $-C_6H_6$, and $-C_{12}H_{26}$, respectively, over Ti_3C_2 [25, 26]. Thus modification of MXene results in enhanced properties along with creating synergistic effect in MXene/polymer composite.

8.3.2 Non-Covalent Interaction

Non-covalent interaction of MXene can be achieved by various methods such as hydrogen bonding, Electrostatic interaction, or Vander Waals interactions. Hydrogen bonding interaction is one of best methods to introduce modification of MXene and it is usually done by introducing oxygen rich functional groups on MXene surface [27]. Presence of oxygen rich group results in the formation of hydrogen bond which actually leads to formation of cross-linking between the filler and the matrix materials [28] and also promotes energy transmission in the composite material. Eventually, H-bonding enhances the thermal and mechanical properties of the MXene nanocomposite [29]. Electrostatic interaction due to positive and negative charges over the components of MXene/polymer composite also expands its application. Presence of negative charges over the MXene makes it to combine with negatively charged components in hybrid [30]. Ti_3C_2 MXene modified with dihexadecyl-dimethylammonium bromide (DDAB) and decyltrimethylammonium bromide (DTAB), to introduce by electrostatic interaction with polystyrene (PS) has led to the enhanced thermal stability of the composites [31].

8.4 Synthesis Methods and Fabrication of MXene-Based Polymer Composites

2D MXene materials have displayed excellent electronic and photo thermal properties, which are vital for the numerous applications in the field of purification, catalysis and sensing. However, the stability of the MXene materials is still a concern. One of the strategies to improve the stability of MXene materials is to introduce polymers as matrix along with MXene fillers, which also reduces the overall cost of the composite and expands their applications.

Polymers being macromolecules have known for their excellent processibility, flexibility, stability, and mechanical properties. MXene are recognized for their excellent electrical and thermal properties. Hence, a composite of polymer with MXene is expected to have synergistic effect which is good for any kind of applications. However, incorporation of MXene in polymer is crucial to get a synergistic effect. The following are the various techniques involved with the incorporation of MXene in polymer to prepare a composite.

Preparing an MXene/polymer composite is accomplished by various methods such as *ex situ* or *in situ* polymerization method due to presence of active functional group on MXene [32, 33]. *Ex situ* polymerization results in a well-formed polymer than *in situ* method, since in *in situ* polymerization, more monomers are chemically bond to the MXene. *Ex situ* mixing creates covalent bonding between the polymer and MXene along with other weak interaction like van der Waals force [34] or hydrogen bonds [35] are also possible.

8.4.1 *Ex Situ* Mixing

Ex situ blending offers an efficient mode of mixing MXenes in the polymer matrices using appropriate solvent which takes full benefit of MXenes surface functional groups that combines with polymers for good intermolecular interactions in the MXene/polymer composite [36]. However, mixing of the polymer and fillers need to be homogenous which can also improve the interaction of the MXene/polymer composite. The $Ti_3C_2T_x$/PDDA (polydiallyldimethylammonium chloride) or PVA (polyvinyl alcohol) composites were synthesized by mixing of an aqueous PDDA/PVA solution into the colloidal solution of MXene, kept stirring/sonication ($Ti_3C_2T_x$/PVA) and centrifuged to get the products. Fabrication of the above prepared composites was done by vacuum-assisted filtration through polypropylene with about 45 mm of diameter [35]. This method has rendered a stronger interaction between the polymer and the MXene flakes which further alleviated the possibility of restacking of the MXene and hence improved the stability of the intercalated composite by precluding the possibility of flakes coming in contact with oxygen [37]. Sobolciak *et al.* reported nanofibers of 2D $Ti_3C_2T_x$/CNC (cellulose nanocrystals)/PVA fabricated by an electro spinning technique. The tensile and conductivity measurements results showed enhanced mechanical and electrical properties [38].

8.4.2 *In Situ* Mixing

The *in situ* method of preparing the composite is by adding the polymer precursor in the mixture of curing agent and the MXene filler followed by *in situ* polymerization. To obtain homogenous mixture is the key component which depends on the mixing of the component before fabrication of the composite. In this technique, the polymer penetrates the MXene layer and the inserted polymer increases the MXene layer spacing leading to increase stability of the material [39]. *In situ* combination of polyacrylamide (PAM) with Ti_3C_2 to get Ti_3C_2/PAM hydrogels exhibited excellent mechanical properties and good drug release activity [40]. The MXene/PANI (polyaniline) composite was prepared from delaminated $Ti_3C_2T_x$ and aniline monomers by *in situ* chemical oxidative polymerization under acidic conditions. The synthesis and fabrication taken place at single step [41]. The hybrids of $Ti_3C_2T_x$/PEDOT were prepared by *in situ* polymerization of 3, 4-ethylenedioxythiophene on $Ti_3C_2T_x$ in water without any oxidant by stirring at RT for 24 h [32]. $Ti_3C_2T_x$/PPy composite was synthesized by stirring the suspension of pyrrole as monomer and $Ti_3C_2T_x$ as substrate in water at RT for 24 h without using any catalyst [4]. The reaction mixture was worked up using water and ethanol in vacuum filter to obtain the composite. The above procedure was followed to prepare NA- $Ti_3C_2T_x$/PPy with 1,5-naphthalene di-sulfonic acid (NA) and NA-CTAB- $Ti_3C_2T_x$/PPy with 1,5-naphthalene di-sulfonic acid (NA) and cetyltrimethylammonium bromide (CTAM) into $Ti_3C_2T_x$ and pyrrole solution mixture [42]. V_2C@PDMAEMA [poly(2-(dimethylamino) ethyl methacrylate)] hybrid material was fabricated by self-initiated photografting and photopolymerization (SIPGP). This smart material is very sensitive/responsive for CO_2 and temperature and also exhibited tunable conductivity in aqueous media [43]. $Ti_3C_2T_x$-polyacrylamide nanocomposite was fabricated by taking both materials in aqueous media by sonication. This mixture was poured into a Teflon tray kept at RT for curing and it was dried in an oven for overnight at 60°C. Material was showed good flexibility and conductivity with 6 wt% MXene in PAM [44]. Certainly, the MXene/polymer composites produced *via in situ* polymerization or *ex situ* mixing) have good processing property and the morphologies composite membrane depends on the methods to fabricate MXene/polymer membranes. MXene/polymer composites are typically synthesized by drop casting (DC), vacuum-assisted filtration (VAF) [45], and hot press (HP) [46].

8.4.3 Fabrication Techniques

8.4.3.1 Drop Casting

Solution casting of a well-mixed solution of polymer with fillers to get a composite material is known as are Drop casting method which is a simple method of fabrication of the composite film simply by solvent evaporation after pouring the composite solution on a mold. A well dispersed mixture of MXene and PVA after stirring and ultrasonic conditions and the$Ti_3C_2T_x$/PVA nanocomposites was obtained by drop casting method. The resultant film was observed by scanning electron microscope (SEM) which showed a homogeneous dispersion of MXene in the polymer matrix [47].

8.4.3.2 Vacuum-Assisted Filtration (VAF)

VAF is another simple method commonly used to fabricate MXene/polymer composite films which involves vacuum assisted removal of the solvents from the composite blend solution. Well dispersed solution of MXene/polymer in a solvent with good dispersion is poured onto a filter membrane and under vacuum, a composite film is obtained which is also has good cross-linking network between the MXene/polymer composite [48]. $Ti_3C_2T_x$ with polyurethane (PU) in water as solvent, was fabricated in to a high ordered nanocomposite films having good interaction between the polymer and MXene like H-bonding which leads to improved mechanical toughness and good electromagnetic interference (EMI) shielding ability of the films [49].

8.4.3.3 *Hot Press (HP)*

Fabrication of MXene/polymer composites at elevated temperature compared to the melting point of polymers into composite film is known as HP method. This technique takes the advantage of the ability of the inorganic MXene to with stand higher temperature stability and hence, the MXene fillers are dispersed in the polymer which is in molten state. This technique alleviates the use of any solvent for the fabrication of composite film. This solvent free techniques are economical and also flexible towards formulation of MXene in polymer. A well dispersed MXene filler in molten polymer blend is processed in to composite film by hot press method and the appropriate shape of the composite is obtained thereafter.

8.5 Properties of MXene/Polymer Composite

MXenes are inorganic compounds that can conduct electricity owing to presence of plenty of dipoles or mobile charge carriers on the surface [50]. The conducting properties of MXenes are enhanced by elimination of polar functional groups by calcination at low temperature [51]. Most of the organic polymers are insulator or have low conductivity, which restricts the application of these materials in electronic industries. However, if the polymers are intercalated into multilayers of the MXenes flake, then, there is interaction between these two materials and the hybrid composite shows a better electron or ion transfer which enhance the polymer conductivity [42]. To understand the influence of MXenes on the conducting property of polymers, lower to higher weight percentage of MXene are added to polymer and recorded the conductivity of polymer/MXene composite. This result revealed that as increase the MXene content, conductivity of composite increases up to certain amount after that addition of MXene into polymer there is no much change in the conductivity [52]. Polymers such as epoxy resin [52], poly (vinylidene fluoride) [53] PVA [54], PS [48], PAM [44], copolyamide matrix [55], PDMS [56], polyvinylchloride (PVC) [57], PU [58] etc. are utilized to make composites with MXenes and their conducting properties are tested. When MXenes are added to polymers, the thermal conductivity of polymers increases. With 1.0 weight % of Ti_3C_2 added to epoxy resin, increased the thermal conductivity by 141.3% [59]. MXenes are blended to various polymers such as PVDF [60], PVA [54], polymethyl methacrylate (PMMA) [39], etc. to fabricated composites and tested their enhancement of thermal conductivity. The improvement in the thermal property is due to the hydrogen bonds were formed between MXenes and polymeric materials which create a channel for heat dissipation and makes it more thermally stable.

8.5.1 Electronic and Dielectric Property

The MXene/polymer composite consists of MXene with excellent electronic property and polymer with poor electrical conductivity. However, intercalation of polymer chains into the MXene multilayers leads to molecular-level coupling, that results in induced conductivity of polymer [61] and dielectric property [34]. With the increase in the MXene filler loading in the polymer, dielectric permittivity increases up to a loading concentration of 15.3 wt % and after which there is a decrease in the dielectric constant which indicates that the highest percolation level reached at the 15.3 wt % loading of the MXene filler (Figure 8.4a). Similarly, dielectric loss also was observed to increase with the filler loading and the increase is five times elevated with 10 wt % loading of filler in the polymer (Figure 8.4b) [34]. The dielectric constant of PVDF based composites reached the maximum (>100 000) by the incorporation of 15.3 wt% of $Ti_3C_2T_x$ filler and further showed no change in the dielectric constant [34]. Also, addition of MXene in higher weight % reduces the space in the filler with improved nano-capacitance and enhanced dielectric constant [53]. The dipole interaction at the interface between MXene surface atoms (F, O) and H and the polymer chain is schematically shown in (Figure 8.4c) [34]. Furthermore, enriched interfacial compatibility such as hydrogen bonding between the MXene and the polymer, improves the permittivity values and also reduces the interfacial leakage current [64]. Due to high dielectric constant and low dielectric loss of the polymers, dielectric property of the polymer has huge applications as energy storage capacitors and flexible electronics [62]. With the incorporation of

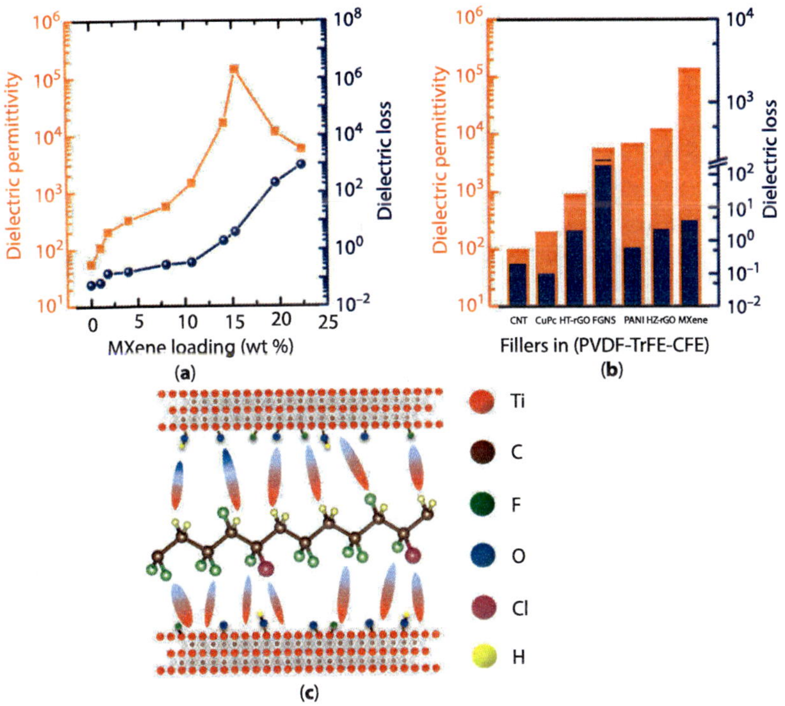

Figure 8.4 (a) Variation of the dielectric permittivity and dielectric loss with change in MXene content wt % in MXene/P(VDF-TrFE-CFE) composite. (b) dielectric permittivity and corresponding dielectric loss represented in the bar charts for the P(VDF-TrFE-CFE) composite with variation of filler concentrate. (c) Schematic representation of dipole interaction at the interface between MXene surface atoms (F, O) and H on the polymeric chain. Reproduced with permission from Ref. [34]. Copyright 2018, American Chemical Society.

MXene, the dielectric property of the polymer is expected to improve since, accumulation of charge occurs at the MXene and polymer interface. These microscopic dipole is amplified due to large surface area of MXene resulting in higher dipole- dipole interaction and hence increased interfacial polarization and enriched dielectric constant [39, 63]. The presence of surface functional group over the MXene fillers also creates a channel for proton transport which is also essential for the electrochemical energy storage system [65]. The proton conductivity (σ) increases with increase in the water uptake because of the swelling of the polymer. Proton migration and proton transportability increases when the polymer is hydrated as the water molecules provides a channel for more proton migration and hence, proton conductivity in increased [66]. Thus, the excellent conductivity of the MXene in the MXene/polymer composite plays a crucial role in the electrical and electronic applications. However, the percentage loading and surface modification to optimize the properties need in-depth exploration.

8.5.2 Dielectric Constant

A recent study emphasis on the enriched dielectric constant ascribed for generating an interfacial polarization for both binary PMMA/MXene & PMMA/ZnO composites and ternary PMMA/MXene/ZnO composite [39]. The dielectric constant value is found to be 27.024 at 25 Hz for the addition of 1 weight percentage of MXene. But a substantial enhancement of dielectric constant value is found to be 66.776 for increasing the MXene concentration up to 4 weight percentages in binary composites. The dielectric performances of polymer composites could be described using Debye model, micro-capacitor/nano-capacitors generation, and percolation theory [39]. Pure polymer matrix was alike dipoles which produces polarization in the similar way for external electric field by bring into line them. In certain direction the relaxation time was unpretentious due to like non-interacting dipoles. In the same way, the relaxation time of trifling overloaded PMMA/MXene leftovers unaffected as the rising minor interfacial dipoles as well aligned themselves in-line with the applied electric field. When increasing the concentration of MXene there is a decreasing trend of distance from PMMA to MXene nano-capacitors observed that leads to insert more numbers of nano-capacitors resulting increased dielectric constant. The effect of MXene concentration could be implicit by means of percolation theory at known concentration level (4 wt %) that is called as percolation threshold. Variation in the frequencies also affected dielectric constant of polymer composites meanwhile ultrahigh frequency created perceptible diminution of dielectric constant. As a result, a dielectric constant of 66.776 was found at 25 Hz that decreased into 36.276 while frequency increased to 1 MHz for binary PMMA/4 weight percentage of MXene composite [39]. Conversely, the dipoles number was decreased at high-frequency charge density through a rapid alteration of applied AC electric field and an increased trend of frequency which lowers external electric field leads an addition of charges in an interfaces that offers lowering of dielectric constant. The ternary nanocomposite (PMMA/MXene/ZnO) displayed a maximum dielectric constant value of 435 than binary (PMMA/4 wt % MXene) composite (66.677) at 25 Hz [67]. The increased dielectric constant of ternary nanocomposite revealed that more number of interfaces formation which gives additional improvement in the interface polarization.

MXene/polymer composites contain MXene with superb electronic performance and polymer with deprived electrical conductivity but intercalating a polymer keen on MXene led a persuaded conductivity and dielectric property. Though, the filler loading in polymer remains indispensable to divulge conductivity and the epoxy resin with conductivity of 10-16 S cm^{-1} was reduced with a conductivity of 4.52×10^{-4} S cm^{-1} meant for EP/$Ti_3C_2T_x$ polymer composites while MXene loading stretched to 0.85 weight percentage [68]. As a result of dielectric characteristics of the fabricated composites including higher dielectric constant and lower dielectric loss of polymers provide innumerable applications in the fields of energy storage devices. In a study by Wu *et al.* [69], MXene surface was functionalized with cetyltrimethylammonium bromide (CTAB) to achieve improved dispersion in the PVDF matrix. After functionalization, the MXene@CTAB dispersed homogeneously in the PVDF matrix, showing good compatibility. Moreover, the incorporation of 7 wt% of MXene@CTAB into pure PVDF has furnished a dielectric permittivity of 82.1 with the dielectric loss of 0.1 at 1KHz [69]. The CTAB acted as an insulating layer that prevented the formation of conductive network thereby suppressing the dielectric loss of the nanocomposite. The obtained dielectric permittivity was about 8.5 times more compared to pure PVDF which is about 9.7. The modification of pure PVDF with MXene@CTAB and the dielectric property variation is depicted in Figure 8.5 [69]. Electrical polarization occurs in the interior of dielectric and charges mount up and an electrical field persuaded by means of mount up charges matches an applied electric field. The charging/discharging patterns can be represented by electric displacement-electric field (D-E) loop.

8.5.3 Dielectric Loss

Dielectric loss of the capacitor is the loss of the energy that is stored in the capacitor. Hence, a better dielectric should possess a low dielectric loss so that it has less power wastage through loss [70–72]. To get low dielectric loss and to rectify the energy loss and storage,

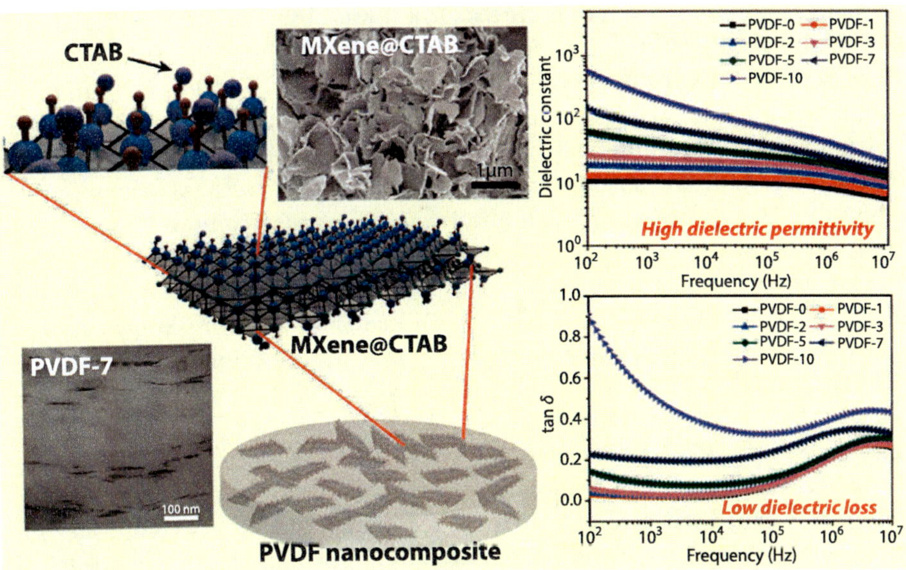

Figure 8.5 TEM image of exfoliated MXene and the measure of dielectric property for the PVDF/MXene@CTAB nanocomposite. Reproduced with permission from Ref. [69]. Copyright 2019, Elsevier.

ternary nanocomposite like PMMA/MXene/ZnO has been utilized with different proportions of MXene and ZnO NPs [39]. The obtained data indicated that the decrease in dielectric loss for both binary and ternary nanocomposites due to increase frequency and MXene concentration. A dielectric loss of PMMA is found to be 0.04 which indicates the intrinsic sequestering environment. The dielectric loss is found to be 0.20, 0.07, and 0.04 for 4 weight percentage of MXene-binary nanocomposite at 25, 10, and 1 MHz frequencies respectively. Whereas, the dielectric loss is found to be 0.37, 0.12, and 0.08 for 4:2 weight percentages of ternary nanocomposite (PMMA/MXene: ZnO NPs) at 25, 10, and 1 MHz frequencies respectively that provides a wide spectrum of capacitor applications. Accordingly, the dielectric loss of binary/ternary nanocomposites mainly depends the concentrations of MXene and the conducting behavior of MXene is more noticeable owing to the existence of ZnO NPs and increasing trend of dielectric loss value was observed because of extra outflow of current.

8.5.4 Breakdown Strength

The performance of the dielectric capacitor is measures with reference to the energy density, which mainly depends on dielectric constant and the breakdown strength. High interfacial leakage leads to reduced breakdown strength. However, breakdown strength can be improves by incorporation of semi-conductive ceramic particles along with the conductive MXene fillers in preparing a composite with the polymer [73]. Hence, a ternary composite of the polymer/MXene-ceramic composite is an interesting strategy to get higher breakdown strength. Ternary composite of polymer with V_2C-CuO (10 wt %) shown high breakdown strength of 204MV m^{-1} [74]. Feng, *et al.* fabricated binary and ternary nanocomposite of PVDF/graphene oxide (GO) and PVDF/GO@MXene for analysis of the dielectric properties of the nanocomposite with the increase in the filler loading [75]. Schematic illustration of synthesis steps of PVDF/GO and PVDF/GO@MXene nanocomposite films is shown in Figure 8.6 [75]. It was observed that the dielectric properties of the binary PVDF/GO composite increased gradually with the filler loading up to 0.12 wt % indicating that the percolation limit reached at 0.12 wt %. Beyond this value of the filler loading the dielectric constant increased abruptly indicating increased network of conducting GO in the matrix (Figure 8.7a). However, in the ternary composite of PVDF/GO@MXene the abrupt increase in the dielectric constant was not perceived indicating that the effect of percolation limit in the ternary composite is not effective (Figure 8.7b). Likewise, the dielectric loss for the binary composite had similar trend of abrupt increase in the dielectric loss (Figure 8.7c) and gradual variation in the ternary composite of PVDF/GO@MXene (Figure 8.7d). Reduce electric breakdown strengths of composites by increase the content of results from the induced conductive network in the matrix (Figure 8.7e). This trend was also seen in the ternary composite however, it was noticed that the breakdown strength loss is relatively lower compared to the binary composite (Figure 8.7f) [75]. In this work, GO/N-doped Ti_3C_2-MXene quantum dots (both binary and ternary polymer/hybrid-particle nanocomposite) were synthesized through H-bonding persuaded self-assemble method. As a result, the ternary composites showed enhanced dielectric constants and breakdown strengths than binary composites due to

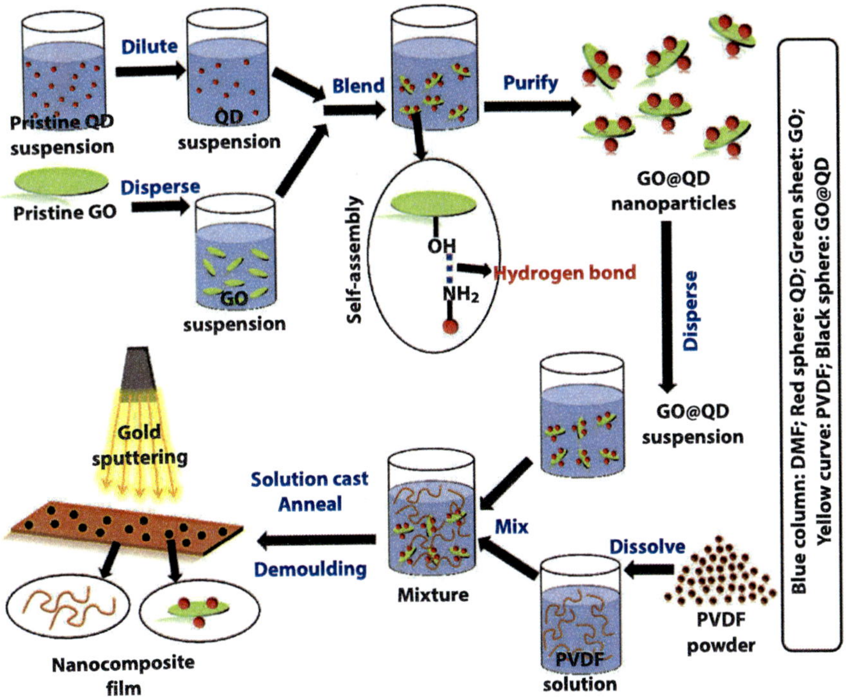

Figure 8.6 Synthesis steps of PVDF/GO and PVDF/GO@MXene nanocomposite films. Reproduced with permission from Ref. [75]. Copyright 2019, Elsevier.

synergistic effect [75]. A dielectric constant of 53 and breakdown strength of 205 MV/m at 1 kHz were recognized using by means of 0.12 weight percentage of MXene in ternary nanocomposites.

8.5.5 AC Electrical Conductivity

To increase the concentration of MXene between 0 and 4 weight percentage in composites led to enhance the AC conductivity between 7×10^{-11} to 1.20×10^{-5} S m^{-1} at 25 Hz. A small change of electrical conductance occurs while increasing the MXene concentration beyond 4 weight percentages subsequently a drastic rise in the electrical conductance. This property could be described based on percolation theory that the known concentration of MXene (4 wt %) as percolation threshold [34]. Thus a precise percolation limit of MXene in PMMA facilitated the production of conductive medium leads for the formation of additional nano-capacitors prolonged all over the entire matrix. Therefore, enhancement of AC conductivity with an augmented quantity of MXene gives indication for the reality of improved interfacial polarization between PMMA and MXene. Alternatively, the ternary Mxene-ZnO NPs composites provide slightly improved AC conductivity as compared to binary composites and exhibited relatively high concentration (PMMA/4:2 weight percentages) MXene: ZnO NPs displayed an electrical conductance of 4.88×10^{-5} S m^{-1} at 25 Hz. But both composites are followed the similar increasing trend of AC conductivity by means of increasing frequency between 25 Hz and 1 MHz where the MXene acted as a primary

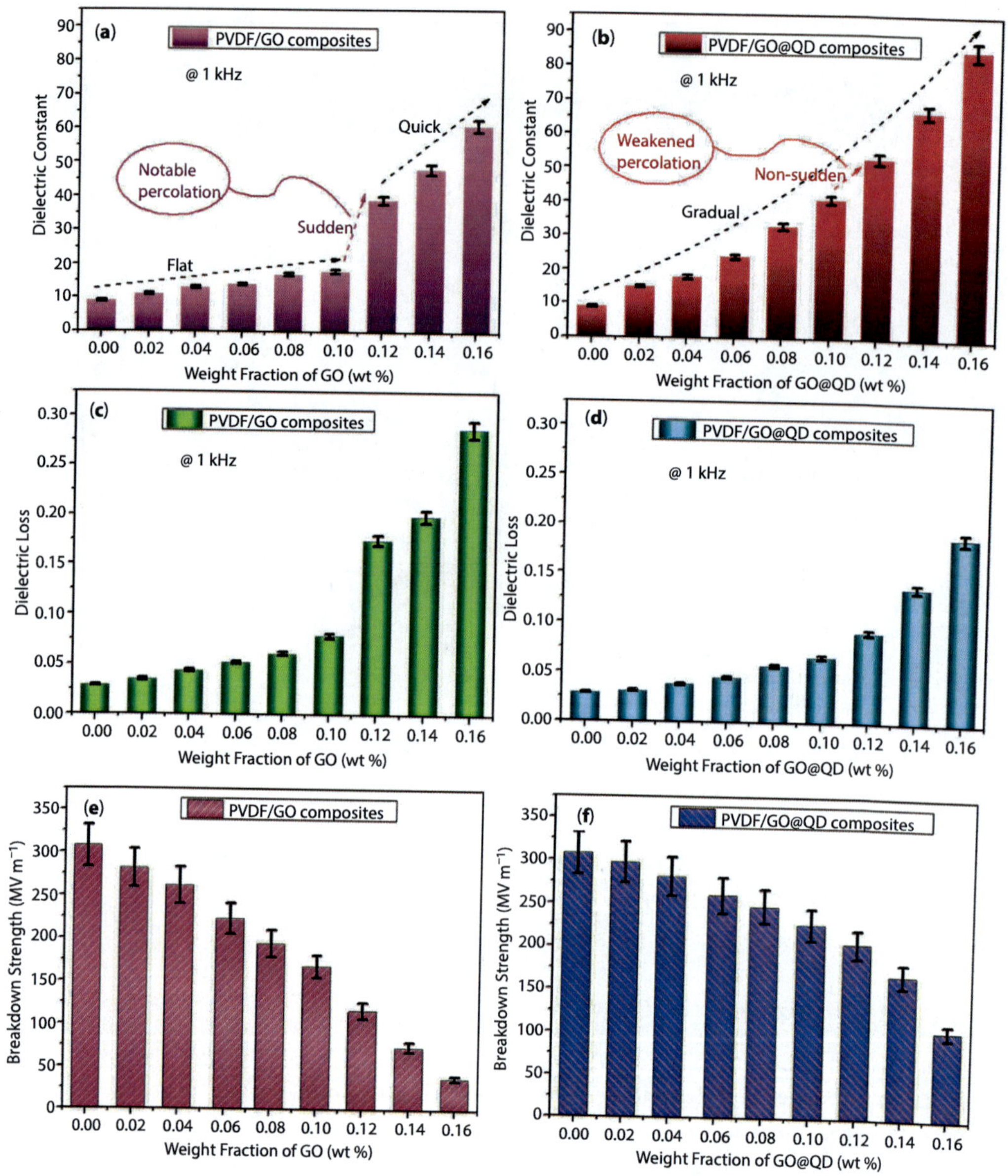

Figure 8.7 (a) Dielectric constant of binary composite of PVDF/GO, (b) Dielectric constant of binary composite of PVDF/GO@MXene nanocomposites, (c) Dielectric loss of binary composite of PVDF/GO, (d) Dielectric loss of ternary composite of PVDF/GO@MXene nanocomposites, (e) Breakdown strength of binary composite of PVDF/GO, (f) Breakdown strength of ternary composite of PVDF/GO@MXene nanocomposites. Reproduced with permission from Ref. [75]. Copyright 2019, Elsevier.

conductive filler owing to its intrinsic high electrical conductivity whereas ZnO NPs acted as a subordinate semi-conductive nanofiller with band gap value of 3.3 eV. The semi conductive performance of ZnO NPs keen on PMMA gave an electrical conductance of around 10^{-9} S m^{-1} for extreme concentration of 2 weight percentage and ZnO NPs acted as linker amongst MXene and PMMA which is responsible for the increased electrical conductivity of ternary composites as compared to binary composites.

$Ti_3C_2T_x$ MXene/PVDF composite with a multilayer-structure and extraordinary dielectric constant and ultra-low dielectric loss was prepared through spin, spray coatings and hot-press approaches. 4-MXene/5-PVDF exhibited a great dielectric constant of 41 and an ultra-low dielectric loss of 0.028 which are lesser than the pure PVDF at 1 kHz. Remarkably, the as-fabricated MXene/PVDF composite showed excellent broad-band dielectric characteristics and dielectric constant of 4-MXene/5-PVDF could be reached to 32.2 at 1 MHz that remained as extraordinary as 78.4% of MXene/PVDF at 1 kHz. The multi-layered structure could be prevented for the formation of a conductive network over the whole film proficiently which leads to decrease in the dielectric loss, relative breakdown strength with pristine PVDF and an extreme discharge energy density of 7.4 J cm^{-3} [53].

Feng *et al.* [75] has used to reduce electric breakdown strengths of MXene composites by increase the content of PVDF/graphene for high-field uses without losing dielectric constants due to interlayer persuaded quantum confinement and Coulomb barrier effects. GO/N-doped Ti_3C_2-MXene quantum dots (both binary and ternary polymer/hybrid-particle nanocomposite) are synthesized through H-bonding persuaded self-assemble method. As a result, the ternary composites showed enhanced dielectric constants and breakdown strengths than binary composites due to synergistic effect. Figure 8.8(a) shows the ac conductivity of PVDF/GO composites taken at 1 kHz. The changes in the ac conductivity was similar to that of the dielectric loss indicating a significant contribution of conductivity in the dielectric loss. At the percolation threshold, the conductivity was significantly improved while before percolation it gradually increased with increase in filler loading. Figure 8.8(b) shows the ac conductivity of PVDF/GO@MXene composites taken at 1 kHz. The ac conductivity trend of PVDF/GO@MXene composite was similar to that of PVDF/GO composites where gradual increase in ac conductivity was observed in consistent with the dielectric loss. Moreover, a decrease

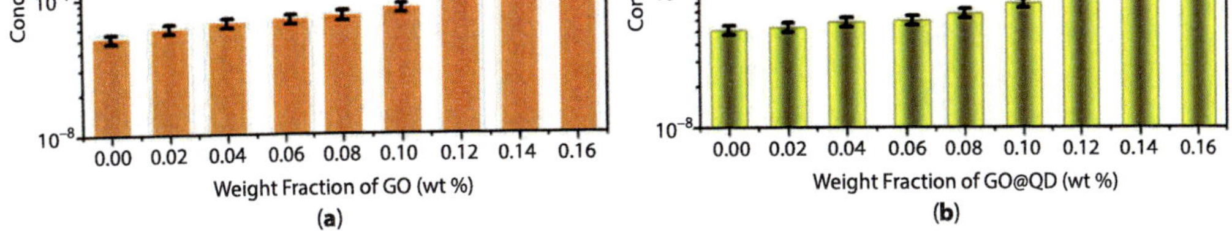

Figure 8.8 (a) AC conductivity of PVDF/GO and (b) AC conductivity of PVDF/GO@MXene composites at 1 kHz. Reproduced with permission from Ref. [75]. Copyright 2019, Elsevier.

in ac conductivity was observed in PVDF/GO@MXene composites as compared with PVDF/GO composites at high filler contents [75].

8.6 Dielectric Applications of MXene/Polymer Composite Materials

The tuneable structural properties of 2D-MXenes, an excellent interaction (chain-chain) for enhancing its performances towards wide-spectrum of applications such as energy, purification, electro-magnetic shielding, sensors, biomedical and so on as depicted in Figure 8.9. General functional properties of 2D-MXenes including conductivity, wealthy active-sites, surface defects, hydrophilicity, low band gap, terminal groups, and semiconducting property showed great potential for gas sensors, stress-strain sensors, biosensors, humidity sensor and so on [76–81]. However, dielectric applications of the MXene/polymer composites are discussed here in this chapter.

MXene/polymer composites are notably used as dielectric capacitors which exhibit excellent power density and good cyclic stability and hence, are used for the energy storage and battery applications, since dielectric capacitors exhibit better long life and excellent working voltage [81]. Intrinsic electrical insulating ability of polymeric materials is desired for dielectric applications and for energy storage devices. Nevertheless, low permittivity of the polymers restricts them for further application. However, the incorporation of conductive MXene material improve the energy density of the composite for the energy storage applications. High energy density of the PVDF polymer incorporated with $Ti_3C_2T_x$ composite was achieved due to interface compatibility existing because of the p–π conjugation and hydrogen bonding between the polymer and the MXene [82]. Hence, high dielectric constant and low dielectric loss of MXene/polymer composite have shown potential dielectric applications. Figure 8.10(a) displays the plots of permittivity vs. frequency for all the composite samples [82]. The permittivity was gradually decreased for all the composite films with an increase in frequency which indicate that dielectric response of PVDF matrix is weak at higher frequencies [82]. The dielectric loss plots of PVDF/MXene composite films are shown in Figure 8.10(b) [82]. It was noted that the dielectric loss values of all the composite films were higher than pristine PVDF film. Furthermore, it was observed that the

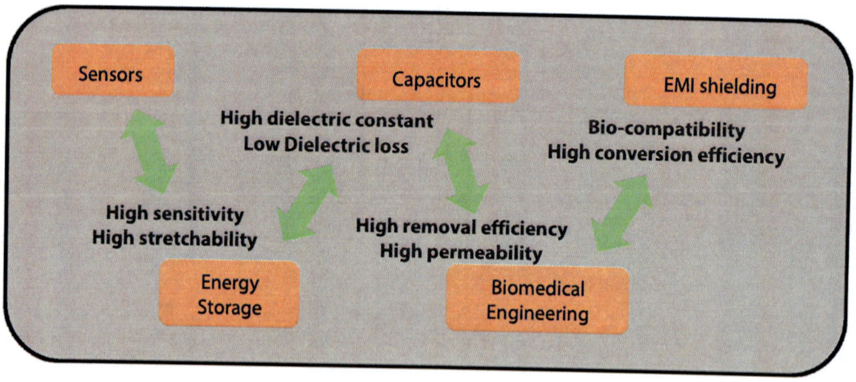

Figure 8.9 Schematic diagram of various applications using MXene/polymer composites.

higher MXene loading resulted in a higher dielectric loss. This may be due to the fact that an increase of the MXene loading resulted in an improved MWS polarization between MXene and PVDF thereby increasing the interphase relaxation induced loss in the composites [82]. Figure 8.10(c, d) depicts the AC conductivity and resistivity of different composite films as a function of frequency. The AC conductivity showed a linear behavior with an increase in frequency for all the composite films indicating their highly insulating nature [82]. As the MXene loading increases, the AC conductivity of composite films also increases which is ascribed to the high conductivity of MXene. Moreover, the electrical resistivity (Figure 8.10d) decreases with an increase in the MXene loading for all the composite films which is in good agreement with the Ac conductivity results [82]. Figure 8.11 demonstrates the source of the high permittivity obtained in the sandwich-structured PVDF/MXene composite films. The high compatibility between MXene/PVDF interface can be well understood from Figure 8.11(a) which was achieved for all the composites. A good adhesion between PVDF and terminal groups of MXene was obtained due to hydrogen bonding interaction between them. Moreover, the P-π interaction between the p electron orbit of MXene and the p electron orbit of PVDF could also be responsible for good adhesion between the

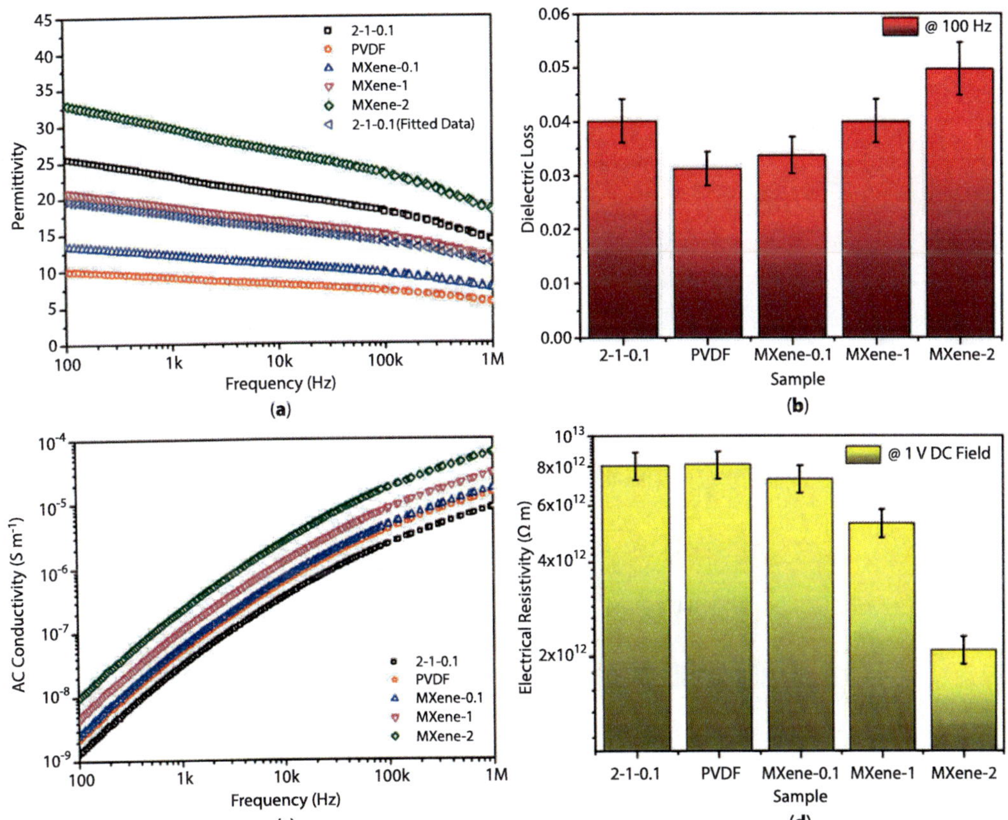

Figure 8.10 (a) Plots of Permittivity vs. frequency, (b) Dielectric loss plots of each sample (at 100 Hz), (c) plot of ac conductivity vs. frequency and (d) plot of electrical resistivity of each sample films (at 1 V dc field). Reproduced with permission from Ref. [82]. Copyright 2018, The American Chemical Society.

PVDF and the MXene [82]. Figure 8.11(b) describes the MWS polarization zone at the interface between PVDF and MXene that has contributed in the enhancement of dielectric constant. The strong MWS polarization in the zone A and B of the interface of the sandwiched composite film ascribing to the permittivity difference between the neighboring sublayers is shown in Figure 8.11(c). The high interface adhesion between the neighboring two sub-layers have helped in the enhancement of MWS polarization thereby playing a key role in the high dielectric properties of the composite films [82]. Figure 8.11(d) schematically represents the proposed micro capacitor mechanism of the composites [82].

Polymers usually possess good flexibility but poor dielectric values which restricts their practical electronic applications. However, incorporation of fillers with admirable dielectric properties in to the polymer matrix improves the dielectric properties of the polymer composite. However, the amount of loading of the filler is very crucial in the polymer matrix as high amount of filler loading (~40 vol%) will make the composite lose its flexibility and also results in high dielectric loss. Hence, optimized loading is necessary to develop MXene/polymer composite [83, 84]. Guo *et al.* [85] reported the enhancement of micro-capacitance of $Ti_3C_2T_x$ sheets in PVDF for piezoelectric application due to interfacial polarization caused by dielectric micro-capacitance. Surface termination of MXene flake strongly affect the

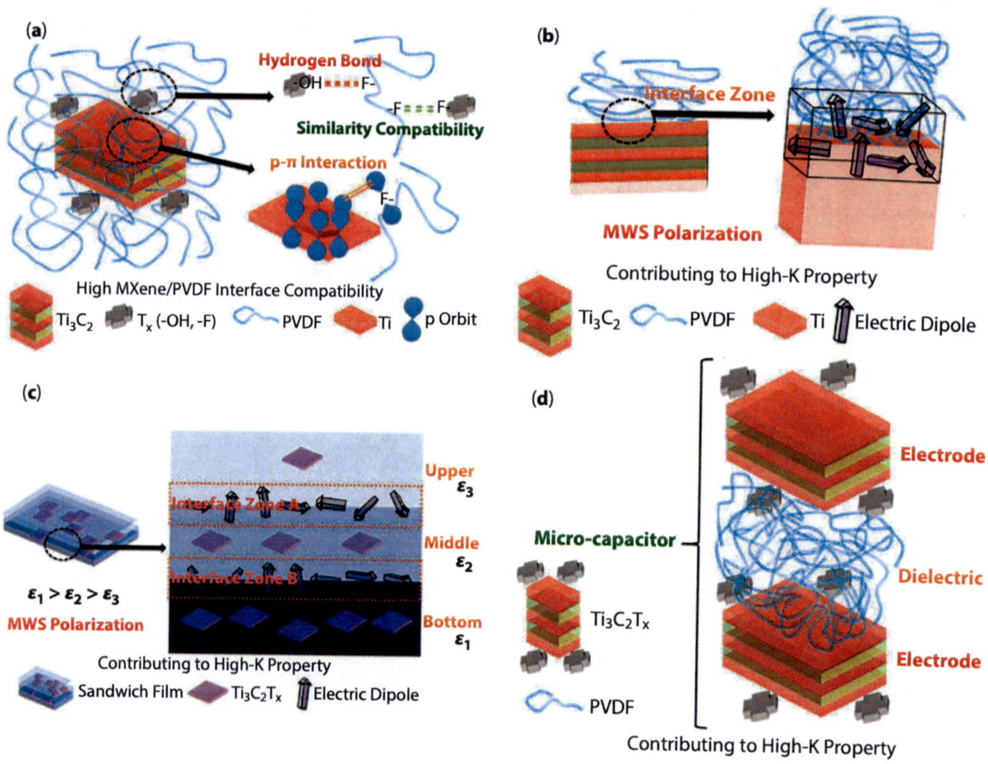

Figure 8.11 (a) Diagram displaying the interaction of MXene/PVDF at interface, (b) schematic diagram illustrating polarization of MXene/PVDF MWS, (c) schematic diagram illustrating polarization of adjacent sub-layer-by-layer MWS (d) schematic diagram illustrating the micro-capacitor mechanism, respectively. Reproduced with permission from Ref. [82]. Copyright 2018, The American Chemical Society.

dielectric permittivity of the MXene/polymer composite. The presence of –O, –F, and –OH terminating groups on $Ti_3C_2T_x$ flakes in poly(vinylidene fluoride-trifluoro-ethylene-chloro-fluoroehylene) (P(VDF-TrFE-CFE)) polymer increase the dielectric permittivity to 10^5 and dielectric constant of 2204 with 15.3 wt % $Ti_3C_2T_x$ loading [86]. In another study, flexible thin nanocomposite of PVC/MXene showed excellent thermal stability, mechanical flexibility at percolation level of 10 wt % MXene in the polymer leading to improved interfacial polarization. The fabricated composite reached a highest dielectric constant of 11800 and a low dielectric loss of 1.31 [57]. Conventionally, insulating materials like polymers are used as dielectric material that exhibit excellent stability and good dielectric constant along with good flexibility. Another strategy for the dielectric fabrication is tri-layer structure in which the loading contents are dispersed in gradient leading to interfacial barrier effect and hence results in micro- capacitor behavior. The dielectric capacitor can be designed based on the sandwiched structure consisting of dielectric material between the two conducing layers (electrodes) as shown in Figure 8.12 [87]. Nevertheless, multilayer dielectrics having more than three layers notably shows enhanced energy storage ability. Hence, multi-layered structures are investigated extensively in the recent works [88].

Even though, MXene/polymer composite has shown promising dielectric applications, their stark leakage loss needed an urgent attention. This problem was addressed by Chen *et al.* [89], by fabricating ternary composites of $CaCu_3Ti_4O_{12}$, $Ti_3C_2T_x$ MXene and silicone rubber (SR) which exhibited low dielectric loss and high dielectric constant. Figure 8.13 depicts the fabrication process of $CaCu_3Ti_4O_{12}/Ti_3C_2T_x$ MXene/SR ternary composites [89]. The incorporation of 1.2 wt % of $Ti_3C_2T_x$ MXene and 12 wt % of CCTO in the silicon rubber matrix resulted in a high dielectric constant of 8.8 and the dielectric loss of less than 0.0015. The optimization of the device properties was attributed to the high polarity in the ceramic hetero-junction. The improved dielectric properties of the ternary composites were ascribed to the homogeneous dispersion and good compatibility of the hybrid nanofillers and the polymer matrix. This suggests that ternary composites could be feasible option for the development of high performance dielectric composites instead of binary composites. In another study, PVC composites based on V_2C MXene and Cu_2O hybrid fillers were prepared which showed high dielectric constant as well as breakdown strength [74]. The ternary nanocomposite of PVC with V_2C/Cu_2O hybrid filler revealed a reduced dielectric loss compared to binary composite of PVC/V_2C [74]. This improvement is perceived due to the existence of high-polarity in the hybrid hetero-junction interface-phase. Figure 8.14(a)

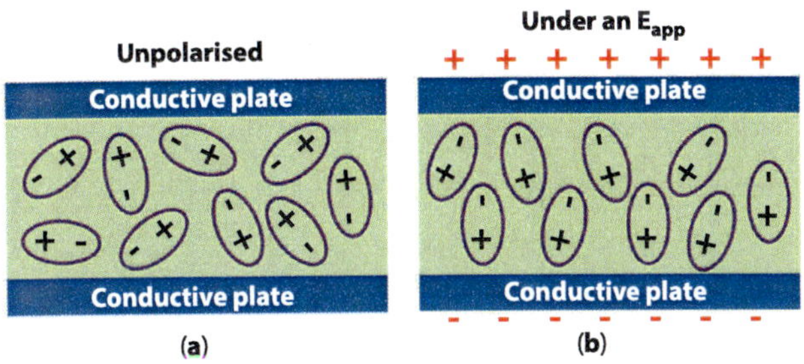

Figure 8.12 Schematic representation of a sandwiched dielectric capacitor (a) Unpolarized and (b) polarized under charging. Tri-layer structure of $Ti_3C_2T_x$/PVDF micro-capacitor mechanism. Reproduced with permission from Ref. [87]. Copyright 2021, Elsevier.

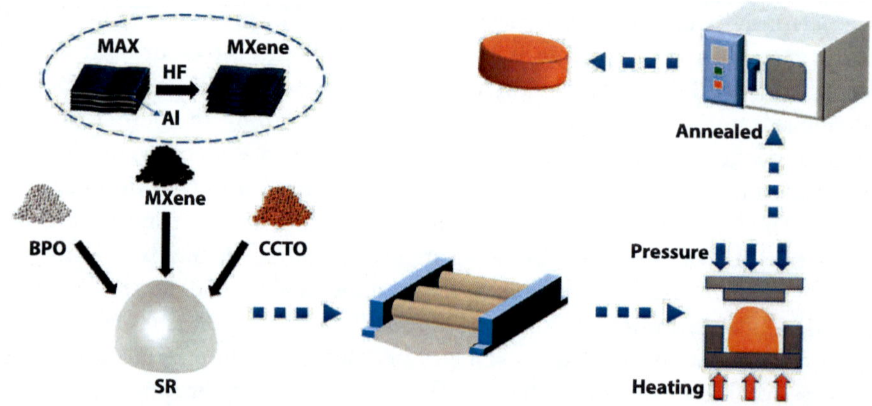

Figure 8.13 Fabrication of $CaCu_3Ti_4O_{12}/Ti_3C_2T_x$ MXene/SR ternary composites. Reproduced with permission from Ref. [89]. Copyright 2022, Elsevier.

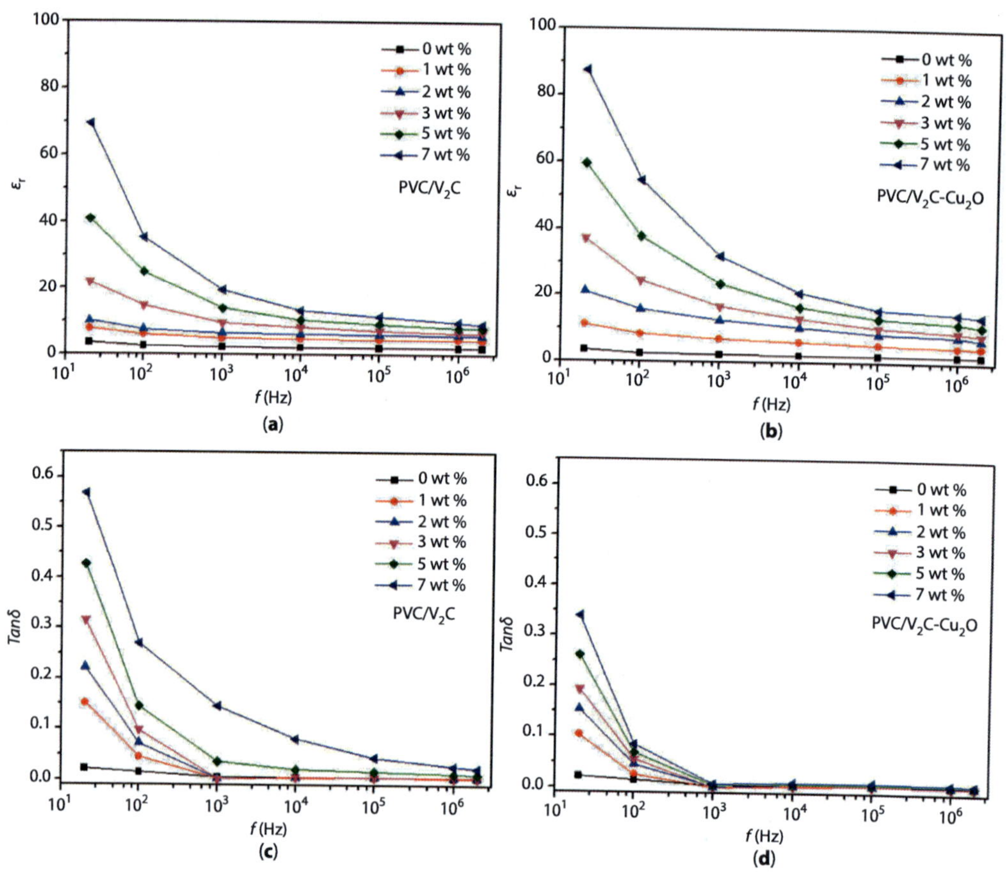

Figure 8.14 (a) Plot of dielectric constant vs. frequency of polymer/MXene binary composite, (b) plot of dielectric constant *vs.* frequency of polymer/MXene@Cu_2O ternary composites, (c) plot of dielectric loss *vs.* frequency of binary composites, (d) plot of dielectric loss *vs.* frequency of ternary composites. Reproduced with permission from Ref. [74]. Copyright 2021, Elsevier.

shows the dielectric constant of PVC/V_2C binary composite film as a function of frequency [74]. Pure PVC film showed low dielectric constant owing to its low polarity and linear dielectric nature [74]. For all the composite films, the dielectric constant decreases when the frequency was increased. However, at the low frequency range, the dielectric constant of PVC/V_2C composites was increased with increase in V_2C loading [74]. Figure 8.14(b) shows the dielectric constant plots of PVC/V_2C/Cu_2O ternary composites which revealed frequency dependent dielectric behavior. For ternary composites also, the dielectric constant decreased with increase in frequency and the trend was similar to that of binary composites. However, at low frequency region, the dielectric constant of ternary composites was higher than that of binary composites which could be due to high polarity of V_2C/Cu_2O interfacial phase and the interfacial effect between the components of ternary composites [74]. Figure 8.14(c, d) shows the dielectric loss plots of binary and ternary composites. Pure PVC film showed low dielectric loss due to its highly insulating nature. In the low frequency region,

Table 8.1 Comparison of dielectric properties of different MXene/polymer composites at different test frequencies.

MXene/polymer composite	Dielectric constant	Dielectric loss	Breakdown strength	Test frequency	Ref.
MXene/P(VDF-TrFE-CFE)	10^5	0.35	–	1 kHz	[34]
PMMA/MXene/ZnO	437	0.36	–	25 Hz	[39]
4MXene/5PVDF	41	0.028	< 270 kV mm^{-1}	1 kHz	[53]
PVC/MXene	11800	1.31	–	25 Hz	[57]
$Ti_3C_2T_x$ MXene@SiO_2/PVA	27.2	0.057	–	100 Hz	[62]
PANI-$Ti_3C_2T_x$/AE	298	0.29	–	100 Hz	[63]
PVA/V_2C MXene	24	0.14	31 MV m^{-1}	1 kHz	[64]
PVDF/MXene@CTAB	82.1	0.2	–	1 kHz	[69]
PVC/V_2C-Cu_2O composites	55	0.085	332 MV m^{-1}	100 Hz	[74]
PVDF/Graphene@MXene	53	0.188	205 MV m^{-1}	1 kHz	[75]
Ti_3C_2/PVDF	26	0.04	350 MV m^{-1}	100 Hz	[82]
Ti_3C_2/PVDF	46.37	–	–	10 Hz	[85]
MXene/P(VDF-TrFE-CFE)	2204	1425	–	1 kHz	[86]
CCTO/$Ti_3C_2T_x$/SR	8.08	0.0015	–	1 kHz	[89]

the dielectric loss of the binary composites was decreased with an increase in frequency. This is because at low frequencies, the leakage conductance at the interface was high owing to the easy transfer of charges that were trapped under low frequencies [74]. In addition, the dielectric loss of binary composites was increased at low frequency with an increase in V_2C loading. At low frequencies, the dielectric loss in ternary composites was less (due to low leakage conductance) as compared to the binary composites. Table 8.1 compares the dielectric properties of different types of MXene/polymer composites at different test frequencies.

8.7 Conclusion

MXene has emerged as an incredible 2D material in the past decade. However, the synthetic route for the preparation of MXene and the surface modification techniques play a crucial role for fabricating MXene@polymer nanocomposite interaction between the hybrid components solely depends on the surface functional groups. In this chapter we have discussed an introduction, synthesis, properties of MXene/polymer composites, and their enhanced utilization as potential dielectric applications. Since, the high electrical conductivity and low current leakage is crucial for the application, it is a challenge to impart the entire necessary requisite for the industrial application. However, optimization in the various processes and techniques to achieve optimum combination of the properties is anticipated and it is essential for industrialization and commercialization of MXene/polymer composite. Many MXene/polymer composite materials are investigated successfully due to inherent conductivity, thermal, mechanical properties and dielectric properties. Various configuration of the dielectric capacitors fabricated with polymer and MXene sandwich, tri-layer, or multilayer configuration reveal the potential exploitation of the dielectric capacitors for energy storage and power application. Finally, incorporation of another ceramic semiconducting material also are gaining momentum due to their superior low dielectric loss characteristics.

References

1. Naguib, M., Kurtoglu, M., Presser, V., Lu, J., Niu, J., Heon, M., Hultman, L., Gogotsi, Y., Barsoum, M.W., Two-dimensional nanocrystals produced by exfoliation of Ti_3AlC_2. *Adv. Mater.*, 23, 4248–4253, 2011.
2. Naguib, M., Mochalin, V.N., Barsoum, M.W., Gogotsi, Y., 25th anniversary article: MXenes: A new family of two-dimensional materials. *Adv. Mater.*, 26, 992–1005, 2014.
3. Benchakar, M., Loupias, L., Garnero, C., Bilyk, T., Morais, C., Canaff, C., Guignard, N., Morisset, S., Pazniak, H., Hurand, S., Chartier, P., Pacaud, J., Mauchamp, V., Barsoum, M.W., Habrioux, A., Célérier, S., One MAX phase, different MXenes: A guideline to understand thecrucial role of etching conditions on $Ti_3C_2T_x$ surface chemistry. *Appl. Surf. Sci.*, 530, 147209, 2020.
4. Boota, M., Anasori, B., Voigt, C., Zhao, M.Q., Barsoum, M.W., Gogotsi, Y., Pseudocapacitive electrodes produced by oxidant-free polymerization of pyrrole between the layers of 2D titanium carbide (MXene). *Adv. Mater.*, 28, 1517–1522, 2016.
5. Qin, L., Tao, Q., El Ghazaly, A., Fernandez-Rodriguez, J., Persson, P.O.Å., Rosen, J., Zhang, F., High-performance ultrathin flexible solid-state supercapacitors based on solution processable Mo1.33C MXene and PEDOT: PSS. *Adv. Funct. Mater.*, 28, 1703808, 2018.

6. Naguib, M., Mashtalir, O., Carle, J., Presser, V., Lu, J., Hultman, L., Gogotsi, Y., Barsoum, M.W., Two-dimensional transition metal carbides. *ACS Nano*, 6, 1322–1331, 2012.
7. Wang, X., Kajiyama, S., Iinuma, H., Hosono, E., Oro, S., Moriguchi, I., Okubo, M., Yamada, A., Pseudocapacitance of MXene nanosheets for high-power sodium-ion hybridcapacitors. *Nat. Commun.*, 6, 6544, 2015.
8. Ghidiu, M., Lukatskaya, M.R., Zhao, M.Q., Gogotsi, Y., Barsoum, M.W., Conductive two-dimensional titanium carbide clay with high volumetric capacitance. *Nature*, 516, 78–81, 2014.
9. Karlsson, L.H., Birch, J., Halim, J., Barsoum, M.W., Persson, P.O.Å., Atomically resolved structural and chemical investigation of single MXene sheets. *Nano Lett.*, 15, 4955–4960, 2015.
10. Munir, S., Rasheed, A., Rasheed, T., Ayman, I., Ajmal, S., Rehman, A., Shakir, I., Agboola, P.O., Warsi, M.F., Exploring the influence of critical parameters for the effective synthesis of high-quality 2D MXene. *ACS Omega*, 5, 26845–26854, 2020.
11. Naguib, M., Halim, J., Lu, J., Cook, K.M., Hultman, L., Gogotsi, Y., Barsoum, M.W., New two-dimensional niobium and vanadium carbides as promising materials for Li-ion batteries. *J. Am. Chem. Soc.*, 135, 15966–15969, 2013.
12. Yang, J., Naguib, M., Ghidiu, M., Pan, L.-M., Gu, J., Nanda, J., Halim, J., Gogotsi, Y., Barsoum, M.W., Two-dimensional Nb-based M4C3 solid solutions (MXenes). *J. Am. Ceram. Soc.*, 99, 660–666, 2016.
13. Anasori, B., Xie, Y., Beidaghi, M., Lu, J., Hosler, B.C., Hultman, L., Kent, P.R.C., Gogotsi, Y., Barsoum, M.W., Two-dimensional, ordered, double transition metals carbides (MXenes). *ACS Nano*, 9, 9507–9516, 2015.
14. Kim, S.J., Naguib, M., Zhao, M., Zhang, C., Jung, H.-T., Barsoum, M.W., Gogotsi, Y., High mass loading, binder-free MXene anodes for high areal capacity Li-ion batteries. *Electrochim. Acta*, 163, 246–251, 2015.
15. Panda, S., Deshmukh, K., Hussain, C.M., Pasha, S.K.K., 2D MXenes for combatting COVID-19 Pandemic: A perspective on latest developments and innovations. *FlatChem*, 33, 100377, 2022.
16. Wang, L., Zhang, H., Wang, B., Shen, C., Zhang, C., Hu, Q., Zhou, A., Liu, B., Synthesis and electrochemical performance of $Ti_3C_2T_x$ with hydrothermal process, Electron. *Mater. Lett.*, 12, 702–710, 2016.
17. Halim, J., Lukatskaya, M.R., Cook, K.M., Lu, J., Smith, C.R., Näslund, L.-Å., May, S.J., Hultman, L., Gogotsi, Y., Eklund, P., Barsoum, M.W., Transparent conductive two-dimensional titanium carbide epitaxial thin films. *Chem. Mater.*, 26, 2374–2381, 2014.
18. Lipatov, A., Alhabeb, M., Lukatskaya, M.R., Boson, A., Gogotsi, Y., Sinitskii, A., Effect of synthesis on quality, electronic properties and environmental stability of individual monolayer Ti_3C_2 MXene flakes. *Adv. Electron. Mater.*, 2, 1600255, 2016.
19. Urbankowski, P., Anasori, B., Makaryan, T., Er, D., Kota, S., Walsh, P.L., Zhao, M., Shenoy, V.B., Barsoum, M.W., Gogotsi, Y., Synthesis of two-dimensional titanium nitride Ti4N3 (MXene). *Nanoscale*, 8, 11385–11391, 2016.
20. Deshmukh, K., Muzaffar, A., Kovarik, T., Ahamed, M.B., Pasha, S.K.K., Chapter 1 Introduction to 2D MXenes: Fundamental aspects, MAX phases and MXene derivatives, current challenges, and future prospects, in: *Micro and Nano Technologies, MXenes and their Composites*, K.K. Sadasivuni, K. Deshmukh, S.K. Khadheer Pasha, T. Kovařik (eds.), pp. 1–47, Elsevier, 2022.
21. Wang, J., Tang, J., Ding, B., Malgras, V., Chang, Z., Hao, X., Wang, Y., Dou, H., Zhang, X., Yamauchi, Y., Hierarchical porous carbons with layer-by-layer motif architectures from confined soft-template self-assembly in layered materials. *Nat. Commun.*, 8, 15717, 2017.
22. Li, T., Yao, L., Liu, Q., Gu, J., Luo, R., Li, J., Yan, X., Wang, W., Liu, P., Chen, B., Zhang, W., Abbas, W., Naz, R., Zhang, D., Fluorine-free synthesis of high-purity $Ti_3C_2T_x$ (T=OH, O) *via* alkali treatment. *Angew. Chem. Int. Ed.*, 57, 6115–6119, 2018.

23. Sun, W., Shah, S.A., Chen, Y., Tan, Z., Gao, H., Habib, T., Radovic, M., Green, M.J., Electrochemical etching of Ti_2AlC to Ti_2CT_x (MXene) in low-concentration hydrochloric acid solution. *J. Mater. Chem. A*, 5, 21663–21668, 2017.
24. Borges, J. and Mano, J.F., Molecular interactions driving the layer-by-layer assembly of multilayers. *Chem. Rev.*, 114, 8883–8942, 2014.
25. Hao, L., Zhang, H., Wu, X., Zhang, J., Wang, J., Li, Y., Novel thin-film nanocompositemembranes filled with multi-functional $Ti_3C_2T_x$ nanosheets for task-specific solvent transport. *Compos. Part A: Appl. Sci. Manuf.*, 100, 139–149, 2017.
26. He, D., Sun, M., Cao, D., He, G., Chen, H., Rational design of nano-Fe_3O_4 encapsulated in3D honeycomb biochar for enhanced lithium storage performance. *Nanotechnology*, 33, 035401, 2021.
27. Lipatov, A., Alhabeb, M., Lu, H., Zhao, S., Loes, M.J., Vorobeva, N.S., Dall, Y., Agnese, Gao, Y., Gruverman, A., Gogotsi, Y., Sinitskii, A., Electrical and elastic properties of individual single-layer $Nb_4C_3T_x$ MXene flakes. *Adv. Electron. Mater.*, 6, 1901382, 2020.
28. Liu, Y., He, Y., Vargun, E., Plachy, T., Saha, P., Cheng, Q., Porous Ti_3C_2 MXene NiCo-MOF composites for enhanced lithium storage. *Nanomaterials*, 10, 4, 2020.
29. Heinz, W.R., Agirrezabal-Telleria, I., Junk, R., Berger, J., Wang, J., Sharapa, D.I., Gil-Calvo, M., Luz, I., Soukri, M., Studt, F., Wang, Y., Wöll, C., Bunzen, H., Drees, M., Fischer, R.A., Thermal defect engineering of precious group metal–organic frameworks: A case study on Ru/Rh-HKUST-1 analogues. *ACS Appl. Mater. Interfaces*, 12, 40635–40647, 2020.
30. Zhu, X., Liu, B., Hou, H., Huang, Z., Zeinu, K.M., Huang, L., Yuan, X., Guo, D., Hu, J., Yang, J., Alkaline intercalation of Ti_3C_2 MXene for simultaneous electrochemical detection of Cd(II), Pb(II), Cu(II) and Hg(II). *Electrochim. Acta*, 248, 46–57, 2017.
31. Si, J.-Y., Twiah, B., Sun, W.-L., Lin, B., Wang, C., Yuen, A.C., Yu, B., Li, A., Yang, W., Lu, H.-D., Chan, Q.N., A.U., Yeoh, G.H., Functionalization of MXene nanosheets for polystyrene towards high thermal stability and flame retardant properties. *Polymers*, 11, 6, 2019.
32. Chen, C., Boota, M., Xie, X., Zhao, M., Anasori, B., Ren, C.E., Miao, L., Jiang, J., Gogotsi, Y., Charge transfer induced polymerization of EDOT confined between 2D titanium carbide layers. *J. Mater. Chem. A*, 5, 5260–5265, 2017.
33. Zhang, Z., Yang, S., Zhang, P., Zhang, J., Chen, G., Feng, X., Mechanically strong MXene/Kevlar nanofiber composite membranes as high-performance nanofluidic osmotic power generators. *Nat. Commun.*, 10, 2920, 2019.
34. Tu, S., Jiang, Q., Zhang, X., Alshareef, H.N., Large dielectric constant enhancement in MXene percolative polymer composites. *ACS Nano*, 12, 3369–3377, 2018.
35. Ling, Z., Ren, C.E., Zhao, M.Q., Yang, J., Giammarco, J.M., Qiu, J., Barsoum, M.W., Gogotsi, Y., Flexible and conductive MXene films and nanocomposites with high capacitance. *Proc. Natl. Acad. Sci. U.S.A.*, 111, 16676–16681, 2014.
36. Khazaei, M., Ranjbar, A., Esfarjani, K., Bogdanovski, D., Dronskowski, R., Yunoki, S., Insights into exfoliation possibility of MAX phases to MXenes. *Phys. Chem. Chem. Phys.*, 20, 8579–8592, 2018.
37. Zhao, X., Vashisth, A., Prehn, E., Sun, W., Shah, S.A., Habib, T., Chen, Y., Tan, Z., Lutkenhaus, J.L., Radovic, M., Green, M.J., Antioxidants unlock shelf-stable $Ti_3C_2T_x$ (MXene) nanosheet dispersions. *Matter*, 1, 513–526, 2019.
38. Sobolčiak, P., Ali, A., Hassan, M.K., Helal, M.I., Tanvir, A., Popelka, A., Al-Maadeed, M.A., Krupa, I., Mahmoud, K.A., 2D $Ti_3C_2T_x$ (MXene)-reinforced polyvinyl alcohol (PVA) nanofibers with enhanced mechanical and electrical properties. *PloS One*, 12, 0183705, 2017.
39. Ul Haq, Y., Murtaza, I., Mazhar, S., Ahmad, N., Qarni, A.A., Ul Haq, Z., Khan, S.A., Iqbal, M., Investigation of improved dielectric and thermal properties of ternary nanocomposite PMMA/MXene/ZnO fabricated by *in-situ* bulk polymerization. *J. Appl. Polym. Sci.*, 137, 49197, 2020.

40. Zhang, P., Yang, X.-J., Li, P., Zhao, Y., Niu, Q.J., Fabrication of novel MXene(Ti_3C_2)/polyacrylamide nanocomposite hydrogels with enhanced mechanical and drug release properties. *Soft Matter*, 16, 162–169, 2020.
41. Xu, H., Zheng, D., Liu, F., Li, W., Lin, J., Synthesis of an MXene/polyaniline composite with excellent electrochemical properties. *J. Mater. Chem. A*, 8, 5853–5858, 2020.
42. Cao, J., Han, Y., Zheng, X., Wang, Q., Preparation and electrochemical performance of modified $Ti_3C_2T_x$/polypyrrole composites. *J. Appl. Polym. Sci.*, 136, 47003, 2019.
43. Chen, J., Chen, K., Tong, D., Huang, Y., Zhang, J., Xue, J., Huang, Q., Chen, T., CO_2 and temperature dual responsive "smart" MXene phases. *Chem. Commun.*, 51, 314–317, 2015.
44. Naguib, M., Saito, T., Lai, S., Rager, M.S., Aytug, T., Parans Paranthaman, M., Zhao, M.-Q., Gogotsi, Y., $Ti_3C_2T_x$ (MXene)–polyacrylamide nanocomposite films. *RSC Adv.*, 6, 72069–72073, 2016.
45. Shahzad, F., Alhabeb, M., Hatter, C.B., Anasori, B., Man Hong, S., Koo, C.M., Gogotsi, Y., Electromagnetic interference shielding with 2D transition metal carbides (MXenes). *Science*, 353, 1137–1140, 2016.
46. Carey, M., Hinton, Z., Sokol, M., Alvarez, N.J., Barsoum, M.W., Nylon-6/$Ti_3C_2T_z$ MXene nanocomposites synthesized by *in situ* ring opening polymerization of ε-caprolactam and their water transport properties. *ACS Appl. Mater. Interfaces*, 11, 20425–20436, 2019.
47. Pan, Y., Fu, L., Zhou, Q., Wen, Z., Lin, C.-T., Yu, J., Wang, W., Zhao, H., Flammability, thermal stability and mechanical properties of polyvinyl alcohol nanocomposites reinforced with delaminated $Ti_3C_2T_x$ (MXene). *Polym. Compos.*, 41, 210–218, 2020.
48. Sun, R., Zhang, H.B., Liu, J., Xie, X., Yang, R., Li, Y., Hong, S., Yu, Z.Z., Highly conductive transition metal carbide/carbonitride (MXene)@polystyrene nanocomposites fabricated by electrostatic assembly for highly efficient electromagnetic interference shielding. *Adv. Funct. Mater.*, 27, 1702807, 2017.
49. Zhou, J., Shi, D., Wang, Y., Chen, M., Dong, W., Bioinspired MXene/polyurethane plastic films with exceptional flexibility and toughness for electromagnetic interference shielding. *Mater. Res. Bull.*, 154, 111939, 2022.
50. Chen, X., Zhao, Y., Li, L., Wang, Y., Wang, J., Xiong, J., Du, S., Zhang, P., Shi, X., Yu, J., MXene/polymer nanocomposites: Preparation, properties, and applications. *Polym. Rev.*, 61, 80–115, 2021.
51. Wang, L., Chen, L., Song, P., Liang, C., Lu, Y., Qiu, H., Zhang, Y., Kong, J., Gu, J., Fabrication on the annealed $Ti_3C_2T_x$ MXene/epoxy nanocomposites for electromagnetic interference shielding application. *Compos. Part B: Eng.*, 171, 111–118, 2019.
52. Feng, A., Hou, T., Jia, Z., Zhang, Y., Zhang, F., Wu, G., Preparation characterization of epoxy resin filled with TiCTx MXene nanosheets with excellent electric conductivity. *Nanomaterials*, 10, 1, 2020.
53. Li, W., Song, Z., Zhong, J., Qian, J., Tan, Z., Wu, X., Chu, H., Nie, W., Ran, X., Multilayer-structured transparent MXene/PVDF film with excellent dielectric and energy storage performance. *J. Mater. Chem. C*, 7, 10371–10378, 2019.
54. Jin, X., Wang, J., Dai, L., Liu, X., Li, L., Yang, Y., Cao, Y., Wang, W., Wu, H., Guo, S., Flame-retardant poly(vinyl alcohol)/MXene multilayered films with outstanding electromagnetic interference shielding and thermal conductive performances. *Chem. Eng. J.*, 380, 122475, 2020.
55. Tanvir, A., Sobolciak, P., Popelka, A., Mrlik, M., Spitalsky, Z., Micusik, M., Prokes, J., Krupa, I., Electrically conductive transparent polymeric nanocomposites modified by 2D $Ti_3C_2T_x$ (MXene). *Polymers*, 11, 8, 2019.
56. Wang, D., Lin, Y., Hu, D., Jiang, P., Huang, X., Multifunctional 3D-MXene/PDMS nanocomposites for electrical, thermal and triboelectric applications. *Compos. Part A: Appl. Sci. Manuf.*, 130, 105754, 2020.

57. Mazhar, S., Qarni, A.A., Ul Haq, Y., Ul Haq, Z., Murtaza, I., Promising PVC/MXene based flexible thin film nanocomposites with excellent dielectric, thermal and mechanical properties. *Ceram. Int.*, 46, 12593–12605, 2020.
58. Seyedin, S., Uzun, S., Levitt, A., Anasori, B., Dion, G., Gogotsi, Y., Razal, J.M., MXene composite and coaxial fibers with high stretchability and conductivity for wearable strain sensing textiles. *Adv. Funct. Mater.*, 30, 1910504, 2020.
59. Kang, R., Zhang, Z., Guo, L., Cui, J., Chen, Y., Hou, X., Wang, B., Lin, C.-T., Jiang, N., Yu, J., Enhanced thermal conductivity of epoxy composites filled with 2D transition metal carbides (MXenes) with ultralow loading. *Sci. Rep.*, 9, 9135, 2019.
60. Cao, Y., Deng, Q., Liu, Z., Shen, D., Wang, T., Huang, Q., Du, S., Jiang, N., Lin, C.-T., Yu, J., Enhanced thermal properties of poly(vinylidene fluoride) composites with ultrathin nanosheets of MXene. *RSC Adv.*, 7, 20494–20501, 2017.
61. Raagulan, K. and Braveenth, R., Electromagnetic shielding by MXene-graphene-PVDF composite with hydrophobic, lightweight and flexible graphene coated fabric. *Materials*, 11, 1803, 2018.
62. Wan, W., Tao, M., Cao, H., Zhao, Y., Luo, J., Yang, J., Qiu, T., Enhanced dielectric properties of homogeneous $Ti_3C_2T_x$ MXene@SiO_2/polyvinyl alcohol composite films. *Ceram. Int.*, 46, 13862–13868, 2020.
63. Gao, X.-H., Wang, J.-W., Liu, D.-N., Wang, X.-Z., Wang, H.-Q., Wei, L., Ren, H., Enhanced dielectric properties of acrylic resin elastomer (AE)-based percolative composite with modified MXene. *Polym. Test.*, 102, 107344, 2021.
64. Deng, Q., Zhou, F., Qin, B., Feng, Y., Xu, Z., Eco-friendly poly(vinyl alcohol)/delaminated V_2C MXene high-k nanocomposites with low dielectric loss enabled by moderatepolarization and charge density at the interface. *Ceram. Int.*, 46, 27326–27335, 2020.
65. Ekinci, K.Z., Çelik, S. Ü., Bozkurt, A., Enhancing the anhydrous proton conductivity of sulfonated polysulfone/polyvinyl phosphonic acid composite membranes with hexagonal boron nitride. *Int. J. Polym. Mater. Polym. Biomater.*, 64, 683–689, 2015.
66. Zhang, J., Liu, Y., Lv, Z., Zhao, T., Li, P., Sun, Y.-l., Wang, J., Sulfonated $Ti_3C_2T_x$ toconstruct proton transfer pathways in polymer electrolyte membrane for enhanced conduction. *Solid State Ionics*, 310, 100–111, 2017.
67. Psarras, G.C., Manolakaki, E., Tsangaris, G.M., Electrical relaxations in polymericparticulate composites of epoxy resin and metal particles. *Compos. Part A: Appl. Sci. Manuf.*, 33, 375–384, 2002.
68. Feng, A., Hou, T., Jia, Z., Zhang, Y., Preparation and characterization of epoxy resin filled with $Ti_{(3)}C_{(2)}T_{(x)}$ MXene nanosheets with excellent electric conductivity. *Nanomaterials*, 10, 162, 2020.
69. Wu, W., Zhao, W., Sun, Q., Yu, B., Yin, X., Cao, X., Feng, Y., Li, R.K.Y., Qu, J., Surface treatment of two dimensional MXene for poly (vinylidene fluoride) nanocomposites with tunable dielectric permittivity. *Compos. Commun.*, 23, 100562, 2021.
70. Zeraati, A.S., Arjmand, M., Sundararaj, U., Silver nanowire/MnO_2 nanowire hybrid polymer nanocomposites: Materials with high dielectric permittivity and low dielectric loss. *ACS Appl. Mater. Interfaces*, 9, 14328–14336, 2017.
71. Ding, Z., Yuan, L., Liang, G., Gu, A., Thermally resistant thermadapt shape memorycrosslinked polymers based on silyl ether dynamic covalent linkages for self-folding and self-deployable smart 3D structures. *J. Mater. Chem. A*, 7, 9736–9747, 2019.
72. Baniecki, J.D., Shioga, T., Kurihara, K., Kamehara, N., Investigation of the importance of interface and bulk limited transport mechanisms on the leakage current of high dielectric constant thin film capacitors. *J. Appl. Phys.*, 94, 6741–6748, 2003.

73. Wu, W., Huang, X., Li, S., Jiang, P., Toshikatsu, T., Novel three-dimensional zinc oxide superstructures for high dielectric constant polymer composites capable of withstanding high electric field. *J. Phys. Chem. C*, 116, 24887–24895, 2012.
74. Feng, Y., Qiu, H., Mao, B., Bo, M., Deng, Q., Preparation of hybrid ceramic/PVC composites showing both high dielectric constant and breakdown strength ascribed to interfacial effect between V_2C MXene and Cu_2O. *Colloids Surf. A: Physicochem. Eng. Asp.*, 630, 127650, 2021.
75. Feng, Y., Zhou, F., Dai, Y., Xu, Z., Deng, Q., Peng, C., High dielectric and breakdown performances achieved in PVDF/graphene@MXene nanocomposites based on quantum confinement strategy. *Ceram. Int.*, 46, 17992–18001, 2020.
76. Hassan, G., Sajid, M., Choi, C., Highly sensitive and full range detectable humidity sensor using PEDOT: PSS, methyl red and graphene oxide materials. *Sci. Rep.*, 9, 15227, 2019.
77. Choi, J., Kim, Y.-J., Cho, S.-Y., Park, K., Kang, H., Kim, S.J., Jung, H.-T., *In situ* formation of multiple schottky barriers in a Ti_3C_2 MXene film and its application in highly sensitive gas sensors. *Adv. Funct. Mater.*, 30, 2003998, 2020.
78. Seroka, N.S. and Mamo, M.A., Application of functionalised MXene-carbon nanoparticle-polymer composites in resistive hydrostatic pressure sensors. *SN Appl. Sci.*, 2, 413, 2020.
79. Han, S., Duan, Z., Meng, X., Zhao, Q., Zhang, L., Ouyang, X., Ma, N., Wei, H., Zhang, X., Tunable fabrication of conductive Ti3C2Tx MXenes *via* inflating a polyurethane balloonfor acute force sensing. *Langmuir*, 36, 1298–1304, 2020.
80. Wang, D., Zhang, D., Li, P., Yang, Z., Mi, Q., Yu, L., Electrospinning of flexible poly (vinyl alcohol)/MXene nanofiber-based humidity sensor self-powered by monolayer molybdenum diselenide piezoelectric nanogenerator. *Nano-Micro Lett.*, 13, 57, 2021.
81. Sun, Z., Wang, Z., Tian, Y., Wang, G., Wang, W., Yang, M., Wang, X., Zhang, F., Pu, Y., Progress, outlook, and challenges in lead-free energy-storage ferroelectrics. *Adv. Electron. Mater.*, 6, 1900698, 2020.
82. Feng, Y., Deng, Q., Peng, C., Hu, J., Li, Y., Wu, Q., Xu, Z., An ultrahigh discharged energydensity achieved in an inhomogeneous PVDF dielectric composite filled with 2D MXene nanosheets *via* interface engineering. *J. Mater. Chem. C*, 6, 13283–13292, 2018.
83. Luo, B., Wang, X., Wang, Y., Li, L., Fabrication, characterization, properties and theoretical analysis of ceramic/PVDF composite flexible films with high dielectric constantand low dielectric loss. *J. Mater. Chem. A*, 2, 510–519, 2014.
84. Dang, Z.-M., Zhou, T., Yao, S.-H., Yuan, J.-K., Zha, J.-W., Song, H.-T., Li, J.-Y., Chen, Q., Yang, W.-T., Bai, J., Advanced calcium copper titanate/polyimide functional hybrid films with high dielectric permittivity. *Adv. Mater.*, 21, 2077–2082, 2009.
85. Tian, G., Deng, W., Xiong, D., Yang, T., Zhang, B., Ren, X., Lan, B., Zhong, S., Jin, L., Zhang, H., Deng, L., Yang, W., Dielectric micro-capacitance for enhancing piezoelectricity via aligning MXene sheets in composites. *Cell Rep. Phys. Sci.*, 3, 100814, 2022.
86. Tu, S., Jiang, Q., Zhang, J., He, X., Hedhili, M.N., Zhang, X., Alshareef, H.N., Enhancement of dielectric permittivity of $Ti_3C_2T_x$ MXene/polymer composites by controlling flake size and surface termination. *ACS Appl. Mater. Interfaces*, 11, 27358–27362, 2019.
87. Hu, J., Zhang, S., Tang, B., 2D filler-reinforced polymer nanocomposite dielectrics for high-k dielectric and energy storage applications. *Energy Storage Mater.*, 34, 260–281, 2021.
88. Jiang, J., Shen, Z., Cai, X., Qian, J., Dan, Z., Lin, Y., Liu, B., Nan, C.-W., Chen, L., Shen, Y., Polymer nanocomposites with interpenetrating gradient structure exhibiting ultrahigh discharge efficiency and energy density. *Adv. Energy Mater.*, 9, 1803411, 2019.
89. Zeng, Y., Rao, S., Xiong, C., Du, G., Fan, Z., Chen, N., Enhanced dielectric and mechanical properties of $CaCu_3Ti_4O_{12}/Ti_3C_2T_x$ MXene/silicone rubber ternary composites. *Ceram. Int.*, 48, 6116–6123, 2022.

9

MXenes-Reinforced Polymer Composites for Microwave Absorption and Electromagnetic Interference Shielding Applications

B. D. S. Deeraj[1], Jitha S. Jayan[2], Asok Aparna[2], Appukuttan Saritha[2] and Kuruvilla Joseph[1]*

[1]*Department of Chemistry, Indian Institute of Space Science and Technology, Thiruvananthapuram, Kerala, India*

[2]*Department of Chemistry, Amrita Vishwa Vidyapeetham, Amritapuri, Kollam, Kerala, India*

Abstract

Two dimensional (2D) nanomaterials have enticed notable consideration among the research fraternity for their outstanding application in various fields. Among 2D materials, MAX phases and its derivatives called MXenes gained enormous interest owing to their easiness in synthesis and their stable nature as colloidal suspensions in aqueous and a wide range of organic solvents. The blends of properties, including large surface area, hydrophobicity, absorption capability, enhanced electrical conductivity, and better surface reactive nature, have rendered MXenes as significant candidates for a diversity of applications like energy storage, microwave absorption, electromagnetic interference (EMI) shielding, conductive electrodes etc. Current research works point out that MXenes outshine the existing two-dimensional nanomaterials. The chapter will cover the structure, properties, and associated applications of MXenes, focusing on their microwave absorption and EMI shielding applications. Additionally, the chapter summarizes the EMI shielding and microwave absorption features of MXene/polymer composites with a special mention on MXene based other materials for the same.

Keywords: 2D MXene, MXene/polymer composites, EMI shielding, microwave absorption, electrospinning

9.1 Introduction to MXenes

Owing to the ridiculous aspect ratios and few atoms layer thickness, 2D materials achieved magnificent interest among researchers from various fields. In the 1970s, the invention of the catalytic behavior of transition metal based carbides and nitrides, that are considered as chemically stable, wear resistant and hard materials paved the way for the investigation of their energy storage applications [1]. After the discovery of in 2D transition metal carbides and nitrides called "MXenes" in 2011 and after numerous research works happened in this

**Corresponding author*: kuruvilla@iist.ac.in

arena, MXenes highly attracted by the entire scientific community [2]. Various MXenes with enhanced chemical diversity as well as structural complexity [3], having different characteristics including surface functional tunability [4], increased spin-orbit coupling [5], and wide elemental composition possibilities, are developed, which provide enormous applications in numerous fields.

9.1.1 Structure and Properties

The exfoliation of these 2D transition metal carbides and nitrides through selective etching of 'A' components from MAX phases produces "MXenes" in which the suffix 'ene' indicates their similarity to graphene [6]. MAX phases are denoted as Mn+1AXn where M represents a transition metal, A denotes a set 13-16 element, X denotes carbon, nitrogen otherwise their mix, and n varies from 1 to 3 [7]. A combination of ionic, metallic, and covalent nature was exhibited by the M-X bond, whereas the metallic nature was shown by the M-A bond [8]. The selective removal of 'A' layers from the MAX phases generates MXenes with a common chemical formula of Mn+1XnTz, where M denotes an early transition metal, X denotes carbon, nitrogen, or their mix, Tz denotes different terminating moieties including fluorine (F), hydroxyl (-OH), or oxygen (O) atoms which are able to provide hydrophilic nature to their surface [9]. Since the n values varies from 1-3, three kinds of lattice structures are known for MXenes, as illustrated in Figure 9.1 [9]. More than one transition metal element can also be found in MXene and are attained in two forms such as ordered phases which forms a sandwich of mono/bilayer of one transition metal between another type layer, and solid solutions in which two distinct transition metals are randomly distributed in the M layers [10]. The structure of MXene was first introduced using density functional theory (DFT) simulations and in common, they have hexagonal close packing (HCP) crystal structure.

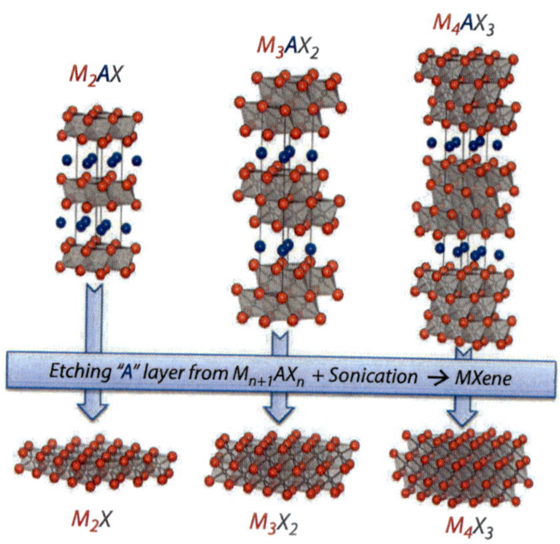

Figure 9.1 MAX phase materials and MXenes [9]. Copyright 2014, Reproduced with permission from Wiley-VCH.

MXenes possess excellent properties which ensure its applications in different fields. They own a high negative zeta potential which provide better stability to MXene colloidal solution which further helps to develop MXene films, and their hydrophilicity helps to make good dispersion in water as well as various organic solvents [11, 12]. Bare MXenes possess metallic character like that of their MAX precursors and they are converted to semiconductors/metals after surface terminations. hydroxyl or fluorine terminations make them semiconducting with a bandgap of 0.05 eV and 0.1 eV correspondingly. Semiconducting property can also be generated by altering the elemental composition of transition metal layers. In a research work by Anasori *et al.* [13], replacement of two layers of Ti in $Ti_3C_2(OH)_2$ with Mo layers was done and noticed the formation of semiconducting nature with a bandgap of 0.05 eV owing to the as-mentioned property of Mo. In addition to these electronic properties, MXenes own excellent mechanical properties. Zha *et al.* [14] evaluated the mechanical characteristics of newly fabricated Mo_2C MXene and noticed an elastic modulus of 312 ± 10 GPa, ideal strength of 20.8 GPa with a critical strain of 0.086 which confirms the outstanding mechanical strength of as-mentioned MXene. They also analyzed the thermal characteristics of MXene and observed a thermal conductivity value of 48.4 $Wm^{-1} K^{-1}$ which can be raised by the introduction of n-type dopants. The enhanced miscibility, electron rich density, ability to absorb electromagnetic waves, flexibility, strength, and better surface area to volume ratio provides unlimited applications to MXenes.

Numerous researchers reported the structural varieties of MXene which showcases their importance in various fields. Different characterization methods including Transmission Electron Microscopy (TEM), Scanning Electron Microscopy (SEM), Atomic Force Microscopy (AFM), Raman spectroscopy etc. helps to unveil the structural diversity of

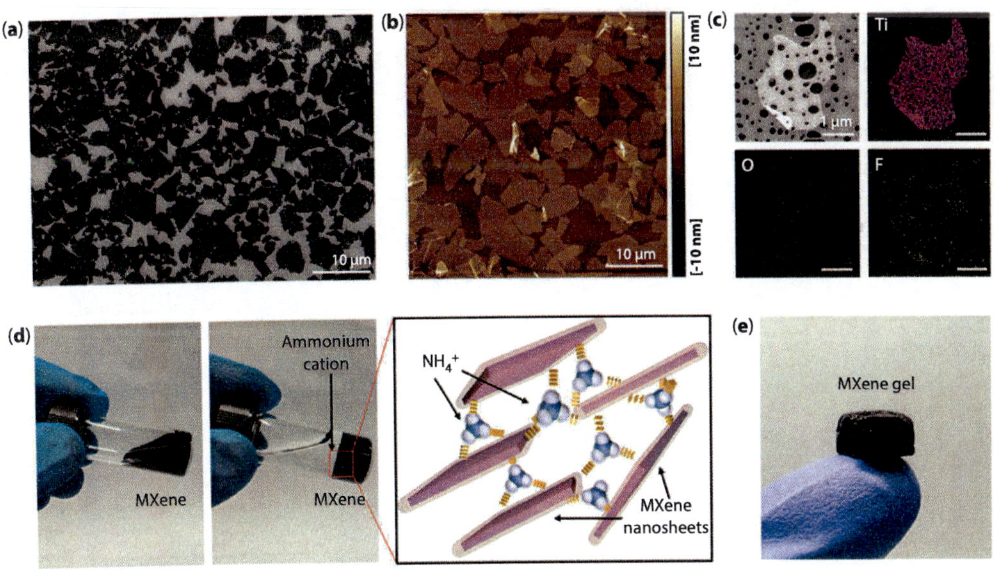

Figure 9.2 MXene and their gelation (a) SEM image (b) AFM image (c) TEM image and elemental mapping. (d) Optical images of dispersion and gelled MXene and schematic representation of the electrostatic interaction (e) Optical image of gel [15]. Copyright 2021, Reproduced with permission from American Chemical Society.

MXenes. The formation of MXene hydrogels from MXene colloid nanosheets and their further transformation into MXene fibers was reported by Shin *et al.* [15]. They showed the morphology of coated MXene as represented in Figure 9.2 and the morphology of MXene fibers as represented in Figure 9.3 [15]. The development of small as well as large crystals of MXene owing to the variations in sources are also reported [16]. Large crystals are generated from graphite and TiC sources where carbon lampblack generates small crystals as depicted in Figure 9.4 [16].

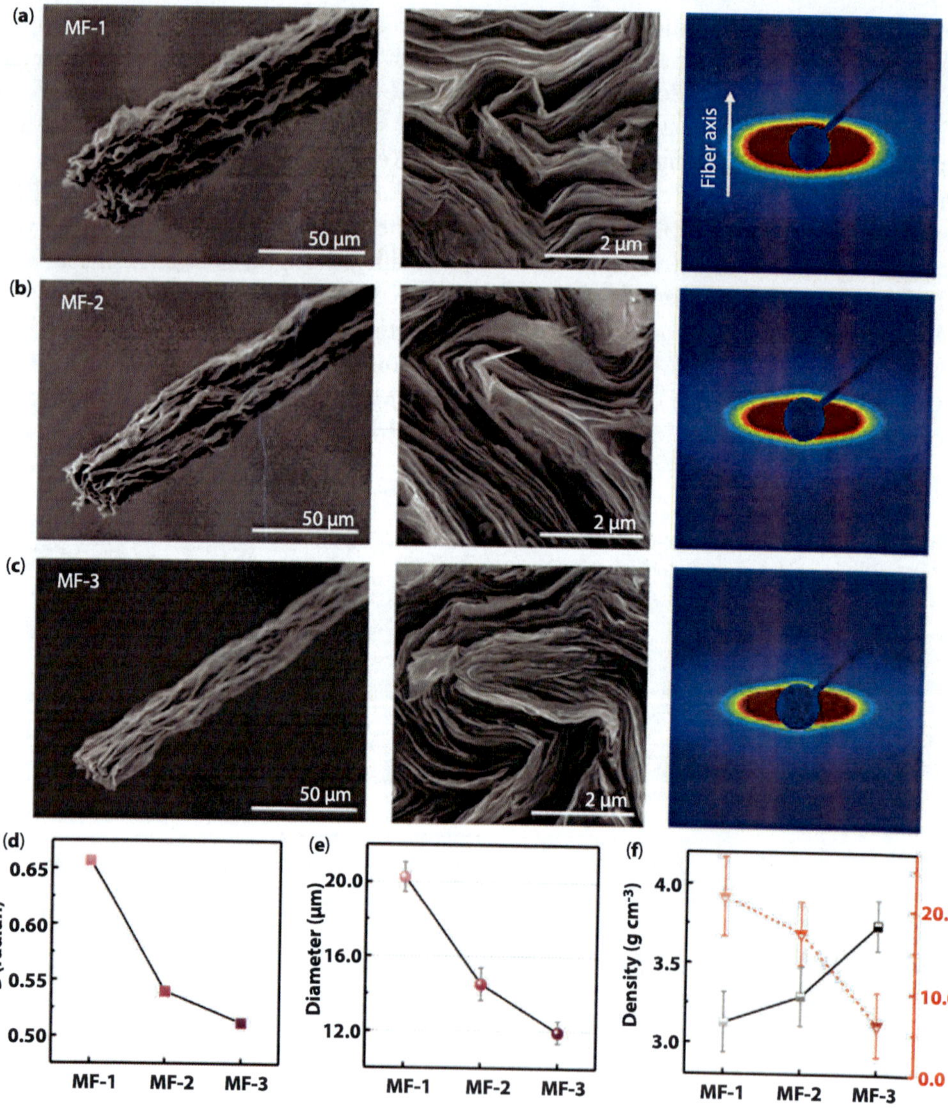

Figure 9.3 SEM image and SAXS of MF-1 (a), MF-2 (b), and MF-3 (c). (d) orientation change of the microvoids, (e) diameter, (f) density, and porosity [15]. Copyright 2021, Reproduced with permission from American Chemical Society.

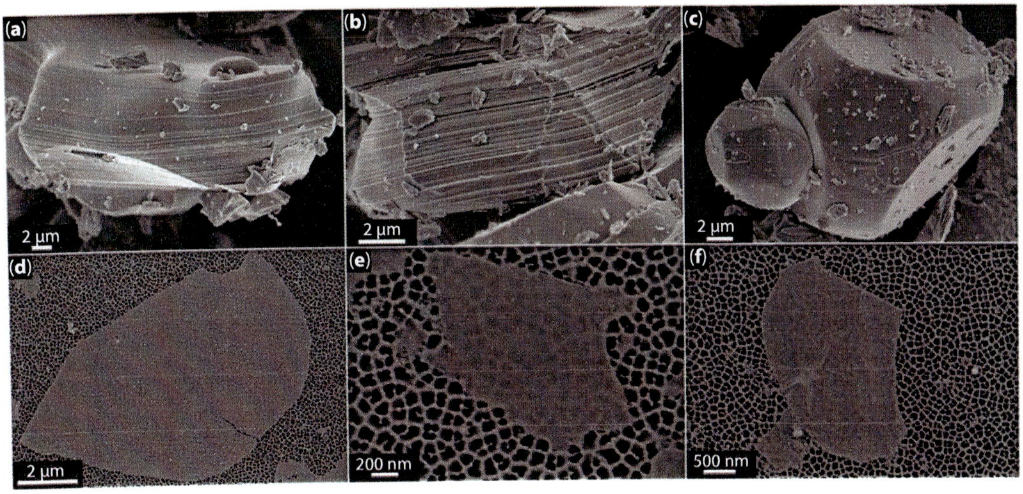

Figure 9.4 SEM micrographs of Ti_3AlC_2 synthesized by (a) graphite, (b) TiC, and (c) carbon lampblack; $Ti_3C_2T_x$ MXene produced by a MILD-like process, using (d) graphite, (e) TiC, and (f) carbon lampblack-derived MAX phases [16]. Copyright 2019, Reproduced with permission from American Chemical Society.

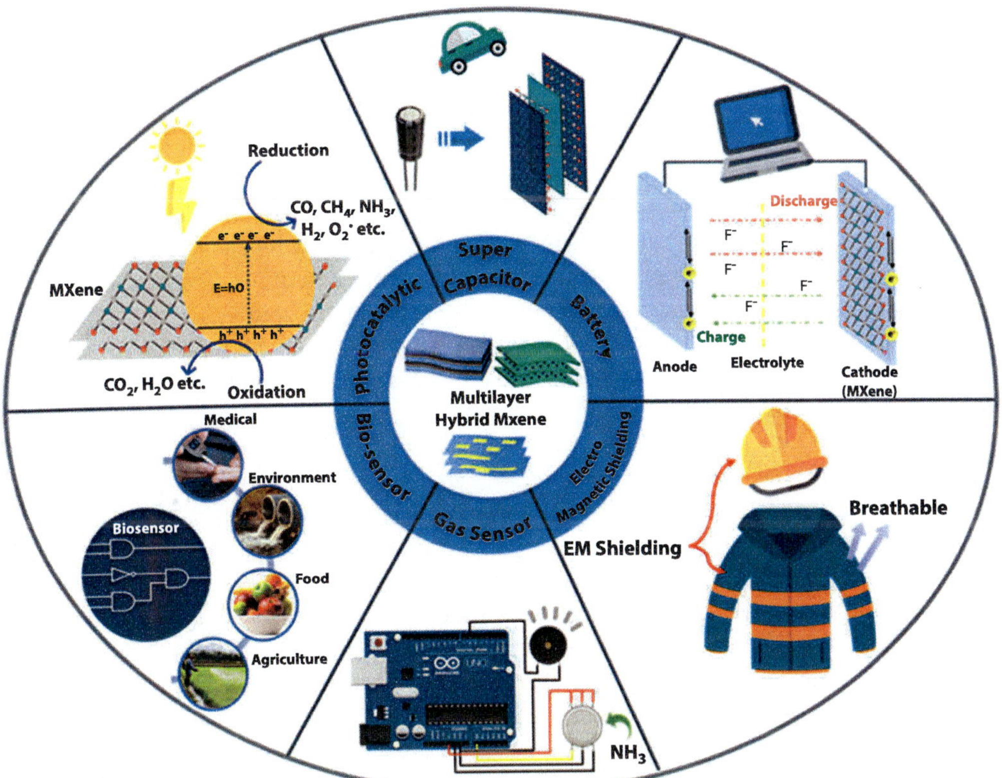

Figure 9.5 Explored properties and applications of MXenes [31]. Copyright 2022, Reproduced with permission from American Chemical Society.

9.1.2 Applications

Since MXene holds excellent properties, their places of usage are countless. They find application in numerous areas including energy storage [17], EMI shielding [18], microwave absorption [19], biosensors [20], water purification [21], nanomedicine [22] in which they are found either as flexible thin films [23], as anchors in hybrid systems [24], as wearable materials [25] or as incorporated in various matrices [26]. They are promising electrodes for lithium, sodium, and aluminium ion batteries owing to their enhanced electrical conductivity and capability to diffuse ions between layers. The mechanism for storing charge in MXene is pseudocapacitive in which surface redox reactions causes the transferring of electron charge [27]. Recently, various works are reported based on the environmental application of MXenes. They can act as an effective absorbent and as a membrane for separating different pollutants such as metal ions, gases, and dyes [28]. Owing to the direct charge transfer mechanism, they are extensively used as gas sensors and researchers found out the enhanced affinity of T_3C_2 for selective CO_2 capturing compared to N_2 capture. In addition, they can sense ethanol, acetone, ammonia and methanol even at room temperature [29, 30]. Among all these applications, EMI shielding got enormous interest owing to the flexibility, ability to form thin films, better conductivity of MXenes. Various applications of MXene are represented in Figure 9.5 [31].

9.2 Materials for EMI Shielding and Microwave Absorption

EMI not only affecting the working of electronic devices but also badly affects the human health [32, 33]. Microwave-absorbing materials with solid absorption over a wide frequency range are needed to overcome the EMI problem. Carbon materials, magnetic materials [34, 35], intrinsic conductive polymer [36], MXenes [37], ceramic materials like silicon carbide and barium titanate [38, 39], transition metal dichalcogenides [40, 41], etc., are used as EMI shielding and/or absorbing materials. Conductive polymer composites possess better corrosion resistance, manageable conductivity, higher temperature resistance, and the presence of localized/bound charges that lead to strong polarization and relaxation effects. These allow them to be used as suitable materials for dissipation of electrostatic charge as well as EMI shielding [42, 43]. The non-redox doping, better environmental stability and low cost of polyaniline gives a distinct position among other conducting polymers.

In case of EMI shielding, the term 'shielding effectiveness' (SE) is used to describe the ability of the material shield to obstruct electromagnetic (EM) waves. It is expressed in decibels (dB). The absorption component contributing to shielding is denoted by SE_A and the reflection component of shielding is denoted by SE_R. The more the SE value the more is the capability of the shielding materials in attenuating the EM waves. We are looking for a SE value of 20 dB or more for commercial applications. The value 20 dB represents 99% attenuation, which means the shield is able to restrict the EM waves up to 99%. The more the value the more is the attenuation of the EM energy. Lakshmi *et al.* [44] studied polyaniline–polyurethane composite for microwave absorption and EMI shielding applications. It is found that the composite's shielding ability enhanced with the sample's thickness and can ideally shield at 2.23 and 8.82 GHz. Carbon materials, like carbon black, carbon nanotubes, graphite, carbon fibers (CFs) and filaments, are also used as ideal

Table 9.1 Details of materials used for EMI shielding and microwave absorption applications.

Dimensionality	Materials	Fabricated system	Preparation method	Ref.
One dimensional (1D)	Bagasse fiber	Bagasse fiber/polyaniline composites	*In-situ* polymerization	[53]
	1D structures of polypyrrole	1D structure of polypyrrole	From dyes	[54]
	Broccoli-like short carbon fibers (SCF)	SCF@TiO_2 composites	One-pot hydrothermal method	[55]
	MnO_2 nanotubes	MnO@N-doped carbon nanotubes	Hydrothermal method and thermal decomposition of pyrrole	[56]
Two dimensional (2D)	$Ti_3C_2T_x$	$Ti_3C_2T_x$@$NiCo_2O_4$ composite	Hydrothermal method followed by annealing process	[57]
	Basalt slabs	two-dimensional plate@ shell basalt@NiO	Water bath method and then after by sintering process	[58]
	CoNi alloys	Cobalt and nickel alloys@mesoporous carbon (CoNi@MC) composites	Pyrolysis	[59]
	MOF	CuHT (HT = 4-hydroxythiophenol)	--	[60]
Three dimensional (3D)	Macroporous polymer scaffold (PS)	TiO_2 Aerogel-PS composites	High internal phase emulsion technique (HIPE) and chemical liquid deposition	[61]
	Graphene nanosheets wrapped melamine foam	Thermoplastic polyurethane (TPU)-graphene nanosheets @melamine foam	Dip-drying method	[62]

microwave absorbers [45–47]. Kumar *et al.* [48] used composites containing nanoscale iron/iron-carbide-graphite particles incorporated in carbon matrix and polyvinylidene fluoride and it is found to effectively enhance the EMI shielding by absorption of electromagnetic wave in the composite coating.

Yang *et al.* [49] studied the shielding performance of methyl vinyl silicon rubber nanocomposite foams with Fe_3O_4 and MWCNT. Fe_3O_4 and MWCNT offered better magnetic and dielectric loss which increased the microwave-absorbing capability of the nanocomposite foams. Flexible vinyl silicon rubber /MWCNTs/Fe_3O_4 foams having a density of 0.48 g·cm^{-3} and 3.95 volume % fillers loading showed EMI shielding efficiency of about 27.5 dB with a conductivity of about 14.6 S/m in the X-band. About 64% absorption efficiency is attained in the range of 8.2 to 12.4 GHz of frequency. Silicone rubber based nanocomposite foams also showed outstanding EMI shielding stability by retaining 96.2% of efficiency afterwards 1000 bending. Wang *et al.* [50] synthesized carbon nanotube (CNT)/polyimide (PI) foam incorporating polyvinyl pyrrolidone as surfactant for EMI shielding and microwave absorption. This showed an effective EMI shielding up to 41.1 dB at a form density of 32.1 mg·cm^{-3}. It also displayed an excellent absorption coefficient of up to 82.3%. In another work Li *et al.* [51] used carbon nanotube (CNT)/chitosan (CS) foams prepared via freeze- drying method for studying EMI shielding and microwave absorption. The composite showed a better effective EMI shielding of 37.6 dB with a density of 17.6 mg·cm^{-3} and showed an absorption coefficient of 81.73% at higher EMI shielding effectiveness of 37.6 dB. Interestingly, 2D MXene materials possess abundant natural defects, larger specific surface area, distinct layered structure and metallic properties [52]. MXenes can also be used as a conductive filler showing excellent performance for EMI shielding applications. Table 9.1 shows the details of various materials used for EMI shielding and Microwave absorption applications [53–62].

9.3 MXenes-Based Materials for EMI Shielding and Microwave Absorption

MXenes falls under the advanced class of materials that are having promising scope in EMI shielding and absorption applications. Since their first report, these materials have proved themselves to be promising materials for a wide variety of applications especially EM wave absorbing/reflecting purposes [63–67]. Especially $Ti_3C_2T_X$ based systems are widely investigated for this application. This section classifies briefly how these MXene based materials can be categorized.

9.3.1 MXenes

In the case of MXenes, studies related to the structure, synthesis route, dielectric properties and electrical conductivity are being studied related to EMI shielding and microwave absorption [52]. There is wide array of studies related to their structural optimization [68, 69], their hybrids and composites for enhanced absorption and shielding.

9.3.2 MXenes/Carbon Composites

For shielding applications carbon materials are the mostly obvious choice. Materials like carbon fibers, carbon nanotubes, GO, graphite, carbon black and rGO are extensively used for adsorption and shielding applications [70, 71]. These materials possess interesting properties like electrical conductivity, stability, good surface area, and light weight. MXenes are often mixed with carbon materials for employing them in advanced shielding applications [72–74].

9.3.3 MXenes/Magnetic Materials

In shielding applications, magnetic materials are used because of their unique permeability. These materials include metals, metal oxides and more. Few researchers combined magnetic materials with MXenes to attenuate electromagnetic waves [75, 76].

9.3.4 MXenes/Polymer Composites

MXenes are often mixed with different polymers and MXene/polymer composites were prepared for multifunctional applications. conducting polymers were used along with MXene materials to selective tailor shielding materials [77, 78]. A section related to this is mentioned in the later stage.

9.3.5 Hybrid Combinations

Different material combinations like semiconductors, metal oxides, dielectric/carbon materials, dielectric/magnetic materials are used with MXenes to developed shielding materials. More details regarding the shielding behavior is presented in the following sections. In Figure 9.6, the conductivities of different MXenes are presented [79].

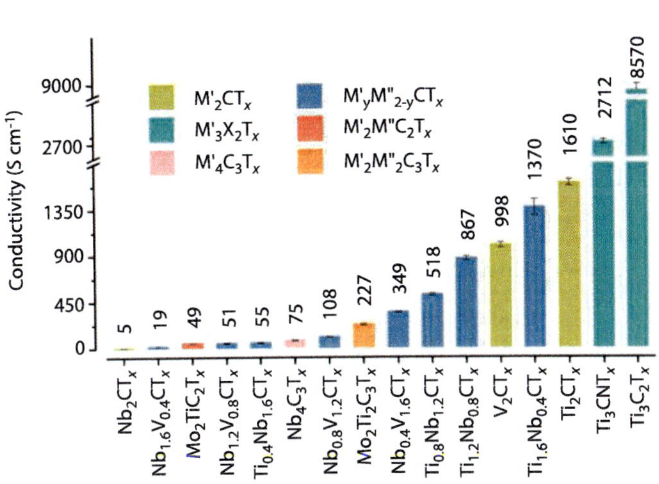

Figure 9.6 Conductivities of different MXenes [79]. Copyright 2020, Reproduced with permission from American Chemical Society.

9.4 EMI Shielding Mechanisms for MXene-Based Materials

Generally, reflection and absorption are the two main modes in which shielding takes place. In Figure 9.7, a diagrammatic representation of EMI shielding is presented [70]. The reflection is because of the impedance discrepancy between the electromagnetic (EM) wave and shield while absorption is because of presence of magnetic and electrical dipoles in the shield material. The complex dielectric and magnetization of the shield materials constitute to EMI shielding either by reflection or absorption. Usually for absorption dominant materials (SE_A > 10 dB), the effect of multiple reflections is not considered. In case of MXenes, different additives are added or MXenes themselves are used as fillers in other materials for improving the shielding property. The dielectric losses and magnetic losses contribute to overall loss of EM wave energy and thereby helps in shielding.

The research interest is on developing wave absorption dominant materials as the reflection of EM waves by the shield possess secondary EMI, which is not the solution for the problem. MXenes constitutes such wave absorption dominant behavior. In MXenes case, here conducting core and electron hopping between the sheets makes them conductive and provides platform for conduction losses. By improving this conductivity with functionalization or additives can make these MXenes better shielding materials. For materials with dominant dielectric properties, the polarization occurs in different modes, namely (a) ionic based, (b) electronic based, (c) interfacial based and (d) dipole based. But only interfacial and dipole polarization plays effective role in dielectric losses in GHz range. So, these two polarizations play major role in dielectric losses in this frequency range. While the ionic and electronic polarizations play a vital role in THz and PHz frequency range. For the materials with dominant magnetic properties, the magnetic loss is owing to hysteresis, natural and domain wall resonance, magnetic aftereffect loss and eddy current losses [80–82]. Cao *et al.*, made significant contributions in understanding the losses mechanisms in MXene based systems [81, 83]. There studies showed that in case of MXenes and their composites, along with the conductivity, the interfacial polarization occurs and it contributes to overall

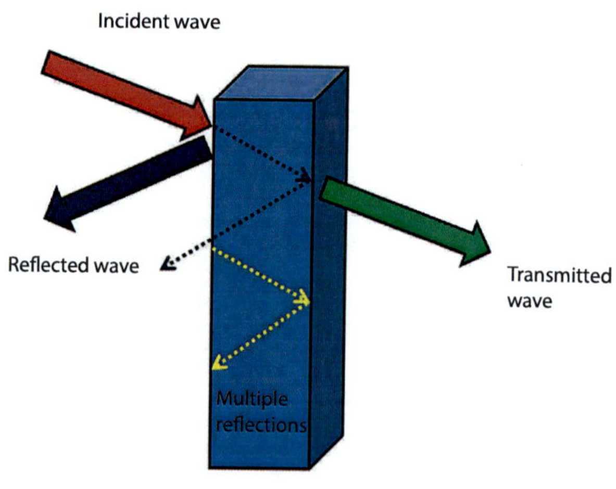

Figure 9.7 Scheme of EMI shielding mechanism [70]. Copyright 2020, Reproduced with permission from Elsevier.

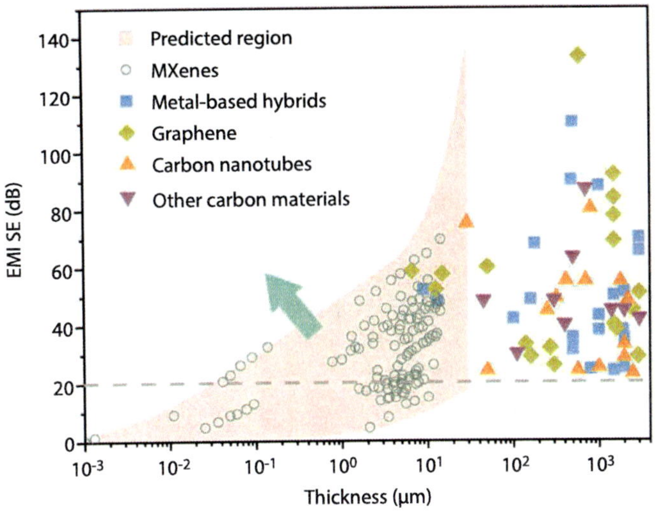

Figure 9.8 Shielding performance with respect to thickness [79]. Copyright 2020, Reproduced with permission from American Chemical Society.

shielding [84]. The testing range of frequency, kind of material, shape, as well as size of materials and their loading in composites are found to have significant effect on the shielding performance. The next section with different MXenes with polymer matrices and their shielding performance. In Figure 9.8, the performance of materials for shielding applications is presented [79].

9.5 MXenes/Polymer Composites for EMI Shielding and Microwave Absorption

For the gadgets to be used for a continued period of time, electronic and electric materials must be protected from EMI. The increased data carrying capacity of 5G networks is anticipated to have an impact on the widespread production of electronic devices [64, 79]. Therefore, there is a strong likelihood that electromagnetic waves (EMW) may be generated in this era of Internet of Things, which could put the safety of electrical devices at risk [79, 85]. Usually polymer composites are ideal materials for functional applications because of low weight to performance and corrosion resistance properties. Many reports highlight the use of polymer composites in microwave absorption and shielding applications [71, 86].

It was reported that a $Ti_3C_2T_x$ film can exhibit a shielding effectiveness of 92 dB owing to the presence of charge carriers present over the surface of $Ti_3C_2T_x$ causing microwave reflection loss; Moreover, the terminal groups induce polarization and which leads to microwave absorption loss [64]. Advanced materials are therefore crucial for eliminating the problems that EMI presents for devices and living things. Strong interfaces allow for recurrent EMR reflection and scattering, resulting in longer conductive routes and higher EMI shielding qualities. These interfaces are present in the 3D structure of MXenes. However, building 3D interconnected networks with superior mechanical performance is thought to be difficult due to the weak interactions between MXenes and the weak gelation qualities [87–89]. MXenes are capable of showing varying electrical properties ranging from metallic to insulating depending on the composition, thickness and surface functionalization [90].

Owing to the advances in the wearable and portable devices, EMI shielding materials with excellent mechanical stability, low density, light weight, and flexibility are highly essential. Owing to the intrinsic electrical conductivity of MXenes, they are greatly utilized for the production of EMI shielding material after CNT and graphene [91, 92]. By combining MXenes with polymer materials it is possible to construct highly interconnected electrical conductive network even with lower loading of MXenes [93–96]. Polymers like polyvinyl alcohols (PVA) are capable of showing better interaction with MXene via H-bonding and can be effectively utilized for the fabrication of shielding materials. PVA/MXene based alternating multilayers can show better EMI SE of about 44.4 dB due to the improved absorption and interfacial reflection [97]. It was observed that the increased interaction in the MXene/polymer composite film lead to the formation of highly interconnected pathways which are highly conductive and inhibit the EM pollution [98]. It is observed that highly conductive MXenes on the interface of polystyrene (PS) are capable of very conductive pathways that are highly efficient and continuous. More interestingly, these materials are capable of showing a low percolation threshold (0.26 vol%) with highly electrical conductivity of 1081 S m^{-1}. The better EMI SE is attributed by the cellular structured network structure which are highly conductive leading to attenuation and multireflection of EMI [99]. It is also observed that the annealing of polymer composites can enhance the electrical conductivity and EMI SE, because of the reflection of EM waves by the mismatch in the impedance and the absorption of EMR by the charge carriers [100]. Moreover, the process of annealing help in enhancing the strength of the conductive pathway, promoting the conduction of electrons. Annealing also enhances the multiple reflection and reabsorption, by the interaction of EMR with abundant dipoles present in the annealed composites [101]. Foams can also be made out of polymer/MXene composites for the EMI shielding applications generally by following the method of freeze drying. Self-assembling Ti_3C_2Tx in polymer matrices can lead to an absorption-based EMI shielding mechanism. The formation of highly porous foam like structures will generally enhance the multiple internal reflection and enlarge absorption, reduce reflection via reducing the conductivity. Figure 9.9 shows the EMI shielding performance of Ti_3C_2Tx/PVA foam over CNT and graphene based foams [100]. In Figure 9.10, the possible shielding mechanism of the composites was proposed and presented [100]. When the incident electromagnetic wave is obstructed by the prepared samples the phenomena of reflection, absorption and transmission takes place. In case of PVA/MXene films the conductivity of the films is as high as 8.0×10^{-4} S/m, the reflection dominant behavior was observed (figure). For the PVA/MXene composites with the foam structure better wave adsorption and less wave reflection takes place. This is because of the better impedance correlation among the electromagnetic wave as well as the foam sample. Here, in the foam structure the electromagnetic waves experience internal reflection because of the porous foam structure and layered MXene structures when it enters into the material. Hence, the energy dissipation is more there by promoting the absorption dominant shielding mechanism.

It is noteworthy that the MXenes with honey comb like network structures are highly effective in enhancing the conductivities and SE of sponges owing to the synergistic of MXenes and silver nanowires [75]. MXene/sodium alginate polymer foams are capable

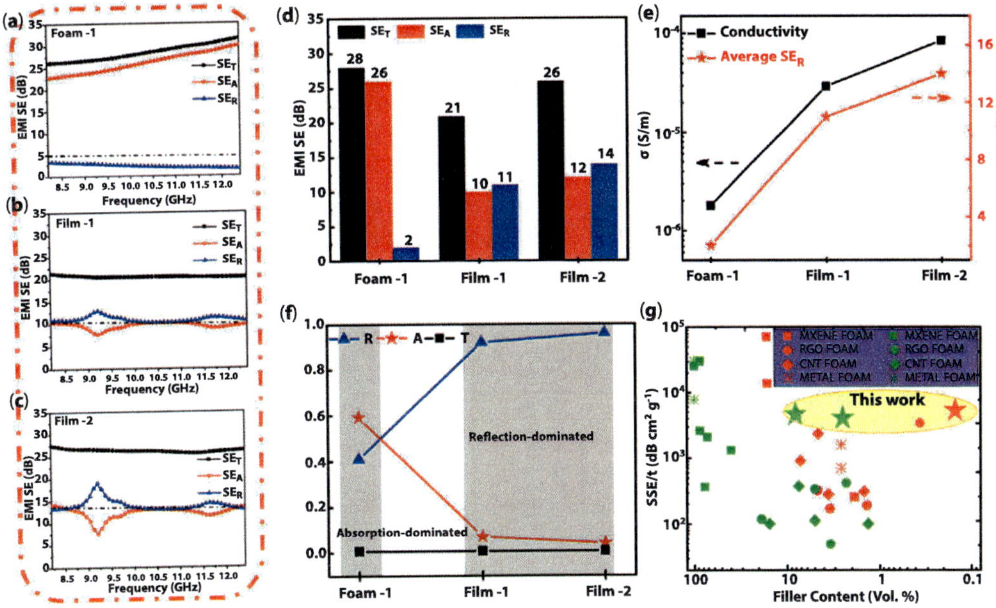

Figure 9.9 (a–c) EMI shielding performances. (d) EMI SE_T, SE_A, and SE_R. (e) Average SE_R and conductivity. (f) Proportion of reflection (R), absorption (A), and transmission (T). (g) SSE/t of different reported materials [100]. Copyright 2019, Reproduced with permission from American Chemical Society.

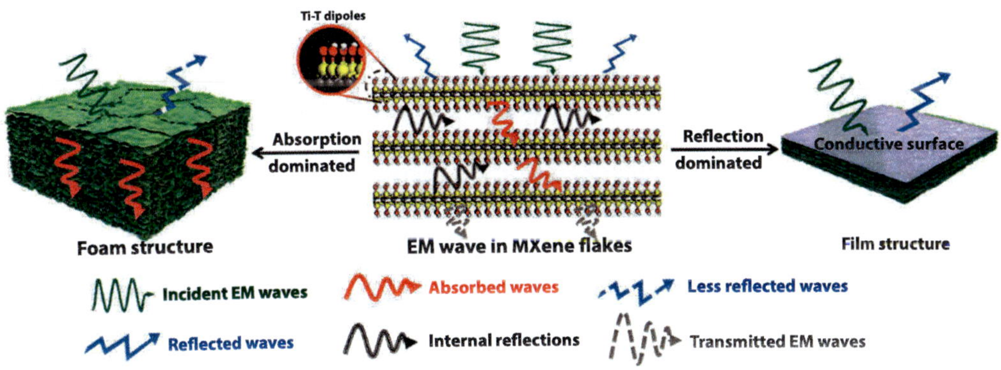

Figure 9.10 Possible shielding mechanism in case of a foam structure and a film structure of MXene/PVA composites [100]. Copyright 2019, Reproduced with permission from American Chemical Society.

of showing better EMI SE and electrical conductivity owing to the exciting layered structure. PDMS layer of coating helps in enhancing the structural stability and durability of the porous structure, thus enhancing the EMI SE [102]. The auxetic structure of the porous foam enhances the surface area and thereby enhances the interfaces for EMI shielding. Owing to the higher electrical conductivity of the MXene over CNT and graphene, it is observed that the porous conductive MXene/polymer composites with auxetic structure showed better SE values. Moreover, these composite materials are highly effective for impact attenuation process [103]. In a study, MXene-decorated polymer foam beads (MPFBs) were fabricated by 3D accumulation of MXene followed encapsulation of PDMS, which ensured higher SE by the absorption dominated mechanism. It is also noticed that the presence of foam beads in the composites helps in decreasing the dielectric property and thus a small amount of EMR is reflected due to frail impedance mismatch. The penetrated waves will interact with MXene conduction networks and lead to conduction loss and thus attenuate energy. Moreover, the dipole and interface polarizations promote the dissipation of EMR which are penetrated [104]. It was observed that the induced polarization of Ti and the terminating groups lead to higher shielding efficiency, moreover the hopping and contacting lead to conduction loss due to increased conductivity and the layers of MXenes lead to conduction loss which in turn enhances the shielding effectiveness of the polymer [105]. It is also important to note that in polymer/MXene composites, the brick-and-mortar structure provides more interfaces for reflection as well as scattering of EMR and lead to absorption attenuation. Massive electrons present on the surface also lead to the reflection of EM waves, remaining waves will travel through the layers of MXene and lead to the decay of energy and still remaining waves will get attenuated [77]. The defects and the functional groups on the surface of MXene enhances the space charge polarization under alternating EMR and thus improve the SE of the composites film [106]. The EMI shielding effectiveness of different MXene/polymer composites are tabulated in Table 9.2.

Studies have shown that the 'brick-and-mortar' structure of MXene polymeric composites are good enough in enhancing the EMI shielding properties by easing absorption attenuation by providing interfaces suitable for scattering and recurrent. Due to the presence of massive electrons present in the highly conductive Ti_3C_2Tx, part of EM radiations is getting reflected back from the composite surface. Remaining waves will interact with the Ti_3C_2Tx having high charge density lead to the reduction in the energy of EM radiations. Remaining radiations will get transferred to the next layers and again get reflected. Moreover, the lamellar structure helps in the absorption of microwave. MXene based composites shows superior EM shielding property compared to other nanofillers like CNT, rGO, carbon aerogel, carbon foam etc. due to its high SSE and ultrathin structure. It is also observed that the introduction of poly (3,4-ethylenedioxythiophene)/poly(styrenesulfonate) (PEDOT/PSS) into MXene and its acid treatment leads to an enhancement in EMI shielding property along with enhancement in tensile strength. From Figure 9.11a and Figure 9.11b it is clear that with the incorporation of PEDOT/PSS, the SE decreases due to inferior electrical conductivity of PEDOT/PSS compared to MXene. But the acid treatment enhances the electrical conductivity of PEDOT/PSS, which helps in enhancing the SE of MXene film as shown in Figures 9.11c–e. The film depth of the materials is presented in Figure 9.11f [108].

Recent study shows that the Ti_3C_2Tx based bacterial cellulose (BC) composites are capable of showing better shielding property (EMI SE of 29141 dB cm^2 g^{-1}) compared to other existing carbon based and MXene based composite materials. Hence this study developed a method for the fabrication of strong, ultrathin, and extremely flexible freestanding composite films of $Ti_3C_2T_x$/BC for competent EMI shielding application. The preparation process is presented in Figure 9.12a and the digital images are presented in Figures 9.12(b–d).

Table 9.2 MXene/Polymer composites in EMI shielding applications.

MXene-Polymer composites	Preparation methods	Electrical conductivity, EMI SE	Mechanism	Ref.
Silver nanowire (AgNW)/MXene	Dip-coating and freeze-drying	24.5 S cm^{-1}, 52.6 dB	Multiple reflections and scattering	[75]
$Ti_3C_2T_x$/poly(3,4-ethylenedioxy-thiophene)–poly-(styrene sulfonate)	Vacuum-assisted filtration	340.5 S/cm, 42.10 dB	Attenuation, reflection	[77]
$Ti_3C_2T_x$/PVA	Multilayered casting	716 S/m, 44.4dB	Microwave absorption	[98]
Ti_3C_2Tx/polyvinylidene fluoride (PVDF)	Solvent assisted mixing followed compression molding	0.98 S/m, 48.47 ± 3.5 dB	Microwave absorption	[99]
MXene/Sodium alginate	Directional freezing and freeze drying	2211 S m–1, 72 dB	Attenuation	[102]
MXene/auxetic polyurethane composite foam	Compression followed annealing	106 S/m, 76.2 dB	Absorption	[103]
MXene@PANI/mPP foam beads	Oxidation treatment, dip coating	~23.5–39.8 dB	Absorption, attenuation	[104]
PVDF/CNT/MXene foams	solution mixing, compression molding, CO_2 based delamination	65.1 dB	Absorption	[105]
rGO-MXene/epoxy	Solution casting	387.1 S/m, (55 dB	Multi-reflection, reabsorption	[106]
MXene-Decorated polyester textiles	Dip coating	1000 S m^{-1}, 90 dB	Reflection, absorption	[107]

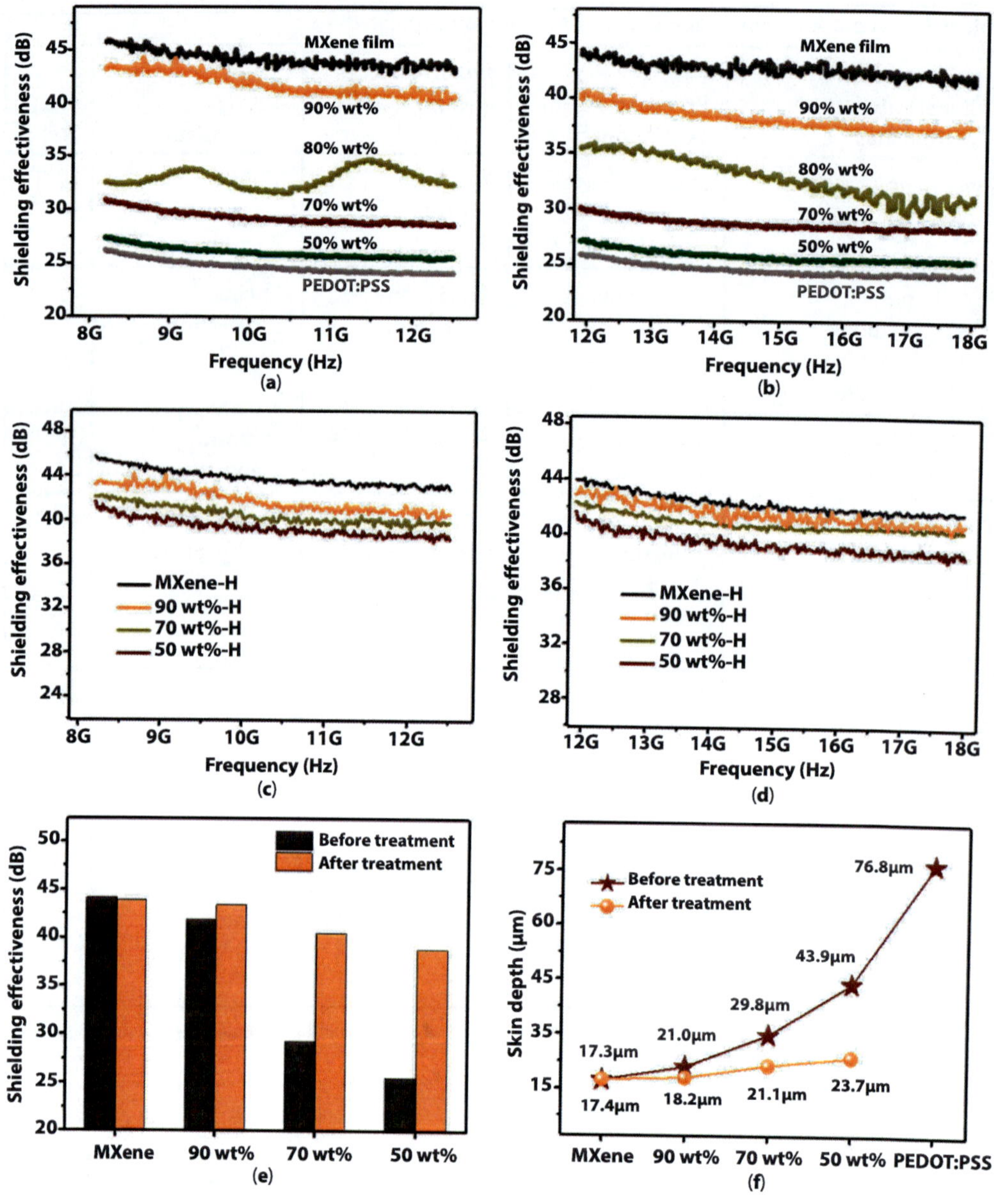

Figure 9.11 Shielding performance of MXene composite films (a), (b) before and (c), (d) after acid treatment; comparison of (e) EMI SE and (f) skin depth of films with and without treatment [108]. Copyright 2020, Reproduced with permission from Elsevier.

SSE of $Ti_3C_2T_x$/BC composites and comparison with the existing composites are portrayed in Figure 9.13a. High intrinsic electrical conductivity offered by the entangled network attributes higher SE. Overall SE (SE_T) is measured in terms of multiple reflections (SEM), absorption (SE_A) and reflection (SE_R) and it is observed that SE_T improves with the filler loading owing to absorption and reflection mechanism (Figure 9.13b). The mechanism

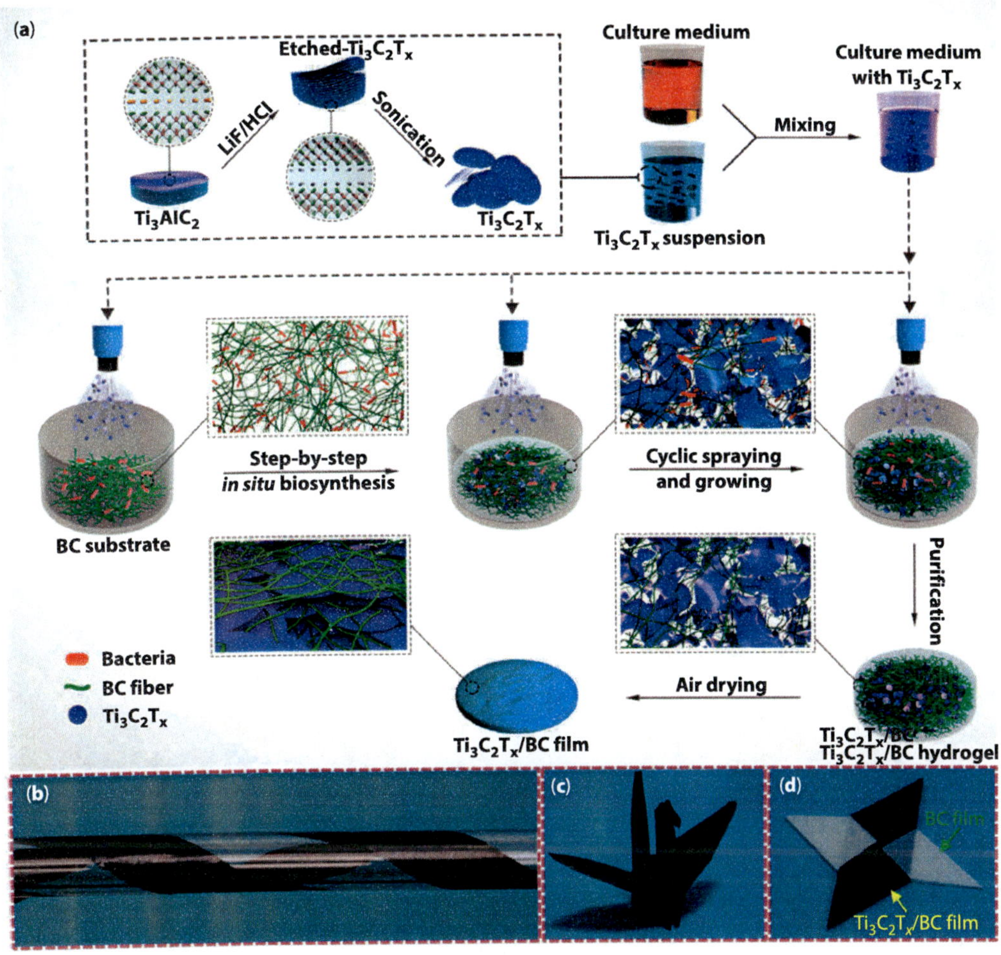

Figure 9.12 Schematic representation of composite preparation (b–d) Digital images of prepared composite films [109]. Copyright 2021, Reproduced with permission from American Chemical Society.

behind EM shielding is portrayed in Figure 9.13c. Owing to the EM absorption ability, the same material can be utilized as light to heat conversion applications. The film is lighted for 240 seconds, during which time the temperature quickly rose to 55.4°C and then stabilized as shown in Figure 9.13d [109]. Ti_2CT_x MXene/poly (vinyl alcohol) composite based foams made using few layered MXene via freeze drying are capable of showing a SE of 5136 dB cm^2 g^{-1} with an ultra-low content of the filler owing to the adsorption dominant behavior. Internal reflection, and polarization effect synergically plays significant role in improving the EMI SE of highly porous structure [100]. Mortar-brick structures of wood-derived porous carbon (WPC)/MXeneene aerogel composite solve the structural instability of MXene aerogel and forms a prolonged transmission path and thus dissipates EMR in the form electricity and heat [110]. Porous carbon foam (PCF)/MXene shows SE and SSE of 75 dB and 216.9 dB cm3/g, respectively because of its cellular structures and multilayer interfaces [111]. Chitosan (CS)/MXene films capable of show an EMI SE of ~ 34.7 ± 0.2 dB

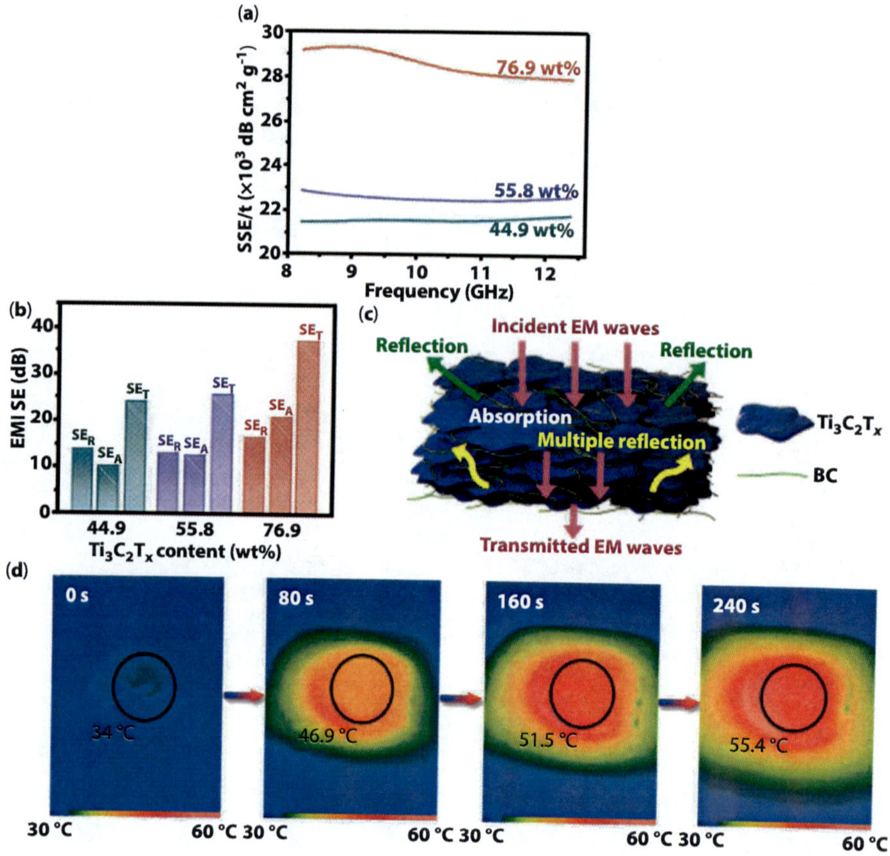

Figure 9.13 (a) EMI SSE/t of $Ti_3C_2T_x$/BC composite films. (b) comparison of SET, SEA, and SER of composite films. (c) schematic of shielding mechanism. (d) infrared images showing light [109]. Copyright 2021, Reproduced with permission from American Chemical Society.

owing to the multiple internal reflections attributed by the MXene flakes and outstanding electrical conductivity [112].

9.6 Electrospun Fibers with MXenes as Additives

Electrospinning is a versatile and well-established technique to prepare continuous polymer fibers [113–116]. These fibers can be incorporated with various functional additives and their combinations for synthesizing composite fibers for multifunctional applications. By selectively optimizing the electrospinning parameters (like voltage, flow rate, distance, collector speed, polymer concentration etc.) we can modify the shape and diameters of the prepared fibers. These nonwoven mats prepared via electrospinning can be used in wide range of applications, especially EMI shielding [116–120]. In Figure 9.14, typical electrospinning setups for fiber preparation is presented [121]. For EMI shielding electrospun fibers in three ways are prepared 1) coating EM wave active materials (metals, metal oxides etc.) on the surface of fiber mats, 2) incorporate EM wave active particles

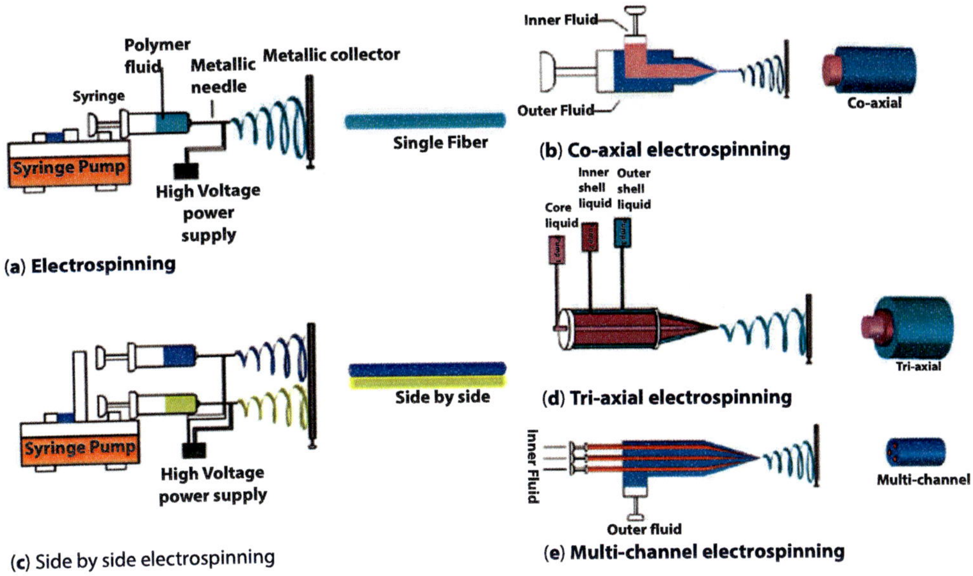

Figure 9.14 Electrospinning setups for getting various fiber morphologies [121]. Copyright 2021, Reproduced with permission from American Chemical Society.

inside the electrospun fibers and preparing nanocomposites and 3) Convert electrospun polymer fibers into electrospun carbon fibers via the stabilization-carbonization route. Among all these the last option to prepare pure carbon fibers or composite carbon fibers with embedded EM wave active materials is gaining a lot of attention. Electrospun carbon fibers are prepared from a suitable polymer precursor, which has the ability to produce stable carbon fibers. Mostly PAN polymer is used for this purpose. Many researchers have incorporated various materials inside PAN polymer and converted them to carbon fibers and used them for EMI shielding and microwave absorbing purposes [120–123]. In Figure 9.15, different electrospun fiber composites form shielding applications are highlighted [121].

MXenes are often used as fillers in electrospun fibers and composites were made. Many works on MXene/carbon fibers prepared via electrospinning process are also reported for many applications. In Figure 9.16, the process of preparation of MXene/carbon fibers is presented [124]. For this PAN is used as precursor polymer and DMF as the organic solvent for dispersing both PAN polymer and MXene ($Ti_3C_2T_x$). The concentration of MXenes are varied are varied from 2w% to 16 w% and electrospun fibers were made. The dependence of fiber diameter on additive loading was studied. From Figure 9.17, we can clearly see the diameter of fibers increased drastically with increase in the additive loading. When no MXenes are added the fibers have a mean diameter of 179±33 nm which is very fine. But as the filler loading increased to 2 w%, 6 w%, 10 w% and 16 w%, the fiber diameter increased to 209±44 nm, 411±84 nm, 499±94 nm and 958±190 nm. In literature, the preparation of hollow carbon fibers with MXenes is presented. There a sacrificing polymer is used to get core and a carbonizing polymer is used to form the shell of electrospun core/shell structure. For example, a polymer X is core and polymer Y is the shell polymer. During carbonization the core polymer X will be removed and hollow structure will be formed [125].

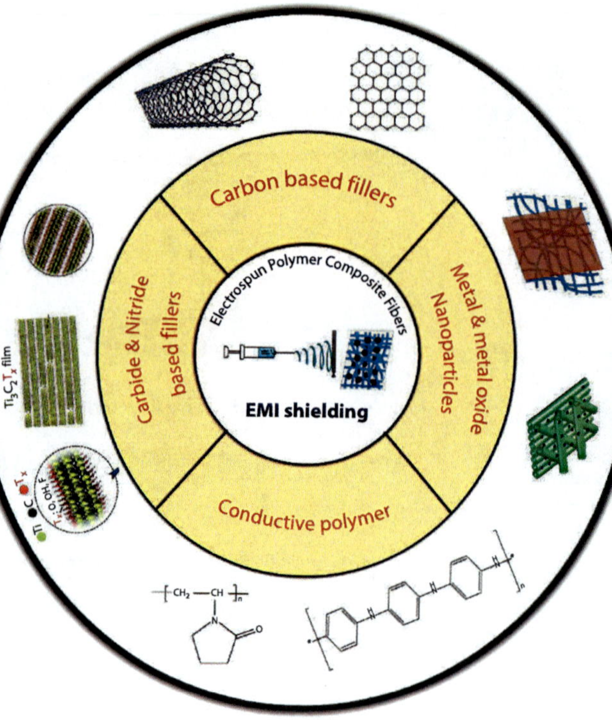

Figure 9.15 Electrospun fiber with different additives for shielding applications [121]. Copyright 2021, Reproduced with permission from American Chemical Society.

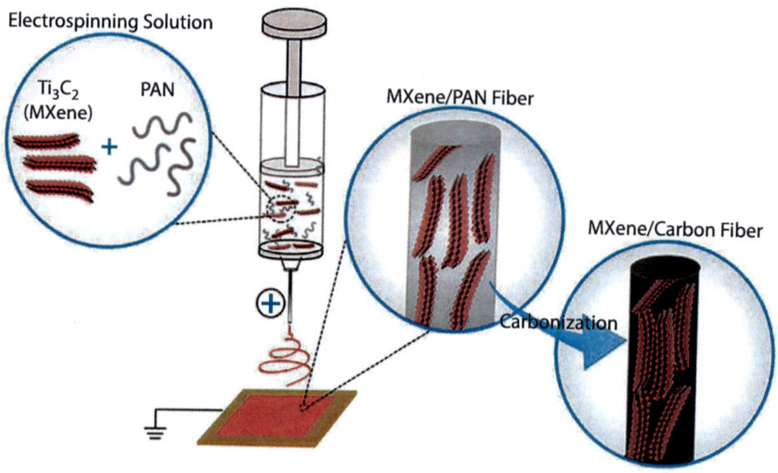

Figure 9.16 Preparation of electrospun/MXene composites [124]. Copyright 2019, Reproduced with permission from Royal Society of Chemistry.

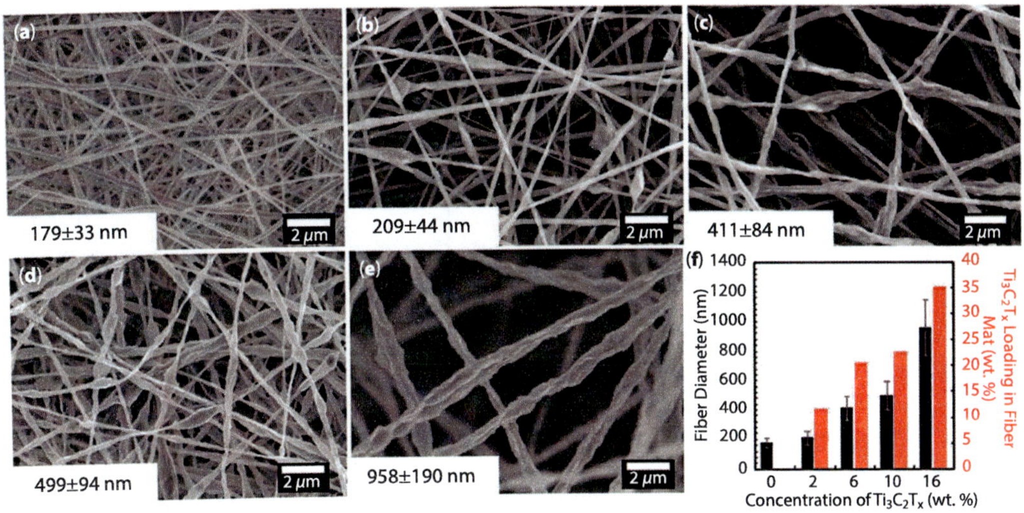

Figure 9.17 (a–e) SEM micrographs of electrospun/MXene fibers with different MXene loading (f) Barchart representation [124]. Copyright 2019, Reproduced with permission from Royal Society of Chemistry.

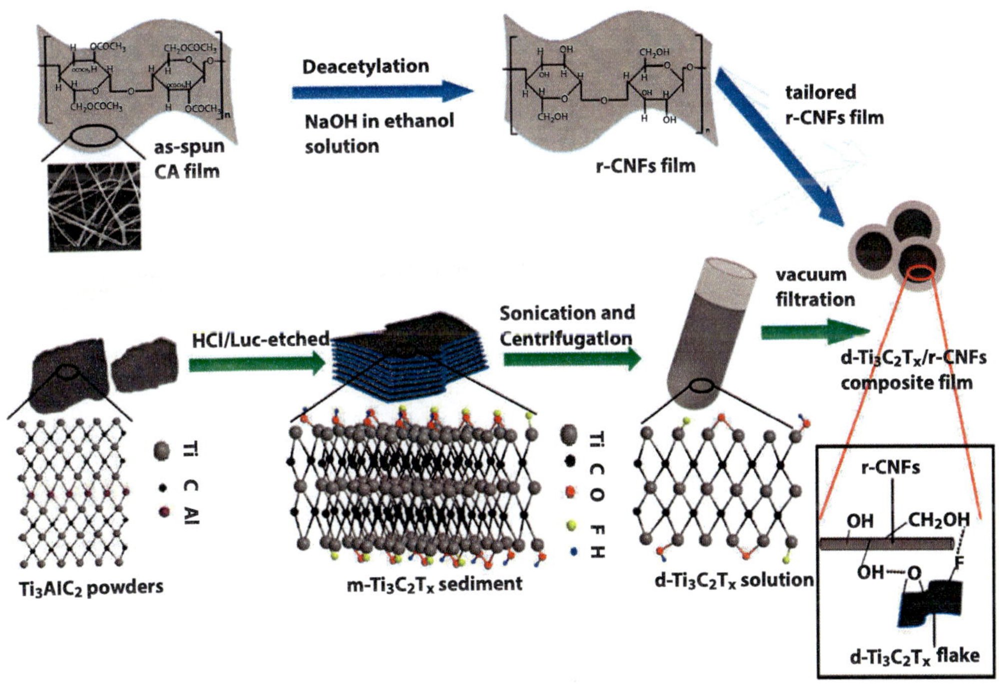

Figure 9.18 Schematic of the preparation process employed for preparing d-Ti3C2Tx/r-CNFs film [129]. Copyright 2019, Reproduced with permission from Elsevier.

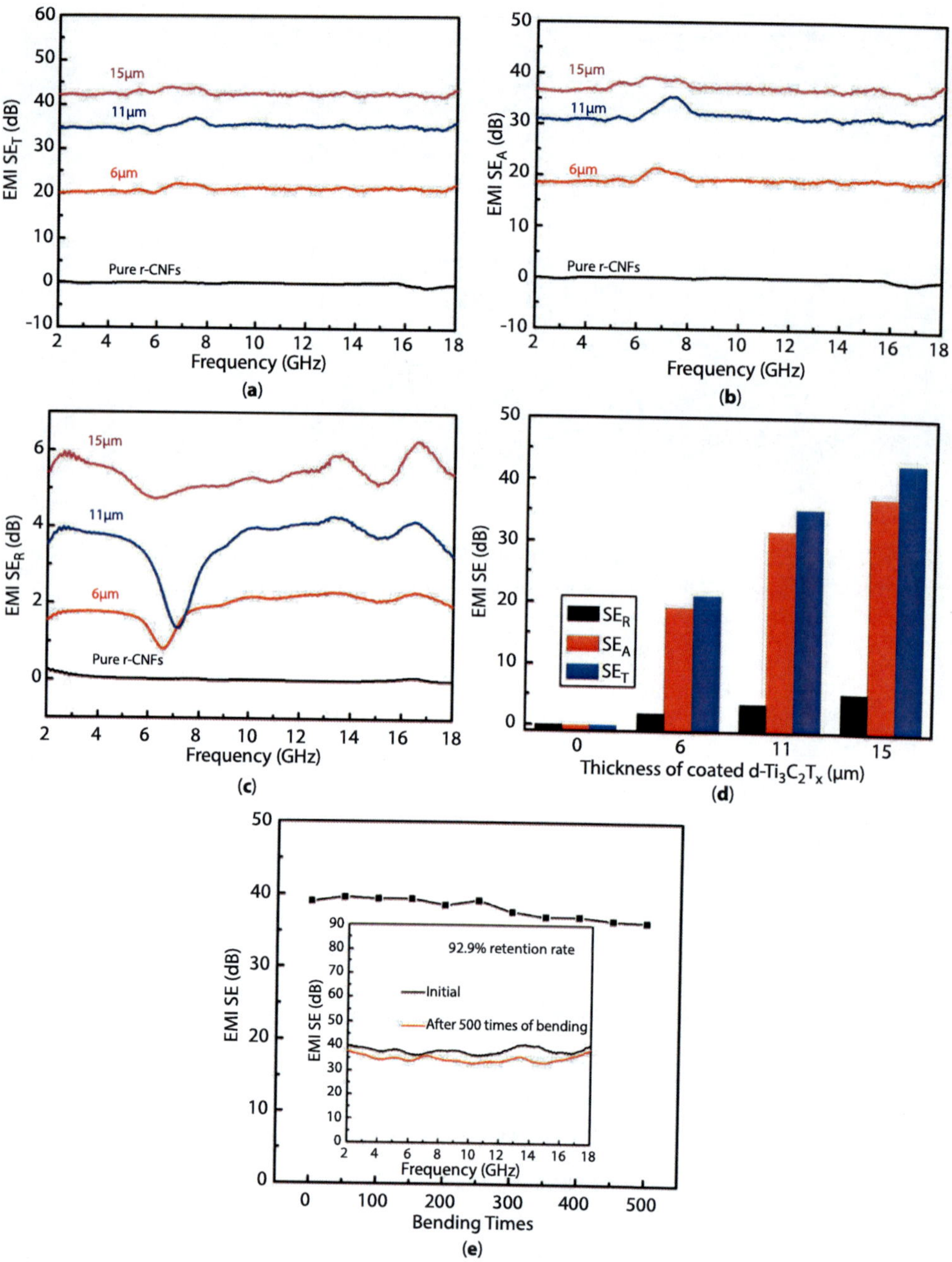

Figure 9.19 (a) EMI SET, (b) EMI SEA, (c) EMI SER of films (d) average EMI SE of films with different thickness of d-$Ti_3C_2T_x$, and (e) EMI SE of composite film after 500 times of bending [129]. Copyright 2019, Reproduced with permission from Elsevier.

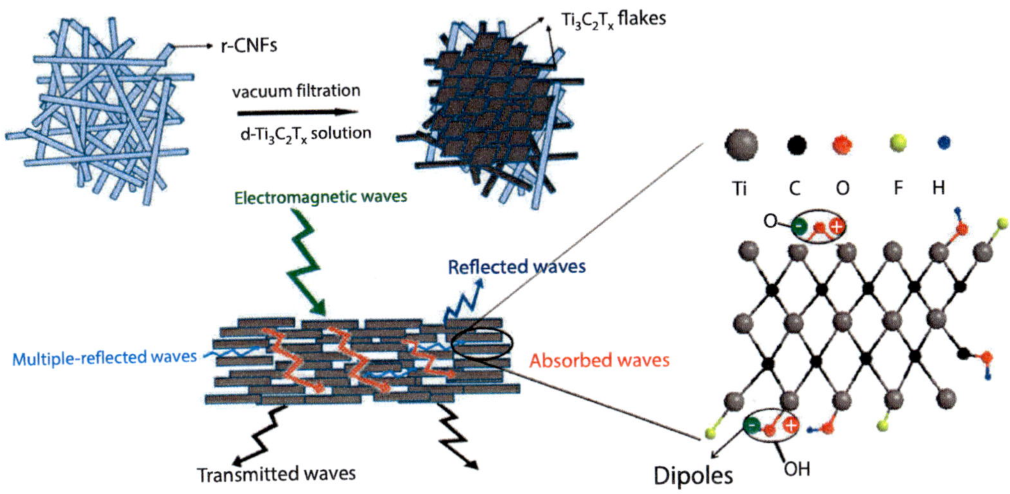

Figure 9.20 Scheme of the mechanism of EMI shielding [129]. Copyright 2019, Reproduced with permission from Elsevier.

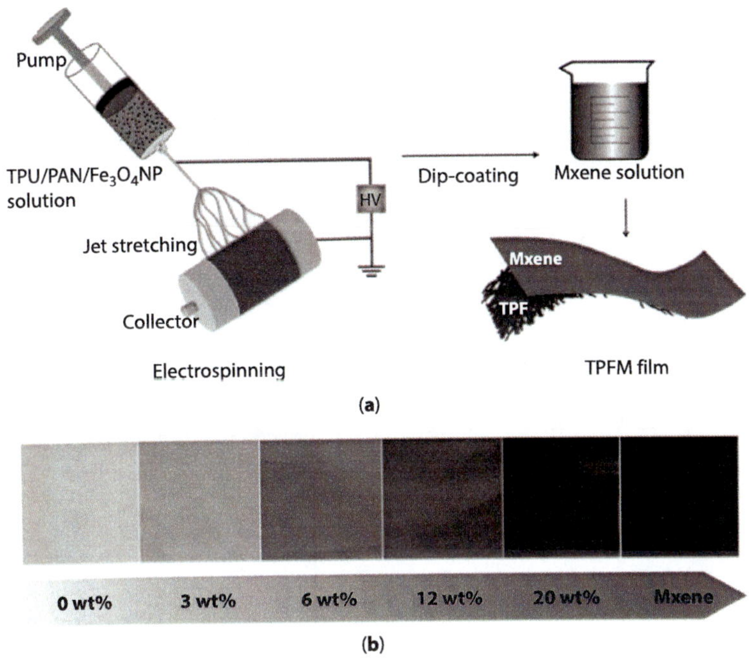

Figure 9.21 (a) Preparation procedure for hybrid fiber, (b) optical image with increase in Fe_3O_4 loading [130]. Copyright 2022, Adapted from MDPI.

MXene based fibers are gaining attention in various applications in material science. These fibers are produced from 1) coating, 2) spinning (wet/electro) and 3) trapping [126]. For EMI shielding, MXenes with different polymer matrices like epoxy, polyurethane, polystyrene, Poly (lactic acid) are being investigated [91, 92, 107, 127, 128]. Cui *et al.* in their work prepared delaminated MXene (d- $Ti_3C_2T_x$)/regenerate cellulose fibers via vacuum assisted filtration route. The detailed schematic for the preparation of the composite films is presented in Figure 9.18 [129]. In Figures 9.19(a–c), the values corresponding to total shielding effectiveness and individual contribution from reflection and absorption is presented. The variation of these values with thickness is presented in Figure 9.19d. The stability of these samples after bending is presented in Figure 9.19e. These films exhibited a SE of 42.7 dB owing to their surface area and conductivity. From the curves, it is very clear that MXene coating has a significant effect on the shielding performance. The proposed shielding mechanism is presented in Figure 9.20 [129]. Miao *et al.* [130] in their work, prepared thermoplastic polyurethane (TPU))/polyacrylonitrile (PAN)/Fe_3O_4 nanoparticle membrane was prepared by varying the content of Fe_3O_4. The schematic for the hybrid mat preparation is presented in Figure 9.21 [130]. After that they dipped the hybrid mat in MXene solution of both 2 mg/ml and 5 mg/ml. They observed the hybrid mats with MXene coating have better SE than the mats without MXene coating (Figures 9.22 a, b). Further the performance increased with Fe_3O_4 nanoparticle loading. The coating concentration MXenes

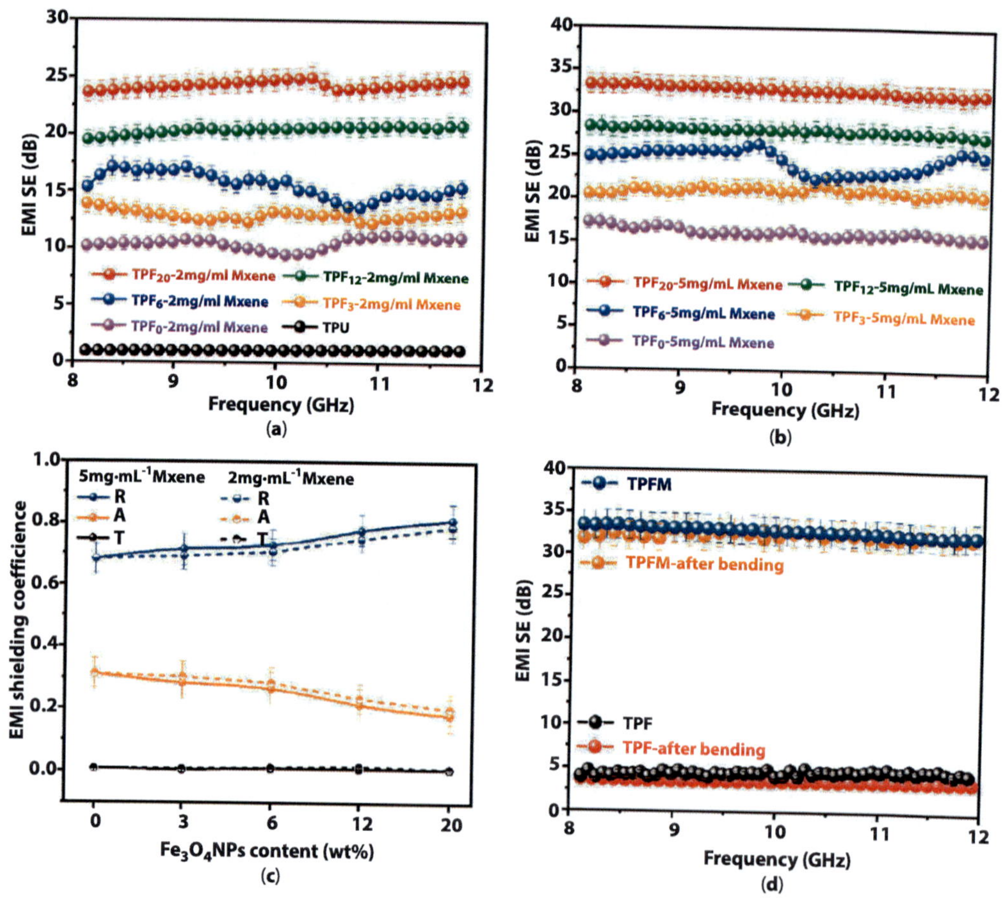

Figure 9.22 (a, b) EMI SE of hybrid fillers in the X band. (c) R, A, and T coefficients with respect to filler loadings (d) EMI SE performance after bending cycles at room temperature [130]. Copyright 2022, Adapted from MDPI.

also play role in improving the shielding performance. We can see, from the figure in all compositions the fiber mats with 5 mg/ml has more SE than fiber mats with 2 mg/ml. They concluded that for the prepared material the EMI SE of more than 30 dB can be achieved for prepared mats because of the hydrogen bonds between the PAN polymer and MXene functional groups. The role of Fe_3O_4 NPs in influencing electromagnetic wave attenuation is presented in Figure 9.22c. The R, A, and T coefficients are tested and it was observed that as the loading of nanoparticle increases the R increases and A decreases slightly. Overall, the nanoparticles are showed to have positive impact on shielding. In Figure 9.22d, the stability of the material is tested at room temperature. After 500 bending tests, the EMI SE of the hybrid material is reduced by 1.02%; this is attributed to the destruction of the MXene networks and the detaching of nanoparticles from the fiber.

9.7 Conclusions and Future Outlook

MXene based shielding materials are the upcoming and promising category for efficient shielding applications. By carefully synthesizing these materials we can design MXenes with different shape, structure and functionality. By mixing these materials with carbon, metals, metal oxides, conducting polymers and metal organic frameworks we can tailor and achieve advanced functional materials. This chapter brings together many such ideas along with MXene/polymer composites for microwave adsorption and EMI shielding purposes. In future, there is a tremendous scope in this area with different dielectric/dielectric combinations, dielectric/conductive combinations, dielectric/magnetic combinations and more for designing materials based on required applications. MXene based aerogels, MXene embedded fibers and MXene-MOFs hybrids are establishing themselves are ideal material combinations for a variety of applications. MXenes based fibers are also a promising material combination that can be promising applications in future. The high aspect ratio, enhanced functionality and flexibility makes them potential material for many applications. Overall, by carefully playing with the structure, size and combination of the MXene with can tune the absorption and reflection properties, thereby tailoring them for shielding applications. when you see literature the case of secondary EMI is not recommended (materials with reflection as the primary shielding mechanism), so absorption dominant materials are preferred. In this regard, MXenes play a vital role in developing such materials. The research in case of MXene/polymer composites is in early stage. So, lot of work needs to be done to address all the disadvantages associated with their preparation methods. The use of MXenes in electrospinning to made flexible MXene mats is one interesting idea to improve the applicability of MXenes.

References

1. Levy, R.B. and Boudart, M., Platinum-like behavior of tungsten carbide in surface catalysis. *Sci. (80-)*, 181, 547–549, 1973.
2. Naguib, M., Kurtoglu, M., Presser, V., Lu, J., Niu, J., Heon, M., Hultmen, L., Gogotsi, Y., Barsoum, M.W., Two-dimensional nanocrystals produced by exfoliation of Ti_3AlC_2. *Adv. Mater.*, 23, 4248–4253, 2011.

3. Zhan, C., Sun, W., Xie, Y., Jiang, D., Kent, P.R.C., Computational discovery and design of MXenes for energy applications: Status, successes, and opportunities. *ACS Appl. Mater. Interfaces*, 11, 24885–24905, 2019.
4. Sun, Q., Fu, Z., Yang, Z., Tunable magnetic and electronic properties of the Cr-based MXene (Cr2C) with functional groups and doping. *J. Magn. Magn. Mater.*, 514, 167141, 2020.
5. Jin, M.-J., Um, D.-S., Ogbeide, O., Kim, C.-I., Yoo, J.-W., Robinson, J.W.A., Crossover from weak anti-localization to weak localization in inkjet-printed $Ti_3C_2T_x$ MXene thin-film. *Adv. Nano Res.*, 14, 259–267, 2022.
6. Naguib, M. and Gogotsi, Y., Synthesis of two-dimensional materials by selective extraction. *Acc. Chem. Res.*, 48, 128–135, 2015.
7. Magnuson, M. and Mattesini, M., Chemical bonding and electronic-structure in MAX phases as viewed by X-ray spectroscopy and density functional theory. *Thin Solid Films*, 621, 108–130, 2017.
8. Sun, Z., Music, D., Ahuja, R., Li, S., Schneider, J.M., Bonding and classification of nanolayered ternary carbides. *Phys. Rev. B*, 092102, 1–3, 2004.
9. Naguib, M., Mochalin, V.N., Barsoum, M.W., Gogotsi, Y., 25th anniversary article: MXenes: A new family of two-dimensional materials. *Adv. Mater.*, 26, 992–1005, 2013.
10. Bae, S., Kang, Y.-G., Khazaei, M., Ohno, K., Kim, Y.-H., Han, M.J., Chang, K.J., Raebiger, H., Electronic and magnetic properties of carbide MXenes—The role of electron correlations. *Mater. Today Adv.*, 9, 100118, 2021.
11. He, Z., Huang, D., Yue, G., Zhu, J., Zhao, P., Ca2+ induced 3D porous MXene gel for continuous removal of phosphate and uranium. *Appl. Surf. Sci.*, 570, 150804, 2021.
12. Guan, Y., Zhang, M., Qin, J., Ma, X., Li, C., Tang, J., Hydrophilicity-dependent distinct frictional behaviors of different modified MXene nanosheets. *J. Phys. Chem. C*, 124, 13664–13671, 2020.
13. Anasori, B., Shi, C., Moon, J., Xie, Y., Voigt, C.A., Gogotsi, Y., Control of electronic properties of 2D carbides (MXenes) by manipulating their transition metal. *Nanoscale Horiz.*, 1, 227–234, 2016.
14. Zha, X.-H., Yin, J., Zhou, Y., Huang, Q., Luo, K., Lang, J., Francisco, J.S., He, J., Du, S., Intrinsic structural, electrical, thermal, and mechanical properties of the promising conductor Mo_2C MXene. *J. Phys. Chem. C*, 120, 15082–15088, 2016.
15. Shin, H., Eom, W., Lee, K.H., Jeong, W., Kang, D.J., Han, T.H., Highly electroconductive and mechanically strong $Ti_3C_2T_x$ MXene fibers using a deformable MXene gel. *ACS Nano*, 15, 3320–3329, 2021.
16. Shuck, C.E., Han, M., Maleski, K., Hantanasirisakul, K., Kim, S.J., Choi, J., Reil, W.E.B., Gogotsi, Y., Effect of Ti_3AlC_2 MAX phase on structure and properties of resultant $Ti_3C_2T_x$ MXene. *ACS Appl. Nano Mater.*, 2, 3368–3376, 2019.
17. Liao, L., Jiang, D., Zheng, K., Zhang, M., Liu, J., Industry-scale and environmentally stable $Ti_3C_2T_x$ MXene based film for flexible energy storage devices. *Adv. Funct. Mater.*, 31, 2103960, 2021.
18. Cheng, H., Pan, Y., Chen, Q., Che, R., Zheng, G., Liu, Q., Li, M., Zhou, Y., Jia, K., Yang, L., Wang, D., Ultrathin flexible poly (vinylidene fluoride)/MXene/silver nanowire film with outstanding specific EMI shielding and high heat dissipation. *Adv. Compos. Hybrid Mater.*, 4, 505–513, 2021.
19. Zhang, Z., Cai, Z., Zhang, Y., Peng, Y., Wang, Z., Xia, L., Ma, S., Yin, Z., Wang, R., Cao, Y., Li, Z., Huang, Y., The recent progress of MXene-Based microwave absorption materials. *Carbon*, 174, 484–899, 2021.
20. Yoon, J., Shin, M., Lim, J., Lee, J.-Y., Choi, J.-W., Recent advances in MXene nanocomposite-based biosensors. *Biosensors*, 10, 185, 2020.

21. Fan, D., Lu, Y., Zhang, H., Xu, H., Lu, C., Tang, Y., Yang, X., Synergy of photocatalysis and photothermal effect in integrated 0D perovskite oxide/2D MXene heterostructures for simultaneous water purification and solar steam generation. *Appl. Catal. B Environ.*, 295, 120285, 2021.
22. Rafieerad, A., Yan, W., Amiri, A., Dhingra, S., Bioactive and trackable MXene quantum dots for subcellular nanomedicine applications. *Mater. Des.*, 196, 109091, 2020.
23. Zhang, C.J., Anasori, B., Seral-ascaso, A., Park, S., Mcevoy, N., Shmeliov, A., Duesberg, G.S., Coleman, J.N., Gogotsi, Y., Nicolosi, V., Transparent, flexible, and conductive 2D titanium carbide (MXene) films with high volumetric capacitance. *Adv. Mater.*, 1702678, 1–9, 2017.
24. Huang, H., Cui, J., Liu, G., Bi, R., Zhang, L., Carbon-coated MoSe2/MXene hybrid nanosheets for superior potassium storage. *ACS Nano*, 13, 3448–56, 2019.
25. Levitt, A., Zhang, J., Dion, G., Gogotsi, Y., Razal, J.M., MXene-based fibers, yarns, and fabrics for wearable energy storage devices. *Adv. Funct. Mater.*, 30, 2000739, 2020.
26. Rakhi, R.B., Nayak, P., Xia, C., Alshareef, H.N., Novel amperometric glucose biosensor based on MXene nanocomposite. *Sci. Rep.*, 6, 1–10, 2016.
27. Xu, J., Hu, X., Wang, X., Wang, X., Ju, Y., Ge, S., Lu, X., Ding, J., Yuan, N., Gogotsi, Y., Low-temperature pseudocapacitive energy storage in $Ti_3C_2T_x$ MXene. *Energy Storage Mater.*, 33, 382–389, 2020.
28. Chen, J., Huang, Q., Huang, H., Mao, L., Liu, M., Zhang, X., Wei, Y., Recent progress and advances in the environmental applications of MXene related materials. *Nanoscale*, 12, 3574–3592, 2020.
29. Liu, G., Cheng, L., Chen, G., Liang, F., Liu, G., Jin, W., Pebax-based membrane filled with two-dimensional Mxene nanosheets for efficient CO_2 capture. *Chem. Asian J.*, 15, 2364–2370, 2020.
30. Chen, W.Y., Lai, S.-N., Yen, C.-C., Jiang, X., Peroulis, D., Stanciu, L.A., Surface functionalization of $Ti_3C_2T_x$ MXene with highly reliable superhydrophobic protection for volatile organic compounds sensing. *ACS Nano*, 14, 11490–11501, 2020.
31. Mahmud, S.T., Hasan, M.M., Bain, S., Rahman, S.T., Rhaman, M., Hossain, M.M., Ordu, M., Multilayer MXene heterostructures and nanohybrids for multifunctional applications: A review. *ACS Mater. Lett.*, 4, 1174–1206, 2022.
32. Cheng, Y., Dai, J., Zhu, X., Wu, D., Sun, Y., Preparation, magnetic and microwave absorption properties of La0. 5Sr0. 5MnO3/La (OH) 3 composites. *Mater. Res. Bull.*, 45, 663–667, 2010.
33. Micheli, D., Pastore, R., Apollo, C., Marchetti, M., Gradoni, G., Primiani, V.M., Moglie, F., Broadband electromagnetic absorbers using carbon nanostructure-based composites. *IEEE Trans. Microw. Theory Tech.*, 59, 2633–2646, 2011.
34. Deng, R., Chen, B., Li, H., Zhang, K., Zhang, T., Yu, Y., Song, L., MXene/Co_3O_4 composite material: Stable synthesis and its enhanced broadband microwave absorption. *Appl. Surf. Sci.*, 488, 921–930, 2019.
35. Gargama, H., Thakur, A.K., Chaturvedi, S.K., Polyvinylidene fluoride/nickel composite materials for charge storing, electromagnetic interference absorption, and shielding applications. *J. Appl. Phys.*, 117, 224903, 2015.
36. Faisal, M. and Khasim, S., Broadband electromagnetic shielding and dielectric properties of polyaniline-stannous oxide composites. *J. Mater. Sci. Mater. Electron.*, 24, 2202–2210, 2013.
37. Verger, L., Natu, V., Carey, M., Barsoum, M.W., MXenes: An introduction of their synthesis, select properties, and applications. *Trends Chem.*, xx, 1–14, 2019.
38. Li, X., Yin, X., Xu, H., Han, M., Li, M., Liang, S., Cheng, L., Zhang, L., Ultralight MXene-coated, interconnected SiCnws three-dimensional lamellar foams for efficient microwave absorption in the X-band. *ACS Appl. Mater. Interfaces*, 10, 34524–34533, 2018.
39. Mei, H., Zhao, X., Gui, X., Lu, D., Han, D., Xiao, S., Cheng, L., SiC encapsulated Fe@ CNT ultra-high absorptive shielding material for high temperature resistant EMI shielding. *Ceram. Int.*, 45, 17144–17151, 2019.

40. Liu, Y., Wu, Y., Li, K., Wang, J., Zhang, C., Ji, J., Wang, W., Amorphous SnS nanosheets/graphene oxide hybrid with efficient dielectric loss to improve the high-frequency electromagnetic wave absorption properties. *Appl. Surf. Sci.*, 486, 344–353, 2019.
41. Wang, H. and Ma, H., The electromagnetic and microwave absorbing properties of MoS2 modified $Ti_3C_2T_x$ nanocomposites. *J. Mater. Sci. Mater. Electron.*, 30, 15250–15256, 2019.
42. Mathew, K.T., Kumar, A.V.P., John, H., Polyaniline and polypyrrole with PVC content for effective EMI shielding. *IEEE Int. Symp. Electromagn. Compat.*, 443–445, 2006. Available at: http://ieeexplore.ieee.org/document/1706344
43. Saib, A., Bednarz, L., Daussin, R., Bailly, C., Lou, X., Thomassin, J.-M., Pagnoule, C., Detrembleur, C., Jerome, R., Huynen, I., Carbon nanotube composites for broadband microwave absorbing materials. *IEEE Trans. Microw. Theory Tech.*, 54, 2745–2754, 2006.
44. Lakshmi, K., John, H., Mathew, K.T., Joseph, R., George, K.E., Microwave absorption, reflection and EMI shielding of PU–PANI composite. *Acta Mater.*, 57, 371–375, 2009.
45. Luo, X. and Chung, D.D.L., Electromagnetic interference shielding reaching 130 dB using flexible graphite. *MRS Online Proc. Libr.*, 445, 235–238, 1996.
46. Song, W., Cao, M., Hou, Z., Yuan, J., Fang, X., High-temperature microwave absorption and evolutionary behavior of multiwalled carbon nanotube nanocomposite. *Scr. Mater.*, 61, 201–204, 2009.
47. Liang, J., Wang, Y., Huang, Y., Ma, Y., Liu, Z., Cai, J., Zhang, C., Gao, H., Chen, Y., Electromagnetic interference shielding of graphene/epoxy composites. *Carbon N. Y.*, 47, 922–925, 2009.
48. Kumar, R., Choudhary, H.K., Pawar, S.P., Bose, S., Sahoo, B., Carbon encapsulated nanoscale iron/iron-carbide/graphite particles for EMI shielding and microwave absorption. *Phys. Chem. Chem. Phys.*, 19, 23268–23279, 2017.
49. Yang, J., Liao, X., Li, J., He, G., Zhang, Y., Tang, W., Wang, G., Li, G., Light-weight and flexible silicone rubber/MWCNTs/Fe3O4 nanocomposite foams for efficient electromagnetic interference shielding and microwave absorption. *Compos. Sci. Technol.*, 181, 107670, 2019.
50. Wang, Y.Y., Zhou, Z.H., Zhou, C.-G., Sun, W.J., Gao, J.F., Dai, K., Yan, D.X., Li, Z.M., Lightweight and robust carbon nanotube/polyimide foam for efficient and heat-resistant electromagnetic interference shielding and microwave absorption. *ACS Appl. Mater. Interfaces*, 12, 8704–8712, 2020.
51. Li, M.-Z., Jia, L.-C., Zhang, X.-P., Yan, D.-X., Zhang, Q.-C., Li, Z.-M., Robust carbon nanotube foam for efficient electromagnetic interference shielding and microwave absorption. *J. Colloid Interface Sci.*, 530, 113–119, 2018.
52. Wang, Z., Cheng, Z., Fang, C., Hou, X., Xie, L., Recent advances in MXenes composites for electromagnetic interference shielding and microwave absorption. *Compos. Part A Appl. Sci. Manuf.*, 136, 105956, 2020.
53. Zhang, Y., Yang, Z., Yu, Y., Wen, B., Liu, Y., Qiu, M., Tunable electromagnetic interference shielding ability in a one-dimensional bagasse fiber/polyaniline heterostructure. *ACS Appl. Polym. Mater.*, 1, 737–745, 2019.
54. Moučka, R., Sedlačík, M., Kasparyan, H., Prokeš, J., Trchová, M., Hassouna, F., Kopecky, D., One-dimensional nanostructures of polypyrrole for shielding of electromagnetic interference in the microwave region. *Int. J. Mol. Sci.*, 21, 8814, 2020.
55. Liang, H., Liu, J., Zhang, Y., Luo, L., Wu, H., Ultra-thin broccoli-like SCFs@TiO_2 one-dimensional electromagnetic wave absorbing material. *Compos. Part B Eng.*, 178, 107507, 2019.
56. Xu, D., Yang, Y., Lyu, L., Ouyang, A., Liu, W., Wang, Z., Wu, L., Yang, F., Liu, J., Wang, F., One-dimensional MnO@ N-doped carbon nanotubes as robust dielectric loss electromagnetic wave absorbers. *Chem. Eng. J.*, 410, 128295, 2021.

57. Hou, T., Wang, B., Ma, M., Feng, A., Huang, Z., Zhang, Y., Jia, Z., Tan, G., Cao, H., Wu, G., Preparation of two-dimensional titanium carbide ($Ti_3C_2T_x$) and $NiCo_2O_4$ composites to achieve excellent microwave absorption properties. *Compos. Part B Eng.*, 180, 107577, 2020.
58. Di, J., Duan, Y., Pang, H., Ma, X., Liu, J., Sintering-regulated two-dimensional plate@shell basalt@NiO heterostructure for enhanced microwave absorption. *Appl. Surf. Sci.*, 574, 151590, 2022.
59. Qiu, Y., Yang, H., Hu, F., Lin, Y., Two-dimensional CoNi@mesoporous carbon composite with heterogeneous structure toward broadband microwave absorber. *Nano Res.*, 15, 7769–7777, 2022.
60. Miao, P., Zhang, T., Wang, T., Chen, J., Gao, T., Wang, Y., Kong, J., Chen, K.J., A two-dimensional semiconductive metal-organic framework for highly efficient microwave absorption. *Chin. J. Chem.*, 40, 467–474, 2022.
61. Phadtare, V.D., Parale, V.G., Kim, T., Lee, K.-Y., Pathak, S., Park, H.-H., Facile synthesis of a lightweight three-dimensional polymer scaffold dip-coated with multiple layers of TiO_2 aerogel for X-band microwave absorption applications. *J. Alloys Compd.*, 823, 153847, 2020.
62. Fu, B., Ren, P., Guo, Z., Du, Y., Jin, Y., Sun, Z., Dai, Z., Ren, F., Construction of three-dimensional interconnected graphene nanosheet network in thermoplastic polyurethane with highly efficient electromagnetic interference shielding. *Compos. Part B Eng.*, 215, 108813, 2021.
63. Zhang, H., Chen, J., Ji, H., Wang, N., Feng, S., Xiao, H., Electromagnetic interference shielding with absorption-dominant performance of $Ti_3C_2T_x$ MXene/non-woven laminated fabrics. *Text. Res. J.*, 91, 2448–2458, 2021.
64. Shahzad, F., Alhabeb, M., Hatter, C.B., Anasori, B., Hong, S.M., Koo, C.M., Gogotsi, Y., Electromagnetic interference shielding with 2D transition metal carbides (MXenes). *Science*, 353, 1137–1140, 2016.
65. Muzaffar, A., Deshmukh, K., Ahamed, M.B., MXene-based multifunctional polymer composites for electromagnetic interference shielding applications, in: *MXene and Their Composites*, pp. 649–688, 2022.
66. He, S., Sun, X., Zhang, H., Yuan, C., Wei, Y., Li, J., Preparation strategies and applications of MXene-polymer composites: A review. *Macromol. Rapid Commun.*, 42, 2021.
67. Xia, X. and Xiao, Q., Electromagnetic interference shielding of 2d transition metal carbide (Mxene)/metal ion composites. *Nanomaterials*, 11, 2929, 2021.
68. Cui, G., Sun, X., Zhang, G., Zhang, Z., Liu, H., Gu, J., Gu, G., Electromagnetic absorption performance of two-dimensional MXene $Ti_3C_2T_x$ exfoliated by HCl + LiF etchant with diverse etching times. *Mater. Lett.*, 252, 8–10, 2019.
69. Bian, R., He, G., Zhi, W., Xiang, S., Wang, T., Cai, D., Ultralight MXene-based aerogels with high electromagnetic interference shielding performance. *J. Mater. Chem. C*, 7, 474–478, 2019.
70. Deeraj, B.D.S., Mathew, M.S., Parameswaranpillai, J., Joseph, K., *EMI shielding materials based on thermosetting polymers*, Elsevier Inc., Amsterdam, The Netherlands, 2020.
71. Deeraj, B.D.S., Jayan, J.S., Raman, A., Saritha, A., Joseph, K., Polymeric blends and nanocomposites for high performance EMI shielding and microwave absorbing applications. *Compos. Interfaces*, 29, 1–43, 2022.
72. Raagulan, K., Braveenth, R., Jang, H.J., Lee, Y.S., Yang, C.M., Kim, B.M., Moon, J.J., Chai, K.Y., Fabrication of nonwetting flexible free-standing MXene-carbon fabric for electromagnetic shielding in S-band region. *Bull. Korean Chem. Soc.*, 39, 1412–1419, 2018.
73. Li, X., Yin, X., Han, M., Song, C., Xu, H., Hou, Z., Zhang, L., Cheng, L., Ti_3C_2 MXenes modified with: *In situ* grown carbon nanotubes for enhanced electromagnetic wave absorption properties. *J. Mater. Chem. C*, 5, 4068–4074, 2017.
74. Zhao, M.Q., Trainor, N., Ren, C.E., Torelli, M., Anasori, B., Gogotsi, Y., Scalable manufacturing of large and flexible sheets of MXene/graphene heterostructures. *Adv. Mater. Technol.*, 4, 1–7, 2019.

75. Weng, C., Wang, G., Dai, Z., Pei, Y., Liu, L., Zhang, Z., Buckled AgNW/MXene hybrid hierarchical sponges for high-performance electromagnetic interference shielding. *Nanoscale*, 11, 22804–22812, 2019.
76. Li, Y., Zhou, X., Wang, J., Deng, Q., Li, M., Du, S., Han, Y.H., Lee, J., Huang, Q., Facile preparation of: *In situ* coated $Ti_3C_2T_x$/Ni0.5Zn0.5Fe2O4 composites and their electromagnetic performance. *RSC Adv.*, 7, 24698–24708, 2017.
77. Liu, R., Miao, M., Li, Y., Zhang, J., Cao, S., Feng, X., Ultrathin biomimetic polymeric Ti3C2Tx MXene composite films for electromagnetic interference shielding. *ACS Appl. Mater. Interfaces*, 10, 44787–44795, 2018.
78. Gao, X., Wang, B., Wang, K., Xu, S., Liu, S., Liu, X., Jia, Z., Wu, G., Design of $Ti_3C_2T_x/TiO_2$/PANI multi-layer composites for excellent electromagnetic wave absorption performance. *J. Colloid Interface Sci.*, 583, 510–521, 2021.
79. Han, M., Shuck, C.E., Rakhmanov, R., Parchment, D., Koo, C.M., Friedman, G., Gogotsi, Y., Beyond $Ti_3C_2T_x$: MXenes for electromagnetic interference shielding. *ACS Nano*, 14, 5008–5016, 2020.
80. Cao, M.S., Cai, Y.Z., He, P., Shu, J.C., Cao, W.Q., Yuan, J., 2D MXenes: Electromagnetic property for microwave absorption and electromagnetic interference shielding. *Chem. Eng. J.*, 359, 1265–1302, 2019.
81. Cao, M., Han, C., Wang, X., Zhang, M., Zhang, Y., Shu, J., Yang, H., Fang, X., Yuan, J., Graphene nanohybrids: Excellent electromagnetic properties for the absorbing and shielding of electromagnetic waves. *J. Mater. Chem. C*, 6, 4586–4602, 2018.
82. Zhang, Y., Wang, X., Cao, M., Confinedly implanted NiFe2O4-rGO: Cluster tailoring and highly tunable electromagnetic properties for selective-frequency microwave absorption. *Nano Res.*, 11, 1426–1436, 2018.
83. Cao, M., Wang, X., Cao, W., Fang, X., Wen, B., Yuan, J., Thermally driven transport and relaxation switching self-powered electromagnetic energy conversion. *Small*, 14, 1–8, 2018.
84. Lu, M.M., Cao, W.Q., Shi, H.L., Fang, X.Y., Yang, J., Hou, Z.L., Jin, H.B., Wang, W.Z., Yuan, J., Cao, M.S., Multi-wall carbon nanotubes decorated with ZnO nanocrystals: Mild solution-process synthesis and highly efficient microwave absorption properties at elevated temperature. *J. Mater. Chem. A*, 2, 10540–10547, 2014.
85. Iqbal, A., Shahzad, F., Hantanasirisakul, K., Kim, M.-K., Kwon, J., Hong, J., Kim, H., Kim, D., Gogotsi, Y., Koo, C.M., Anomalous absorption of electromagnetic waves by 2D transition metal carbonitride Ti_3CNT_x (MXene). *Science*, 369, 446–450, 2020.
86. Yun, T., Kim, H., Iqbal, A., Cho, Y.S., Lee, G.S., Kim, M., Kim, S.J., Kim, D., Gogotsi, Y., Kim, S.O., Koo, C.M., Electromagnetic shielding of monolayer MXene assemblies. *Adv. Mater.*, 32, 1906769, 2020.
87. Sushmita, K., Madras, G., Bose, S., Polymer nanocomposites containing semiconductors as advanced materials for EMI shielding. *ACS Appl. Mater. Interfaces*, 5, 4705–4718, 2020.
88. Liu, J., Zhang, H., Xie, X., Yang, R., Liu, Z., Liu, Y., Yu, Z.Z., Multifunctional, superelastic, and lightweight MXene/polyimide aerogels. *Small*, 14, 1802479, 2018.
89. Sambyal, P., Iqbal, A., Hong, J., Kim, H., Kim, M.-K., Hong, S.M., Han, M., Gogotsi, Y., Koo, C.M., Ultralight and mechanically robust $Ti_3C_2T_x$ hybrid aerogel reinforced by carbon nanotubes for electromagnetic interference shielding. *ACS Appl. Mater. Interfaces*, 11, 38046–38054, 2019.
90. Shang, T., Lin, Z., Qi, C., Liu, X., Li, P., Tao, Y., Wu, Z., Li, D., Simon, P., Yang, Q.H., 3D macroscopic architectures from self-assembled MXene hydrogels. *Adv. Funct. Mater.*, 29, 1903960, 2019.
91. Song, P., Liu, B., Qiu, H., Shi, X., Cao, D., Gu, J., MXenes for polymer matrix electromagnetic interference shielding composites: A review. *Compos. Commun.*, 24, 100653, 2021.

92. Yuan, W., Yang, J., Yin, F., Li, Y., Yuan, Y., Flexible and stretchable MXene/polyurethane fabrics with delicate wrinkle structure design for effective electromagnetic interference shielding at a dynamic stretching process. *Compos. Commun.*, 19, 90–98, 2020.
93. Song, P., Liang, C., Wang, L., Qiu, H., Gu, H., Kong, J., Gu, J., Obviously improved electromagnetic interference shielding performances for epoxy composites via constructing honeycomb structural reduced graphene oxide. *Compos. Sci. Technol.*, 181, 107698, 2019.
94. Liu, Z., Zhang, Y., Zhang, H.-B., Dai, Y., Liu, J., Li, X., Yu, Z.Z., Electrically conductive aluminum ion-reinforced MXene films for efficient electromagnetic interference shielding. *J. Mater. Chem. C*, 8, 1673–1678, 2020.
95. Liu, J., Liu, Z., Zhang, H., Chen, W., Zhao, Z., Wang, Q., Yu, Z.Z., Ultrastrong and highly conductive MXene-based films for high-performance electromagnetic interference shielding. *Adv. Electron. Mater.*, 6, 1901094, 2020.
96. Fan, Z., Wang, D., Yuan, Y., Wang, Y., Cheng, Z., Liu, Y., Xie, Z., A lightweight and conductive MXene/graphene hybrid foam for superior electromagnetic interference shielding. *Chem. Eng. J.*, 381, 122696, 2020.
97. Liang, C., Song, P., Qiu, H., Zhang, Y., Ma, X., Qi, F., Gu, H., Kong, J., Cao, D., Gu, J., Constructing interconnected spherical hollow conductive networks in silver platelets/reduced graphene oxide foam/epoxy nanocomposites for superior electromagnetic interference shielding effectiveness. *Nanoscale*, 11, 22590–22598, 2019.
98. Jin, X., Wang, J., Dai, L., Liu, X., Li, L., Yang, Y., Cao, Y., Wang, W., Wu, H., Guo, S., Flame-retardant poly(vinyl alcohol)/MXene multilayered films with outstanding electromagnetic interference shielding and thermal conductive performances. *Chem. Eng. J.*, 380, 122475, 2020.
99. Rajavel, K., Luo, S., Wan, Y., Yu, X., Hu, Y., Zhu, P., Sun, R., Wong, C., 2D Ti3C2Tx MXene/polyvinylidene fluoride (PVDF) nanocomposites for attenuation of electromagnetic radiation with excellent heat dissipation. *Compos. Part A Appl. Sci. Manuf.*, 129, 105693, 2020.
100. Xu, H., Yin, X., Li, X., Li, M., Liang, S., Zhang, L., Cheng, L., Lightweight Ti_2CT_x MXene/poly (vinyl alcohol) composite foams for electromagnetic wave shielding with absorption-dominated feature. *ACS Appl. Mater. Interfaces*, 11, 10198–10207, 2019.
101. Sun, R., Zhang, H., Liu, J., Xie, X., Yang, R., Li, Y., Hong, S., Yu, Z.Z., Highly conductive transition metal carbide/carbonitride (MXene)@polystyrene nanocomposites fabricated by electrostatic assembly for highly efficient electromagnetic interference shielding. *Adv. Funct. Mater.*, 27, 1702807, 2017.
102. Wu, X., Han, B., Zhang, H.-B., Xie, X., Tu, T., Zhang, Y., Dai, Y., Yang, R., Yu, Z.Z., Compressible, durable and conductive polydimethylsiloxane-coated MXene foams for high-performance electromagnetic interference shielding. *Chem. Eng. J.*, 381, 122622, 2020.
103. Kim, E., Zhang, H., Lee, J.-H., Chen, H., Zhang, H., Javed, M.H., Shen, X., Kim, J.K., MXene/polyurethane auxetic composite foam for electromagnetic interference shielding and impact attenuation. *Compos. Part A Appl. Sci. Manuf.*, 147, 106430, 2021.
104. Jia, X., Shen, B., Zhang, L., Zheng, W., Construction of compressible polymer/MXene composite foams for high-performance absorption-dominated electromagnetic shielding with ultra-low reflectivity. *Carbon*, 173, 932–940, 2021.
105. Li, R., Ding, L., Gao, Q., Zhang, H., Zeng, D., Zhao, B., Fan, B., Zhang, R., Tuning of anisotropic electrical conductivity and enhancement of EMI shielding of polymer composite foam via CO2-assisted delamination and orientation of MXene. *Chem. Eng. J.*, 415, 128930, 2021.
106. Song, P., Qiu, H., Wang, L., Liu, X., Zhang, Y., Zhang, J., Kong, J., Gu, J., Honeycomb structural rGO-MXene/epoxy nanocomposites for superior electromagnetic interference shielding performance. *Sustain. Mater. Technol.*, 24, e00153, 2020.
107. Wang, Q., Zhang, H., Liu, J., Zhao, S., Xie, X., Liu, L., Yang, R., Koratkar, N., Multifunctional and water-resistant MXene-decorated polyester textiles with outstanding electromagnetic interference shielding and joule heating performances. *Adv. Funct. Mater.*, 29, 1806819, 2019.

108. Wan, Y.-J., Li, X.-M., Zhu, P.-L., Sun, R., Wong, C.-P., Liao, W.-H., Lightweight, flexible MXene/polymer film with simultaneously excellent mechanical property and high-performance electromagnetic interference shielding. *Compos. Part A Appl. Sci. Manuf.*, 130, 105764, 2020.
109. Wan, Y., Xiong, P., Liu, J., Feng, F., Xun, X., Gama, F.M., Zhang, Q., Yao, F., Yang, Z., Luo, H., Xu, Y., Ultrathin, strong, and highly flexible $Ti_3C_2T_x$ MXene/bacterial cellulose composite films for high-performance electromagnetic interference shielding. *ACS Nano*, 15, 8439–8449, 2021.
110. Liang, C., Qiu, H., Song, P., Shi, X., Kong, J., Gu, J., Ultra-light MXene aerogel/wood-derived porous carbon composites with wall-like "mortar/brick" structures for electromagnetic interference shielding. *Sci. Bull.*, 65, 616–622, 2020.
111. Qi, F., Wang, L., Zhang, Y., Ma, Z., Qiu, H., Gu, J., Robust Ti3C2Tx MXene/starch derived carbon foam composites for superior EMI shielding and thermal insulation. *Mater. Today Phys.*, 21, 100512, 2021.
112. Liu, F., Li, Y., Hao, S., Cheng, Y., Zhan, Y., Zhang, C., Meng, Y., Xie, Q., Xia, H., Well-aligned MXene/chitosan films with humidity response for high-performance electromagnetic interference shielding. *Carbohydr. Polym.*, 243, 116467, 2020.
113. Liu, S., White, K.L., Reneker, D.H., Electrospinning polymer nanofibers with controlled diameters. *IEEE Trans. Ind. Appl.*, 55, 5239–5243, 2019.
114. Reneker, D.H. and Yarin, A.L., Electrospinning jets and polymer nanofibers. *Polym. (Guildf)*, 49, 2387–2425, 2008.
115. Raman, A., Deeraj, B.D.S., Jayan, J.S., Saritha, A., Joseph, K., A brief review on electrospun lignin nanofibres. *J. Sib. Fed. Univ. – Biol.*, 14, 465–474, 2021.
116. Raman, A., Jayan, J., Deeraj, B.D.S., Saritha, A., Joseph, K., Electrospun nanofibers as effective superhydrophobic surfaces: A brief review. *Surf. Interfaces*, 24, 101140, 2021.
117. Deeraj BDS, R.H., Jayan, J.S., Saritha, A., Joseph, K., Enhanced visco-elastic and rheological behavior of epoxy composites reinforced with polyimide nanofiber. *Nano-Struct. Nano-Objects*, 21, 100421, 2020.
118. Gupta, A., Gurunathan, P., Ramesha, K., Singh, M., Dhakate, S.R., Effect of heat treatment temperature on energy storage performance of PAN co-MMA based carbon nanofibers as free-standing lithium ion batteries anode. *Energy Storage*, 1, e89, 2019.
119. Atchison, J.S., Zeiger, M., Tolosa, A., Funke, L.M., Jäckel, N., Presser, V., Electrospinning of ultrafine metal oxide/carbon and metal carbide/carbon nanocomposite fibers. *RSC Adv.*, 5, 35683–35692, 2015.
120. Sharma, G.K. and James, N.R., Carbon black incorporated carbon nanofiber based polydimethylsiloxane composite for electromagnetic interference shielding. *Carbon Trends*, 8, 100177, 2022.
121. Sharma, G.K. and James, N.R., Progress in electrospun polymer composite fibers for microwave absorption and electromagnetic interference shielding. *ACS Appl. Electron. Mater.*, 3, 4657–4680, 2021.
122. Im, J.S., Kim, J.G., Bae, T.S., Lee, Y.S., Effect of heat treatment on ZrO2-embedded electrospun carbon fibers used for efficient electromagnetic interference shielding. *J. Phys. Chem. Solids*, 72, 1175–1179, 2011.
123. Deeraj, B.D.S., George, G., Dhineshbabu, N.R., Bose, S., Joseph, K., Electrospun ZrO2@carbon nanofiber mats and their epoxy composites as effective EMI shields in Ku band. *Mater. Res. Bull.*, 144, 111477, 2021.
124. Levitt, A.S., Alhabeb, M., Hatter, C.B., Sarycheva, A., Dion, G., Gogotsi, Y., Electrospun MXene/carbon nanofibers as supercapacitor electrodes. *J. Mater. Chem. A*, 7, 269–277, 2019.
125. Seo, D., Kim, M.-R., Kyu Song, J., Kim, E., Koo, J., Kim, K.-C., Han, H., Lee, Y., Ahn, C.W., Hollow Ti_3C_2 MXene/Carbon nanofibers as an advanced anode material for lithium-ion batteries. *ChemElectroChem*, 9, e202101344, 2022.

126. Qin, S., Usman, K.A.S., Hegh, D., Seyedin, S., Gogotsi, Y., Zhang, J., Razal, J.M., Development and applications of MXene-based functional fibers. *ACS Appl. Mater. Interfaces*, 13, 36655–36669, 2021.
127. Du, Z., Chen, K., Zhang, Y., Wang, Y., He, P., Mi, H.Y., Wang, Y., Liu, C., Shen, C., Engineering multilayered MXene/electrospun poly(lactic acid) membrane with increscent electromagnetic interference (EMI) shielding for integrated Joule heating and energy generating. *Compos. Commun.*, 26, 100770, 2021.
128. Lian, Q., Xu, W., Li, Y., Chen, H., Duan, H., Wang, Z., An, D., Zhao, G., Liu, Y., Deposited structure design of epoxy composites with excellent electromagnetic interference shielding performance and balanced mechanical properties. *J. Mater. Chem. C*, 8, 16930–16939, 2020.
129. Cui, C., Xiang, C., Geng, L., Lai, X., Guo, R., Zhang, Y., Xiao, H., Lan, J., Lin, S., Jiang, S., Flexible and ultrathin electrospun regenerate cellulose nanofibers and d-$Ti_3C_2T_x$ (MXene) composite film for electromagnetic interference shielding. *J. Alloys Compd.*, 788, 1246–1255, 2019.
130. Miao, Z., Chen, X., Zhou, H., Liu, P., Fu, S., Yang, J., Gao, Y., Ren, Y., Rong, D., Interfacing mxene flakes on a magnetic fiber network as a stretchable, flexible, electromagnetic shielding fabric. *Nanomaterials*, 12, 20, 2022.

10

Polymer/MXene Composites for Supercapacitor and Electrochemical Double Layer Capacitor Applications

Anju C.

Rajagiri School of Engineering & Technology (Autonomous), Kochi, Kerala, India

Abstract

MXene, a 2D transition metal carbonitrides with exciting possibilities in diverse fields, has sparked tremendous amount of research interest. They are considered as an assuring material for electrode to be used in energy storage applications due to their superior physicochemical properties like feasible charge transfer process due to transition metals and high volumetric specific capacitance. They are suitable for compounding with diversity of polymers with restricted conductivity owing to their characteristic 2D layered structure with metallic conductivity and many active terminations on their surface. The contact between MXene and polymer in such hybrid composites is reinforced by electrostatic interactions, strong covalent bond, hydrogen bonds, and van der Waals forces. Furthermore, accumulation of polymer chains upgrades the stability of MXene by reducing their oxygen interaction and alleviates their restacking tendency. A rationally designed MXene-polymer hybrid composites have great potential as a supercapacitive material as it utilizes both EDLC and psuedocapacitive mechanisms. In addition, a unique combination of good hydrophilicity, flexibility with high mechanical strength of MXene-polymer composites promotes their practical applications in many supercapacitor devices like symmetric, asymmetric and microsupercapacitor. This review summarizes the preparation methods, properties, and application of various M-xene-polymer composites for supercapacitor applications along with the recent advancements, achievements, and challenges.

Keywords: Energy storage, MXene, polymer, composites, supercapacitor, electrochemical double layer, specific capacitance, energy density, power density, cyclic stability

10.1 Introduction

The need for pristine and sustainable energy has risen dramatically in view of environmental pollution and the energy crisis [1]. Because of their quick power delivery, outstanding rate performance, and long cycling life, supercapacitors are ideal electrochemical energy storage devices [2]. The high catalytic activity, high specific surface area, and wide-bandgap

Email: anjuc@rajagiritech.edu.in

modulation capacity of 2D materials are currently a hot topic among scientists, and these materials hold enormous potential for disruptive applications and innovations in varied spheres. Elements including silicene, graphene, borophene, and black phosphorus and the compounds like transition metal oxides and sulphides, MXene and layered double hydroxide make up this large family [3–6].

MXenes are a peculiar conductive, water loving two-dimensional nanomaterial family discovered in 2011. These are structurally 2D sheets of transition metal carbonitrides and so far, more than 30 distinct MXene compounds have been synthesized [7, 8]. $M_{n+1}X_nT_x$ is the generic formula for MXenes, M denotes an early transition metal like Ti where X stands to represent carbon and/or nitrogen, T represents a surface functional group like O, F or OH, and x represents the functional groups number wherein number 'n' is a positive integer from 1 to 4 [9].

MXenes have distinct features, including better specific surface area, strong surface hydrophilicity, electrical conductivity, and appreciable electrochemical activity [10]. MXenes are similar to GO in terms of morphology, since a layer of an MXene has a sheet thickness of about 1 nm and its lateral dimensions ranging from a few 100nm - tens of microns, and their properties are greatly influenced by their shape as well as the chemical moiety that occur amidst their layers [11]. MXenes have good electrochemical properties due to their intrinsically conductive nature and charge transfer feasibility. This is attributed to the transition metal M with changeable oxidation number, as well as their distinctive layered structure, which renders them appealing in energy storage applications. In H_2SO_4 electrolyte, $Ti_3C_2T_x$ exhibits appreciable volumetric capacitance of 1500 F cm^3 as a representative of MXenes [12].

Polymers, because of their outstanding performance and ease of processing, are widely employed in our daily lives and their performance is insufficient to meet wide range applications. Fracture resistance, impact strengths, fatigue, tensile strength, fracture resistance, abrasion, corrosion, and other bulk characteristics are all outstanding in polymers. Their organic solvents solubility improves their congeniality with nanoparticles and makes it easier to incorporate them into two-dimensional material systems. By incorporating diverse fillers like 2D nanoparticles into the matrices of polymer, nanotechnology provides a way for the production of polymers with good characteristics and by altering the type of fillers, it has been possible to affect the physical-chemical properties of polymeric composites [13]. Hybrid materials benefit from both polymer and nanoparticle features, which has prompted the incorporation of 2D nanoparticles into polymer matrices [14, 15].

Two dimensional MXenes are a suitable nanofiller for constructing unique polymer composites which satisfies mechanical, thermal, electrical, flame-retardant, and other performances due to their unique features [16]. MXene can provide nucleation sites for composites when combined with other nanostructured polymers due to its abundant O, OH, and/or F surface functional groups [13–17]. The finely tailored layer spacing and functional groups obtained through etching method, together with MXenes two dimensional morphological structures, facilitates the inclusion of exceptional properties of modifiers to further improve the characteristics of MXene-polymer composites [18].

The capacity of MXene to disperse without usage of dispersing agents within the polymer matrix makes it ideal for use during the earliest steps of polymer synthesis and integration into an existing manufacturing line [9]. In MXene–polymer composites, a substantial association with charge assemblage at the interfaces between the polymer matrix and two dimensional flakes under an externally applied electrical field has also been found [19]. The practical application of MXene-polymer composites is determined by their performance, which is strongly tied to the synthesis method where complete etching and exfoliation and after delamination, adequate modification of MXene can not only increase MXenes performance, but also give MXene composites new properties [18].

The chapter briefly describes the classification, synthesis methods, and properties of MXene-polymer composites and their usage as supercapacitive material with recent advancements and achievements. We also discuss the challenges and opportunities that these nanocomposites face in the future.

10.2 MXene-Polymer Composites

10.2.1 Classification

MXene-polymer composites are mostly divided into two categories: filled composites and complexes. MXene is utilized as a filler in filled composites, where the polymer and MXene are randomly spread. However, in complexes, both MXene and polymer are disseminated in an organized way and are synthesized with appropriate structures such as MXene coated polymeric fiber, MXene-polymer aerogel, foam, and so on. The most often used composites are filled composites, which incorporate both thermoplastic and thermosetting polymers.

10.2.2 Preparation Methods

Diverse preparation methods are explored for MXene-polymer composites where the performance strongly rely on the characteristics of MXene and polymer matrixes chosen, MXene dispersity, MXene to polymer ratio, bonding between polymer and matrix, preparation process etc. [18].

10.2.2.1 Ex Situ Blending (Solvent Processing)

Ex situ blending is a direct approach that improves the performance of water soluble polymers such as PVC and PVDF. The appreciable fluidity of the solution makes this method suitable for the development of films composites and 3D porous functional polymer composites. The elimination of organic solvents and the limiting of MXene dispersion in polar solvents are two of the method's drawbacks.

In this method MXene and polymer are mixed thoroughly in a suitable solvent or solution. Composite materials with excellent properties can be formed through intermolecular interactions between MXene and polymer [18]. The indigenous compatibleness of MXene, makes this highly stable in solvents, with a Hildebrand solubility parameter of ~27 $MPa^{1/2}$. Hence, the choice of solvent is restricted to polar (protic/aprotic) solvents [20]. But surface modifications can greatly enhance the stability in non-polar solvents [21]. Si *et al.* studied the usage of long chain cationic bromide to modify the surface of Ti_3C_2 which improves its dispersibility in DMF, polystyrene etc. [22]. The modification of Ti_3C_2 surface by cetyltrimethylammonium bromide was reported by Yu *et al.* to enhance the dispersion thermoplastic polyurethane (TPU) [23]. MXene is typically introduced to polymer solutions in colloidal form, with the solvent being removed later via vacuum, evaporation, precipitation, and other methods [24].

Many MXene-hydrophilic polymer nanocomposites has been synthesized using water as solvent such as natural rubber [25], sulfonated polyether ether ketone [26], PU [27], PVA [28], PDDA [29], chitosan [30], cellulose nanofibrils [31], PEDOT:PSS [32] etc. Also solution based composite preparation in polar organic solvents was reported for PPy [33], PVA [34], polysulfone [35], epoxy [36] acrylic resin elastomers [37] etc.

Solution casting is a simple procedure in which the polymer and filler mixture is properly mixed, the solvent is evaporated, and the finished composite material is poured into a mold. Li *et al.* studied the synthesis of high dispersible MXene-PVA composite under ultrasonic conditions followed by stirring and pouring into a mold [38]. The same method was reported by to prepare PVDF-MXene, PEDOT:PSS, PVC composite materials [18].

In compression molding technique, the desired shape is obtained after compressing the mixture of polymer and filler in solution under high temperature. Polystyrene, TPU, polyvinylidene fluoride, and poly (lactic acid) have all employed this approach. Rajavel *et al.* studied the compression molding method for preparation of MXene-PVDF composites, in which a homogenous mixture of MXene and PVDF was made by ultrasonication in ethanol, the solvent was removed by heating, and the composite was formed by pressing [39].

Another widely used method is vacuum assisted filtering, which increases the crosslinking network and the performance of MXene-polymer composites [40]. Initially, a homogenous mixed solution of MXene and polymer is made, and solute-solvent are unpaired under vacuum by using a membrane filter [41]. Highly systematized nano-films having nacre-like brick-mortar nanostructured $Ti_3C_2T_x$ MXene with aqueous polyurethane were reported by Liu *et al.* [42].

The synthesis of MXene-polymer composite aerogels has also been reported utilizing the solution casting approach, in which a combination of etched MXene and PVA is poured into a mold and freeze dried and strong hydrogen bonding results in a 3D crosslinked network structure after drying [18].

10.2.2.2 In Situ Polymerization

Due to the intercalation of polymer matrix in between polymer chains, this approach generates stronger connections between polymer and MXene and greater dispersity, and it is extensively employed in the synthesis of EP, PMMA, and PS composites. Solubility of MXenes in good organic solvents is one of the challenge faced by this method.

In situ polymerization is a broadly accepted mode for fabricating MXene-polymer composites, wherein a mixture of monomers, initiators, fillers, and other materials are thoroughly mixed with MXene, and the monomers are polymerized to macromolecules, allowing MXene to be evenly distributed throughout the polymer host [43]. Appropriate stirring is the most significant criteria in producing high-performance polymer composites. Because MXene has sufficient layer spacing, intercalation of monomers followed by polymerization can result in MXene layer exfoliation [24]. This polymerization process has been reported for thermosetting polymers, linear macromolecules, and polymers with cyclic or heterocyclic units that polymerize under mild circumstances [44]. To improve dispersion, solvents such as acetone and DMF have been used and the method has been reported in the case of polypyrrole, $Ti_3C_2T_x$/epoxy resin nanocomposites, polythiophene, polydopamine, polyaniline.

10.2.2.3 *Other Preparation Methods*

Melt blending is a method used for non-polar thermoplastic polymers like PP, TPU and PLA where the polymer is melted at higher temperature and then mixed with M-xenes and fillers with constant stirring. In this method internal mixers and extruder were used as dispersion tools [44]. This method is devoid of organic solvents and few limitations of this method include viscosity control, MXene oxidation, less effective dispersions.

Another effective way for making MXene-polymer composites is coating technology [18]. MXene coating utilizing polyurethane foam to prepare polymer composite was reported by Lin *et al.* [45]. Spray coating is another approach that has been reported, in which MXene is sprayed over polymers such as rigid polyurethane foam and fabrics (Figure 10.1) to generate a conductive coating [46]. Carey-Shah employed a three-roll mill to integrate MXene into bisphenol-A diglycidyl ether, a popular monomer employed in epoxy manufacture [47].

10.2.3 Properties

MXene discovery facilitated improvements in the physical-chemical properties of polymer. In virtue of MXene-polymer unique properties it find applications in diverse fields like battery, supercapacitor electromagnetic interference shielding, water purification, sensors, coating.

10.2.3.1 *Electrical Properties*

The practical application of polymers is always hindered by their low electrical conductivity. MXene exhibited an electrical conductivity of (9880 S/cm) near to graphene [48]. The intercalation of polymer in between the layers of MXene reinforces the dielectric property,

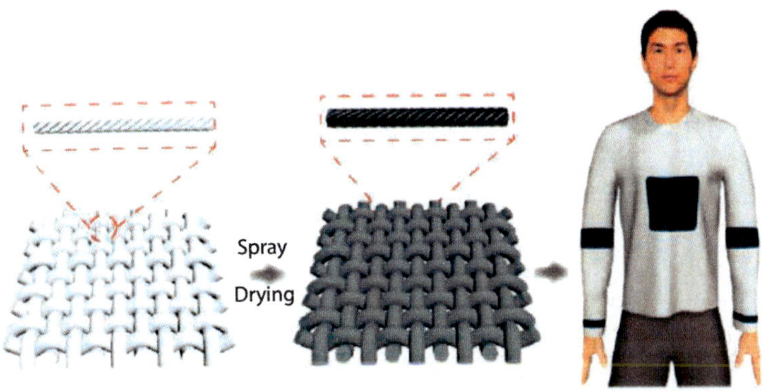

Figure 10.1 Spray coating method for fabricating multifunctional MXene-adorned cotton woven cloth. (Reprinted with permission from ref [46]. Copyright 2020 American Chemical Society).

conductivity and proton transport capacity [18]. The electrical conductivity enhances with the MXene loading and finally attains a plateau and the conductivity of those composites is much greater than other individual components. The conductivity of composites would be impacted by the MXene content and the inherent electrical characteristics [44]. Feng and group reported the preparation of EP/$Ti_3C_2T_x$ composites and maximum conductivity of (4.52×10^{-4} S/m) is obtained [49]. Rajavel *et al.* reported the incorporation of $Ti_3C_2T_x$ into PVDF matrix and was able to attain 0.988 S/m conductivity [39]. Likewise MXene was also added to PS and PVA matrices where an appreciable enhanced in conductivity is observed in all the cases. Thus the combination of MXene with polymer helps in building an effective conductive network for the facile transfer of electrons thereby elevating the electrical properties [18].

The presence of functional groups and good electrical conductivity of MXene creates efficient proton transport channels which endows the employment of MXene-polymer materials for energy devices like supercapacitors [50]. Liu *et al.* as well as Zhang *et al.* reported the combination of sulfonated poly (ether ether ketone) (SPEEK) polymer materials where interaction of sulphonic group of polymer with hydroxyl groups of MXene provide more proton carriers and shortens the proton hopping distance [51, 52]. The in-depth exploration of good conductivity of MXene to polymers is the major difficulty still faced for numerous industrial uses.

10.2.3.2 Thermal Properties

Polymers are considered as the inevitable electronic packaging material where the longevity of the device depends on the efficient heat dissipation. The addition of conductive fillers like MXene to polymer matrices enhances the heat conduction in resulting composites due to its inherent outstanding thermal conductivity. Kang *et al.* added the multilayer Ti_3C_2 MXene embedded into epoxy material to investigate thermal conductivity property of composites [53]. The adding up of MXene reduces the epoxy molecular chain randomness around the fillers and acts as a bridge to enhance transfer of heat. Chen and group [171] investigated thermal conductivities performance of as synthesized $Ti_3C_2T_x$ films which achieved 2.84 W/(mm·K) at 290 K [54].

Excellent thermal conductivity has been achieved in polymers like PU, PVA and PMMA where heat diffusion channels are created via strong H-bonding. Ul Haq *et al.* reported a ternary composite of MXene/PMMA/ZnO, where conductivity 14 times higher than pure PMMA was achieved [55]. High filler content sometimes hinders the performances of matrices due to irregular stacking and to overcome such defects Guo *et al.* fabricated 3D pore structured carbon fiber with MXene-Epoxy composites by freeze drying method where thermal conductivity of 9.68 W/mK was achieved at 30.2% MXene hybrid fillers loading [56]. Presence of carbon fibers minimizes phonon scattering and MXene build bridges in between carbon fibers for improving thermal conductivity. Ji *et al.* reported 3D MXene/Ag nanocomposite with EP where the ternary composite displayed higher temperature than pure EP at the same time indicating higher heat transfer (Figure 10.2) [57]. Thus it can be concluded that nanosheet structure of MXene forms chemical or hydrogen bonds with polymer matrices which acts as bridge in forming heat transfer network and also minimizes polymer chain randomness, scattering of phonons and thermal vibrations [44].

10.2.3.3 *Mechanical Properties*

MXenes are inorganic fillers that have excellent stiffness and flexural characteristics [58]. Ti_3C_2 MXene has a Young's modulus about 333 30 GPa. This is the maximum among solution-processible 2D materials [24]. MXene could substantially enhances the mechanical properties of polymer like tensile strength, Young's modulus, yield strength, hardness, elongation at break, etc. The presence of terminal groups helps in close association of polymer chain with MXene [59]. Surface functional groups, mainly those that are O-terminated and have smaller lattice constants, have a serious impact on the mechanical characteristics of MXenes [60]. By interacting with polar functional groups on polymer chains, these functional groups can attach MXenes to polymer matrices, thus ensuring the mechanical properties of composite materials [61].

Unlike other fillers MXene also improves the toughness of the composite [44]. The researchers found that improving interfacial adhesion, homogenous dispersion with the chemical interaction of MXene with macromolecules are all crucial factors in enhancing the strength and toughness of MXene/polymer nanocomposites [62]. Furthermore, increase in polymer crystallinity generated by MXene filler can donate to the composites' strength [44]. MXenes also have distinct advantages over other fillers, such as a little amount of addition and multifunctional improvements indicating that they have tremendous potential to augument the mechanical properties of polymer composites [18]. Low dosage MXene can assure superior mechanical performance of polymer composites by forming uniform allocation in matrix of polymer and producing strong bonding surfaces with chains of polymer [63]. For example, PLA composites containing only 0.5 wt percent $Ti_3C_2T_x$ MXene altered by stearic acid had much higher break elongation, Young' strength/modulus than those with pure PLA [64].

Because MXene and certain polymers have hydrogen bonds, it is simple to create strong interconnections, which considerably enhances break elongation with tensile strength [65]. Sheng *et al.* found that adding MXene to TPU increased stiffness about 47.1 percent (from 14 to 20 MPa), break elongation by 17.5 percent (from 1580 percent to 1857 percent), and toughness by a significant amount (from 1580 percent to 1857 percent) [66]. The two-dimensional $Ti_3C_2T_x$ can function alone to augument the polymer's elastic modulus, also it can interact synergistically among other fillers for improving the polymer's modulus of elasticity even further. To improve the mechanical strength of PVA nanofibers, Sobolciak *et al.* employed Ti_3C_2 nanosheets and cellulose nanocrystals (CNC) [67].

10.3 Applications of MXene Polymer Composites for Supercapacitor Applications

10.3.1 Introduction to Supercapacitor and Its Classification

Due to fossil fuels scarcity, it is necessary to create energy storage systems that are cost-effective, high-performing, and environmentally friendly in order to satisfy modern energy demand and maintain living standards. This calls for the creation of new materials and specialized architectural designs. In this search, the supercapacitor is unique as an appealing candidate due to its high power density, longer cycle life, faster charge-discharge rate, low

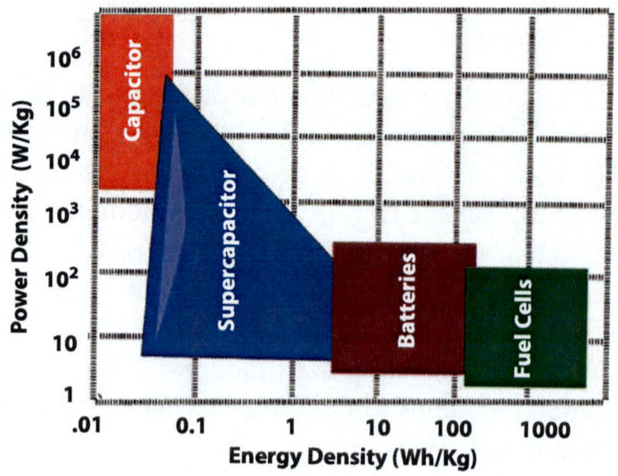

Figure 10.2 Ragone plot.

maintenance cost, and reversibility [68]. Supercapacitors are electrochemical devices that store energy through high-speed electrostatic/Faradaic electrochemical processes, where the charge is primarily stored at the electrode-electrolyte interface [69]. They primarily comprise a positive and negative electrode immersed in an electrolyte which is separated by an electronically insulating, ion-permeable separator [70]. They can store and deliver energy quickly [71] and are intriguing contenders to bridge gaps between rechargeable batteries and traditional capacitors [72]. Supercapacitors store higher energy than ordinary capacitors due to their larger electrode surface area and shortest separation between the two charged layers [73]. They have the ability to distribute the charge in a consistent manner over the discharging potential window [74]. They are found in various systems like electric vehicles, portable electronic gadgets, mobile phone power backup, and air bags [75]. Figure 10.2 shows a comparison of energy storage in batteries, supercapacitors, and capacitors.

10.3.2 Classification of Supercapacitor

Based on their energy storage mechanisms, supercapacitors are divided into three 1) EDLC 2) Pseudocapacitor 3) Hybrid capacitor (Figure 10.3). Carbon-based materials with good surface area, appreciable electrical conductivity, high power density, and stability, like activated carbon, template carbon, carbon aerogel, carbon nanomaterials, mesoporous carbon (and others) are commonly used in EDLCs. They store charges electrostatically via ion absorption at the electrode/electrolyte interface [76]. Pseudocapacitors, known as redox capacitors, are composed primarily of conjugated polymers like polythiophene, polyaniline, polypyrrole, as well as transition metal oxides like Fe_3O_4, NiO, RuO_2, MnO_2, $Ni(OH)_2$, $Co(OH)_2$, where charges are stored via fast and reversible faradaic reactions [77]. Faradaic electrodes' large specific pseudocapacitance outperforms EDLC's double layer charge storage method [78].

The poor capacitance value of EDLCs due to their microstructures and low energy density limits their performance [79]. Pseudocapacitors, on the other hand, have poor cycle stability and a longer response time than EDLCs during charge discharge cycles. The choice of materials has an important role in achieving an efficient supercapacitor with all of the

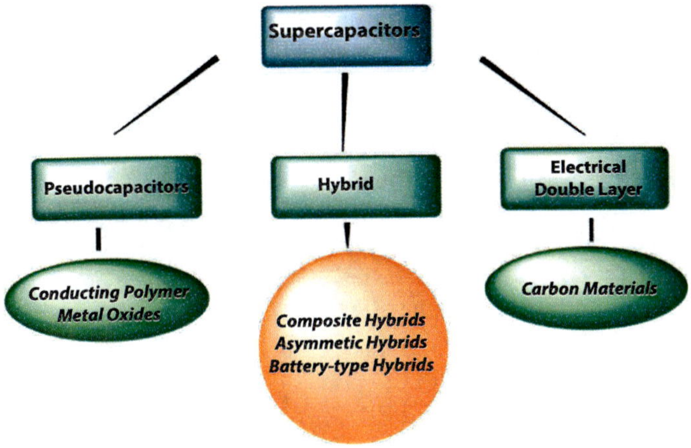

Figure 10.3 Classification of supercapacitor.

needed properties (high capacitance, good energy with power density, capacity retention, good cycle life with stability, manufacturing cost). The development of hybrid materials can help achieve this goal. They are divided into three categories.

1) Composite Hybrids- Here a single electrode comprises of a carbon material and it can be either metal oxides or conjugated polymer. As a result, the best characteristics of an EDLC and a pseudocapacitor are combined in a single electrode, resulting in synergistic effect such as high conductivity, cyclic stability and specific capacitance
2) Asymmetric Hybrid- It combines faradaic and non-faradaic processes by using an EDLC type material as cathode electrode and a conducting polymer/metal oxide as anode electrode. They simultaneously address the power and energy density requirements, and so outperform symmetric supercapacitors.
3) Battery type—Like asymmetric hybrids, it combines two electrodes, one of which is a supercapacitor electrode and the other a battery type electrode. It combines the supercapacitor and battery properties into a single cell.

10.3.3 Recent Advancements and Achievements in Various MXene-Polymer Composites for Supercapacitor Applications

Rechargeable batteries, supercapacitors, and energy conversion are all hot topics in energy storage research. As an electrode, MXenes have good specific area, great conductivity with appreciable capacitance density and large number of redox sites, allowing them to be combined with polymers for energy applications [80]. The capacitance density of a MXene electrode is 900 F cm^3 more than that of conventional carbon electrodes [81]. When MXene is introduced in between polymer chains, the microstructure of the polymer matrix is rebuilt, enabling enhanced ion channel connections and quick proton transfer [18]. Due to their extremely high level of electrical insulation and intrinsic breakdown strength, polymeric materials are crucial dielectrics for energy storage systems. However, their extraordinarily low permittivity, is incompatible with the ultrahigh energy density demands of energy storage devices [80].

The polymer may efficiently diminish MXene stacking and collapse, intercalation resistance and decrease ion diffusion, and increase ion adsorption and reaction sites by stabilizing and intercalating MXene [18]. As a result, MXene/polymer nanomaterials have received a lot of attention for energy applications including photothermal and electrothermal conversion [44]. The MXene/conductive polymer nanocomposites have the following advantages: that they take into account the trade-off between transportation properties and supercapacitor's charge storage capacity [9]. Most commonly used polymeric materials used for the preparation of MXene/polymer nanocomposites for supercapacitor applications involves Polyfluorene derivatives (PFDs), poly (3, 4-ethylenedioxythiophene) (PEDOT) and its derivatives, PPy, PAN!, PVA, and PDAC [82].

Sandwich MXene-Fe_2O_3-C-MoS_2-PEDOT:PSS(FMP)-MXene electrode material having extremely high area capacitance was described by Li *et al.* for use in flexible supercapacitors [83]. Due to the numerous active sites on MXene's surface and its greater hydrophilicity, FMP may be combined with it effectively, thus preventing the development and loss of FMP. This may significantly improve the capacitance value of MXene while reducing performance depletion resulting from the buildup of MXene's own structure. Hence, the good M/FMP/M electrode was able to achieve an area capacitance breakthrough (2700 mF cm^{-2} and 541 F g^{-1}). For symmetric and asymmetric supercapacitor devices, Sharabani *et al.* described a simple, affordable method of synthesizing Polyindole-MXene material (electron transfer mechanism depicted in Figure 10.4) [84]. The MXene-Polyindole composite system with 3:1 MXene-Polyindole ratio has the highest specific surface area, according to surface area analysis. This nanocomposite has a specific capacitance of 226.5 F/g at 2 A/g, and exceptional cyclic stability of 90.5 percent after 8000 cycles. With this composite as an anode, the planned asymmetric device attained a maximum specific capacitance of 117 F/g and an energy density of 65.3 Wh/kg. Notably, the composite material based on

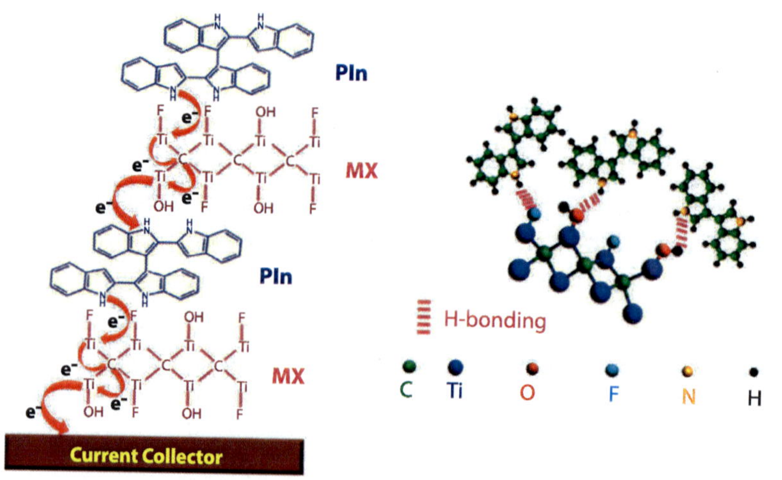

Figure 10.4 MXene and polyindole interaction and electron transport mechanism. (Reprinted with permission from ref [84]. Copyright 2020 American Chemical Society).

MXene performed exceptionally well as both cathode and anode within an asymmetric supercapacitor device. In addition, this device helped power a tiny fan and light up LEDs of various colors.

MXenes are greatly employed as anodes in supercapacitors, however due to the oxidation risk in this electrode, they have a limited use as cathode material. An asymmetric structure using MXene negative electrode and MXene/PANI nanocomposite positive electrode was developed to overcome the problem and produce a high-performance supercapacitor. The volumetric capacitance of 1632 F cm^3 and the rate capability of 827 F cm^3 at 5000 mV/s of this unique positive electrode are the highest reported values [85]. This MXene/PANI positive electrode and pure MXene negative electrode produced an asymmetric supercapacitor with a high power density of 127 kW/L and energy density of 50 Wh/L. Chen *et al.* reported the synthesis of DL tartaric acid-MXene-Polyaniline composite (Synthesis Procedure Figure 10.5) and their performance as supercapacitor electrode materials [86]. TEM images (Figure 10.6) reveals the decoration of PANI on MXene due to the chemical interactions. With the increase in concentration of MXene PANI particles become more agglomerated and hence DLTA can support the ordered growth of PANI on MXene if they are taken in stochiometric proportions. The 3D composite and doping microstructure developed are advantageous for electron and ion transfer, allowing for improved electrochemical performance.

CV curves of samples are shown in Figure 10.7a and GCD curves at Figure 10.7b which shows that specific capacitance of the as-prepared composite material in 1 M H_2SO_4 electrolyte is 452 F/g, which is greater than either composite component (263 F/g for polyaniline and 61 F/g for MXene). It is observed that for DLTA-PANI and TDPs IR drop significantly diminishes. Nyquist curves (Figure 10.7c) shows minimal charge transfer resistance values

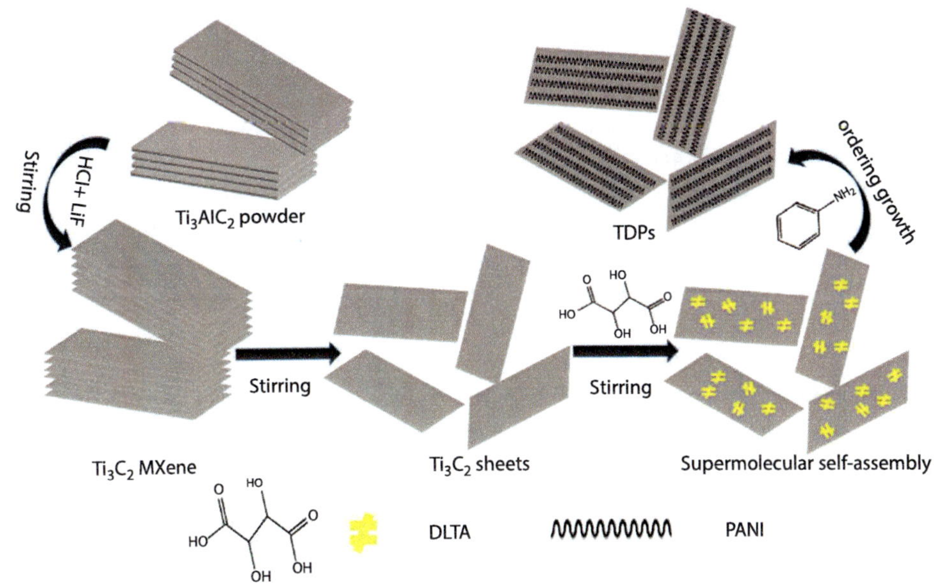

Figure 10.5 An illustration of the TDPs' synthesis process. (Reprinted with permission from ref [86]. Copyright 2020 American Chemical Society).

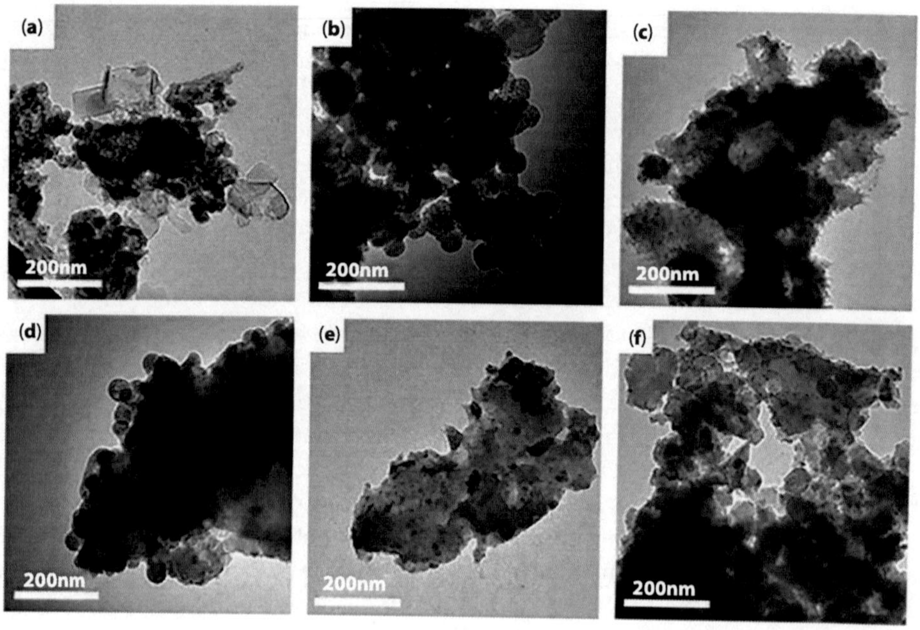

Figure 10.6 TEM images of (a) MXene, (b) pure PANI, (c) DLTAPANI (d) MXene-PANI (1:9) (e) MXene-PANI (2:8) (f) MXene-PANI (3:7). (Reprinted with permission from ref [86]. Copyright 2020 American Chemical Society).

like 0.4 for MXene, 1.4 for pure PANI, 0.3 for DLTA-PANI, 0.2 for TDP1, 0.2 for TDP2, and 0.18 Ω for TDP3. Furthermore, the COOH group on the MXene-DLTA self-assembly effectively dopes PANI, facilitating electron transport.

Fu *et al.* designed and synthesized pouch-type high-performance Asymmetric Supercapacitor based on Graphene-MXene-Polyaniline (GMP) Composite (Figure 10.8) [87]. Due to the utilization of mechanically resistant and densely intercalated conductive polyaniline and chemically inert graphene between the multi-layer MXenes, the resulting GMP electrode had increased electrochemical performance and cycle stability. Electrochemical performance of the samples is shown in Figure 10.9. CV curves of the samples exhibited a rectangular shape with the redox peaks of PANI. At 1 A/g current density, the GMP electrode shows a high gravimetric capacitance of 635 F/g (volumetric capacitance of 1143 F cm^{-3}) with minimum IR drop, and outstanding cycling stability of 97.54 percent after 10,000 cycles. In comparison to the MP and MXene electrodes, whose ESR values are 0.68 Ω and 0.71 Ω, the GMP electrode has the lowest resistance, with a value of 0.33 Ω. The pouch-type asymmetric supercapacitor, which uses graphene as a negative electrode and GMP as a positive electrode, has a power density of 950 W/kg, high energy density of 42.3 Wh/Kg and excellent cycle stability (94.25 percent after 10,000 cycles at 10 A/g) (Figure 10.10).

According to Qin *et al.*, capacitance of electrode was only 568 F cm^3 when MXene was intercalated with PEDOT:PSS to generate supercapacitors, but was more than twice as high when PEDOT:PSS was processed with concentrated H_2SO_4 to boost its conductivity (1310 F cm^3) [81]. This is the best results for the capacitor material's capacitive properties and one of the highest MXene base electrode values. The research on the MXene/PPy composite

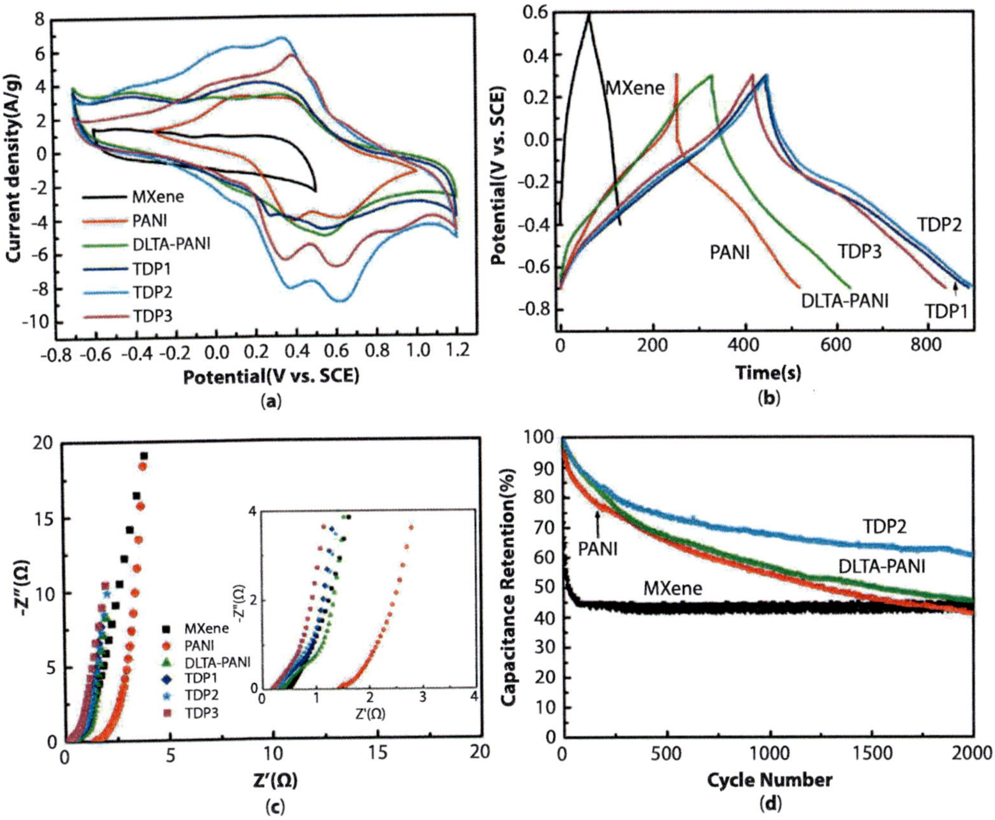

Figure 10.7 (a) CV curves of samples at a 10 mV s^{-1}. (b) GCD curves at [1] A g^{-1} (c) Nyquist plots (d) Specific capacitance retention at 4 A g^{-1}. (Reprinted with permission from ref [86]. Copyright 2020 American Chemical Society).

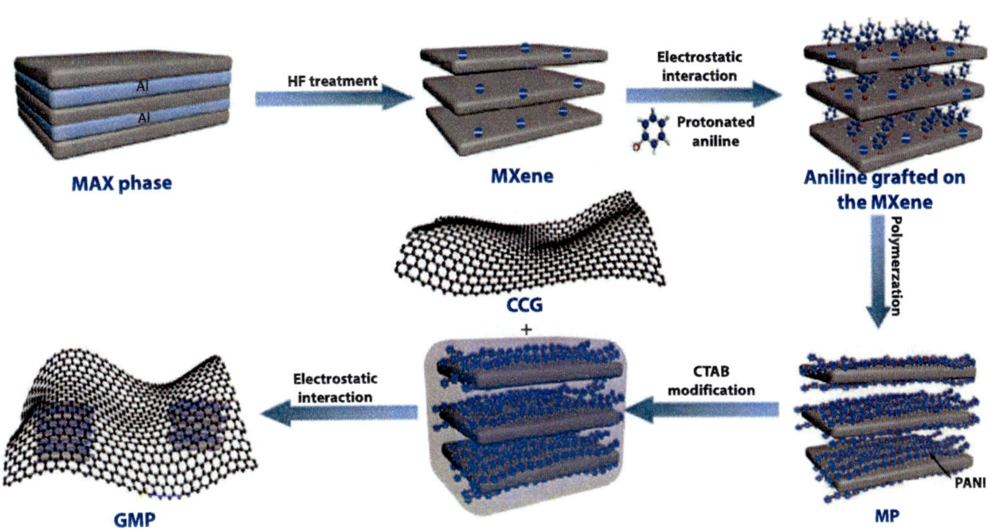

Figure 10.8 Synthesis procedure of GMP. (Reprinted with permission from ref [87]. Copyright 2018 American Chemical Society).

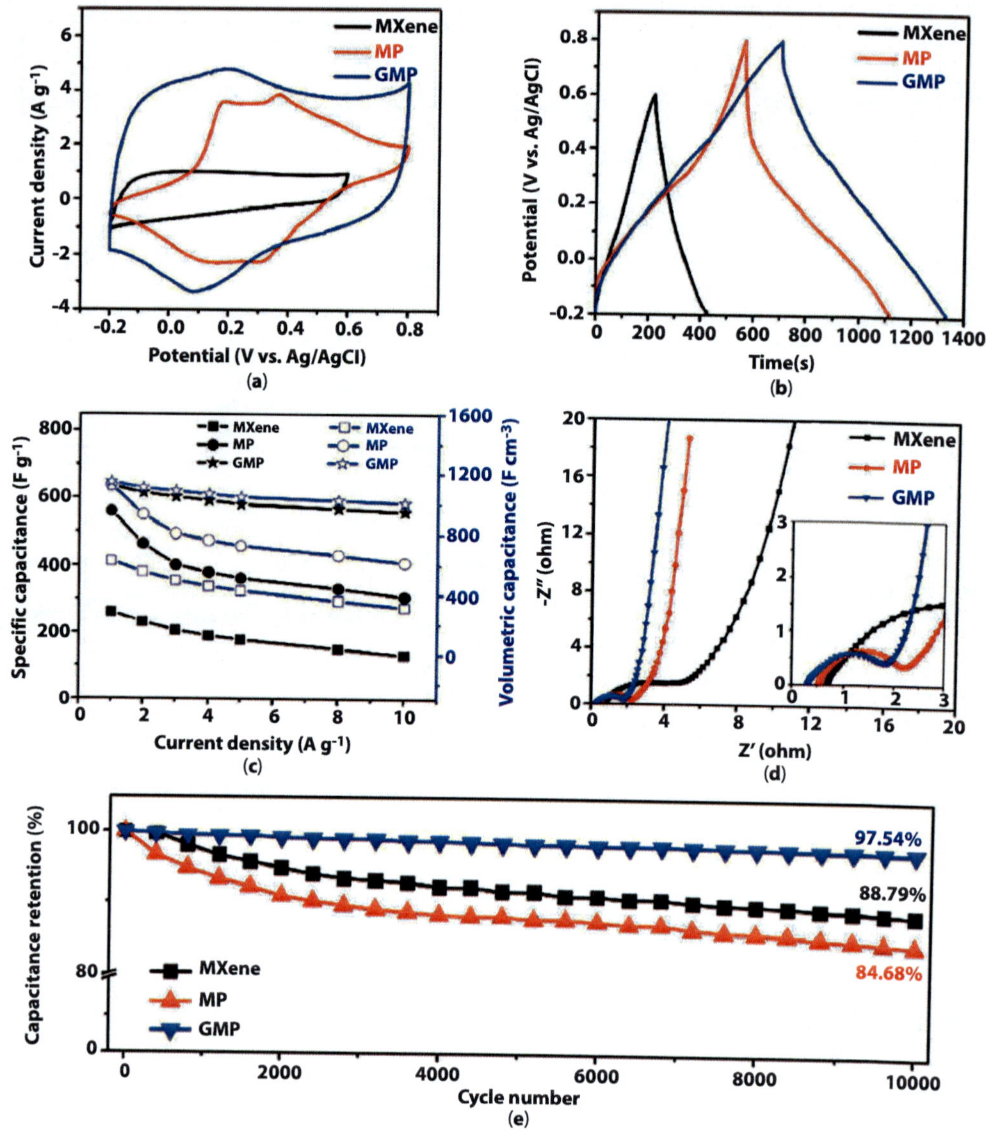

Figure 10.9 (a) CV curves at 5 mV s^{-1}. (b) GCD curves at 1 A g^{-1}. (c) Specific capacitance values at different current densities. (d) EIS curves. (e) Cycling stability at 10 A g^{-1}. (Reprinted with permission from ref [87]. Copyright 2018 American Chemical Society).

electrode provided more proof that higher polymer conductivity was advantageous for higher system capacitance. Boota *et al.* [88] examined the MXene/PFDs nanocomposites' capacitance properties. The researchers discovered that the mass capacitance density of composite electrode made of MXene and charged PFDs polymer was 1.5 times that of MXene and was comparable to that of the composite electrode made of MXene and PPy. According to Boota *et al.*, PFDs increased ability of MXene electrodes to store charge. This improvement was influenced by the polarity, homogeneity of dispersion, and association

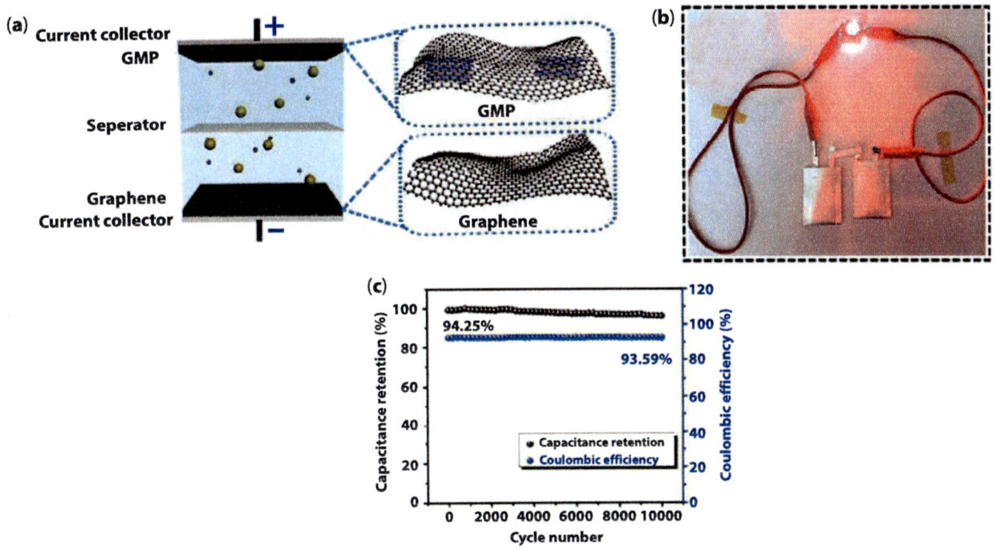

Figure 10.10 (a) Schematic of the GMP//graphene pouch type of ASC device. (b) Digital photo of a red LED lightening by two-series connected p-ASCs. (c) Capacity retention and coulombic efficiency at $10Ag^{-1}$. (Reprinted with permission from ref [87]. Copyright 2018 American Chemical Society).

of PFDs with MXene. The greater the contact with MXene and the obvious change in capacitance effect, the more uniform dispersion will be. Gogotsi's group investigated the capacitance characteristics of MXene/PVA-KOH (10 wt percent) nanocomposites used as electrodes [89]. When compared to the MXene/PDAC composite electrode, the volumetric capacitance of the composites enhanced by 80% (from 296 F/cm^3 for the MXene/PDAC composite to 528 F/cm^3 for the MXene/PVA-KOH composites). The scientists theorized that this rise in volumetric capacitance was due to MXene's increased interlayer spacing, which boosted ion mobility and charge transport capacity due to PVA. Yanchunxiao Qi *et al.* [90] reported the preparation of a binder composite, MXene-clay-polyacrylonitrile-co-butyl acrylate (MXC/PANBA) via *in situ* emulsion polymerization procedure with a superior specific power of 8510 W/kg and energy of 23.6 Wh/kg. The composite further continues to work superbly up to 50 mA cm^2. A better channel for binder film electron transport is provided by the strengthened interaction in the MXC/PANBA. Thus, it can be inferred that conductive MXC is an effectual technique to amend the high-power performance of EDLCs when combined with a nonconducting polymer binder.

Wu *et al.* [91] synthesized Ti_3C_2/PPy, an organ-like nanocomposite (Figure 10.11). SEM image (Figure 10.12) of the as prepared composite shows that with the increase in concentration of PPy, the thickness of the Ti_3C_2 layer increases where PPy itself shows an agglomerated spherical morphology. Figure 10.13 shows that the maximum specific capacitance of 184.36 Fg^{-1} at 2 mVs^{-1} is displayed by the Ti_3C_2/PPy-2 nanocomposite, which also maintains exceptional cycle stability thereby retaining a capacity of 83.33% after 4000 cycles at $1Ag^{-1}$. Notably, the combination of Ti_3C_2 organ shaped nanosheets with an EDLC mechanism and pseudocapacitive behavior of PPy nanoparticles contribute largely to high specific capacitance and also enhance cycle stability. Ti_3C_2/PPy nanocomposite's structural stability is aided by the organ-like Ti_3C_2 framework, which restricts PPy growth, prevents PPy stacking,

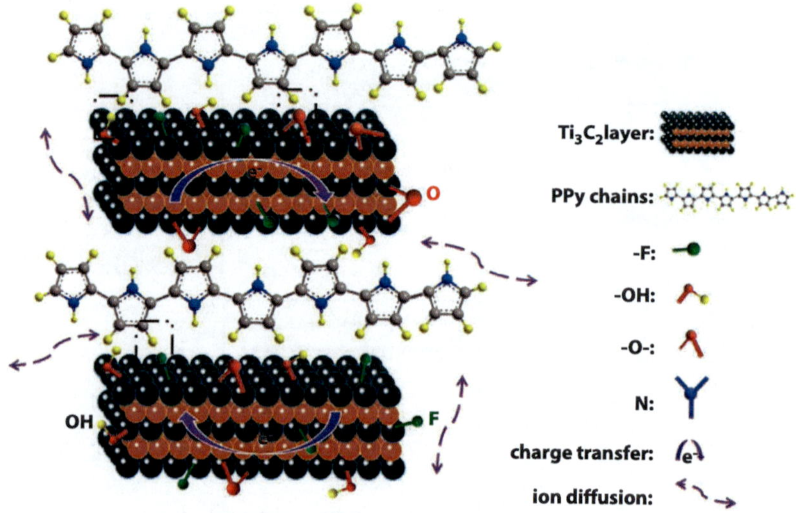

Figure 10.11 Ti_3C_2/PPy preparation is represented schematically. (Reprinted with permission from ref [91]. Copyright 2019 Elsevier).

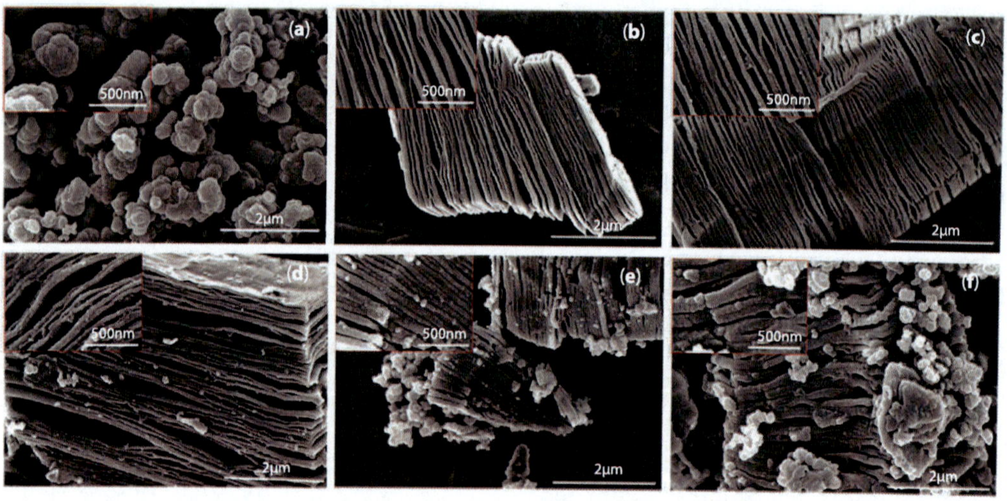

Figure 10.12 SEM images of (a) PPy, (b) Ti_3C_2 MXene, (c) Ti_3C_2/PPy-2, (d) Ti_3C_2/PPy-3, (e) Ti_3C_2/PPy-4, (f) Ti_3C_2/PPy-5. (Reprinted with permission from ref [91]. Copyright 2019 Elsevier).

and improves structural stability. The interlayer spacing of Ti_3C_2 is further increased by the intercalation of homogeneous PPy nanoparticles, and the well-ordered polymer chains provide more opportunities for the diffusion of electrolytic species and transfer of charges, which raises specific capacitance and lowers resistance to charge transfer.

Also the same author [92] reported an organ like high performance supercapacitive material made from Ti_3C_2 MXenes/polyaniline (Figure 10.14) by chemical grafting with a maximum of 228mFcm^{-2} area capacitance and capacitance retention of almost 85% even after 1000 charging/discharging cycles. Ti_3C_2 surface layers via interlamination, N-Ti_3C_2 can

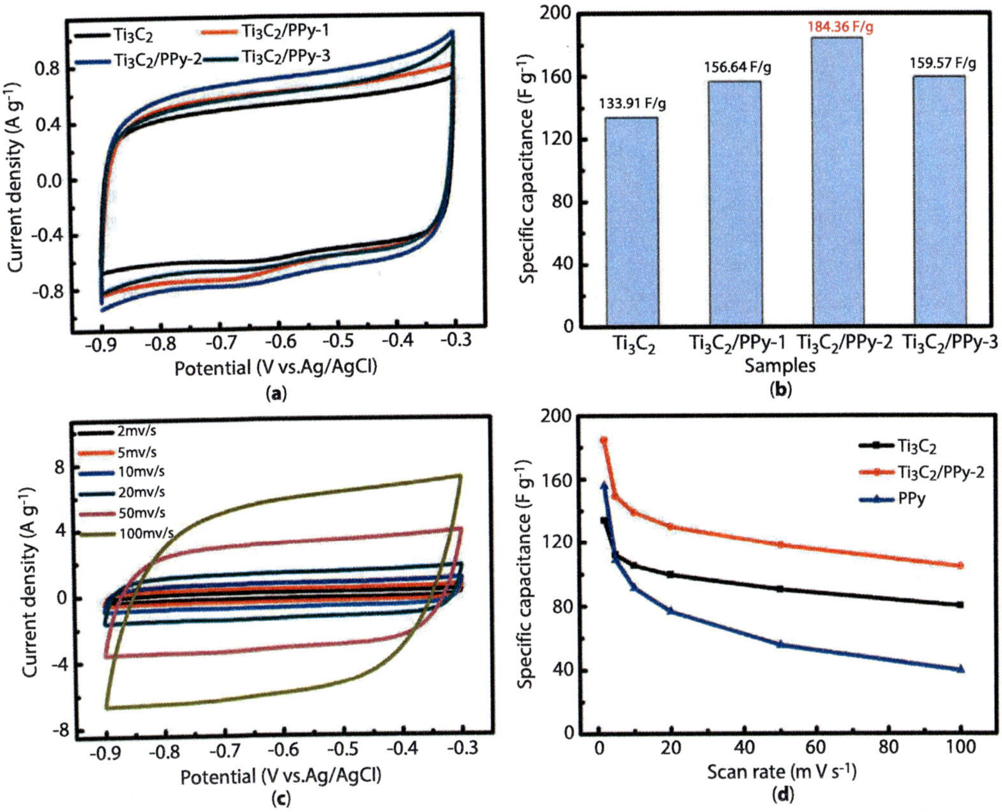

Figure 10.13 (a) Cyclic voltammogram of composites at 10 mVs^{-1}, (b) Specific capacitance of composites at 2 $mV\ s^{-1}$, (c) CV curves of Ti_3C_2/PPy-2 electrode at different scan rates, (d) Specific capacitance of Ti_3C_2 MXene, Ti_3C_2/PPy-2 and PPy electrodes at various scan rates.

function as active sites to react with PANI chains amine nitrogens and encourage the growth monomers of aniline. PANI has good pseudocapacitance properties that can increase the Ti_3C_2 MXenes surface area and interlaminar spacing. SEM images (Figure 10.15) reveal the uniform growth of organ-like N-Ti_3C_2 particles. The presence of PANI in the interlayers of N-Ti_3C_2 can be observed after the electrochemical polymerization.

Polyaniline nanotube incorporated Titanium carbide MXene (Ti_3C_2) nanosheets were synthesized by Wu *et al.* [93] for symmetric supercapacitors. Initially an ultrathinTi_3C_2 nanosheets are exfoliated, followed by aniline *in situ* polymerization. Schematic pathway of synthesis procedure and mechanism is shown in Figure 10.16. CV curves (Figure 10.17a–c) of Ti_3C_2/PANI-NTs show redox peaks arises from the redox transitions between different forms of PANI. The material exhibited a specific capacitance of 596.6 F/g at 0.1A/g shown by Ti_3C_2/PANI-NTs-1 by retaining a capacity of 94.7% after 5000 cycles. GCD curves (Figure 10.17 d-e) further confirmed the above result with highest discharge time and specific capacitance displayed by Ti_3C_2/PANI-NTs-1.

Additionally, in order to assess the practical applicability of the Ti_3C_2/PANI-NTs-1 composite, symmetric supercapacitor device was constructed and the analysis results are shown in Figure 10.18. CV curves pattern exhibits a typical pseudocapacitive nature. The SSC

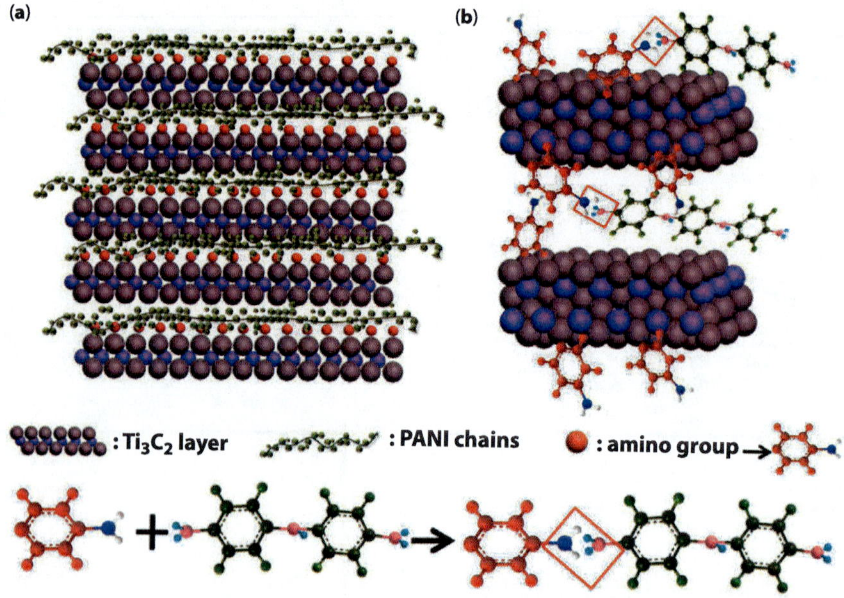

Figure 10.14 Schematic representation for the preparation of Ti_3C_2/Polyaniline. (Reprinted with permission from ref [92]. Copyright 2019 Elsevier).

Figure 10.15 SEM images of (a) N-Ti_3C_2, (b) PANI, (c) N-Ti_3C_2/PANI-240, (d) N-Ti_3C_2/PANI-300, (e) N-Ti_3C_2/PANI-420 and (f) N-Ti_3C_2/PANI-600. (Reprinted with permission from ref [92]. Copyright 2019 Elsevier).

device shows a maximum specific capacitance of about 300.8F/g, Rct value 13.58 Ω and 81.8% capacity retention. Also 25.6 Wh kg^{-1} energy density at 153.2 W kg^{-1} and a remarkable power density of about 1610.8 W kg^{-1} (at 13.2 Wh kg^{-1}) was achieved. This assembled supercapacitor device lighted up the 1.8V LED.

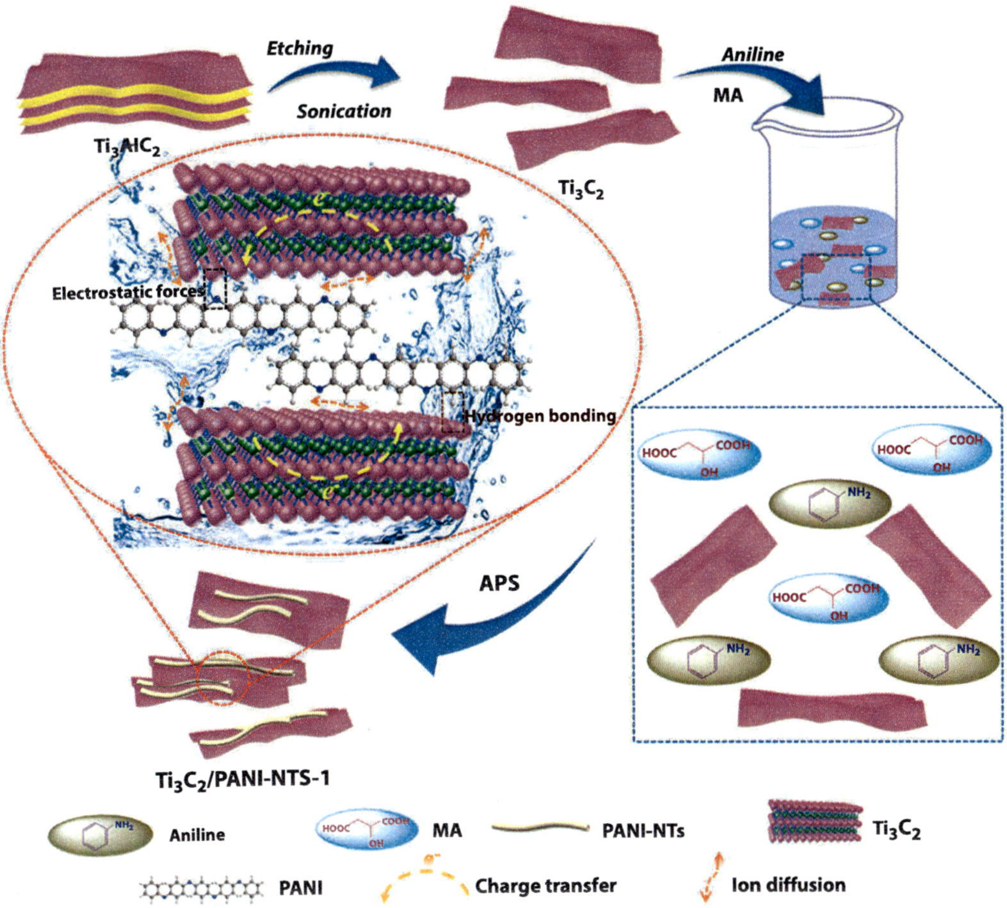

Figure 10.16 Hierarchical Ti_3C_2/PANI-NTs-1 nanocomposite generation and manufacturing process illustrated schematically. (Reprinted with permission from ref [93]. Copyright 2020 Elsevier).

Yue Li *et al.* [94] designed and synthesized an Hierarchical architecture of a hybrid electrode material MXene/PANI for asymmetric supercapacitor. PANI nanofibers were wrapped around MXene layers that had chain-like polyaniline (PANI) surface anchors through to a hydrothermal reaction. Layered MXene offer routes for ion insertion/extraction because it is an active substrate with a large framework and surface area. The charge transfer between various MXene layers might be sped up by the chain-like PANI utilized as conductive chains as well as a path for connecting adjacent layers. Surface morphology studies clearly show the successful exfoliation of MXene layers, additionally, chain-like PANIs were effectively fixed to the MXene layer's surface, enhancing the layer spacing. The material exhibited specific capacitance of 563 F g^{-1} and high capacity retention of 84.72%. An asymmetric supercapacitance fabricated using activated carbon as cathode electrode and MXene/PANI hybrids as anode electrode have reached energy density of 22.67 Wh/kg and power density of 0.217 kW/kg.

An advanced supercapacitor material, with nanospheres of polypyrrole grown on the nanosheets of Ti_3C_2-MXene (procedure shown in Figure 10.19) was developed by Dan

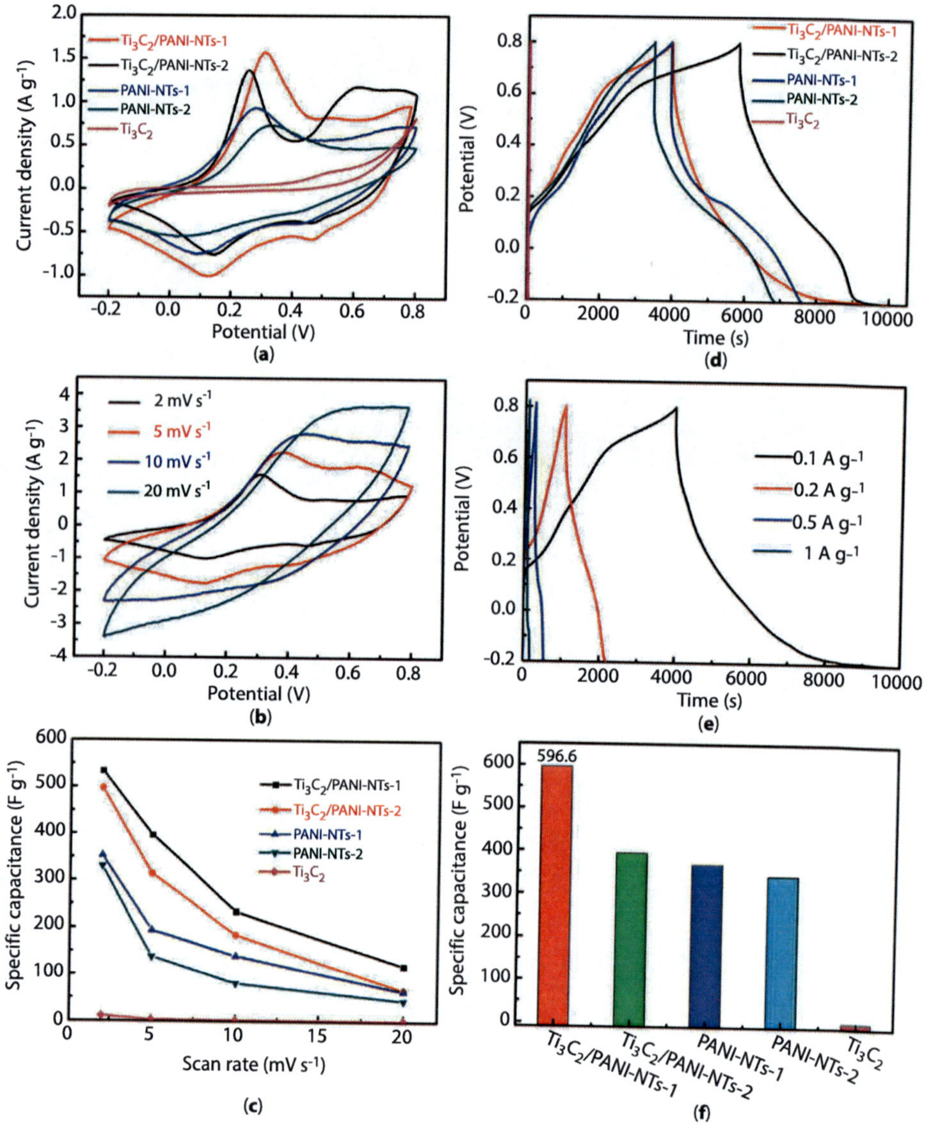

Figure 10.17 (a) CV curves of electrodes at the scan rate 2 mV s^{-1}. (b) CV curves of Ti_3C_2/PANI-NTs-1 at various scan rates, (c) Specific capacitance values of Ti_3C_2/PANI-NTs-1, Ti_3C_2/PANI-NTs-2, PANI-NTs-1, PANI-NTs-2, Ti_3C_2 at various scanning rates, (d) GCD curves of electrodes at 0.1 A g^{-1}, (e) Constant charge–discharge curves of Ti_3C_2/PANI-NTs-1 electrode at current densities 0.1, 0.2, 0.5 and 1 A g^{-1}, (f) Specific capacitance values of electrodes at 0.1 A g^{-1} from the GCD analysis. (Reprinted with permission from ref [93]. Copyright 2020 Elsevier).

Wei *et al.* [95]. By *in situ* polymerization of pyrrole monomers in the proximity of Ti_3C_2 solution, distinct PPy nanospheres having an average diameter of roughly 75 nm were entrenched to ultrathin nanosheets of Ti_3C_2. Ti_3C_2 has a morphology like crumpled paper, as shown in Figure 10.14. It benefits from having a lot of active sites for ion diffusion and charge transport as a matrix material, which helps it perform better electrochemically as a

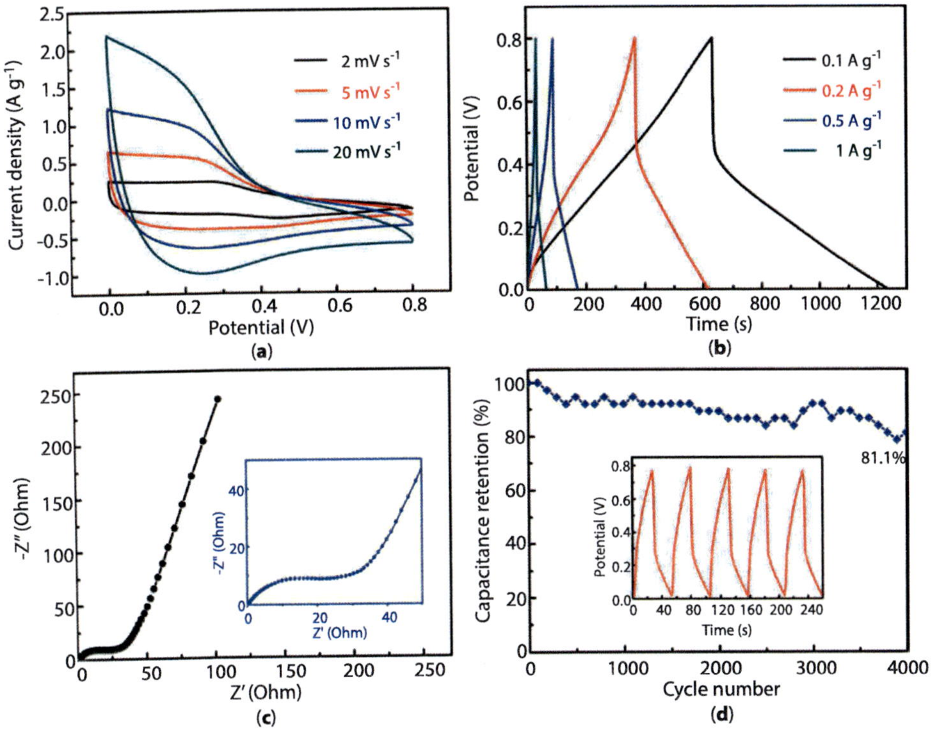

Figure 10.18 CV curves of SSC fabricated using Ti_3C_2/PANI-NTs-1. (a) CV curves from 2 to 20 mV s^{-1} scan rates. (b) GCD curves from 0.1 to 1 A g^{-1} current densities. (c) Nyquist plots of Ti_3C_2/PANI-NTs-1, (d) Capacitance retention during 4000 cycles at 1 A g^{-1}. (Reprinted with permission from ref [93]. Copyright 2020 Elsevier).

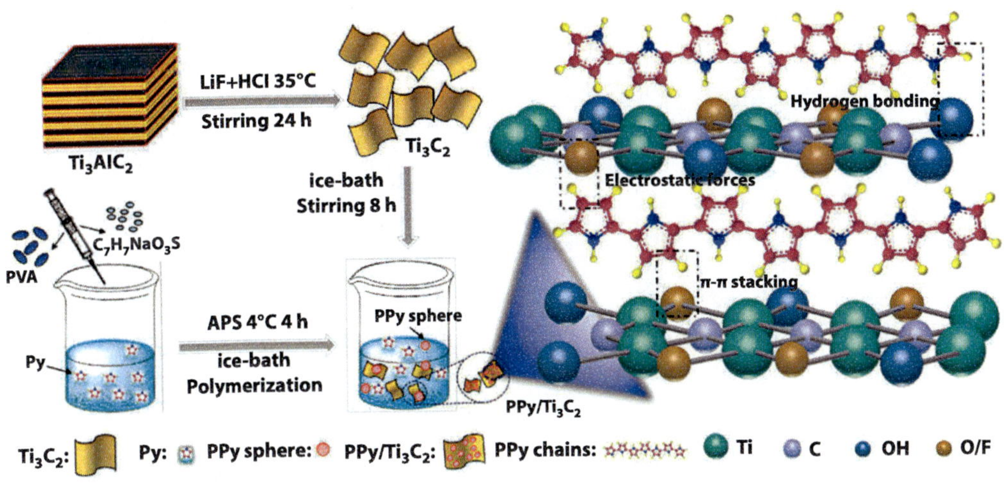

Figure 10.19 Synthesize procedure of PPy/Ti_3C_2 nanocomposite. (Reprinted with permission from ref [95]. Copyright 2020 Elsevier).

supercapacitor electrode. Figure 10.20 (b-d) shows the morphology of the nanocomposite S1, S2, and S3 of the PPy/Ti_3C_2 heterostructure in accordance with varied mass contents of Ti_3C_2 (50 mg, 70 mg, and 90 mg) respectively. With the help of the mutualistic interaction between the various faradaic and non-faradaic electrode materials, the electrode materials of uniform PPy nanospheres and ultrathin Ti_3C_2 nanosheets, each of which has a different energy storage mechanism, were successfully combined and exhibits a specific capacitance of 458 Fg^{-1} at the scan rate of 2 mV s^{-1}. A high energy density of 21.61 $Whkg^{-1}$ was attained by the PPy/Ti_3C_2-S2 symmetric supercapacitor with a power density of 499.94 W kg^{-1} with a cyclic stability of 73.68% at 1A g^{-1} after 4000 cycles.

The planar supercapacitor having interdigital electrodes has received a significant myriad of interest due to its adaptability, portability, and safety. Planar Supercapacitor fabricated using Ti_3C_2/Polypyrrole Composite Film with 109.4 mF/cm^2 areal capacitance in 2M H_2SO_4 was reported by Zhang *et al.* [96] with high cycle stability of 96% after 10000 cycles. A stable two dimensional conductive pathway is provided by Ti_3C_2 nanosheets for ion transport whereas PPy enhances the pseudocapacitive characteristics, which together have a synergistic effect that leads to the good electrochemical performance. In addition, an assimilated self-powered energy storage-conversion system using a supercapacitor and a solar cell that exhibits great practicability was constructed and is shown below in Figure 10.21.

In a study by Wu *et al.* [97], a supercapacitor electrode material that is highly flexible and exhibits very little capacitance loss was created by hybridizing decentralized conducting polymer chains (PDT) with $Ti_3C_2T_x$ (Atomic scale representation shown in Figure 10.22a). The electrode film schematic representation of fabrication is shown in Figure 10.23. Electrochemical analysis results (Figure 10.24) shows that PDT/$Ti_3C_2T_x$ have larger

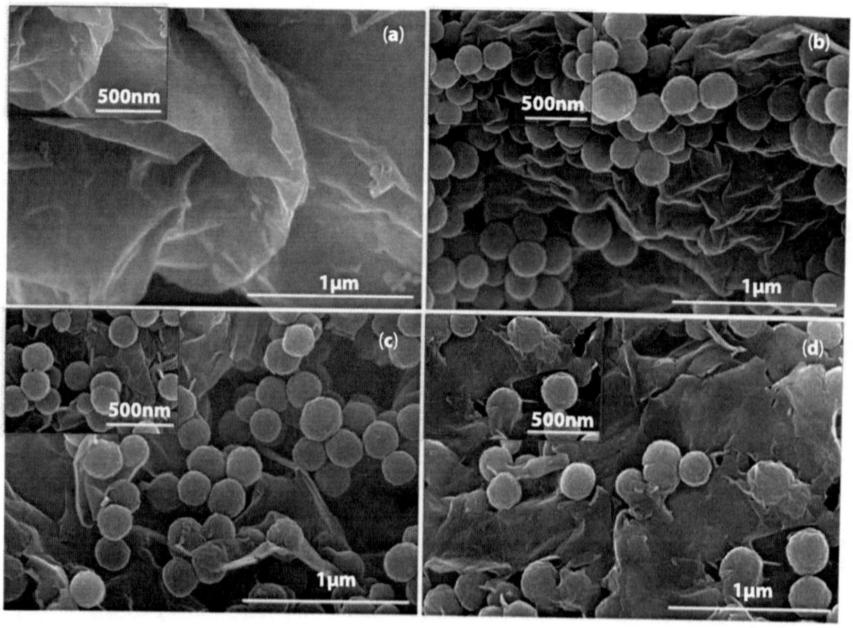

Figure 10.20 SEM images of (a) Ti_3C_2 MXene, (b) S1, (c) S2, (d) S3. (Reprinted with permission from ref [95]. Copyright 2020 Elsevier).

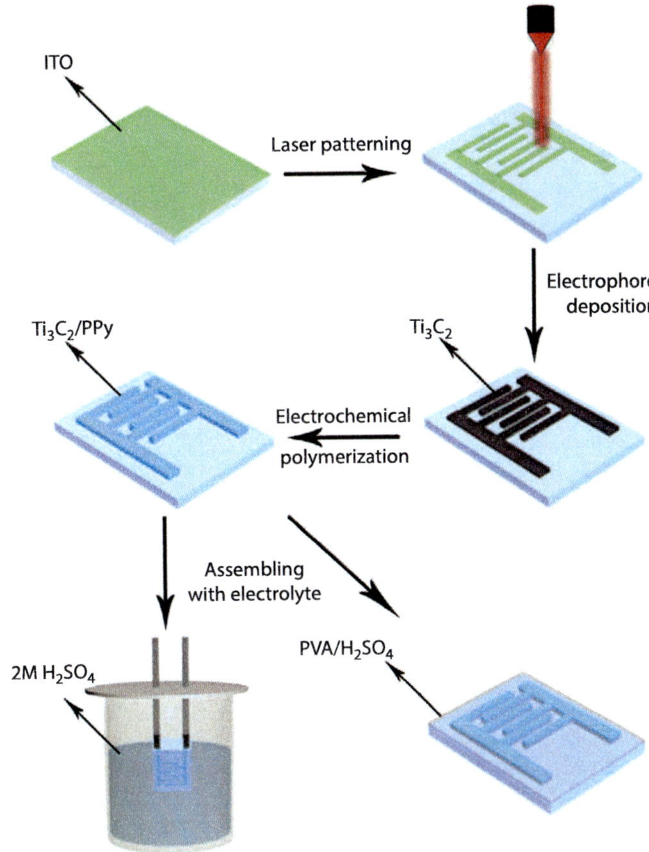

Figure 10.21 A schematic diagram of symmetric planar supercapacitor's production process. (Reprinted with permission from ref [96]. Copyright 2020 Elsevier).

enclosed area and longest discharge time than other individual components. It's interesting to note that the supercapacitor film is highly flexible and was able to withstand the 10,000-cycle 0-90° static bending test. The electrode displays an excellent area capacitance of 284 mFcm^{-2} at 50 mAcm^{-2} and negligible loss of capacitance, even after 10,000 charge-discharge cycles (Figure 10.22c). Strong linkages between the surfaces of $Ti_3C_2T_x$ created by this decentralized chain of PDT are advantageous because they not only offer precise paths for charge transfer but also increase the PDT backbones structural stability.

A new electrochemical polymerization (EP) approach for creating conjugated polymer-MXene composites by utilizing a 2D transition metal carbide was demonstrated by Qin *et al.* [98]. This technique' universality create a variety of unique polymer/MXene composites with regulated thickness with micro-patterns. The as synthesized asymmetric micro super-capacitor have an extremely high volumetric energy density (250.1 mWh cm^{-3}) at 1.6 V, a maximum areal capacitance of 69.5 mF cm^{-2} with magnificent cycle stability up to 10000 cycles. By varying the quantity of PANI nanofibers used as highly performing supercapacitor material for electrode, Luo *et al.* [99] were able to create flexible $Ti_3C_2T_x$/polyaniline films with changeable thickness in this study. The synthesis procedure of the composite

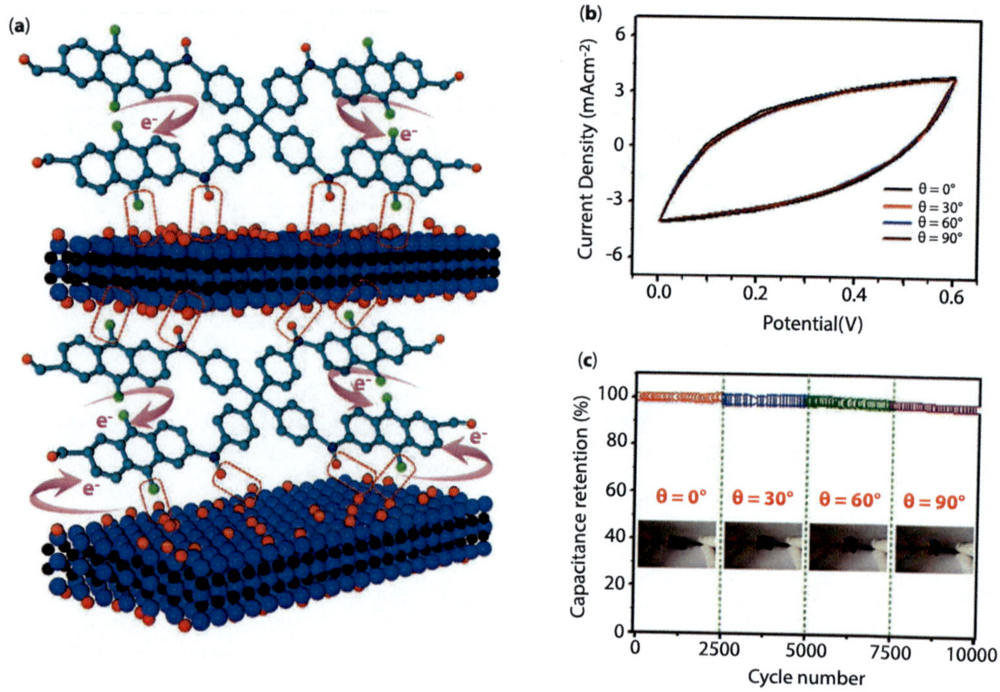

Figure 10.22 (a) Atomic-scale representation of PDT Chains with $Ti_3C_2T_x$ (b) CV under four bending angles (c) Cycling stability after 10,000 cycles under four static bending angles and photograph of the four static bending angles. (Reprinted with permission from ref [97]. Copyright 2019 Elsevier).

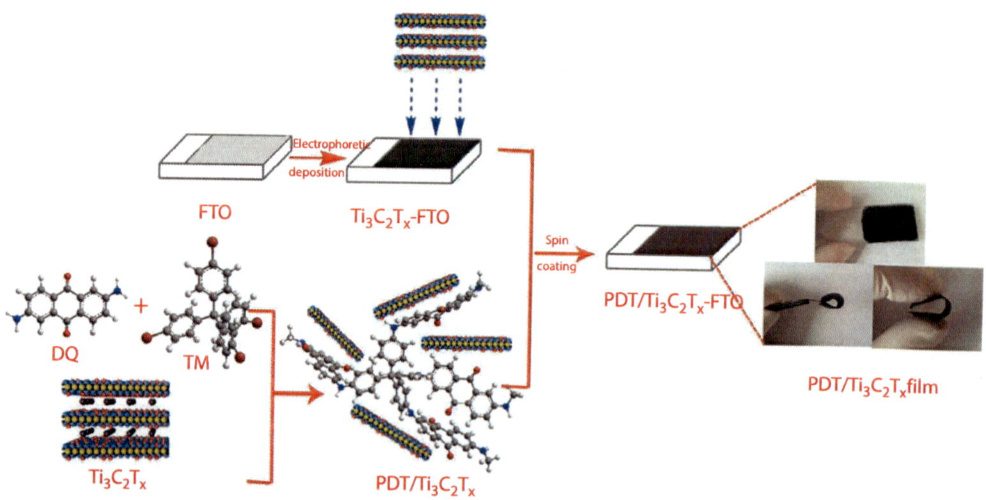

Figure 10.23 A schematic illustration of PDT/$Ti_3C_2T_x$ electrode production process. (Reprinted with permission from ref [97]. Copyright 2019 Elsevier).

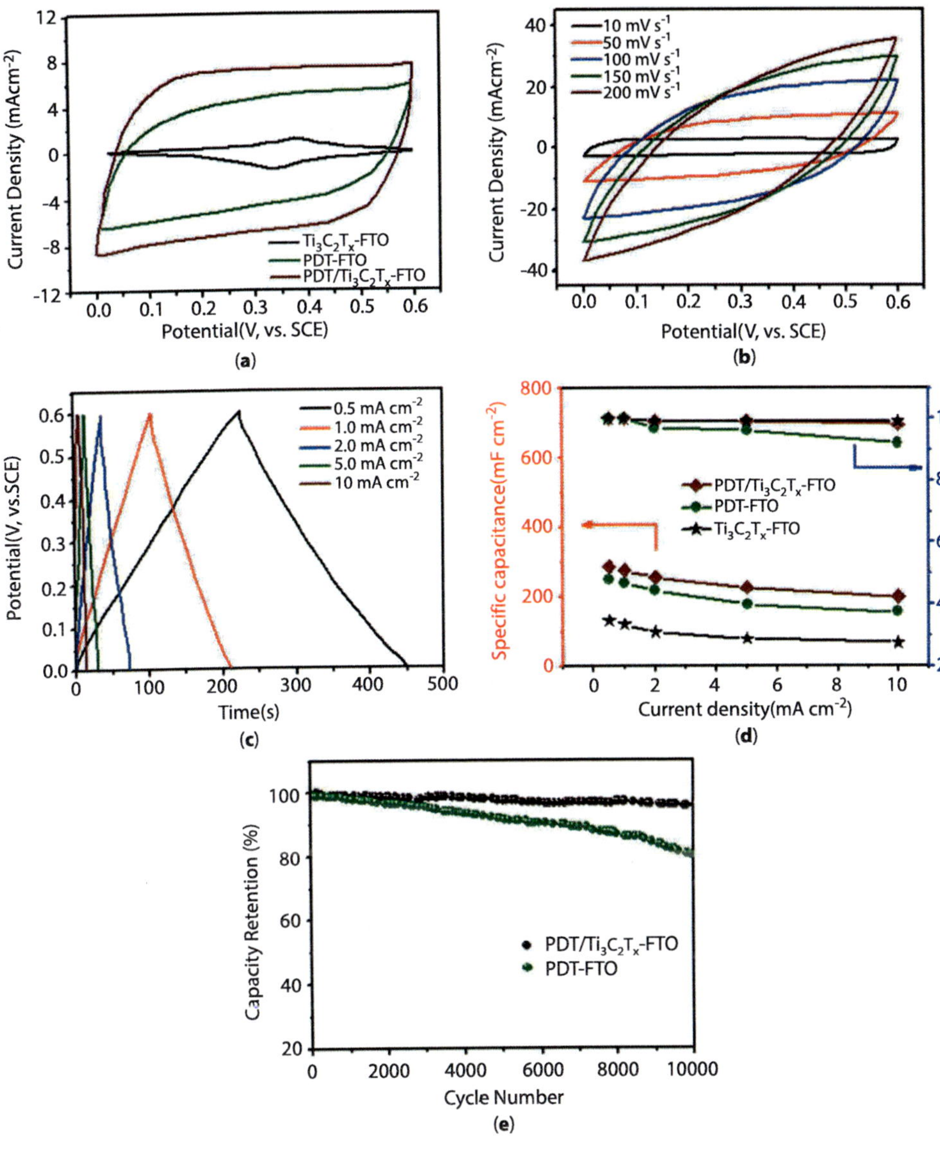

Figure 10.24 (a) CV curves of $Ti_3C_2T_x$, PDT and PDT/$Ti_3C_2T_x$ (b) CV curves of the PDT/$Ti_3C_2T_x$ at varying scan rates (c) GCD curves of PDT/$Ti_3C_2T_x$ electrode at varied current density (d) Specific capacitance and coulombic efficiency at different current density (e) Cycling stability of PDT and PDT/$Ti_3C_2T_x$. (Reprinted with permission from ref [97]. Copyright 2019 Elsevier).

and optical images of the as synthesized films is shown in Figure 10.25. The specific capacitance of the composite film, however, exhibits a tendency of initially increase and to decline with increasing PANI nanofiber content. This is true because too many PANI nanofibers cause the composite layer to thicken, which widens the electron and ion transport pathways. Figure 10.26 shows the device's CV curves at varying scan rates between 10 and 100 mV s^{-1}. Even at 100 mV s^{-1}, there are no polarization phenomena. The as obtained specific

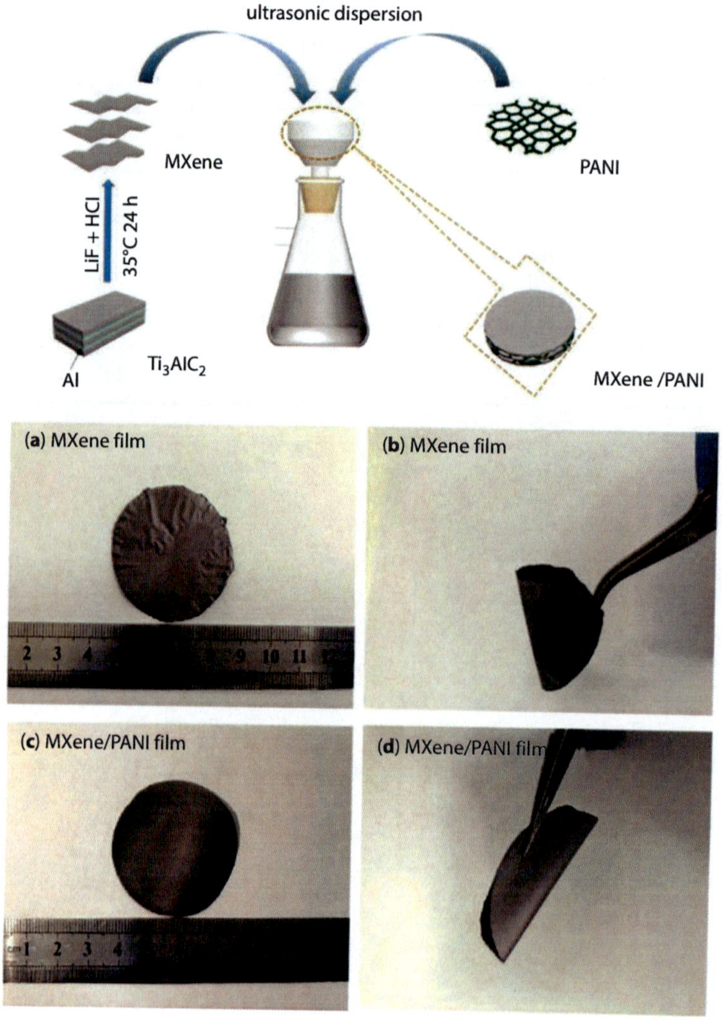

Figure 10.25 Schematic representation of synthesis method of film and optical pictures of (a) & (b) MXene film (MP0) and (c) & (d) MXene/PANI film (MP5). (Reprinted with permission from ref [99]. Copyright 2022 Elsevier).

capacitances of the material are depicted in Figure 10.26 (b), and it is observed that from high to low scan rate, the values obtained are 66.69, 81.57, 133.50, 191.30, and 352.90 F/g. In Figure 10.26 (c), GCD graphs for the same triangular-shaped device are shown for current densities ranging from 0.50 to 2.0 A g^{-1}, with an open circuit voltage of 1.2 V. The capacitance performance of the device declines with current density or scan rate, and the lowered rate exhibits a pattern from fast to slow (Figure 10.26d). The material has an equivalent internal resistance of 7.24 and the contact resistance of 10.74, according to the EIS curve of Figure 10.26 (e). The device can be bent almost 180 degrees in the photograph of Figure 10.19 (f), demonstrating its extraordinary flexibility which demonstrates MXene/PANI film's excellent potential for use in flexible fabric. A 1.8 V LED lamp can be easily lit by two devices linked in series, as seen in the image of Figure 10.26 (f). The device's effective

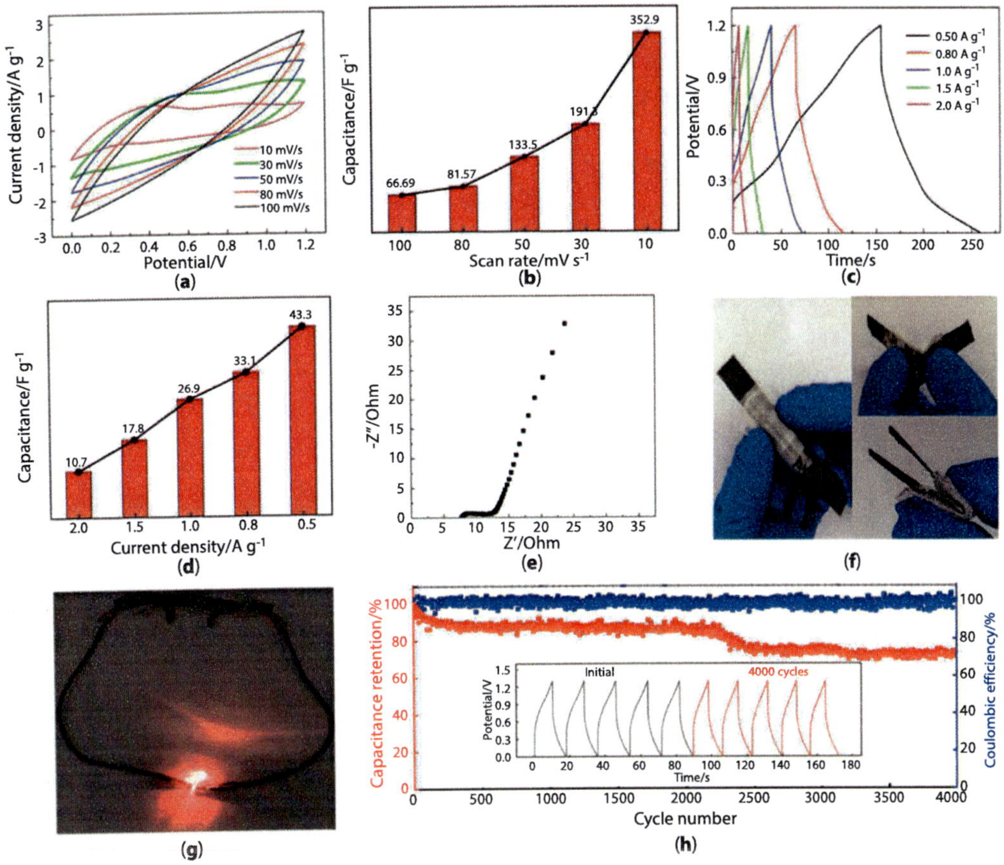

Figure 10.26 (a) CV of device at various scan rates and equaling, (b) Specific capacitances Values, (c) GCD of the device tested at various densities, and the equaling, (d) Specific capacitances values, (e) The device Nyquist plot, (f) Images of the device bended at an angle 90°, (g) Images of two devices connected in series to light a 1.8 V LED, (h) Cyclic stability and coulomb efficiency at 2 A g−1, inset is GCD of first and last five cycles. (Reprinted with permission from ref [99]. Copyright 2022 Elsevier).

lighting indicates that it has a wide voltage window and excellent energy storage capabilities. With an appreciable specific capacitance of 272.5 F g^{-1} at 1 Ag^{-1} and the retention of capacitance of around 71.4% after 4000 cycles at 2 Ag^{-1}, the built device based on the ideal composite film demonstrates impressive performance (Figure 10.26h). The devices' impressive energy storage performance was demonstrated by their high energy density of 31.18 Wh kg^{-1} at a power density of 1079.3 W kg^{-1}.

Shuqing Cao *et al.* [100] utilized Sol-gel and freeze-dried techniques to create three-dimensional porous $Ti_3C_2T_x$/PVA sponges. The porous sponge's pore utilization rate can be increased by the presence of PANI particles, which can also boost the electrode material's specific capacitance. The interlayer spacing of $Ti_3C_2T_x$ nanosheets has been significantly increased by $Ti_3C_2T_x$/PVA porous sponge material having a surface area of 48.8 m^2 g^{-1}, and the flexible supercapacitor based on PANI@$Ti_3C_2T_x$/PVA hydrogel composite shows both faradaic and non-faradaic behavior. A supercapacitor made of PANI@$Ti_3C_2T_x$/PVA has a huge areal energy density of 9.2 μWh/cm^{-2} and an ideal power density of 800 μW/cm^{-2}. After 10,000 charge/discharge cycles, this supercapacitor maintains a 99.5% capacitance retention rate.

A finely crafted solvent-associated self-assembly approach was used to create the flexible free-standing polyaniline@ MXene-carbon nanotubes (CNTs) sheet, done after vacuum filtration was studied by Xing Wu *et al.* [101]. To stop MXene from oxidizing and give high specific capacitance and the surface of 2D MXene nanosheets was coated with 0D polyaniline nanoparticle and is frequently employed as a cathodic electrode material and 0D carbon nanotubes were added to the composite to provide good flexibility and mechanical strength. With 99.4 MPa tensile strength, 463 F g^{-1} of high specific capacitance at 5 mVs^{-1}, and with 92% capacitance retention after 10,000 cycles which has a good cyclic stability, the ternary composite film offers exceptional mechanical performance. The asymmetric supercapacitor assembly with the manufactured free-standing MXene-CNTs as the anode and the ternary composite sheet as the cathode achieves a remarkable power density of 2808 W kg^{-1}.

Tong *et al.* [102] studied the electrochemical deposition of polypyrrole on $Ti_3C_2T_x$ film which was obtained after vacuum-assisted filtration method. The surface studies reveal that polypyrrole successfully intercalates into the gaps as well as on the surface of the $Ti_3C_2T_x$. An exceptional specific capacitance of 420.2 F/g at 1 A/g is demonstrated by an optimized sample with an impressive cyclic stability of 86%. Ke Li *et al.* [103] prepared an ultrafast cathode material by lodging polyaniline homogeneous layer on a 3D porous $Ti_3C_2T_x$ which demonstrates an ultrahigh rate capacity of 827 F cm^{-3} at 5000 mV s^{-1} and a volumetric

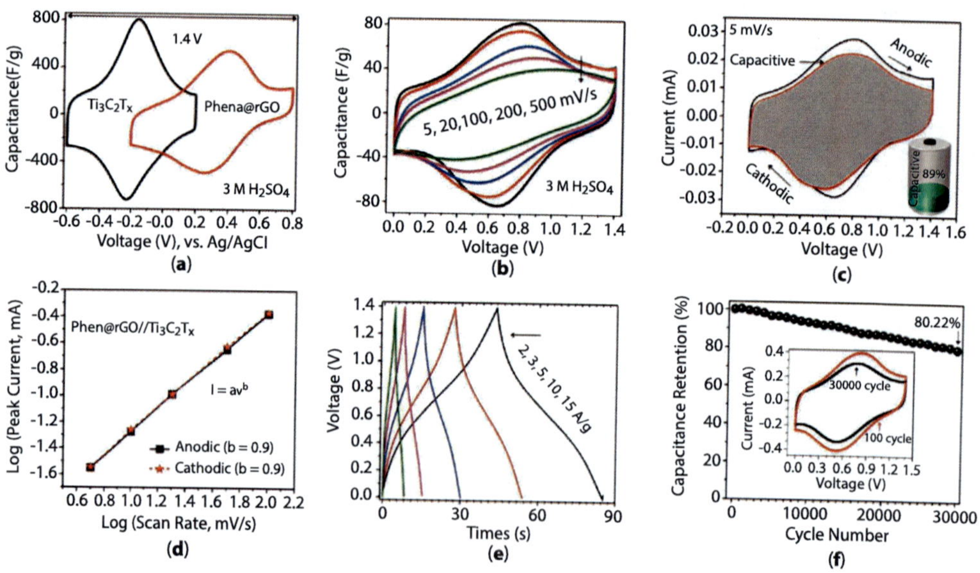

Figure 10.27 (a) CV curves of positive & negative electrode at various potentials (10 mV/s), (b) CV curves at different scan rates of phen@rGO-$Ti_3C_2T_x$ device, (c) Quantification of diffusion limited (unshaded area) and capacitive (shaded area) contributions to total charge storage at 5 mV/s. 89% of the total current was found to be capacitive, (d) Peak current- scan rate relationship, (e) GCD curves at different current densities, (f) Cyclic stability of phe-rGO-$Ti_3C_2T_x$ at 100 mV/s. (Reprinted with permission from ref [105]. Copyright 2020 ACS).

capacitance of 1632 F cm^{-3}. The same author designed an asymmetric pseudocapacitor [104] with a wave like MXene film as a negative electrode and a ternary nanocomposite of graphene/polyaniline/carbon nanotube as anode. The supercapacitor outputs high volumetric power and energy density of 111 kWL^{-1} and of 70 WhL^{-1}. Boota *et al.* [105] designed an asymmetric pseudocapacitor with $Ti_3C_2T_x$ as the negative electrode and a hybrid electrode of phenothiazine coated reduced graphene oxide as the positive electrode which retains a cycle stability of 80% after 30,000 cycles at 100mV/s. The electrochemical analysis of the samples is displayed in the Figure 10.27. After the CV analysis $Ti_3C_2T_x$-MXene electrodes shows a capacitance of 312 F/g at scan rate 5 mV/s. CV curve of the hybrid device has a couple of redox peaks originated from redox reactions of the individual components. About 89% of the total current produced by the device exhibits a capacitive nature. The 'b' value calculated from log peak current Vs log scan rate graph shows a value of 0.9 which confirms surface redox reactions and GCD analysis further confirmed negligible IR drop. The above discussed performance of various MXene-polymer composites are summarized in the Table 10.1.

Table 10.1 Performance chart of various MXene-polymer composites.

Material	Capacitance	Reference
Fe_2O_3-C-MoS_2-PEDOT:PSS-MXene	2700 $mFcm^{-2}$	[83]
Polyindole-MXene	226.5 F/g	[84]
MXene-PANI	1632 Fcm^3	[85]
Dl Tartaric acid-MXene-Polyaniline	452 F/g	[86]
Graphene-MXene-Polyaniline	635 F/g	[87]
MXene-PVA-KOH	528 F/cm^3	[89]
Ti_3C_2-PPy	184.36	[91]
Ti_3C_2-MXene-PANI	228 $mFcm^{-2}$	[92]
MXene-Polyaniline nanotube-Titanium carbide	596.6 F/g	[93]
PANI nanofiber-MXene	563 F/g	[94]
Polypyrrole nanosphere-Ti_3C_2	458 F/g	[95]
Decentralized conjugated polymer chain-$Ti_3C_2T_x$	284 $mFcm^{-2}$	[97]
$Ti_3C_2T_x$-Polyaniline	352.90 F/g	[99]
Free standing PANI-MXene-CNT	463 F/g	[101]
Polyprrrole-$Ti_3C_2T_x$	420.2 F/g	[102]
Polyaniline-3D porous $Ti_3C_2T_x$	1632 Fcm^{-3}	[103]
Hybrid $Ti_3C_2T_x$-Phenothiazine coated rGO	312 F/g	[105]

10.4 Challenges and Future Perspectives

Despite the fact that many of the outcomes described are very astounding, there are still several problems associated with manufacturing these NCs on a large scale. When exposed to air or aqueous solution, MXene can quickly oxidize to TiO_2 and when placed in water it will be totally oxidized in just 2 weeks. This is one of the most critical issues with the material. MXene placed in water will be totally oxidized in just two weeks. The range of oxidization of MXene in individual polymers is unknown, as it will most likely be determined by the polymer's chemistry and/or structure. The vulnerability of MXenes to oxidation and improvement in MXene's storage stability are important considerations.

The following two major technological obstacles must be overcome before MXene/polymer nanocomposites may be used industrially or produced on a large scale. Non-toxic MXene can be produced by inexpensively peeling multilayer MXene into individual MXene flakes. The non-Van der Waals interlayer interaction among the MAX phase layers, however, severely restricts the production of MXene/polymer nanocomposites on a larger scale, at a low cost, and in an environmentally friendly manner. This is because the method of stripping to create MXene is very dependent on harsh and toxic chemical etching routes. MXene has very substantial hydrophilic characteristics as a result of the usage of a strong acid in the etching synthesis, making it ideal for solution blending nanocomposites. The fundamental obstacle to future development and deployment of MXene/polymer nanocomposites will be how to disperse hydrophilic MXene in a hydrophobic polymer with outstanding blending ability.

The availability and cost of MAX phase powders are still the limiting factors, though novel methods to synthesize MXene devoid of HF direct (or indirect) use are more concerning factors in commercial developments of MXene/polymer NCs, due to inherent dangers of handling and storing large amounts of HF. In polymer matrices, MXene nanosheets, like GO, have a propensity to stack, which results in poor dispersion and reduced performance of MXene/polymer composites. Besides, the difficulty in dissolving freeze-dried MXene in water complicates the creation of MXene/polymer composites.

MXene/polymer composites have unique benefits that render a wide range of applications, so it's worth continuing to create multifunctional MXene/polymer composites. A well-functioning composites made from MXene-polymer materials is distinct in a range of composite materials and that MXene will be an essential fillers for study and use in polymer materials to develop unique supercapacitive materials to satisfy growing energy crisis in the near future.

10.5 Conclusion

MXene polymer nanocomposites show great promise for diverse applications, and various fields are still emerging, with further research to follow. The numerous terminations and 2D lamellar structure make it facile to combine MXenes and polymer matrices to make high-performance composites. The combined effects of high conductivity, wide specific area nanosheet structure, and diversified surface morphology are responsible for the remarkable performance advantages of MXene/polymer nanocomposites, according to researchers. Their flexibility has demonstrated intriguing potential and is receiving a lot of attention still for developing unique materials with outstanding performance for energy storage applications.

References

1. Wu, W., Wang, C., Zhao, C., Wei, D., Zhu, J., Xu, Y., Facile strategy of hollow polyaniline nanotubes supported on Ti3C2-MXene nanosheets for high-performance symmetric supercapacitors. *J. Colloid Interface Sci.*, 580, 601–613, 2020.
2. Shao, Y., El-Kady, M.F., Sun, J., Li, Y., Zhang, Q., Zhu, M., Kaner, R.B., Design and mechanisms of asymmetric supercapacitors. *Chem. Rev.*, 118, 9233–9280, 2018.
3. Zhu, J., Ha, E., Zhao, G., Zhou, Y., Huang, D., Yue, G., Hu, L., Sun, N., Wang, Y., Lee, L.Y.S., Recent advance in MXenes: A promising 2D material for catalysis, sensor and chemical adsorption. *Coord. Chem. Rev.*, 352, 306–327, 2017.
4. Huo, C., Yan, Z., Song, X., Zeng, H., 2D Materials via liquid exfoliation: A review on fabrication and applications. *Sci. Bull.*, 60, 1994–2008, 2015, DOI: 10.1007/s11434-015- 0936-3.
5. Nicolosi, V., Chhowalla, M., Kanatzidis, M.G., Strano, M.S., Coleman, J.N., Liquid exfoliation of layered materials. *Science*, 340, 1226419–1221424, 2013, DOI: 10.1126/ science.1226419.
6. Chhowalla, M., Shin, H.S., Eda, G., Li, L.J., Loh, K.P., Hua, Z., The chemistry of two dimensional layered transition metal dichalcogenide nanosheets. *Nat. Chem.*, 5, 263–275, 2013, DOI: 10.1038/nchem.1589.
7. Handoko, A.D., Fredrickson, K.D., Anasori, B., Convey, K.W., Johnson, L.R., Gogotsi, Y., Seh, Z.W., Tuning the basal plane functionalization of two-dimensional metal carbides (MXenes) to control hydrogen evolution activity. *ACS Appl. Energy Mater.*, 1, 1, 173–180, 2017.
8. Jun, B.-M., Kim, S., Heo, J., Park, C.M., Her, N., Jang, M., Yoon, Y., Review of MXenes as new nanomaterials for energy storage/delivery and selected environmental applications. *Nano Res.*, 12, 3, 471–487, 2019.
9. Riazi, H., Nemani, S.K., Grady, M.C., Anasori, B., Soroush, M., Ti3C2MXene-polymer nanocomposites and their applications. *J. Mater. Chem. A*, 9, 8051–8098, 2021, https://doi.org/10.1039/d0ta08023c.
10. He, Z., Xie, H., Wu, H., Chen, J., Ma, S., Duan, X., Chen, A., Kong, Z., Recent advances in MXene/polyaniline-based composites for electrochemical devices and electromagnetic interference shielding applications. *ACS Omega*, 6, 22468–22477, 2021, https://doi.org/10.1021/acsomega.1c02996.
11. Anasori, B., Lukatskaya, M.R., Gogotsi, Y., 2D metal carbides and nitrides (MXenes) for energy storage. *Nat. Rev. Mater.*, 2, 2, 1–17, 2017.
12. Lukatskaya, M.R., Kota, S., Lin, Z., Zhao, M.Q., Shpigel, N., Levi, M.D., Halim, J., Taberna, P.L., Barsoum, M.W., Simon, P., Gogotsi, Y., Ultra-high-rate pseudocapacitive energy storage in two-dimensional transition metal carbides. *Nat. Energy*, 6, 17105, 2017.
13. Jimmy, J. and Kandasubramanian, B., MXene functionalized polymer composites: Synthesis and applications. *Eur. Polym J.*, 122, 109367, 2020.
14. Xia, X., Weng, G.J., Zhang, J., Li, Y., The effect of temperature and graphene concentration on the electrical conductivity and dielectric permittivity of graphene–polymer nanocomposites. *Acta Mech.*, 1–16, 2020.
15. Chen, J., Li, Y., Wang, Y., Dong, J., Xu, X., Yuan, Q., Niu, Y., Wang, Q., Wang, H., Significantly improved breakdown strength and energy density of tri-layered polymer nanocomposites with optimized graphene oxide. *Compos. Sci. Technol.*, 186, 107912, 2020.

16. Li, X., Yin, X., Liang, S., 2D carbide MXene Ti2CTX as a novel high-performance electromagnetic interference shielding material. *Carbon*, 146, 210–7, 2019.
17. Xu, H., Zheng, D., Liu, F., Li, W., Lin, J., Synthesis of an MXene/polyaniline composite with excellent electrochemical properties. *J. Mater. Chem. A*, 8, 12, 5853–58, 2020.
18. Gong, K., Zhou, K., Qian, X., Shi, C., Yu, B., MXene as emerging nanofillers for high-performance polymer composites: A review. *Compos. Part B Eng.*, 217, 108867, 2021, https://doi.org/10.1016/j.compositesb.2021.108867.
19. Tu, S., Jiang, Q., Zhang, J., He, X., Hedhili, M.N., Zhang, X., Alshareef, H.N., Enhancement of dielectric permittivity of Ti3C2Tx MXene/polymer composites by controlling flake size and surface termination. *ACS Appl. Mater. Interfaces*, 11, 30, 27358–27362, 2019.
20. Maleski, K., Mochalin, V.N., Gogotsi, Y., Dispersions of two-dimensional titanium carbide MXene in organic solvents. *Chem. Mater.*, 29, 4, 1632–1640, 2017.
21. Carey, M., Hinton, Z., Natu, V., Pai, R., Sokol, M., Alvarez, N.J., Kalra, V., Barsoum, M.W., Dispersion and stabilization of alkylated 2D MXene in nonpolar solvents and their pseudocapacitive behavior. *Cell Rep. Phys. Sci.*, 1, 100042, 2020.
22. Si, J.-Y., Tawiah, B., Sun, W.-L., Lin, B., Wang, C., Yuen, A.C.Y., Yu, B., Li, A., Yang, W., Lu, H.-D., Functionalization of MXene nanosheets for polystyrene towards high thermal stability and flame retardant properties. *Polymers*, 11, 976, 2019, DOI: 10. 3390/polym11060976.
23. Yu, B., Tawiah, B., Wang, L.-Q., Yin Yuen, A.C., Zhang, Z.-C., Shen, L.-L., Lin, B., Fei, B., Yang, W., Li, A., Interface decoration of exfoliated MXene ultra-thin nanosheets for fire and smoke suppressions of thermoplastic polyurethane elastomer. *J. Hazard. Mater.*, 374, 110–119, 2019, DOI: 10.1016/j.jhazmat.2019.04.026.
24. Carey, M. and Barsoum, M.W., MXene polymer nanocomposites: A review. *Mater. Today Adv.*, 9, 100120, 2021, https://doi.org/10.1016/j.mtadv.2020.100120.
25. Luo, J.Q., Zhao, S., Bin Zhang, H., Deng, Z., Li, L., Yu, Z.Z., Flexible, stretchable and electrically conductive MXene/natural rubber nanocomposite films for efficient electromagnetic interference shielding. *Compos. Sci. Technol.*, 182, 107754, 2019.
26. Liu, Y., Zhang, J., Zhang, X., Li, Y., Wang, J., $Ti_3C_2T_x$ Filler Effect on the proton conduction property of polymer electrolyte membrane. *ACS Appl. Mater. Interfaces*, 8, 31, 20352–20363, 2016.
27. Arabi Shamsabadi, A., Pournaghshband Isfahani, A., Khoshhal Salestan, S., Rahimpour, A., Ghalei, B., Sivaniah, E., Soroush, M., Pushing rubbery polymer membranes to be economic for CO2 separation: Embedment with Ti3C2Tx MXene nanosheets. *ACS Appl. Mater. Interfaces*, 12, 3984e3992, 20192020.
28. Mayerberger, E.A., Urbanek, O., McDaniel, R.M., Street, R.M., Barsoum, M.W., Schauer, C.L., Promising PVC/MXene based flexible thin film nanocomposites with excellent dielectric, thermal and mechanical properties. *J. Appl. Polym. Sci.*, 8, 46, 2020.
29. Ling, Z., Ren, C.E., Zhao, M.Q., Yang, J., Giammarco, J.M., Qiu, J., Barsoum, M.W., Gogots, Y.I., Flexible and conductive MXene films and nanocomposites with high capacitance. *Proc. Natl. Acad. Sci.*, 111, 16676e16681, 2014.
30. Mayerberger, E.A., Street, R.M., McDaniel, R.M., Barsoum, M.W., Schauer, C.L., Antibacterial properties of electrospun Ti3C2Tz MXene/chitosan nanofibers. *RSC Adv.*, 8, 62, 35386–35394, 2018.
31. Zhou, B., Zhang, Z., Li, Y., Han, G., Feng, Y., Wang, B., Liu, C., Flexible, robust and multifunctional electromagnetic interference shielding film with alternating cellulose nanofiber and MXene layers. *ACS Appl. Mater. Interfaces*, 12, 4895e4905, 2020.
32. Liu, Y., Tao, Q., Jin, Y., Liu, X., Sun, H., Ghazaly, A.E., Zhang, F., Mo1.33C MXene-assisted PEDOT:PSS hole transport layer for high performance bulk-heterojunction polymer solar cells. *ACS Appl. Electron. Mater.*, 2, 2020, 163e169, 2019.

33. Seroka, N.S. and Mamo, M.A., Application of functionalised MXene-carbon nanoparticle-polymer composites in resistive hydrostatic pressure sensors. *SN Appl. Sci.*, 2, 1–11, 2020.
34. Mazhar, S., Qarni, A.A., Ul Haq, Y., Ul Haq, Z., Murtaza, I., Promising PVC/MXene based flexible thin film nanocomposites with excellent dielectric, thermal and mechanical properties. *Ceram. Int.*, 46, 12593e12605, 20202020.
35. Zhang, X., Fan, C., Yao, N., Zhang, P., Hong, T., Xu, C., Cheng, J., Quaternary Ti3C2Tx enhanced ionic conduction in quaternized polysulfone membrane for alkaline anion exchange membrane fuel cells. *J. Membr. Sci.*, 563, 882–887, 2018.
36. Kang, R., Zhang, Z., Guo, L., Cui, J., Chen, Y., Hou, X., Wang, B., Te Lin, C., Jiang, N., Yu, J., Enhanced thermal conductivity of epoxy composites filled with 2D transition metal carbides (MXenes) with ultralow loading. *Sci. Rep.*, 9, 1, 24, 2019.
37. Shao, J., Wang, J.W., Liu, D.N., Wei, L., Wu, S.Q., Ren, H., A novel high permittivity percolative composite with modified MXene. *Polymer*, 174, 86–95, 2019.
38. Li, L., Liu, X., Wang, J., New application of MXene in polymer composites toward remarkable anti-dripping performance for flame retardancy. *Compos. Appl. Sci. Manuf.*, 127, 105649, 2019.
39. Rajavel, K., Luo, S., Wan, Y., 2D Ti3C2Tx MXene/polyvinylidene fluoride (PVDF) nanocomposites for attenuation of electromagnetic radiation with excellent heat dissipation. *Compos. Appl. Sci. Manuf.*, 129, 105693, 2020.
40. Tian, W., Vahidmohammadi, A., Reid, M.S., Multifunctional nanocomposites with high strength and capacitance using 2D MXene and 1D nanocellulose. *Adv. Mater.*, 31, 41, 1902977, 2019.
41. Song, G., Kang, R., Guo, L. *et al.*, Highly flexible few-layer Ti3C2 MXene/cellulose nanofiber heat-spreader films with enhanced thermal conductivity. *New J. Chem.*, 44, 17, 7186–93, 2020.
42. Liu, Z., Wang, W., Tan, J. *et al.*, Bioinspired ultra-thin polyurethane/MXene nacrelike nanocomposite films with synergistic mechanical properties for electromagnetic interference shielding. *J. Mater. Chem. C*, 8, 21, 7170–80, 2020.
43. Liu, C., Wu, W., Shi, Y. *et al.*, Creating MXene/reduced graphene oxide hybrid towards highly fire safe thermoplastic polyurethane nanocomposites. *Compos. B. Eng.*, 203, 108486, 2020.
44. Chen, X., Zhao, Y., Li, L., Wang, Y., Wang, J., Chen, X., Zhao, Y., Li, L., Wang, Y., Wang, J., MXene/polymer nanocomposites: Preparation, properties, and applications. *Polym. Rev.*, 61, 1–36, 2020, https://doi.org/10.1080/15583724.2020.1729179.
45. Lin, B., Yuen, A.C.Y., Li, A. *et al.*, MXene/chitosan nanocoating for flexible polyurethane foam towards remarkable fire hazards reductions. *J. Hazard. Mater.*, 381, 120952, 2020.
46. Zhang, X., Wang, X., Lei, Z. *et al.*, Flexible MXene-decorated fabric with interwoven conductive networks for integrated Joule heating, electromagnetic interference shielding, and strain sensing performances. *ACS Appl. Mater. Interfaces*, 12, 12, 14459–67, 2020.
47. Ye, Q., Xiao, P., Liu, W. *et al.*, Exploring the potential of exfoliated ternary ultrathin Ti4AlN3 nanosheets for fabricating hybrid patterned polymer brushes. *RSC Adv.*, 5, 86, 70339–44, 2015.
48. Zheng, S., Zhang, C., Zhou, F., Dong, Y., Shi, X., Nicolosi, V., Wu, Z., Bao, X., Ionic liquid pre-intercalated MXene films for ionogel-based flexible micro-supercapacitors with high volumetric energy density. *J. Mater. Chem. A*, 7, 9478–9485, 2019, DOI: 10. 1039/C9TA02190F.
49. Feng, A., Hou, T., Jia, Z. *et al.*, Preparation and characterization of epoxy resin filled with Ti3C2Tx MXene nanosheets with excellent electric conductivity. *Nanomaterials*, 10, 1, 162, 2020.
50. Ekinci, K.Z., Ünügür Çelik, S., Bozkurt, A., Enhancing the anhydrous proton conductivity of sulfonated polysulfone/polyvinyl phosphonic acid composite membranes with hexagonal boron nitride. *Int. J. Polym. Mater. Polym. Biomater.*, 64, 13, 683–9, 2015.
51. Liu, Y., Zhang, J., Zhang, X. *et al.*, Ti3C2Tx filler effect on the proton conduction property of polymer electrolyte membrane. *ACS Appl. Mater. Interfaces*, 8, 31, 20352–63, 2016.

52. Zhang, J., Liu, Y., Lv, Z. *et al.*, Sulfonated Ti3C2Tx to construct proton transfer pathways in polymer electrolyte membrane for enhanced conduction. *Solid State Ionics*, 310, 100–11, 2017.
53. Kang, R., Zhang, Z., Guo, L. *et al.*, Enhanced thermal conductivity of epoxy composites filled with 2D transition metal carbides (MXenes) with ultralow loading. *Sci. Rep.*, 9, 1, 53, 2019, https://doi.org/10.1038/s41598-019-45664-4.
54. Chen, L., Shi, X., Yu, N. *et al.*, Measurement and analysis of thermal conductivity of Ti3C2Tx MXene films. *Materials*, 11, 9, 1701, 2018.
55. Ul Haq, Y., Murtaza, I., Mazhar, S. *et al.*, Investigation of improved dielectric and thermal properties of ternary nanocomposite PMMA/MXene/ZnO fabricated by *in-situ* bulk polymerization. *J. Appl. Polym. Sci.*, 137, 40, 49197, 2020.
56. Guo, L., Zhang, Z., Li, M. *et al.*, Extremely high thermal conductivity of carbon fiber/epoxy with synergistic effect of MXenes by freeze-drying. *Compos. Commun.*, 19, 134–41, 2020.
57. Ji, C., Wang, Y., Ye, Z. *et al.*, Ice-templated MXene/Ag–epoxy nanocomposites as high-performance thermal management materials. *ACS Appl. Mater. Interfaces*, 12, 21, 24298–307, 2020.
58. Kurtoglu, M., Naguib, M., Gogotsi, Y., Barsoum, M.W., First principles study of two-dimensional early transition metal carbides. *MRS Commun.*, 2, 133–137, 2012, DOI: 10.1557/mrc.2012.25.
59. Borysiuk, V.N., Mochalin, V.N., Gogotsi, Y., Bending rigidity of two-dimensional titanium carbide (MXene) nanoribbons: A molecular dynamics study. *Comput. Mater. Sci.*, 143, 418–424, 2018, DOI: 10.1016/j.commatsci.2017.11.028.
60. Zha, X., Luo, K., Li, Q. *et al.*, Role of the surface effect on the structural, electronic and mechanical properties of the carbide MXenes. *Europhys. Lett.*, 111, 2, 26007, 2015.
61. Gao, L., Li, C., Huang, W. *et al.*, MXene/polymer membranes: Synthesis, properties, and emerging applications. *Chem. Mater.*, 32, 1703–47, 2020.
62. Yu, B., Tawiah, B., Wang, L. *et al.*, Interface decoration of exfoliated MXene ultra-thin nanosheets for fire and smoke suppressions of thermoplastic polyurethane elastomer. *J. Hazard. Mater.*, 374, 110–9, 2019.
63. Hu, J., Li, S., Zhang, J. *et al.*, Mechanical properties and frictional resistance of Al composites reinforced with Ti3C2Tx MXene. *Chin. Chem. Lett.*, 31, 4, 996–9, 2020.
64. Yi, Z., Yang, J., Liu, X. *et al.*, Enhanced mechanical properties of poly(lactic acid) composites with ultrathin nanosheets of MXene modified by stearic acid. *J. Appl. Polym. Sci.*, 137, 17, 48621, 2019.
65. Pan, Y., Fu, L., Zhou, Q. *et al.*, Flammability, thermal stability and mechanical properties of polyvinyl alcohol nanocomposites reinforced with delaminated Ti3C2Tx (MXene). *Polym. Compos.*, 41, 1, 210–8, 2019.
66. Sheng, X., Zhao, Y., Zhang, L., Lu, X., Properties of two-dimensional Ti3C2 MXene/thermoplastic polyurethane nanocomposites with effective reinforcement via melt blending. *Compos. Sci. Technol.*, 181, 107710, 2019, DOI: 10.1016/j.compscitech.2019. 107710.
67. Sobolciak, P., Ali, A., Hassan, M.K., Helal, M.I., Tanvir, A., Popelka, A., Al-Maadeed, M.A., Krupa, I., Mahmoud, K.A., 2D Ti3C2Tx (MXene)-reinforced polyvinyl alcohol (PVA) nanofibers with enhanced mechanical and electrical properties. *PLoS One*, 12, e0183705, 2017, DOI: 10.1371/journal.pone.0183705.
68. Mondal, S., Rana, U., Malik, S., Reduced graphene oxide/Fe3O4/polyaniline nanostructures as electrode materials for an all-solid-state hybrid supercapacitor. *J. Phys. Chem. C*, 121, 7573–7583, 2017, doi: 10.1021/acs.jpcc.6b10978.
69. Chen, W., Rakhi, R.B., Alshareef, H.N., Facile synthesis of polyaniline nanotubes using reactive oxide templates for high energy density pseudocapacitors. *J. Mater. Chem. A*, 1, 3315, 2013, doi: 10.1039/c3ta00499f.

70. Shao, Y., El-Kady, M.F., Sun, J., Li, Y., Zhang, Q., Zhu, M., Wang, H., Dunn, B., Kaner, R.B., Design and mechanisms of asymmetric supercapacitors. *Chem. Rev.*, 118, 9233–9280, 2018, doi: 10.1021/acs.chemrev.8b00252.
71. Zhang, F., Liu, T., Li, M., Yu, M., Luo, Y., Tong, Y., Li, Y., Multiscale pore network boosts capacitance of carbon electrodes for ultrafast charging. *Nano Lett.*, 17, 3097–3104, 2017, doi: 10.1021/acs.nanolett.7b00533.
72. Miao, Y.E., Fan, W., Chen, D., Liu, T.X., High-performance supercapacitors based on hollow polyaniline nanofibers by electrospinning. *ACS Appl. Mater. Interfaces*, 5, 4423–4428, 2013, doi: 10.1021/Am4008352.
73. Wang, H., Lin, J., Xiang, Z., Advanced materials and devices polyaniline (PANi) based electrode materials for energy storage and conversion. *J. Sci. Adv. Mater. Devices*, 1, 225–255, 2106, doi: 10.1016/j.jsamd.2016.08.001.
74. Eftekhari, A., Li, L., Yang, Y., Polyaniline supercapacitors. *J. Power Sources*, 347, 86–107, 2017, doi: 10.1016/j.jpowsour.2017.02.054.
75. Shi, F., Li, L., Wang, X.L., Gu, C.D., Tu, J.P., Metal oxide/hydroxide-based materials for supercapacitors. *RSC Adv.*, 4, 79, 41910–41921, 2014, DOI: 10.1039/C4RA06136E.
76. Miao, F., Shao, C., Li, X., Wang, K., Liu, Y., Flexible solid-state supercapacitors based on freestanding nitrogen-doped porous carbon nanofibers derived from electrospun polyacrylonitrile@polyaniline nanofibers. *J. Mater. Chem. A*, 4, 4180–4187, 2016, doi: 10.1039/c6ta00015k.
77. He, Y., Chen, W., Gao, C., Zhou, J., Li, X., Xie, E., An overview of carbon materials for flexible electrochemical capacitors. *Nanoscale*, 5, 8799–8820, 2013.
78. Wang, Y., Song, Y., Xia, Y., Characterization and applications chemical functional materials. *Chem. Soc. Rev.*, 45, 5925–5950, 2016, doi: 10.1039/C5CS00580A.
79. Zhang, K., Zhang, L.L., Zhao, X.S., Wu, J., Graphene/polyaniline nanofiber composites as supercapacitor electrodes. *Chem. Mater.*, 22, 1392–1401, 2010, doi: 10.1021/cm902876u.
80. Feng, Y., Deng, Q., Peng, C. *et al.*, An ultrahigh discharged energy density achieved in an inhomogeneous PVDF dielectric composite filled with 2D MXene nanosheets via interface engineering. *J. Mater. Chem. C*, 6, 48, 13283–92, 2018.
81. Boota, M., Anasori, B., Voigt, C., Zhao, M.-Q., Barsoum, M.W., Gogotsi, Y., Pseudocapacitive electrodes produced by oxidant-free polymerization of pyrrole between the layers of 2D titanium carbide (MXene). *Adv. Mater.*, 28, 1517–1522, 2016, DOI: 10. 1002/adma.201504705.
82. Ling, Z., Ren, C.E., Zhao, M.-Q., Yang, J., Giammarco, J.M., Qiu, J., Barsoum, M.W., Gogotsi, Y., Flexible and conductive MXene films and nanocomposites with high capacitance. *Proc. Natl. Acad. Sci. U.S.A.*, 111, 16676–16681, 2014, DOI: 10.1073/pnas.1414215111.
83. Chenglong, L., Shuang, W., Yinghe, C., Xiadong, W., Zhipeng, Y., Dan, L., Yuw, C., Zhe, W., Sandwich-like MXene/α-Fe_2O_3–C–MoS_2-PEDOT:PSS/MXene film electrodes with ultrahigh area capacitance for flexible supercapacitors. *ACS Appl. Mater. Interfaces*, 14, 7, 9172–9182, 2022.
84. De, S., Maity, C.K., Saho, S.O., Nayak, G.C., Polyindole booster for Ti3C2TxMXene based symmetric and asymmetric supercapacitor devices. *ACS Appl. Energy Mater.*, 4, 3712–3723, 2021, https://doi.org/10.1021/acsaem.1c00142.
85. Li, K., Wang, X., Li, S., Urbankowski, P., Li, J., Xu, Y., Gogotsi, Y., An ultrafast conducting polymer@ MXene positive electrode with high volumetric capacitance for advanced asymmetric supercapacitors. *Small*, 16, 4, 1906851, 2020.
86. Chen, Z., Wang, Y., Han, J., Wang, T., Leng, Y., Wang, Y., Li, T., Han, Y., Preparation of polyaniline onto dl -tartaric acid assembled MXene surface as an electrode material for supercapacitors. *ACS Appl. Energy Mater.*, 3, 9326–9336, 2020, https://doi.org/10.1021/acsaem.0c01662.
87. Fu, J., Yun, J.M., Wu, S., Li, L., Yu, L., Kim, K., Architecturally robust graphene-encapsulated MXene Ti2CTx @polyaniline composite for high-performance pouch-type asymmetric

supercapacitor. *Appl. Mater. Interfaces*, 10, 40, 34212–34221, 2018, https://doi.org/10.1021/acsami.8b10195.

88. Boota, M., Pasini, M., Galeotti, F., Porzio, W., Zhao, M.-Q., Halim, J., Gogotsi, Y., Interaction of polar and nonpolar polyfluorenes with layers of two-dimensional titanium carbide (MXene): Intercalation and pseudocapacitance. *Chem. Mater.*, 29, 2731–2738, 2017, DOI: 10.1021/acs.chemmater.6b03933.
89. Ling, Z., Ren, C.E., Zhao, M.-Q., Yang, J., Giammarco, J.M., Qiu, J., Barsoum, M.W., Gogotsi, Y., Flexible and conductive MXene films and nanocomposites with high capacitance. *Proc. Natl. Acad. Sci. U.S.A.*, 111, 16676–16681, 2014, DOI: 10.1073/pnas.1414215111.
90. Qi, Y., Sarang, K., Lutkenhaus, J., Oh, E.S., Electronically conductive MXene clay-polymer composite binders for electrochemical double-layer capacitor electrodes. *J. Power Sources*, 506, 230138, 2021, https://doi.org/10.1016/j.jpowsour.2021.230138.
91. Wu, W., Wei, D., Zhu, J., Niu, D., Wang, F., Wang, L., Yang, L., Yang, P., Wang, C., Enhanced electrochemical performances of organ-like Ti 3 C 2 MXenes/polypyrrole composites as supercapacitors electrode materials. *Ceram. Int.*, 45, 7328–7337, 2019, https://doi.org/10.1016/j.ceramint.2019.01.016.
92. Wu, W., Niu, D., Zhu, J., Gao, Y., Wei, D., Liu, X., Wang, F., Wang, L., Yang, L., Organ-like Ti3C2 MXenes/polyaniline composites by chemical grafting as high-performance supercapacitors. *J. Electroanal. Chem.*, 847, 113203, 2019, https://doi.org/10.1016/j.jelechem.2019.113203.
93. Wu, W., Wang, C., Zhao, C., Wei, D., Zhu, J., Xu, Y., Facile strategy of hollow polyaniline nanotubes supported on Ti3C2-MXene nanosheets for High-performance symmetric supercapacitors. *J. Colloid Interface Sci.*, 580, 601–613, 2020, https://doi.org/10.1016/j.jcis.2020.07.052.
94. Li, Y., Kamdem, P., Jin, X.J., Hierarchical architecture of MXene/PANI hybrid electrode for advanced asymmetric supercapacitors. *J. Alloys Compd.*, 850, 156608, 2021, https://doi.org/10.1016/j.jallcom.2020.156608.
95. Wei, D., Wu, W., Zhu, J., Wang, C., Zhao, C., Wang, L., A facile strategy of polypyrrole nanospheres grown on Ti3C2-MXene nanosheets as advanced supercapacitor electrodes. *J. Electroanal. Chem.*, 877, 114538, 2020, https://doi.org/10.1016/j.jelechem.2020.114538.
96. Zhang, C., Xu, S., Cai, D., Cao, J., Wang, L., Han, W., Planar supercapacitor with high areal capacitance based on Ti3C2/polypyrrole composite film. Elsevier Ltd, *Electrochimica Acta,* 330, 135277, 2020, https://doi.org/10.1016/j.electacta.2019.135277.
97. Wu, X., Huang, B., Lv, R., Wang, Q., Wang, Y., Highly flexible and low capacitance loss supercapacitor electrode based on hybridizing decentralized conjugated polymer chains with MXene. *Chem. Eng. J.*, 378, 122246, 2019, https://doi.org/10.1016/j.cej.2019.122246.
98. Qin, L., Tao, Q., Liu, X., Fahlman, M., Halim, J., Persson, P.O.A., Rosen, J., Zhang, F., Polymer-MXene composite films formed by MXene-facilitated electrochemical polymerization for flexible solid-state microsupercapacitors. *Nano Energy*, 60, 734–742, 2019, https://doi.org/10.1016/j.nanoen.2019.04.002.
99. Luo, W., Wei, Y., Zhuang, Z., Lin, Z., Li, X., Hou, C., Li, T., Ma, Y., Fabrication of Ti3C2Tx MXene/polyaniline composite films with adjustable thickness for high-performance flexible all-solid-state symmetric supercapacitors. *Electrochim. Acta*, 406, 139871, 2022, https://doi.org/10.1016/j.electacta.2022.139871.
100. Cao, S., Zhao, T., Li, Y., Yang, L., Ahmad, A., Jiang, T., Shu, Y., Jing, Z., Luo, H., Lu, X., Zhang, H., Fabrication of PANI@Ti3C2Tx/PVA hydrogel composite as flexible supercapacitor electrode with good electrochemical performance. *Ceram. Int.*, 48, 15721–15728, 2022, https://doi.org/10.1016/j.ceramint.2022.02.108.
101. Wu, X., Hu, W., Qiu, J., Geng, B., Du, M., Zheng, Q., Solvent-assisted self-assembly to fabricate a ternary flexible free-standing polyaniline@MXene-CNTs electrode for high-

performance supercapacitors. *J. Alloys Compd.*, 921, 166062, 2022, https://doi.org/10.1016/j.jallcom.2022.166062.

102. Tong, L., Jiang, C., Cai, K., Wei, P., High-performance and freestanding PPy/Ti3C2Tx composite film for flexible all-solid-state supercapacitors. *J. Power Sources, 465*, 228267, 2020, doi: 10.1016/j.jpowsour.2020.
103. Li, K., Wang, X., Li, S., Urbankowski, P., Li, J., Xu, Y., Gogotsi, Y., An ultrafast conducting polymer@MXene positive electrode with high volumetric capacitance for advanced asymmetric supercapacitors. *Small*, 16, 4, 1906851, 2019.
104. Li, K., Wang, X., Wang, X., Liang, M., Nicolosi, V., Xu, Y., Gogotsi, Y., All-pseudocapacitive asymmetric MXene-carbon-conducting polymer supercapacitors. *Nano Energy*, 75, 104971, 2020.
105. Boota, M., Becuwe, M., Gogotsi, Y., Phenothiazine—MXene aqueous asymmetric pseudocapacitors. *ACS Appl. Energy Mater.*, 3, 4, 3144–3149, 2020.

11

MXene-Based Polymer Composites for Hazardous Gas and Volatile Organic Compound Detection

Sachin Karki[1,2], Rajashree Bhuyan[3], Sachin R. Geed[1,2*] and Pravin G. Ingole[1,2†]

[1]*Chemical Engineering Group, Engineering Sciences and Technology Division, CSIR-North East Institute of Science and Technology, Jorhat, India*
[2]*Academy of Scientific and Innovative Research (AcSIR), Ghaziabad, Uttar Pradesh, India*
[3]*Kalinga Institute of Industrial Technology, Patia, Bhubaneswar, Odisha, India*

Abstract
Nowadays new class of two-dimensional materials called MXenes has fascinated enormous attention in a range of applications because of the availability of numerous active sites, metallic conductivity, tunable surface chemistry, and outstanding stability. These characteristics, as mentioned above make MXene as best advanced gas and volatile organic compound (VOC) detection material. Due to the presence of advanced properties like convenient solution processibility, high flexibility, and easy functionalization process, MXenes have opened the door to making their composite with other materials for detection applications. In this chapter, we have focused on the recent research and advancement of MXene-based polymer composites for the detection of hazardous gases, VOCs, and water remediation. Also, in this chapter, we have covered the synthesis of MXenes, the properties of Mxene and finally, their application for detection gases, heavy metals, and biomolecules.

***Keywords*:** MXenes, MXene-based polymer composites, hazardous gases, electrocatalysis

11.1 Introduction

Nowadays, the emission of gaseous pollutants causes serious environmental issues. Auspiciously, public awareness increases due to the impact of the emission of gas on the environment and aquatic and terrestrial animal health [1]. Nitrogen-based gases, NO_2 released from automobiles and fossil fuels cause acid rain which affects human health as well as the environment [2–4]. NH_3 is used in many industries such as fertilizer, explosives, and petroleum industries. NH_3 damage the human respiratory system, eyes, and skin. H_2S gas is highly toxic and flammable in nature and causes serious health issues. After inhalation of concentrations >500 mg/L cause, people lose their lives within a shorter period of exposure. Another class of hazardous gaseous pollutants is Volatile Organic Compounds (VOC).

**Corresponding author*: sachingeed23@gmail.com
†*Corresponding author*: ingolepravin@gmail.com

Primarily VOC pollutants are released from the emission of petrochemical plants, refineries, metal industries, transportation, and other industries. To mitigate such issues gaseous detection/sensing technology has caught attention in many fields like early diagnosis, safety at work, and pollution control [2, 3].

Sensors are an important component to collect huge information about the surrounding environment. For instance, VOCs are significant airborne contaminants with detrimental impacts on human health. Therefore, it is crucial to create a cordless working gas sensor system that can detect volatile organic chemicals at ambient temperature in a selective, sensitive, quick, and reversible manner. Indoor Prolonged exposure to VOCs can irritate eyes respiratory problems and also cause cancer etc. [4, 5]. Hexanal aldehyde is an indoor odorant of wood-based products, is frequently found in indoor spaces as air contaminants and can be harmful to people's health. Longed exposure to hexanal gas damages the eyes and respiratory system [4, 5]. Nevertheless, hexanal gas concentration levels are low in biological samples with a picomolar to the micromolar range and it is difficult to detect. With time gas sensors based on In_2O_3, SnO_2, ZnO, and other metal oxides have been investigated for the detection of toxic gases [6, 7]. Nevertheless, most of the sensors are operated at a higher temperature and have high-power consumption, which restricts their alteration for minimal power utilizing devices like smart wearable and health monitors for the detection of hazardous gaseous [8–11].

Recently, various materials have been developed for the detection of gases such as metal oxides [12–16], metal-organic frameworks (MOF) [17, 18], carbon-based materials [19–21], conductive polymers [22–24], and rare-earth oxides [25–27], etc. Both n-type (TiO_2 and SnO_2) and p-type (CuO and ZrO_2) metal oxides exist. Due to its sensitivity, compactness, and inexpensive price, the majority of studies employed metal oxide for gas detection [28]. Metal oxide sensors caused hazards at a high temperature hence increasing the cost of safe detection of gases. Graphene as well as carbon nanotubes, which are based on carbon, seemed to be utilized at room temperature for gas sensing owing to their conductance, high surface area, mechanical properties, and special structure [25]. Reduced graphene oxide (rGO) film prepared through chemical synthesis and modification was done by high edge functionalization. This reduced graphene oxide gas sensor responded to CO_2 quite well [29]. Polyaniline, polypyrrole, and polythiophene are conductive polymers with high sensing performance due to their low working temperature [22, 23]. Producing polyaniline is simple, and it provides maximum stability and versatile conductance. Therefore, polyaniline is widely used as a conductive polymer for detection [30]. Polypyrrole is generally used as a gas detector because of its controllable conductivity and low cost of manufacture [24]. MOF are the porous structure of metal centers and organic connectors. MOFs are gaining attention due to their high porosity, structural diversity, ultra-high surface area, and good adsorption affinity [31]. Zeolite imidazole ester-based metal frameworks have been used for gas sensitivity [32]. Out of these available detectors, two-dimensional nanomaterials-based gas sensors are well-known due to their high surface area and adsorption sites.

In the last decade' with numerous advancements in two-dimensional material synthesis processes, composite materials were made ahead of graphene [33, 34]. These include the sizable family of two-dimensional transition metals termed as MXenes, which are nitrides and carbides [35–41]. The MXenes-based materials are synthesized by etching of A layers (A element from group IIIA or IVA) from the MAX phases (layered hexagonal structures) of ternary carbides or nitrides with chemical formula $M_{n+1}AX_n$, where M is the early transition

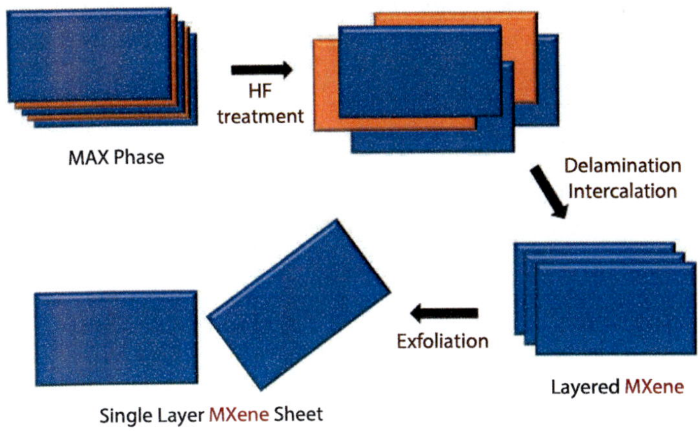

Figure 11.1 Schematic method of preparing single layer MXene sheet.

metal as V, Ti, Sc, Cr and the X is carbon or nitrogen where n = 1,2,3 [42–44]. In $M_{n+1}AX_n$ materials, Because M-X bonds are stronger compared to M-A bonds, A layers could be chemically etched without damaging the M-X bonds, leading to $M_{n+1}X_n$ layers which are poorly linked and easily detached as shown in Figure 11.1. The formed materials after stretching of A layers from the source MAX are called MXenes and they share graphene's characteristics. Figure 11.1 shows a general representation of preparing MXene sheets.

Firstly, MXene is synthesized by Naguib *et al.*, [44] in 2011; it is the 2D multilayer nanostructure of $Ti_3C_2T_x$ material from Ti_3AlC_2 and TiC powder. Currently, more than 200 MAX phases based MXene materials have been synthesized, for example, Mo_2C, Nb_2C, Ta_4C_3, TiCN, and V_2C [44, 45]. Due to improved properties such as large specific surface area, rich organic bonds, surface functional groups, large porosity, high hydrophilicity and conductivity of MXene materials. These materials are used as electrode material, sensor, energy storage, supercapacitors, electromagnetic shielding material, battery materials, and nanofluids [46–49].

Though, limited studies are reported on gas sensing applications. Therefore, an energy-efficient sensor is considered necessary, especially in the application domains of indoor monitoring systems and real-time health monitoring from exhaled breath, to selectively detect low levels of hexanal at ambient temperature. Few articles are focused on gas sensors made from MXene-based materials [50]. This chapter is focused on synthesis methods of MXene, the properties of MXene and the application of MXene in gas sensors for hazardous gases and VOCs.

11.2 Synthesis of MXenes and MXene–Polymer Composites

11.2.1 Synthesis of MXenes

Chemically specific etching of particular layers of atoms from stacked nitride/carbonitride/carbide precursors has yielded over twenty distinct MXenes. Acidic medium consisting of fluoride ions (NH_4HF_2 or HF, a combination of HCl and LiF), as well as salts consisting of fluorine ions (NaF/NH_3F/LiF/KF), are the two principal types of etchants [35]. MXenes are first isolated from MAX phases ($M_{n+1}AX_n$ layered, hexagonal carbides and nitrides) by immersing them in particular acids & breaking the M–A linkages. A particular corrosive period and complete mixing are required for this procedure [51]. By soaking Ti_3AlC_2 particles in 50% concentrated hydrofluoric acid (HF) at room temperature (RT) for 2 hours, Naguib *et al.* created the very first MXene, Ti_3C_2Tx. In the case of Ti_3AlC_2, the primary peak in the X-ray diffraction XRD pattern changes from roughly 40 °C to about 10 °C following HF treatment, showing that Ti_3AlC_2 is substituted with TiC_2T_x [44]. A schematic process of preparation by exfoliation of Ti_3AlC_2 used by Naguib *et al.* is represented in Figure 11.2. In time HF etching is used for other MAX phases, generating plenty of novel MXenes such as $Cr_2TiC_2T_x$, $Mo_2Ti_2C_3T_x$, $V_4C_3T_x$, $Ta_4C_3T_x$, $Nb_4C_3T_x$, $(V_{0.5}Cr_{0.5})_3C_2T_x$, Ti_3CNT_x, $Ti_3C_2T_x$, $W_{4/3}CT_x$, $Nb_{4/3}CT_x$, $Mo_{4/3}CT_x$, $(Ti_{0.5}Nb_{0.5})_2CT_x$, Ti_2NT_x, Nb_2CT_x, $Mo_2TiC_2T_x$, V_2CT_x and Ti_2CT_x [44, 52]. Because HF is dangerous for humans as well as for the environment, it is essential to discover alternative etchants that are safe. To make $Ti_3C_2T_x$, Ghidiu *et al.* utilized a safe and secure combination of lithium fluoride (LiF) & hydrochloric acid (HCl) wherein Ti_3AlC_2 particles were immersed for 24 h at 30 °C. The appearance of $Ti_3C_2T_x$ under the Transmission Electron Microscope reveals some defects in the manufactured material [53]. It is assumed that the LiF & HCl mixed solvent has a similar etching process for the MAX phases with the HF medium as it contains ions of fluoride & hydrogen. MXenes were also generated by combining different kinds of fluorine ion-consisting salts (CaF_2, FeF_3, CsF, KF, NaF) [50, 52]. Several MXenes are synthesized utilizing a combination of salts & acids having fluorine ions, including $(Nb_{0.8}Ti_{0.2})_4C_3T_x$, Ti_2CT_x, $(Nb_{0.8}Zr_{0.2})_4C_3T_x$, Mo_2CT_x, $Cr_2TiC_2T_x$, V_2CT_x, Ti_3CNT_x, $W_{4/3}CT_x$ and $Ti_3C_2T_x$. This combination of solutions does, nevertheless, emit some hazardous HF fumes during the etching procedure [54].

It's worth noting that the etching procedure needs an absolute transformation of the MAX phase to MXene, which necessitates proper control of the etching requirements. The etching parameters typically need to be altered when the M atom and the n are present in the chemical formula $M_{n+1}AX_n$ [55]. The energy M–A bond rises as the atomic number of M grows, hence the concentration of fluoride ion in the etchant fluid and the etching duration must also be raised. A longer etching period and more fluoride ions must be used when the n value is higher [42, 48]. A greater temperature is also required when the etchant turns

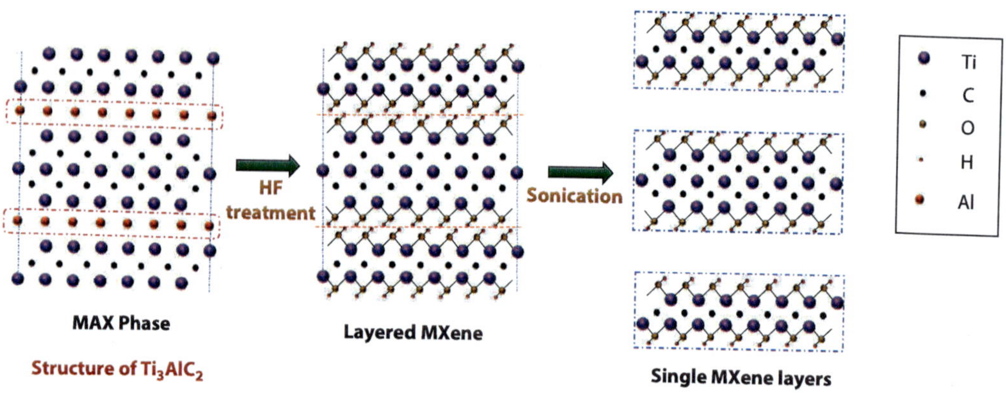

Figure 11.2 The process of schematic exfoliation process of Ti_3AlC_2.

less acidic. As a result, acid medium, as well as salts consisting of fluorine ions can be used as etchants for the production of MXenes, however, later requires a significantly increased etching temperature. Urbankowski *et al.* also succeeded in creating the first MXene based on nitride (Ti_4N_3Tx) by heating a mixture of melted fluoride salts and Ti_4AlN_3 particles in an argon environment at a temperature of 550 °C. The creation of various 2D transition metal nitrides can be modeled after the production of $Ti_4N_3T_x$ [56]. Ghasali *et al.,* [48] prepared V_2C MXene from the V_2AlC MAX phase. For the production of the MAX phase, they heated the V_2O_5-Al-C. Figure 11.3 represents the method of their preparation of V_2C. Figure 11.4 represents the FE-SEM images of the prepared V_2C MXene while treated with HF and after finer treatment.

MXenes that are produced are generally of multiple layers, requiring additional treatment to achieve a single layer MXene. At present Intercalation delamination & manual delamination are the two ways for delaminating multilayered MXenes [35, 47]. MXenes with multiple layers have high interconnections, so the production of single-layer MXenes is quite poor when using the first approach. Few reports have succeeded in delaminating $Ti_3C_2T_x$ having multiple layers using the mechanical layering method till now [35]. As MXenes having multiple layers can tolerate diverse ions and molecules within their layers, the majority of monolayer MXene particles are made by intercalation. Intercalated MXenes can be separated into different layers by sonication in air-treated water because intercalation widens the interfacial gap of MXenes with numerous layers [57].

Ionic and organic materials are the two types of intercalants. Organic materials are commonly used as intercalants for the delamination of HF-synthesized MXenes [58]. Polar organic molecules like dimethyl sulfoxide (DMSO), urea, tetrabutylammonium hydroxide (TBAOH), isopropylamine and hydrazine are examples of organic compounds. TBAOH is being used to effectively delaminate V_2CT_x, Ti_3CNT_x, $Ti_4N_3T_x$, and Mo_2CT_x, while isopropylamine and DMSO are being used to effectively delaminate Nb_2CT_x and $Ti_3C_2T_x$,

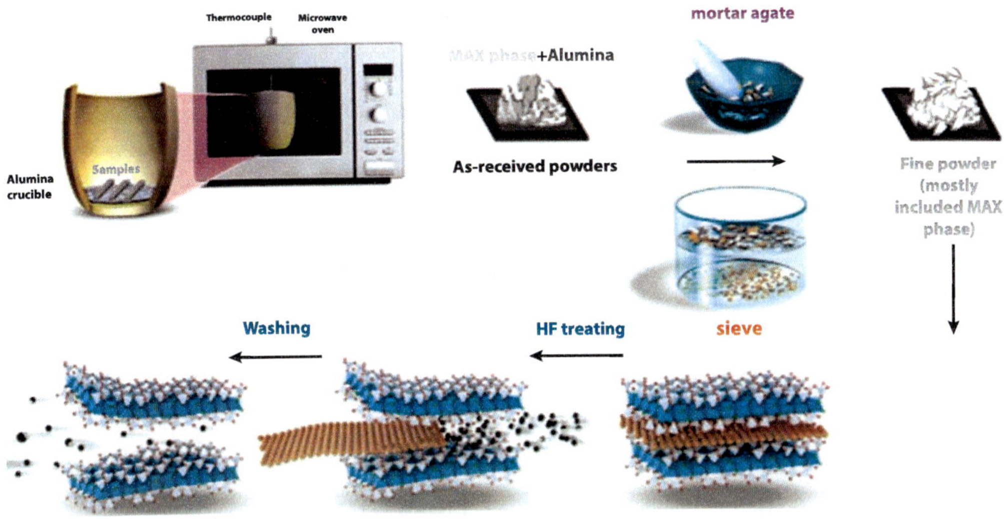

Figure 11.3 Method of preparing V_2C MXene from V_2AlC MAX phase which was obtained by microwave heating of V_2O_5-Al-C. The figure represents the schematic way of preparation along with a graphical diagram. The figure is taken from ref. [48], with copyright © 2020, Elsevier B.V.

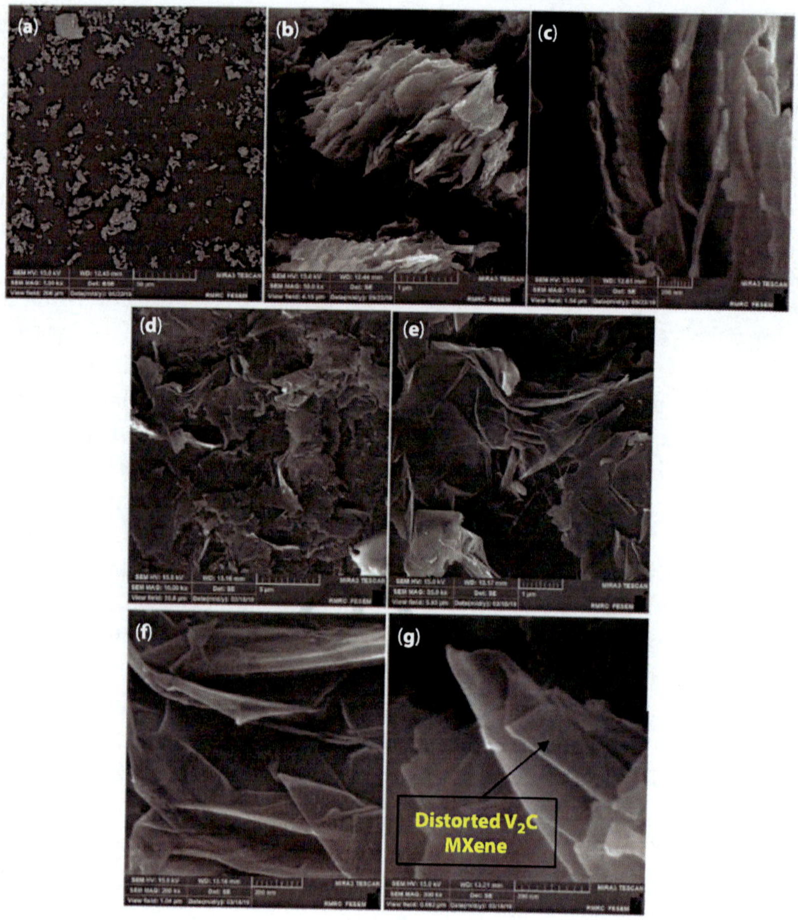

Figure 11.4 FE-SEM images of the MXene V2C obtained by microwave heating. (a, b, c) represent the SEM images of the MXene after HF treatment; (d, e, f, g) are the FE-SEM images of the MXene after the final treatment. The figure is taken from ref. [48], with copyright © 2020, Elsevier B.V.

accordingly [50, 52, 59]. NH_4^+/Li^+ ions can enter between the MXenes sheets in the case of MXenes having multiple layers generated with the $NH_4F/LiF + HCl$ etchant. As a result, these MXenes possess lengthier lattice parameters compared to MXenes generated by HF, and the single-layer MXenes may be created straight via sonication without further intercalation. Furthermore, as opposed to prior approaches, this process produces single-layer MXenes with lesser flaws [52, 59].

11.2.2 Synthesis of MXene-Based Composites

Manufacturing composites has become a popular approach for developing robust and adaptable substances recently [47]. MXenes are regarded as ideal prospects for the production of multipurpose composites owing to their two-dimensional geometry, stacked architectures, as well as remarkable elasticity, which has sparked a rush in research for composites based

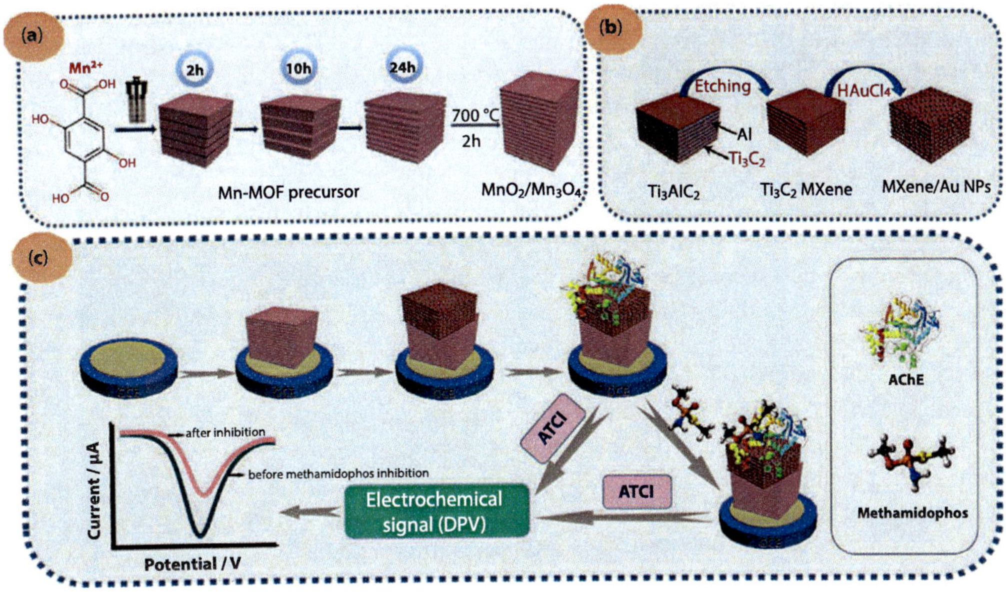

Figure 11.5 (a) represents the formation of the MnO_2/Mn_3O_4 composite. (b) synthesizing MXene/Au-nanoparticle, a composite MXene. (c) fabrication of AChE-Chit/MXene/Au NPs/MnO_2/Mn_3O_4/GCE biosensor. The figure is taken with permission from reference [60], with copyright © 2019, Elsevier B.V.

on MXene. MXenes are being coupled with an array of substances, notably carbon nanotubes, polymers as well as metal oxides to create a number of unique composites [15, 41, 52]. In a work by Song *et al.*, gold nanoparticles incorporated MXenes were prepared and utilized for pesticide detection [60].

They first synthesized the MOF-derived MnO_2/Mn_3O_4 composites by hydrothermal reaction of $MnCl_2{\cdot}4H_2O$ and 2,5-dihydroxyterephthalic acid in the autoclave. After that, they prepared the MXene composite by simply mixing the ultrasonicated aqueous Ti_3C_2 MXene with $HAuCl_4$ with mild shaking. Then a glass carbon electrode (GCE) was taken and MOF-derived MnO_2/Mn_3O_4 dropped on the electrode surface. On the above of it, MXene/Au composite was poured on it and finally a mixture of AChE and Chit was spread which leads to the formation of the biosensor for sensing methamidophos assay. Figure 11.5 displays the schematic representation of preparing the biosensor [60].

11.2.3 MXene–Polymer Composites

MXenes possessing superior hydrophilic surfaces, mechanical properties & metallic conductivity could enhance the thermal and mechanical properties of polymers when building composites with them [56, 59]. Monolayer MXenes possess better polymer suitability & available hydrophilic surfaces than MXenes having multiple layers. As a result, before getting combined with polymers, MXenes are frequently delaminated. By combining the colloidal solution containing $Ti_3C_2T_x$ films and a PVA aqueous solution, Zhang *et al.* [61] proposed the production of monolayered $Ti_3C_2T_x$-polyvinyl alcohol (PVA) composite. PVA is dispersed amongst $Ti_3C_2T_x$ films in the $Ti_3C_2T_x$-PVA composite, which is a thick layer since extremely hydrophilic PVA could establish powerful hydrogen linkages with $Ti_3C_2T_x$, and has great thermal stability. $Ti_3C_2T_x$-PVA possesses superior elasticity, compressive and tensile strengths in comparison to separated PVA & $Ti_3C_2T_x$. $Ti_3C_2T_x$-PVA composite's tensile strength with $Ti_3C_2T_x$ 40 wt.% approaches 91±10 MPa, that's about 4 times greater compared to the $Ti_3C_2T_x$ film. Furthermore, good conductivity is provided to this composite by $Ti_3C_2T_x$. Zhang *et al.* [62] in their other work prepared MXene/PEDOT:PSS hybrid. A right amount of dispersed delaminated MXene ($Ti_3C_2T_x$) and poly(3,4-ethylenedioxythiophene): polystyrene sulfonate (PEDOT:PSS) mixed with each other to prepare spinning dopes. The spinning dope is then injected into the H_2SO_4 bath where the dope coagulates and forms fibers. Then the fibers were rinsed in the water/ethanol bath and dried on a spool.

In a work of synthesizing the stretchable wearable conductometric VOC sensors Tang *et al.*, synthesized MXene/polymer composite [63]. They have prepared micro structured core-sheath fibers of $Ti_3C_3T_x$/polyurethane (PU) composite of three types, which they have named Flat MXene sheath, cracked MXene sheath and wrinkled MXene sheath as seen in Figure 11.6(a). They first cut the PU fiber 10 cm long and then sprayed 10 ml MXene solution under 40 psi air pressure for 45 minutes, resulting in the flat MXene sheath fibers. For making the cracked-MXene sheath fibers the PU fiber is first stretched 5–10% and then sprayed with the MXene solution for developing microcracks. For the wrinkled-MXene sheath, 50% stretching of PU fiber was done first. The prepared composite nanofibers are flexible and stretchable as shown in Figure 11.6(b).

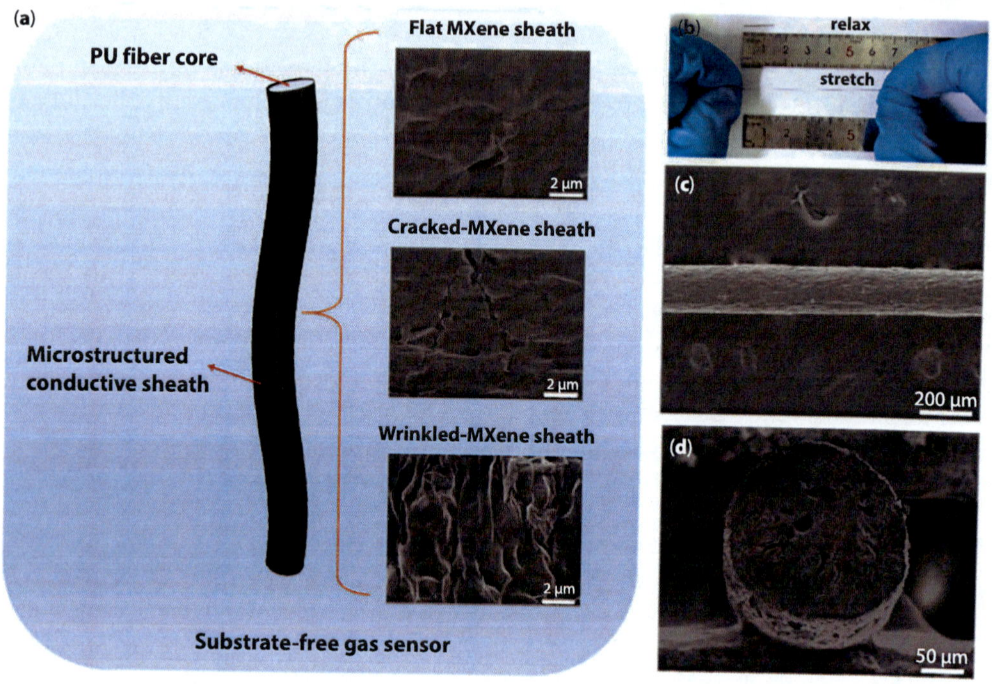

Figure 11.6 (a) representation of the micro structured core-sheath fibers with the FE-SEM images of the three types of MXene-sheaths. (b) image representing the relaxed and stretched state of the prepared MXene/ polyurethane composite. (c, d) outer shell and cross-section FE-SEM images of the prepared fiber, respectively. The figure is taken with permission from reference [63], with copyright © 2021, Elsevier B.V.

11.3 Properties of MXenes and MXene–Polymer Composites

11.3.1 Mechanical Properties

Surface termination strongly influences the mechanical properties of MXenes. While MXenes terminated by the O group have been predicted to possess extremely high stiffness, those OH and F terminated groups exhibit less elastic stiffness. This may be because the lattice parameters in MXenes vary with different terminations [64]. In the case of MXenes with O-terminations, the lattice parameters are smaller than those with F or OH-terminations. MXenes that have been surface-functionalized exhibit greater flexibility than bare MXenes. The functionalization of Ti_2C, for example, reduced Young's modulus, yet functionalized Ti_2C was more able to withstand strain than bare Ti_2C and even graphene [65]. Ti_2C fractures are increased in a critical strain where the groups terminated at the surface act as buffers, slowing the disintegration of Ti layers under tensile deformation. The quantity of layers of atom in MXenes, based on "n" present in their system of chemistry $M_{n+1}X_n$, affects MXene's physical characteristics. In the case of the functionalized MXene, when n decreases, the $M_{n+1}X_nT_x$'s strength and hardness progressively increase. A composite of MXenes with polymers or carbon nanotubes enhanced the mechanical properties of these materials. MXenes can be improved to varying extents in terms of elasticity, hardness, tension and compression force by adding various kinds of polymers [65, 66]. Polyvinyl alcohol compounds have outstanding elasticity as well as large tension and compression force, tensile force is about four times higher in $Ti_3C_2T_x$–PVA composites than in isolated $Ti_3C_2T_x$. Furthermore, high hardness and yield strength have been reported for $Ti_3C_2T_x$–PAM80 and $Ti_3C_2T_x$–UHMWPE81 [66].

11.3.2 Electronic Properties

Because of the enormous variety in composition, different surface functionalization options, and controllable thickness flexibility, MXenes cover a wide range of electrical characteristics, from metallicity and semiconductivity to topological insulative [67]. The estimated work functions (WFs) of metallic MXenes can be any value between 1.8 and 8 eV. For a particular MXene, unlike its revealed surface, the OH (O) design consistently lowers (raises) it's WF, while the F design shows any pattern depending on the particular substance [67]. It is important to note that all the MXenes with OH-terminations exhibited extremely low WFs (2.8 eV), almost as low as Sc, and hence considered the least value of all elemental metals [67]. Certain MXenes with O-terminations, on the contrary, have WFs that are greater compared to those of Pt, the elemental metal with the highest WF. As with OH- and O-terminated compounds, F-terminated MXenes always have a WF between the two [68]. The change in WF following surface functionalization occurs due to functionalization changing the surface dipole moment. Every time the termination of OH (O) produces a minus (plus) surface dipole moment, lowering (increasing) WF, but the F termination can produce a positive or negative surface dipole moment. However, most ongoing studies performed on the combination of O, OH and F groups on the MXene's surface bring back the WFs of MXenes to a value in the intermediate range [67]. Therefore, controlling the kinds of functional groupings on the surface is of great importance to achieving an extraordinarily low or high activity of MXenes with OH-terminations [67, 68].

A high/low WF metallic MXene shows great promise when Schottky-barrier (SB)-free metal gets in touch with two-dimensional semiconductors. MoS_2 has gotten so much attention out of all the existing 2D semiconductors, and it has a lot of potential for replacing silicon in upcoming nanoelectronics. As the MoS_2's ionization energy is high (~6.0 eV), making an SB-free p-type connection to MoS_2 remains a significant difficulty [67, 69]. Analytically when MoS_2 is in touch with metals, it frequently exhibits n-type activity. Interestingly, among the MXene family V_2NO_2, Cr_2NO_2, $V_4C_3O_2$, Mo_2CO_2, Cr_2CO_2, and V_2CO_2 were found MoS_2 with vanishing p-type Schottky hindrances at the interfaces of MoS_2–MXene, prompting exceptionally proficient opening infusions into MoS_2. This will work with the acknowledgement of elite execution MoS_2 field-impact semiconductors with the two polarities [67, 68]. Furthermore, many MXenes have a high electronic conductivity due to their metallicity, rendering them potential in a wide spectrum of uses, including as electrode substances in supercapacitors and batteries. MXenes's conductivity goes from less than 1 to 100 S/cm which relies unequivocally upon its delamination and synthesis techniques. For example, the HF-etched $Ti_3C_2T_x$ has a lower conductivity of 2 S/cm, whereas LiF + HCl-etched $Ti_3C_2T_x$ have a conductivity of 1500 S/cm [70]. A huge contrast like this is conceivable in light of the fact that the previous consists of more deformities and high F-terminations. Delamination of LiF + HCl-etched $Ti_3C_2T_x$ will improve its conductance. For experimental uses, MXenes are generally joined with other favorable electrode materials (for example TMOs) to make composites that show uncommon electrochemical attributes in lithium-sulfur batteries as well as sodium and lithium-particle batteries. MXene–TMO composite's exceptional electrochemical properties are ascribed to the symbiotic impacts of MXenes and TMO: MXene with conductivity works with quick electron transfer between electrolytes and composite materials; in the interim, the high-rich redox reactions of TMOs consisting of various ions which leads to a notably high capacitance of MXene–TMO compounds compared to electrode substances based on carbon. Four compounds with three separate lithium storage methods stand out among hundreds of TMO composites that were produced [68, 70]. TiO_2–$Ti_3C_2T_x$, for example, uses either intercalation or deintercalation process. SnO_2–$Ti_3C_2T_x$ uses either alloying or dealloying process, whereas Co_3O_4–$Ti_3C_2T_x$ and Fe_3O_4-$Ti_3C_2T_x$ use a distinct process involving reactions of conversion wherein the high-valence Co/Fe oxidation state efficiently retains Li-ions [62, 63]. Topological insulators in two dimensions are a novel type of material with a gap of insulation in the core but no gaps at the border. Surprisingly, the side state's distribution is disallowed, resulting in a dissipation less charge or spin transport [63, 64]. On the irregular borders of $Mo_2HfC_2O_2$, a set of geometrical border states joins the valence bands and core conduction, leading to the formation of a mono Dirac cone at the M point. Nevertheless, in MXenes, no semiconducting or topologically insulated states have been produced since all the generated MXenes have mixed terminations due to the current synthesis processes; nonetheless, the expected superconducting and topological conditions need a particular kind of termination on the surfaces of MXene [62]. To obtain semiconducting or topologically insulated MXenes, it would be necessary to have control over their surface chemistry [61, 64].

11.3.3 Magnetic Properties

Most of the 2D materials that have been discovered are nonmagnetic in nature. As a result, achieving controllable magnetic properties in 2D materials always have been a long-term aim. Few of the pure MXenes, including Mn_2C, Cr_2C, Ti_2N, Cr_2N, and Ti_2C, are projected to be materials containing innate magnetism. Ti_2N, Cr_2C and Ti_2C are all ferromagnetic, but Mn_2C and Cr_2N are both antiferromagnetic [71]. MXenes with ½ metallicity that is magnetic, in particular, have lately received much interest. Half-metals have one metallic spinning channel and one semiconducting spinning channel which leads to electrons at the Fermi level. Mxene family's 1st assumed ½ metal is Cr_2C [72]. Furthermore, the half-metallic gap, which is the variation between the Fermi level as well as the highest of the conquered spin-down band, seems to be as wide as 2.9 eV, implying that the 100% spin-filter effectiveness can be maintained throughout a wide biased region. ½ metallicity has been expected in Ti_2N and Ti_2C as a result [24, 35, 62, 73].

Surface functionalization can simply change the magnetic and electrical characteristics of bare MXenes. Since the magnetism of raw MXenes is mainly derived from the unpaired electrons that are present on the metal atoms having a transitioned surface, most MXenes' magnetism will be reduced after surface metallization [73]. The magnetic moments of MXenes consisting of Cr and Mn do not vanish following surface functionalization, which is surprising. Functional groups, on the other hand, would impact their magnetic couplings and electronic configuration. For instance, when organized by OH, Cl, F, or H compounds, the ferromagnetic half-metal Cr_2C is changed into an antiferromagnetic semiconductor with a large energy gap. In comparison to Cr_2C, the Cr_2N turns from an antiferromagnetic metal to a ferromagnetic ½ metal following surface passivation by O [8, 14, 30, 74]. Like symmetrical operations, the asymmetrical operation appears to generate a similar ferromagnetic–antiferromagnetic transition in Cr_2C [73]. Nevertheless, in contrast to symmetrical operationalization, this will cause certain substantial alterations to Cr_2C's electrical framework. Unlike Cr_2CF_2, which has a symmetrical density of states (DOS), Cr_2CFCl has the property of magnetic bipolar semiconductors, in which the CBM (Conduction band minimum) and VBM (valence band maximum) have opposing spinning polarizations. Other Cr_2CXX' also have similar bipolar magnetic characteristics, making them attractive choices for spintronic uses [73, 74].

11.4 Mxene–Polymer Composites Applications

11.4.1 Detection of VOCs and Hazardous Gases

2D materials were employed in sensors for over 10 years owing to the huge specific surface area, outstanding conductive characteristics, and acceptable biocompatibility [13, 24]. MXenes have demonstrated amazing performance in biosensors and gas sensors, which is unsurprising. Because the active protein in a biosensor lacks its bioactivity when it comes into direct contact with the electrode surface, special compounds are used to immobilize active proteins and keep them active [50, 51]. Being protein immobilization matrixes, MXenes could effectively preserve active proteins while simultaneously facilitating the rapid transfer of electrons between the enzyme and the electrode, creating mediator-free biosensors [75]. $Ti_3C_2T_x$ was first selected to immobilize hemoglobin (Hb) in a biosensor for sensing nitrites (Nafion/Hb/$Ti_3C_2T_x$/GCE). Because of $Ti_3C_2T_x$'s comparatively higher conductivity, the GCE /$Ti_3C_2T_x$/Hb/Nafion biosensor exhibits outstanding biosensor activity, consisting of a wider observation range & a lower observation limit. The biosensor's nitrite detection range has been increased, with the minimum limit of detection reduced to 0.12 μM [75, 76]. MXenes can be employed in gas sensors as well as biosensors. The use of MXenes in gas sensors addresses the problem of conventional gas sensors in that they have a shortage of selectivity for specific gases like NH_3. Yu *et al.*, [77] explore the adsorption of gas particles (H_2, NH_3, CO, CH_4, N_2, CO_2, O_2 and NO_2) on single-layered Ti_2CO_2 using first-principles computations. In comparison to other particles, NH_3 is observed to be selectively adsorbed on Ti_2CO_2. The estimated current-voltage relation curve of Ti_2CO_2 following NH_3 adsorption displays a considerable variation, compared to the modest variations of the equivalent graphs of phosphorene and MoS_2 in a similar situation, suggesting Ti_2CO_2's higher sensitivity to NH_3. The gas sensor based on MXenes–$Ti_3C_2T_x$ is built as a result of this experimental analysis. At ambient temperature, this sensor can effectively cover a wide variety of gases like methanol, ethanol, NH_3 and acetone and exhibits p-type sensing characteristics. $Ti_3C_2T_x$ sensor has the strongest NH_3 reaction and the weakest acetone response. The $Ti_3C_2T_x$ sensor's sensing method requires gas desorption or absorption by functional as well as imperfection $Ti_3C_2T_x$ groups [77].

The VOCs sensors with low-power consumption and high selectivity have been predicted for monitoring applications [78]. Kuang *et al.*, [79] have reported that the $Ti_3C_2T_x$-TiO_2 nanocomposites gas sensor is to detect VOCs at ambient temperature. The $Ti_3C_2T_x$-TiO_2 nanocomposites are synthesized by hydrothermal strategy using $Ti_3C_2T_x$ as a titanium source. At ambient temperature, the $Ti_3C_2T_x$-TiO_2 sensor's carrier density modulation displays responses that are 1.5–12.6 times more sensitive than the detection of pure MXene for different VOCs. They observed that the $Ti_3C_2T_x$-TiO_2 sensor's gas response to 10 ppm hexanal is approximately 3.4%. Their study has provided a comparison study between the normal MXene $Ti_3C_2T_x$ and nanocomposite MXene $Ti_3C_2T_x$-TiO_2 for hexanal detection. Figure 11.7 represents the detection of various VOCs carried out in this study and shows the enhancement they have achieved in sensing response after reinforcing the MXene with TiO_2 nanoparticles.

In order to create gas-sensing materials, Chen *et al.*, [80] investigated the hybridization of MXenes with two-dimensional transition metal dichalcogenides. They selected the $Ti_3C_2T_x$ and WSe_2 for the hybridization and the detection of VOCs. The $Ti_3C_2T_x$/WSe_2 hybrid sensor has an extremely quick reaction and faster recovery, minimal noise, and considerable adaptability for detecting different VOCs. They have also proposed a mechanism of sensing ethanol by the $Ti_3C_2T_x$/WSe_2 nanohybrids. Figure 11.8(a) represents the band structure of the $Ti_3C_2T_x$/WSe_2 nanohybrid. The $Ti_3C_2T_x$ nanosheets are highly conductive and readily provide electrons to WSe_2. In the presence of air, the electronegative oxygen adsorbs the electrons thus creating a lag in supplying the electron due to which the depletion layer gets thicker between $Ti_3C_2T_x$ and WSe_2. But when the exposure of oxygen-containing VOC like ethanol occurs, the ethanol molecule reacts with the electron trapped oxygen (as it has become more reactive) forms volatile gases like CO_2 and releases the adsorbed electron to the conduction band which increases the conductivity, decreases the resistance of the

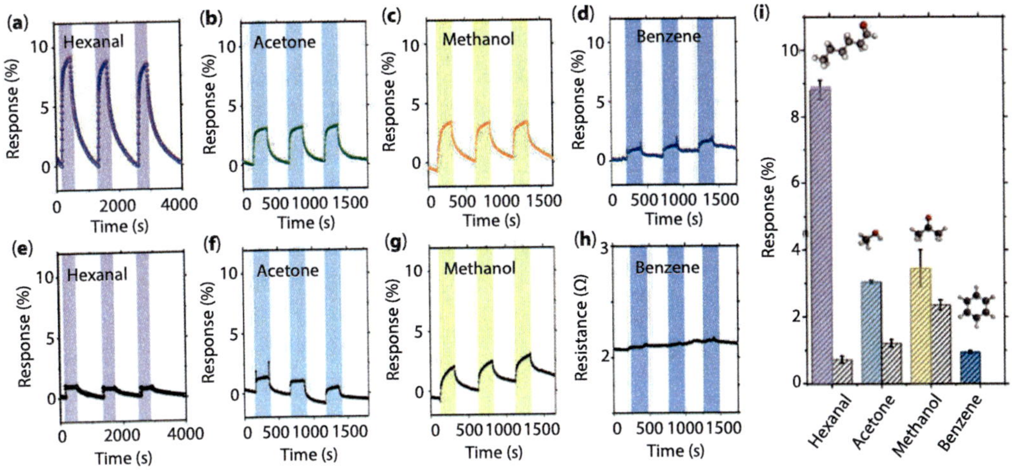

Figure 11.7 Real-time response of $Ti_3C_2T_x$-TiO_2 against (a) hexanal, (b) acetone, (c) methanol and (d) benzene which show enhancement over the real-time responses of $Ti_3C_2T_x$ against (e) hexanal, (f) acetone, (g) methanol and (h) resistance variation of Benzene. All these data were measured in exposure to 100 ppm concentration for three successive cycles. (i) selectivity of the $Ti_3C_2T_x$-TiO_2 and $Ti_3C_2T_x$ sensor to the VOCs shown, where the colored bar represented the $Ti_3C_2T_x$-TiO_2 sensor. Figure is taken from [79], with copyright © 2021, Elsevier B.V.

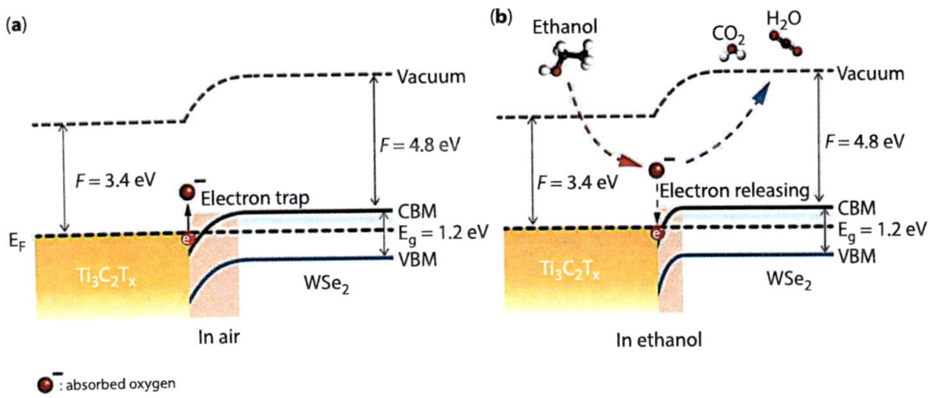

Figure 11.8 Proposed mechanism of sensing VOC (ethanol) with the help of energy-band diagram of MXene $Ti_3C_2T_x/WSe_2$ in (a) air and (b) ethanol. The figure is taken from Ref. [80].

sensor and reduces the depletion layer as shown in Figure 11.8(b). This improves the sensing response and selective detection of VOCs.

Two-dimensional carbides-based transition metals ($Ti_3C_2T_x$ MXene) have been used potentially for gas sensing due to their highest sensitivity and gas selectivity, though these materials have poor stability in a humid environment against hydration and oxidation. Chen *et al.* [81] have used MXene-based composite materials with the integration of a protection layer with MXenes. This composite is a promising deal to overcome in a humid environment. They demonstrated the surface functionalization of $Ti_3C_2T_x$ with fluoroalkyl silane molecules through surface treatment. This surface modification provides the gadget with a superhydrophobic surface, mechanical resilience, and improved sensing capabilities. They showed in their results that high sensitivity, good repeatability, long-term stability, selectivity, and faster response/recovery property were achieved by the functionalization when $Ti_3C_2T_x$ was integrated into chemo resistive sensors sensitive to oxygen-containing VOCs (ethanol, acetone). Fluoroalkyl silane functionalization gives protection to sensing response when the dynamic response of the $Ti_3C_2T_x$-F sensor to 30 ppm of ethanol was measured over in the 5 to 80% relative humidity range [81, 82].

Jin *et al.*, [82] prepared polymeric titanium carbide ($Ti_3C_2T_x$) MXene composite by *in situ* polymerization on $Ti_3C_2T_x$ MXene. They utilized the prepared PEDOT: PSS/MXene material for gas sensor to measure the NH_3 gas at room temperature. Figure 11.9 represents the preparation of the composite and the gas sensor using the MXene. Figure 11.10 represents the selectivity of the sensor at 100 ppm at ambient temperature with different gases. The shown data exhibit the strong sensing of NH_3 (36.6%) by the sensor compare to other gases.

Furthermore, TiO_2 has great biocompatibility and chemical durability, making it a good candidate for creating a protected enzyme microenvironment [83]. The enzyme is mounted on the internal surface of the Ti_3C_2Tx nano-layers which look like organ structures and are next to TiO_2, ensuring its long-term stability and functioning. The GCE/Ti_3C_2Tx/Hb/Nafion biosensor has a quicker sensitivity (<3 s) and a larger measurement range (0.1 to 380 μM) than the (Hb/Nafion/GCE/Ti_3C_2Tx–TiO_2) biosensor, which uses $Ti_3C_2T_x$ as the immobilization matrix. Au nanoparticles are commonly utilized in biosensors to detect glucose to maintain enzyme biological performance by decreasing the insulating impact of the protein coat for immediate transfer of electrons [52, 61, 83]. An MXene-Au nanocomposite with a morphology comparable to TiO2–$Ti_3C_2T_x$ is utilized as an immobilization matrix in a biosensor Nafion/GC/MXene/GOx/Au so that MXene can have outstanding activity in the enzymatic glucose biosensor [59, 76]. The manufactured biosensor has a broad limit of detection for glucose concentrations ranging from 0.1 mM to 18 mM, a lower detecting threshold of 5.9 μM, high durability, as well as outstanding

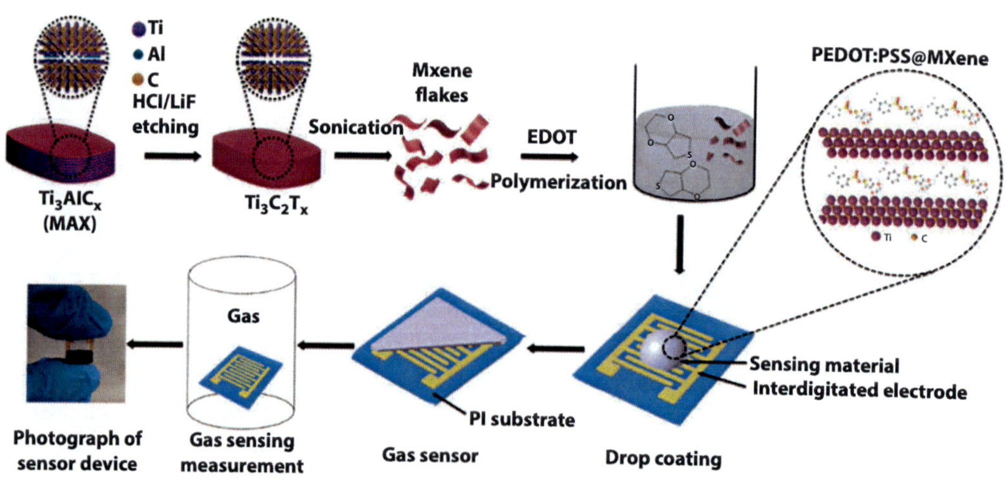

Figure 11.9 Schematic illustration of the preparation of MXene/polymer composite-based gas sensor. Figure adapted from [82], copyright © 2020, American Chemical Society.

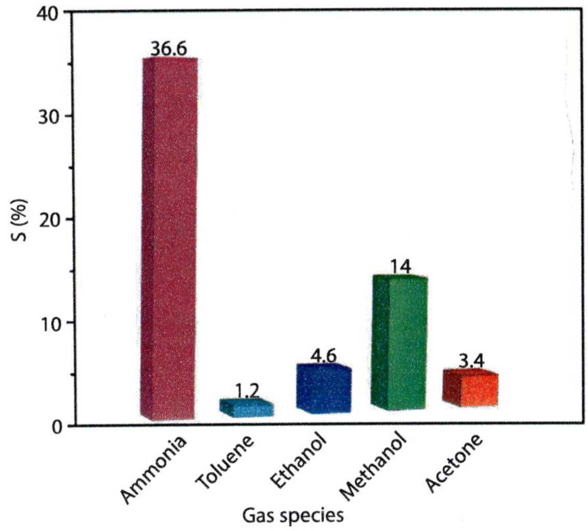

Figure 11.10 Data representing the selectivity % of the gas sensor with different gases at 100 ppm at room temperature. Figure adapted from [82], copyright © 2020, American Chemical Society.

repeatability, which is remarkable [76, 84]. The distribution of Au nanoparticles on the MXene nanosheets surfaces boosts electrical conductivity even further thereby increasing electron exchange between the enzyme's active core and the electrode [84].

11.4.2 Environment-Related Applications

11.4.2.1 Catalysis

Since the worldwide air contamination is continually expanding and energy emergencies keep on being bothered, proficient clean energy change advancements like electrochemical water parting, photocatalytic water parting, and photocatalytic decrease of CO_2 are happening to incredible importance. Catalysts with high efficiency are urgent for the fruitful execution of these advancements. As the uses of customary catalysts as well as valuable metals and oxides based on oxides (IrO_2, RuO_2, and Pt) are restricted by their shortage and significant expense, creating productive catalysts which have elite exhibitions as well as are economical is essential for clean energy transformation advances [85]. Since MXenes, along with their composites are a new group of materials consisting of excellent conductivity of electrons, better surface solubility in water, and a large number of revealed metal locations, they were put to use as highly active electrode materials in electrochemical water splitting and as catalysts or co-catalysts in photocatalytic processes such as water splitting and CO_2 reduction, exhibiting great potential for replacing precious metals [86]. Figure 11.11 represents the catalytic applications of MXene-based catalysts.

11.4.2.2 Electrocatalysis

The water splitting mechanism is of two types: the Oxygen Evolution Reaction (OER) and the Hydrogen Evolution Reaction (HER). Regarding the first one, the latest electrocatalysts,

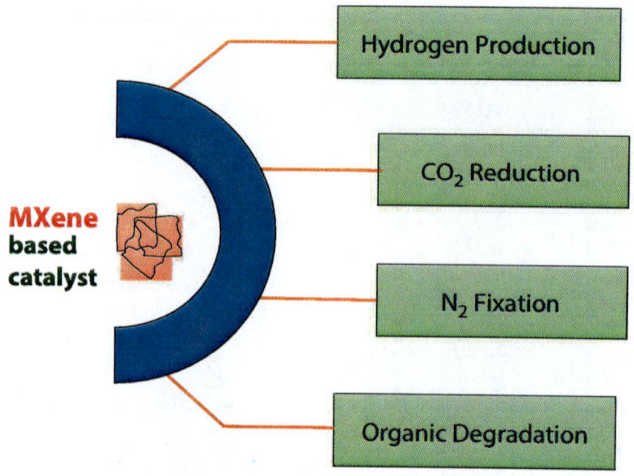

Figure 11.11 Catalyst-based applications of MXene.

for example, metal–natural systems (MOFs), layered double hydroxides (LDHs) & graphitic carbon nitride (g-C_3N_4) are produced to supplant valuable metals [87]. Nevertheless, because of their inferior electrical conductivity, these latest catalysts generally show lower electrocatalytic exhibitions in the oxygen evolution reaction. Consolidating the exceptionally conductive materials with the aforementioned latest electrocatalysts for making composites is a favorable approach for enhancing their electrocatalytic activities. MXenes of higher conductivity happen to be great for boosting conductivity [87].

When g-C_3N_4 and Ti_3C_2Tx combine, a Ti_3C_2Tx-g-C_3N_4 (TCCN) composite layer with surfaces possessing hydrophilicity and conducting structures is generated with enhanced electrocatalytic activity. In the inter-layers of the extremely pliable TCCN layer which comprises a couple of inches of size and 8–10 mm of thickness, numerous wide pores are present which vary from tens to hundreds of nanometers in size [87]. $Ti_3C_2T_x$ produces units of Ti-N_x in addition to g-C_3N_4 and excellent conductivity for the TCCN composite. Anodic current recorded on TCCN increases rapidly at 1.44 V and sometimes surpasses IrO_2/C at potentials over 1.54 V, indicating significant OER sensitivity and catalytic performance possessed by TCCN. Furthermore, the polarization curve of TCCN remains nearly unaltered after 5000 cycles, demonstrating the compound's high stability [86–88]. It's worth noting that TCCN's hierarchical pores and particular Ti–Nx units, in addition to its high conductivity, contribute significantly to the material's exceptional catalytic activity [88]. Reactants use Ti–Nx units as electroactive locations, while the hierarchical pores allow for rapid transit of reactants and reaction products. The group of metal ions, clusters or organic ligands in metal-organic frameworks, or MOFs, is a promising strategy for developing new good electrocatalysts for the oxygen evolution reaction. Usually, MOFs have porosity and a lot of uncovered effective metal sites, however, they have low electronic conductance [89]. As a result, it is expected that combining MOFs along with a conductive material will boost MOF's electrocatalytic activity in the OER. An interdiffusion-mediated method is used to produce a new composite of Mxene–MOF, known as Ti_3C_2Tx–CoBDC (cobalt 1,4-benzene dicarboxylate) [89].

LDHs (Layered Double Hydroxides) having layers of metal hydroxide that are charged positively and charge-adjusting anions (for example, NO_3, etc.) offer tremendous potential in the form of substitutes to native metal electrocatalysts for the oxygen evolution reaction due to their high redox activity [90]. Furthermore, LDH's applicability in the OER is constrained due to its weak conductivity and restricted electrochemically active surfaces. By combining highly conductive MXenes with LDH, Yu *et al.* [77] aim to address this challenge. As a novel OER electrocatalyst, they create a FeNi-LDH/Ti_3C_2Tx composite including an interconnected porous network structure that shows high activity for the OER. The main determinants for effective electrocatalytic catalytic activity are good electronic interaction and contact of interfaces involving Ti_3C_2Tx & FeNi-LDH. Furthermore, by covering an electrode using nickel foam, the electrocatalysts FeNi-LDH/Ti_3C_2Tx's oxygen evolution reaction catalytic activity can be increased much more [77, 90].

MXenes have shown potential as advanced electrocatalysts for the Hydrogen Evolution Reaction, in addition to being employed as conductive materials to increase the performance of current catalysts. The 1st two-dimensional material found as a potential HER catalyst is MoS_2 [91]. The revealed border sites of MoS_2 are particularly active catalytically, while the underlying plane seems inactive, necessitating a sophisticated design to optimize the border sites. Based on both theoretical analysis and observations, many authors claim that MXenes can be employed in the form of electrocatalysts for the Hydrogen Evolution Reaction [90, 91]. They anticipate that Mo_2CT_x is the most effective electrocatalyst by measuring over potential as well as hydrogen adsorption free energy (DGH) M_2XT_x series (M = Mo, V, Hf, Zr, Ti, and Sc) substances [92]. Furthermore, unlike MoS_2, Mo_2CT_x's basal plane contains a considerable number of catalytic sites. A linear-sweep voltammetry measurement employing glassy carbon as the substrate electrode is used to analyze the genuine electrocatalytic capabilities of MXenes for the HER [91, 92]. When compared to Ti_2CT_x, which requires a 609 mV overpotential to obtain the multi-lt density of 10 mA cm^2 throughout the preliminary swept, Mo_2CT_x requires a much lesser overpotential (283 mV), showing that Mo_2CT_x does have greater electrocatalytic performance over Ti_2CT_x [92]. The Mo_2CT_x slope, in contrast, shows a lesser shift following continuous cycling compared to that of Ti_2CT_x, showing that for HER, Mo_2CT_x is a pretty steady catalyst. Because metal atoms occupy a major amount of Mxene surfaces, they have a low tolerance to oxygen in an electrolytic solution that contains oxygen, which has a significant impact on Mxene stability as electrocatalysts and hence limits their applicability [93]. Wu *et al.* [83] use a nano plating technique to effectively produce hierarchical MoS_2–$Ti_3C_2T_x$@C, which improves oxygen resistance. $Ti_3C_2T_x$ ($Ti_3C_2T_x$@C) has a carbon layer on its surface which significantly suppresses surface oxidation, hence improving $Ti_3C_2T_x$ stability. Furthermore, the MoS_2's bifunctional action to both HER and Li storage expands the scope of MoS_2–$Ti_3C_2T_x$@C multi-field uses [94]. This multi-layered arrangement built of $Ti_3C_2T_x$@C & MoS_2 has been discovered to have high electrocatalytic activity for the HER. To obtain a 10 mA/cm^2 current density, $Ti_3C_2T_x$@C needs 135 mV of overpotential, which is smaller compared to MoS_2–rGO@C and $Ti_3C_2T_x$@C. The remarkable electrocatalytic performance of MoS_2–$Ti_3C_2T_x$@C is confirmed by the low overpotential. Furthermore, following cyclic voltammetry consisting of 2000 cycles, the MoS_2 $Ti_3C_2T_x$@C electrode's polarization curve shows no substantial shift, showing that MoS_2–$Ti_3C_2T_x$@C is highly stable. MXenes composites (CoP@3D–$Ti_3C_2T_x$, $Ni_{0.7}Fe_{0.3}PS_3$– $Ti_3C_2T_x$ and Ni_2P–Cr_2CT_x) have just been validated as dual-functional electrocatalysts by scientists for HER and OER [94, 95]. Furthermore, the electrocatalytic activity of the CoP@3D–$Ti_3C_2T_x$ and $Ni_{0.7}Fe_{0.3}PS_3$–$Ti_3C_2T_x$ produced materials exceeds the standard bifunctional electrocatalysts RuO_2 and Pt [95].

11.4.2.3 Photocatalysis

Photocatalytic water splitting is now achievable thanks to the development of photocatalysis technology. Utilizing solar energy to instantly split water using a semiconductor photocatalyst eliminates the use of extra energy sources, as opposed to electrochemical water splitting. Currently, the production of novel photocatalysts that are affordable, stable, and can act successfully under visible light is the main emphasis of photocatalytic water splitting [96].

MXenes have been proposed as a potential catalyst for photocatalytic hydrogen production reactions. Using *ab initio* calculations, Guo *et al.* identified two promising photocatalysts in the MXene class for the photocatalytic hydrogen generation reaction: Zr_2CO_2 and Hf_2CO_2. The size of the bandgap, the locations of the conduction minimum (CBM) as well as the valence band maximum (VBM), and other electrical structural properties of a substance are commonly utilized as characteristics describing the material's viability as a photocatalyst [97]. The photocatalyst bandgap, as well as gap value for the photocatalytic hydrogen production reaction, must be within 1.55 - 3.0 eV and their CBM must also be larger compared to the hydrogen reduction potential of H^+/H_2 while their valence band maximum must be lesser compared to the water oxidation potential of H_2O/O_2. These conditions are met by Zr_2CO_2 and Hf_2CO_2 [97]. In addition, a good photocatalyst must be capable of using solar energy effectively. $Hf_2CO_2$2D and Zr_2CO_2 have high optical absorption in visible light, indicating that both $Hf_2CO_2$2D and Zr_2CO_2 can be utilized as photocatalysts in water-splitting reactions to produce hydrogen.

Aside from generating novel catalysts, researchers have focused on finding innovative strategies to improve the catalytic activity of current photocatalysts [97]. Co-catalysts have exhibited significant promise in enhancing the performance and durability of photocatalysts in the latest years [97, 98]. The basic components must be present in an excellent co-catalyst: (i) a large number of surface functionalities to form powerful linkages amidst the photocatalyst for long-term stabilization and fast interlayer charge transport; (ii) adequate connections with molecules of water; and (iii) an increased conductivity to allow effective charge transfer. Surprisingly, MXenes can satisfy all of the aforementioned criteria regarding co-catalysts. Hydrophilic functional groups (–O and –OH) are present on MXenes' surfaces, allowing easy creation of strong solid contacts with a variety of semiconductor photocatalysts and molecules of water. MXenes' high electrical conductivities provide effective electron transport. As a result, there is now a massive rise in the use of MXenes as photocatalysts [98].

At least two co-catalysts have been coupled to study more effective co-catalytic methods, taking into account the synergic activity of diverse co-catalysts. For instance, when nanoparticles of $Ti_3C_2T_x$ and nanoclusters of Pt are used for photocatalyst g-C_3N_4 in the form of double co-catalysts, the photocatalytic hydrogen generation ability of g-C_3N_4 is raised to 5100 mol/h/g, significantly greater than when Pt or $Ti_3C_2T_x$ are employed individually as co-catalysts for g-C_3N_4. The quick interfacial charge transmission from g-C_3N_4 to $Ti_3C_2T_x$ and Pt, which leads to the successful segregation of photo-generated carriers, is attributed to this extraordinary action in the hybrid of Pt, g-C_3N_4 and $Ti_3C_2T_x$ [99].

The photocatalytic CO_2 reduction reaction can generate hydrocarbon solar biofuels, which can help with energy shortages and air contamination. MXene can also be used in photocatalytic CO_2 reduction as a co-catalyst [98]. For example, $Ti_3C_2T_x$ ($Ti_3C_2T_x$–OH) which is surface alkalinized has been used as a co-catalyst to boost titania (P25) photo-activity in photocatalytic CO_2 reduction. When the molar ratio of $Ti_3C_2T_x$–OH to P25 is increased to 5 percent, the transformation levels of CH_4 and CO surface alkalinized obtain high values of 16.61 and 11.74 μmol/g/h, respectively, which are significantly greater compared to that of $Ti_3C_2T_x$–P25, P25, and Pt–P25, implying that the photocatalytic efficiency of P25 has been significantly increased utilizing $Ti_3C_2T_x$–OH as a co-catalyst. The increased amount of hydroxyl groups on the surface of $Ti_3C_2T_x$ provides ample locations for carbon dioxide to get adsorbed, which explains the outstanding activity of $Ti_3C_2T_x$–OH [97–99].

11.4.3 Water Remediation

Water pollution is a serious environmental and ecological problem. Toxic heavy metals, organic chemicals (dyes), eutrophic substances, and nuclear waste are among the pollutants. For wastewater treatment, many technologies for e.g., biological treatment, adsorption, catalytic degradation and chemical oxidation were established [100]. Adsorption and catalytic breakdown are two of the best strategies for eliminating hazardous compounds from water. MXenes, as novel adsorbents, are predicted to make a significant contribution to water remediation because of their benefits of ease of production, large specific surface area, and excellent reducibility [35]. Cr(VI), a type of heavy metal ion found in a large number of industrial wastewater, is extremely harmful to plants, animals, and humans. As a result, industrialized effluent must be treated with Cr(VI) adsorption [101]. Because $Ti_3C_2T_x$ has been effectively utilized to adsorb hefty Pb metal ions, $Ti_3C_2T_x$ is believed to be able to remove Cr(VI) from wastewater as well. According to Ying *et al.*, the $Ti_3C_2T_x$ has an exceptional elimination capability for Cr(VI) of up to 250 mg/g. The remaining Cr(VI) content in the water after $Ti_3C_2T_x$ processing is below the World Health Organization's standard Cr(VI) concentration in drinking water [102]. There is a significant adsorption capability of $Ti_3C_2T_x$ nanosheets for Cu ions, according to Shahzad *et al.*, and delaminated (DL)-$Ti_3C_2T_x$ has a better adsorption capacity compared to that of multilayered-$Ti_3C_2T_x$. At 78.45 mg/g, DL-$Ti_3C_2T_x$ had the maximum adsorption capability for Cu ions, which is 1.17 and 2.7 times higher than $Ti_3C_2T_x$ having multiple layers as well as generally employed activated carbon, respectively. XRD measurement of DL-$Ti_3C_2T_x$ reveals the adsorption mechanism [98, 102].

The appearance of the precipitates Cu_2O and CuO after Cu ion adsorption can be seen in the XRD pattern, implying that Cu ions combine with DL-$Ti_3C_2T_x$ surface functional groups to produce CuO and Cu_2O. The DL-$Ti_3C_2T_x$ is dissolved in simulated electroplating wastewater consisting of 4 types of heavy metal ions (Cd^{2+}, Cu^{2+}, Cr^{3+} and Pb^{2+}) to demonstrate its actual adsorption capability for Cu ions [103]. MXenes can further be used in dye sewage treatment as possible adsorbents. Mashtalir [52] has studied the $Ti_3C_2T_x$'s adsorption capability for an anionic dye AB80 as well as a cationic dye MB When $Ti_3C_2T_x$ is used in the dark to remove MB and AB80 from the solution, the concentration of AB80 remains unchanged after 20 hours, but the concentration of MB drops dramatically within 8 hours. $Ti_3C_2T_x$ appears to selectively trap the cationic dye, which is likely due to $Ti_3C_2T_x$'s negatively charged surface, which may link effectively to MB [52].

MXenes and their composites could remediate eutrophic chemicals in wastewater besides heavy metal ions and dyes. MXenes can thus be utilized for water eutrophication (phosphate) restoration thanks to manufacturing a sandwich-like $Fe_3O_4/Fe_2O_3/Ti_3C_2T_x$ nanocomposite (MXI). Nano-Fe_2O_3 is dispersed in the inner stacks of MXenes in the MXI, increasing the amount of accessible operated layers of Ti–OH, and Fe_3O_4 fills the spaces among MXene's numerous sheets, generating the Fe–OH molecule, which aids phosphate adsorption selectively and enhances sorption performance [104]. Every kg of MXI could manage wastewater consisting of 2300/2400 kilos of phosphorus in practice, resulting in the utilization of MXI in treating wastewater on a large scale. Furthermore, by immersing the sample in a blend of alkaline and salt solution, the $Ti_3C_2T_x/Fe_3O_4/Fe_2O_3$ can be reproduced multiple times [104].

Researchers have also looked into MXenes' antibacterial capabilities to see if they may be used as biocides in water treatment. Begum *et al.* [105] have found that $Ti_3C_2T_x$, as a prominent material in the family of MXene, has remarkable antibacterial effectiveness to *Escherichia coli (E. coli)* as well as *Bacillus subtilis* (*B. subtilis*), then graphene oxide (GO), which is recognized worldwide as an antibacterial agent, that paves the way for uses of MXenes in antimicrobial layering and membranes for water treatment [105]. MXenes' strong antibacterial property may be due to their hydrophilic surfaces, which might aid in bacterial elimination through immediate contact. High-density flaws and functional groups could boost the antibacterial properties of MXenes. Many chemical interactions among MXenes and bacterial membranes are likely during the bactericidal procedure, but they need to be investigated further [106]. Figure 11.12 represents a general categorized diagram of applications, dimensions and method of preparations of MXenes. As shown in Figure 11.12 the MXenes have materialized as the best alternative for improving photocatalytic performance in the renewable energy field along with the environmental remediation field [107].

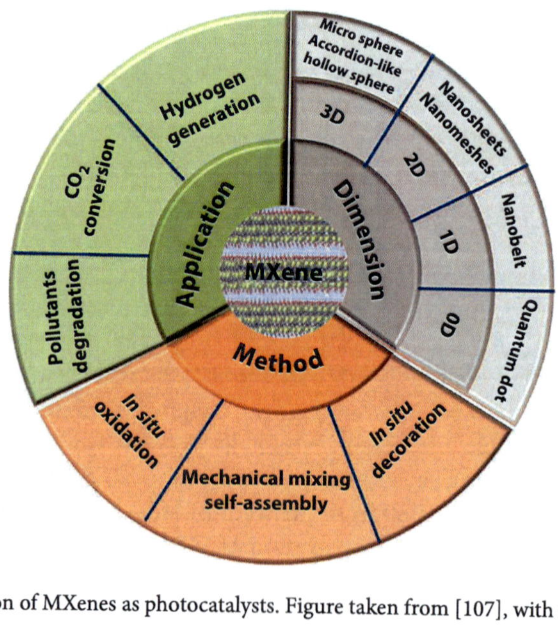

Figure 11.12 Application of MXenes as photocatalysts. Figure taken from [107], with copyright © 2021, Elsevier B.V.

11.5 Future Directions

Due to improved properties like high specific surface area, surface functional groups, abundant organic bonds, large porosity, high hydrophilicity and conductivity, MXenes has a variety of application in future such as detection of hazardous gases, biomedical applications for the medical sector preparation, like antimicrobial, anticancer activity, drug administration, biosensors, photothermal treatment (PTT), diagnostic imaging, and bone regeneration, which are attributed to the 2D planar structure and extraordinary capabilities demonstrated by polymer functionalized MXenes.

11.5.1 Bioimaging

The assumption of the tumor location and the observation of the PTT are both possible using diagnostic imaging. Various diagnostic imaging methods exist. MRI is a powerful imaging technology for studying the anatomical structure of soft tissues by identifying the non-ionizing electromagnetic waves generated by the tissues. Computed tomography is regarded as the best imaging modality because of its great spatial resolution and anatomical tomography. Because the photoacoustic signal produced by NIR irradiation has a great tissue perforation form and exceeds the optical imaging perforation limitation, PA imaging is a real-time method of detecting stress waves in irradiated tissues. An effective way to track the therapy's effectiveness is Imaging-guided synergistic PTT/chemotherapy.

11.5.1.1 Magnetic Resonance Imaging (MRI)

Magnetic functionalization of Ta_4C_3 with iron oxide nanoparticles (IONPs) gave the composite superparamagnetic properties and higher magnetic concentration. Ta_4C_3-IONP-SP nanocomposite has a great transverse relaxivity making it a good contrast enhancer for T2-weighted MR imaging. The disintegrating of Mn-O bonds occurs effortlessly in acidic tumor surroundings, resulting in the production of Mn^{2+} ions and the enhanced connection involving Mn and H_2O molecules. The MnOx component is what gives the MnOx/Ta_4C_3-SP and MnOx/Ti_3C_2-SP Mxene nanocomposite pH-sensitive T1-weighted MR imaging.

11.5.1.2 Photoacoustic (PA) Imaging

Ti_3C_2-SP, MnOx/Ti_3C_2-SP and Ta_4C_3-SPnanoflakes possess excessive NIR-absorptivity and η properly added to PA imaging. With the increase in signal and deterioration with the gradual accumulation and discharge of the contrast upgraded nano agents, the cancer area occurs to be unequivocal.

11.5.2 Computed Tomography (CT)

The contrast increases while CT imaging is due to the nanocomposite's large X-ray attenuation coefficient and Ta's greater atomic number. As a result of the increasing accumulation of Ta_4C_3-SP as well as Ta_4C_3-IONP-SPs in cancer cells, the increase in CT values could be used as contrasting probes for CT imaging. The Hounsfield unit (HU) is proportional to the concentration of Ta in Ta_4C_3/MnOx, and also it has a greater value compared to the commonly utilized Iopromide 300 when presenting contrast-enhanced CT images at a similar concentration.

11.5.3 Bone Regeneration

Guided bone regeneration (GBR) therapy, which is increasingly used for orthodontic surgery and periodontal rejuvenation, is based on the concept of using a GBR barrier to protect the bone from soft tissue interruption. While disintegrating in aerobic conditions, the molecules based on Ti discharged by the MXene integrated polymeric membrane, and water aid in the formation of new bones. A biomaterial and robust membrane for GBR by embedding Ti_3C_2Tz nanosheets in PLA matrix and using n-octyltriethoxy silane (OTES) as an intermediating agent. The solvent casting OTES-Ti_3C_2Tz/PLA nanocomposite membrane had outstanding mechanical capabilities than the pure PLA membrane. OTES-Ti_3C_2Tz and PLA have a close association because mechanical properties are improved with a good load transfer. The Ti_3C_2Tz nanosheets' biocompatibility was superior to GO on the film in promoting preosteoblast adherence, multiplication, and osteogenic differences, and it increased with the concentration of MXene. Many authors used electrospinning to create MXene/polycaprolactone fibers that were more biocompatible with pre-osteoblast cells regeneration than compared to fibroblast cells. MXene percentage in PCL was raised, which improved the composite fiber mat's wettability, promoting cell attachment and growth. The bioactivity of calcium or phosphorus accumulation as well as *in-vitro* biomineralization on the scaffolding is adequate for osteointegration. As a result, they concluded that such composites are being used for bone tissue engineering and dressing of wounds but that more research is needed.

11.6 Conclusion

The emission of gaseous pollutants causes serious environmental issues. Fortunately, the consequences of gas emissions on the environment and the health of aquatic and terrestrial animals are raising public awareness. Sensors are an important component to collect huge information about the surrounding environment. Therefore, there is a pressing need to build a wireless operating gas sensor system that can detect volatile organic chemicals at ambient temperature in a selective, sensitive, rapid, and reversible manner. A newly discovered class of two-dimensional materials called MXenes has fascinated enormous attention in a range of applications because of the availability of numerous active sites, metallic conductance, customizable surface chemistry, and exceptional permanence. Due to improved properties of MXene for example high specific surface area, abundant organic bonds, surface functional groups, large porosity, high hydrophilicity and conductivity MXene materials are widely used for sensing hazardous environments. MXene composite materials can be also used as a sensor, energy storage, electrode materials, supercapacitors, electromagnetic shielding materials, battery materials and nanofluids.

Acknowledgement

The authors are thankful to the Director, CSIR-NEIST, Jorhat for encouragement and continuous support. We also acknowledge the Centre for Petroleum Research, CSIR-NEIST Jorhat along with CSIR, New Delhi for valuable support. PGI acknowledge to the Science and Engineering Research Board (SERB), New Delhi, India, for funding under the CRG/2022/004008 project. SK acknowledges DST, New Delhi, India for DST/INSPIRE fellowship award No. IF190333.

References

1. McMichael, A.J. and McMichael, T., *Planetary overload: Global environmental change and the health of the human species*, p. 352, Cambridge University Press, Cambridge, England, 1993.
2. Turner, A.P. and Magan, N., Electronic noses and disease diagnostics. *Nat. Rev. Microbiol.*, 2, 161–166, 2004.
3. Wetchakun, K., Samerjai, T., Tamaekong, N., Liewhiran, C., Siriwong, C., Kruefu, V., Wisitsoraat, A., Tuantranont, A., Phanichphant, S., Semiconducting metal oxides as sensors for environmentally hazardous gases. *Sens. Actuators B Chem.*, 160, 580–591, 2011.
4. Ernstgard, L., Iregren, A., Sjogren, B., Svedberg, U., Johanson, G., Acute effects of exposure to hexanal vapors in humans. *J. Occup. Environ. Med.*, 48, 573–580, 2006.
5. Chen, F., Wang, C.Q., Zhang, M.J., Zhang, X.L., Liu, Y., Ye, J.N., Chu, Q.C., Sensitive determination of endogenous hexanal and heptanal in urine by hollow fiber liquid-phase microextraction prior to capillary electrophoresis with amperometric detection. *Talanta*, 119, 83–89, 2014.
6. Huang, K., Zhu, C., Yuan, F., Xie, C., Nanoscale SnO_2 flat-type coplanar Hexanal gas sensor arrays at ppb level. *J. Nanosci. Nanotechnol.*, 13, 4370–4374, 2013.
7. Malara, A., Bonaccorsi, L., Donato, A., Frontera, P., Piscopo, A., Poiana, M., Leonardi, S.G., Neri, G., *Sensing properties of indium, tin and zinc oxides for hexanal detection*. Lecture Notes in Electrical Engineering, vol. 539, pp. 39–44, Springer, Cham, 2019.
8. Lou, Z., Wang, L., Shen, G., Recent advances in smart wearable sensing systems. *Adv. Mater. Technol.*, 3, 1800444, 2018.
9. Lim, H.R., Kim, H.S., Qazi, R., Kwon, Y.T., Jeong, J.W., Yeo, W.H., Advanced soft materials, sensor integrations, and applications of wearable flexible hybrid electronics in healthcare, energy, and environment. *Adv. Mater.*, 32, 1901924, 2020.
10. Nikolic, M.V., Milovanovic, V., Vasiljevic, Z.Z., Stamenkovic, Z., Semiconductor gas sensors: Materials, technology, design, and application. *Sensors*, 20, 6694, 2020.
11. Kumar, R., Goel, N., Hojamberdiev, M., Kumar, M., Transition metal dichalcogenides-based flexible gas sensors. *Sens. Actuators A: Phys.*, 303, 111875, 2020.
12. Hung, N.M., Hung, C.M., Van Duy, N., Hoa, N.D., Hong, H.S., Dang, T.K., Viet, N.N., Phuoc, P.H., Van Hieu, N., Significantly enhanced NO_2 gas-sensing performance of nanojunction-networked SnO_2 nanowires by pulsed UV-radiation. *Sens. Actuators A: Phys.*, 327, 112759, 2021.
13. Pi, M., Zheng, L., Luo, H., Duan, S., Li, C., Yang, J., Zhang, D., Chen, S., Improved acetone gas sensing performance based on optimization of a transition metal doped WO_3 system at room temperature. *J. Phys. D: Appl. Phys.*, 54, 155107, 2021.
14. Yang, J., Han, W., Ma, J., Wang, C., Shimanoe, K., Zhang, S., Sun, Y., Cheng, P., Wang, Y., Zhang, H., Lu, G., Sn doping effect on NiO hollow nanofibers based gas sensors about the humidity dependence for triethylamine detection. *Sens. Actuators B Chem.*, 340, 129971, 2021.

15. Xu, T.T., Zhang, X.F., Dong, X., Deng, Z.P., Huo, L.H., Gao, S., Enhanced H_2S gas-sensing performance of Zn_2SnO_4 hierarchical quasi-microspheres constructed from nanosheets and octahedra. *J. Hazard. Mater.*, 361, 49–55, 2019.
16. Yao, D., Dong, C., Bing, Q., Liu, Y., Qu, F., Yang, M., Liu, B., Yang, B., Zhang, H., Oxygen-defective ultrathin $BiVO_4$ nanosheets for enhanced gas sensing. *ACS Appl. Mater. Interfaces*, 11, 23495–23502, 2019.
17. Mi, Q., Zhang, D., Zhang, X., Wang, D., Highly sensitive ammonia gas sensor based on metal-organic frameworks-derived $CoSe_2$@ nitrogen-doped amorphous carbon decorated with multi-walled carbon nanotubes. *J. Alloys Compd.*, 860, 158252, 2021.
18. Ingole, P.G., Sohail, M., Abou-Elanwar, A.M., Baig, M.I., Jeon, J.D., Choi, W.K., Kim, H., Lee, H.K., Water vapor separation from flue gas using MOF incorporated thin film nanocomposite hollow fiber membranes. *Chem. Eng. J.*, 334, 2450–2458, 2018.
19. Kim, J.K., Han, M., Kim, Y., An, H.K., Lee, S., Kong, S.H., Jung, D., Pd-Decorated multi-walled carbon nanotube sensor for hydrogen detection. *J. Nanosci. Nanotechnol.*, 21, 3707–3710, 2021.
20. Ridhi, R., Gautam, S., Saini, G.S.S., Tripathi, S.K., Rawat, J.S., Jha, P., Study of the effect of orbital on interaction behaviour of SWCNT-metal phthalocyanines composites with ammonia gas. *Sens. Actuators B Chem.*, 337, 129767, 2021.
21. Jeong, D.W., Kim, K.H., Kim, B.S., Byun, Y.T., Characteristics of highly sensitive and selective nitric oxide gas sensors using defect-functionalized single-walled carbon nanotubes at room temperature. *Appl. Surf. Sci.*, 550, 149250, 2021.
22. Qu, C., Zhao, P., Wu, C., Zhuang, Y., Liu, J., Li, W., Liu, Z., Liu, J., Electrospun PAN/PANI fiber film with abundant active sites for ultrasensitive trimethylamine detection. *Sens. Actuators B Chem.*, 338, 129822, 2021.
23. Thangamani, G.J., Deshmukh, K., Nambiraj, N.A., Pasha, S.K., Chemiresistive gas sensors based on vanadium pentoxide reinforced polyvinyl alcohol/polypyrrole blend nanocomposites for room temperature LPG sensing. *Synth. Met.*, 273, 116687, 2021.
24. Zhang, Z., Du, H., Yi, W., Sun, Y., Zheng, Y., Wu, Y., Sun, S., Xu, S., Investigation of ammonia-sensing mechanism on polypyrrole gas sensor based on experimental and theoretical evidence. *Sens. Mater.*, 33, 1443–1454, 2021.
25. Michel, C.R., Martínez-Preciado, A.H., Contreras, N.L.L., Gas sensing properties of Nd_2O_3 nanostructured microspheres. *Sens. Actuators B: Chem.*, 184, 8–14, 2013.
26. Michel, C.R. and Martinez-Preciado, A.H., CO sensing properties of novel nanostructured La_2O_3 microspheres. *Sens. Actuators B Chem.*, 208, 355–362, 2015.
27. Bi, H., Zhang, L.-X., Xing, Y., Zhang, P., Chen, J.-J., Yin, J., Bie, L.-J., Morphology-controlled synthesis of CeO_2 nanocrystals and their facet-dependent gas sensing properties. *Sens. Actuators B Chem.*, 330, 129374, 2021.
28. Lei, G., Lou, C., Liu, X., Xie, J., Li, Z., Zheng, W., Zhang, J., Thin films of tungsten oxide materials for advanced gas sensors. *Sens. Actuators B Chem.*, 341, 129996, 2021.
29. Gupta, M., Hawari, H.F., Kumar, P., Burhanudin, Z.A., Tansu, N., Functionalized reduced graphene oxide thin films for ultrahigh CO_2 gas sensing performance at room temperature. *Nanomaterials*, 11, 623, 2021.
30. Yuan, Y., Wu, H., Bu, X., Wu, Q., Wang, X., Han, C., Li, X., Wang, X., Liu, W., Improving ammonia detecting performance of polyaniline decorated rGO composite membrane with GO doping. *Materials*, 14, 2829, 2021.
31. Gohain, M.B., Pawar, R.R., Karki, S., Hazarika, A., Hazarika, S., Ingole, P.G., Development of thin film nanocomposite membrane incorporated with mesoporous synthetic hectorite and MSH@UiO-66-NH_2 nanoparticles for efficient targeted feeds separation, and antibacterial performance. *J. Membr. Sci.*, 609, 118212, 2020.

32. Li, X., Zhang, Y., Cheng, Y., Chen, X., Tan, W., MOF-derived porous hierarchical $ZnCo_2O_4$ microflowers for enhanced performance gas sensor. *Ceram. Int.*, 47, 9214–9224, 2021.
33. Bhimanapati, G.R. *et al.*, Recent advances in two-dimensional materials beyond graphene. *ACS Nano*, 9, 11509–11539, 2015.
34. Ingole, P.G., Jeon, J.D., Hazarika, S., Lee, H.K. Polymeric nanocomposite membranes for diverse applications, in: *Handbook of Polymer Nanocomposites for Industrial Applications, Micro and Nano Technologies*, pp 169–199, 2021.
35. Anasori, B., Lukatskaya, M.R., Gogotsi, Y., 2D metal carbides and nitrides (MXenes) for energy storage. *Nat. Rev. Mater.*, 2, 16098, 2017.
36. Naguib, M., Mochalin, V.N., Barsoum, M.W., Gogotsi, Y., Two-dimensional materials: 25th Anniversary article: MXenes: A new family of two-dimensional materials. *Adv. Mater.*, 26, 992–1005, 2014.
37. Hantanasirisakul, K. and Gogotsi, Y., Electronic and optical properties of 2D transition metal carbides and nitrides (MXenes). *Adv. Mater.*, 30, e1804779, 2018.
38. Khazaei, M., Ranjbar, A., Arai, M., Sasaki, T., Yunoki, S., Electronic properties and applications of MXenes: A theoretical review. *J. Mater. Chem. C*, 5, 2488–2503, 2017.
39. Liu, Y.T., Zhu, X.D., Pan, L., Hybrid architectures based on 2D MXenes and low-dimensional inorganic nanostructures: Methods, synergies, and energy-related applications. *Small*, 14, e1803632, 2018.
40. Nan, J., Guo, X., Xiao, J., Li, X., Chen, W., Wu, W., Liu, H., Wang, Y., Wu, M., Wang, G., Nanoengineering of 2D MXene-based materials for energy storage applications. *Small*, 17, e1902085, 2019.
41. Barsoum, M.W. and El-Raghy, T., The MAX phases: Unique new carbide and nitride materials: Ternary ceramics turn out to be surprisingly soft and machinable, yet also heat-tolerant, strong and lightweight. *Am. Sci.*, 89, 334–343, 2001.
42. Sun, Z.M., Progress in research and development on MAX phases: A family of layered ternary compounds. *Int. Mater. Rev.*, 56, 143–166, 2013.
43. Högberg, H., Hultman, L., Emmerlich, J., Joelsson, T., Eklund, P., Molina-Aldareguia, J.M., Palmquistc, J.-P., Wilhelmsson, U., Growth and characterization of MAX-phase thin films. *Surf. Coat. Technol.*, 193, 6–10, 2005.
44. Naguib, M., Kurtoglu, M., Presser, V., Lu, J., Niu, J., Heon, M., Hultman, L., Gogotsi, Y., Barsoum, M.W., Two-dimensional nanocrystals produced by exfoliation of Ti_3AlC_2. *Adv. Mater.*, 23, 4248–4253, 2011.
45. Wu, X., Ma, P., Sun, Y., Du, F., Song, D., Xu, G., Application of MXene in electrochemical sensors: A review. *Electroanal.*, 33, 1827–1851, 2021.
46. Cao, C., Chang, Q., Qiao, H., Shao, R., Guo, X., Xiao, G., Shi, W., Huang, L., Determination of H plus ion diffusion in Ti_3C_2-rGO glucose sensor. *Sens. Actuators B Chem.*, 340, 129943, 2021.
47. Mei, J., Ayoko, G.A., Hu, C., Bell, J.M., Sun, Z., Two-dimensional fluorine-free mesoporous Mo_2C MXene via UV-induced selective etching of Mo_2Ga_2C for energy storage. *Sustain. Mater. Technol.*, 25, e00156, 2020.
48. Ghasali, E., Orooji, Y., Azarniya, A., Alizadeh, M., Kazem-zad, M., Ebadzadeh, T., Production of V_2C MXene using a repetitive pattern of V_2AlC MAX phase through microwave heating of Al-V_2O_5-C system. *Appl. Surf. Sci.*, 542, 148538, 2021.
49. Zhang, Y., Yu, J., Lu, J., Zhu, C., Qi, D., Facile construction of 2D MXene (Ti_3C_2Tx) based aerogels with effective fire-resistance and electromagnetic interference shielding performance. *J. Alloys Compd.*, 870, 159442, 2021.
50. Huang, J. and Wu, J., Robust and rapid detection of mixed volatile organic compounds in flow through air by a low cost electronic nose. *Chemosensors*, 8, 73, 2020.

51. Qin, Z., Su, T., Ji, H., *MXene-based photocatalysts: Fabrication and applications*, CRC Press, Boca Raton, 2022.
52. Mashtalir, O., *Chemistry of two-dimensional transition metal carbides (MXenes)*, Drexel University, Philadelphia, 2015.
53. Uzun, S., *Multiscale integration of two-dimensional transition metal carbides (MXenes) into textile-based devices*, Drexel University, Philadelphia, 2020.
54. Yang, J., Naguib, M., Ghidiu, M., Pan, L.M., Gu, J., Nanda, J., Halim, J., Gogotsi, Y., Barsoum, M.W., Two dimensional Nb based M_4C_3 solid solutions (MXenes). *J. Am. Ceram. Soc.*, 99, 660–666, 2016.
55. Parajuli, D. and Samatha, K., MXene as topological insulator. *J. Emerg. Technol. Innov. Res.*, 6, 689–706, 2019.
56. Urbankowski, P., Anasori, B., Makaryan, T., Er, D., Kota, S., Walsh, P.L., Zhao, M., Shenoy, V.B., Barsoum, M.W., Gogotsi, Y., Synthesis of two-dimensional titanium nitride Ti_4N_3 (MXene). *Nanoscale*, 8, 11385–11391, 2016.
57. Chaudhari, N.K., Jin, H., Kim, B., San Baek, D., Joo, S.H., Lee, K., MXene: An emerging two-dimensional material for future energy conversion and storage applications. *J. Mater. Chem. A*, 5, 24564–24579, 2017.
58. Alhabeb, M., Maleski, K., Anasori, B., Lelyukh, P., Clark, L., Sin, S., Gogotsi, Y., Guidelines for synthesis and processing of two-dimensional titanium carbide ($Ti_3C_2T_x$ MXene). *Chem. Mater.*, 29, 7633–7644, 2017.
59. Biswas, S. and Alegaonkar, P.S., MXene: Evolutions in chemical synthesis and recent advances in applications. *Surfaces*, 5, 1, 1–34, 2022.
60. Song, D., Jiang, X., Li, Y., Lu, X., Luan, S., Wang, Y., Li, Y., Gao, F., Metal–organic frameworks-derived MnO_2/Mn_3O_4 microcuboids with hierarchically ordered nanosheets and Ti_3C_2 MXene/Au NPs composites for electrochemical pesticide detection. *J. Hazard. Mater.*, 373, 367–376, 2019.
61. Zhang, L., Jiang, Y., Zhou, L., Jiang, Z., Li, L., Che, W., Yu, Y., Mechanical, thermal stability, and flame retardancy performance of transparent wood composite improved with delaminated $Ti_3C_2T_x$ (MXene) nanosheets. *J. Mater. Sci.*, 57, 1–12, 2022.
62. Zhang, J., Seyedin, S., Qin, S., Wang, Z., Moradi, S., Yang, F., Lynch, P.A., Yang, W., Liu, J., Wang, X., Razal, J.M., Highly conductive $Ti_3C_2T_x$ MXene hybrid fibers for flexible and elastic fiber shaped supercapacitors. *Small*, 15, 1804732, 2019.
63. Tang, Y., Xu, Y., Yang, J., Song, Y., Yin, F., Yuan, W., Stretchable and wearable conductometric VOC sensors based on microstructured MXene/polyurethane core-sheath fibers. *Sens. Actuators B: Chem.*, 346, 130500, 2021.
64. Caffrey, N.M., Effect of mixed surface terminations on the structural and electrochemical properties of two-dimensional $Ti_3C_2T_2$ and V_2CT_2 MXenes multilayers. *Nanoscale*, 10, 13520–13530, 2018.
65. Ibragimova, R., Erhart, P., Rinke, P., Komsa, H.P., Surface functionalization of 2D MXenes: Trends in distribution, composition, and electronic properties. *J. Phys. Chem. Lett.*, 12, 2377–2384, 2021.
66. Bayram, V., Unidirectionally freeze-cast MXene aerogels for electrochemical capacitors, Doctoral dissertation, University of Manchester, UK, 2020.
67. Khazaei, M., Arai, M., Sasaki, T., Ranjbar, A., Liang, Y., Yunoki, S., OH-terminated two-dimensional transition metal carbides and nitrides as ultralow work function materials. *Phys. Rev. B*, 92, 075411, 2015.
68. Zhan, X., Si, C., Zhou, J., Sun, Z., MXene and MXene-based composites: Synthesis, properties and environment-related applications. *Nanoscale Horiz.*, 5, 235–258, 2020.

69. Liu, Y., Xiao, H., Goddard III, W.A., Schottky-barrier-free contacts with two-dimensional semiconductors by surface-engineered MXenes. *J. Am. Chem. Soc.*, 138, 15853–15856, 2016.
70. Zeraati, A.S., Mirkhani, S.A., Sun, P., Naguib, M., Braun, P.V., Sundararaj, U., Improved synthesis of $Ti_3C_2T_x$ MXenes resulting in exceptional electrical conductivity, high synthesis yield, and enhanced capacitance. *Nanoscale*, 13, 3572–3580, 2021.
71. Shukla, V., The tunable electric and magnetic properties of 2D MXenes and their potential applications. *Mater. Adv.*, 1, 3104–3121, 2020.
72. Kumar, H., Frey, N.C., Dong, L., Anasori, B., Gogotsi, Y., Shenoy, V.B., Tunable magnetism and transport properties in nitride MXenes. *ACS Nano*, 11, 7648–7655, 2017.
73. Jiang, X., Kuklin, A.V., Baev, A., Ge, Y., Ågren, H., Zhang, H., Prasad, P.N., Two-dimensional MXenes: From morphological to optical, electric, and magnetic properties and applications. *Phys. Rep.*, 848, 1–58, 2020.
74. Wang, G., Theoretical prediction of the intrinsic half-metallicity in surface-oxygen-passivated Cr_2N MXene. *J. Phys. Chem. C*, 120, 18850–18857, 2016.
75. Wang, F., Yang, C., Duan, M., Tang, Y., Zhu, J., TiO_2 nanoparticle modified organ-like Ti_3C_2 MXene nanocomposite encapsulating hemoglobin for a mediator-free biosensor with excellent performances. *Biosens. Bioelectron.*, 74, 1022–1028, 2015.
76. Khan, R. and Andreescu, S., Mxenes-based bioanalytical sensors: Design, characterization, and applications. *Sensors*, 20, 5434, 2020.
77. Yu, X.F., Li, Y.C., Cheng, J.B., Liu, Z.B., Li, Q.Z., Li, W.Z., Yang, X., Xiao, B., Monolayer Ti_2CO_2: A promising candidate for NH_3 sensor or capturer with high sensitivity and selectivity. *ACS Appl. Mater. Interfaces*, 7, 24, 13707–13713, 2015.
78. Bai, H. and Shi, G., Gas sensors based on conducting polymers. *Sensors*, 7, 267–307, 2007.
79. Kuang, D., Wang, L., Guo, X., She, Y., Du, B., Liang, C., Qu, W., Sun, X., Wu, Z., Hu, W., He, Y., Facile hydrothermal synthesis of Ti_3C_2Tx-TiO_2 nanocomposites for gaseous volatile organic compounds detection at room temperature. *J. Hazard. Mater.*, 416, 126171, 2021.
80. Chen, W.Y., Jiang, X., Lai, S.N., Peroulis, D., Stanciu, L., Nanohybrids of a MXene and transition metal dichalcogenide for selective detection of volatile organic compounds. *Nat. Commun.*, 11, 1–10, 2020.
81. Chen, W.Y., Lai, S.N., Yen, C.C., Jiang, X., Peroulis, D., Stanciu, L.A., Surface functionalization of $Ti_3C_2T_x$ MXene with highly reliable superhydrophobic protection for volatile organic compounds sensing. *ACS Nano*, 14, 11490–11501, 2020.
82. Jin, L., Wu, C., Wei, K., He, L., Gao, H., Zhang, H., Zhang, K., Asiri, A.M., Alamry, K.A., Yang, L., Chu, X., Polymeric $Ti_3C_2T_x$ MXene composites for room temperature ammonia sensing. *ACS Appl. Nano Mater.*, 3, 12071–12079, 2020.
83. Wu, S., Weng, Z., Liu, X., Yeung, K.W.K., Chu, P.K., Functionalized TiO_2 based nanomaterials for biomedical applications. *Adv. Funct. Mater.*, 24, 5464–5481, 2014.
84. Gong, K., Zhou, K., Qian, X., Shi, C., Yu, B., MXene as emerging nanofillers for high-performance polymer composites: A review. *Compos. Part B Eng.*, 217, 108867, 2021.
85. Pantoja Suárez, L.F., *Carbon nanotubes grown on stainless steel for supercapacitor applications*, Universitat de Barcelona, 2019.
86. Peng, J., Chen, X., Ong, W.J., Zhao, X., Li, N., Surface and heterointerface engineering of 2D MXenes and their nanocomposites: Insights into electro-and photocatalysis. *Chem*, 5, 18–50, 2019.
87. Abazari, R., Mahjoub, A.R., Sanati, S., Rezvani, Z., Hou, Z., Dai, H., Ni–Ti layered double hydroxide@ graphitic carbon nitride nanosheet: A novel nanocomposite with high and ultra-fast sonophotocatalytic performance for degradation of antibiotics. *Inorg. Chem.*, 58, 1834–1849, 2019.

88. Ma, T.Y., Cao, J.L., Jaroniec, M., Qiao, S.Z., Interacting carbon nitride and titanium carbide nanosheets for high performance oxygen evolution. *Angew. Chem.*, 128, 1150–1154, 2016.
89. Qiu, T., Liang, Z., Guo, W., Tabassum, H., Gao, S., Zou, R., Metal–organic framework-based materials for energy conversion and storage. *ACS Energy Lett.*, 5, 520–532, 2020.
90. Bouali, A.C., Serdechnova, M., Blawert, C., Tedim, J., Ferreira, M.G.S., Zheludkevich, M.L., Layered double hydroxides (LDHs) as functional materials for the corrosion protection of aluminum alloys: A review. *Appl. Mater. Today*, 21, 100857, 2020.
91. Chia, X. and Pumera, M., Characteristics and performance of two-dimensional materials for electrocatalysis. *Nat. Catal.*, 1, 909–921, 2018.
92. Zeng, M. and Li, Y., Recent advances in heterogeneous electrocatalysts for the hydrogen evolution reaction. *J. Mater. Chem. A*, 3, 14942–14962, 2015.
93. Huang, J. and Chen, R., An overview of recent advances on the synthesis and biological activity of α-aminophosphonic acid derivatives. *Heteroat. Chem.: An International Journal of Main Group Elements*, 11, 480–492, 2000.
94. Li, J., Yan, D., Hou, S., Li, Y., Lu, T., Yao, Y., Pan, L., Improved sodium-ion storage performance of $Ti_3C_2T_x$ MXenes by sulfur doping. *J. Mater. Chem. A*, 6, 1234–1243, 2018.
95. Peera, S.G., Liu, C., Sahu, A.K., Selvaraj, M., Rao, M.C., Lee, T.G., Koutavarapu, R., Shim, J., Singh, L., Recent advances on MXene based electrocatalysts toward oxygen reduction reaction: A focused review. *Adv. Mater. Interfaces*, 8, 2100975, 2021.
96. Pan, H., Principles on design and fabrication of nanomaterials as photocatalysts for water-splitting. *Renew. Sust. Energ. Rev.*, 57, 584–601, 2016.
97. Zhu, X.T., Xu, Y., Cao, Y., Zhao, Y.Q., Sheng, W., Nie, G.Z., Ao, Z., Investigation of the electronic structure of two-dimensional GaN/Zr_2CO_2 hetero-junction: Type-II band alignment with tunable bandgap. *Appl. Surf. Sci.*, 542, 148505, 2021.
98. An, X., Wang, W., Wang, J., Duan, H., Shi, J., Yu, X., The synergetic affects of Ti_3C_2 MXene and Pt as co-catalysts for highly efficient photocatalytic hydrogen evolution over gC_3N_4. *Phys. Chem. Chem. Phys.*, 20, 16, 11405–11411, 2018.
99. Feng, X., Yu, Z., Sun, Y., Shan, M., Long, R., Li, X., 3D MXene/Ag_2S material as Schottky junction catalyst with stable and enhanced photocatalytic activity and photocorrosion resistance. *Sep. Purif. Technol.*, 266, 118606, 2021.
100. Crini, G. and Lichtfouse, E., Wastewater treatment: An overview, in: *Green Adsorbents for Pollutant Removal*, pp. 1–21, 2018.
101. Soliman, N.K. and Moustafa, A.F., Industrial solid waste for heavy metals adsorption features and challenges: A review. *J. Mater. Res. Technol.*, 9, 10235–10253, 2020.
102. Ying, Y., Liu, Y., Wang, X., Mao, Y., Cao, W., Hu, P., Peng, X., Two-dimensional titanium carbide for efficiently reductive removal of highly toxic chromium (VI) from water. *ACS Appl. Mater. Interfaces*, 7, 1795–1803, 2015.
103. Yang, J., Hou, B., Wang, J., Tian, B., Bi, J., Wang, N., Li, X., Huang, X., Nanomaterials for the removal of heavy metals from wastewater. *Nanomaterials*, 9, 424, 2019.
104. Zhu, J., Ha, E., Zhao, G., Zhou, Y., Huang, D., Yue, G., Hu, L., Sun, N., Wang, Y., Lee, L.Y.S., Xu, C., Recent advance in MXenes: A promising 2D material for catalysis, sensor and chemical adsorption. *Coord. Chem. Rev.*, 352, 306–327, 2017.
105. Begum, S., Pramanik, A., Davis, D., Patibandla, S., Gates, K., Gao, Y., Ray, P.C., 2D and hetero-structure nanomaterial based strategies for combating drug-resistant bacteria. *ACS Omega*, 5, 3116–3130, 2020.

106. Lim, G.P., Soon, C.F., Morsin, M., Ahmad, M.K., Nayan, N., Tee, K.S., Synthesis, characterization and antifungal property of $Ti_3C_2T_x$ MXene nanosheets. *Ceram. Int.*, 46, 20306–20312, 2020.
107. Zhang, K., Li, D., Cao, H., Zhu, Q., Trapalis, C., Zhu, P., Gao, X., Wang, C., Insights into different dimensional MXenes for photocatalysis. *Chem. Eng. J.*, 424, 130340, 2021.

12

MXene-Reinforced Polymer Composites as Flexible Wearable Sensors

J. Aarthi[1,2], K. Selvaraju[1*], S. Gowri[2], K. Kirubavathi[1] and Ananthakumar Ramadoss[3†]

[1]*PG & Research Department of Physics, Government Arts College (Affiliated to Bharathidasan University), Ariyalur, India*

[2]*PG & Research Department of Physics, Cauvery College for Women (Autonomous), (Affiliated to Bharathidasan University), Tiruchirappalli, Tamil Nadu, India*

[3]*School for Advanced Research in Petrochemicals (SARP): Advanced Research School for Technology and Product Simulation (ARSTPS), Central Institute of Petrochemicals Engineering and Technology (CIPET), Chennai, India*

Abstract

As a result of the COVID-19 pandemic, the healthcare system has undergone a transformation in which patients must be regularly observed without the requirement for hospitalisation. The advancement of wearable sensors has helped accelerate the revolution in personalised healthcare with their ability to accurately sense different body parameters. MXenes, a unique category of two-dimensional materials have risen to prominence as an option for wearable applications in recent years due to their good mechanical properties, hydrophilicity, surface functional groups, electrical conductivity, large and tunable interlayer spacing, and greater specific surface area. However, the limitations of easy stacking, limited flexibility and poor stability in an oxygen environment impede the continued development of pure MXenes. The advantages of polymers which include low cost, easy fabrication, and tunable functionality have made them widely used in the advancement of MXene-based composites as a wearable sensor. A combination of polymer flexibility and MXene sensing ability can be utilized to improve the sensing operation of flexible sensors. This chapter primarily describes the structure, synthesis techniques, characterisitics of MXenes-reinforced polymer composite materials as physical sensors, including pressure and strain sensors for applications in monitoring the human health.

Keywords: Flexible, wearable, hydrophobic, electronic skin, health monitoring, hydrophilicity

12.1 Introduction

The transformation of the global healthcare system will result in the accelerated development of flexible and wearable sensors for personalised healthcare, remote patient observing,

**Corresponding author*: kselvaraju17@gmail.com

†*Corresponding author*: ananth.cipet@gmail.com; ananth@cipet.gov.in

clinical treatment and healthcare internet of things [1, 2]. Wearable sensors are important in the development of personalised medicine because they can always capture information from the human body and record relevant health state fluctuations in time for precautionary action [3]. Various flexible and wearable sensors with excellent mechanical compliance and unprecedented sensitivity in mimicking human skin's sensory function have been developed by converting numerous physical and chemical subtle changes such as humidity, temperature, strain, pressure, and shear force into electrical signals such as capacitance, voltage, current, and resistance [4–8]. Sensing materials for wearable applications ought to be biocompatible, flexible, conductive, electrochemically active, high-sensitivity, and cost-effective [9, 10].

Two-dimensional (2D) materials are renowned as active sensing materials owing to their distinctive physicochemical properties. 2D materials like transition metal dichalcogenides (TMDs), hexagonal boron nitride (h-BN), black phosphorus (BP), MXenes, and graphene have received considerable interest in a variety of domains, most notably detection and sensing methodologies [11, 12]. At first, 2D materials are preferably single-layer materials with a few nanometers in thickness, resulting in a remarkably high surface-area-to-volume ratio over 0D, 1D and 3 D [13]. Consequently, it offers more reactive spots between the material and the analytes making 2D materials more appropriate for sensor applications [14]. These sensors can detect analytes at low concentrations. Also, 2D materials exhibit a broad range of conductivity. Reducing structural imperfections, doping, expanding layers, and post-functionalizing may all be used to control the conductivity of 2D materials [15]. Furthermore, the morphological features of 2D materials enhance their electrical properties and make them more efficient for signal transduction. 2D materials have good compatibility, device miniaturization, exceptional optical transparency and mechanical strength, which are lacking in 0D, 1D and 3D materials [16].

New classes of 2D materials have emerged in recent years that could be used to make wearables, known as MXenes. MXenes are more effective for sensors than typical 2D materials like MoS_2 and graphene with their difficult functionalization, low hydrophilicity, and complex synthesis conditions and procedures [17]. MXenes are 2D layered materials comprised of transition metal nitrides, carbides or carbonitrides. A layer of metal carbide/nitride or carbonitride called MAX is etched selectively to produce MXene which has the general formula $M_{X+1}X_n$ (n = 1, 2, 3). Here, M represents a transition metal, A symbolises an element from groups 13 and 14 and X denotes carbon/nitrogen. Mxene an $M_{X+1}AX_n$ structure is formed by chemically and selectively etching the A layer. Surface moieties of oxygen, hydrogen and fluorine terminate the MXene layer during the synthesis. MXenes are thus denoted as $M_{X+1}X_nT_x$, where T_x denotes the surface functional group. The type and intensity of the surface functional units in MXenes determine its performance in sensor applications [18, 19].

The surface functional units linked to the surface of MXenes substantially impact their electrical conductivity properties and hydrophilicity. The presence of functional groups increased their hydrophilicity, enabling them to be readily dispersed in the aqueous medium without any need for surfactants or additives [20, 21]. As a result, MXenes facilitate processing functional inks for various thin film methods like casting techniques,coating and printing to produce wearable devices. Furthermore, the surface moieties strongly relies on the electronic properties of MXene. The band gap is initiated by modifying the surface moieties and may offset the density states at the fermi level, impacting electrical conductivity and band structure. Thus, their excellent electrical conductivity and hydrophilicity make them useful in sensing. MXenes has a substantial and variable layer spacing, surface functional units, excellent biocompatibility, easy surface functionalization and along with the benefits of conventional 2D materials such as large conductivity, greater mechanical characteristics and a large specific surface area. As a result, MXene's application in various sensors like gas, temperature, strain, pressure, optical/photoelectric, and electrochemical biosensors has been extensively explored [22–26].

Several factors inhibit the development of pristine MXenes, such as poor stability in the oxygen atmosphere, low flexibility, and easy restacking [27]. The intercalation of polymer chains will prevent MXene flakes from restacking and enhance their stability by preventing oxygen contact [28]. As well, the inclusion of polymers into MXenes may also increase mechanical robustness and sensing range [29]. Polymers offer superior impact and tensile strengths, as well as fatigue, abrasion, corrosion, fracture resistance, and other bulk qualities. Their solubility in organic solvents improves their compatibility with nanoparticles and makes their insertion into 2D material systems easier. Furthermore, 2D MXenes are reinforced with polymers, which are frequently utilized to increase the operation of sensing in flexible sensors owing to the combined effect of polymer flexibility and MXenes sensing capability.

A brief synopsis of current developments in the preparation, processing, and applications of MXene and its reinforced polymer derivatives in flexible wearable electronics is presented in this chapter. The human health indications of interest include pulse, respiration and strain caused by muscle actions. The majority of the MXene-based sensors are designed to respond to the mechanical distortion produced by strain and pressure. To date various physical sensors, including strain and pressure sensors, were extensively developed for human health observation. Flexible pressure/strain sensors utilise piezoresistive, capacitive and piezoelectric sensing mechanisms. The chapter focus on the MXenes flexible pressure/strain sensors is to improve sensing performance by material modification and structural design with polymer reinforcement as well as practical applications such as healthcare monitoring, motion tracking and electronic skin.

12.2 Performance Parameter for Flexible Pressure and Strain Sensor

The performance of pressure and strain sensors is evaluated using a variety of parameters like sensitivity, hysteresis, stretchability, response, recovery time, durability and range. Here are some commonly used flexible sensor performance parameters, which can be used to make a preliminary assessment of an individual sensor's performance discussed.

12.2.1 Sensitivity

Sensitivity is a critical parameter that directly impacts a sensor's performance. The capacity of sensor to change external stimuli into electrical signals is called as sensitivity. In order to evaluate sensor performance, a graph is often drawn between the input values of pressure/strain and the electrical output signal. For the sensitivity of pressure/strain sensors, the sensitivity is measured in kPa^{-1} when the imposed stimulus is pressure, whereas the sensitivity is measured in gauge factor (GF) and is dimensionless when the imposed stimulus is strain. The proper choice of MXene synthesis technique, sensor design, and detecting mechanism is critical for obtaining high sensitivity in wearable observations [1].

12.2.2 Stretchability

A sensor with great stretchability is essential for providing health monitoring applications with the ability to continuously detect human health signals without causing discomfort or invasiveness. Stretchability is determined as the highest value of strain/pressure that strain/pressure sensor can withstand while maintaining a stable response. Stretchability is determined by the type of substrate or active material, their interaction, method of fabrication and the aspect of the filler materials. The material added with elastomer [30], use of material as thin film [31], and the porous structure [32] are the ways of achieving stretchability.

12.2.3 Hysteresis

The hysteresis of strain sensors is particularly important in wearable purposes where strain sensors are often subjected to dynamic loading. Under large strain circumstances, the electrical properties of the flexible sensors cannot be recovered to previous value in time. The area difference between the loading and releasing curves induced by energy dissipation during stretching/compression is used to determine the amount of hysteresis. Compared to capacitive sensors, a flexible Resistive sensors are more susceptible to hysteresis [33]. Capacitive strain sensors have a stable overlapping dielectric layer between electrodes because of the viscoelasticity of the polymer, while resistive strain sensors have a connection-disconnection structure. Strain sensor with low or quantified hysteresis may be ignored or accounted in the output response for practical wearable device applications [34].

12.2.4 Durability and Range

Durability is an essential factor to consider when evaluating the protracted performance of flexible sensors. Wearable sensors will be exposed to numerous situations like bending, buckling and stretching which may cause sensor layers to break or detach. The sensor may have mechanical wear and tear along with sweat and distinct climate conditions may occur throughout its usage. Introducing superhydrophobic properties on the material will increase environmental adaptability, durability, corrosion resistance and liquid resistance [35, 36].

The highest and least value of the factor that the sensors can measure is defined as the working range. There is an inverse relationship between sensitivity and range which means that a highly sensitive sensor will possess a restricted detecting range. Furthermore, sensors with a broad detecting range have limited sensitivity. Recent study has focused on obtaining high sensitivity over a broad sensing scale. Human movement includes both minor and massive joint motions. As a result, it is envisaged that the sensor would notice both movements with great sensitivity. The design of flexible sensors can benefit from structural changes that help achieve a wide detection range.

12.3 Design of MXenes/Polymer Composites as Flexible Pressure Sensors

MXenes or MXenes polymer composites are extensively employed as wearable sensors for human-machine interfaces, monitoring human health, and electronic skin. A flexible wearable pressure sensor can be made from a variety of complexes of MXenes and Polymers can be fabricated into fibre, sponge, foam, composite aerogel, flexible polymer composite, and other types. According to the criteria of flexible sensors, polymer materials offer flexibility and stretchability. The flexibility of pure MXene films and MXene-based composites is limited. Furthermore, the hydrophilic characteristic of MXene hybrid adsorbs gases and humidity from the atmosphere, affecting the pressure sensor's reliability and stability. The spontaneous oxidation of MXenes reduces device stability by lowering MXene conductivity. The pressure sensor combined with organic polymers and inorganic MXenes is viewed as a viable solution to the aforementioned issues. The section discusses the structure, fabrication processes, and associated performance of the MXenes-reinforced polymer flexible sensor. Also, the performance of various sensors for practical applications is evaluated using various parameters such as sensitivity, stretchability, range, recovery/response time, hydrophilicity/hydrophobic surface, air permeability, environmental adaptability, electrical conductivity and stability.

A high-performance piezoresistive pressure sensor of micro-protrusion-structured MXene/Polydimethysiloxane (MXene/PDMS) was fabricated using sandpaper as a template [37]. In addition to the appropriate polymer template and conductive substances, proper microstructure morphology is assumed to be critical for achieving overall sensing performance. As a result of the micro-protrusions morphological structure, the sensor has demonstrated a 2.6 kPa^{-1} of sensitivity with 0–30 kPa of pressure range and 40 ms of response and recovery time. The sandpaper with a mesh of #280 for the upper layer of the sensor and the lower layer with the meshes of # 280, #400, #600, #800, #1000 and #1500 was used to obtain seven sensors of MXene/PDMS film, such as 280/P, 280/280, 280/400, 280/600, 280/800, 280/1000, and 280/1500, to study the electromechanical performance. Even at low pressure, the sensors 280/P, 280/280, 280/400, 280/600, and 280/800 exhibited greater sensitivity, but their sensing capability is lost at pressures greater than 2 kPa. The sensor 280/1500 has shown high sensitivity over a wide detection range, but its sensitivity was reduced when compared to the sensor 280/1000, which was chosen for testing applications. To understand sensing mechanisms the structural changes of the 280/400, 280/1000, and 280/P sensors under different pressure is illustrated in Figure 12.1 (a-c). A low pressure causes the layers to deform elastically, which increases the contacted area and provides additional conductive pathways causing a reduction in electrical resistance. Thus, the sensors 280/400 and 280/P exhibit low sensitivity for a wide range of pressure. When a low-height protuberance on MXenes/PDMS 280 comes into contact with multiple small protuberances on MXenes/PDMS 1000 at high pressure, an electrical resistance decreases, resulting in higher conductive routes. The electrical resistance decreases continuously as the pressure rises across a wide range, reflecting the 280/1000 sensor's great sensitivity and broad range of detection. Limb movement, Heart rate, recognition of handwriting and monitoring vocal cord of human motion detected by the sensor. The observation could pave the way for fabricating flexible piezoresistive sensors for wearables, electronic skin and health monitoring.

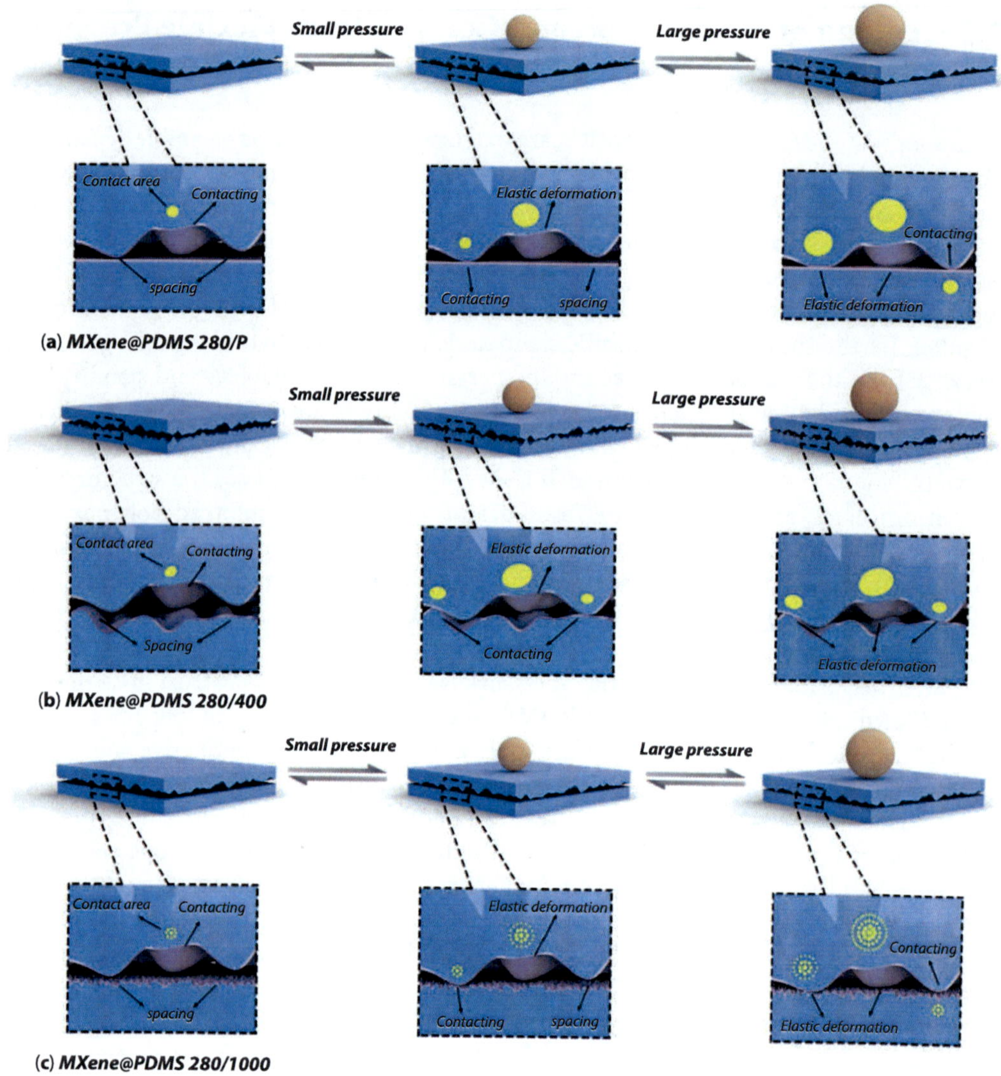

Figure 12.1 Structural and sensing mechanism of MXene@PDMS 280/P (a), MXene@PDMS 280/400 (b) and MXene@PDMS 280/1000 sensor (c) [37]. Copyright 2022. Reproduced with permission from Elsevier.

The material's air permeability and environmental adaptability are important factors in ensuring the comfort of wearable electronics. The sensors lack of air permeability makes them unsuitable for long-term human body wear. The sensor on cotton fabric was designed to address the aforementioned shortcomings. However due to its hydrophilicity, cotton fabric is highly susceptible to chemical degradation caused by human perspiration, impeding the fabrication of sensors on fabric in wearables. Wearable superhydrophobic PPy/MXene pressure sensors were developed with corrosion and liquid resistance [38]. Also, incorporating superhydrophobic properties into materials improves the material's environmental adaptability. A superhydrophobic PPy/MXene sensor is created using a simple chemical oxidation method as well as active materials PPy, MXene, and cotton fabric as a base is shown in Figure 12.2A. The wettability of the fabric served as a defining characteristics of the superhydrophobic properties to explore the consequences of diverse conditions. Figure 12.2B (a-b). shows the wettability test confirmed that the surface has superhydrophobicity for the three pH values of 1, 7, and 13 and has excellent waterproof performance. A study of mechanical durability in the underwater ultrasonic environment showed that superhydrophobic PPy/MXene cotton fabrics decreased slightly until 144° after 5 h, but remained hydrophobic shown in Figure 12.2B (c). Furthermore, the sensor's properties such as low water adhesion, good air permeability, antifouling, self-cleaning, good electrical conductivity and stability were investigated experimentally, as illustrated in Figure 12.2C (a-c). The sensor's electromechanical sensing performance was assessed in dry, wet, acidic and alkaline environments and it was found that it preserved its performance under wet and corrosive conditions, shown in Figure 12.2D (a). Response and recovery time is a key metric to assess the performance of sensing and was observed at 100 and 80 ms. As a result, the sensor is extremely sensitive to pressure changes, which is advantageous for the design of real-time human motion detection materials. The pressure sensor of superhydrophobic PPy/MXene has a detection range of 0–80 kPa, sensitivity of (S_1 = – 20 kPa^{-1} at 0 –2 kPa), quick response and recovery time and favourable cycling stability (>1000 cycles). These findings perfectly integrate the exceptional performance in the applications of water-resistant smart electronic devices, human body movement, and data communication.

A flexible piezoresistive pressure sensor MXene/Polyvinyl alcohol (PVA) with high sensitivity (2320.9 k Pa^{-1}), minimal detection limit (6 Pa), rapid response time (70 ms), and favourable cycle stability (<10000 cycles) was developed [39]. MXene's sensitive layer was made of filter paper, which was chosen for its improved sensitivity due to its surface concave-convex structure and internal porous structure, as displayed in Figure 12.3 (A) (a-b). The active sensing layer's microstructure, interface contact resistance and contact surface roughness may all be modified to meet the sensor's various sensitivity requirements. In a composite film made of MXene and PVA, the resistivity quickly improved with a rise in PVA substance, suggesting a film with adjustable resistivity. Thus, the amount of PVA

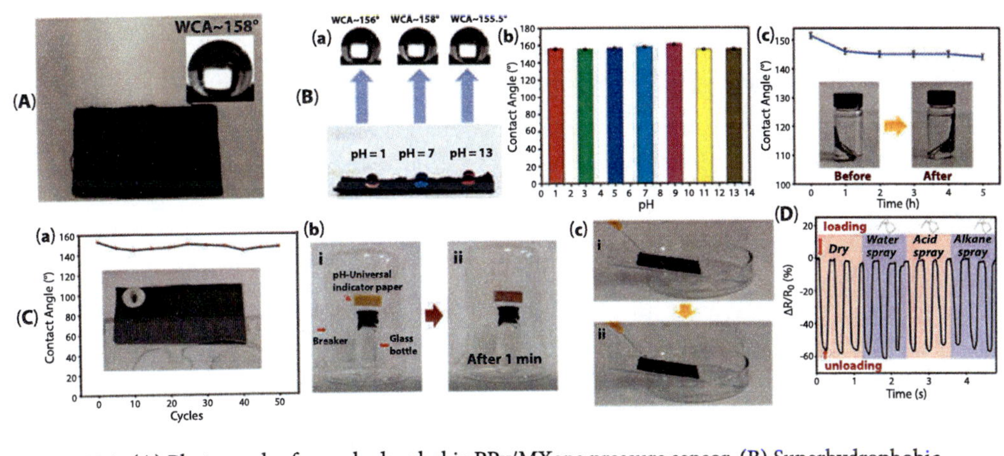

Figure 12.2 (A) Photograph of superhydrophobic PPy/MXene pressure sensor, (B) Superhydrophobic property Testing (a-c), (C) Performance of Abrasion testing (a), air permeability testing (b), self-cleaning testing (c), (D) Assessment of sensing performance in dry, wet, acidic and alkaline environments (a) [38]. Copyright 2022. Reproduced permission from Elsevier.

polymer added to the active layer of MXene is important to achieve adjustable resistivity and higher sensitivity. PVA displays a steady waveform for the current-time (I-T) curve in the pressure range from 65.3 Pa to 294 kPa of the MXene with 0.025 g of composition indicating improved sensing range than sensors using pure MXene and 0.015 g, 0.05 g, 0.035 g PVA of MXene composite film. The current-time (I-T) curve for the MXene composite film with various amounts of PVA addition (0.015g PVA, 0.035 g PVA, 0.025g PVA) implies that the MXene composite film resistance should be appropriate, that is neither too high nor too low which limits the application range. The sensor has a sensitivity in the order of 10^3, which is superior than that of conventional flexible pressure sensors with MXene. The sensor could be used to track changes for the physiological signals of the wrist and neck over time. Figure 12.3(B) (a-d) illustrates the characteristic peaks of the human pulse waveform, viz. the P (percussion), T (tidal), and D (diastolic) waves, which provide the critical evidence for remote monitoring of cardiovascular diseases with a mobile phone. Thus, the findings show that the sensor can identify low pressure and corroborate the excellent sensing response.

For large-scale sensor fabrication, a flexible sensor made of polymer sponge/foams (PU, melamine or cellulose) with minimal density, excellent elasticity, and low cost is preferred. In this study, the polymer sponge was prepared by *in-situ* polymerization. A flexible pressure sensor of 3D microporous MXene/PANI functionalized on a melamine sponge was developed [40] by freeze-drying-assisted depositing, as shown in Figure 12.4A. Flexible sensors must possess good flexibility and compressibility so they can bend/straighten or compress and release upon stimulation without being destroyed. MXene/PANI sponge has good compressibility and structural stability. The MXene/PANI sponge has a sensitivity of 0.08232 kPa^{-1} in the low-pressure range and a greater value 0.3106 kPa^{-1} in the high-pressure range. The response and recovery times of 80 and 130 ms are significantly greater than

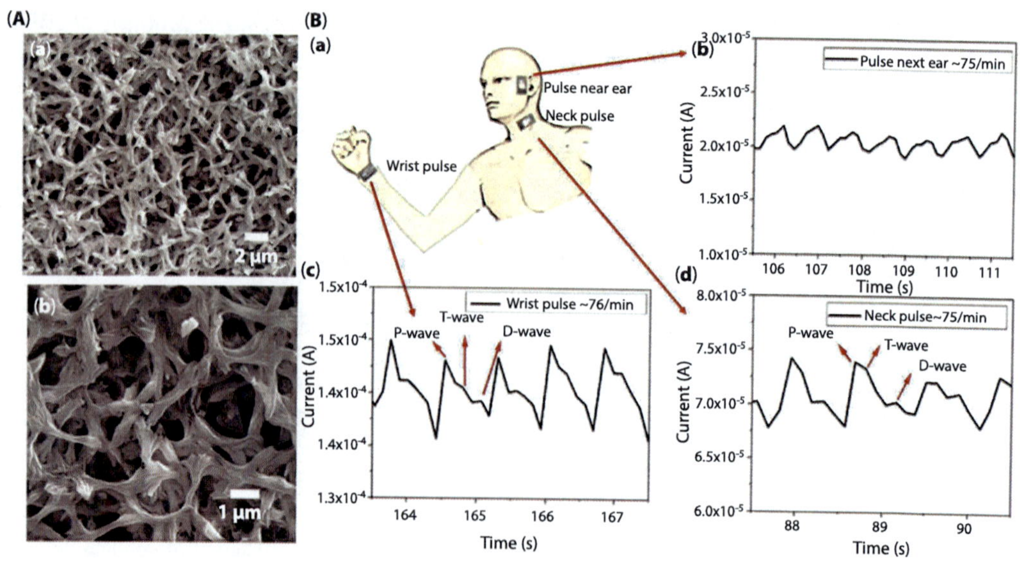

Figure 12.3 (A) Structure of filter paper (a-b), (B) Characteristic peaks can be detected for monitoring cardiovascular diseases (a-d) [39]. Copyright 2022. Reproduced with permission from Elsevier.

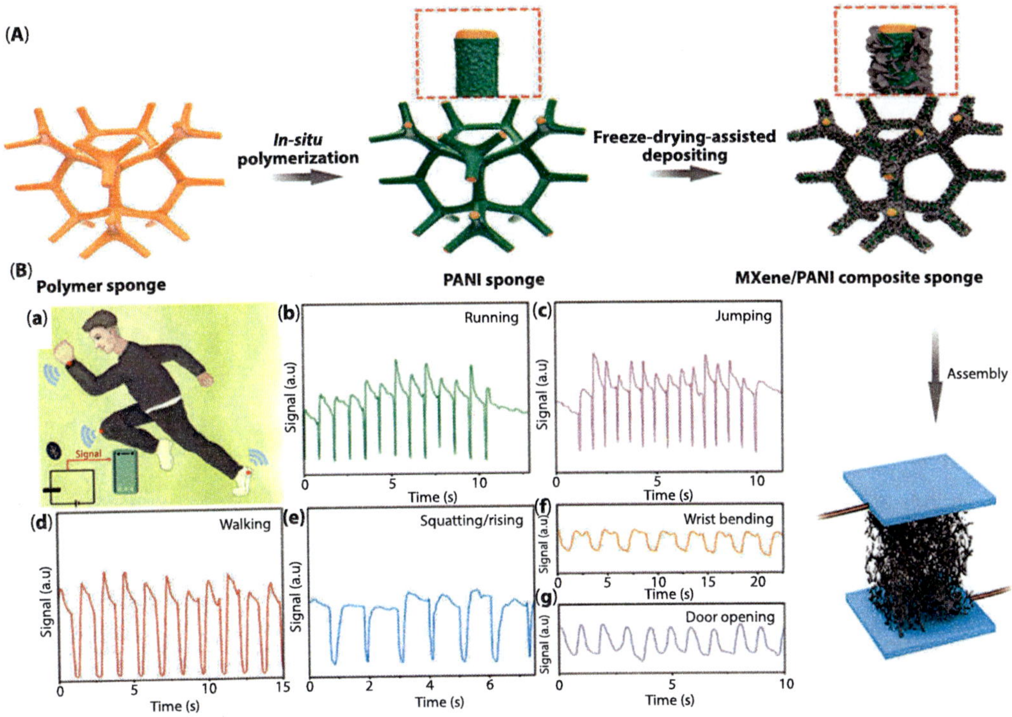

Figure 12.4 (A) Fabrication 3D microporous MXene/PANI functionalized on melamine sponge, (B) MXene/PANI Sponge sensor of wireless real-time human motion recognition (a-g) [40]. Copyright 2021. Reproduced with permission from Elsevier.

the bare PANI sponge's 300/400 ms. It also demonstrates good stability over 5000 cycles of compressing/releasing under 11 kPa corroborating the stability, durability and exceptional sensing performance of the sensor. The characteristics suggest potential applications for wireless real-time human activity recognition, is shown in Figure 12.4(B) (a-f). Thus, the MXene/PANI pressure sensor finds use in wireless wearable devices.

A hydrophobic piezoelectric pressure sensor with exceptional durability and stability of MXene/Poly(vinylidene fluoride) Trifluoroethylene (P VDF-TrFE) hybrid film [41] was assembled using spin-coating is demonstrated in Figure 12.5A (a-b). The wettability test proved the hybrid film's hydrophobic characteristic, as illustrated in Figure 12.5A (c) with a greater water contact angle of 90° than pure MXene, which has a contact angle of 55°. The sensor has a superior GF of 817.4 kPa^{-1} and 2213.68 kPa^{-1} in the ranges from 0.072 to 0.74 kPa and 0.74 to 3.083 kPa, a quick response time and a recovery time of 16 ms. The high level of responsiveness (99%) preserved in ambient air for 20 weeks affirmed the sensor's stability, as evidenced by the similarity in XRD patterns of hybrid films prior to and after 20 weeks of contact to ambient air, which reaffirmed the sensor's stability over 20 weeks. The incorporation of elastomer (P (VDF-TrFE) improves the sensor's stretchability. Figure 12.5B (a-d) depicts the investigation of practical use of pressure sensors in wearable electronics, speech recognition and human activity. The device monitors the real-time current response and waveform in response to the periodic change in cheek and knee movements, indicating that it is an excellent health monitoring device. Further exploration of sensor applications

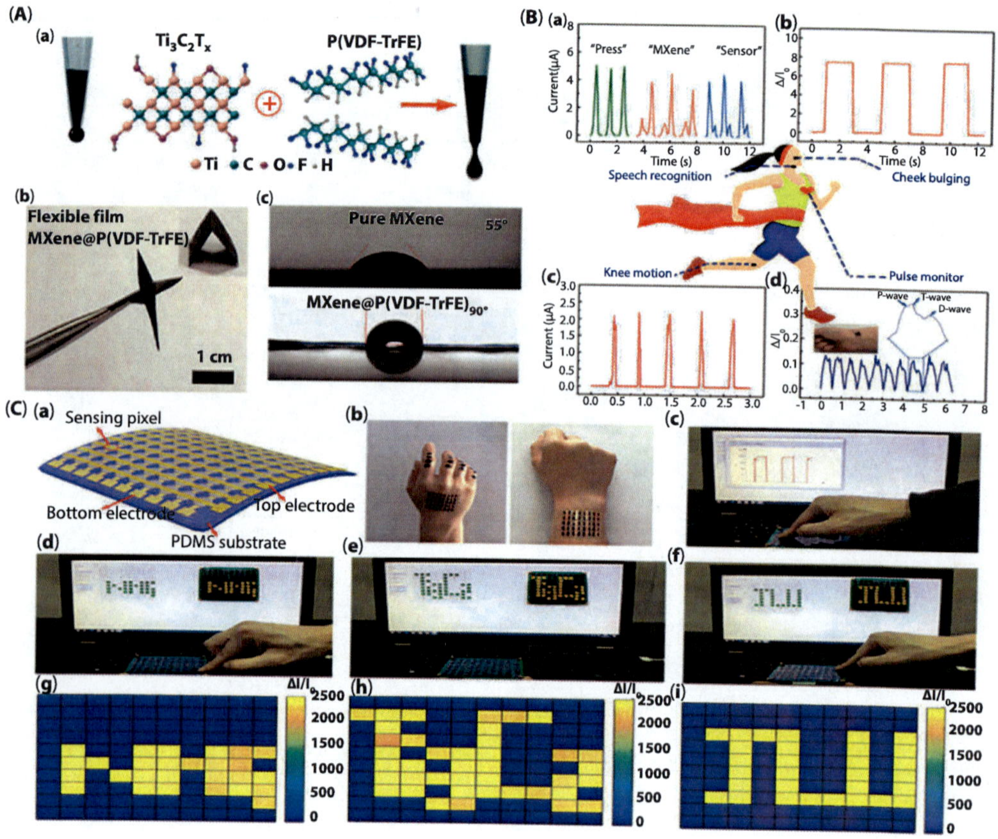

Figure 12.5 (A) Schematic fabrication process of flexible (P VDF-TrFE) hybrid film (a-b), wetting property (c), (B) Speech recognition and human motion monitoring (a-d), (C) 10 × 10 sensor array and affixed to the top of a user's hand and forearm (a-b), Real-time current variation with pressure (c) Real-time pressure distribution in response to a finger with word of NMG, Ti_3C_2, and JLU motions, respectively (d–i). Adapted from Ref [41]. Copyright 2020, American Chemical Society.

led to the advancement of e-skin attached to the human body that is extremely flexible and mechanically stable. Figure 12.5C (a-i) shows a sensor array of pixels with a 10x10 matrix was designed to monitor the current change at the same time as pressure is applied, the response of pressure movement with finger movements. The spin-coating method was used to create a piezoresistive organic-inorganic hybrid sensor made of reduced graphene oxide (rGO), MXene, and P(VDF-TrFE) [42]. The sensor exhibits a great sensitivity of 2298 kPa^{-1} due to the micropores in the organic polymer. The sensor is hydrophobic with a contact angle of 95° which is significant for the wear resistant of pressure sensor. The composite has good stability after being exposed to the air for 16 weeks because the encapsulation of the organic polymer network preserves MXene and rGO from oxidation. It has a fast response/recovery time of 10 ms/9ms, stability after 10000 cycles of testing and a detection limit of 8 Pa, all of which are significant for the practical application of monitoring health activities.

The flexible pressure sensor MXene/Poly (vinylidene fluoride) (PVDF) hybrid film [43] was developed in three steps: electrospinning, hot-pressing, and rolling as illustrated in Figure 12.6 (a-b). Unlike multilayered MXene, the nanosheet forms have more exposed

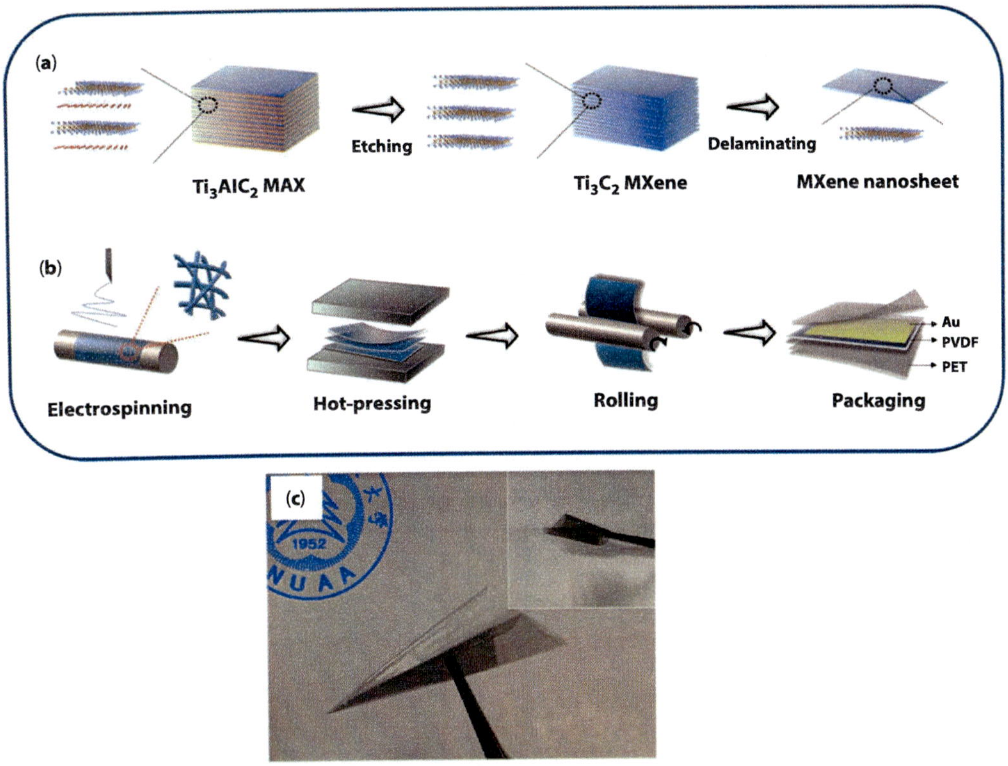

Figure 12.6 (a, b) Schematic design of three-step development of MXene PVDF sensor, (c) Picture of flexible MXene/PVDF hybrid film sensor [43]. Copyright 2021. Reproduced with permission from Elsevier.

surface atoms, enabling larger contact surfaces for the insertion of molecules and charges. MXene nanosheets can be used as conductive fillers. The resultant MXene/PVDF film is shown in Figure 12.6 (c) may preserve PVDF flexibility by allowing it to be folded or rolled. The good homogeneity of the film is attributed to the molecular interaction between OH groups (MXene nanosheets) and C-F (PVDF). High-polarity materials, which are responsible for piezoelectricity, are produced by optimising the crystalline forms of PVDF into β phase. The mechanical characteristics of hybrid films with high tensile strengths are greatly influenced by MXene nanosheet formation and uniform dispersion. Furthermore, the conductive nanosheet ensures a high dielectric constant, superior ferroelectricity and enhanced piezoelectricity. MXene can be employed as an effective reinforcement for improving the output of piezoelectric and the PVDF sensitivity. The sensors can run consistently for over 10,000 cycles without performance degradation, exhibiting their superior durability and stability as long-time effective and durable sensors. Hybrid film sensor has a recovery time of 3.1 ms, which is faster than the PVDF sensor's (2.8 ms), allowing it to detect most physiological phenomena in the 1–3 Hz range.

By electrospinning, a flexible capacitive MXene ($Ti_3C_2T_x$) and polyvinyl pyrrolidone (PVP) nanofiber membranes (CNMs) pressure sensor can be made easily and efficiently [44]. A small amount of MXene added to PVP may reduce its compression modulus and also increase the dielectric constant. The sensing performance can be increased using the

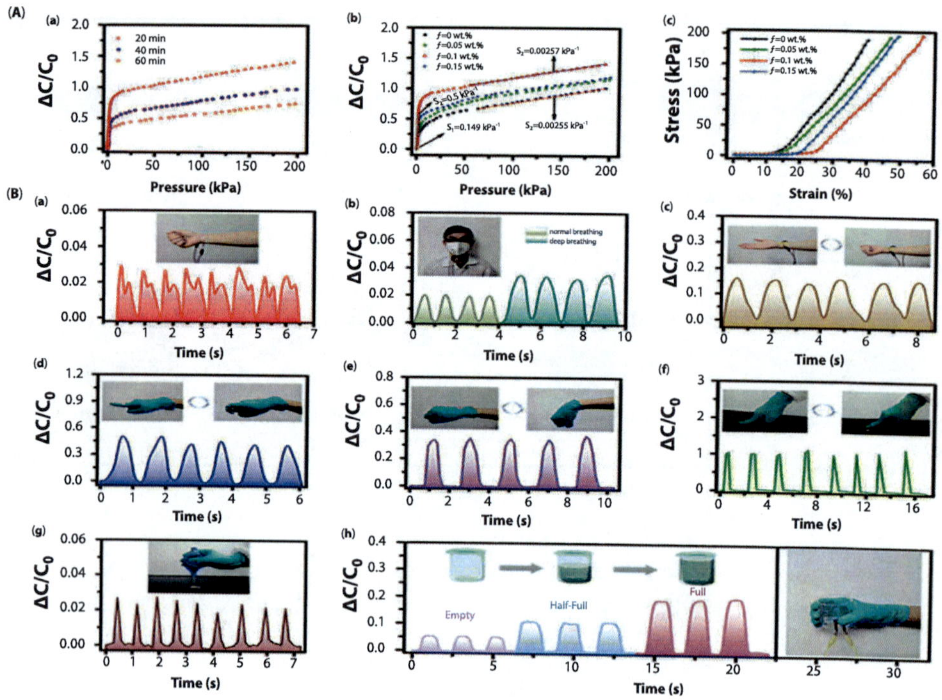

Figure 12.7 (A) Relative capacitance change ($\Delta C/C_0$) at various time periods of MXene/PVP CNM (a), Change in relative capacitance of the pressure sensors depending on f(MXene)/PVP CNM (b), Sensor's Compressive stress–strain performance (c), (B) Practical Functions of the sensor, monitoring the real-time human pulse wave (a), Normal and deep breathing detection (b), Capacitance responses to muscle contraction and relaxation by opening the palm and clenching the fist (c), The capacitance response to bending of finger/wrist (d-e), the capacitance response to repetitive finger pressing (f), sensing of tiny airflow (g), capacitive signal response of holding a beaker empty, half-full, or full (h) [44]. Copyright 2022. Reproduced with permission from the Royal Society of Chemistry.

synergy of the two factors mentioned above. Electrospinning the MXene/PVP CNM at the optimal time is vital to improving the dielectric layer thickness and overall pressure sensor performance. As demonstrated in Figure 12.7A (a), the relative capacitance change ($\Delta C/C_0$) of the MXene/PVP CNM electrospun in 20 minutes is 1.419, which is more than that of the sensors electrospun at 40 minutes (0.977) and 60 minutes (0.750). Figure 12.7A (b) depicts the evaluation of sensing performance on the addition of MXene studied ($\Delta C/C_0$) of the sensor with pressure. The compression modulus can be obtained from the compressive stress-strain curve in Figure 12.7A (c). From the curve, it can be inferred that the PVP chains with a small fraction of MXene can become less fluid, and the PVP nanofiber membranes can become softer leading to lower compression modulus/stiffness, while a high concentration of MXene increases the compression modulus/stiffness, causing agglomeration inside the PVP nanofiber. The device has better sensitivity (0.5 kPa^{-1} at 0–1.5 kPa), broad operating range (0–200 kPa), quick response/recovery time (45/45 ms), and minimal threshold detection (~9Pa). Also, the device demonstrates remarkable mechanical stability, proving its long endurance when exposed to fatigue testing at 8000 intervals. As illustrated in Figure 12.7B (a-h), the proposed sensor can be employed for a plethora of pressure observing purposes, such as respiration, pulse, human joint bending, and the capacitive response of beaker hand-holding. The behaviour of airflow distribution detected on the sensor's surface over a rapid response time will aid in the future determination of atmospheric airflow and wind direction. A pressure mapping experiment was performed on the 4×4 sensing arrays, demonstrating their substantial potential for future wearable medical devices and health monitoring.

A wearable sensor of Polyacrylamide-Gelatin-MXene nanocomplex hydrogels (PGM hydrogels) was developed for human activity monitoring [45]. Mechanical behaviour is an crucial aspect in deciding the applicability and lifespan of hydrogel-based sensors. The solid and dense dual network structure, electrostatic effect, interaction of hydrogen bonding, MXene reinforcement provided better mechanical flexibility to the hydrogels making it appropriate long term and large deformation bearable sensor for practical applications. The sensor exhibits high sensitivity of 100.4 kPa^{-1}, wide sensing range and 5000 cycles of durability due to the spider filament porous structure and high conductive MXene nanosheets. Nanocomposite hydrogels with MXene has good skin affinity as shown in Figure 12.8 (a1–a3). Thus, the hydrogel sensor could detect real-time activity of human motion. Typically, flexible sensors made of a single type of sensitive material cannot display greater sensitivity and a broad range of detection due to physical constraints. According to the preceding discussion, the most effective technique to improve the performance of the sensor is with proper structural design and controlled morphology and material mixing of the sensing materials. The effective ways of structural design include micro protrusion, active sensing layer, and microporous structure. Also, Performance parameters like sensitivity, range and stability are considered important factors for developing flexible sensors and it is provided in the Table 12.1.

12.4 Design of MXenes/Polymer Composites as Flexible Strain Sensors

Metals or semiconductors-based traditional strain sensors have a restricted strain range (often less than 5%) and high stiffness, making them unsuitable for future strain sensors [46]. To meet the rising demands stretchable as well as sensitive strain sensor is especially needed. The flexibility of MXene and MXene composites is restricted. As a consequence,

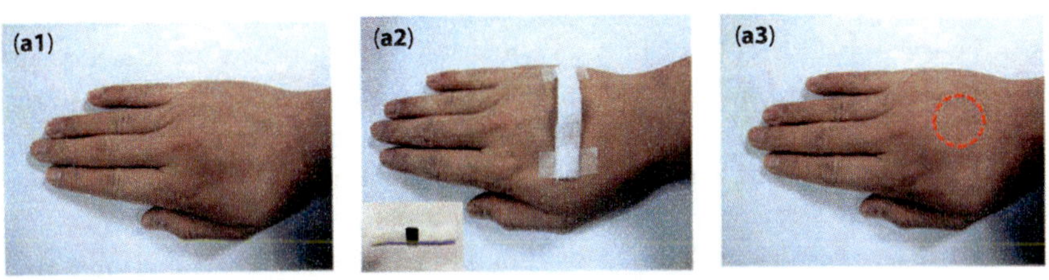

Figure 12.8 Skin affinity test of PGM hydrogel (a1–a3) [45]. Copyright 2022. Reproduced with permission from Elsevier.

Table 12.1 Performance parameters of various MXenes reinforced Polymer pressure sensor.

Sensor	Sensitivity/GF	Response/ recovery time (ms)	Sensing range	Stability (cycles)	Ref
MXene@PDMS	2.6 kPa^{-1}	40/40	0–30 kPa	800	[37]
PPy/MXene	- 20.1 kPa^{-1} at 0–2 kPa	100/80	0–80 kPa	>1000	[38]
MXene/PVA	2320.9 kPa^{-1} (65.3 Pa–6.53 kPa), 1313.6 kPa^{-1} (6.53 kPa –65.3 kPa), 594.2 kPa^{-1} (65.3 kPa–294 kPa)	70 /70	65.3Pa–294 kPa	10000	[39]
MXene/PANI	0.08232 kPa^{-1} & 0.3106 kPa^{-1}	80/130	0–23 kPa	5000	[40]
MXene/(PVDF-TrFE)	817.4 kPa^{-1} 2213.68 kPa^{-1}	16/16	0.072–0.74 kPa 0.74–3.083 kPa	–	[41]
MXene/rGO/ P(VDF-TrFE)	2298 KPa^{-1}	10/9	–	10000	[42]
MXene/PVDF film	0.0480 V/N (Voltage sensitivity)	-/3.1	–	10000	[43]
PVP/MXene nanofiber	0.5 kPa^{-1} (0–1.5 kPa)	45/45	0–200 kPa	8000	[44]
PGM Hydrogel	100.4 kPa^{-1} 4.7 kPa^{-1}	196/409	-	5000	[45]

polymer-based flexible strain sensors have garnered a growing amount of interest. The sensing mechanisms and manufacturing processes used in various MXene-reinforced polymer flexible strain sensors are discussed in this section. The high-performance flexible strain sensors were developed with MXenes and a hydrophobic elastomer of thermoplastic polyurethane (TPU). Polyacrylonitrile (PAN) was introduced into TPU and fabricated into the flexible TPU/PAN mat substrate by the electrospinning method [47]. A stretchable conductive MXene nanosheets/TPU/PAN mat nanofiber made using a facile dip coating method. In this study, the PAN considerably improved the interaction among nanofiber and nanosheet of MXene. The mechanical responses of the sensor under strain are characterized by $\Delta R/R_0$, where R_0 and ΔR signify the sample's initial resistance and the change from R_0, respectively. The sensor was capable of monitoring and distinguishing extremely small tensile strain (1–5%) as evidenced by the intensity of the response signal increased with strain. Moreover, the sensor functions over a broad strain range (from 60% to 80%) with steady sensitivity and no hysteresis. The sensor maintains a stable response at varied stretching rates from 10 to 30 mm/min. Therefore, strain sensor MXene/TPU/PAN offers a broad sensing range (0–80%), a quick response time (<140.6 ms), and a small detection limit (<0.1%). At 10 % strain, the strain sensor shows good durability and stability under loading and unloading cycles (1750). Also, the long-term stability corroborates the output signals for first and last 14 cycles. The wrapping of MXene nanosheets around the TPU/PAN fiber ensures a strong interaction between MXenes/TPU/PAN. The practical applications of the human physiological signals were detected. The strain sensor successfully records the human motion with good reproducibility. A flexible sensor based on micro sheet MXene and a TPU substrate encapsulated with PDMS was developed [48]. The PDMS encapsulation offers electrical insulation, wear resistance and hydrophobicity which in turn promotes stability. The sensor features stable and has no hysteresis, a broad response range (0–90%), quick response time (<120.1 ms), ultra-low detection limit (<0.005%, 0.5 μm) and good durability (>3000 cycles). Because of its hydrophobic, dust-proof and epidermal properties, PDMS has great potential in speech recognition and electronic skin.

Strain sensors with low deformation sensitivity are greatly needed for health monitoring. Cracks are often considered flaws to be eliminated since they are detrimental to mechanical characteristics. Cracks can serve as strain-sensitive active sites and mechanically weak points that are highly sensitive to local strain, leading to significant conductive films. In this study, Conductive MXenes and stretchable polyurethane (PU) fibre with highly porous structures and micropatterns were used to create a MXene/PU fibrous membrane (m-MXene FM) strain sensor [49] by electrospinning method in conjunction with mechanically driven approach with superior sensitivity and wide detection capabilities is displayed in Figure 12.9(A). The m-MXene FM strain sensor has a broad and adjustable sensing range with the introduction of micro wrinkles pattern (0–120% and 120–270%), as well as an ultralow detection limit of 0.05% and 5 μm for small vibrations, indicating a promising approach for monitoring health applications. The piezoresistivity of the strain sensor can be effortlessly altered by varying the fracture-stretching deformation (0–1000%). Figure 12.9(B) (c-f) shows that the increasing microcracks density with fracture stretching deformation improves the strain-sensitive nature. Cracked MXenes-based fibre has significantly higher sensitivity and sensing range than defect-free fiber. The quantity of MXene loaded on the Fibrous membrane is significantly connected to sensing ability, with larger concentrations being preferable. Furthermore, owing to the robust interfacial connection between the MXene coating and PU fibres, which display exceptional flexibility, a steady resistance response with good reversibility is produced even after 500 cycles of stretching and releasing, depicted in Figure 12.9(B) (a,b). Due to its superior sensing qualities, such as rapid and stable sensing and a broad range of sensing capabilities, the sensor was exposed to deformation monitoring and physiological signal detection.

Most traditional strain sensors have a narrow detection range or poor sensitivity, making them unsuitable for extremely full-scale sensitive human motion detection. The conductivity of the microstructure sensing layer is important for subtle deformations; however, the sensing layer is conductive even under large deformations to obtain a broad sensing range. Thus, using layer-by-layer spreading coating method, a wearable strain sensor of highly sensitive and broad sensing range of flexible tile like stacked microstructure Mxene/Polyaniline fibre (PANIF) on an substrate of elastic rubber was designed [50] as presented in Figure 12.10(A) (a-d). A commercial good stretchable and elastic rubber tape was used as substrate. MXenes film was formed by spreading the MXenes suspension

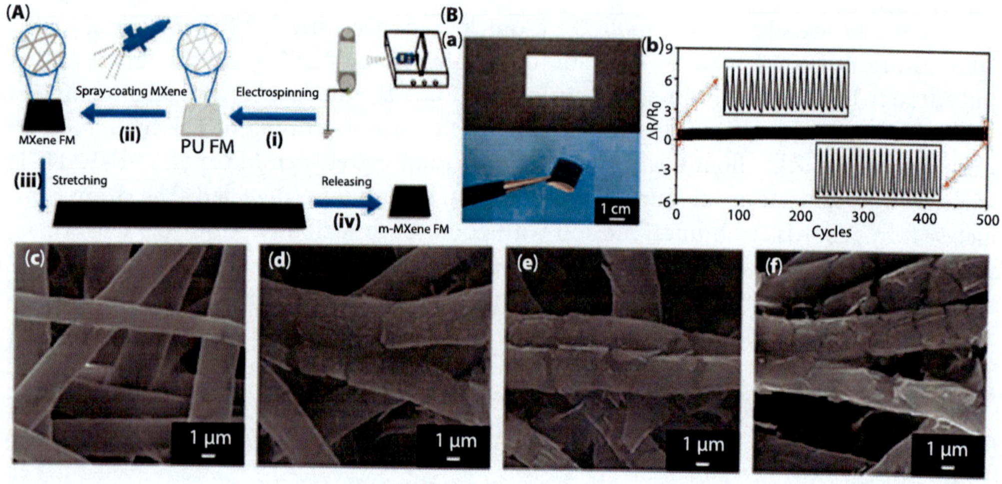

Figure 12.9 (A) Schematic design of strain sensor m-MXene FM, (B) Digital photograph of PU FM (white) and m-MXene FM (black) (a), Testing of Stability under 500 Cycles (b), Microcracks of m-MXene FM under fracture stretching deformation (0–1000%) (c-f) [49]. Copyright 2021. Reproduced with permission from Elsevier.

over the substrate and then PANIF suspension was coated over the MXene film-coated substrate and stretched upto 60% to obtain microcracks. The film coating process was continued and stretched to 100% strain. The strain sensor was created with both ends of the sensitive layer using two copper wire electrodes. To ensure conductivity paths, the PANIF layer is sandwiched between the MXene layer. The presence of strong hydrogen exists among the MXene (O,F O-H group) and PANIF (N-H group) and considerably improves the stability of the sensing structure in Figure 12.10(B) (a). Microcracks are generated even when tiny deformations are applied to rigid MXene films resulting in 20% narrow sensing range. A single MXenes/PANIF layer of strain sensor has a 37% greater sensing range compared to a single MXene (20% strain), which may be attributed to the synergistic action of the PANIF in the MXenes. The addition of PANIF in the strain sensor greatly widens the operating range, but it is still insufficient for practical applications. The strain sensor with an overlapped tile like stacked hierarchical microstructure was developed to widen the sensitivity range and enhance the sensitivity more than single sensing layer. The overlapping structure of the MXene/PANIF microstructure dramatically broadened the range of sensing and better sensitivity when compared with the overlapped microstructure of pure MXene and pure PANIF. The influence of sensing materials mass ratio on the sensing characteristics of the strain sensor was studied. It is noteworthy that increasing the quantity of PANIF loading into the MXenes increases the sensing range by providing more conductive routes even with high deformation, but the highest sensing sensitivity decreases with increasing PANIF content. The strain sensor demonstrated great sensitivity (up to 2369.1 for GF) and a large range of sensing (80% strain). The stability sustained after 2500 cycles of loading/unloading ranging from 0–10% of strain is shown in Figure 12.10(B) (b). The sensor's response and recovery time (0.6s/0.8s) under a minor strain of 1% at a lower stretching rate of 100% min^{-1} meet the standards of human movement finding. The MXene/PANIF strain sensor was evaluated

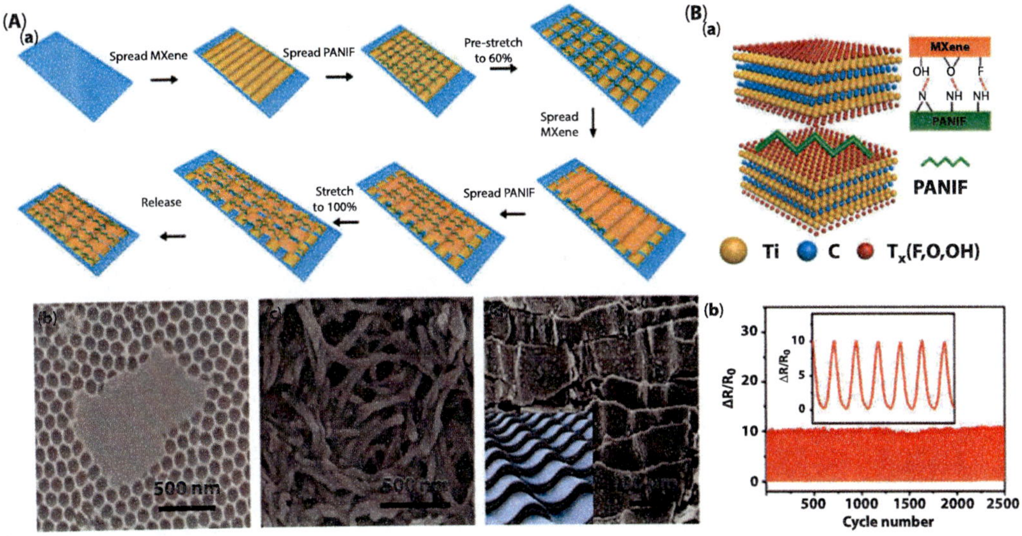

Figure 12.10 (A) Fabrication (a), and SEM images (b-d) of tile-like stacked hierarchical microstructure MXene/PANIF strain sensor, (B) Strong hydrogen bond between MXene and PANIF (a), and stability of MXene/PANIF strain sensor (b) [50]. Copyright 2020. Reproduced with permission from Elsevier.

for practical applications in human physiological and large-scale motion because of its superior sensitivity, wide sensing range, and flexibility of its tie-like stacking structure. Therefore, MXene/PANIF is a potential wearable electronic device.

A strain sensor of composite film MXene/Polyurethane was easily coated on several substrates, including nitrile, natural rubber (NR), polyethylene (PE), and polyvinyl chloride (PVC) [51]. The as-prepared composite films in different proportions (1:14, 1:10, 1:7, 1:5 and 2:7) on NR substrates were cut, and carbon conductive tapes are affixed as electrodes to both ends of the film using silver adhesives. The final composite film has an uneven structure with wrinkles and ridges on the surface, functionally similar to fibrous networks. The composite film fabricated with MXene and PU exhibits adjustable stretchability and conductivity due to electrostatic repulsion and hydrogen bonds. When the MXene: PU = 1:5, the composite conductivity improves dramatically from 0.32 to 9.05 S/m and the sensing range is 0–50% illustrated in Figure 12.11(A) (a). The sensitivity of the film was evaluated and it is denoted with the parameter GF. The GF increases with increasing stretching of the strain and reached the maximum of GF at 3.49×10^4 at 40 % strain, shown in Figure 12.11(A) (b-c).The film has a low hysteresis of 7.21% at 5% strain, suggesting that it has excellent shape stability, illustrated in Figure 12.11(A) (d). The fast response (200 ms at 10% strain) suggests that it might be utilized for real-time deformation monitoring. Stretching the film causes wrinkles and ridges to form as microcracks that create additional conducting channels, displayed in Figure 12.11(B) (a-b). The relative resistance changes of the composite film increased by approximately 0.2 and 0.03 after 5000 cycles of stretching-releasing tests at 5% strain and 1000 cycles of bending/recovering tests (180°), indicating the film's stability in practical applications as revealed in Figure 12.11(C) (a-b). Overall, the strain sensor finds potential applications for preventive measures and constant assessment of cardiovascular disorders.

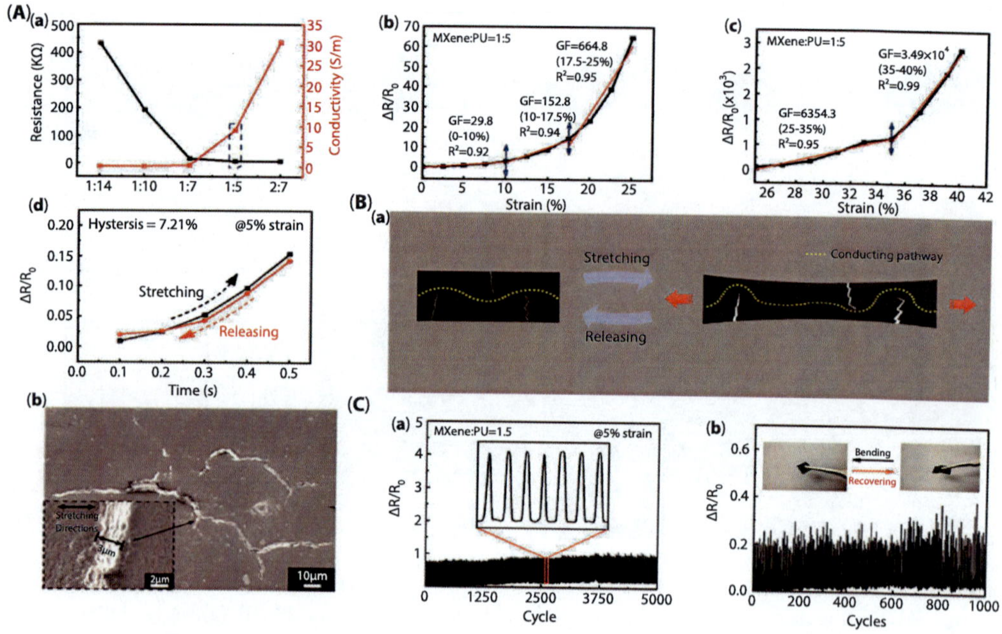

Figure 12.11 (A) Initial resistance, conductivity with varied ratios of MXene for MXene/PU (a), GF in the strain range of 0–25% and 25–40% (b-c), Low hysteresis of the composite film (d), (B) concept of resistance change during stretching of the composite film (a), formation of microcracks after stretching (b), (C) ΔR/R0 response of the MXene/PU composite film subjected to 5000 cyclic stretching-releasing tests and 1000 cyclic bending and recovering tests (a-b) Adapted from ref [51], Copyright 2022, Elsevier.

Due to their outstanding flexibility, biocompatibility, and mechanical properties that are comparable to those of human tissue, conductive polymer hydrogels (CPHs) are considered as appealing substrates for skin sensors [52]. CPHs are susceptible to a loss of sensory responsiveness, stretchability, and lifespan, which limits their widespread application of wearable sensors. To address this difficulty, the materials with electrical capabilities and nanoscale reinforcement have been combined with hydrogels to fabricate highly sensitive sensors [53, 54]. A one pot and two step method was used to create an MXene/Polyacrylamide/Sodium Alginate-Ca^{2+} Double Network (DN) organohydrogel as a multifunctional strain sensor [55]. Figure 12.12 (a) shows the high stretchability and mechanical resilience of organohydrogel under external force. Further, mechanical characteristics such as stress and toughness were enhanced with the addition of Ca^{2+} and glycerol due to the formation of hydrogen bonds and coordination bonds. The synergy action of Ca^{2+} and MXene in the DN organohydrogel could increase its wide linearity and high sensitivity. The DN organohydrogel retained its flexibility after 5 days at -20°C in the presence of glycerol and hardened in the absence of glycerol. As shown in Figure 12.12 (b), at low temperatures (-20°C), organohydrogel exhibits a similar stable electrical signal to external stimuli as it does at room temperature (25°C). Considering the stretchability of the sensor, MXene concentration is critical. MXene may aggregate within the gel at increasing concentrations, resulting in a non-uniform gel structure which decreases stretchability. Another study found that, the surface functional groups of MXene act as a cross-linking agent to

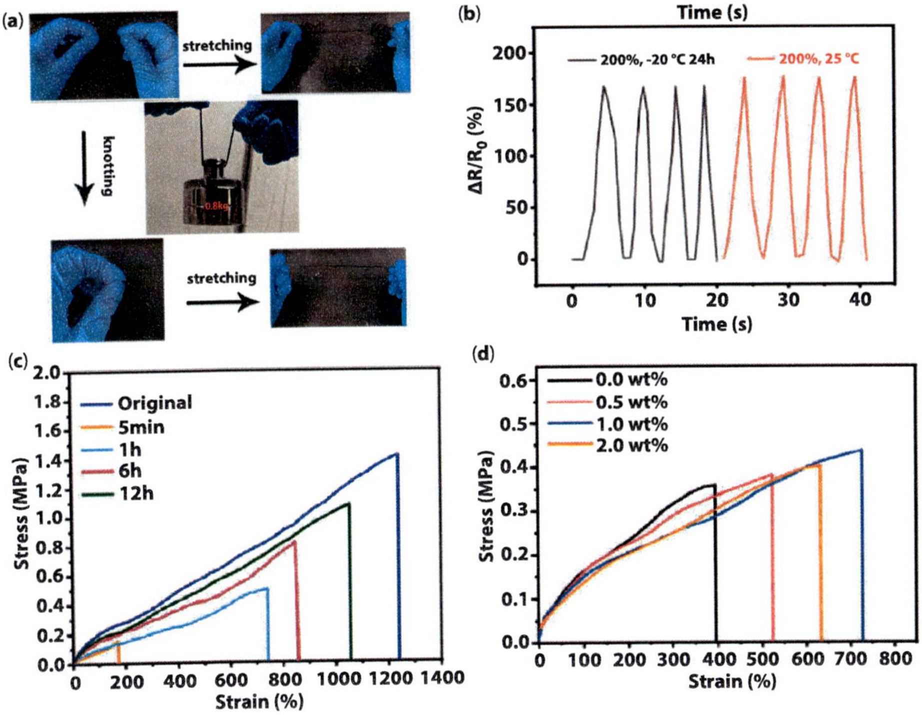

Figure 12.12 Mechanical Characteristics of MXene-PAM/SA-Ca^{2+} organohydrogel (a), Stable electrical signal at – 20°C after 24 h and at room temperature (b) [55]. Copyright 2022. Reproduced with permission from Elsevier. Self healing efficiency with increase of healing time (c) and increase of MXene Content (d) [57]. Copyright 2022. Reproduced with permission from Elsevier.

bond with the polymer matrix potentially improving the mechanical characteristics of the MXene-PAM/SA gel [56]. Upon increasing the MXene content in the double network hydrogel strain sensor made with Ti_3C_2 MXene with acrylamide (AM) and sodium alginate (SA) lowers stretchability. The developed hydrogel has stretchability of 3150 %, a Youngs modulus of 2.03 kPa^{-1}, a rapid response time of 74 ms, negligible hysteresis and anti-fatigue performances. The highly conductive MXene nanosheet matrix further enhanced sensitivity to tension (GF = 18.15) and compression stimuli (S = 0.383 kPa^{-1}) by dynamically changing contact angles and areas between adjacent nanosheets. A good combination of electrical and mechanical performance of the MXene/polyampholytes (MXene/PMN) composite hydrogel strain sensor was developed [57]. It can withstand a large stretching ratio and has a maximum uniaxial tensile strain of 1200% to monitor human body activity. The nano-reinforcement effect of highly dispersed MXene as well as the toughening effect of the weak hydrogen bonds between the functional groups of MXene and the polyampholyte network are two likely causes for mechanical performance improvement. The efficiency of self-healing improved with increasing healing time, and the efficiency increased with increasing MXene concentration owing to hydrogen interactions between MXene and PMN as revealed in Figure 12.12 (c-d). The MXene/PMN hydrogel sensor's high self-adhesion makes it ideal for monitoring human activities. Furthermore, the developed sensor was employed to collect real-time behavioural data from ADHD patients.

A temperature-tolerant, self-healing, self-adhesive, mechanically resilient strain sensor was built into a Poly(acrylamide-acrylic acid)/Chitosan/MXene (PACG-M) hydrogel [58]. The inclusion of a small proportion of MXene sheets resulted in substantial electrical and mechanical performance. The hydrogen bonds exist among the nanosheets of MXene and polymer chains as shown in Figure 12.13 (a-b) allowing the hydrogel to self-heal. The hydrogel's use of polyol glycerol has anti-freezing and anti-drying properties. Outstanding temperature tolerance could be owing to the hydrogen bond between the polyol/water molecule and the nanosheet MXene/polymer chains. Mechanical characteristics with temperature tolerance were evaluated using tensile tests, tensile strength, strain, and work of fracture essentially remained unchanged in extreme temperature environments. The sensor adheres well to a variety of surfaces, including steel, glass, rubber, plastic, and skin as represented in Figure 12.13 (c). As a significance of the sensor's great sensitivity, wide detection range, and temperature range flexibility, it is well suited for customised healthcare. The swelling ultrasonic method was used to design a MXene/CNTs/EVA microcrack structure fibre strain sensor using $Ti_3C_2T_x$ (MXene), carbon nanotubes (CNTs), and ethylene vinyl acetate polymer (EVA) [59]. The addition of CNT's may enhance the effective strain range

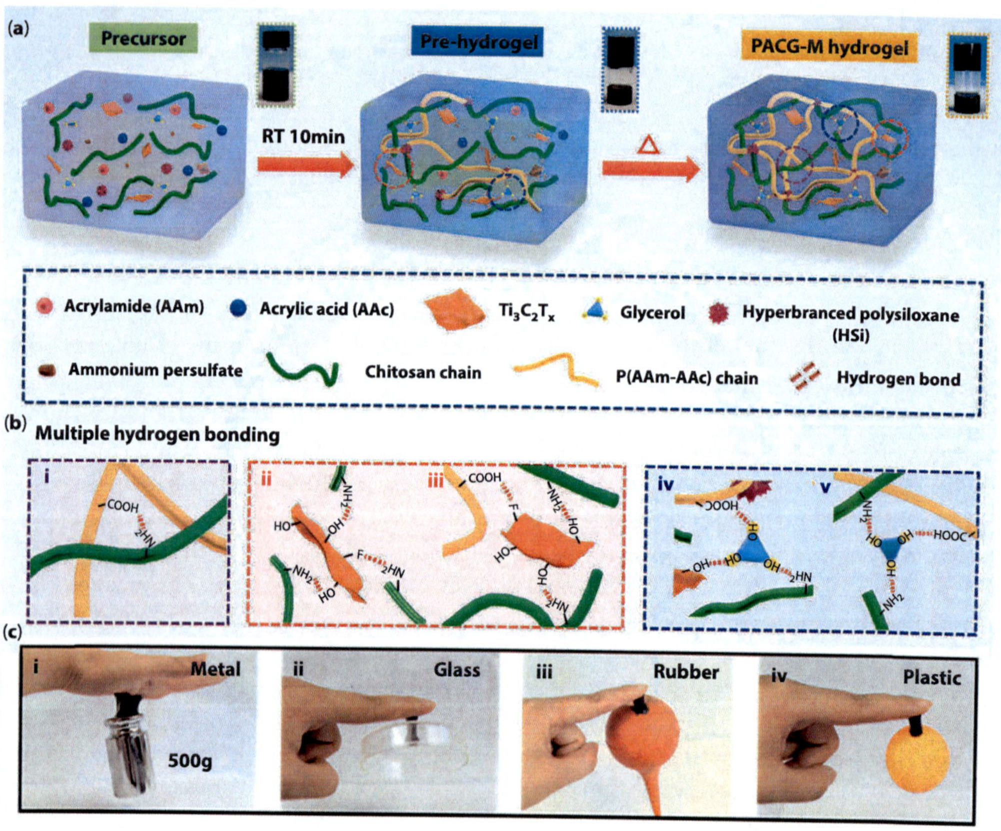

Figure 12.13 Synthesis process and mechanism of PACG-M hydrogel and hydrogen exist in the PACG-M hydrogel (a-b) Adhesive performance in different surfaces (c) [58]. Copyright 2022. Reproduced with permission from Elsevier.

Table 12.2 Performance parameters of various MXene/polymer composite strain sensor.

Material	Sample type	Sensing range	GF	Stability (cycles)	Detection limit (%)	Response time (ms)	Ref
MXene nanosheets/TPU/ PAN mat nanofiber	Composite fiber	80%	9.69	>1750	<0.1	<140.6	[47]
PDMS-MXene/TPU	Composite film	0–90%	-	>3000	<0.005%	<120.1	[48]
m-MXene FM	Fibrous membrane	0–120% to 0–270% (tunable)	1000	500 (stretching-releasing) 1000 (bending-recovering)	0.05	-	[49]
tile-like stacked MXene/ PANIF strain sensor	Composite film	80%	97.6 (0–10%) 563.6 (10–40%) 2369.1 (40%-80%)	2500	0.1538	600	[50]
MXene/PU	Composite film	50% (MXene: PU= 1:5)	3.49×10^4	5000 (stretching-releasing) 1000 (bending-recovering)	0.5–3	200	[51]
MXene/Polyacrylamide/ Sodium Alginate-Ca^{2+} DN	Hydrogel	0–600%	1.21	200	-	600	[55]

(*Continued*)

Table 12.2 Performance parameters of various MXene/polymer composite strain sensor. (*Continued*)

Material	Sample type	Sensing range	GF	Stability (cycles)	Detection limit (%)	Response time (ms)	Ref
MXene-PAM/SA gel	Hydrogel	-	18.15	1000	-	74	[56]
MXene/PMN	Hydrogel	-	2.34 (0–200%) 6.31 (200–1000%)	200	-	-	[57]
PACG-M hydrogel	Hydrogel	600%	3.93	600	-	-	[58]
MXene/CNTs/EVA fiber	Composite fiber	190%	82.5	>10000	-	178.9	[59]

and linear strain range but it is detrimental to its sensitivity of the MXene/CNTs/EVA fibre strain sensor. The anchoring effect of the 2D sheet structure of MXene and 1D linear structure of CNTs on the EVA surface generated the recoverable deformable conductive network of strain sensor which guarantee its durability of the sensor in real-world wearable electronics applications. Also, the sensor has excellent thermal repair property by heating and infrared light. The sensor with strain sensing range (190%), high sensitivity (82.5), response time (178.9 ms) and excellent stability and durability (>10000 cycles) finds applications in human motion monitoring. Table 12.2 displays the performance parameters of several MXene/polymer composite strain sensors.

In summary, the MXene polymeric composites structural crack formation enabled a higher sensitivity and sensing range of the sensor, which greatly increases their practical applications. While cracks are often regarded as defects to be eliminated, crack-based strain sensors are capable of higher sensitivity because they take advantage of the characteristics crack structures. The advancement of crack-based strain sensors with great stretchability is an effective method for improving high-performance strain sensors. Further, the mass ratio of sensing materials has also been shown to have an impact on the strain sensors sensing performance. Temperature tolerance is lacking in MXene-based hydrogels. Polyols like ethylene glycol and glycerol were included into the hydrogel to ensure long-term environmental stability. It is notable that, the DN structure as well as the concentration of MXene in the polymer network must be considered for the mechanical characteristics of the strain sensors. In order to be employed in individualised health-care devices, strain sensors must also have an appropriate GF or sensitivity, low hysteresis loss, stability and fast response, mechanical durability.

12.5 Design of MXenes/Biopolymer Composites as a Flexible Pressure Sensor

Pressure sensors that are flexible and biodegradable have the potential to be used in the applications of health monitoring wearables and electronic skins. Chitin, chitosan, cellulose, alginate and other biopolymers may be used in the design of wearable electronics. They are extremely biocompatible, biodegradable, can be stretched and bent, can be constructed into many shapes and designs, and is mass produced at a minimum cost [60, 61]. Due to the richness of functional sites and reactive centres, they have strong mechanical properties and ability of chemical modifications, making them promising materials for the improvement of wearable electronics. Sensing materials are often difficult to degrade, leading to electronic waste. The exploration of high-performance flexible sensors reinforced with biopolymers to aid degradability is critically important in wearable electronic devices. Furthermore, the hydrophobicity and biocompatibility of the flexible pressure sensor must be ensured prior to direct application to human skin.

MXene nanosheets were embedded in tissue papers and laminated with a biodegradable PLA sheet and designed by an interdigitated conducting electrode to fabricate the flexible pressure sensor [62]. The fabricated flexible transient pressure sensors demonstrated great sensitivity, a low detection limit (10.2 Pa), quick response (11 ms), low power consumption (10^{-8} W), outstanding repeatability upto 10000 cycles, consistent bio-compatibility, and eco-friendly degradation. The sensor showed a steady sensing response and a wide detection range (23 Pa -30 kPa). The sensitivities of the pressure sensor S_1, S_2 and S_3 are 0.55 kPa^{-1}, 3.81 kPa^{-1}, and 2.52 kPa^{-1} in the pressure range of 23 to 982 Pa, 982 to 10 kPa and 10 to 30 kPa. Furthermore, the sensitivities with varied PLA encapsulation layers of the MXene/tissue paper are shown in Figure 12.14(A) (a) indicating that a thinner encapsulation layer of PLA has superior sensing sensitivity and the bottom layer of PLA was 80 μm during the test. The sensor has been evaluated for real-time detection of human motion due to its outstanding sensitivity and wide sensing range. In addition, the pressure sensor could be knocked on at a characteristic frequency to mimic initial stage Parkinson's disease with static tremors as shown in Figure 12.14(A) (b-c). The Figure 12.14(B) (a-e) represents the sensors degradable performance and SEM images prior to and following 14 days suggesting that the sensor was deteriorated after 14 days owing to the strong degradation rate of tissue paper and PLA and could not cause significant environmental harm. Also, the biocompatibility of the sensor was examined and its applicability in the human body was confirmed by cytotoxicity test. With good reproducibility, timely wireless communication, and stability, wearable pressure sensor MXene/tissue-paper is capable of monitoring human healthcare and human-machine interfaces.

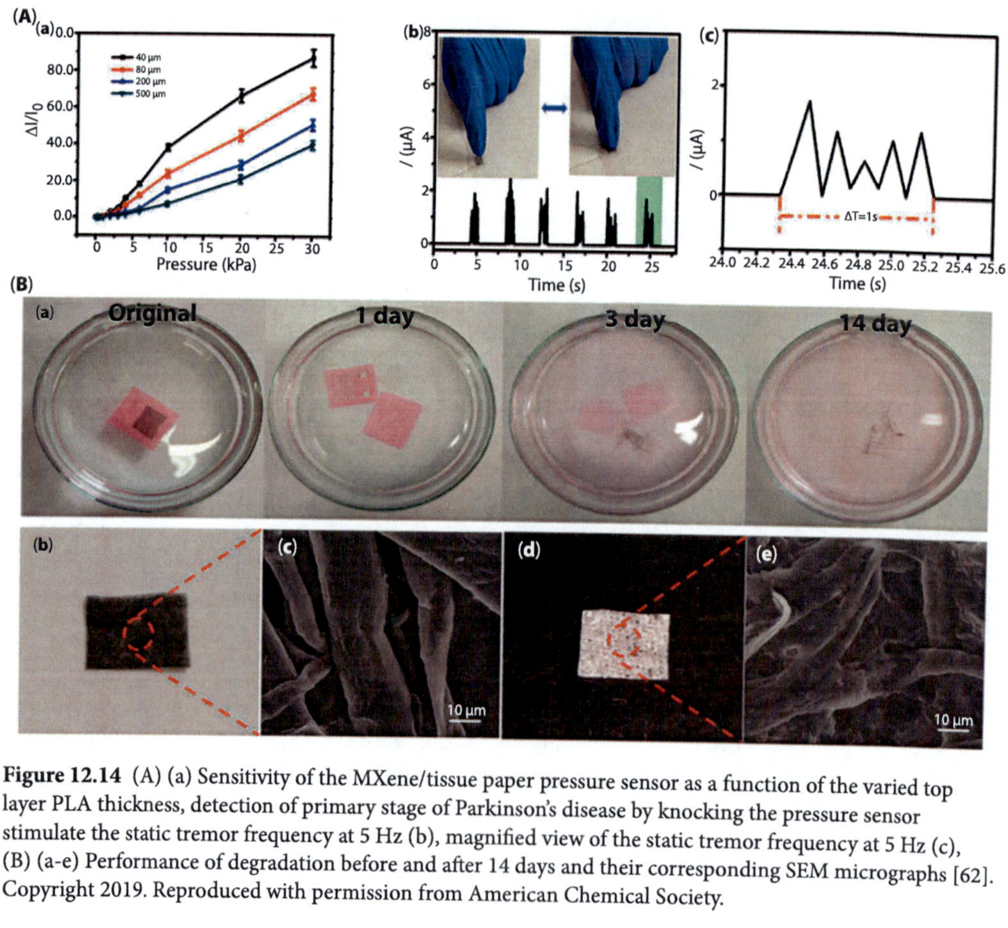

Figure 12.14 (A) (a) Sensitivity of the MXene/tissue paper pressure sensor as a function of the varied top layer PLA thickness, detection of primary stage of Parkinson's disease by knocking the pressure sensor stimulate the static tremor frequency at 5 Hz (b), magnified view of the static tremor frequency at 5 Hz (c), (B) (a-e) Performance of degradation before and after 14 days and their corresponding SEM micrographs [62]. Copyright 2019. Reproduced with permission from American Chemical Society.

A multifunctional wearable sensor fabricated with hybrid chitosan/MXene (CTS/MX) film is based on continuous laminated macrostructures formed by chitosan [63]. This proposed CTS/MX film-based electronic skin is capable of not only identifying pressure quickly but also monitoring humidity continuously for up to 20 days. The cyclic stress-strain test with ε = 30%, as shown in Figure 12.15 (a) displays excellent mechanical stability, demonstrating its promising possibility in a flexible pressure sensor. Four various thickness of bionic CTS/MX were developed by varying MXene and CTS ratio (specified as CTS/MX 1 to CTS/MX 4). Figure 12.15 (b-c) displays SEM micrograph of the various thicknesses of the CTS/MX hybrid films as well as electrical resistance variations. The sensing performance at different pressure levels was studied in Figure 12.15 (d-e) and found that the maximum pressure-sensitive response occurs when the MXene to CTS ratio is 4:1. Benefitting from the film's abundance of conductive channel, rough contact surface on the top and bottom electrodes, the sensor has a good sensitivity of 14.1 and 3.3 kPa^{-1} in the pressure regions 0–200 Pa and 0.2–5 kPa respectively. The developed sensor displays quick response and recovery time (143 and 139 ms), excellent stability (5000 cycles) and a detection limit of 3 Pa. The good stability, great sensitivity and biocompatibility find capabilities in real-time human monitoring.

A superhydrophobic pressure sensor made of MXene-Sodium alginate sponge (SMSS) with exceptional sensitivity, a large range of detection, and moisture resistance can detect varied human movements in all-weather circumstances [64]. The superhydrophobic surface of the SMSS can successfully resist different surface tension of various fluids (water, perspiration, milk, coffee, tea, and juice), as illustrated by the spherical droplet in Figure 12.16(A) (a). Also, the superhydrophobic surface is resistant to tape peeling, corrosion, and

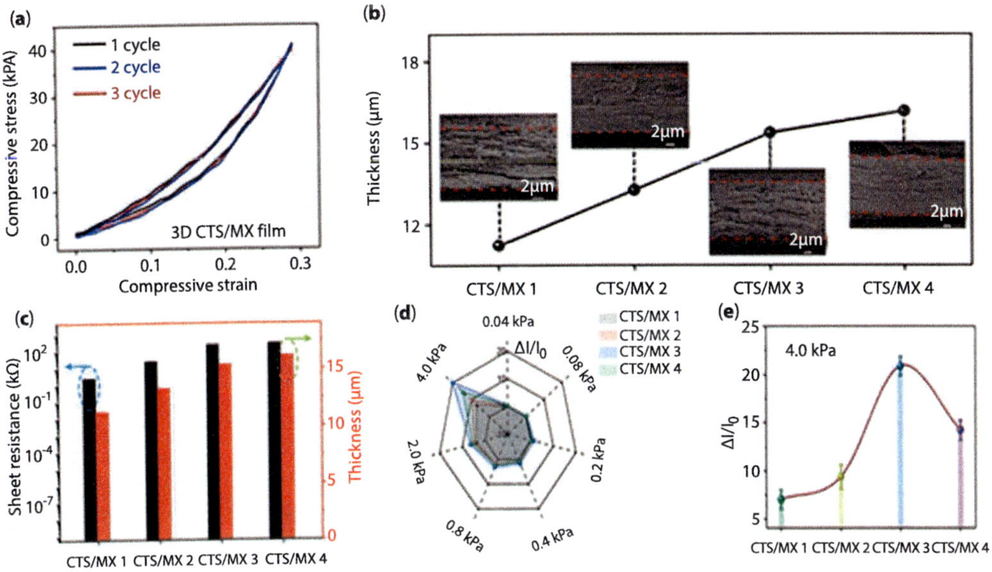

Figure 12.15 Compressive stress-strain curve of the CTS/MX (a), SEM images of the different thicknesses of CTS/MX hybrid film (b), Electrical resistance variations with different thicknesses of the film (c), Sensing performance of the film with different ratio of MXene and CTS (d), Maximum pressure sensitive response at 4.0 kPa (e) [63] Copyright 2021. Reproduced with permission from American Chemical Society.

extrusion, as revealed in Figure 12.16(A) (b-d). When SMSS is immersed in acid, neutral and alkaline (pH = 3,7,11) environments, its contact angle remains essentially unchanged. The effective low-voltage Joule heating, which lowers water molecule adsorption throughout static and dynamic pressure cycling provided by the highly conductive MXene-SA coating, results in persistent superhydrophobicity. The sensor's superhydrophobicity decreasing with increasing exposure duration without the joule heating effect. A special electrode design with self-adhesive rubber reduces the adverse thermal effects on the skin shown in Figure 12.16(B) (a). In addition, the adhesive layer allows for more precise signal detection when compared to no adhesive layer, as revealed in Figure 12.16(B) (b). The pulse detection of the SMSS sensor on different body parts like wrist, elbow, finger and knee are depicted in Figure 12.16(B) (c-f). The resulting sensor featured great sensitivity, a wide detection range, outstanding flexibility and sustained superhydrophobicity indicating its ability to identify human actions in all environmental climate situations.

A vacuum filtration method was used in conjunction with hydrazine-induced foaming process to design a piezoresistive MXene/cellulose nanofiber (CNF) foam [65] as

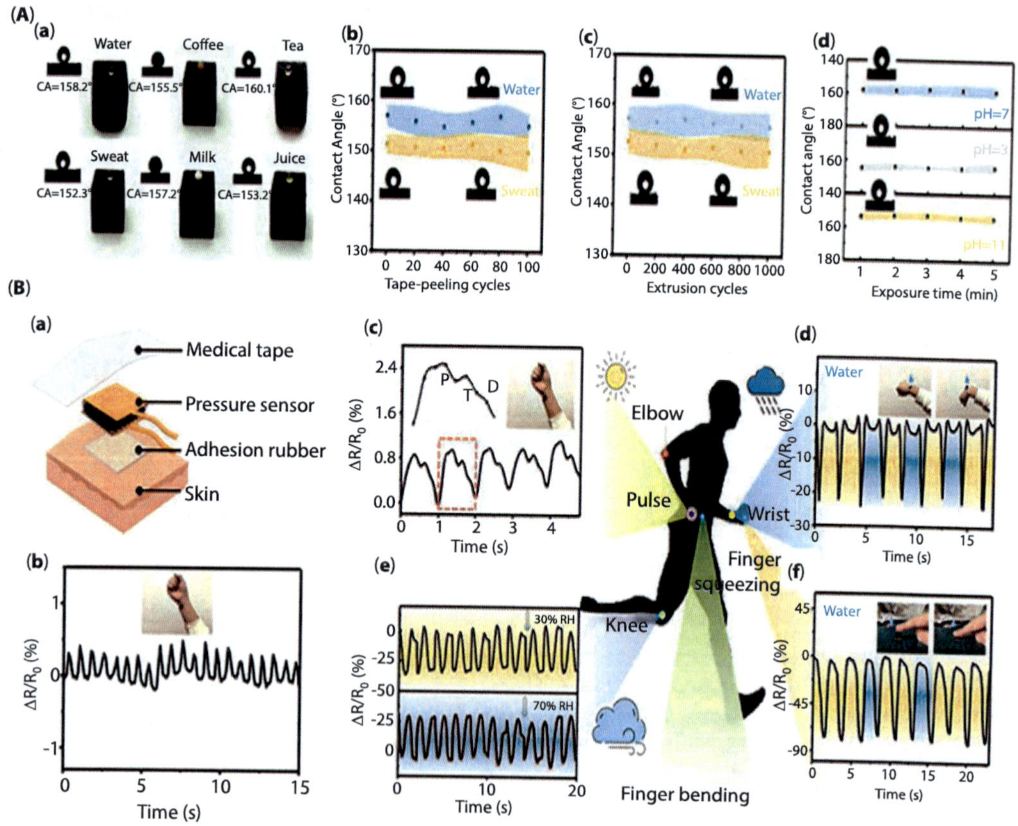

Figure 12.16 (A) superhydrophobic performance of SMSS sensor on the different surface tension of liquids (a), superhydrophobic surface resistance to tape peeling, extrusion, and corrosion cycles (b-d), (B) Self-adhesive SMSS pressure sensor (a), pulse detection without adhesive layer of SMSS sensor (b), Pulse detection of the SMSS sensor to body motions (c-f) [64]. Copyright 2022. Reproduced with permission from Elsevier.

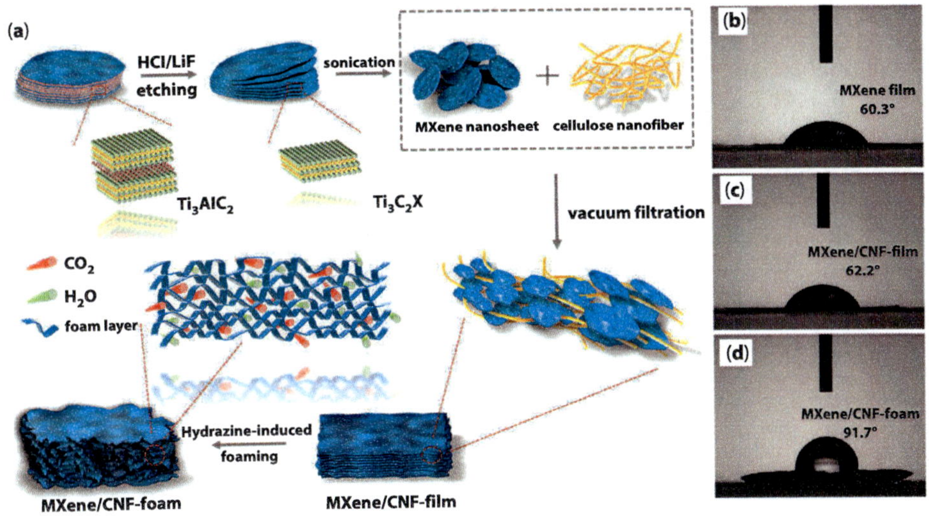

Figure 12.17 Schematic preparation of MXene/cellulose nanofiber (CNF)-foam (a), surface wettability of MXene (b), MXene/cellulose nanofiber (CNF) (c), MXene/cellulose nanofiber (CNF) (d) [65]. Copyright 2021. Reproduced with permission from Elsevier.

Table 12.3 Sensing performances of various MXene/biopolymer composite pressure sensor.

Sensor	Sensitivity	Response time (ms)	Detection limit (Pa)	Cycling stability (Cycles)	Ref
Mxene/tissue Paper with PLA	0.55 kPa^{-1} (23–982 Pa), 3.81 kPa^{-1} (982–10 kPa) 2.52 kPa^{-1} (10–30 kPa)	11	10.2	10000	[62]
Chitosan/MXene	14.1 kPa^{-1} (0–200 Pa) 3.3kPa^{-1} (0.2–5 kPa)	143	3	5000	[63]
Mxene-Sodium alginate sponge	2.4 kPa^{-1} (0–1.85 kPa) -3.56 kPa^{-1} (1.85–20 kPa) -0.5 kPa^{-1} (20–66.6 kPa)	94	-	1000	[64]
MXene/cellulose nanofiber(CNF)-foam	419.7 kPa^{-1} (below 8.04 kPa), 649.3 kPa^{-1} (8.04 kPa and 20.55 kPa)	123	4	10000	[65]

illustrated in Figure 12.17(a). The hydrazine-induced foaming approach may solve the MXene constraints by increasing layer spacing, generating a desirable porous structure, and avoiding MXene oxidation. The maximum contact angle of the MXene/cellulose nanofiber (CNF)-foam is (91.7°) compared to MXene/cellulose nanofiber (CNF) film (62.2°) and MXene film (60.3°) confirming that the surface wettability transitions from hydrophilic to hydrophobic as shown in Figure 12.17(b–d). The pressure sensor offers a broad linear range of sensitivity 419.7 kPa^{-1} for pressures less than 8.04 kPa, and 649.3 kPa^{-1} for pressures between 8.04 kPa and 20.55 kPa region), 4 Pa of detection limit, a fast response time/recovery time of 123/139 ms, 68.9 MPa of tensile strength and outstanding stability up to 10,000 cycles. Moreover, the practical applications of sensing human actions, capturing sound spectrum and discriminating small grains. The degradation of MXene/CNF-foam was also investigated. As a result of the above discussion, it is envisaged that the synergy of MXene with biopolymers would result in the development of a suitable aspirant for a high-performance biocompatible and biodegradable flexible sensors as demonstrated in Table 12.3.

12.6 Conclusions and Future Perspectives

MXene has distinct physicochemical properties, including hydrophilicity, conductivity, specific surface area, high ion migration, ease of functionalization, and excellent biocompatibility, which have enormous potential in wearable sensing applications. Considering its promising characteristics, MXene wearable applications are still in a rudimentary stage. The hydrophilicity of surface groups in Mxene, such as -O, -OH, and -F, may readily adsorb humidity and gases, potentially leading to oxidation. More significantly, MXene-coated electronic devices are easily wet and may be severely damaged in humid or harsh conditions. As a result, although MXene is a potential active material for sensors, its hydrophilicity limits its use in a variety of electrical devices. The polymer reinforcement with MXene will address the aforementioned problem. MXenes have high conductivity and hydrophilic groups, which improve the many interactions with flexible polymers and allow MXene to be easily dispersed in aqueous solutions. As a result, they are often utilised in flexible sensors as a conductive filler in flexible sensors. MXene polymer compositions inhibit surface oxidation and improve mechanical properties owing to cost-effective and simple modification and processing, making them a potential choice for wearable sensor. The sensor's capacity to regulate and manage oxidation will offer stability, extending its shelf life.

Also, the development of MXene devices has not received as much attention as graphene-based sensors. MXene's limited stretchability may be treated by investigating polymer composites. The enhanced stretchability and sensing range provided by the polymer reinforcement will permit it to be used in wearable sensing devices. Mechanical and electrically sensitive polymers, as well as their great flexibility and stretchability, have all contributed to the recent increase in the number of research papers addressing polymers in wearable and skin-attachable devices. Polymer incorporation in MXene materials is among the most extensively used strategies for increasing the stability and decreasing the total cost of MXene materials, which further increases their applicability in sensing. Thus, all performance criteria, such as sensitivity, stretchability, and durability, are challenging to integrate within a single sensor material. The polymer reinforced MXenes-based flexible will serve a growing role in promoting the advancement of healthcare applications.

Wearable sensors for long-term motion monitoring that are made to be worn on the body must also be comfortable. As a result, superhydrophobic properties must be introduced to maintain stability, environmental adaptability and increase the material's durability. In this context, it is important to develop wearable materials with corrosion resistance and mechanical damage resistance. Importantly Electronic waste (E-waste) generation is a growing problem due to the growing production of electronic devices. E-waste is certainly a challenge in the current era that must be tackled. In this context, biocompatible and biodegradable sensors need to be developed to lower the effect on human health and the environment. In terms of real-world applications, more research and discoveries are needed to create high-performance wearable sensing materials integrated with the Internet of Things for enhanced health monitoring systems that may extend our lives.

Acknowledgement

The author J. Aarthi acknowledges DST-CURIE Core grant for Women PG colleges (Reference No. DST/CURIE-PG/2022/8 (G)) for the financial support. The author R. A acknowledges DST-SYST, New Delhi (Ref. No: SP/YO/2019/1432) and UGC-DAE-CSR, Indore (Ref. No: CRS/2021-22/04/614) for the research grant.

References

1. Kedambaimoole, V., Harsh, K., Rajanna, K., Sen, P., Nayak, M.M., Kumar, S., MXene wearables: Properties, fabrication strategies, sensing mechanism and applications. *Mater. Adv.*, 3, 3784–3808, 2022.
2. Mukhopadhyay, S.C., Wearable sensors for human activity monitoring: A review. *IEEE Sens. J.*, 15, 3, 1321–1330, 2015.
3. Reeder, J.T., Xue, Y., Franklin, D., Deng, Y., Choi, J., Prado, O., Kim, R., Liu, C., Hanson, J., Ciraldo, J., Bandodkar, A.J., Krishnan, S., Johnson, A., Patnaude, E., Avila, R., Huang, Y., Rogers, J.A., Resettable skin interfaced microfluidic sweat collection devices with chemesthetic hydration feedback. *Nat. Commun.*, 10, 1, 5513, 2019.
4. Wang, Q., Ling, S., Liang, X., Wang, H., Lu, H., Zhang, Y., Self-healable multifunctional electronic tattoos based on silk and graphene. *Adv. Funct. Mater.*, 29, 16, 1808695, 2019.
5. Wang, Y., Chao, M., Wan, P., Zhang, L., A wearable breathable pressure sensor from metal-organic framework derived nanocomposites for highly sensitive broad-range healthcare monitoring. *Nano Energy*, 70, 104560, 2020.
6. Qi, K., He, J., Wang, H., Zhou, Y., You, X., Nan, N., Shao, W., Wang, L., Ding, B., Cui, S., A highly stretchable nanofiber-based electronic skin with pressure-, strain-, and flexion-sensitive properties for health and motion monitoring. *ACS Appl. Mater. Interfaces*, 9, 49, 42951–42960, 2017.
7. Liu, Z., Li, H., Shi, B., Fan, Y., Wang, Z.L., Li, Z., Wearable and implantable triboelectric nanogenerators. *Adv. Funct. Mater.*, 29, 20, 1808695, 2019.
8. Hou, C., Xu, Z., Qiu, W., Wu, R., Wang, Y., Xu, Q., Liu, X.Y., Guo, W., A biodegradable and stretchable protein-based sensor as artificial electronic skin for human motion detection. *Small*, 15, 11, 1805084, 2019.

9. Sharma, A., Ansari, M.Z., Cho, C., Ultrasensitive flexible wearable pressure/strain sensors: Parameters, materials, mechanisms and applications. *Sens. Actuators: A Phys.*, 347, 113934, 2022.
10. Mathew, M., Radhakrishnan, S., Vaidyanathan, A., Chakraborty, B., Rout, C.S., Flexible and wearable electrochemical biosensors based on two-dimensional materials: Recent developments. *Anal. Bioanal. Chem.*, 413, 727–762, 2021.
11. Sulleiro, M.V., Dominguez-Alfaro, A., Alegret, N., Silvestri, A., Gómez, I.J., 2D Materials towards sensing technology: From fundamentals to applications. *Sens. Bio-Sens. Res.*, 38, 100540, 2022.
12. Radhakrishnan, S., Mathew, M., Rout, C.S., Microfluidic sensors based on two-dimensional materials for chemical and biological assessments. *Mater. Adv.*, 3, 4, 1874–1904, 2022.
13. Rao, C.N.R., Sood, A.K., Subrahmanyam, K.S., Govindaraj, A., Graphene: The new two-dimensional nanomaterial. *Angew. Chem. – Int. Ed.*, 48, 42, 7752–7777, 2009.
14. Varghese, S.S., Varghese, S.H., Swaminathan, S., Singh, K.K., Mittal, V., Two-dimensional materials for sensing: Graphene and beyond. *Electronics*, 4, 3, 651–687, 2015.
15. Late, D.J., Huang, Y.K., Liu, B., Acharya, J., Shirodkar, S.N., Luo, J., Yan, A., Charles, D., Waghmare, U.V., Dravid, V.P., Rao, C.N.R., Sensing behavior of atomically thin-layered MoS_2 Should be MoS_2 transistors. *ACS Nano*, 7, 6, 4879–4891, 2013.
16. Bolotsky, A., Butler, D., Dong, C., Gerace, K., Glavin, N.R., Muratore, C., Robinson, J.A., Ebrahimi, A., Two-dimensional materials in biosensing and healthcare: From *in vitro* diagnostics to optogenetics and beyond. *ACS Nano*, 13, 9, 9781–9810, 2019.
17. Zhu, J., Ha, E., Zhao, G., Zhou, Y., Huang, D., Yue, G., Hu, L., Sun, N., Wang, Y., Lee, L.Y.S., Xu, C., Wong, K.Y., Astruc, D., Zhao, P., Recent advance in MXenes: A promising 2D material for catalysis, sensor and chemical adsorption. *Coord. Chem. Rev.*, 352, 306–327, 2017.
18. Xiao, B., Li, Y.C., Yu, X.F., Cheng, J.B., MXenes: Reusable materials for NH_3 sensor or capturer by controlling the charge injection. *Sens. Actuators B Chem.*, 235, 103–109, 2016.
19. Zhao, L., Wang, K., Wei, W., Wang, L., Han, W., High-performance flexible sensing devices based on polyaniline/MXene nanocomposites. *InfoMat*, 1, 3, 407–416, 2019.
20. Zhang, Y.Z., Wang, Y., Jiang, Q., El-Demellawi, J.K., Kim, H., Alshareef, H.N., MXene printing and patterned coating for device applications. *Adv. Mater.*, 32, 21, 1908486, 2020.
21. Li, N., Peng, J., Ong, W.J., Ma, T., Arramel, Zhang, P., Jiang, J., Yuan, X., Zhang, C., MXenes: An emerging platform for wearable electronics and looking beyond. *Matter*, 4, 2, 377–407, 2021.
22. Li, T., Chen, L., Yang, X., Chen, X., Zhang, Z., Zhao, T., Li, X., Zhang, J., A flexible pressure sensor based on an MXene-textile network structure. *J. Mater. Chem. C*, 7, 4, 1022–1027, 2019.
23. Yang, H., Xiao, X., Li, Z., Li, K., Cheng, N., Li, S., Low, J.H., Jing, L., Fu, X., Achavananthadith, S., Low, F., Wang, Q., Yeh, P.L., Ren, H., Ho, J.S., Yeow, C.H., Chen, P.Y., Wireless $Ti_3C_2T_x$ MXene strain sensor with ultrahigh sensitivity and designated working windows for soft exoskeletons. *ACS Nano*, 14, 9, 11860–11875, 2020.
24. Kim, S.J., Koh, H.J., Ren, C.E., Kwon, O., Maleski, K., Cho, S.Y., Anasori, B., Kim, C.K., Choi, Y.K., Kim, J., Gogotsi, Y., Jung, H.T., Metallic $Ti_3C_2T_x$ MXene gas sensors with ultrahigh signal-to-noise ratio. *ACS Nano*, 12, 2, 986–993, 2018.
25. Zuo, Y., Gao, Y., Qin, S., Wang, Z., Zhou, D., Li, Z., Yu, Y., Shao, M., Zhang, X., Broadband multi-wavelength optical sensing based on photothermal effect of 2D MXene films. *Nanophotonics*, 9, 1, 123–131, 2020.
26. Rhouati, A., Berkani, M., Vasseghian, Y., Golzadeh, N., MXene-based electrochemical sensors for detection of environmental pollutants: A comprehensive review. *Chemosphere*, 291, 132921, 2022.

27. Mashtalir, O., Cook, K.M., Mochalin, V.N., Crowe, M., Barsoum, M.W., Gogotsi, Y., Dye adsorption and decomposition on two-dimensional titanium carbide in aqueous media. *J. Mater. Chem. A*, 2, 35, 14334–14338, 2014.
28. Zhao, X., Vashisth, A., Prehn, E., Sun, W., Shah, S.A., Habib, T., Chen, Y., Tan, Z., Lutkenhaus, J.L., Radovic, M., Green, M.J., Antioxidants unlock shelf-stable $Ti_3C_2T_x$ (MXene) nanosheet dispersions. *Matter*, 1, 2, 513–526, 2019.
29. Ling, Z., Ren, C.E., Zhao, M.Q., Yang, J., Giammarco, J.M., Qiu, J., Barsoum, M.W., Gogotsi, Y., Flexible and conductive MXene films and nanocomposites with high capacitance. *Proc. Natl. Acad. Sci.*, 111, 47, 16676–16681, 2014.
30. Bouhamed, A., Müller, C., Choura, S., Kanoun, O., Processing and characterization of MWCNTs/epoxy nanocomposites thin films for strain sensing applications. *Sens. Actuators A: Phys.*, 257, 65–72, 2017.
31. Yao, S., Ren, P., Song, R., Liu, Y., Huang, Q., Dong, J., O'Connor, B.T., Zhu, Y., Nanomaterial-enabled flexible and stretchable sensing systems: Processing, integration, and applications. *Adv. Mater.*, 32, 15, 1902343, 2020.
32. Zhang, P., Chen, Y., Li, Y., Zhang, Y., Zhang, J., Huang, L., A flexible strain sensor based on the porous structure of a carbon black/carbon nanotube conducting network for human motion detection. *Sensors*, 20, 4, 1154, 2020.
33. Khalid, M.A.U. and Chang, S.H., Flexible strain sensors for wearable applications fabricated using novel functional nanocomposites: A review. *Compos. Struct.*, 284, 115214, 2022.
34. Lin, Y., Liu, S., Chen, S., Wei, Y., Dong, X., Liu, L., A highly stretchable and sensitive strain sensor based on graphene–elastomer composites with a novel double-interconnected network. *J. Mater. Chem. C*, 4, 6345–6352, 2016.
35. Wang, L., Wang, D., Wu, Z., Luo, J., Huang, X., Gao, Q., Lai, X., Tang, L.C., Xue, H., Gao, J., Self-derived superhydrophobic and multifunctional polymer sponge composite with excellent joule heating and photothermal performance for strain/pressure sensors. *ACS Appl. Mater. Interfaces*, 12, 11, 13316–13326, 2020.
36. Dai, Z., Ding, S., Lei, M., Li, S., Xu, Y., Zhou, Y., Zhou, B., A superhydrophobic and anti-corrosion strain sensor for robust underwater applications. *J. Mater. Chem. A*, 9, 27, 15282–15293, 2021.
37. Chen, B., Zhang, L., Li, H., Lai, X., Zeng, X., Skin-inspired flexible and high-performance MXene@polydimethylsiloxane piezoresistive pressure sensor for human motion detection. *J. Colloid Interface Sci.*, 617, 478–488, 2022.
38. He, J., Shi, F., Liu, Q., Pang, Y., He, D., Sun, W., Peng, L., Yang, J., Qu, M., Wearable superhydrophobic PPy/MXene pressure sensor based on cotton fabric with superior sensitivity for human detection and information transmission. *Colloids Surf. A: Physicochem. Eng. Asp.*, 642, 128676, 2022.
39. Qin, R., Li, X., Hu, M., Shan, G., Seeram, R., Yin, M., Preparation of high-performance MXene/PVA-based flexible pressure sensors with adjustable sensitivity and sensing range. *Sens. Actuators A. Phys.*, 338, 113458, 2022.
40. Chang, K., Li, L., Zhang, C., Ma, P., Dong, W., Huang, Y., Liu, T., Compressible and robust PANI sponge anchored with erected MXene flakes for human motion detection. *Compos. Part A*, 151, 106671, 2021.
41. Li, L., Fu, X., Chen, S., Uzun, S., Levitt, A.S., Shuck, C.E., Han, W., Gogotsi, Y., Hydrophobic and stable MXene-polymer pressure sensors for wearable electronics. *ACS Appl. Mater. Interfaces*, 12, 13, 15362–15369, 2020.
42. Peng, Z., Zhang, X., Zhao, C., Gan, C., Zhu, C., Hydrophobic and stable MXene/reduced graphene oxide/polymer hybrid materials pressure sensors with an ultrahigh sensitive and

rapid response speed pressure sensor for health monitoring. *Mater. Chem. Phys.*, 271, 124729, 2021.

43. Zhao, Q., Yang, L., Ma, Y., Huang, H., He, H., Ji, H., Wang, Z., Qiu, J., Highly sensitive, reliable and flexible pressure sensor based on piezoelectric PVDF hybrid film using MXene nanosheet reinforcement. *J. Alloys Compd.*, 886, 161069, 2021.
44. Ren, M., Sun, Z., Zhang, M., Yang, X., Guo, D., Dong, S., Dhakal, R., Yao, Z., Li, Y., Kim, N.Y., A high-performance wearable pressure sensor based on an MXene/PVP composite nanofiber membrane for health monitoring. *Nanoscale Adv.*, 4, 18, 3987–3995, 2022.
45. Wang, H., Xiang, J., Wen, X., Du, X., Wang, Y., Du, Z., Cheng, X., Wang, S., Multifunctional skin-inspired resilient MXene-embedded nanocomposite hydrogels for wireless wearable electronics. *Compos. Part A: Appl. Sci. Manuf.*, 155, 106835, 2022.
46. Hempel, M., Nezich, D., Kong, J., Hofmann, M., A novel class of strain gauges based on layered percolative films of 2D materials. *Nano Lett.*, 12, 11, 5714–5718, 2012.
47. Jia, Z., Li, Z., Ma, S., Zhang, W., Chen, Y., Luo, Y., Jia, D., Zhong, B., Razal, J.M., Wang, X., Kong, L., Constructing conductive titanium carbide nanosheet (MXene) network on polyurethane/polyacrylonitrile fibre framework for flexible strain sensor. *J. Colloid Interface Sci.*, 584, 1–10, 2021.
48. Zhang, Z., Weng, L., Guo, K., Guan, L., Wang, X., Wu, Z., Durable and highly sensitive flexible sensors for wearable electronic devices with PDMS-MXene/TPU composite films. *Ceram. Int.*, 48, 4, 4977–4985, 2022.
49. Li, X., Yang, J., Yuan, W., Ji, P., Xu, Z., Shi, S., Han, X., Niu, W., Yin., F., Microstructured MXene/polyurethane fibrous membrane for highly sensitive strain sensing with ultra-wide and tunable sensing range. *Compos. Commun.*, 23, 100586, 2021.
50. Chao, M., Wang, Y., Ma, D., Wu, X., Zhang, W., Zhang, L., Wan, P., Wearable MXene nanocomposites-based strain sensor with tile-like stacked hierarchical microstructure for broad-range ultrasensitive sensing. *Nano Energy*, 78, 105187, 2020.
51. Fu, W., Wang, Q., Liu, H., Cao, J., Xu, J., Yuan, X., Zhang, Y., Ren, X., Highly stretchable and conductive MXene/polyurethane composite film coated on various flexible substrates for ultrasensitive strain sensors. *Mater. Lett.*, 320, 132328, 2022.
52. Shao, Z., Hu, X., Cheng, W., Zhao, Y., Hou, J., Wu, M., Xue, D., Wang, Y., Degradable self-adhesive epidermal sensors prepared from conductive nanocomposite hydrogel. *Nanoscale*, 12, 36, 18771–18781, 2020.
53. Zheng, C., Yue, Y., Gan, L., Xu, X., Mei, C., Han, J., Highly stretchable and self-healing strain sensors based on nanocellulose-supported graphene dispersed in electro-conductive hydrogels. *Nanomaterials*, 9, 7, 937, 2019.
54. An, R., Zhang, B., Han, L., Wang, X., Zhang, Y., Shi, L., Ran, R., Strain-sensitivity conductive MWCNTs composite hydrogel for wearable device and near-infrared photosensor. *J. Mater Sci.*, 54, 11, 8515–8530, 2019.
55. Huang, C., Luo, Q., Miao, Q., He, Z., Fan, P., Chen, Y., Zhang, Q., He, X., Li, L., Liu, X., MXene-based double-network organohydrogel with excellent stretchability, high toughness, anti-drying and wide sensing linearity for strain sensor. *Polymer*, 253, 124993, 2022.
56. Yuan, W., Qu, X., Lu, Y., Zhao, W., Ren, Y., Wang, Q., Wang, W., Dong, X., MXene-composited highly stretchable, sensitive and durable hydrogel for flexible strain sensors. *Chin. Chem. Lett.*, 32, 6, 2021–2026, 2021.
57. Chen, K., Hu, Y., Wang, F., Liu, M., Liu, P., Li, C., Yu, Y., Xiao, X., Feng, Q., Ultra-stretchable, adhesive, and self-healing MXene/polyampholytes hydrogel as flexible and wearable epidermal sensors. *Colloids Surf. A: Physicochem. Eng. Asp.*, 645, 128897, 2022.

58. Li, S.N., Yu, Z.R., Guo, B.F., Guo, K.Y., Li, Y., Gong, L.X., Zhao, L., Bae, J., Tang, L.C., Environmentally stable, mechanically flexible, self-adhesive, and electrically conductive $Ti_3C_2T_x$ MXene hydrogels for wide-temperature strain sensing. *Nano Energy*, 90, 106502, 2021.
59. Xie, M., Li, S., Qi, X., Chi, Z., Shen, L., Islam, Z., Dong, Y., Thermal and infrared light self-repairing, high sensitivity, and large strain sensing range shape memory MXene/CNTs/EVA composites fiber strain sensor for human motion monitoring. *Sens. Actuators A: Phys.*, 347, 113939, 2022.
60. Divya, Mahapatra, S., Srivastava, V.R., Chandra, P., Nanobioengineered sensing technologies based on cellulose matrices for detection of small molecules, macromolecules, and cells. *Biosensors*, 11, 6, 168, 2021.
61. Suginta, W., Khunkaewla, P., Schulte, A., Electrochemical biosensor applications of polysaccharides chitin and chitosan. *Chem. Rev.*, 113, 7, 5458–5479, 2013.
62. Guo, Y., Zhong, M., Fang, Z., Wan, P., Yu, G., A wearable transient pressure sensor made with MXene nanosheets for sensitive broad-range human-machine interfacing. *Nano Lett.*, 19, 2, 1143–1150, 2019.
63. Wang, L., Wang, D.P., Wang, K., Jiang, K., Shen, G., Biocompatible MXene/chitosan-based flexible bimodal devices for real-time pulse and respiratory rate monitoring. *ACS Mater. Lett.*, 3, 7, 921–929, 2021.
64. Liu, Y., Sheng, Z., Huang, J., Liu, W., Ding, H., Peng, J., Zhong, B., Sun, Y., Ouyang, X., Cheng, H., Wang, X., Moisture-resistant MXene-sodium alginate sponges with sustained superhydrophobicity for monitoring human activities. *Chem. Eng. J.*, 432, 134370, 2022.
65. Su, T., Liu, N., Gao, Y., Lei, D., Wang, L., Ren, Z., Zhang, Q., Su, J., Zhang, Z., MXene/cellulose nanofiber-foam based high performance degradable piezoresistive sensor with greatly expanded interlayer distances. *Nano Energy*, 87, 106151, 2021.

13

MXene-Based Polymer Composites for Various Biomedical Applications

Jamuna Bai Aswathanarayan[1]*, SubbaRao V. Madhunapantula[2] and Ravishankar Rai Vittal[3]

[1]*Department of Microbiology, School of Life Sciences, JSS Academy of Higher Education and Research, Mysore, India*

[2]*Centre of Excellence in Molecular Biology & Regenerative Medicine, Department of Biochemistry, JSS Medical College, JSS Academy of Higher Education and Research, Mysore, India*

[3]*Department of Microbiology, University of Mysore, Mysore, India*

Abstract

MXene are 2D nanomaterials that were discovered in the year 2011. These are hydrophilic and conductive, and structurally metal carbides, nitrides, and carbonitrides sheets. The formula of MXene is written as $M_{n+1}X_nT_z$, here 'M' belongs to transition metal, while 'X' could be either carbon and/or nitrogen, and T_z are surface functional groups which could include a combination of oxygen, hydroxyl, and fluoride groups, and value of n >1 and <3. Mxene are derived from the multilayered ternary $M_{n+1}AX_n$, or MAX phases. Here 'A' is a group IV–V element and etching this layer leads to synthesis of MXenes. Certain functional groups impart hydrophilicity to MXenes and aid in the formation of stable aqueous suspensions, while polar solvents confer neutral pH to the suspensions. The functional groups are also attributed for their reactivity with PVA, PDDA and other such hydrophilic polymers. MXenes have unique structure and due to their biocompatibility, ability to convert light to heat and mechanical flexibility are finding an increased biomedical application. MXenes and their polymer composites are used for engineering sensors, in bioimaging and regeneration of tissues/scaffolds. The therapeutic applications of MXenes composites include photothermal therapy in tumor treatment, as antimicrobials and in drug delivery.

***Keywords*:** MXenes, MXene/polymer composites, biomedical applications, photothermal therapy, antimicrobial activity, tissue engineering

13.1 Introduction to MXenes

The 2D nanomaterial MXene is gaining immense research interest as its structure can be easily modified and it possesses unique physico-chemical characteristics. Mxene was for the first time synthesized in Drexel University, Philadelphia during exfoliation of 3D titanium aluminium carbide (Ti_3AlC_2) with hydrofluoric acid (HF). It resulted in the production of layers of 2D titanium-carbide (Ti_3C_2). Currently, MXenes are synthesized using

**Corresponding author*: jamunabhounsle@gmail.com

transition metal ternary carbides and nitrides called MAX phase [1]. These MAX phases are 3D precursors represented as $M_{n+1}AX_n$. Here, the metal M belongs to transition group, while the element A belongs to III A or IV A, while X is either carbon or nitrogen and $n = 1, 2$ or 3 [2]. In the MAX precursors, M–X is stronger in comparison to M-A bond. The A layer is highly active. Thus, by using a strong acid such as HF, A layers are selectively eliminated by etching resulting in the production of $M_{n+1}X_n$ layers. On sonication these layers, easily separate and further due to their high surface energy are terminated with functional groups of fluorine, hydroxide, and oxygen [2]. The MXene formula is $M_{n+1}X_nT_x$, with T_x as the functional groups [3]. MXenes used for biomedical application are generally Ti_3C_2 and Ti_2C. MXenes have a broad technological and biomedical application due to its unique and specific fundamental structural, optical, electronic, and biological properties [4, 5]. MXenes have large surface area which makes it suitable for drug delivery, therapeutics, and theranostics [6]. The hydrophilic functional groups on surfaces impart it with rigidness and flexibility required to prepare thin films for biosensors [7]. The high atomic numbers and paramagnetic behavior of transition metals make them useful in diagnostics and imaging as in X-ray or CT and as contrasting agents in MRI [8]. Thus, by modifying or functionalizing the biocompatibility of MXenes can be enhanced for biological and medical applications.

In this chapter, the various applications of MXenes and MXene based polymer nanocomposites in health and biomedicine will be discussed. MXenes are used in designing biosensor platforms, as these have enormous surface area with desired electrical and optical properties, hydrophilicity, and numerous surface functional groups. These have biocompatibility and also can stimulate adhesion and proliferation of cells. We will focus on their preparation as composite hydrogels and further use in repair and regeneration of bone, nerve, and myocardial tissue. In this chapter, we will review the mechanism, function, and application of MXenes and their biocomposites as anti-cancer and anti-microbial therapeutics. MXenes such as $Ti_3C_2T_x$ nanosheets exhibit antibacterial activity due to their direct contact with pathogen cell membrane through formation of H bonds with the O group on MXene surface and the LPS in bacterial cell membrane. The small size helps the MXene to enter cells, disrupt cellular structures and membranes resulting in cell death [9]. Similarly, MXenes are being explored in minimally invasive treatments including PTT and PDT to destroy tumor cells. MXenes absorb near-infrared light in the tumor cells and exhibit photothermal conversion which helps in targeted killing of cancer cells. In cancer therapy, modified Ti_3C_2 nanosheets MXenes are also used as a drug carrier and as contrast agents in bioimaging. Similarly, with their use as photosensitizers, they can be applied in PDT to activate ROS based damage to tumor cells. Though they cause EPR effect, the surface modifications can help in designing a controlled release nanoplatform for subsequent use as drug delivery agents [10].

13.2 Synthesis of MXenes and Their Physicochemical Properties

MXenes of varying size and morphology are manufactured using both bottom-up as well as top-down methods based on the required physico-chemical properties [5]. MXenes surface is modified to make them biocompatibile and decrease the cytotoxicity. One of the most commonly used top-down synthesis method is the acid exfoliation of the MAX phase to remove the A layered solid (Figure 13.1). The resulting MXene is structurally similar to its 3D template [11]. The synthesis involves two steps with MAX phase being etched or cleaved, followed by separation of MXene layers. MAX phase is synthesized by mixing pure M, A, and X powders at a specific ratio and subjecting it to temperatures as high as ~1200–1600 °C. Further, applying pressure of ~25–45 MPa by hot or cold press adds to density, removes microvoids or cracks, and imparts grain growth [12]. High quantity of the A element, Si or Al is added to make up for its evaporation in high temperature vacuum/argon atmosphere [13]. Pressure less sintering can also be used to synthesize highly oriented MAX phase. It is a simple method which involves heating MAX powders to specific temperatures. Only limitation is that the sintering has an effect on the orientation of MAX phase [14].

Once the MAX phase is ready, it is etched using a strong etchant, HF. The M–A bonds are weak and fluoride ions strongly bind to A. Therefore, it is easy to selectively etch out A layers. Further, addition of acid creates surface having the functional groups –OH or –O or –F [15]. The MAX phase etching with varying HF concentrations, duration of immersion, and temperature has been tested. MXene for the first time was synthesized by using 50 % HF for immersing the compound Ti_3AlC_2 for 2 h at room temperature resulting in complete MAX phase dissociation [16]. Similarly, MXene has been synthesized by immersing Ti_3C_2 in 48 % HF-water and 40 % HF at RT and 45 °C, respectively. The latter resulted in the production of multi-layered nanosheets [17].

In the recent years, *in situ* HF formation is used for etching. It involves an acid such as HCl and fluoride salt reaction to form HF. MAX phases of Nb_2AlC and Ti_2AlC have been generated using this technique. Nanosheets of Ti_3C_2 and Ti_2C have been prepared using etching fluoride salts of lithium, sodium, potassium, and ammonium in HCl [18]. Though HF is the commonly employed MAX phase etchant, it is highly cytotoxic and induces cell

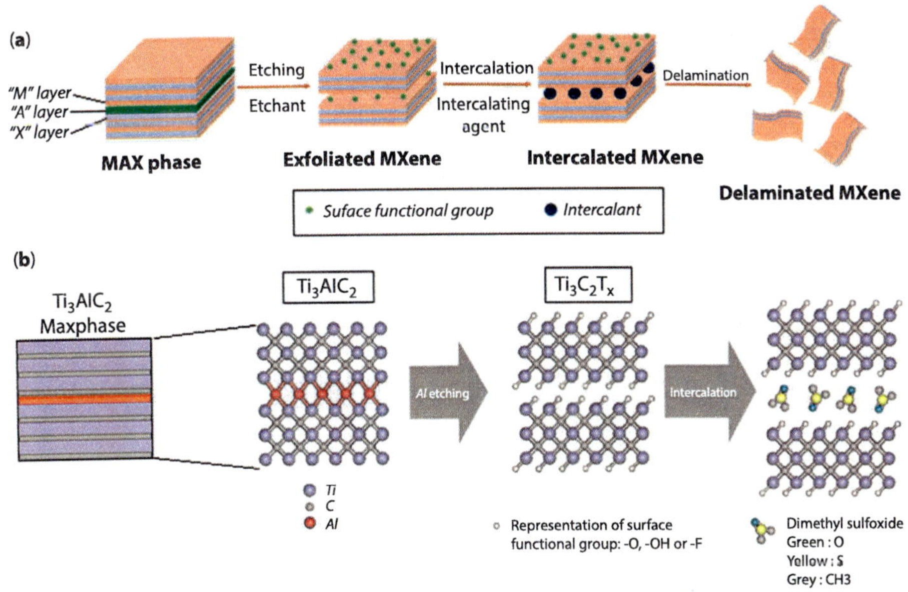

Figure 13.1 (a) MXene synthesis by top-down acid exfoliation method. (b) Etching and intercalation of Ti_3AlC_2 to form Ti_3C_2 MXene. Adapted from Ref. [5]. Copyright 2021. Springer Nature.

death. Therefore, its toxicity is a major impediment in the MXenes' biological application [19]. The fluoride ions (F^-) present on the surface of MXene limits the addition of functional groups and conjugate fluoride ions [20]. Thus, fluorine-free MXenes synthesis is preferable for biomedical applications as it allows for functional surface termination. It is also important that for biomedical uses, MXene nanosheets should be in nano size range. The multilayered MXenes are large sized and not suitable due to low therapeutic efficacy and biosafety issues. If MXene sheets are not in nano-sized range, they cannot be used for intravenous administration, due to their poor transport in circulatory system [21]. In top-down approach, it is essential to optimize suitable etchant, etching time and temperature, and type of intercalant.

MXene can also be synthesized by bottom-up approach [6]. The commonly used method is the chemical vapor deposition (CVD) where multi-layered thin films can be easily obtained as explained in Figure 13.2 [5]. CVD is used to produce high quality MXene of ultrathin Mo_2C with methane and a Cu/Mo foil substrate at 1085°C. These films are defect-free, have high crystallinity, and lack surface functional groups [22, 23]. However, biomedical application requires functionalized MXenes in nanosize range for cell permeability. The bottom-up approaches used for MXene synthesis also include template and pulsed laser deposition enhanced by plasma [24]. The limitation of these methods is that the MXene synthesized have biocompatibility issues [25].

13.3 Biomedical Applications of MXenes

MXenes due to their unique physico-chemical properties, biocompatibility, amenability to surface functionalization and modification have led to their potential application as biosensors and diagnostics, cancer theranostics, drug delivery, and antimicrobials.

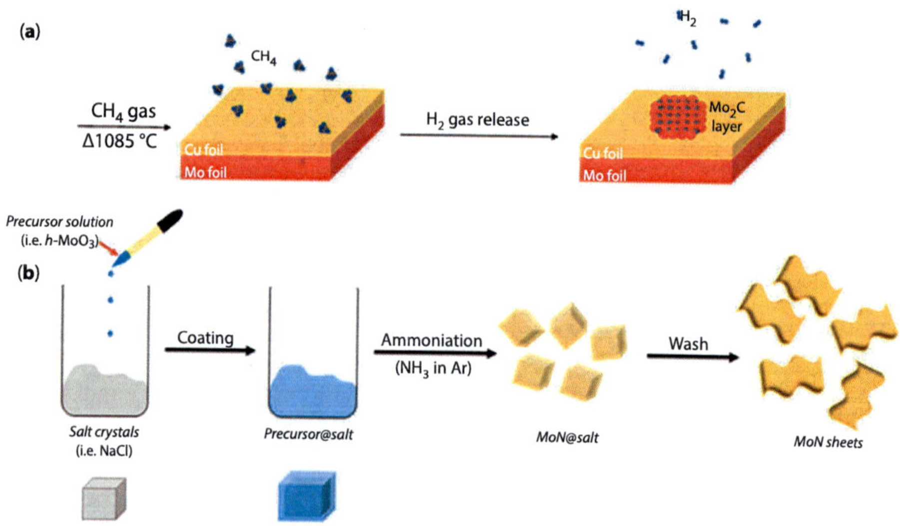

Figure 13.2 (a) Bottom-up approach for MXene synthesis by chemical vapor deposition of Mo and C to form Mo_2C thin film in gas chamber. (b) Salt-template synthesis of MoN nanosheets. Adapted from Ref. [5]. Copyright 2021. Springer Nature.

13.3.1 Engineering Biosensors and Bioimaging Platforms

The MXene composites are attractive as biosensors and for imaging applications as they are amenable to surface modification as well as functionalization. They have a surface negative charge due to the functional groups including hydroxyl, oxy and fluoride. Hence, positively charged particles can selectively adsorb to MXenes by electrostatic interaction. Therefore, biocompatible and biodegradable polymers with large surface area can be used to prepare Mxene composites enabled with sensing capabilities for application in biomedical settings. TiO_2-Ti_3C_2 MXene and nafion nanocomposite was used in developing a mediator-free biosensor to immobilize hemoglobin and detect hydrogen peroxide (Figure 13.3) [26]. The nanocomposite is a good immobilization matrix and has biocompatibility for protein (Figure 13.4) [26]. The nanocomposite facilitates transfer of electron in Hb. This biosensor can detect H_2O_2 between 0.1 and 380 micromoles and has a LOD of 14 nanomoles. MXenes sense H_2O_2 as it can oxidize the surface functional groups, which in turn increases oxygen density and results in electron transfer to Hb without affecting its bioactivity. The performance of biosensor for hydrogen peroxide detection was assessed by amperometry which in presence of H_2O_2 showed a response time less than 3 s. The sensitivity was 447.3 $\mu A\ mM^{-1}\ cm^{-2}$ with a LOD of 14 nM. Nafion/Hb/Ti3C2/GCE had lower current response to the substrate but a detectable and weaker sensitivity than Nafion/Hb/TiO2-Ti3C2/GCE (Figure 13.5) [26]. TiO_2 nanoparticles due to their biocompatibility retain Hb bioactivity and confer long-term stability of up to 94.6 % even after 60 days of storage in the biosensors [26].

MXenes are used in developing organic electrochemical transistors (MOECTs) biosensors. These have been used for detecting fPSA/tPSA. The presence of fPSA/tPSA between

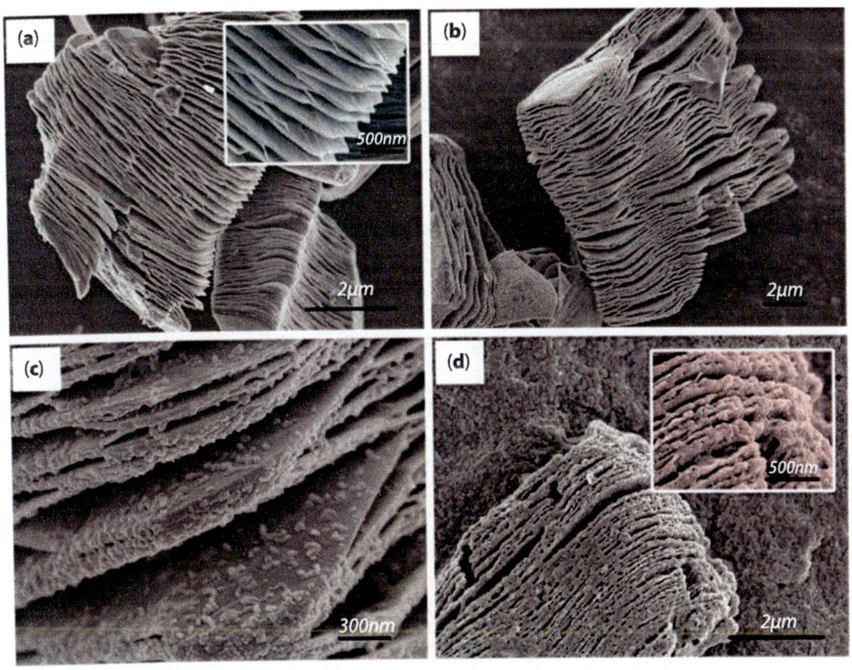

Figure 13.3 SEM images of Ti_3C_2 (a), TiO_2-Ti_3C_2 nanocomposite (b)-(c), and Nafion/Hb/TiO_2-Ti_3C_2 composite film (d). Reproduced with permission from Ref. [26]. Copyright 2015, Elsevier.

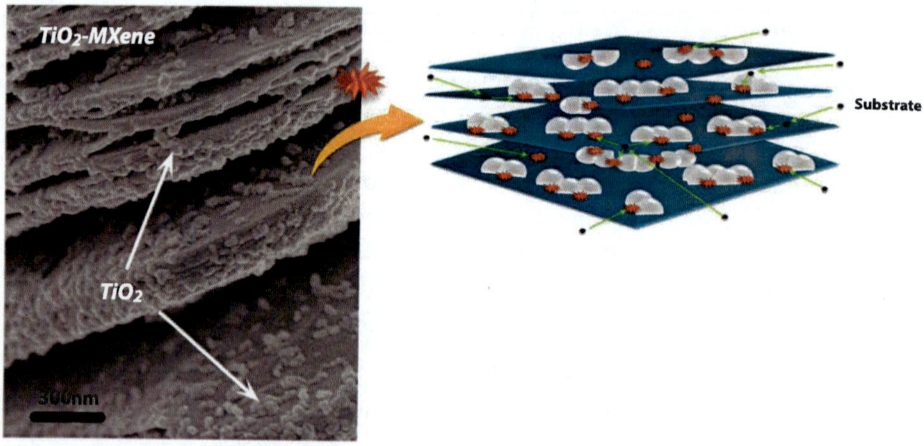

Figure 13.4 TiO_2-Ti_3C_2 MXene nanocomposite for hemoglobin immobilization and detection of hydrogen peroxide. Reproduced with permission from Ref. [26]. Copyright 2015, Elsevier.

4 and 10 ng/ml can be used as a biomarker for diagnosing prostate cancer. However, these are present in trace quantity in peripheral blood which also has biological complexity making it difficult to detect them. The interdigitated spiral MXene biosensor is useful in sensitive detection of fPSA/tPSA. MXene having multiple spiral design causes amplification of amperometric signal. The functionalization with both anti-tPSA and –fPSA, increases sensitivity and improves detection limit to 0.01 pg/ml. Therefore, MOECTs can be used in clinics for ultrasensitive detection and quantification of fPSA/tPSA and in other similar point-of-care testing applications [27]. Electrochemical biosensor made of MXene based ultrathin micro-patterned PEDOT: PSS amplifies signal and enhances transistor's performance. It is used for the detection of survivin protein in osteosarcoma in children. The use of intercalated MXene increases the capacitance as well as mobility of biosensor. The biosensor had a LOD for surviving at 10 pg mL^{-1}. The MXene based OECTs biosensor are non-invasive and portable and therefore, have potential for the point-of-care applications [28].

$Ti_3C_2T_x$ MXene nanosheets are used for fabrication of biosensor for detection of glucose based on enzymatic reaction. It was developed by depositing 32 glucose oxidase enzyme (GOx) units on nanocomposite of Nafion Au and MXene. Au NPs of crystalline nature are fixed to the surface MXene sheets using chemical reduction. The Au/MXene nanocomposite is further used as a matrix for immobilization GOx. Dropcasting technique can be used for depositing GOx on Au/MXene/Nafion/ in GCE. On attaching gold nanoparticles on MXene nanosheets, the efficiency of transfer of electron between GCE and GOx is enhanced. Due to its high electrocatalytic activity, the biosensor detected glucose between 0.1 to 18 mM, exhibited 4.2 μA mM^{-1} cm^{-2} of sensitivity, and its LOD was 5.9 μM. Thus, this Mxene GCE biosensor can detect glucose at ultrasensitive concentrations in biological samples for diagnosing Diabetes Mellitus [29].

GOx has been immobilized on Ti_3C_2 MXene nanosheets for glucose oxidation and hydrogen peroxide catalysis for glucose detection. Poly-l-lysine (PLL) is an amino-rich polypeptide and is used to functionalize the ultra-thin Ti_3C_2 MXene nanosheets. This makes the MXene surface compatible for GOx immobilization as they now possess a positive charge

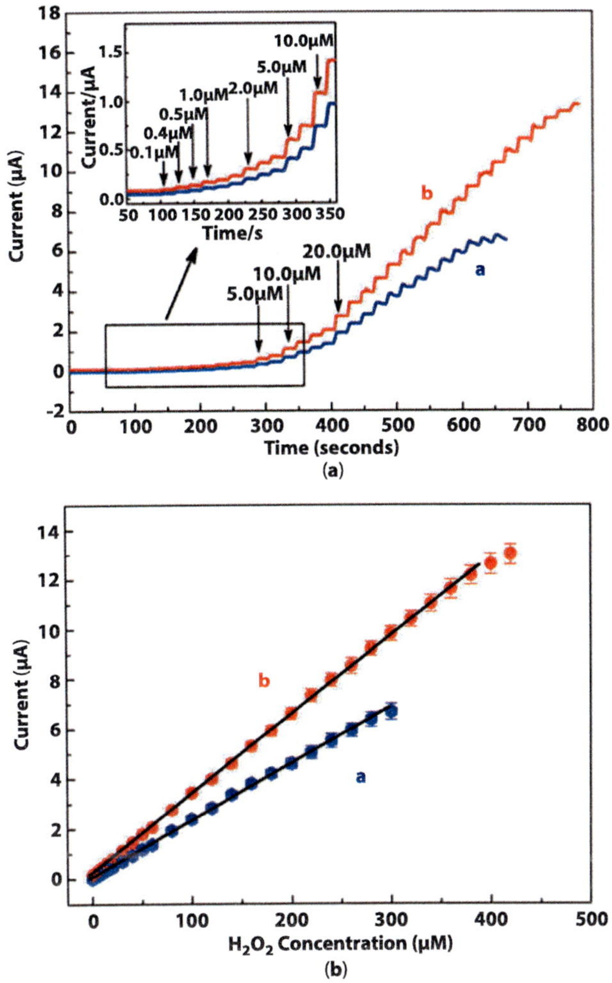

Figure 13.5 (a) Current and time response of GCE composed of Nafion, Hb & Ti_3C_2 (curve a) and Nafion, Hb & TiO_2-Ti_3C_2 (curve b) at -0.35V with increasing concentrations of H_2O_2 (b) Steady state current vs. H_2O_2 concentration: GCE composed of Nafion, Hb & Ti_3C_2 (curve a) and Nafion, Hb & TiO_2-Ti_3C_2 (curve b). Reproduced with permission from Ref. [26]. Copyright 2015, Elsevier.

on the surface. Such a modification increased GOx deposition on MXene nanosheets by 50 %. GOx cross-linked to the PLL amine groups can be observed as GOx-PLL complex on Mxene nanosheets. Ti_3C_2-PLL-GOx nanosheets exhibit higher catalytic activity than GOx immobilized on porous silica. Ti_3C_2 nanosheets catalyze H_2O_2 hydrolysis that is produced as a result of glucose oxidation. The removal of this toxic intermediate enhances the enzymatic activity. As Ti_3C_2 MXene have high electric conductivity, the Ti_3C_2-PLL-GOx nanosheets can be combined with GCE to design biosensor with a LOD of 2.6 μM for glucose [30]. pH-responsive point-of-care immunoassay biosensor functionalized using GOx enzyme and secondary antibody on Ti_3C_2-MXene nanosheets can quantify H1N1 virus. The immunoassay involves sandwich reaction in the antibody-coated microplate. The nanosheets of Ti_3C_2-MXene offer sufficient surface area for loading GOx molecules and thus improve signal amplification. The GOx converts glucose to hydrogen peroxide and gluconic acid. The resulting gluconic acid reduces pH which is detected using a pH meter. This immunoassay platform has a high sensitivity and can detect H1N1 virus between 0.01 to 100 μg mL^{-1} and a LOD of 1.3 ng mL^{-1}. The biosensor has high sensitivity, specificity, reproducibility, reliability, and accuracy in detecting H1N1 virus from clinical samples [31].

Hydrothermal technique has been used to develop peroxidase containing Ti_3C_2/CuS MXene nanocomposites. The synergism between MXene-Ti_3C_2 nanosheets and CuS NPs increased peroxidase activity based on ping-pong mechanism. The peroxidase in presence of H_2O_2 catalyzes the 3,3,5,5-tetramethylbenzidine into a blue color product. The cost effective colorimetric biosensor platform can be used for the detecting cholesterol (10–100 μM). As LOD is 1.9 μM, the biosensor is highly sensitive and can be used for clinical diagnostics [32]. MXenes are used in the food industry for developing high-performance biosensors for taste sensing. The MXenes due to their features including high conductivity, electron transport, and amenability to functionalization have been used to design biomimetic tongues [33]. $Ti_3C_2T_x$ MXene nanosheets with β-hydroxybutyrate (BHB) dehydrogenase are used to develop amperometric biosensor for detecting BHB. The MXene due to its layered structure and biocompatibility is a good immobilization matrix for BHB dehydrogenase. The MXene-based biosensor with - 0.35 V can detect even 0.36 mM of BHB and has a LOD of 45 μM [34].

Ti_3C_2-MXene nanosheets are uniformly coated with aminosilane and immobilized with bio-receptor anti-carcinoembryonic antigen (anti-CEA). This label free biosensor showed cancer biomarker CEA detection with high sensitivity. The f-Ti_3C_2-MXene biosensor used a hexa-amine-ruthenium probe and had a LOD and sensitivity for CEA at 0.0001 to 2000 ng/mL and 37.9 μA ng/mL/cm^{-2}, respectively. The wide detection ranges of these fabricated MXene biosensors is greater than 2D and hybrid 2D nanosensors and thus can find a wide range of biomedical application [35]. SPR biosensor has been used for ultrasensitive recognition of CEA. It was developed using Ti_3C_2-MXene multi-walled carbon nanotube-polydopamine as sensing platform and Ag nanoparticle for enhancing signals. Ti_3C_2-MXene surface area is an ideal hydrophilic-biocompatible matrix for SPR and in combination with AuNP is used to prepare a nanocomposite for attaching staphylococcal protein A and immobilizing monoclonal anti-CEA Ab. Further, multi-walled carbon nanotube-polydopamine conjugated AgNPs linked with polyclonal anti-CEA antibody is added. The biosensor has high specificity and detects CEA in serum between 2×10^{-16} to 2×10^{-8} M at a LOD of 0.07 fM. It can be used for rapid diagnosis and routine monitoring of various types of cancer [36]. Similarly, another SPR based biosensor, Ti_3C_2-MXene (N-Ti_3C_2-MXene) functionalized with amino groups has been used to detect CEA. Here, the ultrathin N-Ti_3C_2-MXene is used as sensor and hollow gold nanoparticles (HGNPs) is used for signal enhancing. The aminosilane functionalized MXene nanosheets are hydrophilic and biocompatible for attaching SPA and covalently immobilizing anti-CEA capture antibody. The nanohybrids with SPA and immobilized with anti-CEA detection antibody forms a sandwiched nanocomplex. This nanocomplex of MXene-based SPR chip biosensor can detect 0.001-1000 pM of CEA with 0.15 fM as LOD [37].

The presence of osteopontin (OPN) protein in patient samples is used for the diagnosis of osteosarcoma. $Ti_3C_2T_x$ with phosphomolybdic acid (PMo_{12}) and polypyrrole nanohybrid is used to attach OPN aptamer. This impedimetric aptamer biosensor was used to detect OPN. It has high electrochemical activity, good stability, biocompatibility, and binds strongly to OPN aptamer. PPy@$Ti_3C_2T_x$/PMo_{12} based electrochemical aptasensors had a

LOD of 0.98 fg/mL for OPN. Thus, MXene based aptasensors can be used for sensitive detection of OPN in clinical diagnosis of osteosarcoma [38]. The limitations of using the conventional fluorescent probes are photobleaching and non-biodegradable nature resulting in unsteady fluorescence signal and toxicity. Therefore, photostable and biodegradable fluorescent probes have potential application in biomedicine. Novel Nb_2C quantum dots (Nb_2C QDs) were synthesized using ultrasound technique. Nb_2C QDs exhibited chemical- and photostable fluorescence emission and high biocompatibility and biodegradation by human myeloperoxidase making them safe for *in vivo* applications. The MXene based Nb_2C QDs can be used for imaging and sensing application in biomedical diagnosis [39].

Exosomes are effective biomarkers for early cancer detection by liquid biopsy. An electrochemical biosensor was developed using Prussian Blue and MXenes to detect exosome and its surface proteins. *In situ* Fe_4 $[Fe(CN)_6]_3$ was generated on Ti_3C_2 MXene as nanoprobes. Further it was modified with a CD63 aptamer altered with PAMAM and gold nanoparticle for specific interaction with CD63 on OVCAR exosomes. A CD63-modified Ti_3C_2 MXene nanocarrier can be used to load numerous aptamers to bind the exosomes. High loading of Prussian blue onto Ti_3C_2 MXene can amplify low potential electrochemical signal and avoid usage of electrochemical active species. The electrochemical biosensor has a LOD of 229 particles/μL. It can also detect exosomes of HeLa and BT474, and can be used for early detection of cancers [40]. Similarly, hierarchical gold nanoarrays have been added to Ti_2CT_x MXene building blocks to generate a 2D membranes as sensors for detection of exosomes. This biosensor has enhanced electrocatalytic activity and numerous aptamer immobilization as a result of good conductivity of gold nanoarrays. Such MXene composite membrane functionalized with aptamer can bind specifically to EpCAM and exosomes. CD63 on exosomes interact with aptamers and enhance the sensitivity and accuracy. Thus, the biosensor has high sensitivity for exosome with 58 particles/μL and demonstrated high detection efficacy in real serum samples [41]. Gold nanoparticles and Ti_3C_2 MXenes hybrid functionalized with CD63 aptamers have been used to design electrogenerated chemiluminescence biosensor for detection of exosomes. Here, MXenes have a dual role and is used as both a reductant and stabilizer. The ultrasensitive detection of exosomes is due to synergism of large surface area of the composite and its functional properties including good conductivity and catalytic activity. In case of exosomes from HeLa cell line, it had a LOD of 30 particles/μL and similar selectivity to exosomes from OVCAR cells well as HepG2 cells [42].

Aminoglycoside antibiotics can be detected using an electrochemical aptasensor on a screen-printed carbon electrode. On modifying Ti_3C_2 MXene with a mesoporous carbon, there was an improvement in the stability of the aptasensor and signal amplification. The MXene nanocomposites acts as carrier for aptamers. As the aptamers and antibiotics interact, the electron transfer on the electrode generates signals. The aptasensor has a LOD of 3.51 nM and is effective in detecting residual antibiotics in foods including milk [43]. Long range SPR designed using PDA, Ti_3C_2-MXene and gold nanoparticles have enhanced sensing and detection activity in comparison to conventional SPR based sensing platforms. The design involves sandwich of Ti_3C_2-MXene gold nanoparticles film between polydopamine films, wherein the substrate in composite acts and signal is enhanced using gold nanoparticles. This platform can bind to high concentrations of rabbit anti-human IgG due to its surface area and further amplification due to sandwich design increases biosensor sensitivity (from 0.075 to 40 μg mL^{-1}) [44].

Label free biocompatible biosensors with high conductive interfaces have been developed using MXenes. It can be loaded with immobilizing Ab or Ag and used for ultrasensitive detection of CEA with an impressive LOD of 0.0429 ng/mL. This platform was constructed by functionalizing MXene with polyaniline and assembling gold nanoparticle decorated β-Cyclodextrin on its surface. The former provides high electron conductivity, while the latter is used for immobilizing antibody. The MXene nanocomposite sensor platform has stability and efficient performance in clinic serum samples [45]. MicroRNAs when expressed abnormally are considered diagnostic and prognostic biomarkers for several cancer types. Label free electrochemical biosensor have been designed using MXene-MoS_2 heterostructure for miRNA detection. Here, MXene composite is linked to a catalytic hairpin assembly. The heterostructures of nanosize offers desired specific area, high electroconductivity with confinement effect resulting in a recycling reaction. miR-21 detection is possible due to anchoring of thionine and gold nanoparticles to the MXene-MoS_2 heterostructure surface, which acts as capture probes. This sensor platform has high selectivity and stability for miR-21 detection at femtomolar concentrations with satisfactory performance in clinical settings. The only limitation is the generation of electronegative ds DNA which can disrupt electron transfer and reduce signal [46].

13.3.2 Photothermal and Photodynamic Therapy of Tumor Cells

Superparamagnetic Ti_3C_2 MXenes have been used in cancer theranostics (Figure 13.6) [47]. The synthesized Ti_3C_2 MXenes exhibited sheets like morphology as depicted in Figure 13.7a. Superparamagnetic Fe_3O_4 nanocrystals have been allowed to grow on surface of Ti_3C_2 MXenes demonstrating high dispersibility (Figure 13.7b) while the highly ordered lattice structure of MXene nanosheets can be observed in Figure 13.7c. The anchoring of Fe_3O_4 nanoparticles on MXene nanosheets resulted in nanocomposites with high T2 relaxivity [47]. This has led to contrast-enhanced MRI of tumors. The superparamagnetic nanoparticles in the nanosheet composite are the contrast agents for T2-weighted MR imaging. Concentration dependent and contrast improved negative T2 signal is evident on imaging the superparamagnetic Ti_3C_2 MXenes composite solution (Figure 13.7b) [47]. The transverse relaxivity r2 is higher in comparison to conventional T2 contrast agents and hence, was investigated in nude mice 4T1 breast cancer xenograft. Significant negatively contrast enhanced MRI-T2 signal was observed in tumor tissue (Figure 13.7c). Quantification of MRI-T2 negative signal confirmed the potential of contrast enhanced MR imaging of Ti3C2-supermagnetic nanocomposites for imaging of tumor cells (Figure 13.7d). Further, the superparamagnetic MXenes due to high PCE cause toxicity and ablation of cancer cells and tumor tissues as seen in CLSM images (Figure 13.7e). As Ti_3C_2-based MXene magnetic composites have high biocompatibility, they can be used in theranostic nanomedicine [47].

MXene nanocomposite of Tantalum carbide (Ta_4C_3) sheets have been used in imaging-guided photothermal tumor ablation. An *in situ* growth of MnO nanoparticles occurs due to activation of redox reaction. Tantalum in MnO_x/Ta_4C_3 is a high contrast agent and results in enhanced CT, while the MnO acts as a contrast agent in the tumor microenvironment for T_1-weighted MRI. MnO_x/Ta_4C_3 led photothermal conversion results in contrast-enhanced photoacoustic imaging, while photothermal hyperthermia leads to suppression of tumor growth. Thus, biocompatible MXenes with functionalization can be used for cancer theranostics [48].

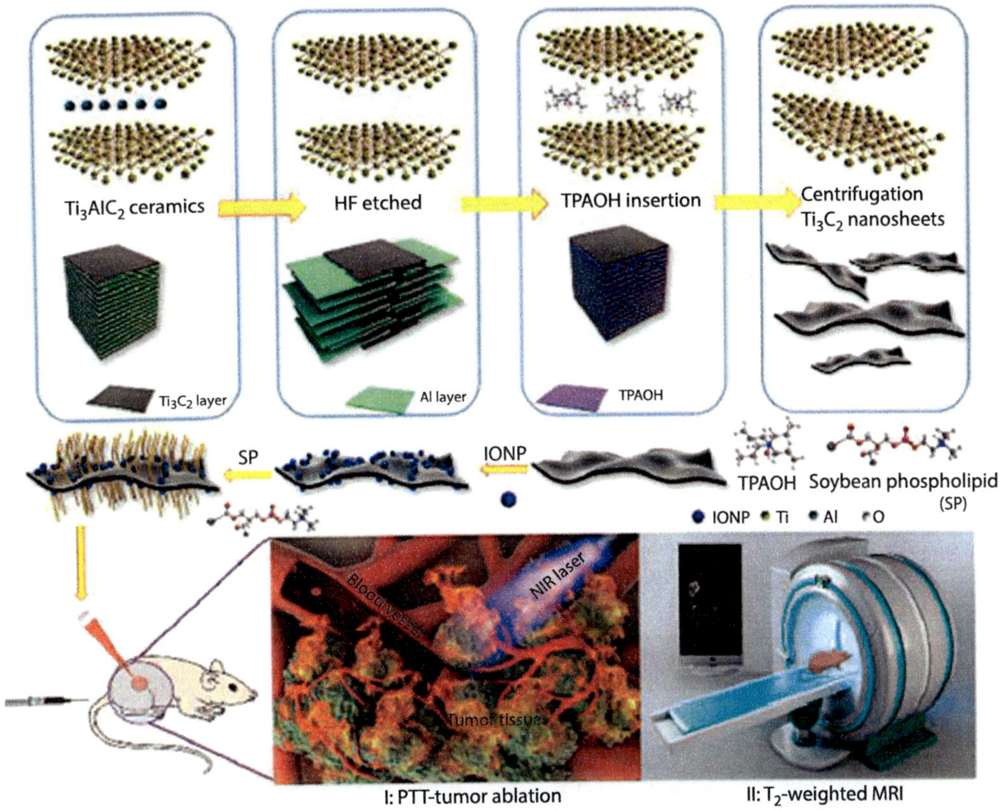

Figure 13.6 Synthesis of magnetic 2D Ti_3C_2 -IONPs-SPs nanocomposites by exfoliation. Surface engineering makes them multifunctional as in theranostic application. Reproduced with permission from Ref. [47]. Copyright 2018, The Royal Society of Chemistry.

Ti_3C_2-based MXene ($MnOx/Ti_3C_2$) are designed for cancer theranostics. MXenes of MnOx/Ti3C2 composite was constructed using MnOx *in situ* growth on the Ti_3C_2 surface due to redox reaction. $MnOx/Ti_3C_2$ MXenes are multifunctional as theranostic agents and can find application in MR as well as photoacoustic/imaging guided therapy. MnOx in $MnOx/Ti_3C_2$ composite creates TME responsive T1 weighted MRI. $MnOx/Ti_3C_2$ increases photoacoustic imaging and tumor ablation. The biocompatibility of $MnOx/Ti_3C_2$ MXene with multifunctionalization increases their potentiality in theranostic [49].

A novel niobium carbide (Nb_2C) MXene is developed with high photothermal ablation efficiency in a mouse derived tumor xenograft in NIR-I and -II window. Ultrathin MXene were synthesized using liquid exfoliation, delamination and intercalation. These nanosheets showed high PCE and good photothermal property. Due to their reduced phototoxicity, biocompatibility and tumor ablation and suppression in both NIR biowindows, they can be used in phototherapy of cancer [50]. Niobium carbide MXenes have also been used as nanoplatforms in planar theranostics. Sol-gel chemistry has been used for surface engineering to improve their surface chemistry and create therapeutic mesopore layer on Nb_2C MXene surface. CTAB was used for *in situ* self-assembly to form mesopore. Nb_2C MXene was coated with mesoporous silica. The surface engineered PEGylated MXenes were with

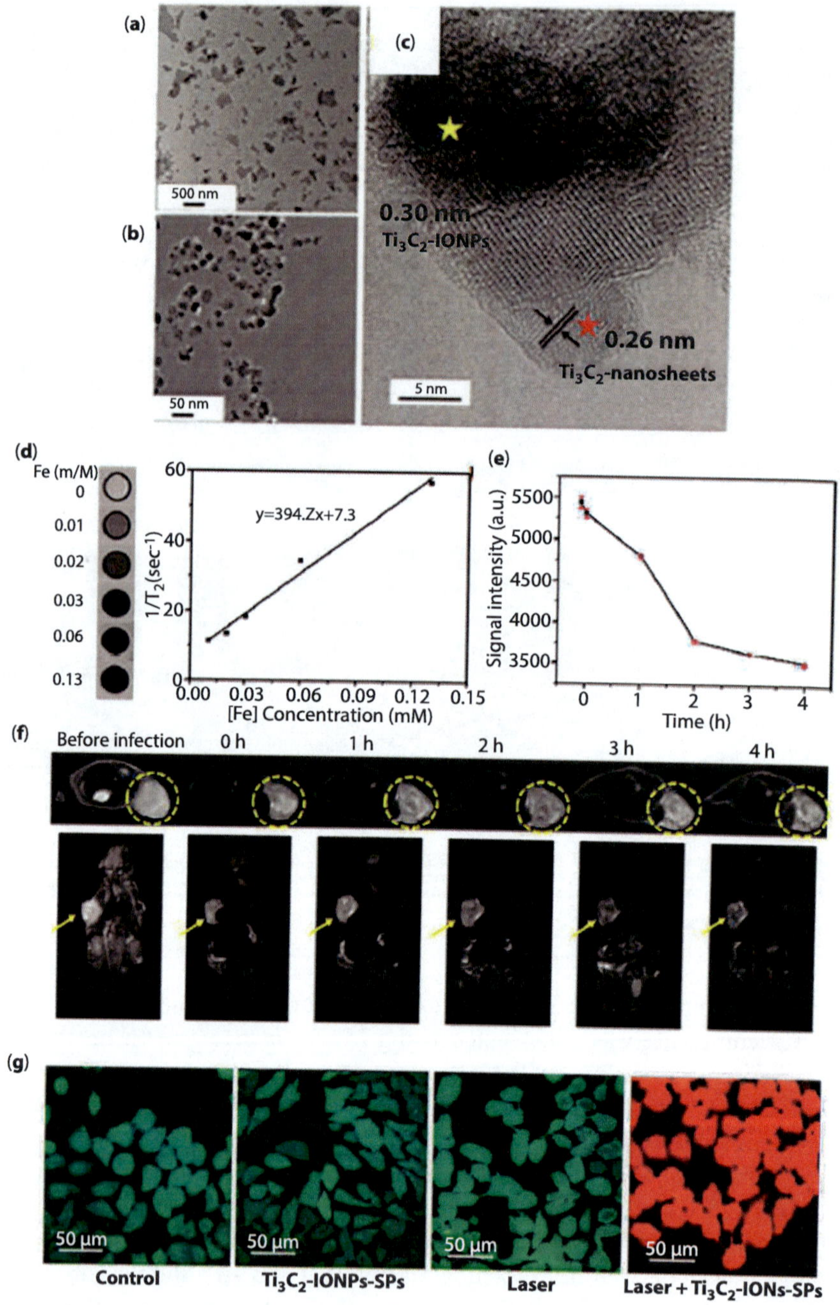

Figure 13.7 (a) TEM images of synthesized MXene, (b) Ti_3C_2–IONP composite, (c) HRTEM of Ti_3C_2-supermagnetic nanoparticle composite nanosheets, (d, e) T2-weighted MR imaging and T2 relaxivity with the corresponding T2-weighted MRI signal intensities at different time intervals on injection of Ti_3C_2 composite nanosheets at tumor sites. (f) TS and CS of T2-weighted MR imaging of mice 4T1 tumor xenograft prior and after injection of Ti_3C_2-supermagnetic at specific time intervals and (g) Confocal fluorescence of 4T1 cells on treatment with Ti_3C_2-IONPs-SPs followed by different methods. Reproduced with permission from Ref. [47]. Copyright 2018, The Royal Society of Chemistry.

cyclic pentapeptide of RGD. These MXenes had good photothermal conversion in NIR II biowindow and was used for photothermal hyperthermia. Mesopore-coated Nb_2C MXenes exhibited targeted hyperthermia in U87 cells and its tumor xenograft at a very high efficacy of 92.37 % in the NIR-II range [51].

Nb_2C nanosheets of 50 nm had high photothermal conversion in NIR-II biowindow and deep tissue penetration. Polydopamine was layered on nanosheets surface resulting in increased stability and loading of R837. The R837 alteration also increased the immune response. Nanoplatforms were coated with red blood cell to avoid blood clearance. The resulting nanoplatforms (Nb_2C@PDA-R837@RBC NPs) had a prolonged blood circulation and could eliminate primary tumors. Further, secondary tumors were suppressed owing to good immune response and thus confirming their ability to inhibit tumor recurrence. Thus, nanoplatform consisting of Nb_2C@PDA-R837@RBC can be used in photothermal and immune therapy in tumor treatment [52]. Nb_2C MXene nanosheets were incorporated into a polyvinylpyrrolidone microneedle. It was used for implanting and photothermally ablating tumors in the NIR-II. The Mxenes are efficient photothermal nanoagents, while PVP are biocompatible matrix required to maintain the structural integrity of the needles. The microneedles have good skin penetration and dissolution ability. The microneedles are injected at the tumor site, and these dissolve and release Nb_2C nanosheets. In NIR-II biowindow, irradiation leads to rise in tumor temperature to 70 °C leading to photonic tumor ablation. It was observed that after two weeks of treatment in a mice model, the tumor growth was suppressed and the survival rate improved. Therefore, MXene microneedle system with good biocompatibility, low toxicity have potential application in non-invasive and localized superficial cancer treatment [53].

MXene quantum dots (MQDs) were prepared using photoluminescent Ti_3C_2 by a hydrothermal method. These MQDs had an excitation-dependent photoluminescence spectrum with an increased quantum yield. These could find application as cellular imaging probes and these have the advantage of biocompatibility and multicolor labeling [54]. Similar, Ti_3C_2 MXene QDs were prepared using fluorine free technique. This safe and simple method resulted in the development of a biocompatible nanoplatform for efficient photothermal therapy. Al oxoanions found on MXene QD surface enhanced NIR absorption and the QDs exhibited a PCE of 52.2 % and significantly inhibited tumor cells. Thus, MXene QDs have good biocompatibility and negligible toxicity with high prospects for clinical applications [55]. QD of Ti_3C_2 MXene encapsulated in liposomes exhibited significant photothermal property. These have been used for detecting and quantifying prostate-specific antigen based on immunoreaction. MXene based detection platform was designed using simple microplate and 3D printed device. The QDs encapsulated using liposomes served as signals and labels for secondary antibody. On NIR irradiation, QDs convert light to heat and this temperature change was measured based on the analyte concentration using thermometer. The immunoassay detected 1.0 to 50 ng/mL of prostate-specific antigen. The visual thermal data was captured using portable NIR imaging camera for quantifying PSA within 3 min. This MXene based photothermal immunoassay is highly specific, reproducible and accurate and hence can be used as biomarker protein for point-of-care diagnostics [56].

Sonodynamic therapy is of interest as noninvasive therapeutic modality as it uses low intensity ultrasound. Its limitation is the inability of sonosensitizers to generate sufficient ROS for effective sonodynamic therapy. Therefore, a photosono responsive nanoplatform was developed using Ti_3C_2 MXene which can be used as sonodynamic therapy to kill tumor cells. MXenes have oxygen vacancies and size of 10 nm and the inherent TiO_{2-x} nanoparticle have narrow band gap and high surface area, useful in trapping charge and preventing charge recombination in presence of ultrasound. This results in sonodynamic ROS generation. The MXenes induce hyperthermia reduces hypoxia by increasing flow of blood and oxygen in tumor microenvironment. On activation by ultrasound and NIR-II stimuli, the nanoplatform kills tumor cells and completely eliminates tumor without recurrence and insignificant systemic toxicity. Thus, combination of photosono responsive nanoplatform can be used for cancer nanotheranostics [57].

MXene Titanium nitride QDs (Ti_2N QDs) of 5 nm size exhibited significant photothermal conversion efficiency by 48.62 % in NIR-I and by 45.51 % in NIR-II. Ti_2N QDs are biocompatible and have remarkable photoacoustic effect, high photothermal efficiency and biodegradation. After 4 h of injecting, Ti_2N QDs aggregated in tumors which aided in photothermal therapy. The Ti_2N QDs degraded and could be excreted from the body proving their stability during circulation and easy elimination from the body after degradation. Thus, the biodegradable Ti_2N QDs can be used for precision cancer therapy based on photoacoustic imaging guided photothermal therapy [58].

Transition metal oxide nanomaterials generate ROS and thus modify tumor microenvironment. A multimodal transition metal oxide based nanoplatform were developed using Ti_3C_2 nanosheets conjugated to nanoparticles of $MnFe_2O_4$. Chitosan was used as a cross-linker resulting in formation of self-assembled Ti_3C_2@Chitosan-$MnFe_2O_4$. The modified MXene nanoplatform showed better ROS generation and biocompatibility. The heterojunction catalyzed H_2O_2 resulting in O_2 production and depletion of high glutathione concentration in the hypoxic tumor microenvironment. Here chemodynamic therapy occurs due to the cyclized Fenton reaction on NIR irradiation and excitation. The nanoplatform due to presence of the high efficiency photothermal Ti_3C_2 exhibited potential for photothermal therapy. It could also be used as contrasting agent for visualization of T_1- and T_2-weighting MRI. As it has no cytotoxicity to normal tissue cells, it has high potential for biological applications as photothermal agent, bioimaging sensor and for loading drug [59].

The biocompatibility of inorganic MXene nanoplatforms for anticancer photothermal therapy was improved by using cellulose to prepare a hydrogel. The MXene cellulose composite hydrogels on NIR stimulation had an immediate response and a continuous dynamic process in water, which helped in sustained release of the anticancer drug doxorubicin hydrochloride. It was possible due to the pores that could expand in the 3D cellulose-based networks on NIR stimulation at 808 nm light. This MXene nanoplatform is multifunctional as it has photothermal property and can be used for controlled and sustained release of DOX. Intratumoral injection of the nanocomposite with PTT and prolonged chemotherapy showed instant tumor killing and suppression of relapse, proving its potential for usage in cancer therapy. The excellent photothermal functionality, biocompatibility, stable behavior, efficient targeted and sustained drug release makes inorganic 2D MXene-based composites attractive for cancer therapy and other applications in biomedicine [60].

TiO_2-based nanosonosensitizers are used for sonodynamic based tumor therapy. However, due to low ROS generation as a result of wide bandgaps, their clinical applications are limited. This limitation has been overcome by the usage of the 2D Ti_3C MXene nanosheets which have a narrow bandgap (~0.94 eV). This was possible due to surface terminations by –F and –OH groups. Further, stability and SDT was enhanced by introducing

0D/2D hybrid heterojunctions between nanosheets of Ti_3C_2 and N doped carbon dots. In comparison to commercial TiO_2 nanoparticles, carbon dot/MXene heterojunctions exhibited ROS generation efficiency by 3.07 and enhanced photothermal conversion by 64.5 % in the NIR-II region. The photothermal effect reduced hypoxia and tumor ablation was achieved using PTT in NIR-II which enhanced sonodynamic therapy in deep tissues. Thus, MXene nanocomposites equipped with nanosonosensitizers can find application in non-invasive therapy of tumors [61].

A biomimetic photoinduced plasmonic assembly was designed using Nb_2C MXene based plasmon, nanozymes of platinum, anticancer drug doxorubicin and tumor cytomembrane for tumor-penetrating photothermal therapy. On internalization into the tumor cells, MXene on irradiation with NIR-II releases electrons in the excited state which stimulates Pt nanozymes to exhibit catalase and oxidase activities resulting in oxygen and ROS production. In hyperpyrexia and acidic environment, doxorubicin is released and its efflux is prevented due to P-glycoprotein inhibition by oxygen and the free radicals. MXene based nanozyme complex reduced viability by 38.67 % in HeLa cell in comparison to treatment with nanozyme. Biocatalytic plasmonic nanoassembly exhibited tumor regression under *in vivo* conditions. Thus, plasmonic hot electrons can be used for improving biocatalysis and therapeutic effect on tumor [62].

Composites of MXenes and hydroxyapatite have been used as pH and NIR based drug delivery platform. Chitosan & hyaluronic acid multilayers present on hydroxyapatite prevent early release of drug DOX. While MXenes and gold nanorods increase the PCE. In acidic conditions of TME, electrostatic force between chitosan and hyaluronic acid decreases and hydroxyapatite dissolves resulting in drug delivery. On exposure to NIR, the photothermal effect of MXene and gold nanorods is observed. Thus, synergistic chemo and phothermal therapy is observed. Thus, MXene nanocomposites with large surface area, NIR responsiveness, PCE and biocompatibility is used for sustained drug release to cancer cells [63]. MXene nanosheets with high PCE and photothermal stability with doxorubicin were incorporated into agarose to obtain hydrogel which on NIR exposure increases the temperature to 60 °C causing agarose to melt and releases the drugs. The release and kinetic parameters are modified by varying the concentration of agarose and MXene, as well as light intensity and exposure. Thus, NIR-responsive hydrogel prepared using MXene can be used for localized cancer treatment [64].

Similarly, hepatic growth factor has been incorporated into the Ti_3C_2 MXene and agarose hydrogel, which can initiate the c-Met-mediated signaling on exposure to NIR light. The platform involves Ti_3C_2 MXene-agarose hydrogel containing protein which is injectable and NIR responsive. *In vivo* studies showed that on NIR light exposure, diffusion and proliferation required for angiogenesis and wound healing was observed. Further, TNF-α was incorporated in the MXene hydrogel composite which stimulated pro-apoptotic signaling pathway resulting in NIR induced apoptosis of tumor spheroids. As NIR light can penetrate deep tissue, the hydrogel MXene composite could eradicate tumors in xenografts. The advantages of using MXene-hydrogel is the light responsive drug release and ability to reach deep and thick tissue [65].

Iron metabolism is an interesting therapeutic target for cancer, however, current strategies for usage of iron in therapeutics is limited due to low efficacy and high off-target toxicity. MXene platform has been used to deliver prodrug DOXjade containing doxorubicin and deferasirox, wherein the former is a topoisomerase inhibitor and the latter chelates iron. This combination of drug incorporated in MXene polyvinylypyrrolidone nanocomposite is used for cancer treatment (Figure 13.8). The Ti_3C_2 MXene nanosheets has a PCE of 40 % on photoactivation. The results have been observed both in *in vitro* and *in vivo* conditions emphasizing success of novel anticancer treatment strategy based on pH-responsive iron chelation and chemotherapy by photothermal approach [66].

Ceramic biomaterials due to challenges in their material functionality such as large dimension, rigid bulk and sintering have been less explored in cancer therapy. However, their functionality can be modulated when prepared as MAX ceramic composites, and can be used for photothermal ablation. Biocompatible MXenes of Ti_3C_2 nanosheets were used to prepare a phase-changeable organic-inorganic hybrid composite using PLGA copolymer. Functionalized biomaterial had high PCE and was successful in photothermal tumor ablation when administered intravenously administration (Figure 13.9) [67]. Thus, MXene based ceramic photothermal composites can be used for cancer treatment and other biomedical applications [67]. Similar MAX ceramic biomaterials using 2D ultrathin tantalum carbide nanosheets have been developed with functionalities for both photoacoustic CT imaging and tumor ablation. The biocompatible, superior PCE and photothermal tumor ablation was achieved by using soybean phospholipid-modified Ta_4C_3

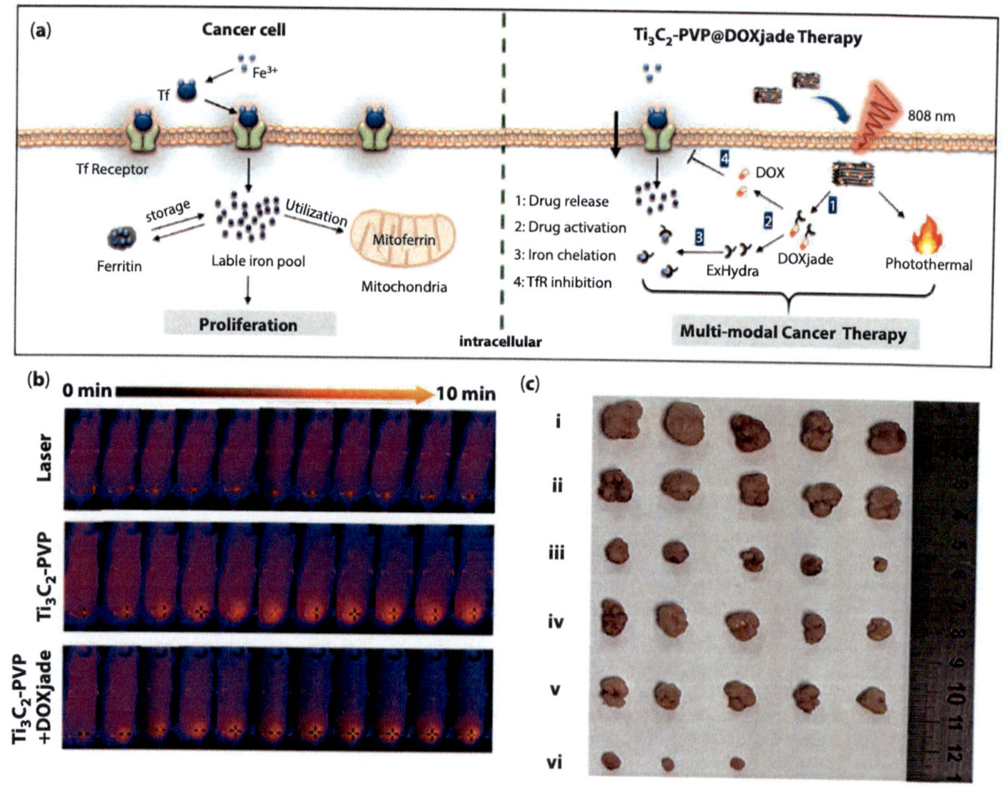

Figure 13.8 (a) Mechanism of Ti_3C_2-PVP@DOXjade-based multi-modal cancer treatment. (b) NIR imaging of tumor on treatment with Ti_3C_2-PVP or Ti_3C_2-PVP@DOXjade and 808 nm laser irradiation (c) Tumors removed in sacrificed mice Control; ii: ExHydra; iii: DOX; iv: DOXjade; v: Ti_3C_2-PVP + NIR; vi: Ti_3C_2-PVP@DOXjade + NIR of 808 nm, 1.0 W cm^{-2} for 10 min. Adapted from Ref. [66]. Copyright 2022, Elsevier.

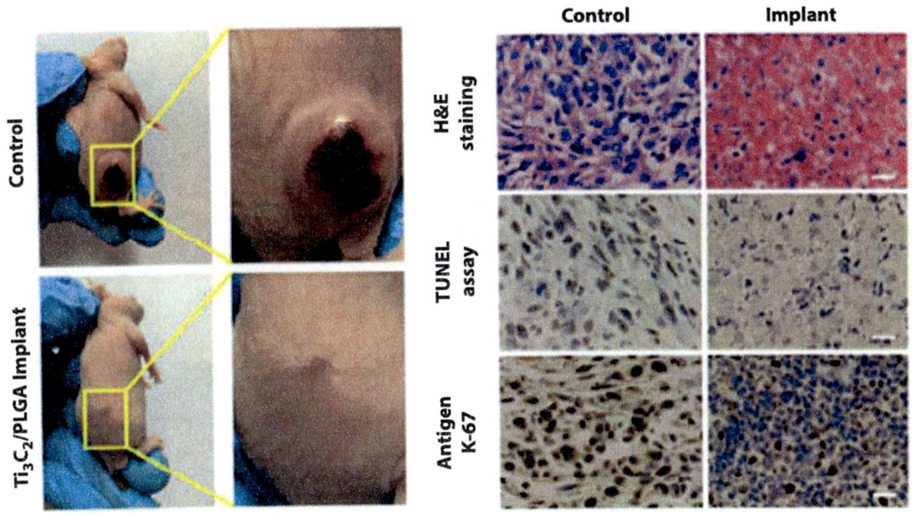

Figure 13.9 Injection and implantation of Ti_3C_2-SP nanosheets and phase-changeable PLGA/Ti_3C_2. Accumulation of nanosheets in tumors is due to EPR effect. Photothermal ablation of cancer cell on Ti_3C_2-SP exposed to NIR laser after 16 days of treatment. H&E staining of tumor tissues confirm *in vivo* efficacy of PTT by PLGA/ Ti_3C_2-SP nanosheets implants. Apoptosis and cellular proliferation is observed by TUNEL and Antigen Ki-67 immunofluorescence staining of tumor sections, respectively. Reproduced with permission from [67]. Copyright 2016, The American Chemical Society.

nanosheets [68]. Highly biocompatible magnetic Ti_3C_2-based MXene composites have been used for cancer theranostics. Superparamagnetic MXenes is constructed by allowing formation of superparamagnetic Fe_3O_4 nanocrystals on Ti_3C_2 MXenes. These have T_2 relaxivity of 394.2 mM^{-1} s^{-1} and high contrast and enhanced MR tumor imaging. These have high PCE and can be used for efficient photothermal killing and tumor ablation. The unique chemistry of MXenes surface allows it to be functionalized for theranostics and other novel nanomedicine applications [69].

13.3.3 Drug Carrier/Delivery Agents

Ti_3C_2 MXenes have high drug loading capacity of around 211.8 %, drug delivery and high therapeutic efficiency in elimination of tumor cells. MXenes show both pH-responsive and near infra-red stimulated drug release. The MXenes show efficient tumor killing due to high PCE and by synergistic photothermal ablation and chemotherapy. The Ti_3C_2 MXenes are used in photoacoustic imaging as contrast agents. They are used in diagnosis and monitoring of progress in cancer treatment. Ti_3C_2 MXenes show histocompatibility and easy excretion, implying their safety in clinical application [70].

Tantalum carbide (Ta_4C_3) MXene surface functionalized with superparamagnetic iron-oxide nanoplatform have been developed for breast cancer theranostics. Ultrathin Ta_4C_3 nanosheet surface synthesized by exfoliation was used for growth of superparamagnetic iron oxide nanoparticles aided by the redox reaction of MXene. Its stability was increased by modification with soybean phospholipid. The resulting nanoplatform when tested in mice breast tumor allograft exhibited high PCE of 32.5 %. Tantalum is used in CT imaging while the iron-oxide are contrast agents in T_2-weighted MRI. Overall, the composite nanosheets had high complete breast-tumor photo-ablation resulting in tumor eradication without reoccurrence. Thus, magnetic functionalization of MXenes can be used as nanoplatform for efficient breast cancer theranostics [71]. MXene composite nanoplatform have been developed for chemotherapy and hyperthermia of hepatocellular carcinoma. MXenes surface functionalization is carried out by surface-nanopore engineering resulting in the formation of thin mesoporous-silica layer on MXene surface. The mesopores improve release and delivery of drug by increasing its hydrophilicity and dispersity. The surface of MXene is engineered with the drug RGD. Now, there is a synergistic anti-tumor action due to mesoporous shell aiding chemotherapy and Ti_3C_2 core causing photothermal hyperthermia. Thus, MXene-based composite nanoplatforms with surface-nanopore engineering can be used to target hepatocellular carcinoma [72].

The enzymes GOx and chloroperoxidase have been conjugated on Ti_3C_2 nanosheets to develop a bionic cascaded enzyme nanoreactor for cancer therapy. Further, the modified nanosheets was loaded with deoxygenation-activated drug tirapazamine. These were embedded into nanovesicles derived from cancer cell membranes that expressed CD47. This helped in overcoming immune response and homologus targeting resulting in internalization in preferred tumor cells. Within the tumor cells, GOx and chloroperoxidase enzyme action produces HClO for enzyme dynamic therapy. At the same time, irradiation increases both catalysis and generation of singlet oxygen. The enzyme dynamic therapy caused depletion of oxygen and local hypoxia environment which activates the deoxygenation-sensitive prodrug tirapazamine used in chemotherapy. The enzyme nanoreactor thus exhibits increased anticancer therapeutic efficacy due to synergistic phototherapy, enzyme action and chemotherapy. Thus, enzyme catalyzed generation of ROS at the tumor sites modulates intracellular redox condition required for anticancer treatment [73].

Though MXenes have photothermal property and could be used in photothermal therapy, they lack biocompatibility and stability. This could decrease their targeting and drug release. Fabrication of composite hydrogels with cellulose and Ti_3C_2 MXene, causes rapid response on NIR stimulation. The cellulose MXene hydrogels on loading with anticancer drug doxorubicin hydrochloride accelerated drug release as a result of increase of the pore size in the cellulose on exposure to light at 808 nm. Intratumoral injection with cellulose MXene hydrogels doxorubicin multifunctional nanoplatform proved the synergistic action of photothermal and chemotherapy in rapidly destroying tumor cells and suppressed tumor relapse [74].

Ti_3C_2 MXene based on good photothermal effect is used for multimodal tumor therapy. Ultrathin nanosheets of Ti_3C_2 MXene of 100 nm size exhibited photo-thermal and -dynamic activity in cancer therapy. The nanosheets on NIR irradiation (808 nm) exhibited PCE by 58 % and produced singlet oxygen. Further, these Ti_3C_2 nanosheets were surface modified with layers of doxorubicin and hyaluronic acid to develop a multifunctional nanoplatform. Ti_3C_2-DOX drug exhibited required biocompatibility, bioaccumulation, and stimulus response release. It killed cancer cells and tumor tissue by synergistic action of photothermal, photodynamic and chemotherapy [75]. To improve the controlled drug release in MXene carriers, heterostructured titanium carbide-cobalt nanowires nanocarriers were developed. These photothermal anticancer agents could be controlled based on their magnetic stimuli-responsive property. The nanocarrier had high PCE on irradiation at 808 nm and with up to 225 % drug loading capacity. The doxorubicin release could be stimulated by both acidic pH of 4-6 or NIR irradiation. It had strong lethal effect on cancer cells than chemo and photothermal therapy [76].

Nanocomposite hydrogels were synthesized by free radical PAM polymerization. Ti_3C_2 MXene nanosheet was used instead of organic conventional crosslinkers. In comparison to acrylamide PAM hydrogels, the Ti_3C_2/PAM hydrogels had greater mechanical properties due to honeycomb assembly with even pores and bendable polymer chains. Further, it had high loading capacity and exhibited good sustained-release performance of loaded drugs i.e. 97 to 127.7 mg/g and releases by 62 to 81 % in comparison to regular hydrogel whose loading capacity and release were 46.4 mg g^{-1} and 45 %, respectively. Thus, MXene fabrication of nanocomposite hydrogels increases their mechanical property and drug loading and release behaviors [77]. The Ti_3C_2 MXene nanosheets incorporated with the antioxidants sodium ascorbate as well as dopamine were stable in both the oxidative and hydrated environments. The modification also increased the number of surface functional groups and it aided in conjugation of GOx and photosensitizer Ce6. It resulted in photodynamic MXene nanosheets causing glucose starvation and properties resulting in killing of cancer cells. These had a photothermal effect with 45.1% conversion efficiency and photodynamic properties resulting in ROS generation on irradiation at 808 and 671 nm. The activity of GOx increased on irradiation due to increase in temperature of the solution. The MXenes had no toxicity to HePG2 and HeLa cancer cells in absence of stimulants. However, they showed high killing with 90 % efficiency on exposure to glucose and irradiation, which is attributed to the due to starvation and photothermal effect. The MXenes with simple modification could be used to treat cancers with minimal side effects [78].

13.3.4 Antimicrobial Therapeutics

Electrospinning process was used to encapsulate delaminated $Ti_3C_2T_z$ MXene in chitosan nanofibers. This was used for application as antibacterial wound dressings. The crosslinked $Ti_3C_2T_z$/chitosan nanocomposite exhibited broad spectrum antimicrobial activity. The Ti_3C_2T nanofibers on 4 h of treatment inhibited *Escherichia coli* and *Staphylococcus aureus* by 95 % and 62 %, respectively (Figure 13.10) [79]. MXene-chitosan nanofibers exhibited antimicrobial activity with no cytotoxicity. Thus, the low cost electrospun $Ti_3C_2T_z$/chitosan nanofibers can be prepared as polymeric bandages and used in wound healing based on their antimicrobial activity, biodegradability and biocompatibility [80].

MAX phases Ti_2AlC and Ti_3AlC_2 were used to synthesize Ti_2C and Ti_3C_2 MXene, respectively, and their antimicrobial activity were compared. Ti_2C MXenes had no effect on *Escherichia coli*, whereas Ti_3C_2 MXenes significantly inhibited bacterial growth. The variation in their antimicrobial activity were attributed to their structures as there was no difference in the surface chemistry. The atomic scale between Ti-C in Ti_2C and in Ti_3C_2 was 9.76 Å and 9.23 Å, respectively. Thus, structure at the atomic scale influences bioactivity in MXenes having similar chemical composition and different stoichiometry [81]. Ti_3C_2Tx MXene have exhibited broad spectrum antibacterial activity. MXene nanosheets even with 0.09 μm thickness have exhibited antibacterial activity in *E. coli* and *B. subtilis* until 3-8 h on incubation in dark conditions. This confirmed that the antimicrobial activity depends on both size and time of exposure to nanosheets. Small sized nanosheets had higher antibacterial activity. It was observed that sharp edges of MXene nanosheets affected bacteria membrane surfaces. The MXenes within a span of 3 h damaged the bacterial cells causing DNA release from the cytosol [82].

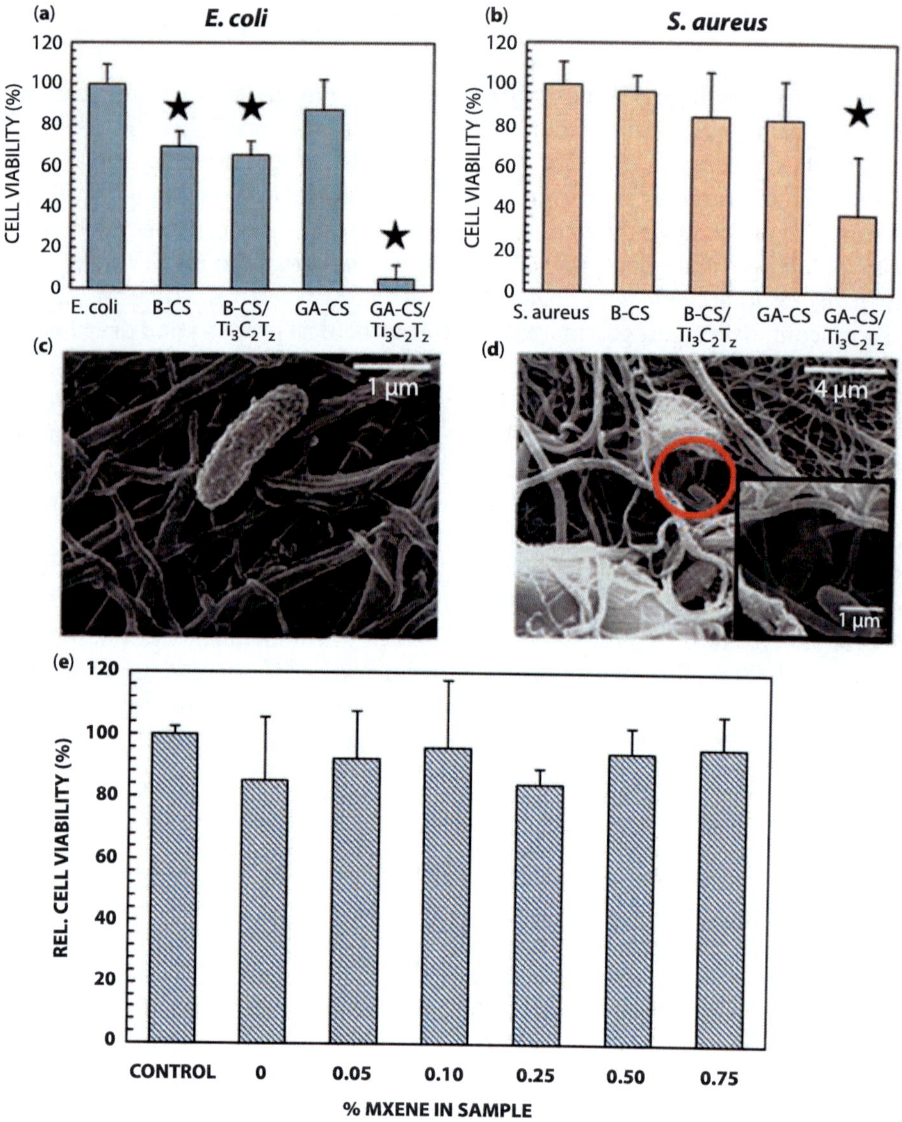

Figure 13.10 Antibacterial activity of basified chitosan, glutaraldehyde linked chitosan and $Ti_3C_2T_z$ MXene chitosan nanofibers in (a) *E. coli* and (b) *S. aureus*. SEM micrographs shows an, (c) intact and, (d) destroyed *E. coli* bacteria on the 0.75 wt% Ti3C2Tz/CS nanofiber mat. (e) Biocompatibility of Ti_3C_2Tz/CS nanofibers as assessed by *in vitro* cytotoxicity assay. Adapted from Ref. [79]. Copyright 2018, The Royal Society of Chemistry.

The potential difference in $Ti_3C_2T_x$ and Bi_2S_3 has been exploited to develop an interfacial Schottky junction which is represented as $Bi_2S_3/Ti_3C_2T_x$. Due to the presence of electrophilic/nucleophilic localization, charge is transferred which results in high electron density on $Ti_3C_2T_x$ surface. Further, Schottky junction prevents the electron backflow and enhances charge transfer. $Bi_2S_3/Ti_3C_2T_x$ possesses photocatalytic activity which increases ROS generation on exposure to NIR of 808 nm. The MXenes exhibited broad spectrum antimicrobial activity. On exposure to hyperthermia for 10 min, it inhibited *S. aureus* and *E. coli* by 99 %. Thus, photoresponsive MXene based Schottky junction can be used to develop antimicrobials [83]. Ti_3C_2Tx and Ag/Ti_3C_2Tx were used to design long-term high efficiency antimicrobials. The antimicrobial activity of the MXenes was increased on exposure to near-infrared light. Ti_3C_2Tx MXene has intrinsic bactericidal activity and the photothermal bactericidal effect and hence can be used to develop near-infrared light-mediated antibacterial nanoplatforms. Wound dressings were prepared from hydrogels of Ag/Ti_3C_2Tx. The hydrogels exhibited growth inhibitory activity and enhanced wound healing on NIR exposure. Thus, MXene-based photothermal platforms can be used as antimicrobials [84].

Titanium carbide MXene are used as antimicrobials as they can damage bacterial membrane and induce oxidative stress. MXene conjugated with nanomaterials exhibit enhanced antibacterial activity. Gold nanoclusters were conjugated on MXene nanosheets, where the MXene disrupted cell membrane and delivered the gold nanoclusters into the cytosol resulting in localized ROS generation which at high concentration oxidized membrane lipids and fragmented DNA. Thus, the synergistic antimicrobial action of MXene and gold nanoclusters result in cell death. These nanoconjugates exhibit broad spectrum activity and IC_{50} of 11.7 μg/mL. The MXene-gold nanocluster could inhibit biofilm formation as the hydrophobic surface prevents bacterial attachment, while the large surface area can be coated with high density of bactericidals [85].

$Ti_3C_2T_x$ MXene nanosheets have also exhibited antifungal activity. The delaminated MXene could inhibit the fungi *Trichoderma reesei*. The MXenes caused cell damage, inhibited hyphal growth and mycelial development. Further, it could inhibit spore germination. MXene treated fungal spores did not germinate after 11 days of incubation. It also disrupted fungal colony structure. Thus, the d-$Ti_3C_2T_x$ MXene nanosheets could be used as anti-fungal agent [86].

MXene polymers based on their antibacterial and antifouling property are having significant potential for commercial application in water purification and desalination plants [87]. MXenes metal carbides and nitrides based on their antimicrobial activity are used in water and wastewater treatment. Membranes of $Ti_3C_2T_x$ MXene were synthesized on a polyvinylidene fluoride matrix. These modified membranes prevented growth of *Escherichia coli* and *Bacillus subtilis* their membrane surface. The freshly prepared titanium carbide inhibited growth of *B. subtilis* by 73 % and *E. coli* by 67 %, respectively. The aged $Ti_3C_2T_x$ membrane inhibited bacterial growth by 99 %. After 24 h of treatment with MXenes, 70 % of microbes were killed due to the damage to the cell surfaces. As the MXenes exhibit antibacterial activity, they can be used as coatings to prevent biofouling of membranes in both water and sewage treatment plants [88]. Similarly, single- and layered MXenes of Ti_3C_2Tx were tested for anti-fouling property. In comparison to graphene oxide, Ti_3C_2Tx exhibited higher broad spectrum antibacterial efficacy in *E. coli* and *B. subtilis*. At 200 μg/mL, Ti_3C_2Tx inhibited cell viability by 98 % within 4 h of exposure. Generation of ROS and stress induction by MXenes resulted in bacterial cell membrane damage and leakage of cytoplasm [89].

A hydrothermal method was used to develop nanocomposites of $Ti_3C_2T_x$ MXenes copper sulfide nanoparticles as a photocatalytic antibacterial material. CuS nanoparticles were anchored on the surface of MXene to improve the photocatalytic activity. The MXene nanocomposite inhibited *E. coli* and *S. aureus* by 99 %. The $Ti_3C_2T_x$@CuS heterojunction supported the electrons and holes to separate leading in an increase in electron transport and generation of ROS. The stronger photothermal effect of the composite in comparison to CuS alone generates more heat and hence increased antibacterial activity. The $Ti_3C_2T_x$@CuS MXene composite can be used as disinfectant based on its antibacterial property [90].

A NIR activated heterostructure MXene composite was developed using 2D Ti_3C_2 MXene nanosheets and 1D cobalt nanowires for antibacterial applications. On NIR illumination at 808 nm, the intercalated nanowires collect the electrons from MXene nanosheets, resulting in an increased ROS production. The synergistic action of plasmonic nanowires and MXene semiconductor increased the photothermal efficacy of the heterostructure which on application as coatings in orthopedic implant on NIR trigger exhibited broad spectrum antimicrobial activity with up to 90% killing efficiency in 20 min due to generation of ROS and hyperthermia. The cytocompatibility of the heterojunction makes it promising as antimicrobial agent and in reducing the usage of antibiotics [91].

Nanoscale catalytic membrane was designed for preventing wound associated infections. The membrane was prepared using electrospun PLGA scaffolds, MXenes with silver sulphide bio-heterojunctions, and the enzyme lactate oxidase. On NIR light exposure, the MXenes heterojunctions exhibited killing of bacteria based on synergistic photothermal effect, ROS generation, increased hydroxyl radicals production through Fenton-like reactions. The photothermal conversion was significantly improved as observed by increase in temperature from 22.5 to 61.2 °C in 10 min. The nanoscale catalytic membrane based on its ability to kill bacteria reversed the chronic wounds into regenerative one and also promoted wound healing through increased angiogenesis, collagen deposition and epithelialization [92].

MXene@PVA hydrogels with excellent mechanical functionality have been prepared inspired by the muscle with its hierarchical assembly of anisotropic structures. The hydrogel due to photothermal nature of MXene on exposure to NIR light at 808 nm exhibited hyperthermia at infection sites in wounds. It had broad-spectrum antibacterial activity against *E. coli* and *S. aureus* by 98 and 95 %, respectively. In *in vivo* studies using animal wound models, it inhibited infection and enhanced healing by 98 %. Thus, MXene@PVA hydrogel composites due to improved mechanical strengths of toughness and anisotropy, and antimicrobial activity can find application as wound dressing [93]. Titanium carbide MXenes nanosheets have been coated onto silk fabric to achieve functional material with antibacterial property. MXene nanosheets dispersion were deposited on silk fiber by dip coating method. The functionally stable MXene and silk fabric composite had high conductivity, PCE and quick thermal response. The fabric had UV protection property and antimicrobial activity. It exhibited remarkable contact inhibition of *E. coli* in 20 min of exposure. Such multifunctional MXene silk fabrics can be used to design smart textiles useful in clinical settings [94].

Smart wearable cotton fabric with multifunctional properties including breathing monitoring and photothermal antibacterial has been designed. The composite of fiber surface with ordered nanosheets contains bark microstructure from *Cinnamomum camphora*, chitosan, and MXene nanosheets self-assembled by electrostatic force. The fabric retains its original features but is conferred with conductivity and stability. The MXene multilayer structures can adsorb and release water which helps the smart fabric to monitor the physiological health and breathing. The local surface plasmon resonance led to increased encouraging photothermal conversion and stability on prolonged irradiation. On irradiation at 400 mW/cm^2 for 5 min, the fabric exhibited 100 % broad spectrum antimicrobial activity as tested with *E. coli* and *S. aureus*. It also had high washing resistance and cycle stability as even after 10 washes, the antibacterial efficacy was 99%. This multifunctional MXene composite based cotton fabric can be used as smart wearable microbreathing sensor and as protective gear against pathogenic bacteria [95].

13.3.5 Regeneration of Tissue/Tissue Engineering

2D Ti_3C_2 MXene have been integrated with 3D-printing glass scaffolds. This MXene biocomposite has dual biological action as it can kill bone tumor due to hyperthermia and also regenerate bone-tissue by means of bioactive scaffolds. The 2D Ti_3C_2 MXene on stimulation with NIR, causes photothermal conversion resulting in photothermal hyperthermia and bone tumor ablation. *In vivo* studies using bone tumor xenografts show complete tumor eradication and accelerated growth of new tissue due to the dual functionality of Ti_3C_2 MXene-glass composite scaffolds [96]. Similarly, 2D composites of ultrathin MXene nanosheets and niobium carbide have been integrated to obtain a 3D bone mimetic scaffolds which can be used to treat osteosarcoma and for bone regeneration. The MXene composite shows NIR-II response and ability to penetrate into deep tissue and thus killing cancer cells. Another interesting feature is the release of Nb from MXene which can enhance neogenesis, migration of blood vessels, transport oxygen and growth factors, recruitment of immune cells at the damaged site for the regeneration process. The scaffold degradation results in Nb products creating space for bone remodeling, while the release of calcium and phosphate increases the mineralization of bone tissue. Composite scaffolds of niobium carbide MXene are intrinsically multifunctional as they can kill tumor cells, stimulate angiogenesis and aid in regeneration of bones. Therefore, these can be used in tissue regeneration and as implant biomaterials [97].

Nanocomposite membranes of ultralong hydroxyapatite nanowires and titanium carbide MXene have been used in bone regeneration. The MXenes exhibited adhesion behavior, improved cell proliferation and differentiation of osteogenic cells, whereas the hydroxyapatite nanowires enhanced mechanical function and hydrophilicity. Increase in concentration of the nanocomposite constituents increased the mechanical and biological properties. *In vivo* studies confirmed the bone tissue regeneration activity of nanocomposite membranes as observed in rat calvarial bone defect. The incorporation of hydroxyapatite nanowires into MXene improved mechanical strength, biocompatibility and osteoinduction. Therefore, MXene based inorganic composites can be used for regeneration of bone tissue [98]. Similar composite scaffolds with uniform structure and mechanical strength were developed using 3D printing. These scaffolds composed of Ti_3C_2 MXene, hydroxyapatite and sodium alginate have exhibited bone regeneration. The scaffold under *in vitro* studies using bone mesenchymal stem cells exhibited biocompatibility and increased cell proliferation, osteogenic gene expression, and mineralization. In the *in vivo* studies using rat calvarial defect model, the scaffolds confirmed their osteogenic activity, bone regeneration and bone healing significantly in comparison to MXene usage alone. Thus, MXene composite 3D-printed tissue-engineered bone scaffolds with their excellent biocompatibitlity could prove to be the next promising treatment option for bone defects [99].

A novel composite was fabricated with 3D printing using titanium carbide MXene on polyethylene glycol hydrogels at cellular resolution. These exhibited no cytotoxicity to induced pluripotent stem cell derived cardiomyocytes even after one week of exposure.

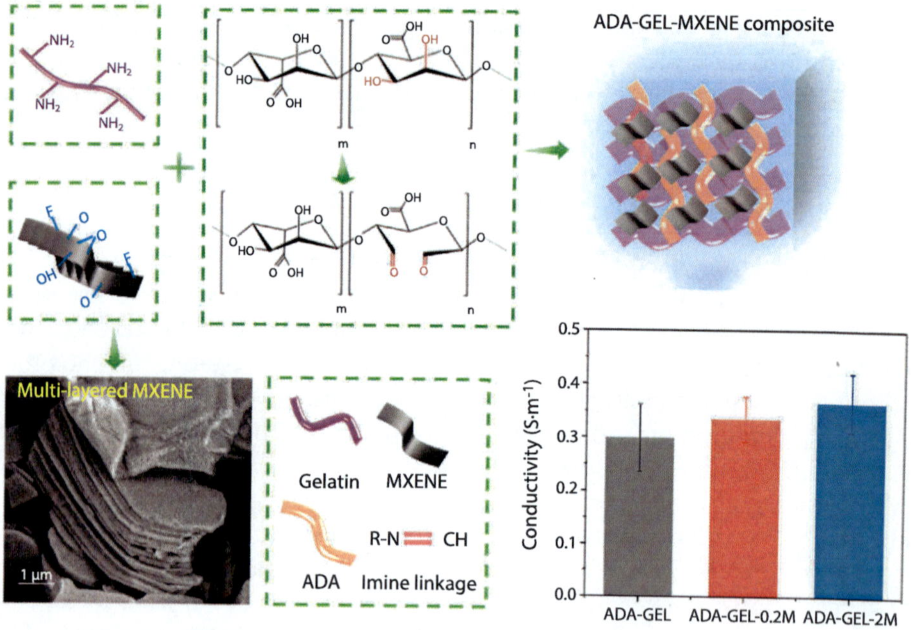

Figure 13.11 Fabrication of MXene nano-flakes composite with ADA and gelatin. Conductivity of different hydrogels increased from 0.299 to 0.366 S/m on addition of MXenes (2 w/v %) to ADA-GEL. Adapted from Ref. [101]. Copyright 2022, MDPI, Basel, Switzerland.

The construct showed physiologically appropriate conductive and topographical cues required for cardiac tissue engineering. Its applications resulted in increased expression of MYH7, SERCA2, and TNNT2. There was also an improvement in synchronous beating and conduction velocity. Thus, 3D printed composites of MXene have potential application in

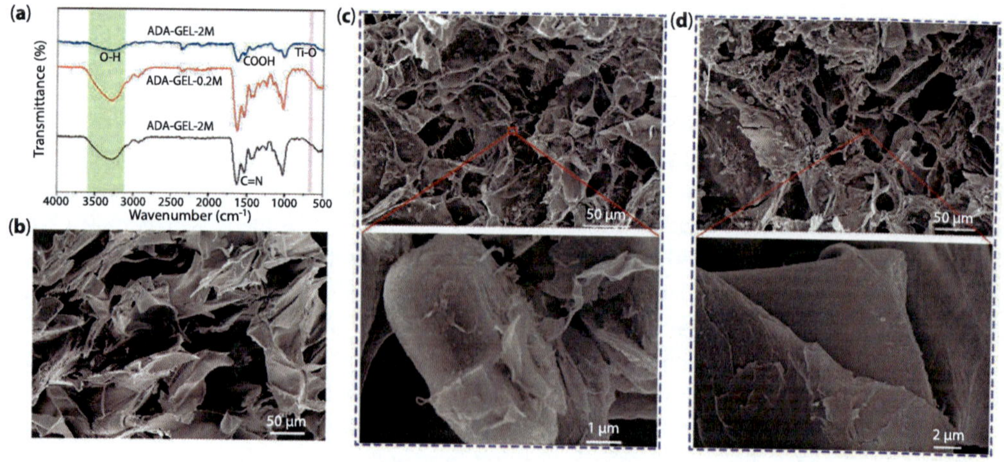

Figure 13.12 (a) FTIR spectra of lyophilized ADA-GEL, ADA-GEL-0.2M and ADA-GEL-2M dry samples. SEM images of the freeze-dried (b) ADA-GEL, (c) ADA-GEL-0.2M and (d) ADA-GEL-2M. Adapted from Ref. [101]. Copyright 2022, MDPI, Basel, Switzerland.

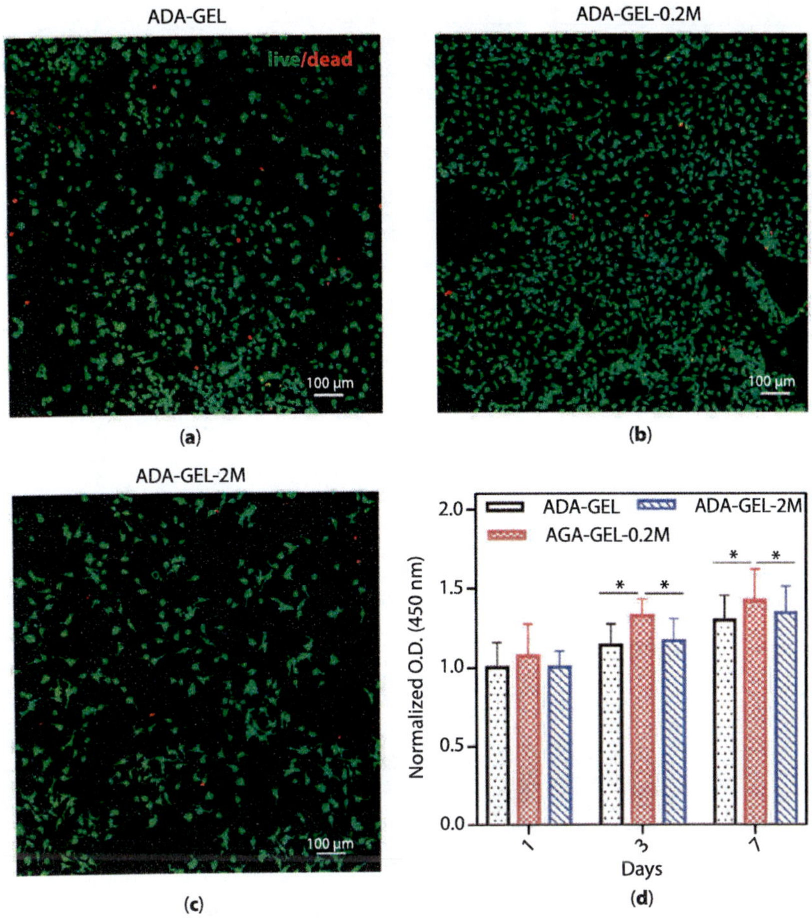

Figure 13.13 (a–c) LIVE (green)/DEAD (red) cell staining of NH3T3s on (a) ADA-GEL, (b) ADA-GEL-0.2M and (c) ADA-GEL-2M. (d) NH3T3 viability cultured on different hydrogels for 1 week tested by CCK-8 assay. Statistically significant differences are indicated as: * $p < 0$. Adapted from Ref. [101]. Copyright 2022, MDPI, Basel, Switzerland.

constructing cardiac cells for myocardial management [100]. In another study, multifunctional composite hydrogels have been prepared by "one-pot" method using 2D MXenes multilayered nanoflakes for fillers within alginate and gelatin oxidized hydrogels (Figure 13.11 and Figure 13.12) [101]. MXene provides large surface groups, excellent conductivity, mechanical property and electroactivity. The composite hydrogel had good self-healing property and promoted attachment to mouse fibroblast and migration probably due to integration of MXenes which could have stimulated conductivity and provided surface hydrophilicity (Figure 13.13 and Figure 13.14) [101]. Thus, electroactive MXene composite hydrogels can find potential application in wound healing [101].

Multifunctional hydrogels composed of bacterial cellulose and titanium carbide MXene have exhibited electrically stimulated cell behaviors which resulted in an increase in skin wound healing rate. The biodegradable composite flexible hydrogel had high electrical conductivity, biocompatibility, required mechanical functions and water absorption capacity.

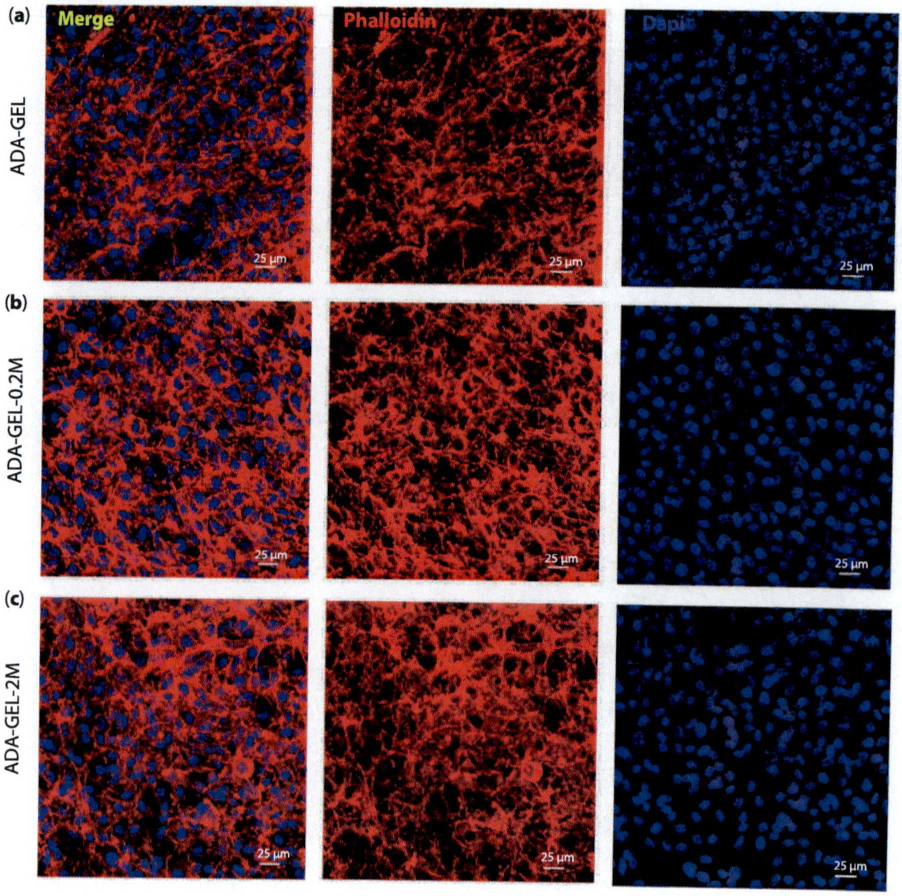

Figure 13.14 Fluorescence microscopy images of NH3T3s stained with F-actin (red) and cell nuclei (blue) after incubation for 7 days on (a) ADA-GELs, (b) ADA-GEL-0.2M and (c) ADA-GEL-2M. Adapted from Ref. [101]. Copyright 2022, MDPI, Basel, Switzerland.

MXene hydrogel composite had improved therapeutic effect in comparison to Tegaderm film in a rat thickness defect model. On electrical simulation, the hydrogel significantly increased fibroblast proliferation and accelerated wound healing. Thus, biodegradable MXene hydrogel composite coupled with electrical simulation can be used as dressing for skin wound healing based on its rapid wound repair property and electroactivity [101]. The MXenes have high potential for biomedical application but there are not many studies on their cytotoxicity and biocompatibility. Their toxicity has been assessed in a zebrafish embryo model and concentration/accumulation is determined by inductively coupled plasma mass spectrometry. Further, investigations are required to elucidate their cellular uptake, impact of surface functionalization/ modification, long-term toxicity, biocompatibility, pharmacokinetics and biodistribution for their safe and numerous biomedical applications. Table 13.1 summarizes various biomedical application of MXenes based polymer composites.

Table 13.1 Biological applications of MXenes and their biocomposites in biosensors/bioimaging, cancer theranostics and antimicrobials.

MXene composites	Biomedical application	References
TiO_2–Ti_3C_2 functionalized with Hemoglobin and Nafion	Biosensor used for detection of hydrogen peroxide (0.1–380 μM) with LOD of 14 nM and nitrite (0.1–380 μM) with LOD of 0.12 μM	[26]
Au/Ti_3C_2 functionalized with GOx	Biosensor used for detection of hydrogen peroxide (0.1–18 mM) with LOD of 5.9 μM	[29]
Ti_3C_2 QDs	Multicolor cellular imaging	[54]
Ti_3C_2 functionalized with tyrosinase and chitosan	Biosensor used for detection of phenol in water samples with LOD of 12 nM	[30]
Ti_3C_2 functionalized with (3-Aminopropyl) triethoxysilane, BSA and anti-CEA	Detection of carcinoembryonic antigen (CEA) with LOD of 1.8×10^{-5} ng/ml	[35]
Ti_3C_2 functionalized with AuNPs, staphylococcal protein A, anti-CEA	Biosensor used for detection of CEA via surface plasmon resonance with LOD of 0.7 fM	[36, 37]
Ti_3C_2 functionalized with Osteopontin aptamer, phosphomolybdic acid and polypyrrole	Biosensor used for detection of osteopontin and LOD of 0.98 fg/ml	[38]
Ti_3C_2 functionalized with β-hydroxybutarate dehydrogenase, BSA and glutarate	Biosensor used for detection of β-hydroxybutarate with LOD of 44.5 μM	[34]
Ti_3C_2 functionalized with Poly-L-lysine and GOx	Biosensor used for detection of hydrogen peroxide with LOD of 2.6 μM	[36]
Ti_3C_2 functionalized with AuNPs, APTES, SPA and anti-CEA	Biosensor used for detection of CEA by SPR with LOD of 0.15 fM	[37]
Nb_2C QDs	Fluorescence imaging and metal ions sensing	[39]
Ta_4C_3	Cancer theranostics and imaging for MRI, CT and PAI; and PTT	[48]
Ti_3C_2	MRI-guided PTT	[49]
Ti_3C_2 QDs	PTT in NIR biowindow	[55]
Ti_3C_2, SP, IONPs	MRI-guided PTT	[68]
Nb_2C with polyvinylpyrrolidone	PAI-guided PTT	[69]

(*Continued*)

Table 13.1 Biological applications of MXenes and their biocomposites in biosensors/bioimaging, cancer theranostics and antimicrobials. (*Continued*)

MXene composites	Biomedical application	References
Nb_2C Cetanecyltrimethylammonium chloride (CTAC), APTES, polyethylene glycol (PEG)	PAI-guided PTT	[72]
Ta_4C_3 and supermagnetic iron oxide nanoparticles	MRI-guided PTT	[73]
Ti_2N QDs	PAI-guided PTT	[58]
Ti_3C_2 loaded with doxorubicin	Drug delivery agent, chemotherapy and PTT	[51]
Ti_3C_2 with doxorubicin and cellulose	Drug delivery agent, chemotherapy and PTT	[60]
Ti_3C_2 with Polyacrylamide	Drug release studies	[57]
Ti_3C_2 loaded with doxorubicin	Drug release studies, chemotherapy and PTT	[70]
Ti_3C_2	Broad spectrum antimicrobial activity	[90]
Ti_3C_2 LiF+HCl etching PVDF	Broad spectrum antimicrobial activity for water purification	[89]
Ti_3C_2 with chitosan nanofibers	Broad spectrum antimicrobial activity	[81]
Ti_3C_2	Broad spectrum antimicrobial activity	[83]
Ti_2AlC and Ti_3AlC_2	Growth inhibitory activity against *Escherichia coli*	[82]
Ti_3C_2	Antifungal activity *Trichoderma reesei*	[87]
Ti_3C_2 MXene gold nanoclusters	Broad spectrum antimicrobial and biofilm inhibitory activity	[86]
Ag/Ti_3C_2Tx hydrogels	Wound healing application	[85]
$Ti_3C_2T_x$ and Bi_2S_3	Antibacterial activity against *S. aureus* and *E. coli*	[84]

13.4 Conclusion and Future Perspectives

In the recent years, various top down and bottom up approaches with or without fluorine etching strategies have been explored to synthesize biocompatible 2D MXenes. The functionalization of MXenes makes them amenable to modification which enhances biocompatibility and reduces their cytotoxicity. Thus, the functionalized and modified MXenes can be used for bio-imaging, biosensing, delivery of drugs, theranostics and as antimicrobials. The NIR absorption in both the biowindows I and II, photodynamic/photo thermal property makes them desirable and promising nanomaterials for biological applications especially in cancer diagnostics and therapy. Due to the antimicrobial activity, MXenes can be used as anti-biofouling agents in medical implants and for purification of water. However, biocompatibility, cell uptake behavior and cytotoxicity of MXenes should be evaluated systematic and environmental toxicity determined before translating their application from bench to bedside. MXene behaviors in biological and physiological conditions, elucidation of their clearance pathway, eco-friendly synthesis strategies should be part of future prospective studies as this would prove their safety in clinical application. Similarly, systematic guidelines, regulations and risk assessment should be strategized for the research and biomedical applications of MXene.

References

1. Tan, C., Cao, X., Wu, X.J., He, Q., Yang, J., Zhang, X., Chen, J., Zhao, W., Han, S., Nam, G.H. *et al.*, Recent advances in ultrathin two-dimensional nanomaterials. *Chem. Rev.*, 117, 6225–6331, 2017.
2. Kang, M.H., Lee, D., Sung, J., Kim, J., Kim, B.H., Park, J., Structure and chemistry of 2D materials, in: *Comprehensive Nanoscience and Nanotechnology*, vol. 1–5, pp. 55–90, 2019.
3. Panda, S., Deshmukh, K., Hussain, C.M., Pasha, S.K.K., 2D MXenes for combatting Covid-19 pandemic: A perspective on latest developments and innovations. *FlatChem*, 33, 100377, 2022.
4. Huang, H., Jiang, R., Feng, Y., Ouyang, H., Zhou, N., Zhang, X., Wei, Y., Recent development and prospects of surface modification and biomedical applications of MXenes. *Nanoscale*, 12, 1325–1338, 2020.
5. Zamhuri, A., Lim, G.P., Ma, N.L., Tee, K.S., Soon, C.F., MXene in the lens of biomedical engineering: Synthesis, applications and future outlook. *Biomed. Eng. Online*, 20, 1, 33, 2021.
6. Wang, Y., Feng, W., Chen, Y., Chemistry of two-dimensional MXene nanosheets in theranostic nanomedicine. *Chin. Chem. Lett.*, 31, 937–946, 2020.
7. Ibrahim, Y., Mohamed, A., Abdelgawad, A.M., Eid, K., Abdullah, A.M., Elzatahry, A., The recent advances in the mechanical properties of self-standing two-dimensional MXene-based nanostructures: Deep insights into the supercapacitor. *Nanomaterials*, 10, 10, 2020.
8. Bai, L., Yi, W., Sun, T., Tian, Y., Zhang, P., Si, J., Hou, X., Hou, J., Surface modification engineering of two-dimensional titanium carbide for efficient synergistic multitherapy of breast cancer. *J. Mater. Chem. B*, 8, 6402–6417, 2020.
9. Maleki, A., Ghomi, M., Nikfarjam, N., Akbari, M., Sharifi, E., Shahbazi, M.A., Kermanian, M., Seyedhamzeh, M., Zare, E.N., Mehrali, M., Moradi, O., Biomedical applications of MXene-integrated composites: Regenerative medicine, infection therapy, cancer treatment, and biosensing. *Adv. Funct. Mater.*, 32, 34, 2203430, 2022.
10. Iravani, P., Iravani, S., Varma, R.S., MXene-chitosan composites and their biomedical potentials. *Micromachines (Basel)*, 13, 1383, 2022.
11. Naguib, M., Tang, W., Browning, K.L., Veith, G.M., Maliekkal, V., Neurock, M., Villa, A., Catalytic activity of Ti-based MXenes for the hydrogenation of furfural. *ChemCatChem*, 12, 22, 5733–5742, 2020.
12. Zhou, A., Methods of MAX-phase synthesis and densification–II, in: *Advances in Science and Technology of Mn+ 1axn Phases*, pp. 21–46, Elsevier, New York, 2012.

13. Shao, J., Zhang, J., Jiang, C., Lin, J., Huang, P., Biodegradable titanium nitride MXene quantum dots for cancer phototheranostics in NIR-I/II biowindows. *Chem. Eng. J.*, 400, 126009, 2020.
14. Li, Z., Wang, L., Sun, D., Zhang, Y., Liu, B., Hu, Q., Zhou, A., Synthesis and thermal stability of two-dimensional carbide MXene Ti3C2. *Mater. Sci. Eng. B*, 191, 33–40, 2015.
15. Halim, J., Lukatskaya, M.R., Cook, K.M., Lu, J., Smith, C.R., Näslund, L.Å., May, S.J., Hultman, L., Gogotsi, Y., Eklund, P., Barsoum, M.W., Transparent conductive two-dimensional titanium carbide epitaxial thin films. *Chem. Mater.*, 26, 2374–2381, 2014.
16. Naguib, M., Kurtoglu, M., Presser, V., Lu, J., Niu, J., Heon, M., Hultman, L., Gogotsi, Y., Barsoum, M.W., Two-dimensional nanocrystals produced by exfoliation of Ti_3AlC_2. *Adv. Mater.*, 23, 4248–4253, 2011.
17. Rozmysłowska-Wojciechowska, A., Mitrzak, J., Szuplewska, A., Chudy, M., Woźniak, J., Petrus, M., Wojciechowski, T., Vasilchenko, A.S., Jastrzebska, A.M., Engineering of 2D Ti_3C_2 MXene surface charge and its influence on biological properties. *Materials*, 13, 1–18, 2020.
18. Liu, F., Zhou, A., Chen, J., Jia, J., Zhou, W., Wang, L., Hu, Q., Preparation of Ti3C2 and Ti2C MXenes by fluoride salts etching and methane adsorptive properties. *Appl. Surf. Sci.*, 416, 781–789, 2017.
19. Pang, S.Y., Wong, Y.T., Yuan, S., Liu, Y., Tsang, M.K., Yang, Z., Huang, H., Wong, W.T., Hao, J., Universal strategy for HF-free facile and rapid synthesis of two-dimensional MXenes as multifunctional energy materials. *J. Am. Chem. Soc.*, 141, 9610–9616, 2019.
20. Huang, K., Li, Z., Lin, J., Han, G., Huang, P., Two-dimensional transition metal carbides and nitrides (MXenes) for biomedical applications. *Chem. Soc. Rev.*, 47, 14, 5109–5124, 2018.
21. Yang, B., Chen, Y., Shi, J., Material chemistry of two-dimensional inorganic nanosheets in cancer theranostics. *Chem*, 4, 1284–1313, 2018.
22. Verger, L., Xu, C., Natu, V., Cheng, H.M., Ren, W., Barsoum, M.W., Overview of the synthesis of MXenes and other ultrathin 2D transition metal carbides and nitrides. *Curr. Opin. Solid State Mater. Sci.*, 23, 149–163, 2019.
23. Xu, C., Wang, L., Liu, Z., Chen, L., Guo, J., Kang, N., Ma, X.L., Cheng, H.M., Ren, W., Large-area high-quality 2D ultrathin Mo2C superconducting crystals. *Nat. Mater.*, 14, 1135–1141, 2015.
24. Xiao, X., Yu, H., Jin, H., Wu, M., Fang, Y., Sun, J., Hu, Z., Li, T., Wu, J., Huang, L., Gogotsi, Y., Zhou, J., Salt-templated synthesis of 2D metallic MoN and other nitrides. *ACS Nano*, 11, 2180–2186, 2017.
25. Soleymaniha, M., Shahbazi, M.A., Rafieerad, A.R., Maleki, A., Amiri, A., Promoting role of MXene nanosheets in biomedical sciences: Therapeutic and biosensing innovations. *Adv. Healthc. Mater.*, 8, 1–26, 2019.
26. Wang, F., Yang, C., Duan, M., Tang, Y., Zhu, J., TiO_2 nanoparticle modified organ-like Ti_3C_2 MXene nanocomposite encapsulating hemoglobin for a mediator-free biosensor with excellent performances. *Biosens. Bioelectron.*, 74, 1022–1028, 2015.
27. Zhu, Y.C., Cai, B., Jiang, Q., Zhang, Y., Sha, J., Xie, S., MXene-assisted organic electrochemical transistor biosensor with multiple spiral interdigitated electrodes for sensitive quantification of fPSA/tPSA. *J. Nanobiotechnology*, 19, 386, 2021.
28. Xu, P., Lu, C., Wang, D., Fu, D., Combination of ultrathin micro-patterned MXene and PEDOT: Poly(styrenesulfonate) enables organic electrochemical transistor for amperometric determination of survivin protein in children osteosarcoma. *Microchim. Acta*, 188, 9, 301, 2021.
29. Rakhi, R.B., Nayak, P., Xia, C., Alshareef, H.N., Novel amperometric glucose biosensor based on MXene nanocomposite. *Sci. Rep.*, 6, 36422, 2016.
30. Wu, M., Zhang, Q., Fang, Y., Deng, C., Zhou, F., Zhang, Y., Wang, X., Tang, Y., Wang, Y., Polylysine-modified MXene nanosheets with highly loaded glucose oxidase as cascade

nanoreactor for glucose decomposition and electrochemical sensing. *J. Colloid Interface Sci.*, 586, 20–29, 2021.

31. Lin, S., Hu, X., Lin, J., Wang, S., Xu, J., Cai, F., Lin, J., Sensitive pH-responsive point-of-care electrochemical immunoassay for influenza A (H1N1) virus using glucose oxidase-functionalized Ti_3C_2-MXene nanosheets. *Analyst*, 146, 13, 4391–4399, 2021.
32. Li, Y., Kang, Z., Kong, L., Shi, H., Zhang, Y., Cui, M., Yang, D.P., MXene-Ti3C2/CuS nanocomposites: Enhanced peroxidase-like activity and sensitive colorimetric cholesterol detection. *Mater. Sci. Eng. C Mater. Biol. Appl.*, 104, 110000, 2019.
33. Do, H.H., Cho, J.H., Han, S.M., Ahn, S.H., Kim, S.Y., Metal-organic-framework- and MXene-based taste sensors and glucose detection. *Sensors*, 21, 7423, 2021.
34. Koyappayil, A., Chavan, S.G., Mohammadniaei, M., Go, A., Hwang, S.Y., Lee, M.H., β-hydroxybutyrate dehydrogenase decorated MXene nanosheets for the amperometric determination of β-hydroxybutyrate. *Microchim. Acta*, 187, 5, 277, 2020.
35. Kumar, S., Lei, Y., Alshareef, N.H., Quevedo-Lopez, M.A., Salama, K.N., Biofunctionalized two-dimensional Ti_3C_2 MXenes for ultrasensitive detection of cancer biomarker. *Biosens. Bioelectron.*, 121, 243–249, 2018.
36. Wu, Q., Li, N., Wang, Y., Liu, Y., Xu, Y., Wei, S., Wu, J., Jia, G., Fang, X., Chen, F., Cui, X., A 2D transition metal carbide MXene-based SPR biosensor for ultrasensitive carcinoembryonic antigen detection. *Biosens. Bioelectron.*, 144, 111697, 2019.
37. Wu, Q., Li, N., Wang, Y., Xu, Y., Wu, J., Jia, G., Ji, F., Fang, X., Chen, F., Cui, X., Ultrasensitive and selective determination of carcinoembryonic antigen using multifunctional ultrathin amino-functionalized Ti_3C_2-MXene nanosheets. *Anal. Chem.*, 92, 4, 3354–3360, 2020.
38. Zhou, S., Gu, C., Li, Z., Yang, L., He, L., Wang, M., Huang, X., Zhou, N., Zhang, Z., Ti3C2Tx MXene and polyoxometalate nanohybrid embedded with polypyrrole: Ultra-sensitive platform for the detection of osteopontin. *Appl. Surf. Sci.*, 498, 143889, 2019.
39. Yang, G., Zhao, J., Yi, S., Wan, X., Tang, J., Biodegradable and photostable Nb2C MXene quantum dots as promising nanofluorophores for metal ions sensing and fluorescence imaging. *Sens. Actuators B Chem.*, 309, 127735, 2020.
40. Zhang, H., Wang, Z., Wang, F., Zhang, Y., Wang, H., Liu, Y., Ti_3C_2 MXene mediated Prussian blue *in situ* hybridization and electrochemical signal amplification for the detection of exosomes. *Talanta*, 224, 121879, 2021.
41. You, Q., Zhuang, L., Chang, Z., Ge, M., Mei, Q., Yang, L., Dong, W.F., Hierarchical Au nanoarrays functionalized 2D Ti_2CT_x MXene membranes for the detection of exosomes isolated from human lung carcinoma cells. *Biosens. Bioelectron.*, 216, 114647, 2022.
42. Zhang, H., Wang, Z., Wang, F., Zhang, Y., Wang, H., Liu, Y., *In situ* formation of gold nanoparticles decorated Ti_3C_2 MXenes nanoprobe for highly sensitive electrogenerated chemiluminescence detection of exosomes and their surface proteins. *Anal. Chem.*, 92, 5546–5553, 2020.
43. Yue, F., Liu, M., Bai, M., Hu, M., Li, F., Guo, Y., Vrublevsky, I., Sun, X., Novel electrochemical aptasensor based on ordered mesoporous carbon/2D Ti_3C_2 MXene as nanocarrier for simultaneous detection of aminoglycoside antibiotics in milk. *Biosensors*, 12, 8, 626, 2022.
44. Zhao, X., Zhang, Y., Wang, X., Ma, P., Song, D., Sun, Y., A Ti_3C_2-MXene-functionalized LRSPR biosensor based on sandwich amplification for human IgG detection. *Anal. Bioanal. Chem.*, 414, 7, 2355–2362, 2022.
45. Wang, Q., Xin, H., Wang, Z., Label-free immunosensor based on polyaniline-loaded MXene and gold-decorated β-cyclodextrin for efficient detection of carcinoembryonic antigen. *Biosensors*, 12, 8, 657, 2022.
46. Zhao, J., He, C., Wu, W., Yang, H., Dong, J., Wen, L., Hu, Z., Yang, M., Hou, C., Huo, D., MXene-MoS_2 heterostructure collaborated with catalyzed hairpin assembly for label-free electrochemical detection of microRNA-21. *Talanta*, 237, 122927, 2022.

47. Liu, Z., Zhao, M., Lin, H., Dai, C., Ren, C., Zhang, S., Peng, W., Chen, Y., 2D magnetic titanium carbide MXene for cancer theranostics. *J. Mater. Chem. B*, 6, 21, 3541–3548, 2018.
48. Dai, C., Chen, Y., Jing, X., Xiang, L., Yang, D., Lin, H., Liu, Z., Han, X., Wu, R., Two-dimensional tantalum carbide (MXenes) composite nanosheets for multiple imaging-guided photothermal tumor ablation. *ACS Nano*, 11, 12, 12696–12712, 2017.
49. Dai, C., Lin, H., Xu, G., Liu, Z., Wu, R., Chen, Y., Biocompatible 2D titanium carbide (MXenes) composite nanosheets for pH-responsive MRI-guided tumor hyperthermia. *Chem. Mater.*, 29, 8637–8652, 2017.
50. Lin, H., Gao, S., Dai, C., Chen, Y., Shi, J., A two-dimensional biodegradable niobium carbide (MXene) for photothermal tumor eradication in NIR-I and NIR-II biowindows. *J. Am. Chem. Soc.*, 139, 45, 16235–16247, 2017.
51. Han, X., Jing, X., Yang, D., Lin, H., Wang, Z., Ran, H., Li, P., Chen, Y., Therapeutic mesopore construction on 2D Nb2C MXenes for targeted and enhanced chemo-photothermal cancer therapy in NIR-II biowindow. *Theranostics*, 8, 16, 4491–4508, 2018.
52. Lu, Y., Zhang, X., Hou, X., Feng, M., Cao, Z., Liu, J., Functionalized 2D Nb_2C nanosheets for primary and recurrent cancer photothermal/immune-therapy in the NIR-II biowindow. *Nanoscale*, 13, 42, 17822–17836, 2021.
53. Lin, S., Lin, H., Yang, M., Ge, M., Chen, Y., Zhu, Y., A two-dimensional MXene potentiates a therapeutic microneedle patch for photonic implantable medicine in the second NIR biowindow. *Nanoscale*, 12, 18, 10265–10276, 2020.
54. Xue, Q., Zhang, H., Zhu, M., Pei, Z., Li, H., Wang, Z., Huang, Y., Huang, Y., Deng, Q., Zhou, J., Du, S., Huang, Q., Zhi, C., Photoluminescent Ti_3C_2 MXene quantum dots for multicolor cellular imaging. *Adv. Mater.*, 29, 15, 1604847, 2017.
55. Yu, X., Cai, X., Cui, H., Lee, S.W., Yu, X.F., Liu, B., Fluorine-free preparation of titanium carbide MXene quantum dots with high near-infrared photothermal performances for cancer therapy. *Nanoscale*, 9, 45, 17859–17864, 2017.
56. Cai, G., Yu, Z., Tong, P., Tang, D., Ti_3C_2 MXene quantum dot-encapsulated liposomes for photothermal immunoassays using a portable near-infrared imaging camera on a smartphone. *Nanoscale*, 11, 33, 15659–15667, 2019.
57. Zhang, D.Y., Liu, H., Younis, M.R., Lei, S., Chen, Y., Huang, P., Lin, J., *In-situ* TiO_{2-x} decoration of titanium carbide MXene for photo/sono-responsive antitumor theranostics. *J. Nanobiotechnology*, 20, 1, 53, 2022.
58. Shao, B., Liu, Z., Zeng, G., Wang, H., Liang, Q., He, Q., Cheng, M., Zhou, C., Jiang, L., Song, B., Two-dimensional transition metal carbide and nitride (MXene) derived quantum dots (QDs): Synthesis, properties, applications and prospects. *J. Mater. Chem. A*, 8, 7508–7535, 2020.
59. Wu, Y., Xiong, W., Wang, Z., Wang, Y., Sun, K.Y., Song, X., Lv, Z., Xu, W., Zhong, W., Zou, X., Cai, H.L., Self-assembled MXene-based Schottky-junction upon transition metal oxide for regulated tumor microenvironment and enhanced CDT/PTT/MRI activated by NIR irradiation. *Chem. Eng. J.*, 427, 131925, 2022.
60. Xing, C., Chen, S., Liang, X., Liu, Q., Qu, M., Zou, Q., Li, J., Tan, H., Liu, L., Fan, D., Zhang, H., Two-dimensional MXene (Ti_3C_2)-integrated cellulose hydrogels: Toward smart three-dimensional network nanoplatforms exhibiting light-induced swelling and bimodal photothermal/chemotherapy anticancer activity. *ACS Appl. Mater. Interfaces*, 10, 33, 27631–27643, 2018.
61. Geng, B., Xu, S., Shen, L., Fang, F., Shi, W., Pan, D., Multifunctional carbon dot/MXene heterojunctions for alleviation of tumor hypoxia and enhanced sonodynamic therapy. *Carbon*, 179, 493–504, 2021.
62. Hao, Z., Li, Y., Liu, X., Jiang, T., He, Y., Zhang, X., Cong, C., Wang, D., Liu, Z., Gao, D., Enhancing biocatalysis of a MXene-based biomimetic plasmonic assembly for targeted cancer treatments in NIR-II biowindow. *Chem. Eng. J.*, 425, 130639, 2021.

63. Wu, Z., Shi, J., Song, P., Li, J., Cao, S., Chitosan/hyaluronic acid based hollow microcapsules equipped with MXene/gold nanorods for synergistically enhanced near infrared responsive drug delivery. *Int. J. Biol. Macromol.*, 183, 870–879, 2021.
64. Dong, Y., Li, S., Li, X., Wang, X., Smart MXene/agarose hydrogel with photothermal property for controlled drug release. *Int. J. Biol. Macromol.*, 190, 693–699, 2021.
65. Wang, S., Zhang, Z., Wei, S., He, F., Li, Z., Wang, H.H., Huang, Y., Nie, Z., Near-infrared light-controllable MXene hydrogel for tunable on-demand release of therapeutic proteins. *Acta Biomater.*, 130, 138–148, 2021.
66. Xu, Y., Wang, Y., An, J., Sedgwick, A.C., Li, M., Xie, J., Hu, W., Kang, J., Sen, S., Steinbrueck, A., Zhang, B., Qiao, L., Wageh, S., Arambula, J.F., Liu, L., Zhang, H., Sessler, J.L., Kim, J.S., 2D-ultrathin MXene/DOXjade platform for iron chelation chemo-photothermal therapy. *Bioact. Mater.*, 14, 76–85, 2021.
67. Lin, H., Wang, X., Yu, L., Chen, Y., Shi, J., Two-dimensional ultrathin mxene ceramic nanosheets for photothermal conversion. *Nano Lett.*, 17, 1, 384–391, 2017.
68. Lin, H., Wang, Y., Gao, S., Chen, Y., Shi, J., Theranostic 2D tantalum carbide (MXene). *Adv. Mater.*, 32, 42, 2003085, 2020.
69. Liu, Z., Zhao, M., Lin, H., Dai, C., Ren, C., Zhang, S., Peng, W., Chen, Y., 2D magnetic titanium carbide MXene for cancer theranostics. *J. Mater. Chem. B*, 6, 21, 3541–3548, 2018.
70. Han, X., Huang, J., Lin, H., Wang, Z., Li, P., Chen, Y., 2D ultrathin MXene-based drug-delivery nanoplatform for synergistic photothermal ablation and chemotherapy of cancer. *Adv. Healthc. Mater.*, 7, 9, 1701394, 2018.
71. Liu, Z., Lin, H., Zhao, M., Dai, C., Zhang, S., Peng, W., Chen, Y., 2D superparamagnetic tantalum carbide composite MXenes for efficient breast-cancer theranostics. *Theranostics*, 8, 6, 1648–1664, 2018.
72. Li, Z., Zhang, H., Han, J., Chen, Y., Lin, H., Yang, T., Surface nanopore engineering of 2D MXenes for targeted and synergistic multitherapies of hepatocellular carcinoma. *Adv. Mater.*, 30, 25, 1706981, 2018.
73. Zhang, X., Cheng, L., Lu, Y., Tang, J., Lv, Q., Chen, X., Chen, Y., Liu, J., A MXene-based bionic cascaded-enzyme nanoreactor for tumor phototherapy/enzyme dynamic therapy and hypoxia-activated chemotherapy. *Nanomicro Lett.*, 14, 1, 22, 2021.
74. Xing, C., Chen, S., Liang, X., Liu, Q., Qu, M., Zou, Q., Li, J., Tan, H., Liu, L., Fan, D., Zhang, H., Two-dimensional MXene (Ti_3C_2)-integrated cellulose hydrogels: Toward smart three-dimensional network nanoplatforms exhibiting light-induced swelling and bimodal photothermal/chemotherapy anticancer activity. *ACS Appl. Mater. Interfaces*, 10, 27631–43, 2018.
75. Liu, G., Zou, J., Tang, Q., Yang, X., Zhang, Y., Zhang, Q., Huang, W., Chen, P., Shao, J., Dong, X., Surface modified Ti_3C_2 MXene nanosheets for tumor targeting photothermal/photodynamic/chemo synergistic therapy. *ACS Appl. Mater. Interfaces*, 9, 46, 40077–40086, 2017.
76. Liu, Y., Han, Q., Yang, W., Gan, X., Yang, Y., Xie, K., Xie, L., Deng, Y., Two-dimensional MXene/cobalt nanowire heterojunction for controlled drug delivery and chemo-photothermal therapy. *Mater. Sci. Eng. C Mater. Biol. Appl.*, 116, 111212, 2020.
77. Zhang, P., Yang, X.J., Li, P., Zhao, Y., Niu, Q.J., Fabrication of novel MXene (Ti_3C_2)/polyacrylamide nanocomposite hydrogels with enhanced mechanical and drug release properties. *Soft Matter*, 16, 1, 162–169, 2020.
78. Korupalli, C., You, K.L., Getachew, G., Rasal, A.S., Dirersa, W.B., Fahmi, M.Z., Chang, J.Y., Engineering the surface of Ti_3C_2 MXene nanosheets for high stability and multimodal anticancer therapy. *Pharmaceutics*, 14, 2, 304, 2022.
79. Mayerberger, E.A., Street, R.M., McDaniel, R.M., Barsoum, M.W., Schauer, C.L., Antibacterial properties of electrospun Ti3C2Tz (MXene)/chitosan nanofibers. *RSC Adv.*, 8, 62, 35386–35394, 2018.

80. Jastrzębska, A.M., Karwowska, E., Wojciechowski, T., Ziemkowska, W., Rozmysłowska, A., Chlubny, L., Olszyna, A., The atomic structure of Ti_2C and Ti_3C_2 MXenes is responsible for their antibacterial activity toward E. *coli bacteria. J. Mater. Eng Perform.*, 28, 1272–1277, 2019.
81. Shamsabadi, A.A., Gh, M.S., Anasori, B., Soroush, M., Antimicrobial mode-of-action of colloidal $Ti_3C_2T_x$ MXene nanosheets. *ACS Sustain. Chem. Eng.*, 6, 12, 16586–16596, 2018.
82. Li, J., Li, Z., Liu, X., Li, C., Zheng, Y., Yeung, K.W.K., Cui, Z., Liang, Y., Zhu, S., Hu, W., Qi, Y., Zhang, T., Wang, X., Wu, S., Interfacial engineering of $Bi_2S_3/Ti_3C_2T_x$ MXene based on work function for rapid photo-excited bacteria-killing. *Nat. Commun.*, 12, 1224, 2021.
83. Zhu, X., Zhu, Y., Jia, K., Abraha, B.S., Li, Y., Peng, W., Zhang., F., Fan, X., Zhang, L., A near-infrared light-mediated antimicrobial based on Ag/Ti_3C_2Tx for effective synergetic antibacterial applications. *Nanoscale*, 12, 37, 19129–19141, 2020.
84. Zheng, K., Li, S., Jing, L., Chen, P.Y., Xie, J., Synergistic antimicrobial titanium carbide (MXene) conjugated with gold nanoclusters. *Adv. Healthc. Mater.*, 9, 19, 2001007, 2020.
85. Lim, G.P., Soon, C.F., Morsin, M., Ahmad, M.K., Nayan, N., Tee, K.S., Synthesis, characterization and antifungal property of $Ti_3C_2T_x$ MXene nanosheets. *Ceram. Int.*, 46, 12, 20306–20312, 2020.
86. Mahar, I., Memon, F.H., Lee, J.W., Kim, K.H., Ahmed, R., Soomro, F., Rehman, F., Memon, A.A., Thebo, K.H., Choi, K.H., Two-dimensional transition metal carbides and nitrides (MXenes) for water purification and antibacterial applications. *Membranes*, 11, 869, 2021.
87. Rasool, K., Mahmoud, K.A., Johnson, D.J., Helal, M., Berdiyorov, G.R., Gogotsi, Y., Efficient antibacterial membrane based on two-dimensional $Ti_3C_2T_x$ (MXene) nanosheets. *Sci. Rep.*, 7, 1, 1598, 2017.
88. Rasool, K., Helal, M., Ali, A., Ren, C.E., Gogotsi, Y., Mahmoud, K.A., Antibacterial activity of Ti_3C_2Tx MXene. *ACS Nano*, 10, 3, 3674–3684, 2016.
89. Li, Q., Wang, W., Feng, H., Cao, L., Wang, H., Wang, D., Chen, S., NIR-triggered photocatalytic and photothermal performance for sterilization based on copper sulfide nanoparticles anchored on $Ti_3C_2T_x$ MXene. *J. Colloid Interface Sci.*, 604, 810–822, 2021.
90. Liu, Y., Tian, Y., Han, Q., Yin, J., Zhang, J., Yu, Y., Yang, W., Deng, Y., Synergism of 2D/1D MXene/cobalt nanowire heterojunctions for boosted photo-activated antibacterial application. *Chem. Eng. J.*, 410, 128209, 2021.
91. Zhou, X., Wang, Z., Chan, Y.K., Yang, Y., Jiao, Z., Li, L., Li, J., Liang, K., Deng, Y., Infection micromilieu-activated nanocatalytic membrane for orchestrating rapid sterilization and stalled chronic wound regeneration. *Adv. Funct. Mater.*, 32, 7, 2109469, 2022.
92. Li, Y., Han, M., Cai, Y., Jiang, B., Zhang, Y., Yuan, B., Zhou, F., Cao, C., Muscle-inspired MXene/PVA hydrogel with high toughness and photothermal therapy for promoting bacteria-infected wound healing. *Biomater. Sci.*, 10, 4, 1068–82, 2022.
93. Yan, B., Zhou, M., Liao, X., Wang, P., Yu, Y., Yuan, J., Wang, Q., Developing a multifunctional silk fabric with dual-driven heating and rapid photothermal antibacterial abilities using high-yield MXene dispersions. *ACS Appl. Mater. Interfaces*, 13, 36, 43414–43425, 2021.
94. Yan, B., Bao, X., Liao, X., Wang, P., Zhou, M., Yu, Y., Yuan, J., Cui, L., Wang, Q., Sensitive micro-breathing sensing and highly-effective photothermal antibacterial Cinnamomum camphora bark micro-structural cotton fabric via electrostatic self-assembly of MXene/HACC. *ACS Appl. Mater. Interfaces*, 14, 1, 2132–2145, 2021.
95. Jiang, Y., Pan, X., Yao, M., Han, L., Zhang, X., Jia, Z., Weng, J., Chen, W., Fang, L., Wang, X., Zhang, Y., Bioinspired adhesive and tumor microenvironment responsive nano MOFs assembled 3D-printed scaffold for anti-tumor therapy and bone regeneration. *Nano Today*, 39, 101182, 2021.

96. Yin, J., Pan, S., Guo, X., Gao, Y., Zhu, D., Yang, Q., Gao, J., Zhang, C., Chen, Y., Nb2C MXene-functionalized scaffolds enables osteosarcoma phototherapy and angiogenesis/osteogenesis of bone defects. *Nano-Micro Lett.*, 13, 1, 1–8, 2021.
97. Fu, Y., Zhang, J., Lin, H., Mo, A., 2D titanium carbide(MXene) nanosheets and 1D hydroxyapatite nanowires into free standing nanocomposite membrane: *In vitro* and *in vivo* evaluations for bone regeneration. *Mater. Sci. Eng. C Mater. Biol. Appl.*, 118, 111367, 2021.
98. Mi, X., Su, Z., Fu, Y., Li, S., Mo, A., 3D printing of Ti3C2-MXene-incorporated composite scaffolds for accelerated bone regeneration. *Biomed. Mater.*, 17, 3, 035002, 2022.
99. Basara, G., Saeidi-Javash, M., Ren, X., Bahcecioglu, G., Wyatt, B.C., Anasori, B., Zhang, Y., Zorlutuna, P., Electrically conductive 3D printed $Ti_3C_2T_x$ MXene-PEG composite constructs for cardiac tissue engineering. *Acta Biomater.*, 139, 179–189, 2022.
100. Zhu, H., Dai, W., Wang, L., Yao, C., Wang, C., Gu, B., Li, D., He, J., Electroactive oxidized alginate/gelatin/MXene ($Ti_3C_2T_x$) composite hydrogel with improved biocompatibility and self-healing property. *Polymers*, 14, 3908, 2022.
101. Mao, L., Hu, S., Gao, Y., Wang, L., Zhao, W., Fu, L., Cheng, H., Xia, L., Xie, S., Ye, W., Shi, Z., Yang, G., Biodegradable and electroactive regenerated bacterial cellulose/MXene ($Ti_3C_2T_x$) composite hydrogel as wound dressing for accelerating skin wound healing under electrical stimulation. *Adv. Healthc. Mater.*, 9, 19, 2000872, 2020.

14

MXene-Reinforced Polymer Composite Membranes for Water Desalination and Wastewater Treatment

Anjana Sreekumar[1], Ajil R. Nair[1], Akhila Raman[1], Akhil Sivan[1*], Mayank Pandey[2†], Kalim Deshmukh[3] and Saritha Appukuttan[1‡]

[1]*Department of Chemistry, Amrita Vishwa Vidyapeetham, Amritapuri, Kollam, Kerala, India*
[2]*Department of Physics and Electronics, Kristu Jayanti College, Bangalore, India*
[3]*Chemical Processes and Biomaterials, New Technologies - Research Centre, University of West Bohemia, Univerzitní 8, Pilsen, Czech Republic*

Abstract
Water pollution has grown to be a severe concern due to unethical human deeds and improper management of water resources. The malignancy of organic wastes and contaminants in aquatic systems can adversely affect the flora and fauna. Traditional and modern water treatment techniques that require liquid separation have been widely used to address this problem, and membrane separation has emerged as the most potential solution to meet the issue. Membrane separation offers excellent industrial uniformity and a low carbon footprint as its key advantages. Unfortunately, water permeance or selectivity interferes with this membrane separation procedure. The hydrophobic properties of commonly used membrane materials like polysulfone and polypropylene have an unfavorable influence on water permeance and separation efficiency. Numerous studies have discussed the usage of novel membranes incorporated with nanomaterials, such as metal-organic frameworks, graphene oxide, carbon nanotubes, and MXenes, in order to solve this issue. MXenes are rising to prominence as a potential material for the purification of water and membrane separation. This chapter throws light on the potentiality of MXene-reinforced polymer composite membranes in the purification of wastewater and desalination.

Keywords: MXene, polymer composites, membranes, water pollution, water treatment, water desalination

14.1 Introduction

Even though water covers 2/3 of the earth's surface, a lack of freshwater worldwide has seriously troubled global social and economic growth [1]. The entire population appreciates the positive dimensions of industrialization, the developments in society, and their living

**Corresponding author*: akhilsivan@am.amrita.edu
†*Corresponding author*: mayank01081989@gmail.com
‡*Corresponding author*: sarithatvla@gmail.com

conditions. The other side of the picture is quite vulnerable, some adverse effects result from certain societal developments, like air pollution, noise pollution, water pollution, soil pollution, etc. The growing global population demands increased water consumption; unfortunately, incautious human actions result in water pollution and have created water insufficiency. The shortage of pure water and the fact 'how to survive in future?' is considered as universal problems that remains unsolved downright. Meeting the requirements of pure and usable water at present and to cater to the shortage expected in near future have forced the researchers to rely on viable alternative water sources like seawater/highly mineralized water and wastewater [2]. There are copious ongoing research investigations in the area of wastewater treatment and desalination of water to solve this universal problem [3–5]; thus, it helps to wipe out the thirst and hydrate our future generations.

Treating wastewater and heavily mineralized water in an effective manner for useful anthropological purposes is very important. Manufacturing industries and factories are the key generators of a significant fraction of worldwide water pollutants [6, 7]. Heavy metal ions, radioactive metal ions, several organic and inorganic pollutants, oily water, dyes, harmful microorganisms, etc., are the crucial concerns to be addressed in wastewater treatment/purification [8–10]. Similarly, the water desalination/desalting process is a widely accepted notion passed down since ancient times to earn pure water from seawater and highly mineralized groundwater [11, 12]. So far, numerous strategies have been used for wastewater treatment and water desalination which includes adsorption methods, bioremediation, coagulation, sedimentation, ultrasonication, chlorination, filtration, distillation, evaporation, membrane-assisted processes, etc. [13–17]. These versatile techniques are primarily conventional or considered outdated and demand substantial energy requirements, generally in the form of fossil fuels, consequently making these processes expensive and less viable [18, 19]. Hence the membrane-assisted processes (filtration and adsorption) have secured the utmost attention besides appreciation and have been widely used/adopted in wastewater treatment and desalting in recent times [20–22]. The cost-effectiveness and low energy consumption of membrane-assisted water treatment encouraged its recognition among the other conventional methods [23, 24].

Membrane-assisted liquid separation has been widely used in the pharmaceutical and chemical industries for water purification [25]. The structure, content, and material properties have a direct influence in the separation effectiveness of membranes in water purification applications. They offer several advantages, including a smaller carbon footprint, better dependability, ease of use, and a lower level of secondary pollution [26, 27]. Low selectivity, diminished mechanical stability, low membrane flux, etc., are some drawbacks of common commercially available polymer-based membrane materials like polypropylene and polysulfone [28]. Their hydrophobic nature affects water permeance and separation performance. Hence, resolving this problem has become a hot research area and consequently boosted the research in this arena. As a result, various studies have demonstrated the fabrication of innovative membranes incorporated with different two-dimensional (2D) nanomaterial's such as carbon nanotubes (CNT) [29], graphene [30], graphene oxides [31], metal-organic frameworks (MOFs) [32], MXenes [33], etc., can resolve these problems effectively.

The features, such as excellent water affinity, controllable functionalities, high specific area, etc., make the aforementioned 2D materials desirable for making modified membranes/thin film nanocomposites for water filtration. Graphene and graphene oxide manifest considerable efficiency as membrane reinforcing materials as they possess outstanding chemical stability, ease of functionalization and molecule-sieving capability [34, 35]. The most extensively investigated materials for desalination are those based on graphene [36]. MOFs are also one of the most omnipresent 2D materials, that have many benefits, including their diverse morphology (polyhedrons, microns and nanoparticles) [37, 38]. The hydrophilic carbon nanotubes (CNT) have also showed good performance in terms of water desalination application [29]. Unfortunately, agglomeration and poor compatibility of CNTs, MOF's dispersibility in organic phase results in low stability in long-term applications. The minor interlayer distance between the graphene nanosheets and the decomposition of graphene oxide in water factors limit its industrial application in membrane technology [39–42]. Apart from the nanomaterials mentioned above as polymer reinforcement candidates, MXenes ($Ti_3C_2T_x$) are a novel two-dimensional nanostructures that researchers from Drexel University have produced in 2011 [43]. This class of materials have gained more attention in recent times because of its versatile characteristics, such as vast surface area, hydrophilicity, outstanding stability, good oxidation resistance, great electrical/thermal conductivity, etc. [44]. Many reports demonstrate that different MXenes find applications in adsorption [45], catalysis [46], membrane applications etc. Modified MXenes are used in polymer composite materials to create new membranes with exceptional properties for water purification and desalination purposes. The fabricated MXene-based membranes were discovered to be highly successful in separating the O/W mixture [5] and removing different salts [47] and impurities [48, 49] from water. These membranes also display great antifouling properties [50, 51].

In this chapter, the MXene-reinforced polymer nanocomposite membranes for targeted applications such as wastewater treatment and water desalination, which is novel and promising in terms of application perspective, are summarized. By considering since 2011, till date reports on MXene-reinforced polymer nanocomposites for desalination and wastewater treatment have exponentially increased in the past five years. Hence, it is necessary to evaluate the present-day condition of research on this topic. The properties, key synthesis methods and different MXenes in terms of application uniqueness are also discussed in detail. By spotlighting the research gaps in this field, we intend to be a torchlight for the preceding researchers and promote the research in this area.

14.2 Preparation

MXene/polymer composites can be developed through various methods, including solution blending [52], *in situ* polymerization [53], emulsion mixing [54], lamination stacking [55], dry mixing [56] etc. The hydrophilic nature of MXene is attributed to the presence of highly polar -F, -O, and -OH terminal groups [57]. Thus, solution blending utilizing water as a solvent is the most common strategy used for MXene/polymer composite preparation. Solution blending process is an efficient and facile strategy to fabricate MXene/polymer composites. In this process, the component ratios of the composites can be easily tuned, and the MXene–polymer dispersion can be enhanced [58]. Usually, polar polymers like polyvinyl alcohol (PVA) [59], polybenzimidazole (PBI) [60], polyacrylamide (PAM) [61], chitosan (CS) [62], polyvinylidene fluoride (PVDF) [63], polylactic acid (PLA) [64], cellulose [65] etc. are involved in this process. The chosen solvent should be able to dissolve both MXene and polymer, or the selected solvents should be miscible, therefore, polar dispersants such as dimethylsulfoxide (DMSO) [66], dimethylacetamide (DMAc) [67], *N, N*-dimethylformamide (DMF) [63] and water [59] are preferred for the process. Coming to MXene/polymer composite membranes, vacuum-assisted filtration (VAF) [68], layer-by-layer (LBL) [69], casting [70], electrospinning [71] etc. are the potential methods for their fabrication.

14.2.1 Vacuum-Assisted Filtration

Vacuum-assisted filtration (VAF) is a simple approach that can be used to congregate MXene to create membranes because of the distinct characteristics of ultrathin flakes. Laminar membrane fabrication using the stacking of MXene/polymer composites has received significant attention. Normally, polar solvents are used to blend polymers and MXenes. This mixture is then rapidly agitated or sonicated before being filtered through a vacuum [72]. Luo *et al.* [73] fabricated pliable and supremely conducting $Ti_3C_2T_x$ MXene/natural rubber (NR) composite films with excellent electromagnetic interference (EMI) shielding efficacy *via* the VAF technique. $Ti_3C_2T_x$/NR membranes were developed by blending MXene suspension with NR latex followed by filtering under vacuum (Figure 14.1) [73]. The negative charges in NR latex and MXene cause an electrostatic repulsion force which permits the selective distribution of MXene sheets in the NR interfaces. Thus, an interlinked network of MXene is formed in the matrix of NR, which encourages load transfer and electron transport even at low MXene concentrations.

Through the same facile VAF strategy, a flexible cellulose nanofiber (CNF) incorporated Ti_3C_2 composite membrane was developed by G. Song *et al.* [74]. The prepared films can be pleated in different shapes as it possesses excellent flexibility, thus it can be functionally employed in pliable electronic devices. The few-layer MXene powder and CNF were disseminated in distilled water through ultra-sonication. The homogenous solution was then treated using vacuum-assisted filtration to produce a composite membrane with nitrogen protection. *In situ* polymerization can also be employed to create an MXene/polymer membrane with a VAF approach besides the *ex situ* blending method. Numerous hydrophilic terminations on MXene flakes make them easily dispersible in polar solvents, opening up the possibility of polymerizing various monomers. Normally, the polymerization is finished before VAF by aggressively stirring the monomers and MXene-containing solution.

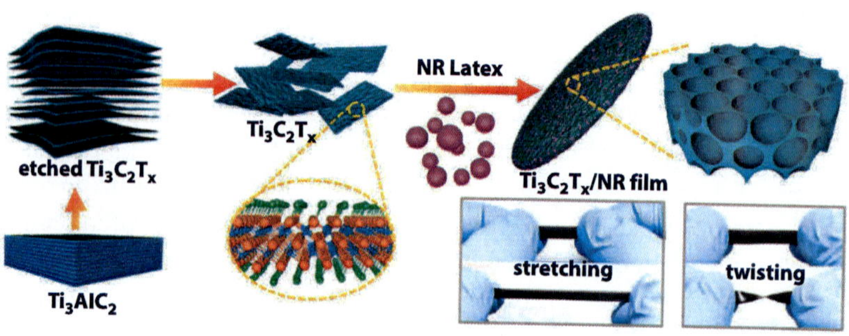

Figure 14.1 Diagrammatic illustration of the synthesis process of $Ti_3C_2T_x$/NR membranes. The fabricated films can be stretched and twisted by applying huge strains [73]. Copyright 2019. Reproduced with permission from Elsevier.

Although the MXene functionalities need to be altered to accommodate different types of monomers, *in situ* polymerization makes it easier for polymers to be homogenously distributed throughout the fabricated membranes. Boota *et al.* [75] publicized the first *in situ* polymerization with VAF method of an electrochemically active polymer, polypyrrole (PPy), using pyrrole as the monomer, with the assistance of MXene terminations.

14.2.2 Layer-by-Layer Assembly

In order to create functionalized membranes with precise thickness and electrostatic or hydrogen bonding interactions, the Layer-By-Layer (LbL) assembly approach is typically used. Traditionally, the LbL technique may deposit oppositely charged species to produce multilayer films with good uniformity. As the MXene materials possess negatively charged surfaces, an MXene-incorporated polymer membrane can be fabricated using the dip-coating LbL method. Additionally, these membranes can also be developed using the spin-spray LbL process, which exhibits superior mechanical and conductivity attributes [76]. Z. Tan *et al.* [77] utilized the electrostatic force between the opposite charges of MXene and chitosan (CS) to fabricate CS/MXene layered membranes *via* LbL assembly method. The chitosan solution was spin-coated on a substrate of polyethylene terephthalate and dried for 20 minutes at 45 °C. Through the same process, 11 layers were coated, of which 5 were MXene and 6 were CS layers. The prepared film should be peeled off from the substrate with utmost care (Figure 14.2) [77]. The developed film displayed an excellent EMI shielding efficiency of 40.8 dB and outstanding Joule heating capacity. Thus, this work paves a novel way to attain excellent heat dissipation capacity and EMI shielding, which can be potentially utilized in next-generation electronic devices. Textiles with EMI shielding efficiency are necessary for various industries and applications. G. Yin *et al.* [78] fabricated polyaniline (PANI) nanowires/MXene nanosheets incorporated fabrics of carbon fiber through LbL

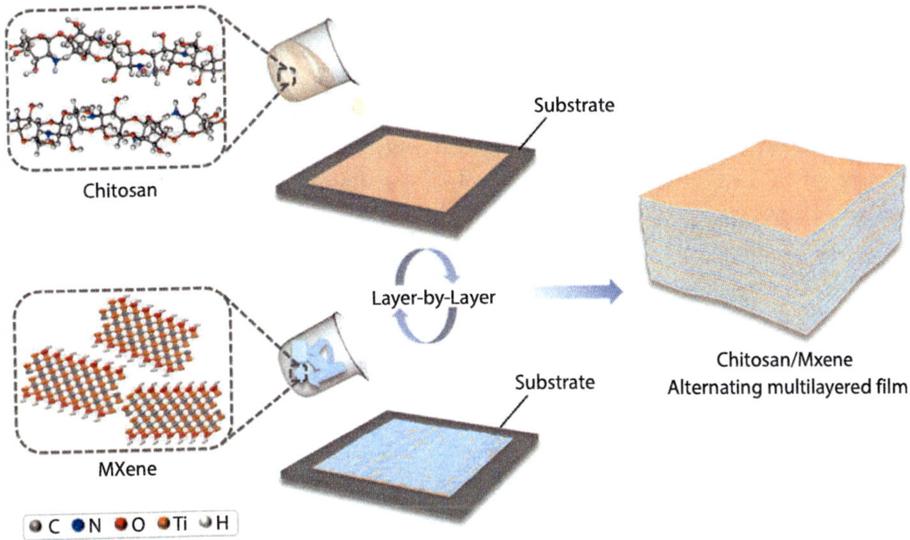

Figure 14.2 Diagrammatic demonstration of the LBL assembly technique for fabricating CS/MXene multi-layered film [77]. Copyright 2022. Reproduced with permission from Elsevier.

assembly method. The coated fabrics were again coated by a layer of polydimethylsiloxane (PDMS). These prepared fabrics displayed excellent EMI shielding efficiency and electrical conductivity. Moreover, the PDMS coating provides self-cleaning ability to the fabric with good hydrophobicity. The developed fabrics also possess beneficial thermal characteristics, air permeability and flexibility, extending their practical applications.

14.2.3 Electrospinning

The well-known electrospinning process uses the force of an electric field applied to a polymer solution to create continuous polymeric fibers. Using this method, ultrafine fibers with enormous surface area can be made. These fiber surfaces have excellent mechanical qualities and are easily functionalizable. As a result, the electrospinning technique is a special and adaptable technique for developing nanofibers quickly [79]. MXene is a 2D inorganic filler that has recently gained extensive recognition for its efficiency in altering electrospun nanofiber surfaces for biomedical applications [80]. Awasthi *et al.* [81] fabricated electrospun polycaprolactone incorporated MXene (PCL-MXene) composite fibers for the first time. The developed fibrous membranes were analyzed to study their biomedical applications. MXene at various concentrations of 0.2, 0.5, 1, and 2 wt% were homogeneously dispersed in 1:8 v/v DMF:chloroform solution. 10 wt% of PCL was added to the dispersion and subjected to stirring to make the solution spinnable. The prepared solutions were electrospun at optimized parameters of 1 mL/h flow rate, applied electric voltage of 15 kV and 15 cm tip-collector distance. The biocompatibility analysis of developed composite membranes proved their cell friendliness and can be employed for advanced biomedical applications. Multilayered MXene/poly(lactic acid) (PLA) membranes with attractive EMI shielding efficiency were synthesized by Du *et al.* [82]. The composite membrane was prepared by arranging MXene layer alternatively onto electrospun PLA nanofibrous films. Multiple-reflection-absorption effect and high conduction of MXene layer contribute to the great EMI shielding efficiency (SE) (55.4 dB) of the developed membrane. The process of wet spinning involves passing a polymer solution through a spinneret before it enters a coagulation bath that contains polymer non-solvent. Polymer fibers are created while this solvent/non-solvent exchange process takes place. The polymer's solvent and the MXene's dispersion medium should be comparable to prepare an MXene/polymer mixture in wet spinning. If they are not comparable, the solvent exchange method enables the use of a similar solvent for both the dissolution of polymers and the dispersion of Ti3C2 [83]. Figure 14.3 (a) gives a pictorial representation of wet-spinning process [84] and Figure 14.3 (b) shows diagrammatic illustration of $Ti_3C_2T_x$ MXene/carbon nanofibers fabrication via electrospinning [85].

14.2.4 Casting

Casting process is frequently used to create MXene/polymer membrane with definite compositions. By continuously stirring or sonicating, specified concentrations of polymer and MXene are typically solubilized in the solvent to form a uniform solution. The prepared solution is then casted on a sterile substrate to enable the solvent to evaporate under high temperatures or in an inert atmosphere. The MXene incorporated polymer membranes can then be detached from the stratum and used for further studies. Only Ti_3C_2/PPy film

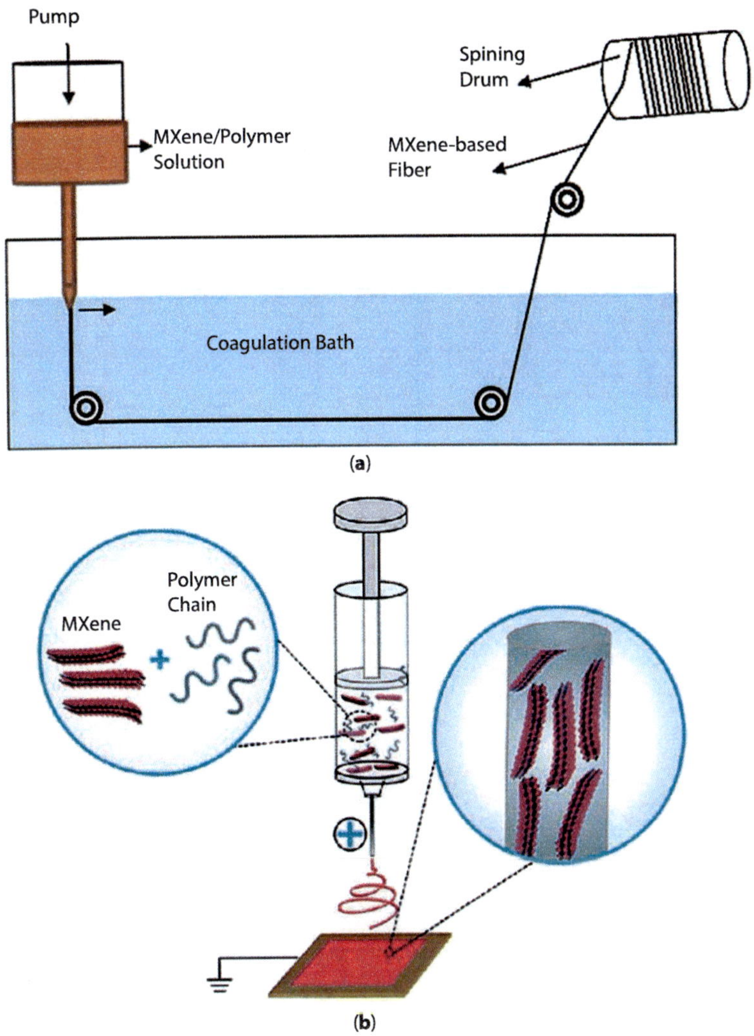

Figure 14.3 (a) The pictorial representation of wet-spinning process [84]. Copyright 2015. Reproduced with permission from Elsevier. (b) Diagrammatic illustration of $Ti_3C_2T_x$ MXene/carbon nanofibers fabrication via electrospinning [85]. Copyright 2018. Reproduced with permission from the Royal Society of Chemistry.

was produced using an *in situ* polymerization, in contrast to the VAF method, where all MXene consolidated polymer membranes were produced using pre-existing polymers [86]. Shen *et al.* [87] fabricated hydrophilic polysulphone/MXene composite membranes with outstanding antifouling characteristics. A solution containing 17 wt% polysulphone (PSF) and 3 wt% polyvinylpyrrolidone (PVP) were degassed and casted onto a clean glass plate to develop a film. The formed film is immersed in a water bath containing MXene nanosheets for MXene embedment (Figure 14.4) [87]. Mirkhani *et al.* [88] fabricated PVA/MXene composite membrane *via* 2 methods of VAF and solution casting. In solution casting method, the required amount of PVA and MXene were blended together in water for 1 hour at 1000 rpm. The solution was kept for sonication for 15 min. in the presence of nitrogen to block

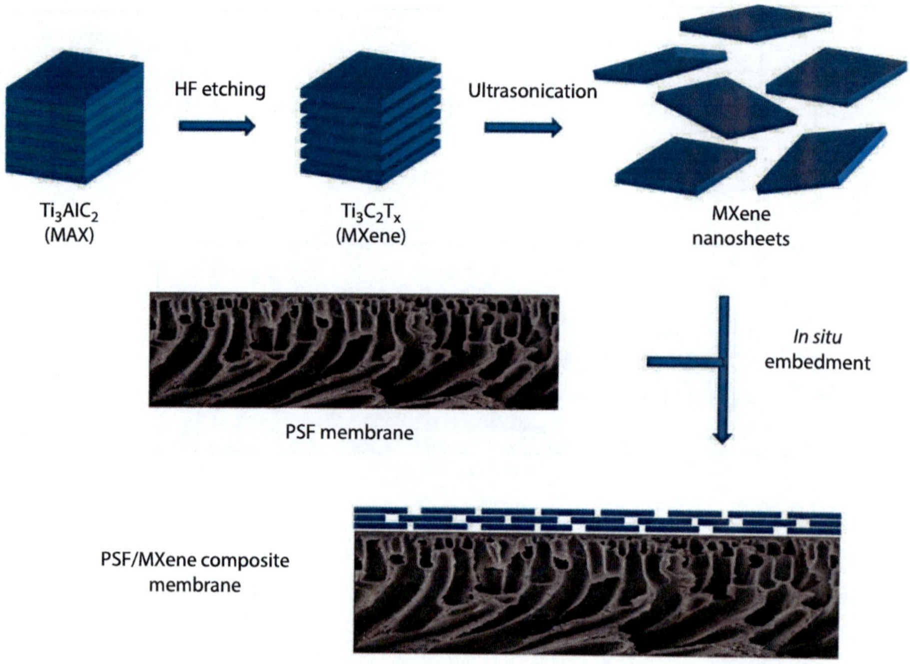

Figure 14.4 Pictorial representation of the development and embedment of the MXene nanosheets [87]. Copyright 2019. Adapted from MDPI. (Open access)

plausible oxidation. The developed solution was casted onto a petridish and kept overnight in vacuum at 40 °C for drying. The solution casted films and VAF films have shown excellent dielectric constant/loss factor values. This work proves the potentiality of MXene to be used in advanced future electronic devices.

Because of the special characteristics of MXene/polymer membranes, researchers are still trying to discover novel methods for producing them. For instance, as MXenes possess exceptional thermal stability, unlike polymers, MXene/polymer composites can be created at temperatures over the polymer melting point. Melt blending is significantly more popular in industrial production than organic solution blending due to its benefits of solvent-free, customizable formulation, affordability, and environmental friendliness. As demonstrated, by melt blending with the proper operating temperature and stirring amplitude, MXene flakes may equally spread in the polymer matrix, allowing the construction of uniform MXene-incorporated polymer membranes by following hot pressing. The developed membranes produced using the hot press process can also be trimmed into the proper sizes and shapes for specific testing and applications [56].

Substantial efforts are needed to improve MXene flakes-polymers interaction, such as by surface-modifying MXene flakes, which will enhance nanofiller dispersion in the composite membranes. In order to create MXene/polymer membranes, it is essential to investigate additional multiple molecular interactions in addition to covalent, hydrogen, electrostatic and van der Waal bonds. Examples of these interactions include biologically specific bonds, complexation bonds, coordination bonds, supramolecular interactions, and others. MXene/polymer membranes can incorporate polymers with a variety of functional groups with fabrication technique advances, expanding the range of uses for these membranes [89].

14.3 Properties of MXene/Polymer Composites

Since certain materials can be used for applications, the specificity of composite membranes' properties is often considered their biggest advantage. MXenes can be considered as the best nanofillers for giving MXene consolidated polymer membranes with a variety of features, including excellent thermal properties, greater conductivities, increased mechanical strength, etc.

14.3.1 Mechanical Property

One of the most crucial aspects of polymer-based composites is their mechanical performance, which can be effectively enhanced by using inorganic fillers. Nanomaterials are desirable as fillers to strengthen polymer-based membranes due to their distinctive characteristics and nanostructures, as well as their reactive nature [86]. The mechanical characteristics, strong adhesion, and flexibility of MXenes, in combination to their electrochemical activity and high electrical conductivity, are critical in practically expanding all their applications. MXenes are an excellent solution for enhancing the mechanical efficiency of thermosetting polymers owing to their structural characteristics of being highly layered and the inclusion of functional groups, including fluorine, hydroxyl, and oxygen groups [90]. In a work reported by Luo *et al.* [73], the strong and 3D MXene network furnishes considerable strength to NR matrix by improving its modulus and tensile strength.

Polysulphone (PSFO) membranes incorporated with 2D functionalized MXene nanosheets with high efficiency and strength were fabricated by Dashtbozorg *et al.* [91]. By increasing MXene concentration to 3 wt%, the mechanical properties, including elastic modulus, tensile strength and elongation at break increases constantly. At 3 wt% MXene concentration, maximum values for elastic modulus, tensile strength and elongation at break are shown. This can be due to the homogenous distribution of MXene nanosheets in the Polysulphone matrix. The stress given on the film will get passed on to the uniformly distributed nanofiller, which enhances their mechanical properties. Mechanical properties of fabricated membranes are

Table 14.1 Mechanical properties of PSFO/MXene composite membranes [91].

Membrane	Elastic modulus (MPa)	Tensile strength at break (MPa)	Elongation at break (%)
PSF0	3.021 ± 0.011	6.831 ± 0.324	2.111 ± 0.440
PSF1	3.213 ± 0.184	8.246 ± 0.412	2.462 ± 0.781
PSF2	3.457 ± 0.023	10.325 ± 0.283	3.262 ± 0.871
PSF3	3.633 ± 0.843	12.517 ± 0.386	4.124 ± 0.183
PSF4	3.066 ± 0.317	7.144 ± 0.216	2.141 ± 0.019

shown in Table 14.1. As the filler concentration increases more than 3 wt%, the mechanical properties of the membranes considerably decrease inversely proportional to the porosity. Thus, an increase in porosity can cause a drop in mechanical properties.

14.3.2 Morphological Properties

MXene based polymer composite membranes generally exhibit porous morphological structures. The SEM images detailing the cross-sectional view of PSF membranes incorporated with MXene are shown in Figure 14.5 [91]. The SEM images manifest that the membranes have an asymmetric morphology comprising of a thick selective layer at the top and a porous sub-layer at the bottom. The membranes thickened and their pores grew larger as the nanofiller concentration raised. Increased thickness suggests greater membrane porosity provided that the polymer weight remained unchanged and that the nanosheets were introduced in small quantity. Moreover, from the figure it is clear that, as the nanosheets loadings increased, membrane holes grew and became more close to the top surface. Similar findings were also described in previous publications that employed hydrophilic nanoparticles in polymeric membranes [91]. In a work done by Ghanbari *et al.* [39], the results of AFM analysis (Figure 14.6) illustrated that the surface roughness of polymer membranes increases significantly with the consolidation of MXene. They have synthesized modified MXene, i.e., MXene@MOF through a solvothermal method and incorporated as a nanofiller in PVDF to prepare polymer nanocomposite membranes through the non-induced phase separation (NIPS) process. 0.00, 0.05, 0.10, 0.15, 0.20 wt% MXene@MOF incorporated PVDF composites were denoted as M_0, M_1, M_2, M_3, and M_4 respectively. The surface roughness increases as the concentration of filler increases, but in M_4 agglomeration of nanofiller causes a decrease in surface roughness.

14.3.3 Thermal Property

In a broad sense, thermal properties, such as thermal conductivity, crystallization, degradation etc., are strongly connected to the processing and application of materials. The fundamental characteristic of thermal transport is thermal conductivity, which is heavily influenced by the mobility of charge carriers [86]. Due to their ability to effectively dissipate heat while maintaining a safe operating temperature as well as their fire retardancy in the event of fire outbreaks, materials with flame retardancy and thermal conductivities are becoming vital components in electronic devices and other engineering materials. The low thermal conductivities and high pervasiveness of flammability of polymeric nanocomposites (PNC) prevent them from being suitable for high-performance devices, despite their benefits in heat dissipation applications, such as easy processing, electrical insulation, low weight, and low cost. It is possible to employ MXene as an additive to increase the thermal conductivity of polymer composites because, its great thermal conductivities are superior than those of most semiconducting and metal nanomaterials [92]. In a work done by Wang *et al.* [93], epoxy composite incorporated with silver (Ag) (epoxy/Ag) have shown a maximum thermal conductivity of 58.3 W/m·K. By incorporating a very small quantity (0.12 vol%) of MXene into epoxy/Ag composite (epoxy/Ag/MXene), the thermal conductivity has shown a significant hike to 72.7 W/m·K, that is 24.7% greater than epoxy/Ag composites (Figure 14.7) [93].

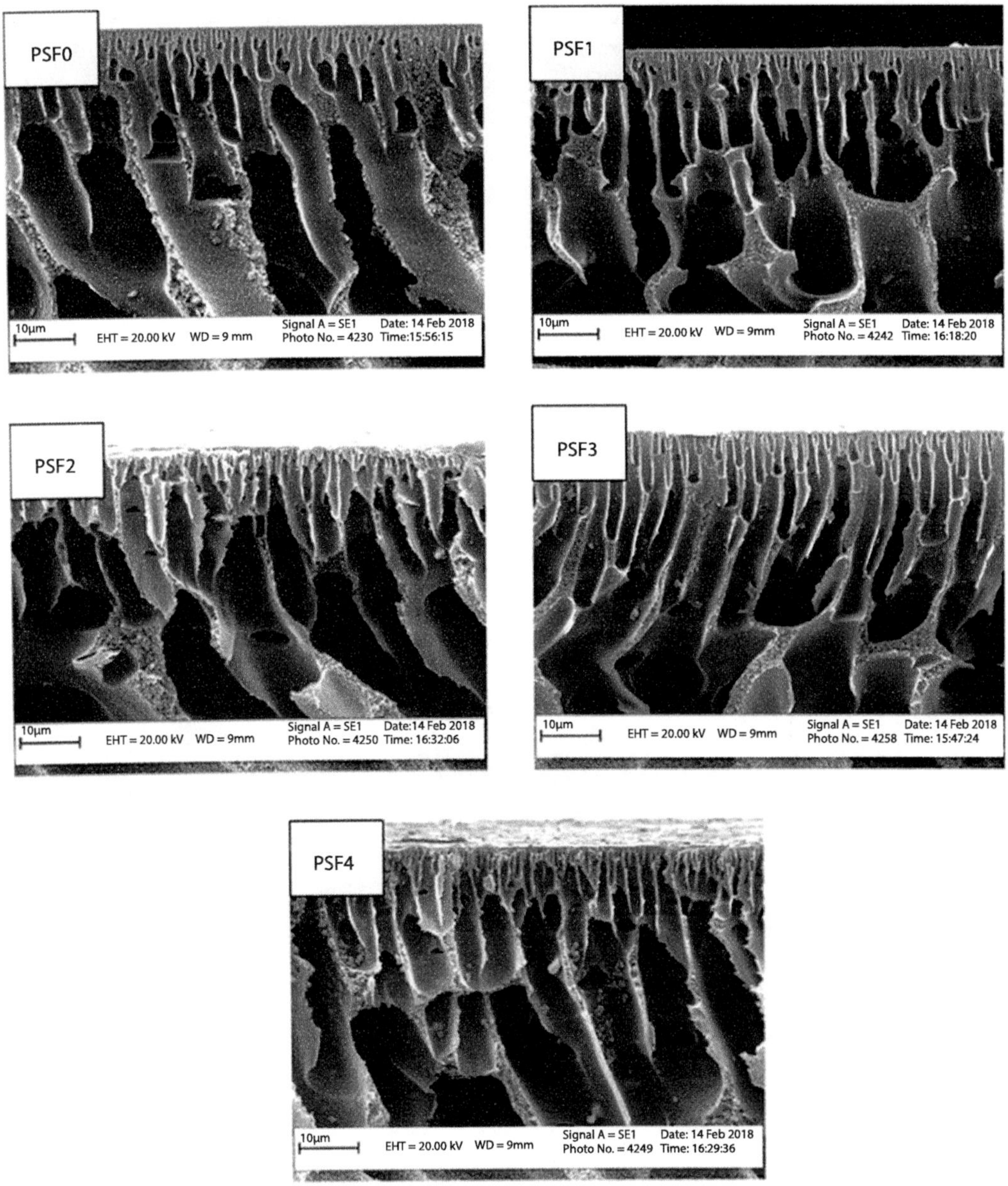

Figure 14.5 SEM images detailing the cross-sectional view of the PSF incorporated MXene membranes [91]. (PSF0 – 0 wt% MXene, PSF1 – 1 wt% MXene, PSF2 – 2 wt% MXene, PSF3 – 3 wt% MXene, PSF4 – 4 wt% MXene). Copyright 2022. Reproduced with permission from Elsevier.

Thermally conductive elastomers having MXene nanosheets at interfaces are fabricated by Aakyiir *et al.* [94]. They have employed unmodified MXenes (un-MXene) and allylamine (m-MXene) as reinforcing agents. The thermal conductivity of fabricated composite increases from 0.23 $W{\cdot}m^{-1}K^{-1}$ to 0.69 $W{\cdot}m^{-1}K^{-1}$ when un-MXene concentration in the nitrile butadiene rubber (NBR) composite increases from 0 wt% to 19.6 wt%. Moreover, a

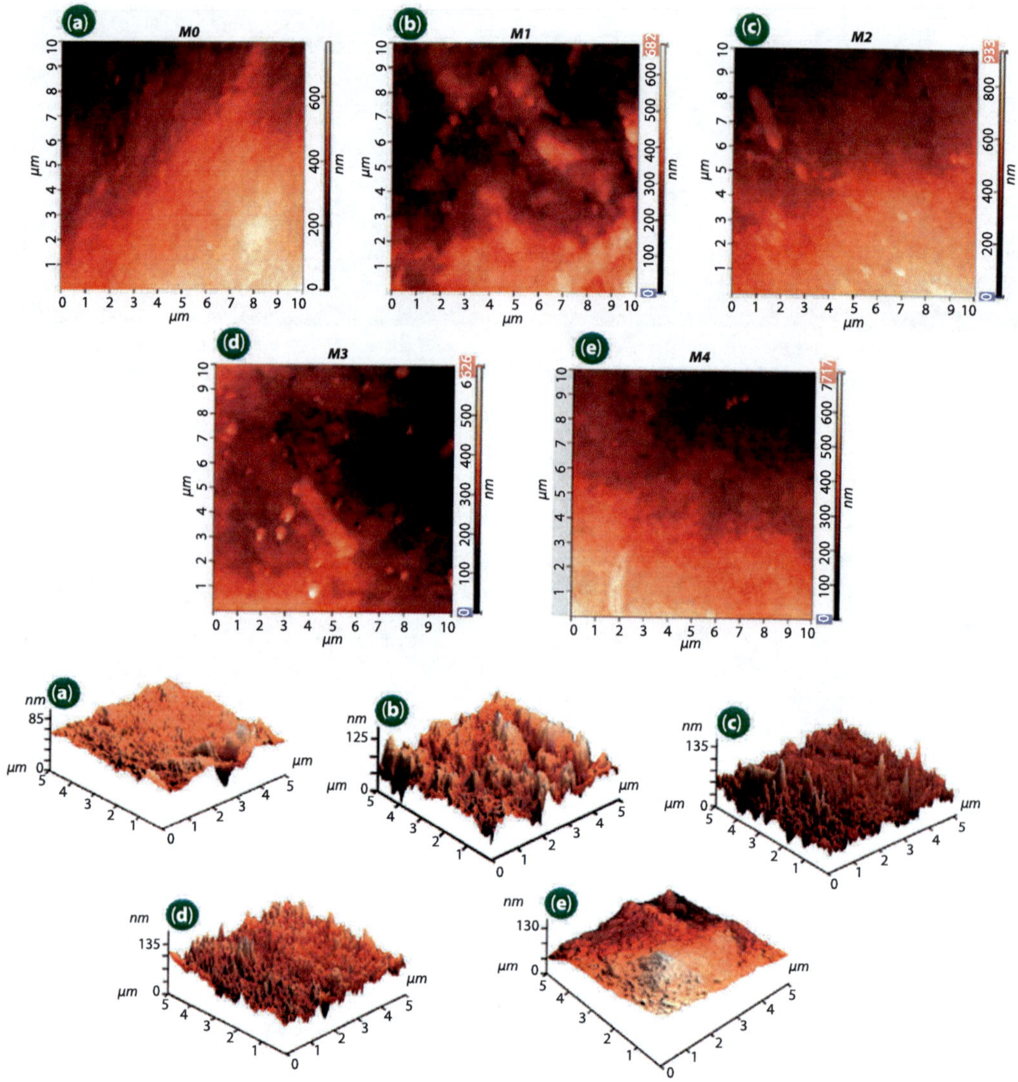

Figure 14.6 The morphological characteristics of (a) M0, (b) M1, (c) M2, (d) M3, and (e) M4 as suggested by AFM studies. The first five being 2-D and the next five being 3-D images [39]. Copyright 2023. Reproduced with permission from Elsevier.

far higher thermal conductivity (1.01 $W{\cdot}m^{-1}K^{-1}$) is exhibited by m-MXene incorporated nitrile butadiene rubber (NBR) (Figure 14.8) [94]. This substantial enhancement in thermal conductivity for m-MXene incorporated NBR is greater than all published thermal conductivity values for elastomeric composites incorporated with alumina, zinc oxide, boron nitride etc. Due to the distance between polymer chains, phonons that scatter through polymers are used to transmit heat. The nanofiller network enhances heat dissipation when nanofillers are introduced to polymers. Additionally, the polymer nanocomposite's efficient heat transmission property is caused by the interfacial interaction between the matrix and the nanofillers, which substantially minimizes interfacial phonon scattering. The allylamine

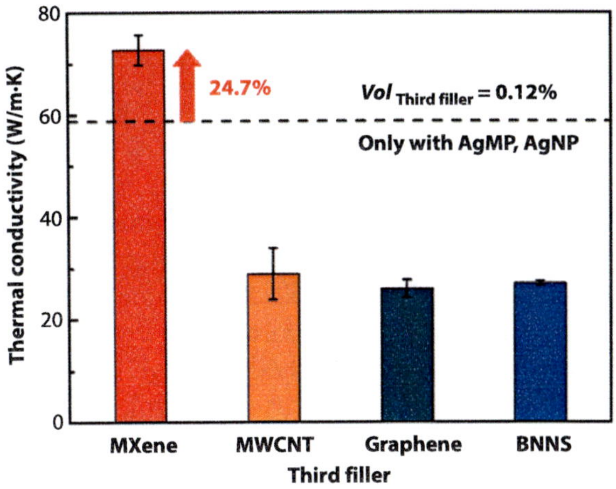

Figure 14.7 Thermal conductivities of Ag/Tertiary filler/Epoxy composites [93]. Copyright 2021. Reproduced with permission from Elsevier.

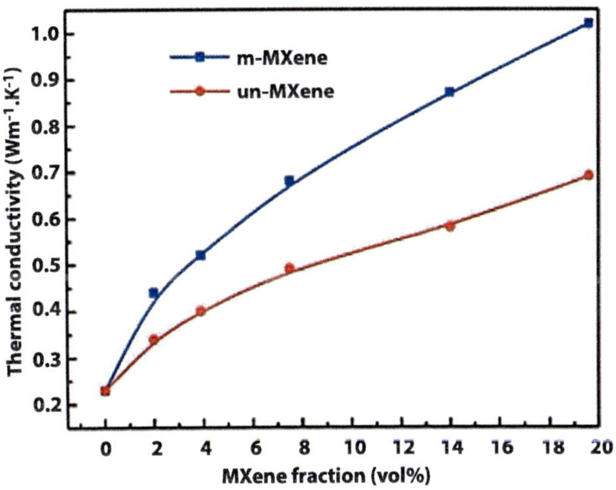

Figure 14.8 Thermal conductivity of m-MXene and un-MXene nanocomposites [94]. Copyright 2020. Reproduced with permission from Elsevier.

modification on MXene help to minimize the interfacial barrier, thus it makes the composites to have low filler-filler resistance and a strong interface.

14.3.4 Electrical Property

MXenes have been found to be effective candidates to be employed in electrodes as they can spontaneously adsorb appropriate cations with possible redox reactions on active sites, enabling relatively significant performance of associated batteries [95]. Even though the majority of polymers are insulators, the addition of MXene flakes can increase their electrical conductivities. In addition to aiding in the delamination of multilayer MXene flakes, the polymer intercalation into MXene layers will also encourage the molecular interaction between polymer and MXene [96]. A recent development involved in the area of $Ti_3C_2T_x$/polymer composite membranes with excellent volumetric capacitance is $Ti_3C_2T_x$/PVA membranes produced *via* simple blending. The pliable $Ti_3C_2T_x$/PVA composite film demonstrated an improvement in mechanical strength, high conductivity, and an exceptional volumetric capacitance as high as 528 Fcm^{-3} at 2 mV/s^{-1} [97]. T. N. Kołtunowicz *et al.* [98] synthesized $Ti_3C_2T_x$ MXene and casted up on an electrospun polycaprolactone (PCL) membrane. The capacitance, inductance, and phase shift angle analysis of fabricated composites are depicted in this work. The frequency dependence of the capacitance displayed a sharp pointed minima in the range of 50 Hz to 10^4 Hz. At the same time, for inductance, the frequency dependence curve exhibits a sharp maximum. These results prove the existence of parallel resonances. Considering the frequency range of 10^4 Hz - 10^5 Hz, a broad minimum peak can be seen in inductance–frequency curve, and this peak coincide precisely with the phase shift angle maximum, which has an amplitude of about 90°. The fabricated composite exhibited an inductance of approximately 1 Henry. Additionally, certain additional characteristics of MXene/polymer membranes, including antiflame, proton conductivity, and gas barrier qualities, have been discovered. Ti_3C_2 has shown to be an effective flame retardant by minimizing the production of toxicity risks and smoke products as a result of its catalytic and barrier effects. Numerous initiatives have been undertaken to unearth the outstanding qualities of MXene/polymer membrane, which expands their range of applications. To enable the designs of MXene/polymer membranes with exceptional potentiality, it is still necessary to thoroughly examine the precise relationship between microstructure and physicochemical properties of the material [99].

14.4 MXene Composite Membranes: Potentiality in Wastewater Treatment and Water Desalination

The past few decades have witnessed indiscriminate and unethical disposal of contaminated water from industries as well as households into global waterways thereby creating health, environmental and climate-associated hazards. Since then, there have been several methodologies implemented for the treatment of wastewater including, chemical precipitation, biochemical oxidation, electrocoagulation, ion-exchange method, and solvent extraction, but the major disadvantages with these methods lies in their lower efficiencies in removing impurities, large emission of greenhouse gases, release of toxic by-products and tedious operating methodology. Also, even after having a wide range of such wastewater treatment methods, 70% of the wastewater produced is left untreated in most of the developing and underdeveloped countries [100]. In this context, 2D materials like MXenes play a pivotal role, as these are found to be promising candidates in wastewater treatment and other environmental remediation applications. Throughout these years we have witnessed a rapid increase in the rate of evaluating the potentiality of MXene based membranes in processes such as water treatment and desalination, wherein there are humongous reports suggesting that both the pristine as well as structurally functionalized MXene-based polymer membranes can act as potential candidates in the effective removal of various pollutants from water. The high selectivity of the structurally modified membranes and water permeance are the major factors that give MXene/polymer composite membranes excellent potential in helping water purification [101].

14.4.1 MXene Composite Membranes in Removing Dyes

MXene-based polymer membranes are efficient in removing toxic dyes from wastewater as per literature reports. A study by Lui *et al.* [102] detailed the synthesis and application of Graphene oxide (GO) incorporated MXene polymer composite membrane, which could help in removing organic impurities including dyes from wastewater with high flux and rejection rate. The introduction of MXenes into the laminar structure of GO result in producing a number of nano-capillary channels with low friction which helps in increasing the water permeability. Additionally, the GO/MXene membranes were also found to possess excellent efficiency of nearly 100% rejection rate in removing small molecule dyes, as well as an almost near to 90% rejection of different types of organic materials like solvents present in raw water. This excellent efficiency of modified MXene was mainly attributed to the mutual effects of molecular sieving and electrostatic interaction taking place in between the different layers of composite membranes. In yet another study, Wei *et al.* [103] also portrayed the application of a similar 2D GO/Ti3C2TxMXene-based lamellar membrane for the separation of organic dyes from wastewater with a rejection of above 90%. Another study by Kim *et al.* [104] reported the use of a dead-end setup for the removal of certain organic dyes like brilliant blue, methylene blue, methyl red, and rose bengal from wastewater, with the help of slot-die-coated and vacuum-filtered $Ti_3C_2T_x$ membranes (Figure 14.9) [104]. The approach is also found to be highly effective in controlling the thickness and enhancing the alignment and stacking of the $Ti_3C_2T_x$ nanosheets which is particularly induced by the force

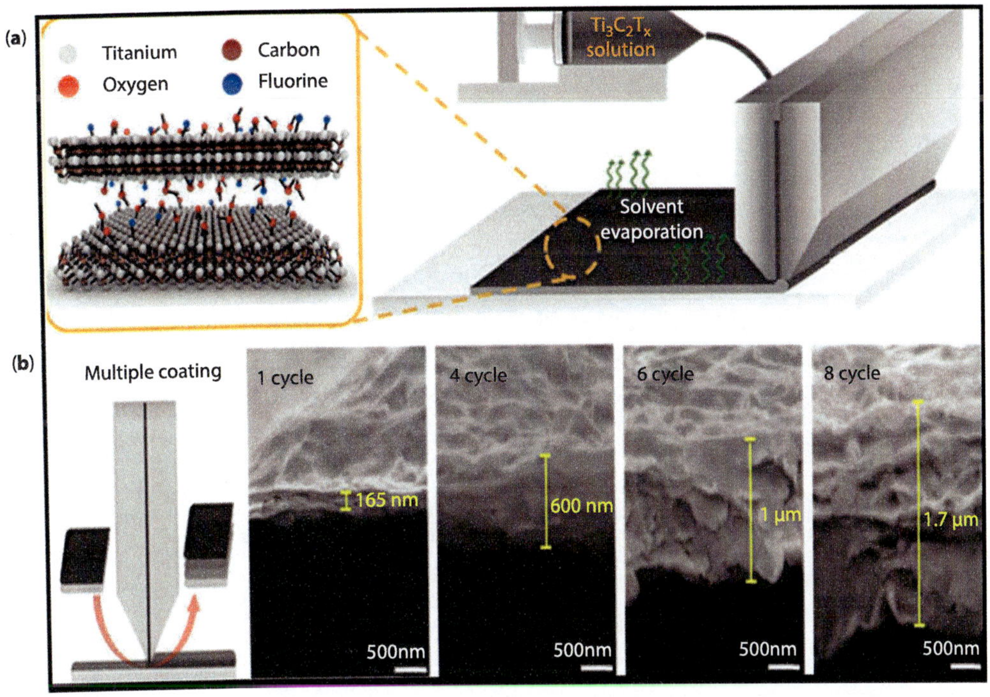

Figure 14.9 (a) Schematic illustration of the synthesized MXene membrane produced using slot-die coater and (b) thickness control of the MXene film by multiple coating cycles and SEM images on multiple coating [104]. Copyright 2021. Reproduced with permission from the American Chemical Society.

produced on the movement of the slot-die head. This coating method can thus be considered scalable and industrially adaptable for the fabrication of large-area MXene films applicable for nanofiltration. Further, it was also identified that the extraordinary mechanical and thermal stability of $Ti_3C_2T_x$ in aqueous solutions which are highly oxidizing in the presence of small organic molecules, thereby making them applicable for long-term nanofiltration operations [104]. This long-term applicability of membranes was confirmed by conducting an immersion test of various dyes, including that of positively-charged, negatively-charged and neutral dyes in aqueous solution for about 1 week, as represented in Figure 14.10 [104]. Usually, the MXene film get oxidized when immersed in deionized water, but the presence of dyes largely helped in lowering the rate of oxidation and degradation of films, which was further confirmed by Ti 2p spectra of XPS. It was also interesting to note that, positively charged dyes played a vital role in hindering the oxidation of MXene films, especially in the presence of polyethylene glycol molecules, thereby making these membranes efficient for long-term application even in oxidizing solvents.

GO/MXene-derived polymer composite membranes were found to possess greater advantages in the nanofiltration process over those membranes prepared using traditional methods. In a study by Han *et al.* [105], a typical MXene was employed as a sacrificial template in order to create continuous and uninterrupted nanochannels within the graphene

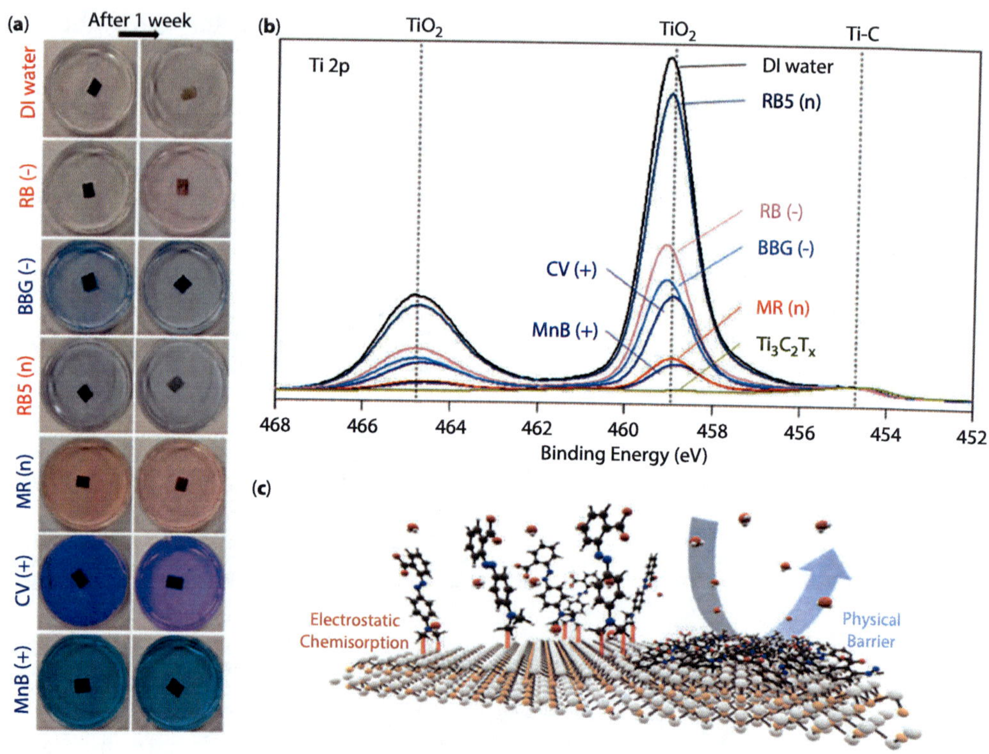

Figure 14.10 The stability of MXene membrane on long term absorption by dyes. (a) color change after immersing different dyes in aqueous solution for about a week. (b) Ti 2p spectra of membrane after immersing them in the dye solutions. (c) schematic illustration of a stabilized MXene nanosheet by dye absorption [104]. Copyright 2021. Reproduced with permission from the American Chemical Society.

oxide-incorporated MXene membranes. Initially, MXene sheets were used as templates to prepare the GO-based MXene membranes which were produced using the vacuum filtration technique, to which continuous nanochannels were created with the help of TiO_2 nanocrystals produced by *in situ* oxidation (Figure 14.11) [105]. The newly produced films were fabricated on a mixed cellulose ester, denoted as HGMR (R= 10,20,30,40, depending on the mass ratio of MXene). The membrane thus developed was found to possess a water permeability rate of about 7 times higher than that shown by the pure GO membrane and about 2-2.5 times higher than that shown by a typical GO/TiO_2 polymer membrane. Further, the membrane also exhibited an excellent rejection rate of over 97% for various organic dyes without a significant detriment to the separation performance compared to the pure GO membrane, as represented in Figure 14.12 [105].

Tong *et al.* [106] reported the applicability of a two-dimensional MXene polymer membrane coated with a tannic-acid (TA)-metal complex in enhancing the membrane contaminant rejection. The newly modified MXene membrane showed low salt ion retention, excellent water permeability, and an extremely high rejection rate for various contaminants including organic dyes like Rhodamine B and methyl blue from water, the rejection rate of which is represented in the graph in Figure 14.13 [106]. The rejection rate of Rhodamine B was almost negligible by means of pristine MXene, which could be increased to about almost complete rejection, by making use of the newly modified composite MXene. The membrane was found to be helpful in the complete rejection of Rhodamine B, but not that of Methyl Blue. This was attributed to the fact that the shorter side of Methyl Blue on a particular orientation had a smaller size compared to the mean pore size of modified MXene, thereby enabling the partial transport of these molecules through them possible, which reduced the rejection rate of this dye (Figure 14.14) [106].

In a study by Long *et al.* [107], the fabrication of a GO/MXene/PPS(polyphenylene sulphide) based polymer composite membrane via vacuum filtration method was reported, which can be utilized for dye removal even in harsh environmental conditions [107]. An interesting fact noted was that the composite membrane exhibited excellent rejection rate

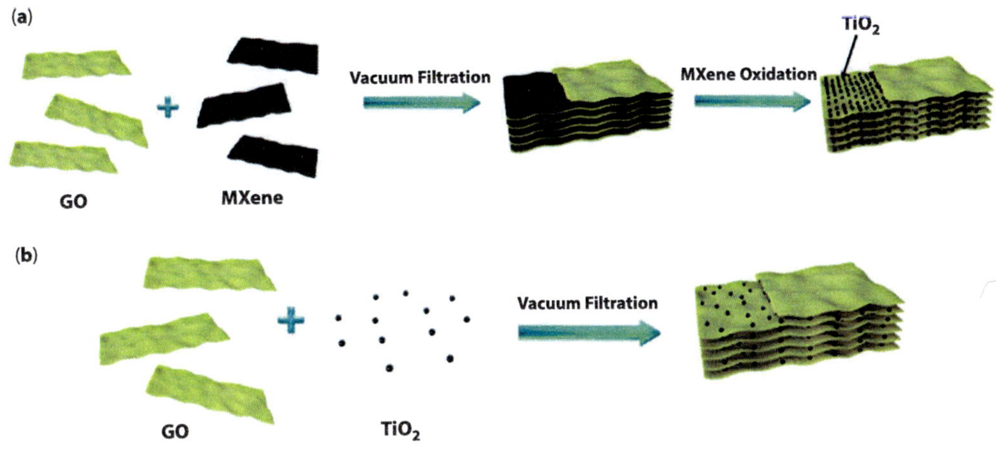

Figure 14.11 Schematic representation of fabricating (a) GO/MXene-derived 2D membranes HGMR, and (b) GO/TiO2 membranes prepared by traditional method [105]. Copyright 2019. Reproduced with permission from The Royal Society of Chemistry.

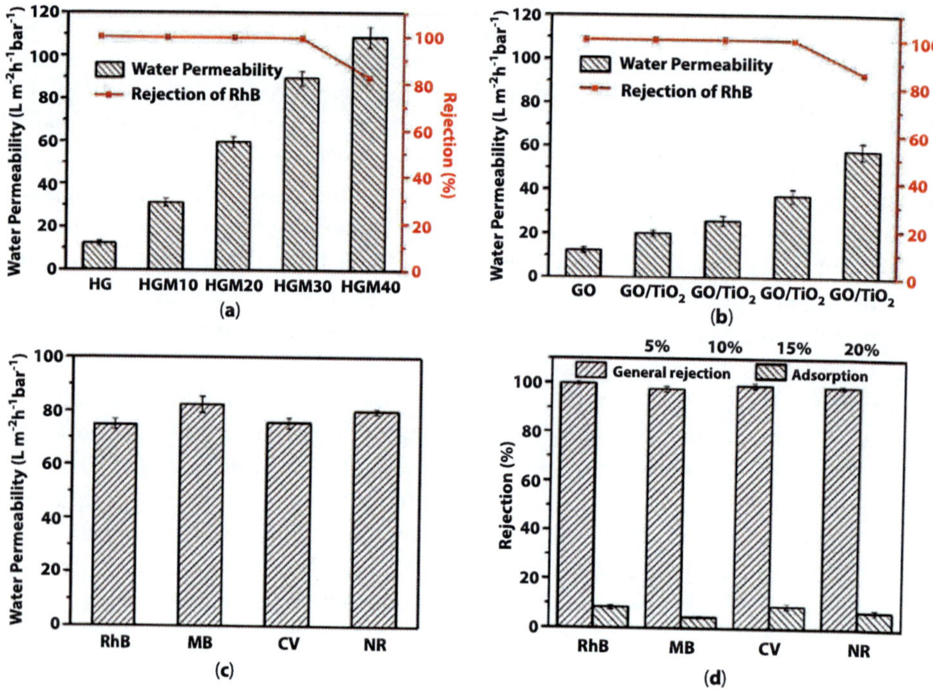

Figure 14.12 Water permeability and removal rate of the dye rhodamine B by (a) the newly synthesized HGMR membranes, (b) typical GO/MXene membrane in pure water, (c) water permeability of different dyes, and (d) general rejection and adsorption rejection of different dyes by HGMR (30w%) membrane [105]. Copyright 2019. Reproduced with permission from The Royal Society of Chemistry.

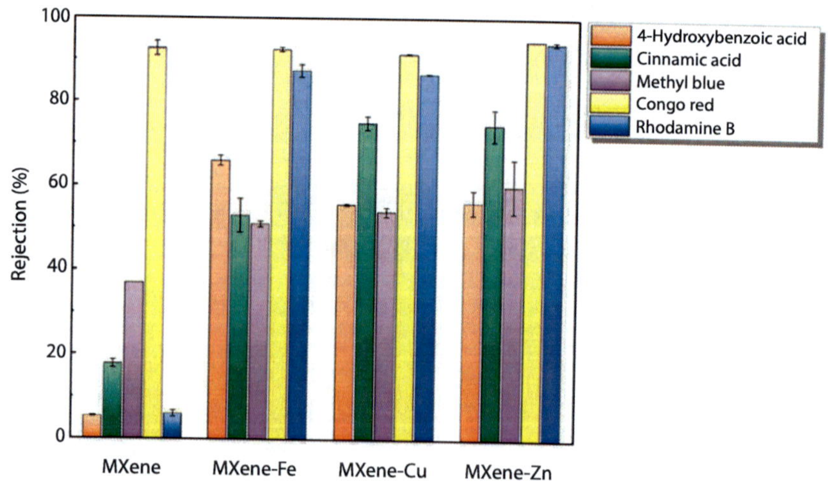

Figure 14.13 Rejection rate of various dyes by different MXene membranes [106]. Copyright 2021. Reproduced with permission from Elsevier.

for dye molecules, and when the loading mass of MXene increased, this efficiency started to decline. Therein, they could conclude that the rejection rate of pristine MXene membrane to various dyes was considerably lower than that of the MXene/PPS composite membrane. The mechanism of action of separation is shown in Figure 14.15 [107], from which it could

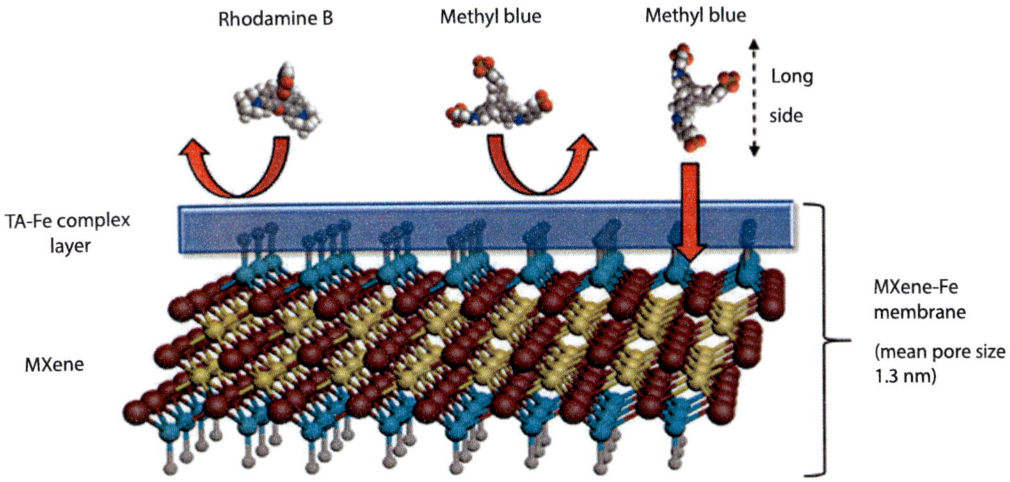

Figure 14.14 Schematic representation of the MXene-Fe composite membranes in the rejection of dyes [106]. Copyright 2021. Reproduced with permission from Elsevier.

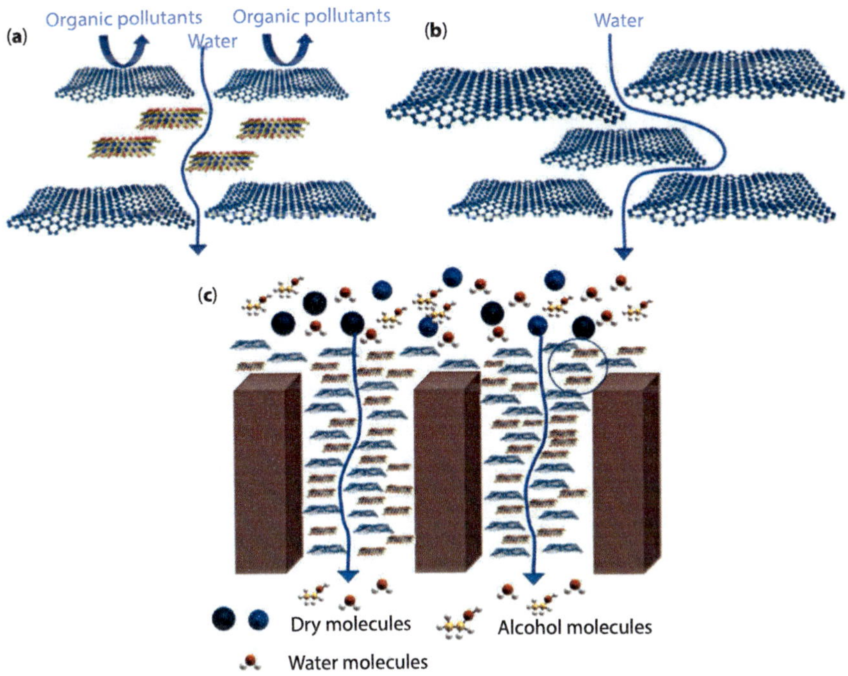

Figure 14.15 Schematic illustration of the mechanism of separation by GO reinforced MXene/PPS composite membrane [107]. Copyright 2022. Reproduced with permission from Elsevier.

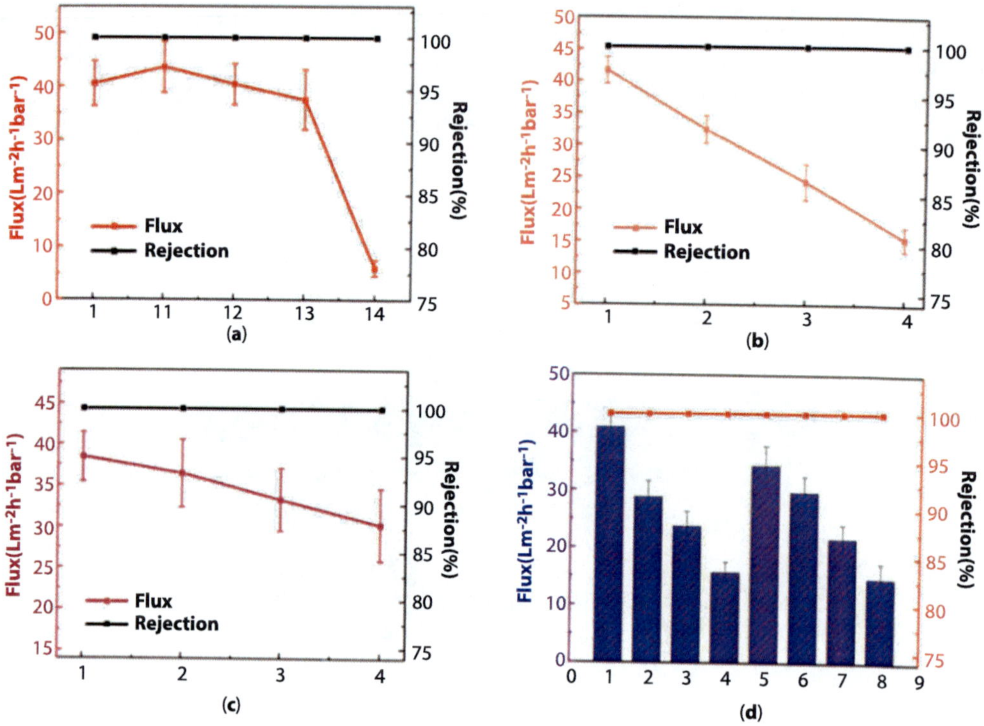

Figure 14.16 Study showing the tolerance of composite to harsh conditions, (a)long time study in acidic and alkaline conditions, (b) strongly acidic condition, (c) strongly alkaline condition, (d) recyclability test [107]. Copyright 2022. Reproduced with permission from Elsevier.

be interpreted that as the mass content of MXene increases, the functional groups containing oxygen moieties on the surface of GO were masked, thereby reducing the π-π bond interactions, which in turn leads to a reduction in separation efficiency. On comparing with the pristine MXene and GO membranes, the composites were also found to possess long term applicability, in normal as well as in harsh pH conditions, with appreciable flux rate at about pH 13 in alkaline conditions and below pH 4 in acidic conditions. The rejection rate was almost unaltered in all kind of pH conditions (Figure 14.16) [107].

14.4.2 MXene Composite Membranes in Removing Radioactive Wastes

The rapid increase in nuclear energy utilization is posing a greater challenge in nuclear waste treatment since the contamination caused by some of the actinides and other radioactive heavy metals with longer half-life periods can create significant hazards to the environment. Even if they are in trace amounts, it very much important to develop new efficient and economical methods for their proper treatment [108]. Surface-functionalized MXenes are found to be excellent candidates for the removal of radioactive isotopes from wastewater. There are several reports suggesting the proper disposal of uranium, thorium, cesium, and other similar radioactive metals using functionalized MXene membranes [100]. Uranium being one of the major constituents in nuclear wastes, so is it important to find a suitable method for their rejection. Wang *et al.* [108] were among the first group of scientists to develop multi-layered Vanadium Carbide based MXene, V_2CT_x /MXene composite membrane to capture Uranium from wastewater. In another report, Wang *et al.* [109] demonstrated the simultaneous adsorption and reduction of U(VI) from water by a 2D Ti_2CT_x MXene membrane which is highly reactive. The synthesized MXene also possessed potential in the treatment of highly acidic wastewater from mines and other related industries wherein the reduced U(IV) species exhibited long-term immobilization under anaerobic conditions. A study by Zhang *et al.* [110] revealed that introducing diazonium salts to carboxylate the surface of $Ti_3C_2T_x$MXenes could highly functionalize the 2D membranes, thereby, helping them remove more than 95% of uranyl ions in wastewater within a short span of time (Figure 14.17) [110]. The functionalization of MXene membrane with carboxyl groups enable them to efficiently enrich radionuclides and the incorporation of the aryl diazonium salt moiety helps in improving the stability of the material in water. Many radioactive isotopes of halogens are found to have an alarming impact on the environment and pose high risks when released into water bodies, for example, radioactive iodine isotopes with half-lives ranging from 8 days to 1.57×10^7 years create serious issues if they are not treated properly. Taking this into consideration, Sha *et al.* [111] developed an MXene polymer composite in which nanosized bismuth oxide particles were restrained on the surface of a polydopamine (PDA) modified MXene owing to the excellent capability of bismuth oxide nanoparticles in adsorbing iodide ions, the preparation of which is as in Figure 14.18 [111].

14.4.3 MXene Composite Membranes in Removing Metals

MXene-based polymer membranes are found to possess the extensive capability in removing metal ions from water. Fan *et al.* [112] established the possibility of ion sieving of K^+/ Pb^{2+} ions and rejection of heavy metal ions like Pb^{2+} at a high rate by making use of a

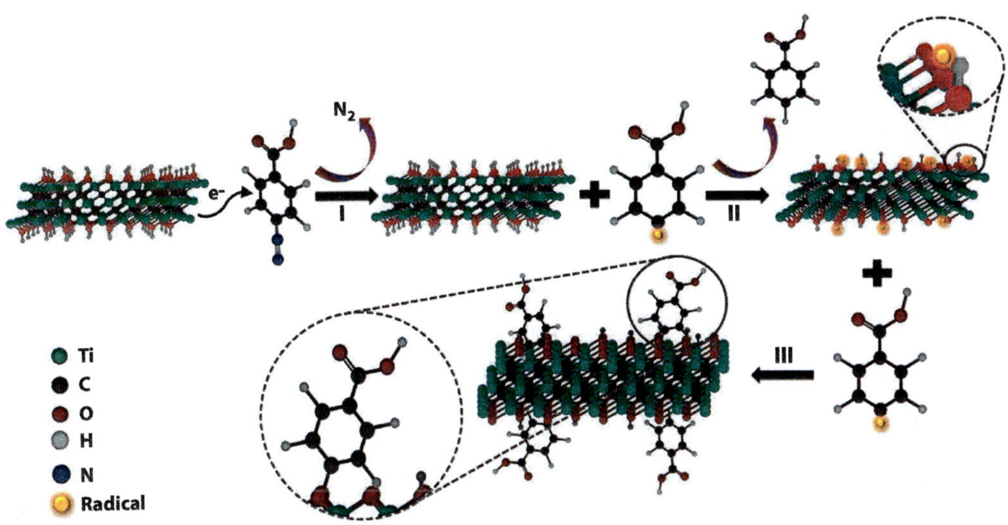

Figure 14.17 Schematic illustration showing the reaction mechanism of functionalizing MXene membrane with diazonium salt [110]. Copyright 2020. Reproduced with permission from Elsevier.

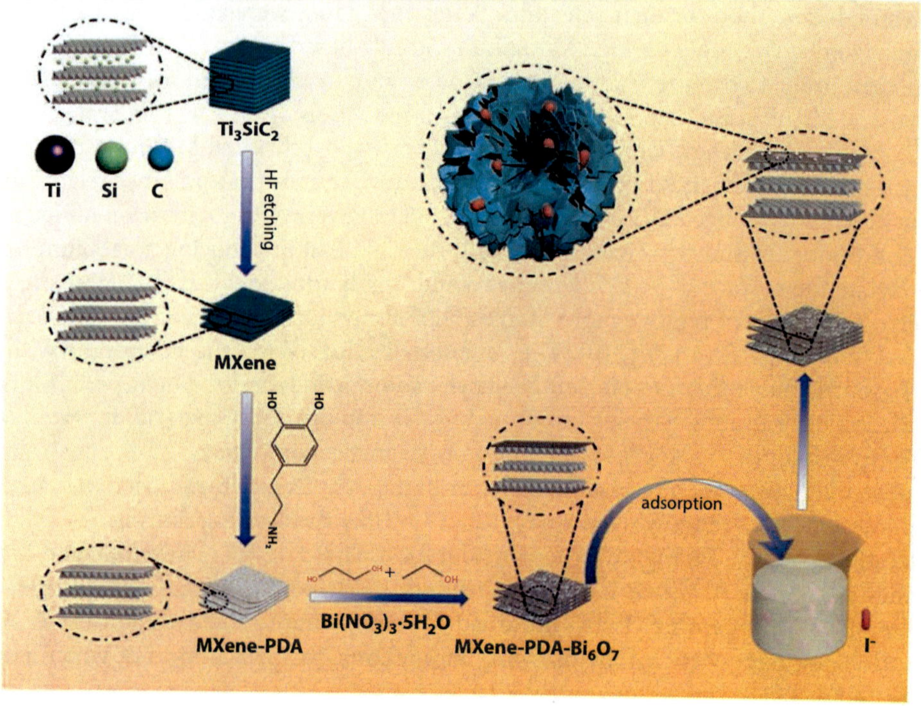

Figure 14.18 Schematic illustration showing the fabrication of MXene-PDA-Bi_6O_7 polymer composite [111]. Copyright 2020. Reproduced with permission from Elsevier.

two-dimensional MXene membrane which is thermally cross-linked under the application of an external voltage. In another report, Jin *et al.* [113] detailed the preparation of 2D $Ti_3C_2T_x$ functionalized by Poly(m-phenylenediamine) (PmPD) for the trapping and reduction of Cr(VI) metal ions (Figure 14.19) [113]. The mechanism of adsorption of chromium

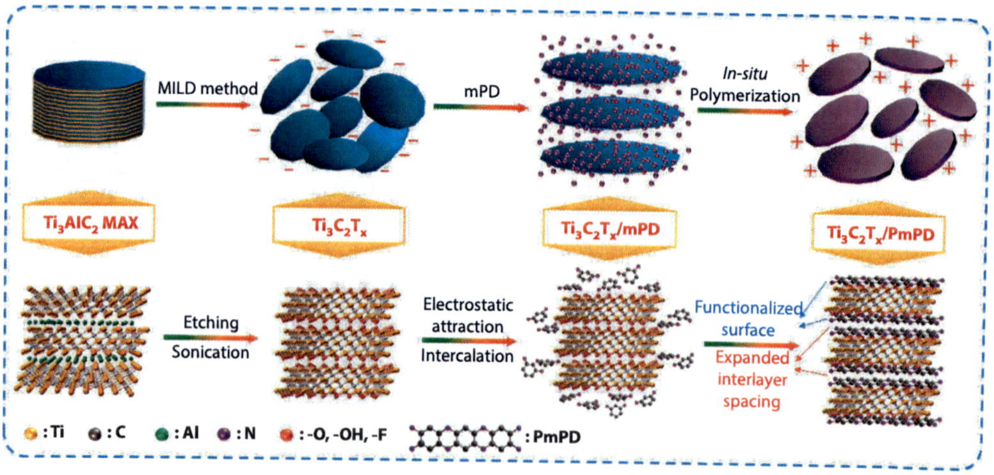

Figure 14.19 Schematic illustration for the preparation of PmPD/MXene membrane [113]. Copyright 2020. Adapted from MDPI. (Open access)

ions on membrane surface involves a series of adsorption, reduction and chelation interactions, as shown in Figure 14.20 [113]. The long-term applicability and recyclability of membrane was also evaluated by adsorption-desorption experiments (Figure 14.21) [113], wherein it was revealed that even after 5 recycle rounds of a 100ppm Cr (VI) ion concentration, the removal efficiency of membrane was still about 90%. A study by Wang *et al.* [114] reported the applicability of $Ti_3C_2T_x$ MXene/CNT hybrid polymer membranes for recovering Au^{3+} even from trace-level solution with minute concentration. The rejection of Au3+ via the aid of the composite membranes takes place in such a way that when the metal ions interact with the MXene (Ti3C2Tx) surface a redox reaction occurs, thereby transferring electrons from the MXene surface to the metal ion, thus enhancing the reduction of Au(III)

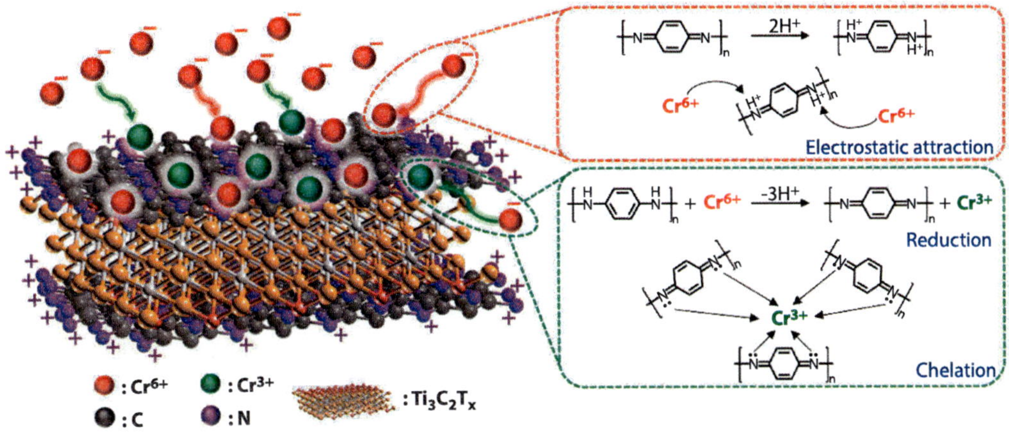

Figure 14.20 Mechanism of adsorption of Cr(VI) on PmPD/$Ti_3C_2T_x$ MXene membrane [113]. Copyright 2020. Adapted from MDPI. (Open access)

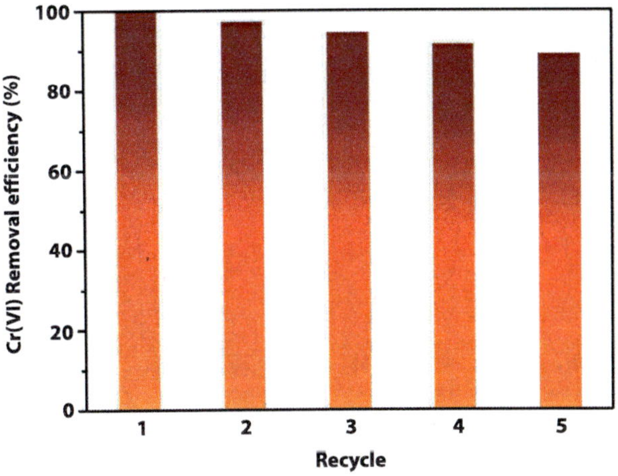

Figure 14.21 Recyclability of PmPD/$Ti_3C_2T_x$ MXene membrane [113]. Copyright 2020. Adapted from MDPI.

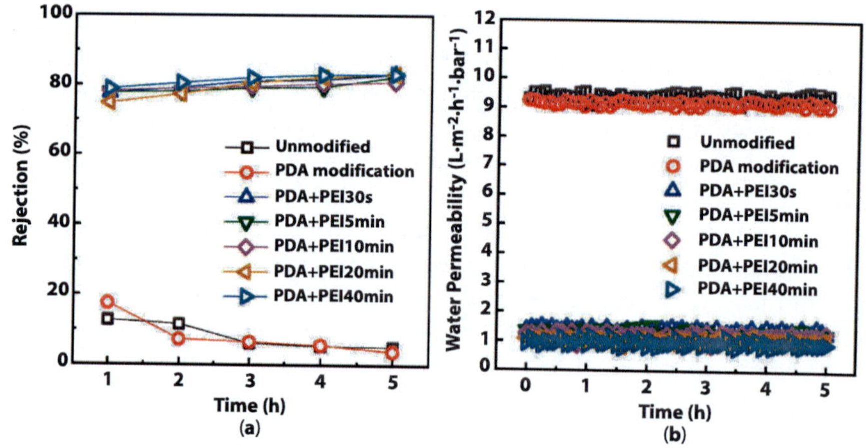

Figure 14.22 (a) Rejection rate and, (b) water permeability of Ca^{2+} ions by different composites [115]. Copyright 2021. Reproduced with permission from Elsevier.

to Au(0), thereby helping in the removal of precious metal ions from untreated water and proper recovery of them.

Zhao *et al.* [115] illustrated the fabrication of polyethyleneimine-modified MXene-based composite membranes incorporated with GO for nanofiltration applications. The unique structural features of the composite membranes helped them in providing efficient rates of water permeability and removing metal ions with high efficiency, which was found to vary upon altering the mass ratio of GO and MXene. The composite membranes fabricated with a GO/MXene mass ratio of 1:4 exhibited higher coefficient of water permeability which was over 7 times higher than that of unmodified GO membranes. The polyethyleneimine-modified (PEI) composite membranes were also found to have an outstanding improvement in retaining the bivalent metal cations. The rate at which Ca^{2+} and Mg^{2+} ions were

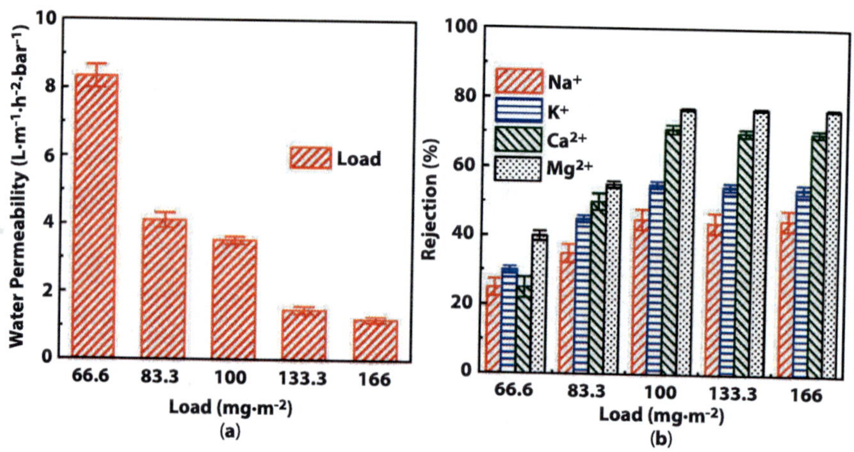

Figure 14.23 (a) Water permeability and (b) ion rejection rate of various ions by GO/MXene/PEI modified composite membranes with different loadings [115]. Copyright 2021. Reproduced with permission from Elsevier.

rejected was approximately about 7 times higher than that shown by a typical pristine GO membrane, thereby making this modified PEI/MXene-based polymer nanocomposites efficient in wastewater treatment applications. The removal efficiency of Ca^{2+} ions by the GO/MXene/PEI membranes was found to increase to about 8 times higher than that shown by pristine composite membranes without any PEI modification. But, the rejection rate of ions and water permeability coefficient of membranes are inversely proportional to each other, thereby making the modified PEI based membranes having lower water permeability, as shown in Figure 14.22 [115]. Further, the separation efficiency of the PEI incorporated GO/MXene membranes in rejecting metal ions was also evaluated based on the loading added. It was identified that as the load of the membranes was reduced, the water permeability of the GO/MXene/PEI membranes gradually increased, but, on the other hand, the rejection rate of metal ions decreased (Figure 14.23) [115].

14.4.4 MXene Composite Membranes in Promoting Oil/Water Separation

A large quantity of oily wastewater produced by various oil industries is being dumped into waterbodies on a daily basis creating huge contamination in freshwater, making it important to develop innovative technologies for treating oily wastewater. Several methods like gravity settlement, and air rotation, are being practiced generally, but the small surface are equipped by the oil droplets in emulsions make these methods less efficient in separating emulsified oil/water mixtures. Later on, studies revealed that size-sieving mechanisms are found to possess excellent potential in removing these tiny oil droplets from emulsified water bodies, of which membrane separation methods are of utmost importance. But, most of the commercial membrane materials available, like that of polyvinylidene fluoride (PVDF) are not hydrophilic, thereby creating serious issues of membrane fouling, which in turn would cause a rapid decline in separation efficiency and permeation flux [116]. The efficient and economic 2D MXene membranes by coating Ti3C2Tx (MXenes) on commercial print paper, for the separation of emulsified oil/water mixtures. The newly fabricated 2D MXene nanoflakes were highly hydrophilic and densely packed on the substrate, thereby acting as a functional layer for the effective separation of emulsified oils, whereas the print paper provides excellent strength and mechanical flexibility. This particular MXene membrane exhibited a good lifetime with less membrane fouling propensity and possesses an easy membrane clean-up procedure [116]. It was also identified that the permeation flux and oil content in filtrates decreases as the mass loadings of MXene composite increases, thereby enhancing the rejection rate to about 98%. The permeation flux was also tested for different oil/water emulsions wherein the highest flux rate was obtained for hexane/water emulsion, and the lowest for sunflower oil/water emulsion filtrate (Figure 14.24) [116]. It is also worth mentioning that the composite membranes possessed long-term applicability, that the operational flux decreased only slightly, with an oil rejection rate of about 99% even after carrying out 8 different operational cycles (Figure 14.25) [116].

Li *et al.* [117] reported the fabrication of an ultra-thin MXene membrane reinforced on a polyethersulfone substrate produced using the VFA method. The membranes produced possessed outstanding stability, fouling resistance, and oil rejection capacity from the emulsion on comparing with the normal separation membranes. In another study, He *et al.* [9] developed MXene/UIO-66-$(COOH)_2$ composite lamellar membranes with excellent stability and separation capacity by making use of a nylon 66 microporous substrate. The membrane was

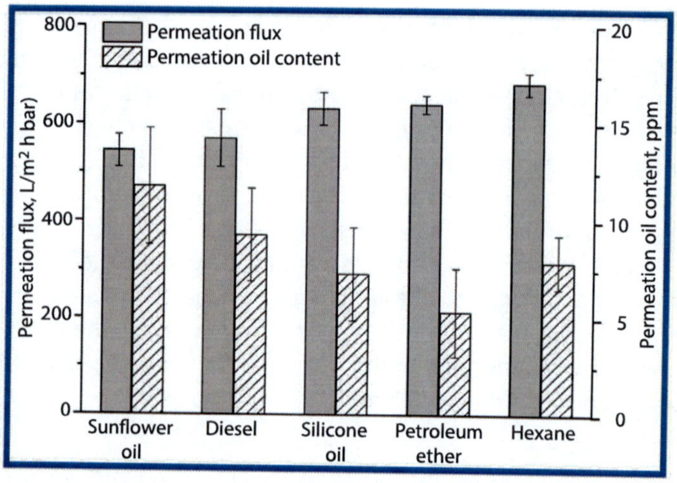

Figure 14.24 Permeation fluxes and oil content in the filtrates of different oil/water emulsions filtrated by composite membranes of MXene [116]. Copyright 2019. Reproduced with permission from the Royal Society of Chemistry.

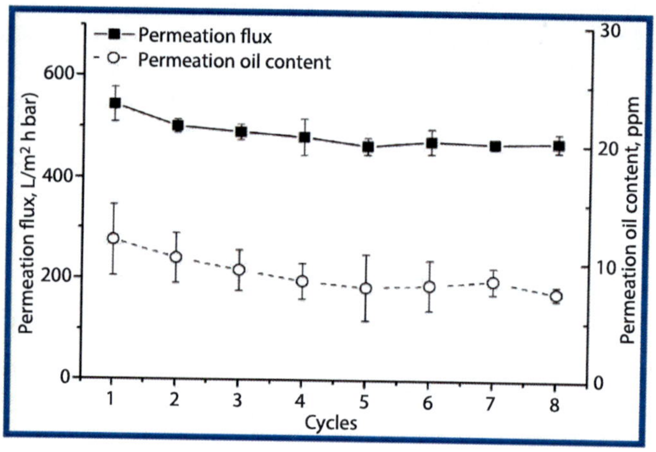

Figure 14.25 Recyclability of MXene composite [116]. Copyright 2019. Reproduced with permission from the Royal Society of Chemistry.

also found to be efficient in the treatment of multi-component pollutant oil/water emulsion. In yet another study, Lui *et al.* [118] fabricated an L-lysine functionalized $Ti_3C_2T_x$ ($Ti_3C_2T_x$-Ly) MXene by arraying L-lysine on $Ti_3C_2T_x$ with the aid of electrostatic interaction and hydrogen bonding, which was further coated with polyurethane (PU) sponge, thereby developing a super hydrophilic $Ti_3C_2T_x$-Ly/PU polymeric sponge. The MXene exhibited excellent underwater superoleophobicity as a result of the presence of hydrophilic functionalities like –OH, and $-NH_3$ groups on the surface of $Ti_3C_2T_x$-Ly nanosheets and the enlarged interlayer spacing of $Ti_3C_2T_x$, the $Ti_3C_2T_x$-Ly/PU sponges, because of which they showed high separation performance for a series of oil–water mixtures. Hou *et al.* [16] prepared modified $Ti_3C_2T_x$ MXene membranes by their intercalation with sodium alginate with the aid of Ca^{2+}-crosslinking as represented in Figure 14.26 [16]. The introduction of functionality like sodium alginate could help in improving the antifouling performance and transport rate of MXene in nanofiltration applications. On investigating the oil rejection performance, it was identified that octane/water emulsion filtrate had the best rejection rate, while that of toluene/water emulsion was comparatively lower, which could be due to its solubility in water (Figure 14.27(a)) [16]. Further the composite membrane was also found to possess excellent recyclability, as it was identified that its separation performance increased after a few cycles, unlike the pristine MXene membranes which drastically loses its rejection rate after 5–6 cycles (Figure 14.27(b)) [16].

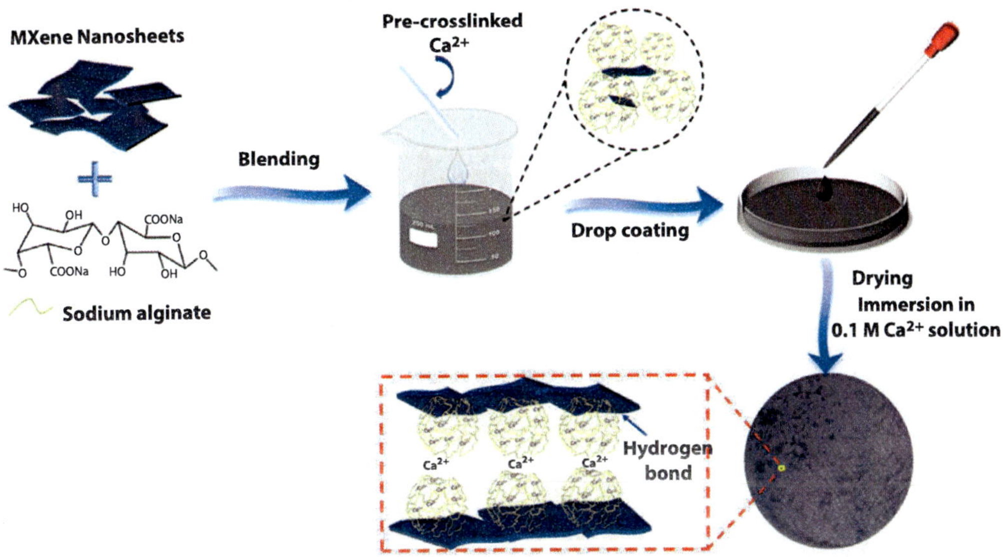

Figure 14.26 Schematic illustration showing the preparation of modified MXene/sodium alginate membrane [16]. Copyright 2023. Reproduced with permission from Elsevier.

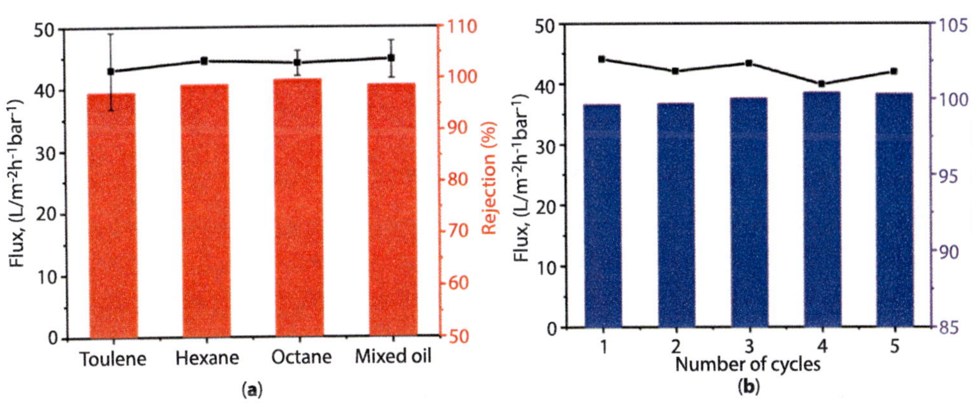

Figure 14.27 (a) Rejection performance of various oil/water emulsions by MXene/sodium alginate membrane and, (b) recyclability of the modified membrane [16]. Copyright 2023. Reproduced with permission from Elsevier.

14.4.5 MXene Composite Membranes in Removing Microbes

Nanoparticles possessing anti-microbial activities could be employed in wastewater treatment, for the removal of hazardous microbes. Recent research has identified that the outstanding characteristics of MXene such as hydrophilicity, nanometer-sized sharp edges, and generation of oxidative stress help them to act as potential candidates having anti-microbial activities [119]. The comparatively high hydrophilicity of MXene membranes aid the easy binding of a bacterial strain to the membrane surface, causing bacterial inactivation. Further, the hydrogen bonds between the functional moieties present on the MXene surface and lipopolysaccharide moieties present on the cell membrane increases, thereby increasing the area of contact between the MXene surface and bacterial cell membrane eventually leading to bacterial death. Ahmad *et al.* [119] investigated the anti-microbial activities of an MXene nanosheet by treating them with bacteria strains of *B. subtilis* and *E. coli* and identified them to be efficient anti-microbial agents, thereby helping remove these bacterial strains from wastewater. Zha *et al.* [120] prepared a highly flexible and anti-biofouling MXene/cellulose fibrous photothermal polymer membrane through a dip-coating method for highly efficient solar-driven water purification. The MXene membrane showed an anti-bacterial efficiency of up to 99% which could generate transparently condensed and pure water from a wastewater suspension contaminated with *E. coli* and *S. aureus* bacterial strains without any bacterial presence. Solar energy can be efficiently utilized to generate steam, which can be considered a sustainable wastewater purification and sterilization method. From the SEM images it is evident that the bacterial fouling on MXene/cellulose membrane surface is less than that on pristine cellulose, rGO/cellulose membrane (Figure 14.28) [120]. It also specifies that the rough and pointed edges of MXene sheets have the ability to effectively limit bacterial growth and prevent the formation of biofilms in addition to physically stressing bacteria and damaging their membranes. The better anti-fouling performance MXene/cellulose membrane with other two are noticeable from statistical data as well. Wei *et al.* [121] proposed the fabrication of a multifunctional Au/Ti_3C_2 photothermal membrane possessing anti-bacterial activities which makes efficient utilization of solar energy for wastewater purification. Studies carried out by the research group revealed that the synthesized Au/Ti_3C_2 photothermal membrane exhibited excellent photothermal anti-bacterial properties, which could completely inactivate the bacterial strain of *E. coli* K^{-12} within 20 minutes. Further research is also being focused on making use of low-cost transition metal nanoparticles, thereby paving the way to provide a much more economical method for water sterilization using solar energy.

14.4.6 MXene Composite Membranes in Water Desalination

Many studies have revealed the potential of MXene membrane (MXM) nanosheets for developing selectively sieving membranes with high flux for water desalination [122]. The abundant surface functionalization possible for MXene membranes provide them with superior advantages for controlling the d-spacing between the membrane layers and tuning them enabling the suppression of swelling, thereby enhancing the rejection rate of monovalent and bivalent ions present in saline water [123]. Ding *et al.* [1] reported the fabrication of an MXene membrane intercalated with Al^{3+} ions for the long-term sieving of small hydrated ions present in saline water. *Via* concentration diffusion technique using $AlCl_3$

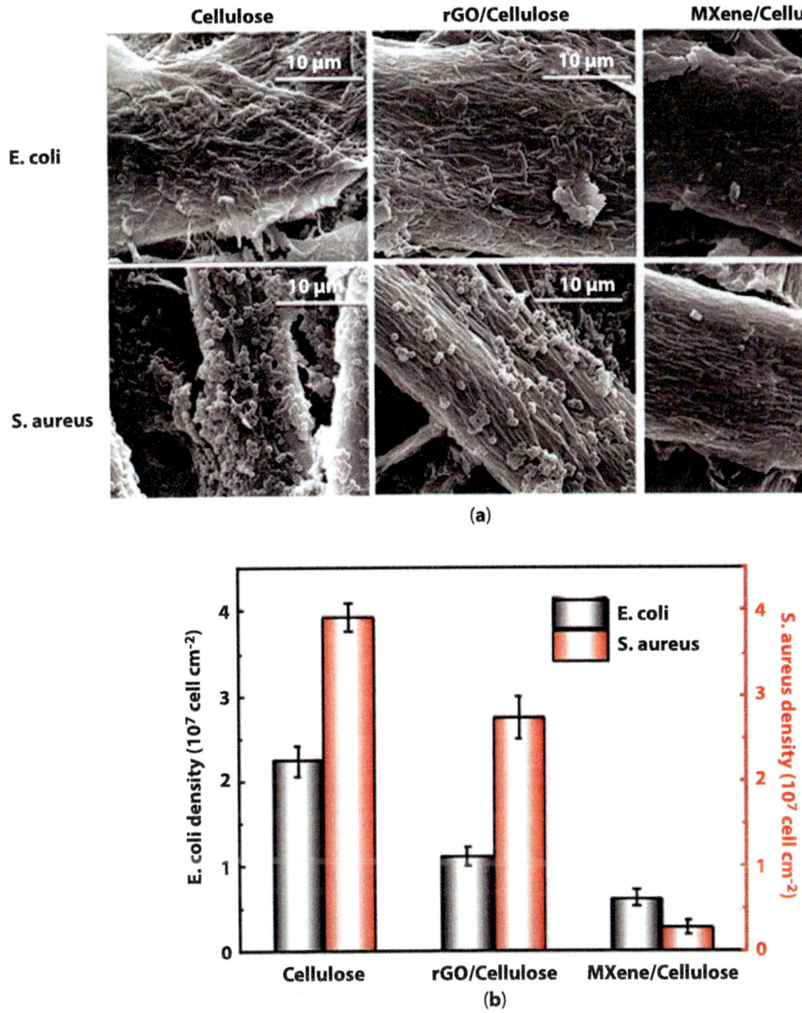

Figure 14.28 (a) SEM images displaying the adhesion of bacteria on membrane surfaces. (b) concentration of *S. aureus* and *E. coli* on the different types of membranes [120]. Copyright 2019. Reproduced with permission from the American Chemical Society.

solution, the pure MXMs were vacuum filtered from MXene dispersions onto a polyethersulfone (PES) substrate to obtain Al^{3+} intercalated MXMs. MD simulations revealed that the newly synthesized Al^{3+} intercalated MXMs could reject both multivalent and monovalent ions with excellent salt (NaCl) rejection rates and high water flux, when compared to the untreated MXMs. Further, the MXMs were also found to exhibit high tolerance against chloride ions, thereby providing the membrane with a longer lifetime. The same group investigated the swelling property of unmodified MXMs and modified Al^{3+} intercalated MXMs, since higher d-spacing results in poor ion sieving performance and reduced stability. The d-spacing rose from 13.5±0.1 Å to 16.5±0.5 Å when the untreated MXMs were soaked in water and various salt solutions. This result is explained by the weak interaction between the MXene layers, which exhibit significant swelling in water-based solutions.

When the Al^{3+} intercalated MXMs were soaked in salt or water solutions, however, the d-spacing only rose slightly, from 14.7±0.1 Å to 15.1±0.2 Å. The effective constriction of Al^{3+} between the MXene nanosheets prevent the swelling of Al^{3+} intercalated MXMs in the aqueous environment, exhibiting the outstanding non-swelling property in aqueous solutions. The proof for this is given by the d-spacing values calculated from the X-ray diffraction (XRD) data (Figure 14.29) [1].

In a study by Ding *et al.* [124] an ultrathin MXene membrane cross-linked with maleic acid supported on a microporous nylon substrate was fabricated for application in pervaporation desalination. Each layer of the MXene nanosheets is affixed strongly by the covalent interactions formed as a result of the esterification reaction between the hydroxyl groups on MXene surface and carboxyl groups in maleic acid. The crosslinked MXene-maleic acid membrane thus formed contains large and well-structurally arranged nanochannels with large slip lengths, thereby promoting excellent desalination performance with high permeance of water and high rejection rates of salt. Moreover, the strong covalent and hydrogen bonds between the MXene layers help to increase the swelling resistance of the newly fabricated membrane providing them with high mechanical and thermal stability. Meng *et al.* [125] developed an ultra-thin surface-charged MXene (SCMX) by coating a layer of polyethyleneimine on the membrane surface. Coating the membrane surface with this polyethyleneimine moiety helped in tuning its properties by enhancing the membrane hydrophilicity, thereby enabling the SCMX membrane to exhibit outstanding water permeance ability and high salt rejection capacity in both nanofiltration and forward osmosis. This is due to the synergistic effect of size sieving and surface electrostatic repulsion behaviors. Zhang *et al.* developed an MXene/PVDF composite membrane (PVDF-MP) with remarkable antifouling and antiwetting properties *via* chemical grafting method [126]. The endurance of the PVDF and PVDF-MP hollow fiber membranes was tested with MHDD under actual desalination conditions for 120 hours of continuous saltwater treatment (Figure 14.30) [126]. For the course of the operation, both membranes' salt rejection rates stayed relatively consistent at >99.8%. Due to its remarkable antifouling capabilities,

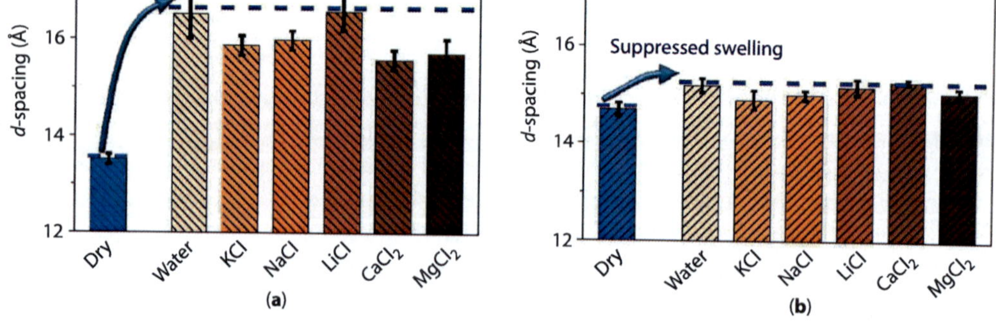

Figure 14.29 The d-spacings for the (a) unmodified MXMs and (b) modified-Al^{3+} intercalated MXMs in the dry state, soaked in deionized water, and exposed to various salt solutions (0.2 M), estimated from the XRD data [1]. Copyright 2020. Reproduced with permission from Springer.

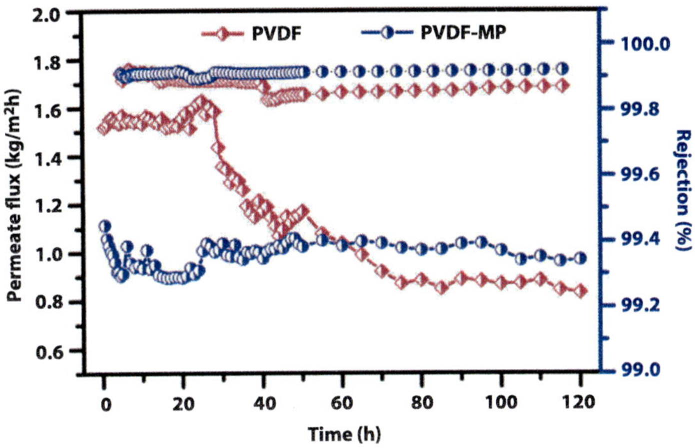

Figure 14.30 Rejection efficiency and permeate flux of the membranes [126]. Copyright 2022. Reproduced with permission from Elsevier.

the PVDF-MP membrane had a salt rejection rate that was higher after 40 hours than the pure PVDF membrane. After treating seawater for 28 hours, the permeate flux of the PVDF hollow fiber membrane decreased gradually from 1.5 to 0.8 kg/(m^2·h). Table 14.2 demonstrate various MXene/polymer composite systems for water purification and desalination applications.

Table 14.2 Various MXene/polymer composite systems for water purification and desalination.

S. no.	Material	Preparation method	Application	Reference
1	2D lamellar Titanium carbide ($Ti_3C_2T_x$)/poly (arylene ether nitrile) (PEN) nanofibrous membranes	Electrospinning followed by vacuum-assisted self-assembly	Oil/water separation	[127]
2	Chitosan/$Ti_3C_2T_x$/ Polypropylene fabric	LbL assembly	Water desalination, Oil/water separation	[41]
3	$Ti_3C_2T_x$-polyvinylidene fluoride (PVDF) janus membrane	VAF	Low surface tension wastewater (LSTW) treatment	[128]
4	Polysulfone (PSF)/$Ti_3C_2T_x$ membrane	Solution casting	Nanofiltration membranes	[91]

(*Continued*)

Table 14.2 Various MXene/polymer composite systems for water purification and desalination. (*Continued*)

S. no.	Material	Preparation method	Application	Reference
5	$Ti_3C_2T_x$@MOF (MIL-101(Al)-NH_2) decorated PVDF membranes	LbL assembly	Removal of heavy metals ions and water desalination	[39]
6	Halloysite nanotubes@ MXene- polydopamine (Hal@MXene-PDA) composite membranes	VAF	Oil/water separation	[129]
7	$Ti_3C_2T_x$ embedded polyamide (PA) membrane	*In situ* polymerization	Water desalination	[130]
8	$Ti_3C_2T_x$-PVDF membrane	VAF	Seawater desalination	[131]
9	MXene/PVDF composite hollow fiber membrane	Chemical grafting	Seawater desalination	[126]
10	Poly(acrylic acid) (PAA)- $Ti_3C_2T_x$/ polyacrylonitrile (PAN) composite membrane	VAF	Wastewater treatment	[132]
11	MXene/ZnO/poly(arylene ether nitrile) fibrous membrane	Electrospinning followed by VAF	Oil/water separation	[133]
12	$Ti_3C_2T_x$ @P84 polyimide	Nonsolvent induced phase inversion method	Organic solvent nanofiltration	[134]
13	$Ti_3C_2T_X$/PVDF cation exchange membrane	Chemical grafting	Electrodialysis	[135]
14	ZIF-67 modified $Ti_3C_2T_X$/ polydopamine/sepiolite (MXene/PDA/Sep@ZIF-67) composite membrane	VAF	Heavy metal removal and oil–water separation	[136]
15	Ag@ $Ti_3C_2T_X$ /PEN membrane	Electrospinning followed by VAF	Fast purification of emulsified oily wastewater	[137]

(*Continued*)

Table 14.2 Various MXene/polymer composite systems for water purification and desalination. (*Continued*)

S. no.	Material	Preparation method	Application	Reference
16	Nanocellulose-intercalated $Ti_3C_2T_X$ membrane	VAF	Seawater desalination, industrial wastewater treatment and remediation of aquatic environment	[138]
17	2D $Ti_3C_2T_X$-PAN membrane	Electrospinning followed by VAF	Efficient oil-water separation and adsorption of organic dyes in wastewater	[139]
18	Na-Bentonite@MXene/ PVDF membrane	VAF	Oil/water separation	[140]
19	GO/MXene/Polyphenylene sulfide (PPS) composite membrane	VAF	Dye wastewater purification treatment	[107]
20	Silane-Grafted $Ti_3C_2T_X$/ polyvinyl alcohol (PVA) membrane	VAF	Water Purification	[141]

14.5 Conclusion and Future Outlook

The MXene/polymer nanocomposite membranes have attained considerable attention during the past five years because of its enhanced thermal, electrical, and mechanical characteristics. The potentiality of MXenes in water purification and desalination has not been completely explored, as only a few publications can be found in this area of research. MXene-based membranes manifest great stability and efficiency in removing different salts, organic solutions, dyes, etc. from water. Moreover, the alluring properties of MXenes like innate negative charge and hydrophilicity help to embody particular functionalities which can enhance selectivity and membrane fouling resistance properties on account of repulsive and attractive interactions, adsorption, etc. The contemporary MXene-based membranes are only concentrated on $Ti_3C_2T_x$ MXenes. Thus, it is necessary to employ other types of MXenes for developing novel materials with excellent efficacy. Furthermore, the studies on the efficiency of MXene-based membranes for water purification are still in lab scale, thus lacking systematic analysis for their potentiality. Therefore, detailed studies are required to make these membranes suitable for commercial applications. Moreover, there is a possibility of combining the synergistic effect of MXenes with other 2D and 1D fillers to create hybrid fillers and integrate them with polymers in the form of polymer nanocomposite membranes for water purification. Thus the entire spectrum of MXene based research on water purification membrane needs to be explored in an effective manner to reap the potential benefits of MXenes.

References

1. Ding, L., Li, L., Liu, Y., Wu, Y., Lu, Z., Deng, J., Wei, Y., Caro, J., Wang, H., Effective ion sieving with $Ti_3C_2T_x$ MXene membranes for production of drinking water from seawater. *Nat. Sustain.*, 3, 296–302, 2020.
2. Afsari, M., Shon, H.K., Tijing, L.D., Janus membranes for membrane distillation: Recent advances and challenges. *Adv. Colloid Interface Sci.*, 289, 102362, 2021.
3. Ayodhya, D., A review of recent progress in 2D MXenes: Synthesis, properties, and applications. *Diam. Relat. Mater.*, 132, 109634, 2023.
4. Jatoi, A.S., Mubarak, N.M., Hashmi, Z., Solangi, N.H., Karri, R.R., Hua, T.Y., Mazari, S.A., Koduru, J.R., Alfantazi, A., New insights into MXene applications for sustainable environmental remediation. *Chemosphere*, 313, 137497, 2023.
5. Wang, G., Xu, Y., Zhang, R., Gai, S., Zhao, Y., Yang, F., Cheng, K., Fire-resistant MXene composite aerogels for effective oil/water separation. *J. Environ. Chem. Eng.*, 11, 109127, 2023.
6. Chaudhry, F.N. and Malik, M.F., Factors affecting water pollution: A review. *J. Ecosystem Ecography*, 7, 1–3, 2017.
7. Arif, A., Malik, M.F., Liaqat, S., Aslam, A., Mumtaz, K., Afzal, A., Ch, D.M., Nisa, K., Khurshid, F., Arif, F., Khalid, M.S.Z., Water pollution and industries. *Pure Appl. Biol.*, 9, 2214–2224, 2020.
8. Yang, H., Zheng, Q., Zhang, P., Nie, G., Ali, T., Raza, S., Fabrication of MXene (Ti2C3Tx) based conducting polymer materials and their applications as anticancer and metal ions removal from wastewater. *Surf. Interfaces*, 36, 102493, 2023.
9. He, S., Zhan, Y., Hu, J., Zhang, G., Zhao, S., Feng, Q., Yang, W., Chemically stable two-dimensional MXene@UIO-66-(COOH)2 composite lamellar membrane for multi-component pollutant-oil-water emulsion separation. *Compos. Part B Eng.*, 197, 108188, 2020.
10. Kumar, J.A., Prakash, P., Krithiga, T., Amarnath, D.J., Premkumar, J., Rajamohan, N., Vasseghian, Y., Saravanan, P., Rajasimman, M., Methods of synthesis, characteristics, and environmental applications of MXene: A comprehensive review. *Chemosphere*, 286, 131607, 2022.
11. Al-Shammiri, M. and Safar, M., Multi-effect distillation plants: State of the art. *Desalination*, 126, 45–59, 1999.
12. Ebrahimi, A., Najafpour, G.D., Yousefi Kebria, D., Performance of microbial desalination cell for salt removal and energy generation using different catholyte solutions. *Desalination*, 432, 1–9, 2018.
13. Karim, A., Islam, M.A., Yousuf, A., Khan, M.M.R., Faizal, C.K.M., Microbial lipid accumulation through bioremediation of palm oil mill wastewater by Bacillus cereus. *ACS Sustain. Chem. Eng.*, 7, 14500–14508, 2019.
14. Shen, Z.Y., Chen, C.Y., Lee, M.T., Recovery of cutting fluids and silicon carbide from slurry waste. *J. Hazard. Mater.*, 362, 115–123, 2019.
15. Gong, C., Lao, J., Wang, B., Li, X., Li, G., Gao, J., Wan, Y., Sun, X., Guo, R., Luo, J., Fast and all-weather cleanup of viscous crude-oil spills with Ti3C2TX MXene wrapped sponge. *J. Mater. Chem. A.*, 8, 20162–20167, 2020.

16. Hou, K., Zhou, H., Zhu, K., Xie, C., Liu, S., He, Y., Zhu, X., Wu, M., Huang, T., Superwetting Ti3C2TX MXene membranes intercalated with sodium alginate for oil/water separation. *Carbohydr. Polym. Technol. Appl.*, 5, 100278, 2023.
17. Song, J., Lu, Y., Luo, J., Huang, S., Wang, L., Xu, W., Parkin, I.P., Barrel-shaped oil skimmer designed for collection of oil from spills. *Adv. Mater. Interfaces*, 2, 1500350, 2015.
18. Crini, G., Lichtfouse, E., Wilson, L.D., Morin-Crini, N., Conventional and non-conventional adsorbents for wastewater treatment. *Environ. Chem. Lett.*, 17, 195–213, 2019.
19. Saththasivam, J., Loganathan, K., Sarp, S., An overview of oil–water separation using gas flotation systems. *Chemosphere*, 144, 671–680, 2016.
20. Li, X., Liu, Y., Wang, J., Gascon, J., Li, J., Van der Bruggen, B., Metal–organic frameworks based membranes for liquid separation. *Chem. Soc Rev.*, 46, 7124–7144, 2017.
21. Kiani, S., Mousavi, S.M., Saljoughi, E., Shahtahmassebi, N., Preparation and characterization of modified polyphenylsulfone membranes with hydrophilic property for filtration of aqueous media. *Polym. Adv. Technol.*, 29, 1632–1648, 2018.
22. Nunes, S.P., Culfaz-Emecen, P.Z., Ramon, G.Z., Visser, T., Koops, G.H., Jin, W., Ulbricht, M., Thinking the future of membranes: Perspectives for advanced and new membrane materials and manufacturing processes. *J. Membr. Sci.*, 598, 117761, 2020.
23. Bayrami, A., Bagherzadeh, M., Navi, H., Nikkhoo, M., Amini, M., Zwitterion-functionalized MIL-125-NH2-based thin-film nanocomposite forward osmosis membranes: Towards improved performance for salt rejection and heavy metal removal. *New J. Chem.*, 46, 15205–15218, 2022.
24. Ghanbari, R. and Amanat, N., Approaches of membrane modification for water treatment. *Mater. Chem. Horizons.*, 1, 153–167, 2022.
25. Liu, P., Hou, J., Zhang, Y., Li, L., Lu, X., Tang, Z., Two-dimensional material membranes for critical separations. *Inorg. Chem. Front.*, 7, 2560–2581, 2020.
26. Ihsanullah, Carbon nanotube membranes for water purification: Developments, challenges, and prospects for the future. *Sep. Purif. Technol.*, 209, 307–337, 2019.
27. Ang, W.L., Mohammad, A.W., Hilal, N., Leo, C.P., A review on the applicability of integrated/hybrid membrane processes in water treatment and desalination plants. *Desalination*, 363, 2–18, 2015.
28. Crini, G. and Lichtfouse, E., Advantages and disadvantages of techniques used for wastewater treatment. *Environ. Chem. Lett.*, 17, 145–155, 2019.
29. Kim, H.J., Choi, K., Baek, Y., Kim, D.G., Shim, J., Yoon, J., Lee, J.C., High-performance reverse osmosis CNT/polyamide nanocomposite membrane by controlled interfacial interactions. *ACS Appl. Mater. Interfaces.*, 6, 2819–2829, 2014.
30. Mousavinezhad, S.A., Mousavi, S.M., Saljoughi, E., Preparation and characterization of styrene-butadiene-styrene membrane incorporated with graphene nanosheets for pervaporative removal of 1, 2, 4-trimethylbenzene from water. *J. Hazard. Mater.*, 378, 120689, 2019.
31. Chu, K.H., Huang, Y., Yu, M., Her, N., Flora, J.R., Park, C.M., Kim, S., Cho, J., Yoon, Y., Evaluation of humic acid and tannic acid fouling in graphene oxide-coated ultrafiltration membranes. *ACS Appl. Mater. Interfaces*, 8, 22270–22279, 2016.
32. Bayrami, A., Bagherzadeh, M., Navi, H., Chegeni, M., Hosseinifard, M., Amini, M., Thin-film nanocomposite membranes containing aspartic acid-modified MIL-53-NH2 (Al) for boosting desalination and anti-fouling performance. *Desalination*, 521, 115386, 2022.
33. Han, R., Ma, X., Xie, Y., Teng, D., Zhang, S., Preparation of a new 2D MXene/PES composite membrane with excellent hydrophilicity and high flux. *RSC Adv.*, 7, 56204–56210, 2017.
34. Surwade, S.P., Smirnov, S.N., Vlassiouk, I.V., Unocic, R.R., Veith, G.M., Dai, S., Mahurin, S.M., Water desalination using nanoporous single-layer grapheme. *Nat. Nanotechnol.*, 10, 459–464, 2015.

35. Chen, L., Shi, G., Shen, J., Peng, B., Zhang, B., Wang, Y., Bian, F., Wang, J., Li, D., Qian, Z., Xu, G., Ion sieving in graphene oxide membranes via cationic control of interlayer spacing. *Nature*, 550, 380–383, 2017.
36. Cohen-Tanugi, D. and Grossman, J.C., Mechanical strength of nanoporous graphene as a desalination membrane. *Nano Lett.*, 14, 6171–6178, 2014.
37. Tan, Y., Sun, Z., Meng, H., Han, Y., Wu, J., Xu, J., Xu, Y., Zhang, X., Aminated metal-organic framework (NH2-MIL-101(Cr)) incorporated polyvinylidene (PVDF) hybrid membranes: Synthesis and application in efficient removal of Congo red from aqueous solution. *Appl. Organomet. Chem.*, 34, e5281, 2020.
38. Li, Y., He, Y., Zhuang, J., Shi, H., Hierarchical microsphere encapsulated in graphene oxide composite for durable synergetic membrane separation and Fenton-like degradation. *Chem. Eng. J.*, 430, 133124, 2022.
39. Ghanbari, R., Zare, E.N., Paiva-Santos, A.C., Rabiee, N., Ti3C2Tx MXene@MOF decorated polyvinylidene fluoride membrane for the remediation of heavy metals ions and desalination. *Chemosphere.*, 311, 137191, 2023.
40. Kang, D., Shao, H., Chen, G., Dong, X., Qin, S., Fabrication of highly permeable PVDF loose nanofiltration composite membranes for the effective separation of dye/salt mixtures. *J. Membr. Sci.*, 621, 118951, 2021.
41. Wu, Z., Li, J., Zhang, S., Yan, J., Gao, J., Zheng, N., Xue, H., Chitosan assisted MXene decoration onto polymer fabric for high efficiency solar driven interfacial evaporation of oil contaminated seawater. *J. Colloid Interface Sci.*, 622, 169–180, 2022.
42. Lim, G.P., Soon, C.F., Al-Gheethi, A.A., Morsin, M., Tee, K.S., Recent progress and new perspective of MXene-based membranes for water purification: A review. *Ceram. Int.*, 48, 16477–16491, 2022.
43. Naguib, M., Kurtoglu, M., Presser, V., Lu, J., Niu, J., Heon, M., Hultman, L., Gogotsi, Y., Barsoum, M.W., Two-dimensional nanocrystals produced by exfoliation of Ti3AlC2. *Adv. Mater.*, 23, 4248–4253, 2011.
44. Ihsanullah, I., Potential of MXenes in water desalination: Current status and perspectives. *Nano-Micro Lett.*, 12, 72, 2020.
45. Sheth, Y., Dharaskar, S., Chaudhary, V., Khalid, M., Walvekar, R., Prospects of titanium carbide-based MXene in heavy metal ion and radionuclide adsorption for wastewater remediation: A review. *Chemosphere*, 293, 133563, 2022.
46. Ihsanullah, I. and Ali, H., Technological challenges in the environmental applications of MXenes and future outlook. *Case Stud. Chem. Environ. Eng.*, 2, 100034, 2020.
47. Wang, Y., Xu, H., Ding, M., Zhang, L., Chen, G., Fu, J., Wang, A., Chen, J., Liu, B., Yang, W., MXene-regulation polyamide membrane featuring with bubble-like nodule for efficient dye/salt separation and antifouling performance. *RSC Adv.*, 12, 10267–10279, 2022.
48. Khatami, M. and Iravani, S., MXenes and MXene-based materials for the removal of water pollutants: Challenges and opportunities. *Comments Inorg. Chem.*, 41, 213–248, 2021.
49. Kim, S., Gholamirad, F., Shin, B., Taheri-Qazvini, N., Cho, J., Yu, M., Park, C.M., Heo, J., Yoon, Y., Application of a Ti3C2TX MXene-coated membrane for removal of selected natural organic matter and pharmaceuticals. *ACS ES&T Water*, 1, 2164–2173, 2021.
50. Hu, J., Zhan, Y., Zhang, G., Feng, Q., Yang, W., Chiao, Y.H., Zhang, S., Sun, A., Durable and super-hydrophilic/underwater super-oleophobic two-dimensional MXene composite lamellar membrane with photocatalytic self-cleaning property for efficient oil/water separation in harsh environments. *J. Membr. Sci.*, 637, 119627, 2021.
51. Huang, Z., Liu, J., Liu, Y., Xu, Y., Li, R., Hong, H., Shen, L., Lin, H., Liao, B.Q., Enhanced permeability and antifouling performance of polyether sulfone (PES) membrane via elevating

magnetic Ni@MXene nanoparticles to upper layer in phase inversion process. *J. Membr. Sci.*, 623, 119080, 2021.

52. Feng, A., Hou, T., Jia, Z., Zhang, Y., Zhang, F., Wu, G., Preparation and characterization of epoxy resin filled with Ti3C2Tx MXene nanosheets with excellent electric conductivity. *Nanomaterials*, 10, 162, 2020.
53. Xu, L., Wang, W., Liu, Y., Liang, D., Nanocellulose-linked Mxene/polyaniline aerogel films for flexible supercapacitors. *Gels*, 8, 798, 2022.
54. Zhi, W., Xiang, S., Bian, R., Lin, R., Wu, K., Wang, T., Cai, D., Study of MXene-filled polyurethane nanocomposites prepared via an emulsion method. *Compos. Sci. Technol.*, 168, 404–411, 2018.
55. Song, P., Qiu, H., Wang, L., Liu, X., Zhang, Y., Zhang, J., Kong, J., Gu, J., Honeycomb structural rGO-MXene/epoxy nanocomposites for superior electromagnetic interference shielding performance. *Sustain. Mater. Technol.*, 24, e00153, 2020.
56. Sheng, X., Zhao, Y., Zhang, L., Lu, X., Properties of two-dimensional Ti3C2 MXene/thermoplastic polyurethane nanocomposites with effective reinforcement via melt blending. *Compos. Sci. Technol.*, 181, 107710, 2019.
57. Zhang, Y., Wang, L., Zhang, N., Zhou, Z., Adsorptive environmental applications of MXene nanomaterials: A review. *RSC Adv.*, 8, 19895–19905, 2018.
58. Zhang, C., Ma, Y., Zhang, X., Abdolhosseinzadeh, S., Sheng, H., Lan, W., Pakdel, A., Heier, J., Nüesch, F., Two-dimensional transition metal carbides and nitrides (MXenes): Synthesis, properties, and electrochemical energy storage applications. *Energy Environ. Mater.*, 3, 29–55, 2020.
59. Liu, R. and Li, W., High-thermal-stability and high-thermal-conductivity Ti3C2T x MXene/ poly (vinyl alcohol) (PVA) composites. *ACS Omega*, 3, 2609–2617, 2018.
60. Fei, M., Lin, R., Deng, Y., Xian, H., Bian, R., Zhang, X., Cheng, J., Xu, C., Cai, D., Polybenzimidazole/Mxene composite membranes for intermediate temperature polymer electrolyte membrane fuel cells. *Nanotechnology*, 29, 35403, 2017.
61. Naguib, M., Saito, T., Lai, S., Rager, M.S., Aytug, T., Paranthaman, M.P., Zhao, M.Q., Gogotsi, Y., Ti 3 C 2 T x (MXene)–polyacrylamide nanocomposite films. *RSC Adv.*, 6, 72069–72073, 2016.
62. Mayerberger, E.A., Street, R.M., McDaniel, R.M., Barsoum, M.W., Schauer, C.L., Antibacterial properties of electrospun $Ti_3C_2T_z$ (MXene)/chitosan nanofibers. *RSC Adv.*, 8, 35386–35394, 2018.
63. Cao, Y., Deng, Q., Liu, Z., Shen, D., Wang, T., Huang, Q., Du, S., Jiang, N., Lin, C.T., Yu, J., Enhanced thermal properties of poly (vinylidene fluoride) composites with ultrathin nanosheets of MXene. *RSC Adv.*, 7, 20494–20501, 2017.
64. Yi, Z., Yang, J., Liu, X., Mao, L., Cui, L., Liu, Y., Enhanced mechanical properties of poly (lactic acid) composites with ultrathin nanosheets of MXene modified by stearic acid. *J. Appl. Polym. Sci.*, 137, 48621, 2020.
65. Ma, M., Liao, X., Chu, Q., Chen, S., Shi, Y., He, H., Wang, X., Construction of gradient conductivity cellulose nanofiber/MXene composites with efficient electromagnetic interference shielding and excellent mechanical properties. *Compos. Sci. Technol.*, 226, 109540, 2022.
66. Zhang, Z., Yang, S., Zhang, P., Zhang, J., Chen, G., Feng, X., Mechanically strong MXene/Kevlar nanofiber composite membranes as high-performance nanofluidic osmotic power generators. *Nat. Commun.*, 10, 1–9, 2019.
67. Han, R., Xie, Y., Ma, X., Crosslinked P84 copolyimide/MXene mixed matrix membrane with excellent solvent resistance and permselectivity. *Chin. J. Chem. Eng.*, 27, 877–883, 2019.
68. Weng, C., Xing, T., Jin, H., Wang, G., Dai, Z., Pei, Y., Liu, L., Zhang, Z., Mechanically robust ANF/MXene composite films with tunable electromagnetic interference shielding performance. *Compos. Part A Appl. Sci. Manuf.*, 135, 105927, 2020.

69. Yin, G., Wang, Y., Wang, W., Yu, D., Multilayer structured PANI/MXene/CF fabric for electromagnetic interference shielding constructed by layer-by-layer strategy. *Colloids Surf. A Physicochem. Eng. Asp.*, 601, 125047, 2020.
70. Waribam, P., Jaiyen, K., Samart, C., Ogawa, M., Guan, G., Kongparakul, S., MXene-copper oxide/sulfonated polyether ether ketone as a hybrid composite proton exchange membrane in electrochemical water electrolysis. *Catal. Today*, 407, 96–106, 2023.
71. Likitaporn, C., Okhawilai, M., Kasemsiri, P., Qin, J., Potiyaraj, P., Uyama, H., High electrolyte uptake of MXene integrated membrane separators for Zn-ion batteries. *Sci. Rep.*, 12, 19915, 2022.
72. Wang, J., He, J., Kan, D., Chen, K., Song, M., Huo, W., MXene film prepared by vacuum-assisted filtration: Properties and applications. *Crystals*, 12, 1034, 2022.
73. Luo, J.Q., Zhao, S., Zhang, H.B., Deng, Z., Li, L., Yu, Z.Z., Flexible, stretchable and electrically conductive MXene/natural rubber nanocomposite films for efficient electromagnetic interference shielding. *Compos. Sci. Technol.*, 182, 107754, 2019.
74. Song, G., Kang, R., Guo, L., Ali, Z., Chen, X., Zhang, Z., Yan, C., Lin, C.T., Jiang, N., Yu, J., Highly flexible few-layer Ti 3 C 2 MXene/cellulose nanofiber heat-spreader films with enhanced thermal conductivity. *New J. Chem.*, 44, 7186–7193, 2020.
75. Boota, M., MXene–organic hybrid materials, in: *2D Metal Carbides and Nitrides (MXenes)*, pp. 221–251, Springer, Springer Nature Switzerland AG, 2019.
76. Lipton, J., Weng, G.M., Röhr, J.A., Wang, H., Taylor, A.D., Layer-by-layer assembly of two-dimensional materials: Meticulous control on the nanoscale. *Matter*, 2, 1148–1165, 2020.
77. Tan, Z., Zhao, H., Sun, F., Ran, L., Yi, L., Zhao, L., Wu, J., Fabrication of chitosan/MXene multilayered film based on layer-by-layer assembly: Toward enhanced electromagnetic interference shielding and thermal management capacity. *Compos. Part A Appl. Sci. Manuf.*, 155, 106809, 2022.
78. Yin, G., Wang, Y., Wang, W., Qu, Z., Yu, D., A flexible electromagnetic interference shielding fabric prepared by construction of PANI/MXene conductive network via layer-by-layer assembly. *Adv. Mater. Interfaces*, 8, 2001893, 2021.
79. Raman, A., Jayan, J.S., Deeraj, B.D.S., Saritha, A., Joseph, K., Electrospun nanofibers as effective superhydrophobic surfaces: A brief review. *Surf. Interfaces*, 24, 101140, 2021.
80. Zhang, Y.R., Wang, B.C., Gao, S.L., Qiu, L.P., Zheng, Q.H., Cheng, G.T., Han, W.P., Ramakrishna, S., Long, Y.Z., Electrospun MXene nanosheet/polymer composites for electromagnetic shielding and microwave absorption: A review. *ACS Appl. Nano Mater.*, 5, 12320–12342, 2022.
81. Awasthi, G.P., Maharjan, B., Shrestha, S., Bhattarai, D.P., Yoon, D., Park, C.H., Kim, C.S., Synthesis, characterizations, and biocompatibility evaluation of polycaprolactone–MXene electrospun fibers. *Colloids Surf. A Physicochem. Eng. Asp.*, 586, 124282, 2020.
82. Du, Z., Chen, K., Zhang, Y., Wang, Y., He, P., Mi, H.Y., Wang, Y., Liu, C., Shen, C., Engineering multilayered MXene/electrospun poly(lactic acid) membrane with increscent electromagnetic interference (EMI) shielding for integrated Joule heating and energy generating. *Compos. Commun.*, 26, 100770, 2021.
83. Riazi, H., Nemani, S.K., Grady, M.C., Anasori, B., Soroush, M., Ti3C2 MXene–polymer nanocomposites and their applications. *J. Mater. Chem. A*, 9, 8051–8098, 2021.
84. Xu, Z. and Gao, C., Graphene fiber: A new trend in carbon fibers. *Mater. Today*, 18, 480–492, 2015.
85. Levitt, A.S., Alhabeb, M., Hatter, C.B., Sarycheva, A., Dion, G., Gogotsi, Y., Electrospun MXene/carbon nanofibers as supercapacitor electrodes. *J. Mater. Chem. A*, 7, 269–277, 2019.
86. Chen, X., Zhao, Y., Li, L., Wang, Y., Wang, J., Xiong, J., Du, S., Zhang, P., Shi, X., Yu, J., MXene/polymer nanocomposites: Preparation, properties, and applications. *Polym. Rev.*, 61, 80–115, 2021.

87. Shen, Z., Chen, W., Xu, H., Yang, W., Kong, Q., Wang, A., Ding, M., Shang, J., Fabrication of a novel antifouling polysulfone membrane with *in situ* embedment of Mxene nanosheets. *Int. J. Environ. Res. Public Health*, 16, 4659, 2019.
88. Mirkhani, S.A., Zeraati, A.S., Aliabadian, E., Naguib, M., Sundararaj, U., High dielectric constant and low dielectric loss via poly (vinyl alcohol)/Ti3C2T x MXene nanocomposites. *ACS Appl. Mater. Interfaces*, 11, 18599–18608, 2019.
89. Karahan, H.E., Goh, K., Zhang, K., Yang, E., Yıldırım, C., Chuah, C.Y., Ahunbay, M.G., Lee, J., Tantekin-Ersolmaz, Ş. B., Chen, Y., Bae, T.H., MXene materials for designing advanced separation membranes. *Adv. Mater.*, 32, 1906697, 2020.
90. Aghamohammadi, H., Amousa, N., Eslami-Farsani, R., Recent advances in developing the MXene/polymer nanocomposites with multiple properties: A review study. *Synth. Met.*, 273, 116695, 2021.
91. Dashtbozorg, A., Saljoughi, E., Mousavi, S.M., Kiani, S., High-performance and robust polysulfone nanocomposite membrane containing 2D functionalized MXene nanosheets for the nanofiltration of salt and dye solutions. *Desalination*, 527, 115600, 2022.
92. Idumah, C., II, Recently emerging advancements in thermal conductivity and flame retardancy of MXene polymeric nanoarchitectures. *Polym. Technol. Mater.*, 62, 510–546, 2022.
93. Wang, D., Wei, H., Lin, Y., Jiang, P., Bao, H., Huang, X., Achieving ultrahigh thermal conductivity in Ag/MXene/epoxy nanocomposites via filler-filler interface engineering. *Compos. Sci. Technol.*, 213, 108953, 2021.
94. Aakyiir, M., Yu, H., Araby, S., Ruoyu, W., Michelmore, A., Meng, Q., Losic, D., Choudhury, N.R., Ma, J., Electrically and thermally conductive elastomer by using MXene nanosheets with interface modification. *Chem. Eng. J.*, 397, 125439, 2020.
95. Anasori, B., Lukatskaya, M.R., Gogotsi, Y., 2D metal carbides and nitrides (MXenes) for energy storage. *Nat. Rev. Mater.*, 2, 1–17, 2017.
96. Ng, V.M.H., Huang, H., Zhou, K., Lee, P.S., Que, W., Xu, J.Z., Kong, L.B., Recent progress in layered transition metal carbides and/or nitrides (MXenes) and their composites: Synthesis and applications. *J. Mater. Chem. A*, 5, 3039–3068, 2017.
97. Ling, Z., Ren, C.E., Zhao, M.Q., Yang, J., Giammarco, J.M., Qiu, J., Barsoum, M.W., Gogotsi, Y., Flexible and conductive MXene films and nanocomposites with high capacitance. *Proc. Natl. Acad. Sci.*, 111, 16676–16681, 2014.
98. Kołtunowicz, T.N., Gałaszkiewicz, P., Kierczyński, K., Rogalski, P., Okal, P., Pogrebnjak, A.D., Buranich, V., Pogorielov, M., Diedkova, K., Zahorodna, V., Balitskyi, V., Investigation of AC electrical properties of Mxene-PCL nanocomposites for application in small and medium power generation. *Energies*, 14, 7123, 2021.
99. Gao, L., Li, C., Huang, W., Mei, S., Lin, H., Ou, Q., Zhang, Y., Guo, J., Zhang, F., Xu, S., Zhang, H., MXene/polymer membranes: Synthesis, properties, and emerging applications. *Chem. Mater.*, 32, 1703–1747, 2020.
100. Damptey, L., Jaato, B.N., Ribeiro, C.S., Varagnolo, S., Power, N.P., Selvaraj, V., Dodoo-Arhin, D., Kumar, R.V., Sreenilayam, S.P., Brabazon, D., Kumar Thakur, V., Surface functionalized mxenes for wastewater treatment—A comprehensive review. *Global Chall.*, 6, 2100120, 2022.
101. Ihsanullah, I. and Bilal, M., Potential of MXene-based membranes in water treatment and desalination: A critical review. *Chemosphere*, 303, 135234, 2022.
102. Liu, T., Liu, X., Graham, N., Yu, W., Sun, K., Two-dimensional MXene incorporated graphene oxide composite membrane with enhanced water purification performance. *J. Membr. Sci.*, 593, 117431, 2020.
103. Wei, S., Xie, Y., Xing, Y., Wang, L., Ye, H., Xiong, X., Wang, S., Han, K., Two-dimensional graphene oxide/MXene composite lamellar membranes for efficient solvent permeation and molecular separation. *J. Membr. Sci.*, 582, 414–422, 2019.

104. Kim, J.H., Park, G.S., Kim, Y.J., Choi, E., Kang, J., Kwon, O., Kim, S.J., Cho, J.H., Kim, D.W., Large-area Ti3C2Tx-Mxene coating: Toward industrial-scale fabrication and molecular separation. *ACS Nano*, 15, 8860–8869, 2021.
105. Han, R. and Wu, P., High-performance graphene oxide nanofiltration membrane with continuous nanochannels prepared by the *in situ* oxidation of MXene. *J. Mater. Chem. A*, 7, 6475–6481, 2019.
106. Tong, X., Liu, S., Qu, D., Gao, H., Yan, L., Chen, Y., Crittenden, J., Tannic acid-metal complex modified MXene membrane for contaminants removal from water. *J. Membr. Sci.*, 622, 119042, 2021.
107. Ma, X.Y., Fan, T.T., Wang, G., Li, Z.H., Lin, J.H., Long, Y.Z., High performance GO/MXene/PPS composite filtration membrane for dye wastewater treatment under harsh environmental conditions. *Compos. Commun.*, 29, 101017, 2022.
108. Wang, L., Yuan, L., Chen, K., Zhang, Y., Deng, Q., Du, S., Huang, Q., Zheng, L., Zhang, J., Chai, Z., Barsoum, M.W., Wang, X., Shi, W., Loading actinides in multilayered structures for nuclear waste treatment: The first case study of uranium capture with vanadium carbide MXene. *ACS Appl. Mater. Interfaces*, 8, 16396–16403, 2016.
109. Wang, L., Song, H., Yuan, L., Li, Z., Zhang, Y., Gibson, J.K., Zheng, L., Chai, Z., Shi, W., Efficient U(VI) reduction and sequestration by Ti2CTx MXene. *Environ. Sci. Technol.*, 52, 10748–10756, 2018.
110. Zhang, P., Wang, L., Du, K., Wang, S., Huang, Z., Yuan, L., Li, Z., Wang, H., Zheng, L., Chai, Z., Shi, W., Effective removal of U(VI) and Eu(III) by carboxyl functionalized MXene nanosheets. *J. Hazard. Mater.*, 396, 122731, 2020.
111. Sha, X., Huang, H., Sun, S., Huang, H., Huang, Q., He, Z., Liu, M., Zhou, N., Zhang, X., Wei, Y., Mussel-inspired preparation of MXene-PDA-Bi6O7 composites for efficient adsorptive removal of iodide ions. *J. Environ. Chem. Eng.*, 8, 104261, 2020.
112. Fan, Y., Li, J., Wang, S., Meng, X., Zhang, W., Jin, Y., Yang, N., Tan, X., Li, J., Liu, S., Voltage-enhanced ion sieving and rejection of Pb2+ through a thermally cross-linked two-dimensional MXene membrane. *Chem. Eng. J.*, 401, 126073, 2020.
113. Jin, L., Chai, L., Yang, W., Wang, H., Zhang, L., Two-dimensional titanium carbides (Ti3C2Tx) functionalized by poly(m-phenylenediamine) for efficient adsorption and reduction of hexavalent chromium. *Int. J. Environ. Res. Public Health*, 17, 167, 2019.
114. Wang, C., Cheng, R., Hou, P.X., Ma, Y., Majeed, A., Wang, X., Liu, C., MXene-carbon nanotube hybrid membrane for robust recovery of Au from trace-level solution. *ACS Appl. Mater. Interfaces*, 12, 43032–43041, 2020.
115. Zhao, X., Che, Y., Mo, Y., Huang, W., Wang, C., Fabrication of PEI modified GO/MXene composite membrane and its application in removing metal cations from water. *J. Membr. Sci.*, 640, 119847, 2021.
116. Saththasivam, J., Wang, K., Yiming, W., Liu, Z., Mahmoud, K.A., A flexible Ti3C2Tx (MXene)/paper membrane for efficient oil/water separation. *RSC Adv.*, 9, 16296–16304, 2019.
117. Li, Z.K., Liu, Y., Li, L., Wei, Y., Caro, J., Wang, H., Ultra-thin titanium carbide (MXene) sheet membranes for high-efficient oil/water emulsions separation. *J. Membr. Sci.*, 592, 117361, 2019.
118. Liu, Z.H., Wang, Q.M., Lü, Q.F., Wu, J., L-lysine functionalized Ti3C2Tx coated polyurethane sponge for high-throughput oil–water separation. *Colloids Surf. A Physicochem. Eng. Asp.*, 640, 128396, 2022.
119. Arabi Shamsabadi, A., Sharifian Gh, M., Anasori, B., Soroush, M., Antimicrobial mode-of-action of colloidal Ti3C2Tx Mxene nanosheets. *ACS Sustain. Chem. Eng.*, 6, 16586–16596, 2018.
120. Zha, X.J., Zhao, X., Pu, J.H., Tang, L.S., Ke, K., Bao, R.Y., Bai, L., Liu, Z.Y., Yang, M.B., Yang, W., Flexible anti-biofouling Mxene/cellulose fibrous membrane for sustainable solar-driven water purification. *ACS Appl. Mater. Interfaces*, 11, 36589–36597, 2019.

121. Qu, W., Zhao, H., Zhang, Q., Xia, D., Tang, Z., Chen, Q., He, C., Shu, D., Multifunctional Au/Ti3C2 photothermal membrane with antibacterial ability for stable and efficient solar water purification under the full spectrum. *ACS Sustain. Chem. Eng.*, 9, 11372–11387, 2021.
122. Ren, C.E., Alhabeb, M., Byles, B.W., Zhao, M.Q., Anasori, B., Pomerantseva, E., Mahmoud, K.A., Gogotsi, Y., Voltage-gated ions sieving through 2D Mxene Ti3C2Tx membranes. *ACS Appl. Nano Mater.*, 1, 3644–3652, 2018.
123. Berdiyorov, G.R. and Mahmoud, K.A., Effect of surface termination onion intercalation selectivity of bilayer Ti3C2T2 (T=F, O and OH) MXene. *Appl. Surf. Sci.*, 416, 725–730, 2017.
124. Ding, M., Xu, H., Chen, W., Yang, G., Kong, Q., Ng, D., Lin, T., Xie, Z., 2D laminar maleic acid-crosslinked MXene membrane with tunable nanochannels for efficient and stable pervaporation desalination. *J. Membr. Sci.*, 600, 117871, 2020.
125. Meng, B., Liu, G., Mao, Y., Liang, F., Liu, G., Jin, W., Fabrication of surface-charged MXene membrane and its application for water desalination. *J. Membr. Sci.*, 623, 119076, 2021.
126. Zhang, T., Guo, X., Solomon, B., Sharifpur, M., Zhang, L.Z., A hydrophobic-hydrophilic MXene/PVDF composite hollow fiber membrane with enhanced antifouling properties for seawater desalination. *J. Membr. Sci.*, 644, 120146, 2022.
127. Feng, Q., Zhan, Y., Yang, W., Sun, A., Dong, H., Chiao, Y.H., Liu, Y., Chen, X., Chen, Y., Bi-functional super-hydrophilic/underwater super-oleophobic 2D lamellar Ti3C2Tx MXene/poly (arylene ether nitrile) fibrous composite membrane for the fast purification of emulsified oil and photodegradation of hazardous organics. *J. Colloid Interface Sci.*, 612, 156–170, 2022.
128. Yan, X., Yang, C., Ma, C., Tao, H., Cheng, S., Chen, L., Wang, G., Lin, X., Yao, C., A novel janus membrane modified by MXene for enhanced anti-fouling and anti-wetting in direct contact membrane distillation. *Chemosphere*, 307, 136114, 2022.
129. Zeng, G., Wei, K., Zhang, H., Zhang, J., Lin, Q., Cheng, X., Sengupta, A., Chiao, Y.H., Ultra-high oil-water separation membrane based on two-dimensional MXene (Ti3C2Tx) by co-incorporation of halloysite nanotubes and polydopamine. *Appl. Clay Sci.*, 211, 106177, 2021.
130. Wang, X., Li, Q., Zhang, J., Huang, H., Wu, S., Yang, Y., Novel thin-film reverse osmosis membrane with MXene Ti3C2Tx embedded in polyamide to enhance the water flux, anti-fouling and chlorine resistance for water desalination. *J. Membr. Sci.*, 603, 11803, 2020.
131. Zhang, T. and Zhang, L.Z., Development of a MXene-based membrane with excellent anti-fouling for air humidification-dehumidification type desalination. *J. Membr. Sci.*, 641, 119907, 2022.
132. Zhang, Y., Li, S., Huang, R., He, J., Sun, Y., Qin, Y., Shen, L., Stabilizing MXene-based nanofiltration membrane by forming analogous semi-interpenetrating network architecture using flexible poly (acrylic acid) for effective wastewater treatment. *J. Membr. Sci.*, 648, 120360, 2022.
133. Sun, A., Zhan, Y., Feng, Q., Yang, W., Dong, H., Liu, Y., Chen, X., Chen, Y., Assembly of MXene/ZnO heterojunction onto electrospun poly (arylene ether nitrile) fibrous membrane for favorable oil/water separation with high permeability and synergetic antifouling performance. *J. Membr. Sci.*, 663, 120933, 2022.
134. Shen, Y., Yao, A., Li, J., Hua, D., Tan, K.B., Zhan, G., Rao, X., Dispersive two-dimensional MXene via potassium fulvic acid for mixed matrix membranes with enhanced organic solvent nanofiltration performance. *J. Membr. Sci.*, 666, 121168, 2023.
135. Zheng, Y., Jin, Y., Zhang, N., Wang, D., Yang, Y., Zhang, M., Wang, G., Qu, W., Wu, Y., Preparation and characterization of Ti3C2TX MXene/PVDF cation exchange membrane for electrodialysis. *Colloids Surf. A Physicochem. Eng. Asp.*, 650, 129556, 2022.
136. Wang, Q., Yu, Z., Zhu, X., Xiang, Q., Chen, H., Pang, Y., ZIF-67 modified MXene/sepiolite composite membrane for oil–water separation and heavy metal removal. *J. Ind. Eng. Chem.*, 115, 314–328, 2022.

137. Feng, Q., Zhan, Y., Yang, W., Dong, H., Sun, A., Li, L., Chen, X., Chen, Y., Ultra-high flux and synergistically enhanced anti-fouling Ag@ MXene lamellar membrane for the fast purification of oily wastewater through nano-intercalation, photocatalytic self-cleaning and antibacterial effect. *Sep. Purif. Technol.*, 298, 121635, 2022.
138. Zhang, H., Zheng, Y., Zhou, H., Zhu, S., Yang, J., Nanocellulose-intercalated MXene NF membrane with enhanced swelling resistance for highly efficient antibiotics separation. *Sep. Purif. Technol.*, 305, 122425, 2023.
139. Imsong, R. and Purkayastha, D.D., Dual-functional superhydrophilic/underwater superoleophobic 2D Ti3C2TX MXene-PAN membrane for efficient oil-water separation and adsorption of organic dyes in wastewater. *Sep. Purif. Technol.*, 306, 122636, 2023.
140. Zhang, L., Liu, Y., Zeng, G., Yang, Z., Lin, Q., Wang, Y., Wang, X., Pu, S., Two-dimensional Na-bentonite@MXene composite membrane with switchable wettability for selective oil/water separation. *Sep. Purif. Technol.*, 306, 122677, 2023.
141. Yousaf, T., Areeb, A., Murtaza, M., Munir, A., Khan, Y., Waseem, A., Silane-grafted Mxene (Ti3C2Tx) membranes for enhanced water purification performance. *ACS Omega*, 7, 19502–19512, 2022.

15

MXene-Based Polymer Composite Membranes for Pervaporation and Gas Separation

S. Manobalan and T. P. Sumangala*

Department of Physics, School of Advanced Sciences, Vellore Institute of Technology, Vellore, Tamil Nadu, India

Abstract

Membrane separation systems are unique in the field of gas separation due to their low cost, outstanding performance, and environmental benefits. MXene membranes have piqued the interest of layered membrane specialists in the fields of gas separation, water treatment, desalination, and solvent extraction due to their remarkable properties, which include great mechanical strength, a high surface area, strong chemical and thermal stability, good electrical conductivity, easiness in functionalization, and hydrophilic nature. MXene based polymer composites makes the membrane preparation easy and straight forward. In addition to this, the development of polyelectrolyte functionalized MXene has also led to excellent separation. Several methods, such as cross-linking MXene nanosheets, adding extra nanochannels, constructing MXene hybrid membranes, and employing MXene as 2D scaffolds, are used to fabricate MXene-based polymer nanocomposite. In this book chapter, we will be systematically discussing these manufacturing processes. The performance of separation is affected by various factors such as stability, hydrophobicity/hydrophilicity, surface functionality, and porous size of the membrane. The mechanism and factors affecting pervaporation and gas separation is discussed in details. The measurement procedure and selectivity is also discussed. Efforts are made to compare the selectivity of various MXene-based composites. This chapter will provide a clear and concise summary of the various MXene based composites in a nut shell and it aims to enlighten the future research in this field.

Keywords: MXene, gas separation, pervaporation, Mxene-based composites

15.1 Introduction

Pervaporation and gas separation have come from the evolution of membrane technology over the last several decades, which has generated a significant amount of curiosity in response to the desire for sustainable and ecologically friendly solutions [1]. Due to their recent use and potential for mass production, two-dimensional (2D) materials were advocated as a suitable replacement [2]. High-performance two-dimensional (2D) materials have been manufactured for a variety of applications, including molecular sieving, including

**Corresponding author*: sumampoornima@gmail.com

graphene oxide (GO) [3], nanosheets of transition-metal dichalcogenides (TMDSs) [4], zeolite membranes [5, 6], metal-organic frameworks (MOFs) [5], and covalent organic frameworks (COFs) [7]. However, attaining consistent pore size distribution across large membrane areas is technically difficult, which restricts their application. Zeolite and a 2D MOF with designed nanopores have been combined to form membranes [8]. However, the formation of crystals without agglomeration and existence of exfoliated sheets still remains a challenge. However, this issue had been sought to tackle using the freeze-thaw exfoliation process wherein crystalline 2D porous nanosheets with better exfoliation, can be obtained. As an alternative, laminar membranes composed of 2D materials and with channels that are horizontally oriented provide a more useful method of utilizing 2D materials for separation [3]. Laminar membranes have been made using 2D graphene or transition metal dichalcogenides (TMDs). Unfortunately, because to their very crystalline stacking and lack of interlayer free space for molecule transport, they demonstrated poor molecular sieving characteristics [9]. Yet another explored material is molybdenum disulfide (MoS_2) membranes. Mi *et al.* have investigated both the filtering potential and the stability of these membranes in aqueous media. They have noticed that their interlayer distance has increased to as much as 1.2 nm because to a significant water permeability. However, these were found more useful for separating molecules in nanosize range, below which it was challenging. In contrast, graphene oxide (GO), the oxidized type of graphene, has exhibited promising results in the sub-nano size ranges [10]. The functional groups of oxygen on the GO planes may be to blame for this, which prohibit the stacking of additional layers. The exfoliation and the functional groups lead to an increase in the interlayer spacing. This can be manipulated so as to allow molecular diffusion through the membranes using these nanochannels. These 2D materials enable selective gas transportation using electrostatic repulsion and/or size sieving [2]. The gas separation membrane selectivity factor, which defines its effectiveness, is thought to be great when bigger molecules are confined while smaller molecules flow by using the available free channels in the membrane.

Due to its impressive electrical, thermal, mechanical, and optical properties, MXene, a new class of 2D transition metal carbides and carbonitrides, has shown great promise in a wide range of uses in recent years [11]. The general formula for MXenes is $M_{n+1}X_nT_x$, where M is an early transition metal, X is either carbon or nitrogen, and n is a number between 1 and 3 [12]. A typical synthesis uses MAX phase as the precursor, wherein 'A' stands for elements from group IIIA or IVA. For synthesizing MXene, the A layer is etched off using a suitable solvent or its combination. Based on the solvent system used for etching, the functional groups (T_x), could vary. The commonly found ones are H, O, and F. This chemical functionalization has seen to have profound role in the performance of MXene. In a manner similar to GO, the functional groups in MXene nanosheet aid in the formation of MXene layers with sufficient interlayer spacing. As a result, these gaps may cause MXene membranes to selectively pass certain molecules. The hydrophilic properties of MXene also help in the production of scalable membranes. As a result, MXene is novel material for creating membranes made of separable 2D materials. There are, however, not many reports on MXene based separation membranes. The different functional groups present can also lead to the orderly arrangement of nanosheets. Such arrangements are instrumental for achieving sieving using membranes [13]. Most important feature of this membrane is its thickness, which controls the selectivity and permeability towards smaller gas molecules. This can also be used to filter out the larger molecules via their inhibiting effect [2]. The permeability and selectivity properties that determine separation performance are significantly influenced by nanochannel size [14]. This chapter covers the development of polymer composite membranes utilizing MXene, the mechanism of pervaporation and gas separation, factors and/or parameters that affect membrane performance, and their subsequent characterization for these processes.

15.2 Development of MXene-Based Polymer Composite Membrane

MXene is a new 2D layered material that has generated a lot of interest from the membrane technology community [15, 16]. 2D MXene membranes differ from conventional polymer and ceramic membranes due to their properties including hydrophilicity, flexibility, and mechanical strength [17–19]. Various preparations methods of MXene have been explored. This includes employing MXene as a 2D scaffold structure, MXene hybrid membranes, cross-linking MXene nanosheets, and creating extra nanochannels. A few fabrication techniques such as Mixed-Matrix membranes, Thin-Film nanocomposites and Nanolaminate membranes are also explored.

15.2.1 Cross-Linking MXene Nanosheets

To achieve the ideal balance between selectivity and permeability, water swelling must be managed. The fundamental problem with MXene stacks is their tendency to expand when wet, which results in increased permeation but unstable behavior. These in turn are stemmed from the higher values of d-spacing due to etched out layers. This could also lead to degradation of sieving performance [20]. This behavior of MXene stacks need to be addressed to enhance the sieving and stability of MXene membranes. This has been attempted by many co-workers using a variety of chemical cross-linking techniques, including molecular cross-linking [21, 22], ionic cross linking [23] and self-cross linking [24].

15.2.1.1 Self-Crosslinking Technique

Use of self-crosslinked MXene membrane is one method for addressing the problem of water swelling (SCMM). Such a work has been reported by Lu *et al.* [24] wherein membranes with exceptional stability and ion-rejection capability were fabricated. During a simple thermal process, hydroxyl functional groups (OH) on surrounding MXene nanosheets suffered dehydration reactions given by Equation (15.1).

$$-OH + -OH \rightarrow -O- + H_2O \tag{15.1}$$

This led to the production of interlocking structures (Figure 15.1) [24]. An "O" bridge is created between neighboring nanosheets. Thus, MXene membrane is prevented from swelling by self-crosslinking, providing the required long-term stability. In addition to this, these sheets also exhibit a capacity to reject monovalent ions. It was shown that this rejection or exclusion capability of SCMMs has enhanced 30 to 70-fold following this treatment.

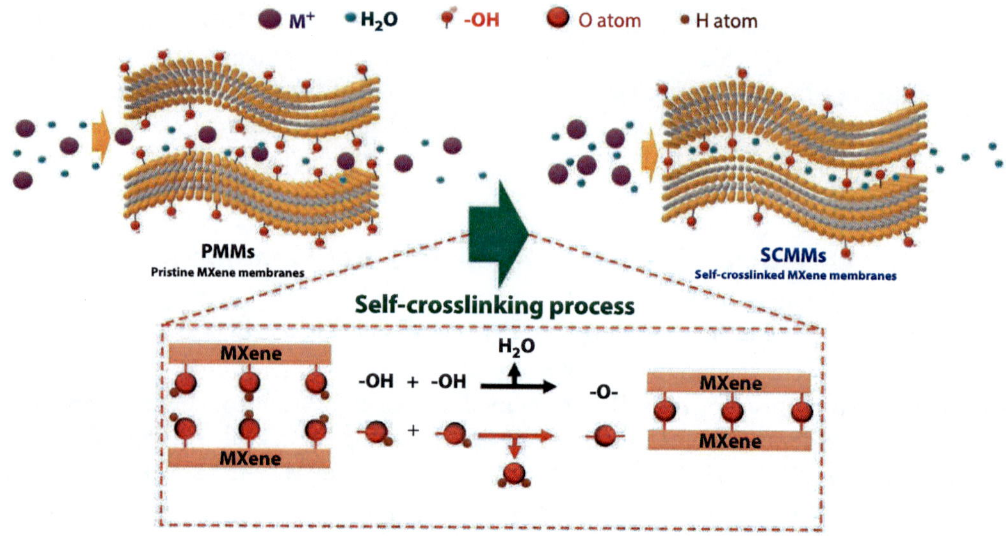

Figure 15.1 MXene membranes prepared by self-crosslinking using hydroxyl to oxygen conversion [24]. Copyright 2019. Reproduced with permission from American Chemical Society.

This is a result of the penetration rate of different ions with single valency, like K^+, Na^+, and Li^+ which has decreased from a value of roughly 10^{-1} to 10^{-3} mol h^{-1} m^{-2}. This straightforward and efficient self-crosslinking method can be used so that many 2D materials with adjustable functional groups can have improved anti-swelling capabilities.

15.2.1.2 Molecular Crosslinking Technique

The molecular crosslinking approach can be utilized to add additional features to MXene membranes to improve separation flexibility in addition to forming interlocked structures that increase structural stability [21]. Liu *et al.* [22] reported the use of isopropanol which was dehydrated by pervaporation using MXene nanosheets of negative charge. A positive charged compound polyethylenimine was intercalated to these sheets. Electrostatic interactions allowed the assembly of highly organized 2D stacked nanochannels. They also demonstrated that the permeate had substantial fluxes of 1069 ± 47 $g.m^{-2}.h^{-1}$ and a large amount of water (>99%). Shen *et al.* [25] showed how adaptable and flexible the molecular crosslinking technique is. They intercalated MXene sheets using CO_2-philic molecules (namely borate and polyethylenimine). These were useful at nanochannels with tunable properties for CO_2 capture. Large d-spacing pristine MXene membranes were unable to distinguish between CO_2, CH_4 and N_2 having d-spacing of 0.33 nm, 0.38 nm, 0.36 nm respectively. The process of CO_2's reversible adsorption and emission in the nanochannels of MXene as a result of crosslinking with CO_2-philic molecules increased selectivity towards CO_2.

15.2.1.3 Ionic Crosslinking Technique

Given the abundance of functional groups that really can react with metal ions to generate coordination bonds on the surfaces of the nanosheets, these methods may be used to enhance the anti-swelling capabilities of MXene membranes [9]. Ding *et al.* [23] have utilized the incorporation of Al^{3+} ions on to MXene membrane for achieving exceptional anti-swelling characteristics. Concentration diffusion allowed Al^{3+} to be intercalated into the pure MXene membrane, and the creation of Aluminium–Oxygen bonds locked neighboring nanosheets. The MXene membrane's swelling was greatly reduced by this interaction (Al-O). In-water tests for long-term stability revealed stability up to 400 hours. Thus Al^{3+} intercalated MXene membrane has exhibited exceptional anti-swelling properties. The rates are two orders lower than they would be for a pristine membrane. Thus, the three crosslinking approaches stated above offer suggestions for creating superior nanosheet membranes.

15.2.2 Construction of Additional Nanochannels

Among the two applications discussed here, MXene has been widely explored for separation. However, for permeation performance, the structure has to be modified. This includes the less fluidic channels, present and the slow diffusion rate. This issue can be tackled by adding more nanochannels wherein high rejection rates can be maintained while improving permeability [26–28]. Xing *et al.* [27] developed a straightforward method of freeze-drying to produce reinforced crumpled MXene lamellar membranes. The strategy was to enhance the number of transport channels formed by the wrinkles. These provided path for the rapid permeation of water and organic solvent (Figure 15.2 (a)) [27]. In addition to wrinkles, nanoparticles (NPs) were also utilized to increase the number of nanochannels or vacancies. Thus, water could travel through shorter path. Pandey *et al.* [28] utilized *in-situ* reduction of silver nitrate and its further intercalation to MXene membrane (Figure 15.2 (b)). They accomplished a water flux of 21%, which is four times more than the pristine membrane. Numerous big compounds, including methyl green (MG), bovine serum albumin, and rhodamine B, can be rejected by these Ag@MXene (RhB). Because Ag nanoparticles are antimicrobial, this technique may also enhance the water quality that goes through the membrane. Etching the NPs in the MXene membranes can reduce transport resistance. Ding *et al.* [26] demonstrated the use of $Fe(OH)_3$ NPs as sacrifice templates to create remarkable nanochanneled MXene membranes (Figure 15.2 (c)). The simple technique of etching with HCl could yield tremendous enhancement in water flux (5 – 10 times) compared to pristine membrane. This was achieved without compromising the rejection rate. This was possible due to presence of additional nanochannels and a shorter route for the water transport. Moreover, the new nanochannels can be created between neighboring MXene nanosheets using functional groups with affinity for particular molecules. Wu *et al.* [29] reported chemical grafting of hydrophobic (C_6H_5, $C_{12}H_{25}$) and hydrophilic (NH_2) groups to produce MXene membranes (Figure 15.2 (d)) [29]. Polar molecules can be synthesized by manipulating the ordering of aggregates in hydrophobic ad hydrophilic channel where in the former, arrangement was disordered and in the latter it was in ordered manner. As a result, polar molecules exhibited an unusually high transmittance value of > 3000 L/m^2 h^1 bar^1; which was thrice larger in comparison to the hydrophobic nanochannels.

15.2.3 MXene Hybrid Membranes

Polymer chains can be chemically bound to the MXene surfaces due to the presence of various functional groups, which is why they are extensively employed as additives to make

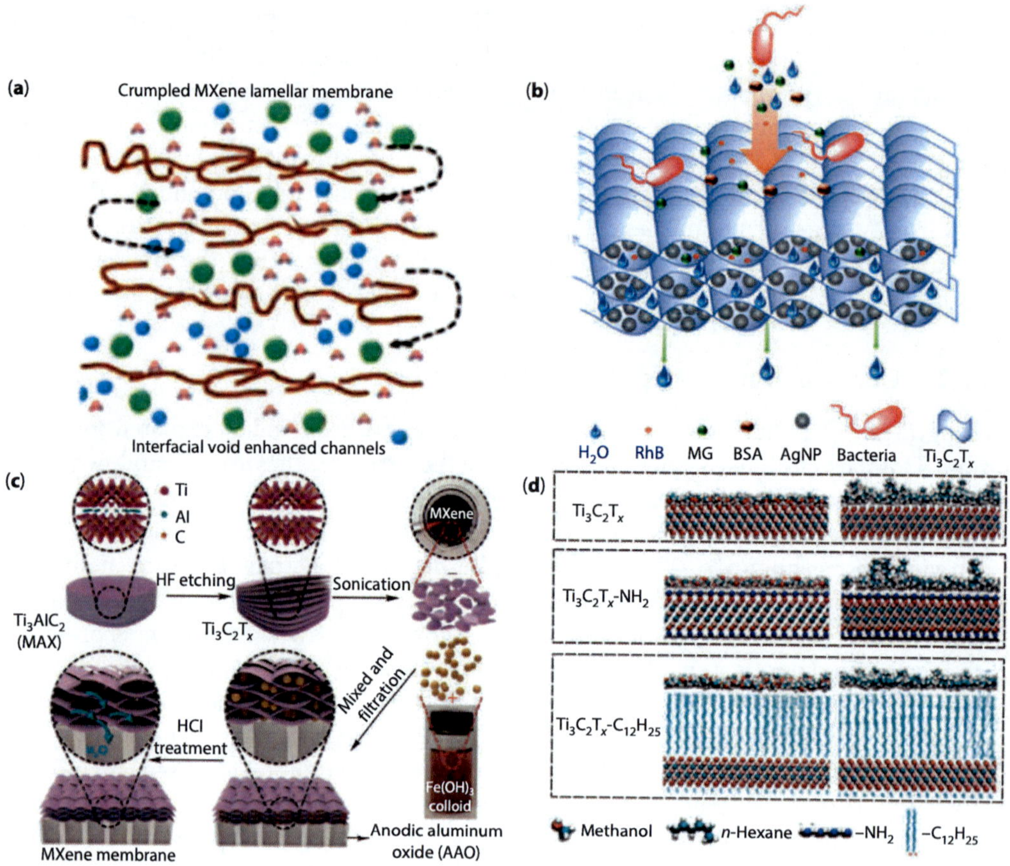

Figure 15.2 (a) Diagram of a creased MXene lamellar membrane [27]. Copyright 2020. Reproduced with permission from American chemical society. (b) Representation of an *in-situ* produced Ag NPs-based ultrahigh-flux and fouling-resistant membrane [28]. Copyright 2018. Reproduced with permission from Royal society of chemistry. (c) Representation of the template-sacrifice approach for creating a porous MXene membrane [26]. Copyright 2017. Reproduced with permission from John Wiley and Sons. (d) Diagram depicting the configuration of molecules inside nanochannels (MD simulations) [29]. Copyright 2019. Reproduced with permission from John Wiley and Sons.

hybrid MXene membranes. This connection prevents the formation of interfacial defects. A finite number of MXene nanosheets can be joined to produce molecular sieving-specific multilayered MXene stacks with interlayer spacing and linear diffusion routes. It has been demonstrated that an increase in the separation performance can be accomplished by adding MXenes to polymer membranes [30–35]. Shamsabadi *et al.* [30] prepared embedding MXene ($Ti_3C_2T_x$) nanosheets in pebax 1657 to develop MXene hybrid membranes for CO_2 capture (Figure 15.3 (a)). The membrane permitted quick and selective CO_2 transport that exceeded Robeson's upper bound. High-speed, selective $Ti_3C_2T_x$ nanochannels are thought to be responsible for this in the membrane. The MXene hybrid membranes fabricated of polyurethane and pebax 2533 shared the same transport properties, demonstrating the usefulness of this strategy [30]. Chitosan (CS) and MXene nanosheets were mixed by Xu *et al.* to create a hybrid membrane for pervaporation dehydration [34]. With the inclusion of

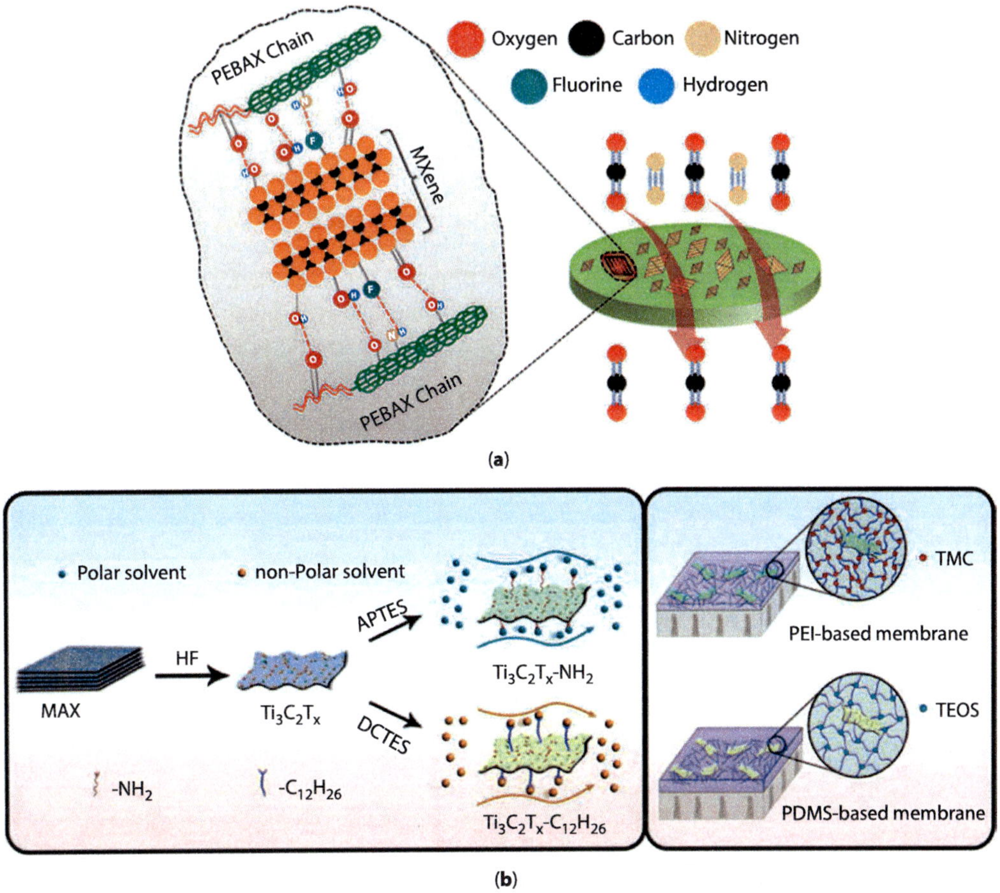

Figure 15.3 (a) Schematic diagram illustrating the hydrogen bonding among pebax chains and MXene ($Ti_3C_2T_x$) nanosheets [30]. Copyright 2020. Reproduced with permission from American chemical society. (b) Synthesis process of $Ti_3C_2T_x$ and $Ti_3C_2T_x$-M and the microstructure of composite membranes [31]. Copyright 2017. Reproduced with permission from Elsevier.

MXene nanosheets, they have observed an enhancement in the membrane's functionality. It was discovered that H_2O diffusion had a better selectivity than ethyl acetate, ethanol and dimethyl carbonate. The MXene/CS hybrid membrane produced separation factors of 1421, 4898, and 906 for the dehydration of ethanol, ethyl acetate, and dimethyl carbonate at 50°C, respectively. The flux in this case was 1.4–1.5 kg.m^{-2}h^{-1}. Additionally, MXene nanosheets dramatically increased the ability of hybrid membranes to separate gases. In addition to functional groups, which improve the performance of hybrids based on polymers, MXene nanosheets can also be employed as a bridge to provide further functionalities to polymer membranes. MXene nanosheets were reportedly added, according to Hao *et al.* [31] with multiple functions to hybrid membranes for enabling specialized solvent transport properties (Figure 15.3 (b)). The solvent transport behavior of hybrid membranes incorporating MXene nanosheets was dependent on the functional groups attached to MXene. Polar molecules can travel more easily through membranes NH_2 and COOR groups, while these membranes inhibit non-polar solvent molecules because of their affinity for polar molecules, such as isopropanol.

15.2.4 MXene as 2D Scaffolds

Inorganic membranes can be made using robust 2D MXene nanosheet assemblies as supporting structures. A lot of work has been done to alter the microstructure using various functional groups of MXene nanosheets. For instance, inorganic mesoporous membranes are frequently prepared using the sol-gel process [36, 37]. However, due to the inherent porous nature of the support, substantial penetration problems persist which leads to the flaws in production. In their work, Xia *et al.* have demonstrated the strategy to lessen gel penetration by the addition of 2D nanosheets to a TiO_2 sol, wherein a toughened network-structured gel was created [38, 39]. Xu *et al.* successfully assisted the sol-coating process in this regard by utilizing the lamellar structures of MXenes [40]. An MXene dispersion was quickly blended with a TiO_2 sol to produce a TiO_2@MXene sol. The 2D MXene nanosheets with the embedded TiO_2 NPs were put together to construct a 3D support framework as depicted in Figure 15.4 [40]. Defects in the TiO_2 sol-based membrane surface were reduced due to the "floor tiles" functionality of MXene nanosheets. Outstanding separation performance was demonstrated by the developed 2D MXene framework supporting a mesoporous membrane. It had 22000 Da as the cut-off molecular weight and a water flux of at least 90 $L{\cdot}m^{-2} \cdot h^{-1} \cdot bar^{-1}$. The efficacy of membrane separation is reportedly improved by the recently developed 2D MXene scaffolds. Sun *et al.* [41] examined the interaction of Mxene and TiO_2 mesoporous membrane using MXene nanosheets. The inclusion of MXene nanosheets had two-fold benefits. In one hand, it aided in the prevention of the penetration of sol during membrane formation. In the other hand, it created paths for dextran to travel both longitudinally and laterally between the MXene nanosheets and TiO_2 NPs. To find the optimum mesoporous membranes for separation, a molecular weight cut-off of 14854 Da and a water flux of more than 102 $Lm^{-2}\ h^{-1}\ bar^{-1}$ were utilized.

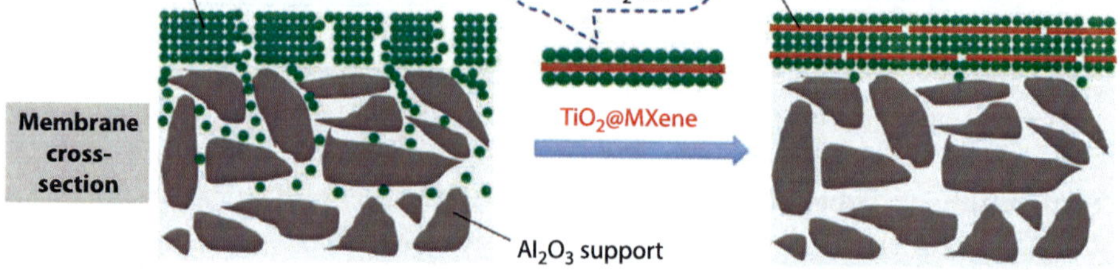

Figure 15.4 Development of TiO_2 and MXene composite membranes to remove probable defects [40]. Copyright 2018. Reproduced with permission from Elsevier.

15.2.5 Mixed Matrix Membrane (MMM)

Composite membranes made of nanomaterials and polymers exhibit synergetic effects of both phases. Nanomaterials are incorporated into the polymeric phase to form these membranes as the dispersed phase (Figure 15.5) [42]. By combining the positive aspects of organic and inorganic substances, this method creates great permeance and selectivity, thereby enhancing separation performance. These membranes were prepared using the following steps. Firstly, the filler materials were dissolved in polymer dope solution. This step is followed by membrane casting. Further to it, a phase inversion was achieved either by following solvent evaporation or by using nonsolvent-induced coagulation. The important points to be considered when creating an MMM are: (i) membrane homogeneity and aggregate state, (ii) chemical compatibility amongst the filler and the matrix of the polymer, particularly at the filler/polymer interface and (iii) the dispersion stability of the filler material in the polymer dope solution's organic solvent system.

Materials with MXene display strong interactions with many solvents because of their hydrophilicity, polarity, and complex surface chemistry. They also provide a wide range of functionalization choices. It is remarkable that $Ti_3C_2T_x$ nanosheets are found to have sufficient stability and dispersion properties in polar organic solvents like N,N-dimethylformamide (DMF) and N-methyl-2-pyrrolidone (NMP), which are essential for the production of MMM [43]. The aqueous solution of PEG was added to 2 w% MXene solution (5.0 mg/ml). Further to it, membrane was prepared by mixed matrix method to form MXene/PEG mixed matrix membrane. At filler/polymer interfaces, $Ti_3C_2T_x$ nanosheets can be surface functionalized to improve their chemical compatibility because they have OH terminations, which can be achieved by silanol chemistry. One of the most popular chemistries for reducing undesirable interfacial morphologies is silanol chemistry. Consequently, successful $Ti_3C_2T_x$-based MMM manufacturing using various polymers was achieved (as a matrix)

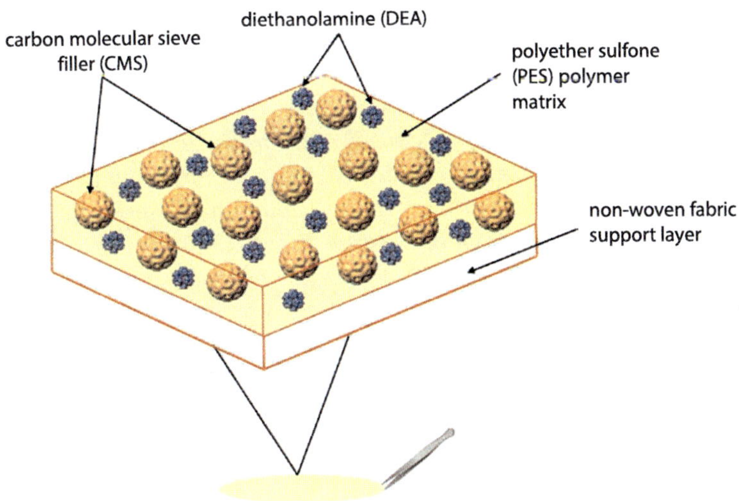

Figure 15.5 Diagrammatic representation amine composite mixed matrix membranes. Adapted from Ref. [42]. Copyright 2020. The Royal Society of Chemistry.

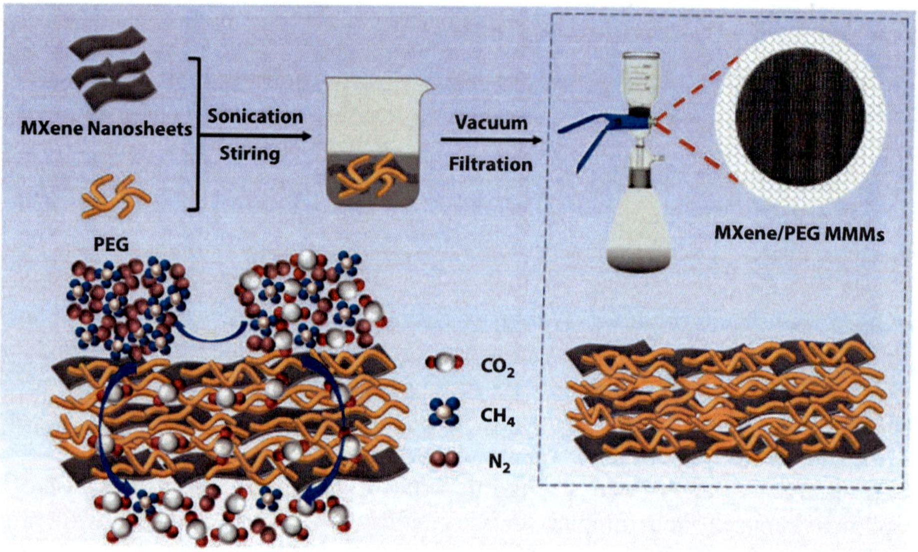

Figure 15.6 The preparation strategies for MXene/PEG mixed matrix membrane via. vacuum filtration [45]. Copyright 2022. Reproduced with permission from Elsevier.

for effective organic solvent and water purification. This is reported by many coworkers [31, 34, 44]. Luo *et al.* [45] have demonstrated the synthesis and usage of mixed matrix membrane of MXene with poly(ethylene glycol). For the preparation of membrane, MXene was dispersed to a solution to which PEG was added. This was stirred magnetically and sonicated under argon atmosphere. This solution was later vacuum filtered using PTFE membrane, dried in room temperature and further transferred to vacuum oven to obtain mixed matrix membrane with MXene and PEG. The synthesis strategies adopted is represented in Figure 15.6 [45].

15.2.6 Thin Film Nanocomposite

TFCs are substantially modified versions of thin film nanocomposite (TFN) membranes, which incorporate the ideas of MMMs and TFCs. The common methods of production and structures of TFC PA FO membrane are represented in (Figure 15.7) [46]. TFNs are interesting for research with a focus on performance since TFC membranes continue to still be the membranes that are most frequently used for desalination and wastewater treatment. Indeed, TFNs have shown a wide range of applications which includes pressure-retarded osmosis (PRO), reverse osmosis (RO), and forward osmosis (FO), as well as nanofiltration (NF). In order to provide an asymmetric structure, TFN membranes is typically comprised of a substrate with pores covered in a thin layer of polyamide (PA)-selective material applied to its (support layer), similar to regular TFCs. The primary methodology for creating these membranes is interfacial polymerization (IP), which has not yet been thoroughly investigated for MXene-filled membranes [47].

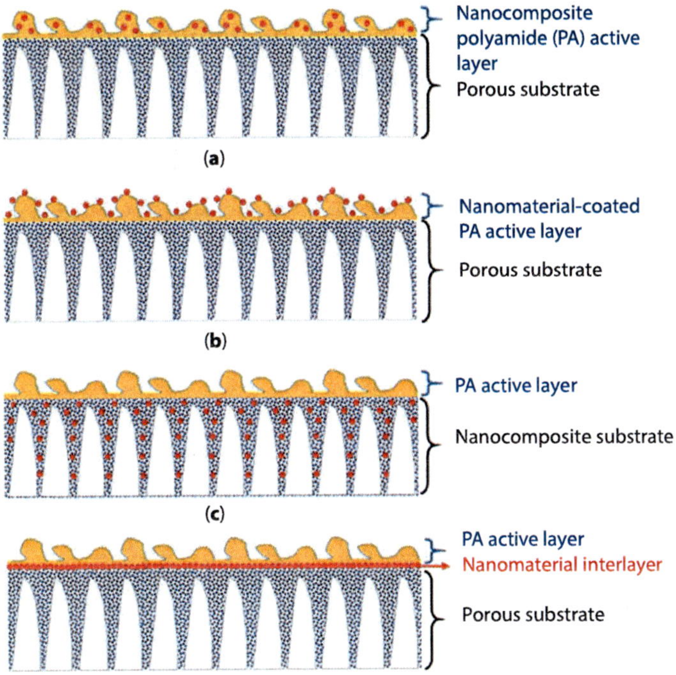

Figure 15.7 Diagram showing the usual topologies of PA TFC membranes with integrated nanomaterials: (a) TFN membrane, (b) TFC membrane with nanomaterial-coated PA layer surface, (c) TFC membrane with nanocomposite substrate and (d) TFC membrane with nanomaterial interlayer [46]. Copyright 2019. Reproduced with permission from Elsevier.

15.2.7 Nanolaminate Membranes

The nanolaminate membranes (NLMs) are produced by the hybridization of H_2TiF_6 and octylamine. This leads to the formation of lamellar hybrid nanosheets. Further to a carbonization step, a sandwich-like carbon-anchored ultrathin TiO_2 nanosheets are obtained. The formation of the sandwich structure is represented in Figure 15.8 [48]. The initial stage in creating NLMs is dispersing MXenes in solvents (often water at the proper pH range), producing a homogenous and stable colloidal dispersion [25, 26, 44, 49]. Important factors affecting how efficiently these membranes function includes the flakes' lateral dimensions

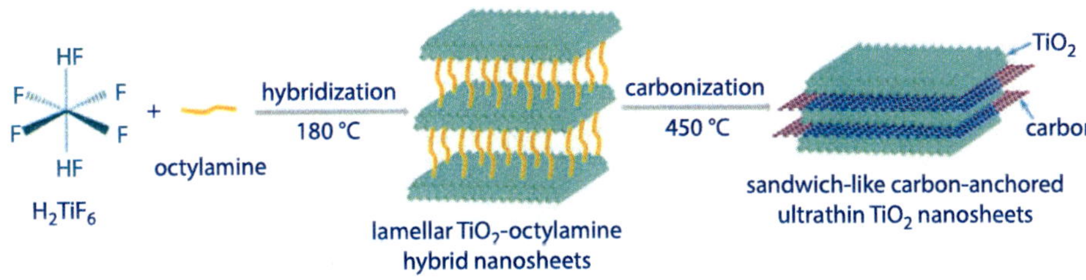

Figure 15.8 A schematic showing formation procedure of nanolaminate carbon-anchored ultrathin TiO_2 [48]. Copyright 2013. Reproduced with permission from Royal society of chemistry.

and the choice of supporting membranes. In addition, various factors including volume, pressure and the concentration of colloidal dispersions will immediately impact the membrane's thickness, compactness, and integrity, among other qualities [50]. It is worth noting that there are alternate ways to make NLMs that are similar to those made using filtration methods. Xu *et al.* reported that on flat-sheet polyacrylonitrile (PAN) substrates, an MXene membrane can be built using the spin-coating process. A four-channel ceramic Al_2O_3 hollow fiber support with an MXene dip-coating was also on display [40]. The ability of MXene-based materials to be processed in solutions, which depends on the rheological characteristics of dispersions, is another crucial component in the development of NLMs. Colloidal dispersions of MXene have been demonstrated to exhibit viscoelastic behaviors at different concentrations, depending on the exfoliation rate. At lower concentration levels of 0.36 mg/mL, the behavior of single-layer $Ti_3C_2T_x$ dispersions is viscoelastic. However, multilayer $Ti_3C_2T_x$ dispersions have higher elastic moduli at higher loadings and display significant viscoelasticity [51]. The above mentioned studies highlight how crucial is the dispersion of $Ti_3C_2T_x$, rheological adaptability for membrane production, even when NLM preparation is not involved. In addition to being a facile technique, NLMs provide the added advantage of being useful for revealing the effects of chemical functionalization and intercalation. Various additives such as PEI, borate ions, silver (Ag), titania (TiO_2), ferric hydroxide (Fe $(OH)_3$), and GO nanosheets have all reportedly improved the efficiency of MXene-based NLMs [25].

15.3 Pervaporation

For industrial-scale applications of purifying organic solvents, Pervaporation (PV) dehydration is a practically and energetically viable separation technique [52, 53]. The creation of high-performance dehydration systems for azeotrope mixtures, chemical compounds with high temperature sensitivity, and organic-organic blends has been carried forward by the investigation of numerous 2D materials for construction of PV membranes [54].

15.3.1 Mechanism for Pervaporation

Pervaporation is a molecular separation technique which can remove any constituents from a feed solution. For this method, a sweep gas or vacuum is applied on one side of a membrane and the liquid feed will be touching the other. This is influenced by the two sides of a membrane's different chemical potential, as was established and demonstrated in Figure 15.9 [55]. The word "permeate" describes the molecules that make up the film as they pass through it, while "retentate" refers to the remaining solution molecules in the feed. This technique is particularly useful while separating azeotropic mixtures or thermally sensitive components that are extremely difficult using conventional separation processes. PV offers the benefits of higher separation efficiency and less energy consumption.

Typically, two factors are used to describe pervaporation membranes: fluxes and separation factor. In accordance with the subsequent equation (15.2), the flow (J) through a membrane is used to calculate the productivity of the membrane.

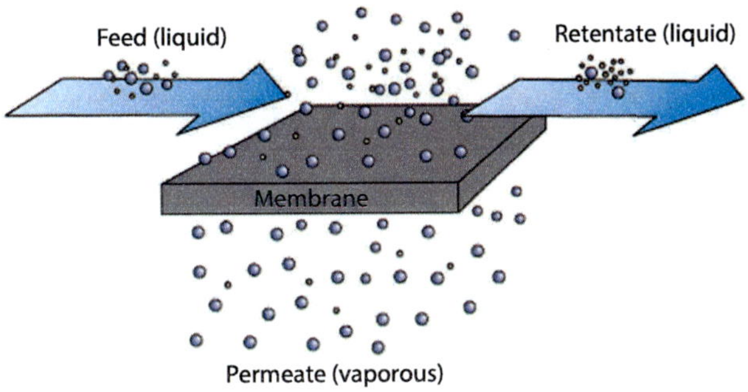

Figure 15.9 Schematic diagram of PV process. Adapted from Ref. [55]. Copyright 2008, Hindawi.

$$J = \frac{Q}{A \times T} \tag{15.2}$$

where Q is the permeate volume (L), t is the interval time, and A is the effective area (m^2) of the membrane (h). Calculating the separation factor (β_{ij}) from Equation (15.3).

$$\beta_{ij} = \frac{\frac{C_{ip}}{C_{jp}}}{\frac{C_{if}}{C_{jf}}} \tag{15.3}$$

where C_{if} and C_{jf} correspond to the component i's and component j's molar concentrations on the feed side surface of the membrane, and C_{ip} and C_{jp} are the permeate side components, respectively. Numerous factors, such as the inherent properties of the materials, temperature, pressure, permeate, and feed concentration, commonly affect the membrane permeability and separation properties. Consequently, comparing membrane flux and parameter separation is challenging in a variety of operating conditions. The thickness of the thick selective layer and the commonalities among GS and PV membranes are directly connected to those of PV membranes. Certain parameters, such as permeability (P_i) (equation (15.4)), selectivity (α_{ij}), and permeances (P_i/l) may aid in fair comparison [56].

$$P_i = \frac{J_{il}}{\gamma_i \gamma_{ip} \times i - y_{ip}} \tag{15.4}$$

where Pi represents component i's permeation, which is provided through diffusivity and solubility. The relative mole proportions of constituent i in the upstream and downstream, respectively, are represented by the letters xi and yi; l stands for membrane's thickness; Ji, γi, and pi* represents, respectively, the component i flow, activity coefficient, and saturation vapor; p is the total pressure at downstream. The Antoine equation and the Wilson equation, respectively, controlled pi*and γi, respectively.

It is possible to determine the selectivity a_{ij} of PV, based on the ratio of permeability between components i and j, equation (15.5).

$$a_{ij} = P_i/P_j \tag{15.5}$$

15.3.2 Theory of Pervaporation

The solution desorption model, defines how a component is transported through a membrane during pervaporation. The processes are as follows: (i) Component diffusion onto the boundary with the liquid and onto the surface of the membrane (ii) Membrane sorption or diffusion, (iii) transfer via the membrane and (iv) diffusion into the majority of the permeance through the vapor phase boundary layer. Because of the low vacuum that is retained on the permeant end, the solute concentration there is regarded as zero. At the vapour phase, the boundary layer offers very little resistance while modeling the transport model for pervaporation. However, the transport resistance on the vapour side will rise and become significant at higher permeant pressures.

The membrane's liquid feed side offers conveyance through the boundary layer and can be modeled by (15.6)

$$J_a^{bl} = K_{lp}(x_f - x_{il}) \tag{15.6}$$

where $x_{i,l}$ stands for the organic concentrations at the liquid-membrane interface and x_f is the same for feed side. On the liquid side, the boundary layer mass transfer coefficient is k_l.

Transport through the membrane is given by equation (15.7),

$$J_o^m = K_{ip}(x_{m,l} - x_{m,g}) \tag{15.7}$$

where the liquid phase's membrane's organic concentration is represented by $x_{m,l}$, the membrane's thickness, t, and its mass transfer coefficient, k_i and $x_{m,g}$ is that in the vapor phase, respectively.

In this instance, the entire flux is described by equation (15.8).

$$J_i = K_{OV_p}(x_f - x_p) \tag{15.8}$$

where the x_f and x_p are the organic concentrations corresponding to feed and permeance respectively [57, 58].

Equation (15.9) and (15.10) provides information on i's partial vapour pressure in both its gaseous and liquid forms. As can be seen from (15.9) and (15.10), partial pressure depends on the concentration of the respective side, $C_{l,i}$ for the liquid side and $C_{g,i}$ for the gaseous side.

$$P_{l,i} = \frac{C_{l,i}\gamma_{iP_i^0}}{\rho_l} \tag{15.9}$$

$$P_{g,i} = \frac{C_{g,i} P_T}{\rho_g} \tag{15.10}$$

In addition to that, it also depends on the liquid side's activity coefficient (γ_i) the liquid side's total pressure (P_T), and the liquid side's saturation vapour pressure (P^0_i). it also depends on ρl and ρg which are the molar density of the liquid and gas [59].

Diffusion through, desorption from, and sorption to the membrane are all included in the membrane mass transfer coefficient (k*i*) (equation (15.11)). It is analogous to the relationship seen between thickness of the membrane's active layer (l) and its permeability coefficient (P*i*). This is demonstrated in equation (15.12).

$$k_i = \frac{P_i}{l}, \tag{15.11}$$

$$P_i = S_i D_i, \tag{15.12}$$

where S_i is the solubility coefficient of membrane permeability and D_i the diffusion coefficient of membrane permeability. They are multiplied by each other to get the permeability coefficient (Pi).

Using a composite membrane, 1,1,1-trichloroethane (TCA) and trichloroethylene (TCE) were separated from an aqueous solution in diluted form according to the literature [60]. The permeability using the aforesaid method was reported as 5.7×10^{-9} and 1.1×10^{-8} m^2/s for TCA and TCE, respectively.

15.3.2.1 *Solution Diffusion Theory*

When a dense nonporous membrane is used to carry a gas, vapor, or liquid, it is characterized by

$$\text{Permeability [P]} = \text{solubility [S]} \times \text{diffusivity [D]}$$

Henry's law describes how a gas can be dissolved. A linear sorption isotherm and a pressure-dependent concentration within the polymer, result from the concentration-independent nature of solubility in ideal systems. Since liquids are not perfect, Henry's law does not apply to them. Additionally, there is a significantly stronger affinity between liquid and polymer; as a result, cross-linking occasionally is necessary to prevent polymer disintegration. Diffusivity is impacted by permeability because it renders more flexibility to polymer chains. Hence a high solubility denotes a high permeability. The nonlinear behavior of the sorption isotherm is described using Flory Huggins thermodynamics and free volume models. The coefficients of diffusion rise with temperature, whereas sorption is typically exothermic. Regarding whether or not the penetration of organic molecules across a membrane is influenced by temperature, many different conclusions have been drawn from the literature. In some circumstances, permeability drops as the temperature rises [61, 62]. Some demonstrate a temperature-dependent increase in permeability [58].

Additionally, there are also instances where temperature within the research range has no impact on permeability [57, 63].

Diffusivity is influenced by the geometry of the penetrant (diffusion coefficient decreases with increasing molecule size) and concentration. The Stokes-Einstein equation (equation (15.13)), describes how the radius (r) of the diffusing component relates to frictional (*f*) resistance, and aids in establishing the dependence on size.

$$f = 6\pi\eta r. \tag{15.13}$$

The relationship between frictional resistance and diffusion (*D*) coefficient is inverse and is stated in Equation (15.14). Here k is the Boltzmann constant and T is the temperature in K.

$$D = \frac{kT}{f} \tag{15.14}$$

A large diffusion coefficient can be produced by large molecules with the capacity to swell polymers. Diffusivity can be determined using the permeation method or the time lag methodology. Equation (15.15) gives the amount of penetrant (Q_l) that is going through the membrane at time t.

$$Q_l = \frac{DC_i}{l}\left[\mathrm{t} - \frac{l^2}{2D}\right] \tag{15.15}$$

Pressure (*P*) can also be measured as an equivalent to Q_l. While plotting *P* versus t, the steady-state portion of the permeation experiment will show permeability as the slope and time lag as the intercept. Since $P = S \times D$, hence *S* can be found once *D* and *P* are known. Diffusivity can also be measured by retaining exerting pressure on the chamber while containing the polymer in a confined volume. As a result of sorption, the pressure gradually drops until equilibrium is attained.

15.3.3 Characterization of Pervaporation Membranes

Anisotropic (asymmetric) dense membranes and isotropic (symmetric) homogeneous membranes are the two membrane types that are most frequently utilized in PV. Due to their simplicity in preparation, symmetric membranes are generally produced in laboratories. A polymer solution is typically placed onto a suitable substrate and is permitted to evaporate in an inert environment. As a result, a dense symmetric membrane is created. However, the membranes are fabricated at an industrial level using an asymmetrical or composite design [64]. A dense top layer of the same material is attached to a porous structure to form symmetric membranes. They combine the increased mechanical resistance offered by the porous support with the higher permeation flux caused by the incredibly thin separating membrane. Typically, a phase inversion method is utilized to make them, in which a nonsolvent bath or atmosphere is used to expose the cast polymer solution.

Composite membranes, which combine a porous foundation with a thin, dense top layer, are made from alternative materials. The capacity of composite membranes to employ a particular thin membrane coating in accordance with the division of interests is their primary advantage (great affinity towards the target molecule). The supported membrane is strengthened mechanically by a highly porous structure. The following section provides a full explanation of the most common characterization tests, which are frequently used to characterize dense membranes.

15.3.3.1 Swelling Tests

An enormous expansion of the polymer membrane is referred to as a swelling, is responsible for the delayed solvent diffusion into the polymer chains [65]. A few factors that have varied degrees of influence on swelling include the intramolecular forces that are present, the crystallinity, the solvent affinity, and the polymer's cross-linking. Additionally, the existence of a porous support affects the swelling of composite membranes [66]. To prevent a significant fall in selectivity in PV, it's necessary that a membrane does not inflate excessively. Most of the time, flux and selectivity are trade-offs, and it is crucial to find a balance between the two to decrease the negative effects of swelling. An extremely low flux also exhibits a very low level of swelling. Therefore, ideal swelling values should typically fall between 5 and 25 wt% of sorption [67]. With PV, the free volume of the membrane can change over time and display various degrees of swelling depending on the feed concentration [66]. Numerous investigations established the impact of membrane swelling on membrane structure and function [68]. Because of this, precise measurements of swelling of the membrane, sorption and solvent absorption are required for the production of PV membranes that are acceptable [69].

15.3.3.2 Contact Angle

A contact angle analyzer can assess the hydrophilicity and hydrophobicity of a membrane. Hydrophilicity and hydrophobicity are specific surface-level characteristics which can affect their overall performance. This technique involves measuring the angle that a liquid creates with the solid surface of the membrane. According to conventional wisdom, surfaces are either hydrophilic or hydrophobic depending on the angle at which water contacts them. Surfaces that have a contact angle of greater than 90 are said to be hydrophobic. How wettable a membrane surface is determined by its contact angle, which mostly depends upon the polymer chemistry and properties of the membrane surface. The type of membrane employed in pervaporation may be hydrophilic or hydrophobic depending on the objective. For operations where it is necessary to promote the entry of water or other polar molecules, hydrophilic membranes are used. They are mostly used for breaking aqueous or organic azeotropes and solvent dehydration [70, 71]. In contrast, aromatic chemicals are recovered from water using hydrophobic membranes in order to remove organic solvents from it [72, 73]. The water contact angle was used to gauge the hydrophilic nature of the membrane surface. As shown in Figure 15.10, membranes made of MXene ($Ti_3C_2T_x$), pure cellulose acetate (CA), cross-linked CA (CCA), 10%MXene/CA, and 10%MXene@CA all have diverse water contact angles [59]. The angle of water contact decreased when MXene

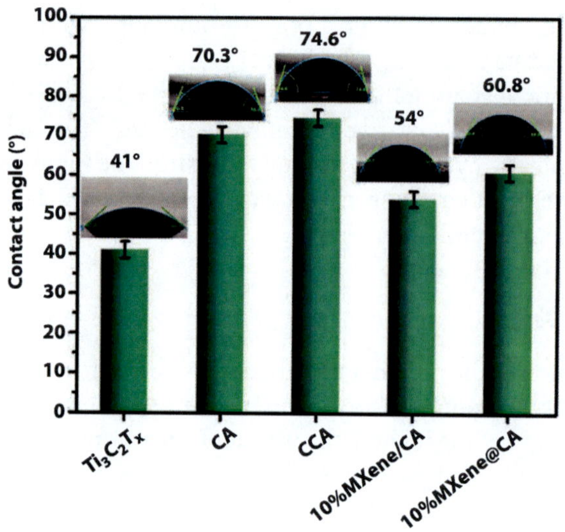

Figure 15.10 Contact angle of water measured for pure cellulose acetate (CA), cross-linked CA (CCA), 10%MXene/CA, and 10%MXene@CA membranes. Adapted from Ref. [74]. Copyright 2020, Elsevier.

was added, going from 70.3° for pure CA to 54.0° for 10%MXene/CA. This confirmed the higher hydrophilicity of the membrane surface and is due to the exceptional hydrophilicity of MXene sheets [74].

15.3.3.3 *Surface Analyses*

To visually analyze membrane structural features, microscopic methods such as atomic force microscopy (AFM), scanning electron microscopy (SEM), and transmission electron microscopy (TEM) can be used. Understanding the mechanism of membrane separation and observing the finer characteristics of the top layer require a thorough understanding of membrane morphology. One often used method for characterizing surfaces is AFM. The sub nanometer to a few micrometers range encompasses the resolution of AFM. It is mostly employed in the membrane field for 3D imaging of materials, which could provide insights on pore size distribution and membrane roughness. Moreover, AFM enables the measurement of adhesion forces among surfaces and various materials to assess the probability of contamination [75]. In SEM examination, secondary electron images are frequently utilized to evaluate membrane morphology. The backscattered electron image mode and elemental studies can also be used to assess the quantity of usable components or elements. SEM resolution is often just 0.6–3 nm [76]. The sample preparation for TEM might involve a difficult process and has a resolution of up to 1 nm. Additionally, due to the requirement that the specimen be extremely thin, this method is mostly limited to membranes without support and with mixed matrix (hundreds of nanometers). Figure 15.11 illustrates SEM, TEM, HR-TEM, AFM, and the height profiles of a representative sheet with a thick lateral dimension of above 3 mm and a thickness of 4.5 nm [77].

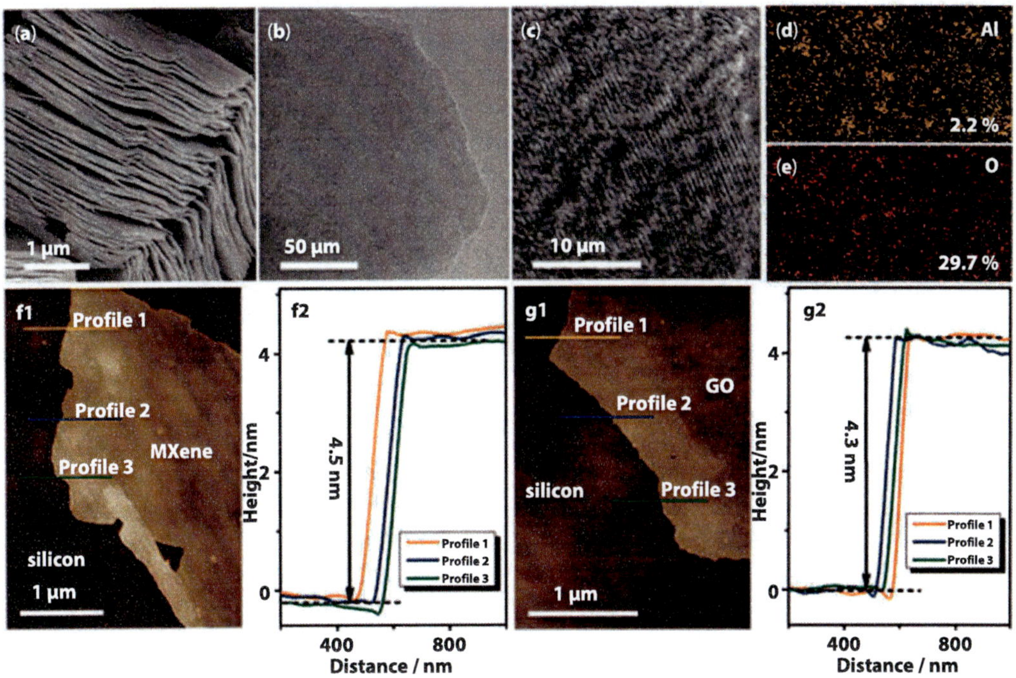

Figure 15.11 (a) SEM image, (b) TEM image, and (c) high-resolution TEM image of $Ti_3C_2T_x$. Elemental maps of (d) Al and (e) O for $Ti_3C_2T_x$. AFM images and the corresponding height profiles of (f) MXene and (g) GO sheets [77]. Copyright 2018. Reproduced with permission from John Wiley & Sons.

15.3.3.4 Positron Annihilation Lifetime Spectroscopy

The solution-diffusion model controls the mechanics of mass transfer over impermeable barriers. According to this theory, the target compounds' permeability through PV membranes determines their solubility and diffusivity. Additionally, the components of free volume found in the gaps between polymer chains continuously contribute to molecule permeation in a PV process [78]. Positron annihilation lifetime spectroscopy (PALS) has attracted a lot of interest recently as a useful strategy for characterizing polymeric materials and determining their free volume size. The antimatter of electrons is represented by positrons (same mass and opposite charge). Positron-electron interactions result in annihilation, which releases energy in the form of 511 KeV g-rays. Positronium is an atom that resembles hydrogen formed when certain positrons interact with electrons. The positronium atoms are entrapped in polymeric materials' imperfections and vacant space preferentially. Measurement of the lifetime of positronium atoms can be used to assess the sample's shape and pore size, which is the amount of time that passes after positrons from a source of the radioisotope ^{22}Na are released and annihilate. Throughout use, the material's pores grow larger. The free volumes of polymeric materials within 2–10 Å can be measured using the PALS method [79–81].

15.3.4 Parameters in Membrane Performance

The two fundamental variables which can be utilized to evaluate the efficiency of a membrane's pervaporation are the permeant flow and the separation factor. The permeant mass is weighed, multiplied by the interval time and area of the membrane to determine the flow. The products of D_iS_i for each portion, where Di represents the diffusion coefficient within the membrane material and Si represents the sorption coefficient (equation (15.16)), determine the fluxes in diluted solutions.

$$S_i = \frac{W_{i,Memb}}{W_{i,feed}} \tag{15.16}$$

Where, $W_{i,Memb}$ and $W_{i,feed}$ represents the component i's weight percentages in the membrane at a soluble equilibrium amongst the membrane phase and the liquid feed, respectively.

The quantity sorbed determines a chemical's solubility. The diffusivity (Di), a kinetic metric which measures the speed of membrane through-permeation, is zero when the membrane is in equilibrium, is replaced by a thermodynamic variable. Di is depending on the membrane's shape and texture (glassy, rubbery, swollen, etc.). A membrane may accommodate smaller molecules more readily than larger ones.

The separation factor controls how well a membrane system separates in its whole. The separation factor's value might range from one to infinity. A high separation factor number indicates increased selectivity. For a specific compound *i*, the separation factor α is described by equation (15.17).

$$\alpha_i = \frac{w_i^{permeate} / (1 - w_i^{permeate})}{w_i^{feed} / (1 - w_i^{feed})} \tag{15.17}$$

Where the superscript shows if it is of permeate or feed.

The following is a representation of the basic pervaporation transport equation (15.18) after the solution-diffusion mechanism.

$$J_i = \left[\frac{P_i}{l}\right](\chi_i \gamma_i P_i^{sat} - y_i P^P) \tag{15.18}$$

Here P_i stands for the membrane permeability, the permeant pressure is P^P, the permeant mole fraction is y_i, and the membrane thickness is *l*. Permeance is given by $[P_i/l]$.

The ideal membrane selectivity ($\beta_{i/j}$) is related to the permeability coefficients or permeances and is represented in equation (15.19).

$$\beta_{i/j} = \frac{P_i}{P_j} \tag{15.19}$$

Pervaporation Separation Index (PSI), which is presented below (equation 15.20), can be used to gauge the membrane's overall performance.

$$PSI = J\ (\alpha - 1) \tag{15.20}$$

where J is permeation flux (kg/m^2 h); when α=1, there is no separation; a PSI of zero indicates no flux or separation.

15.3.4.1 *Effects of Membrane Thickness*

The solution-diffusion model and Fick's equation both suggest that a penetrant's permeability through a membrane must be independent of its thickness despite the fact that the flow is inversely linked to the thickness of the membrane [82]. Hasanoglu *et al.* [83] used polydimethylsiloxane (PDMS) to study the impact of membrane thickness on the hydrolysis reaction of ethyl acetate, water, ethanol (EOH), and acetic acid (AsAc) (PDMS). Two different membrane thicknesses of 250 μm and 300 μm were used. Fluxes through 250 μm membranes were determined to be greater than those across membranes of 300 μm thickness. Hyder *et al.* [84] has conducted their study on dehydration properties of a compound made of ethanol and water using composite membrane of poly(vinyl alcohol) and poly(sulfone) (PVA-PSf). The thickness of PVA layer alone was varied systematically in this cross-linked membrane. The effect of this thickness on the dehydration properties was reported. In comparison to the variation in flow (193%), with a reduction in thickness from 52 to 4 m (92% drop), researchers concluded that the there was a minimal change in selectivity (only 17% decrease). Raisi and Aroujalian [85] carried out investigation on how performance is impacted by membrane thickness behavior in polyoctylmethylsiloxane (POMS) and PDMS pervaporation membrane systems utilizing pomegranate aroma compounds in binary and ternary models. As the correlation involving the penetration rate and thickness is inverse, the findings were used to support the idea that permeation resistance increases with membrane thickness.

15.3.4.2 *Effect of Temperature*

The rate of transportation is significantly impacted by temperature. An Arrhenius type function can be used to represent the temperature effects on a system. Since $P = S{\times}D$, both S and D increase the impact of temperature. D is more dominant in small, noninteractive gases because S is relatively insensitive to temperature changes. The problem is more difficult for larger molecules since solubility and diffusion, two effects that both depend on concentration, are incompatible.

Dependence of S on temperature can be described as equation (15.21),

$$S = S_o \exp\left(-\frac{\Delta H_s}{RT}\right) \tag{15.21}$$

where S_0 is a constant independent of temperature and ΔH_s is the heat of solution. Likewise, equation describes how temperature affects diffusion (15.22),

$$D = D_o \exp\left(-\frac{E_d}{RT}\right) \tag{15.22}$$

where D_o is a constant independent of temperature and E_d is referring to the pre-exponential factor or the activation energy for diffusion.

Using $P=S\times D$ and by applying the previous expression to S and D, we obtain equation (15.23)

$$P = D_o S_o \exp-\left(\frac{\Delta H_s + E_d}{RT}\right) = P_o \exp-\left(\frac{E_P}{RT}\right) \tag{15.23}$$

Given that the distinction among feed and permeance's chemical potentials throughout the pervaporation process is what drives transport, the equation for how permeant pressure affects water flux was developed [86].

Burshe *et al.* [87] reported using membranes made of polyimide, cellulose triacetate, carboxylated polyvinyl chloride, Nafion (NA), and polyethersulfone (PES). They looked into how temperature affected the flow and selectivity towards water in a mixture of glycerine and water. The variation of the temperature was found to be from 30 to 70 °C. An increase in flux with temperature was reported. Although the amount sorbed dropped as the temperature is raised, this was countered by a rise in the solutes diffusion coefficient. NA efficiently transports water due to its low activation energy for diffusion. However, selectivity falls as temperature rises. This happens as a result of the relative motion of polymer chains due to temperature effects which results in the formation of free volumes, which allows penetrating molecules to scatter. Due to increased thermal agitation, the diffusive free volume increases with temperature. This allows for more solute diffusion through the membrane. At all temperatures, there was very little variation in selectivity at water contents > 60%. According to Domingues *et al.*, [88] for slight variations in water content, the flux is nearly linear. They claimed that it rises with temperature and that the 96% water selectivity of the membrane didn't change. Ma *et al.* [89] demonstrated how the CS-TEOS membrane's pervaporation characteristics changed at operation temperatures between 60 and 80°C. Lactic acid and ethanol were present during the reaction, while Amberlyst 15 served as the catalyst. Although the total flux considerably increased, the separation factor decreased as temperature rose. The operational temperature rises as a result, the components' vapour pressure that penetrate the membranes' upstream side increases. Differential increase of the vapour pressure of this side in comparison with the downstream sides provides the transport driving force. Additionally, higher molecule diffusivity was a result of rising temperature.

15.3.4.3 *Effect of Feed Concentration*

The feed concentration, D, depends on the penetrant's volume percent, ϕ. It is observed that D increases exponentially with an exponential constant, γ. This aspect is connected to the plasticizing constant, reflecting the penetrant's plasticizing effect on segmental motion. D is an exponential function as shown in equation (15.24).

$$D = D_o \exp(\gamma \phi) \tag{15.24}$$

Here, D_o is the concentration of diffusion at zero concentration. Depending on the molecule's size, D_o changes inversely. Because of their exponential nature, ϕ and γ are the two most important components. D= D_o because γ=0 applies to gases that seldom interact with polymers.

The concentration-dependent nature of the diffusion coefficient is explained by the free volume hypothesis. Glass and rubber are the two states of polymer, respectively. Below the glass transition temperature, a glassy state persists with limited chain segment mobility. The number of parameters varies above T_g, mobility rises, frozen micro voids are eliminated, and one of those factors is the specific volume, shown as equation (15.25).

$$V_f = V_T - V_o \tag{15.25}$$

At 0 K, molecules occupy a volume called V_0 (estimated by group contribution method).

15.3.5 Reported Works on Pervaporation Using MXene-Based Membranes

The pervaporation characteristics of membranes based on MXene have been investigated by a number of authors. In this section, we summarize a few of the reported works on this area. Wu *et al.* reported a composite NF membrane for solvent resistance made of polyacrylonitrile (PAN)/-PEI-$Ti_3C_2T_x$ [90]. Cao *et al.* developed g-C_3N_4 nanosheet membrane for ethanol/water separation. Approximately 2469 g $m^{-2}h^{-1}$ of good permeate flow and a separation factor of 1663 were displayed by as-prepared membranes [91]. Later, Cui *et al.* showed how to create 2D membrane composites based on MXene that are compatible with PV by combining $Ti_3C_2T_x$ with CS. They used three different kinds of organic solvent, including ethyl alcohol, ethyl acetate, and dimethyl carbonate, to assess the membrane's ability to dehydrate. With a total flow of 1.4 to 1.5 $Kgm^{-2}h^{-1}$, as-prepared membranes produced remarkable results for the dehydration of ethyl alcohol, ethyl acetate, and dimethyl carbonate, with separation factors of 1421, 4898, and 906 correspondingly [34]. These analyses suggest that the permeability of water molecules is considerably enhanced by MXene laminates with controlled nanochannels. Thus, by adding MXene sheets, the flow and separation factors of mixed membranes can be simultaneously enhanced. 2D Ti_2CT_x MXene membranes have been created by Liu *et al.* [92] for pervaporation applications (Figure 15.12 (a, b)). Crosslinking was accomplished using electrostatic interactions involving the positively and negatively charged HPEI molecule and the MXene nanosheets. Trimesoyl chloride interfacial polymerization was then utilized to further strengthen the crosslinks Figure 15.12 (a). With enhanced water flux of 2237+ 292.6 g $m^{-2}h^{-1}$, as-prepared Ti_2CT_x-HPEI/TMC laminates have shown superior methyl alcohol dehydration separation capabilities alongside a reduced water content of 12.59 wt% in permeate (Figure 15.12 (c, d)) [92]. Because of the solvent's larger molecular size, it is feasible to dehydrate alcohol, which increases the permeate's water content and decreases water flux approximately to 1434 + 119.9 g $m^{-2}h^{-1}$ Figure 15.12 (d) [92]. The membranes' performance in the separation of water and isopropyl alcohol was excellent with a value of 7560 + 408.5 g m^{-2} h^{-1} at 100 weight percent water content and 4.7 Å diameter. In comparison to often researched $Ti_3C_2T_x$ MXene, these

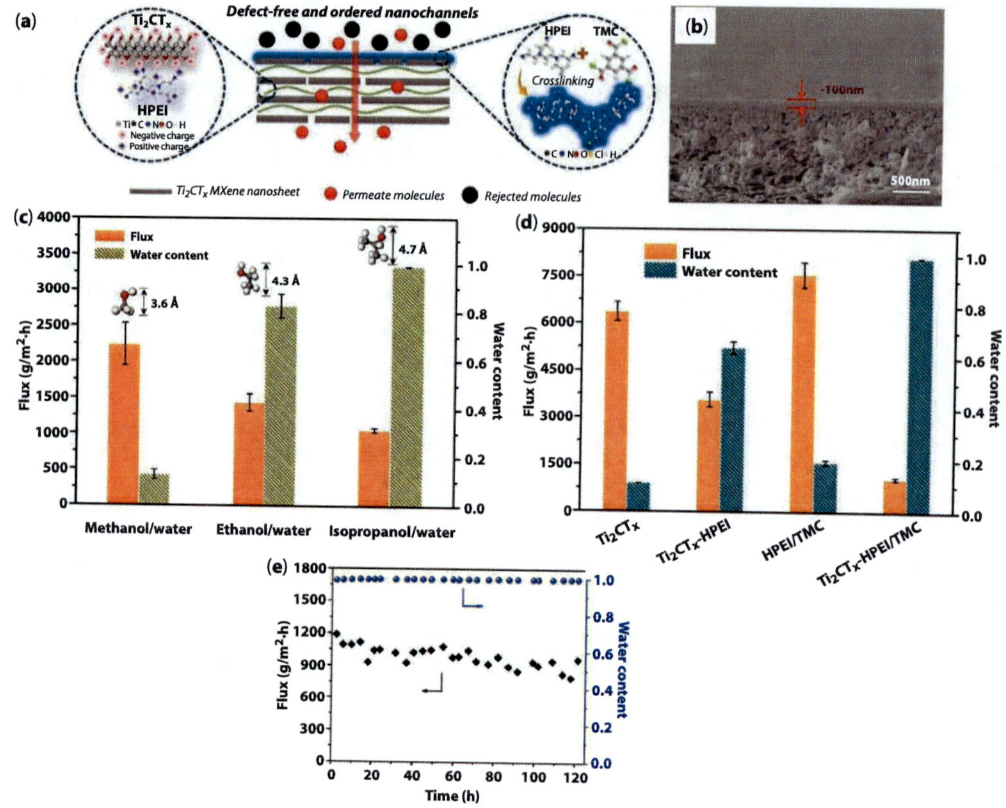

Figure 15.12 (a) Schematic of the interfacial polymerization using TMC and HPEI molecules that produce defect-free, ordered MXene nanochannels. (b) SEM image of Ti_2CT_x-HPEI/TMC membrane. (c) Illustration of the Ti_2CT_x-HPEI/TMC membrane performance for dehydration in various alcohol/water schemes. (d) Performance of the membrane based on Ti_2CT_x-towards isopropanol pervaporation dehydration using a feed of 10% water/isopropyl alcohol. (e) Illustration of the membrane's enduring stability in the dehydration of isopropyl alcohol [92]. Copyright 2019. Reproduced with permission from Royal society of Chemistry.

membranes demonstrated greater selectivity and flow. The authors also assessed the working time of a created membrane for the dehydration of isopropyl alcohol at 50°C and a 72.21 mg m^{-2} MXene loading. It is really remarkable that for 120 hours of nonstop operation, both the flow and the amount of water in the permeate were consistent (Figure 15.12 (e)) [92].

The MXene membrane's functionality during pervaporation desalination was examined in relation to the influence of MXene loading on the polyacrylonitrile (PAN) substrate. As feed loading amounted to 3.662×10^{-6} gm/cm^2 a maximum salt rejection rate up to 99.5% was attained. A very high-water flux of 85.4 Lm^{-2}h^{-1} was produced in this process. This is depicted in Figure 15.13(a) [93]. Depending on the influence of MXene lateral dimension, the manufactured membrane's pervaporation desalination effectiveness was assessed. Smaller MXene nanosheets exhibit a higher water flux than larger MXene nanosheets, although both exhibit the same level of ion rejection. This result is represented in Figure 15.13(b) [93]. The long term stability of MXene based pervaporation membrane was explored and it was noted that after 100 hours of operation, both the water flux and ion

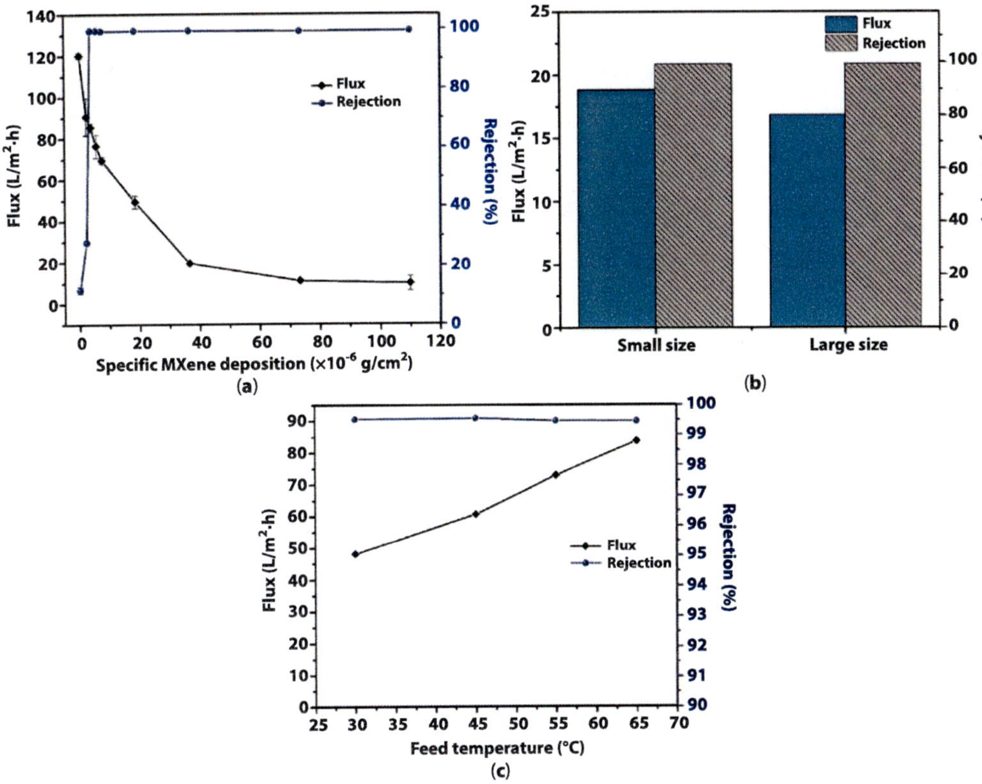

Figure 15.13 (a) Representation of the influence of quantity of MXene on membrane efficiency for pervaporation desalination of 3.5 wt% NaCl solution at 65°C. (b) Illustration of the influence of MXene lateral dimension on MXene membrane capability during pervaporation desalination with MXene loading 36.62 × 10–6 g/cm² for 3.5 wt% NaCl solution at 30°C, (c) Test of MXene membrane's long-term pervaporation desalination with MXene loading of 109.8 × 10^{-6} g/cm² for 3.5 wt% NaCl solution at 30°C [93]. Copyright 2017. Reproduced with permission from Elsevier.

rejection was unchanged suggesting high separation stability of MXene composite membrane (Figure 15.13(c)) [93]. Several common industrial solvents were used to assess the MXene membrane's transport characteristics, including isopropanol, ethanol, methanol, water, acetone, acetonitrile, and dimethylformamide. As shown in Figure 15.14(a), an exceptionally high permeance was obtained for acetonitrile as well as acetone with a value more than 5000 $Lm^{-2}h^{-1}bar^{-1}$ using MXene-230 (it is sample with water permeance of 2302 $Lm^{-2}h^{-1}bar^{-1}$) [77]. Figure 15.14(b) presents the UV-vis spectra of AY79's UV/VIS absorption in 2-propanol both before and after filtering via MXene-230 [77]. When the molecule size is dramatically raised to 2.0 nm, the channels are able to reject 96% of reactive black (RB). The rejection approaches 100% for the probably larger AY79 channels. AY79 are channels with 2.8 nm size. The rejection using various channels are represented Figure 15.14(c) [77]. In particular, the permeance values for organics and water are approximately 5000 and 2300 $Lm^{-2}h^{-1}bar^{-1}$, respectively as shown in Figure 15.14(d) [77]. The channels are capable of rejecting 96% of reactive black when the molecule size is substantially increased to 2.0 nm (RB) [77]. In another study, 3.5% NaCl solution was used to study the pervaporation

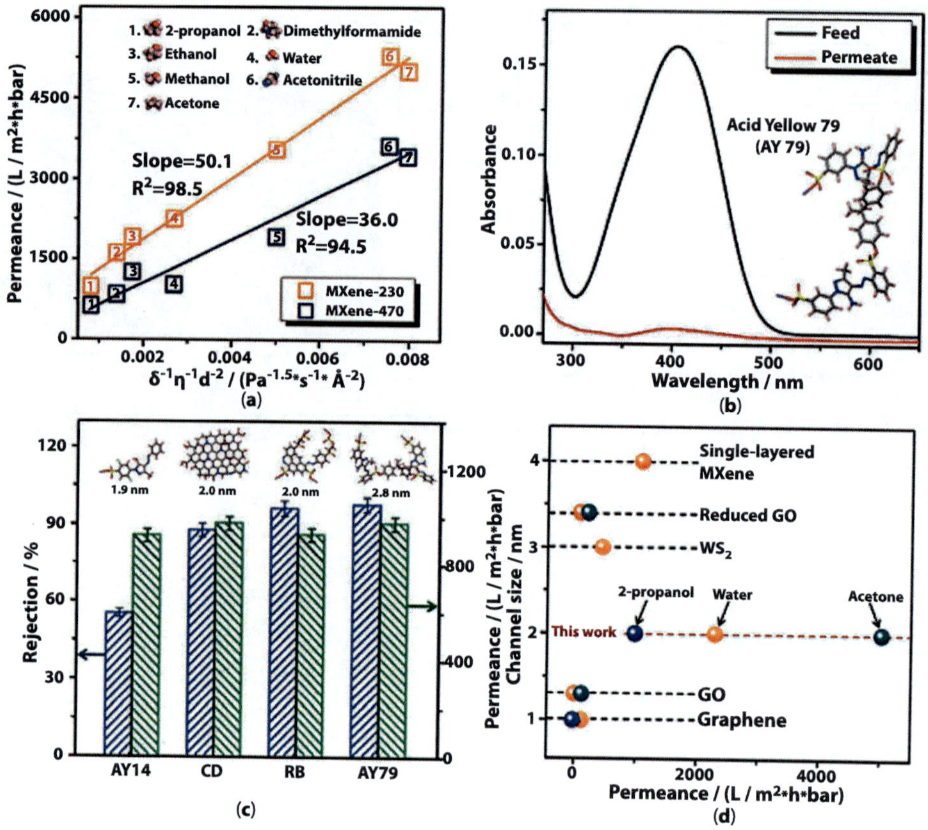

Figure 15.14 (a) Comparison of the solvent permeance versus the combined solvent property for MXene membranes of various thicknesses. (b) Spectra of AY79's UV/VIS absorption in 2-propanol both before and after filtering via MXene-230. (c) Separation performance of MXene-230 for molecules with various sizes. (d) Comparison of the solvent permeance for lamellar membranes with different channel widths (for simplification, the average channel sizes are shown, Single-layered MXene = 2–5 nm; Reduced GO = 2–3.4 nm; WS2 = 2– 4 nm; GO = 1.3 nm; graphene ca. 1 nm) [77]. Copyright 2018. Reproduced with permission from John Wiley and Sons.

behavior of maleic acid cross linked MXene (MXMA) with varied thicknesses at feed temperature 30°C and the outcomes are shown in Figure 15.15 (a) [21]. A value of 22.8 kg m^{-2} h^{-1} was shown for 30 nm, which represents a progressive surge in the water flux with thickness from 370 to 30 nm. When comparing the pervaporation efficiency of MXMA and MX using the identical conditions, MXMA showed a maximum water flux value given by 22.8 kg m^{-2} h^{-1} compared to MX's which was 17.3 kg m^{-2} h^{-1}. This is shown in Figure 15.15(b) [21]. A maximum water permeance value of 79.3 kg m^{-2} h^{-1} has been reached at 60°C by adjusting the feed temperature in the range 30–60°C for the sample MXMA which was having a thickness of 30 nm. This variation is represented in Figure 15.15 (c) [21]. The MXene-based composites for pervaporation separation compiled from literature is represented in Table 15.1.

As per investigation of Chang *et al.* [94], on the ethanol pervaporation in the pilot plant, a significant amount of water is still present in the permeance even though the feed has

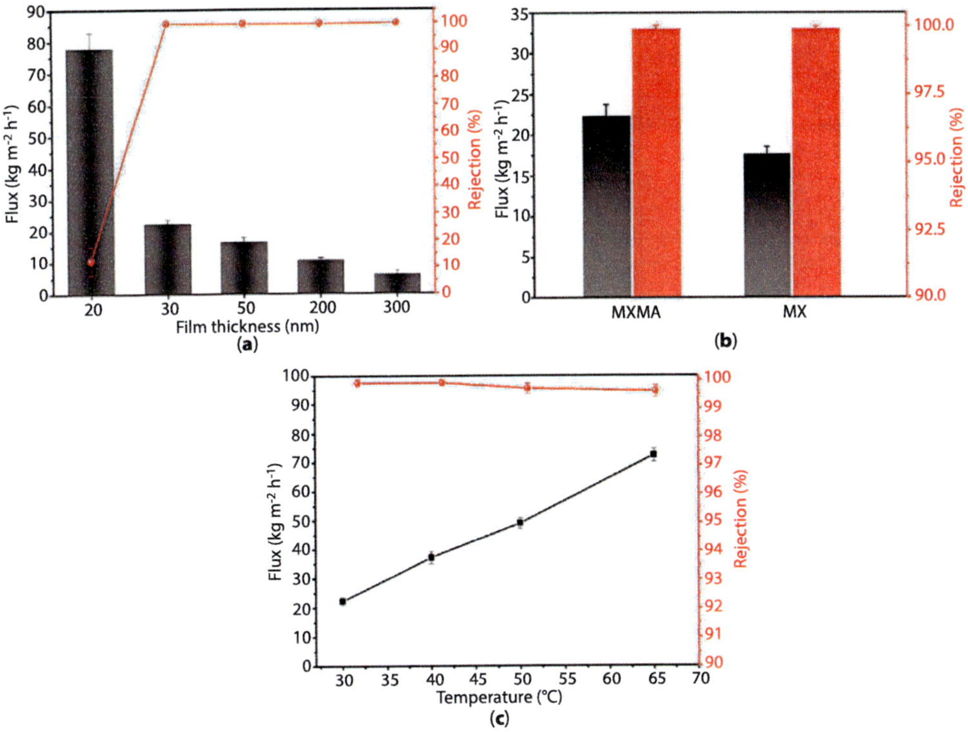

Figure 15.15 (a) MXMA membrane water penetration and ion rejection through different thickness at 30°C for pervaporative desalination of 3.5 wt% NaCl solution (b) MXMA and MX membranes using identical thicknesses of ~ 30 nm at 30°C for pervaporative desalination of 3.5 wt% NaCl solution tested for water permeation and ion rejection (c) Desalination of 3.5 wt% NaCl solution through pervaporation: water permeability and ion rejection of MXMA membrane with thickness of 30nm as a function of temperature [21]. Copyright 2020. Reproduced with permission from Elsevier.

a low water content. In their assessment on the pilot plant's ethanol pervaporation, they noted that despite the feedstock's low water content, there was still a considerable amount of water present in the permeance. The study indicated that changes in permeant pressure hardly ever affect membrane selectivity. According to the amount of water in the feed, permeation fluxes rise linearly, and they reduce as permeant pressure increases. The Arrhenius law describes whether temperature affects a system, and it was found that the activation energy is $E_w = 7.84$ kcal/mol. Guo and Hu [95] discovered that ladder-like phenylsilsesquioxane segments and PDMS segments affected the novel silicone copolymer's ethanol-water transport. Additionally, they demonstrated that the permeance frequently had higher ethanol concentrations than feed. As the feed's ethanol content increases, the separation factor drops. High separation factors can be attained while separating an ethanol-diluted aqueous solution using the membrane. While the flux of water remains essentially constant throughout a range of feed compositions, the flow of ethanol increases proportionally as the feeds ethanol content increases. For methanol, ethanol, 2-propanol, acetone, and THF, the separation factor and flow for pervaporation increase in that particular order. Only the declining characteristics of these organic components' solubility are reflected in this order.

Table 15.1 The selectivity and permeability of MXenes and other pervaporation methods.

Membrane material	Feed condition	Temperature	Separation factor	Permeate flux ($kg\,m^{-2}\,h^{-1}$)	Pressure	Ref.
MXENE/CS (3%wt.)	H_2O/ethanol 10% wt. of water	50°C	1421	1.4	200 Pa	[98]
Mxene/PVA (3% wt.)	H_2O/ethanol 7 %wt. of water	37°C	585	0.0742	200 Pa	[98]
PVA	NaCl 3.5% wt.	25°C	-	7.36 ($L\,m^{-2}\,h^{-1}$)	-	[93]
Mxene/PAN	NaCl 3.5% wt.	65°C	-	85.7 ($L\,m^{-2}\,h^{-1}$)	-	[93]
Mxene/CS	Water/ethyl acetate Water 2.0%wt.	323K	4898	1.471	-	[99]
Graphene/CS	Water/ethanol Water 10%wt.	303k	1093	10g $m^{-2}\,h^{-1}$	-	[99]
MXene/CS	Water/ dimethyl carbonatel Water 2.0%wt.	323k	906	1.428	-	[99]

Polyester-based polyurethanes (PUU-PEs-1, PUU-PEs-2, and PU-PEt-1) membranes were employed by Wolinska-Grabczyk *et al.* [96] to separate benzene in a benzene/cyclohexane combination at high and low concentrations. When employed to separate the benzene/cyclohexane azeotropic combination, PUU-PEs-1 and PUU-PEs-2 demonstrated outstanding selectivity, permeability, and stability. The membrane material (PU-PEt-1) that was created by switching out the polyester macrodiol for a polyether one showed a significantly higher permeability while also having a lower selectivity. The decrease of membrane selectivity is frequently observed in conjunction with an increase in the flux value. The MTBE/MeOH model combination that included 15% (mass) more MeOH than MTBE was created using membranes that were more permeable to MeOH than MTBE. The polyether-based polyurethane-based PU-PEt membrane demonstrates good separation performance (meOH permeance is 100%). In a polyurethane soft sector, polyester took the position of polyether, forming a membrane that was more permeable but having slightly less separation ability. PU-PD-1 polyurethanes have remarkable selectivity for removing benzene and chloroform from water. The administration of the polyether macrodiol (PU-PEt-1 vs. PU-PD-2) instead of the polydiene macrodiol caused structural alterations that reduced the membrane's selectivity.

According to Guo *et al.* [97], the partial water flux increases quickly as feed water concentration rises. PVA with hydroxyl substituted groups makes up the selective layer of the composite membrane. It has a high degree of polarity as a result of it interacts strongly with water via hydrogen bonding. More water molecules come into touch with the membrane's selective layer at increasing feed water concentrations. As a consequence, the membrane's top layer swells up as more water molecules are taken up by it. As the amount of water in the feed grows, the water-swollen membrane can accommodate more water molecules, increasing the water penetration flux. Water permeance rises slightly more slowly than water flow with a rise in feed water concentration. Because the driving factor behind total membrane transport, viz. the fugacity difference, has been significantly reduced, it represents more precise permeant-specific transport capabilities with feed concentration. As feed water concentration increases, butanol concentration in the permeance drops for the 2-butanol, isobutanol, and tert-butanol/water systems, whereas opposite is seen for the 1-butanol/water system. Among all the alcohols, 1-butanol has the greatest flow due to its strong affinity for membranes and extremely linear molecular structure. From the above discussion the potential for MXene based membranes for pervapotation application is confirmed. This is greatly attributed to the hydrophilicity of the MXene membranes. The major challenge still existing with the MXene based membranes are its swelling nature. In spite of the various techniques used to tackle it, this still leads to increase in permeation water flux. Yet another parameter is temperature. Some researchers have reported the poor selectivity which could make it difficult for commercial application. The removal of water content from the upstream remains a challenge. Further, the molecule's structure has a significant impact on the mechanism and hence specific channels need to be designed as per the permeate.

15.4 Gas Separation

Membrane technology is utilized for a variety of gas mixture separation techniques in conjunction with the purification and desalination of water [100–102]. In recent years, the local market and industry have paid close attention to membrane technology for gas separation. However, the main difficulties include using membranes in a lot of situations, with heavy fouling, and in a variety of pH values, particularly in the petrochemical industry. Therefore, it is vitally necessary to develop innovative, stable, and high-performance materials for the aforementioned applications. Owing to their layered structure, optimal thickness, strong thermal characteristics, and remarkably low mass transfer barrier 2D MXenes has recently come to be recognized as a superior material for gas separation membranes [103]. This section explains the usage of MXene-based membrane for gas separation applications and sheds light on the present problems and technical challenges that prevent their widespread use.

15.4.1 Mechanism for Gas Separation

Gas mixtures are separated using a membrane in a continuous process called as "gas separation," where one side of the membrane's feed source is under a higher pressure than the other, and vice versa. The pressure differential between the upstream and downstream acts as the driving power for the GS membrane [104].

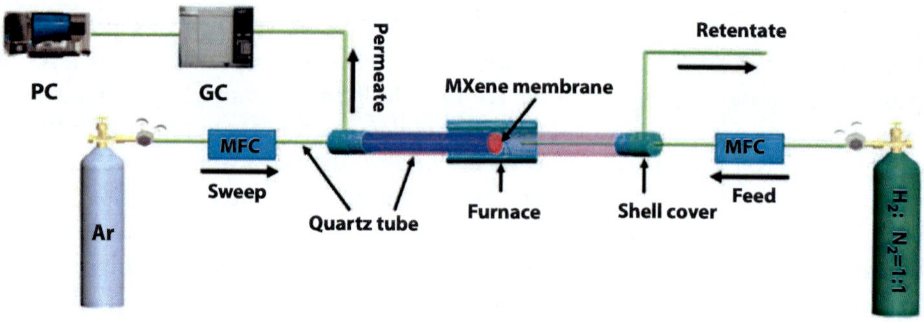

Figure 15.16 Schematic representing gas mixture permeance of MXene membrane [105]. Copyright 2019. Reproduced with permission from Elsevier.

The two main terms of gas membranes are permeability and selectivity. The general pace at which individual gas travels through a membrane is measured by its permeability.

Permeation of pure gas can calculate by following equation (15.26),

$$P = \frac{273.15\times10^{10}}{760}\frac{Vl}{AT\dfrac{P_0\times76}{14.7}}\frac{dp}{dt} \tag{15.26}$$

Where P represents the permeation coefficient of the Barrier, the downstream chamber's volume is V (in cm^3), A is the membrane's effective area (in cm^2), l is the dense layer's thickness (in cm), T is the absolute temperature (in K), dp/dt is the downstream chamber's rate of pressure change (in mmHg/s), and P_0 is the upstream applied pressure. Equation (15.27) describes the appropriate selectivity for two different gas molecules in a membrane. This is the ratio of the two-gas molecule's permeabilities (15.27),

$$\alpha_{A/B} = P_A/P_B \tag{15.27}$$

In order to widely commercialize the GS membrane technology, numerous studies focus on creating membranes with excellent selectivity and enhanced permeability. Figure 15.16 depicts the experimental setup for the permeance test [105]. This homemade instrument was used to test the permeance of a gas mixture. For a typical measurement, a mixture of feed gas with 50 vol% H_2 and 50 vol% N_2 was supplied with flow rate equal to 50 mL min^{-1}. During the measurement cycle, at a specific flow rate of 40 mL min^{-1}, sweep gas; argon was employed under standard pressure condition. Additionally, in order to determine their composition, the sweep side tail gas must be moved to an inline gas chamber (6890N, Agilent) with TCD.

15.4.2 Types of Membrane for Gas Permeation

The membrane material's selectivity and permeability determine how well the gas separation process works. A membrane can be categorized as either porous or nonporous depending on flow and selectivity.

15.4.2.1 *Porous Membrane*

A rigid, hollow porous membrane has pores that are arbitrarily placed and linked. A porous membrane's capacity to separate different components is influenced by both its permeant nature and other features, such as the molecular size, pore size, and its distribution of the membrane polymer. A porous membrane's structure and operation are strikingly similar to those of a typical filter. Typically, microporous membranes are only successful in effectively separating molecules that have a significant difference in size. Porous membranes for gas separation inherit low selectivity values despite having very high flux levels. The tortuosity, porosity, and average pore diameter (d) of microporous membranes can be used to identify them. Porous membranes can be used to segregate gases. The pore diameter needs to be smaller than the usual gas molecule free path. Diameter typically is around 50 nm (100 kPa, 300 K). The square root of the molecular mass, or the velocity of the molecules, or both, determines the inverse relationship between the gas flow through the pore (Knudsen diffusion) [106]. A porous membrane allows for three to five orders of magnitude more flux than a nonporous one. The efficiency of separation is low; oxygen passes more quickly than hydrogen by a factor of 4. The various diffusion types existing in MXene sheets are depicted in Figure 15.17 [107].

i. Knudsen (or Free Molecule) Diffusion

The fraction of the mean free path (usual spacing between collisions) (λ) to a typical physical size range (for instance, the pore radius) (r) of the gas molecules is referred to as the Knudsen number (Kn) and is given in equation (15.28).

$$K_n = \frac{\lambda}{r} \tag{15.28}$$

Membranes with pore diameters less than 50 nm can achieve Knudsen separation. The appropriate separation factors for several groups of gases are shown based on Knudsen flow. However, it is demonstrated that the real separation factor is lower. This is accounted for by the presence of various diffusion processes like back and non-separative diffusion. Viscous

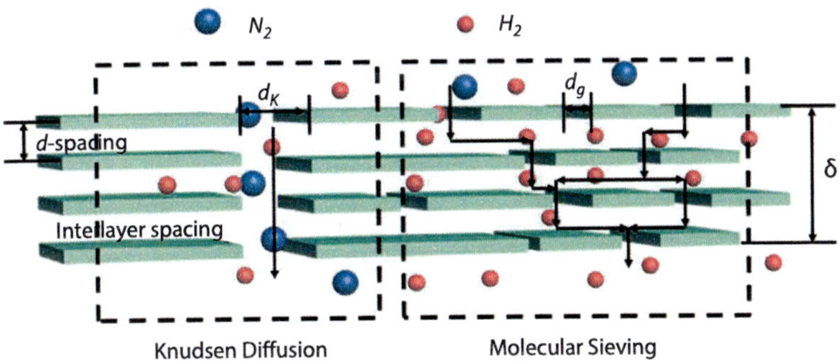

Figure 15.17 The different diffusion mechanism in porous membrane, (1) Knudsen (or free molecule) diffusion, (2) Surface diffusion, (3) Capillary condensation and (4) Molecular sieving and (5) Solution diffusion in non-porous membrane. Adapted from Ref. [107]. Copyright 2019. MDPI.

flow (in large pores) and concentration polarity on the feed/permeate end are additional factors that contribute to the flow.

ii. Surface Diffusion

This can be placed along with Knudsen diffusion. After adsorbing to the membrane pore walls, gas molecules move over the surface. Surface diffusion makes substances more permeable to membrane pores so they can adsorb more successfully. The effective pore diameter is decreased concurrently. Selectivity is increased and non-adsorbing component transport is decreased as a result. Only specific temperature ranges and pore sizes allow for this beneficial effect of surface diffusion.

iii. Capillary Condensation

This occurs when the membrane pores are (partially) covered by a condensed phase. Once condensed phase has completely sealed the pores, only species that can dissolve in the membrane will travel through them. Fluxes and selectivities for capillary condensation are frequently high. But the nature of the gas, the dimensions of the pores, and the regularity of its widths all have a substantial impact on how capillary condensation appears.

iv. Molecular Sieving

When pore sizes (3.0–5.2 Å) are small enough to separate molecules with various kinetic widths, this process is known as molecular sieving. Only the smallest gas molecules are able to cross the membrane at this time. A number of techniques, such as solution casting, sintering, stretching, track etching, and phase separation, can be used to create porous polymeric membranes. The characteristics of the materials and the process parameters used will have a significant impact on the final shape of the membrane generated. Transport via bulk materials or thick membranes, surface diffusion over pore walls, Knudsen diffusion in fine pores, viscous flow in large pores, and other mechanisms can all be utilized to understand the transport in membranes.

15.4.2.2 Non-Porous Membrane

The fundamental transport mechanism in porous polymeric membranes is generally recognized to be solution diffusion. The typical perception of this mechanism is that it involves three steps. The first stage begins with the gas molecules being trapped onto the membrane surface. The next step is the dispersion of the gas molecules throughout the polymer matrix. In the final phase, the gas molecules evaporate on the downstream [108]. Sir Thomas graham developed a solution-diffusion-based method for transporting gases in dense, non-porous polymers [109]. Penetrating molecules disintegrate in the membrane's upstream (or high pressure) face, permeate across that one, and then diffuse from the membrane downstream (or low pressure) face whenever the membrane experiences a pressure gradient. Therefore, the diffusivity and solubility coefficient are the two crucial variables that govern how much of a module permeates the environment, owing to the solution-diffusion model (S). The penetrant permeation phase that regulates rate is diffusion. The thermally driven, random local segmental polymer dynamics that creates holes in the polymer matrix

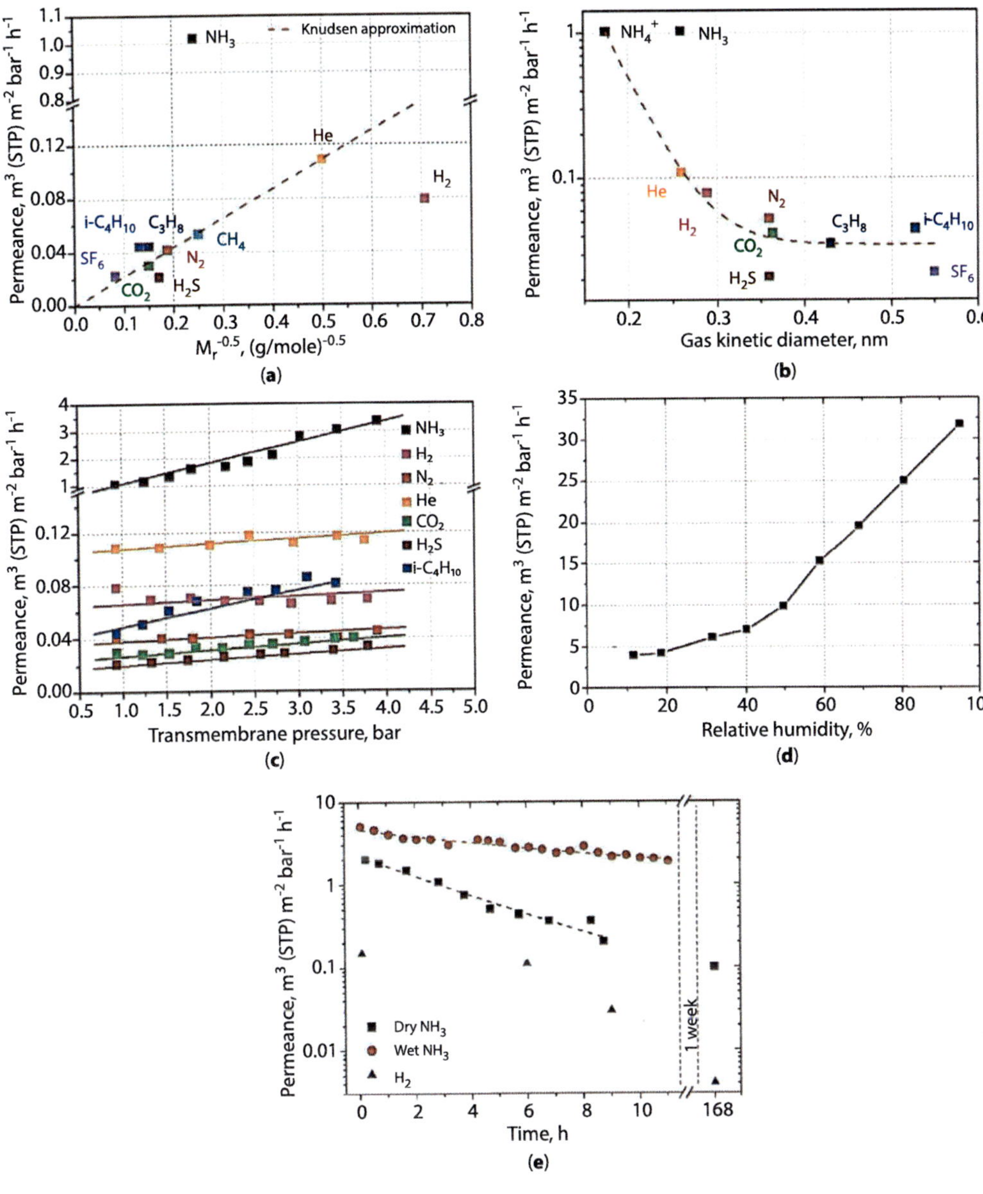

Figure 15.18 The variation of permeance for various gas with (a) the inverse square root of molecular weight, (b) gas kinetic diameter, (c) transmembrane pressure, (d) relative humidity, and (e) effect of time on the permeance of dry and wet ammonia and hydrogen gas [112]. Copyright 2021. Reproduced with permission from Elsevier.

big enough to permit penetrant molecules is the rate-controlling stage in diffusion [110]. Sorption-diffusion theory may be used to explain that a gas permeates a thick polymer (membrane) [111].

15.4.3 Factors Affecting Gas Permeation

Gas permeation is dependent on various factors including pressure and kinetic diameter of gas. However, in addition to these factors, there are few others like water vapour permeance, relative humidity and the type of gases. Petukhov *et al.* [112] have documented the outcome of several variables on the $Ti_3C_2T_x$ nanosheets' gas permeability. The dependence of these factors are represented in Figure 15.18. Figure 15.18 (a) demonstrates that the permeance almost increases linearly with inverse square root of molecular weight [112]. This highlights the idea that it is in agreement with the theory of Knudsen diffusion kinetics. However, gases like ammonia and hydrogen has shown a variation from the kinetic representation. The study on the effect of kinetic diameter of gas (Figure 15.18 (b)) emphasizes the fact that the permeance is higher for lower sizes and decreases almost exponentially with increase in diameter. However, permeance showed a minimum dependence on transmembrane pressure except for ammonia (Figure 15.18 (c)) [112]. Humidity has a major contribution as is evident from Figure 15.18 (d) [112]. The permeance of wet ammonia was more compared to dry ammonia and hydrogen. The permeance however showed a

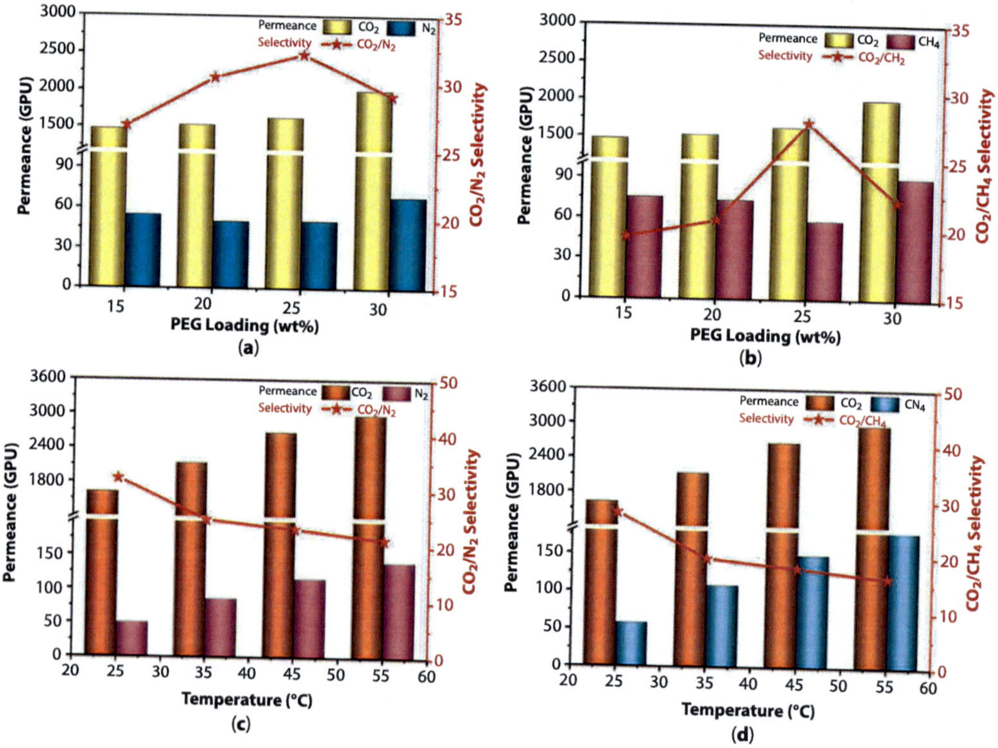

Figure 15.19 (a) Variation of CO_2 permeance with PEG (600) loadings; selectivity of the MXene/PEG (600) MMMs towards CO_2/N_2 is also represented (b) Impact of PEG (600) addition on the CO_2 permeance and MXene/PEG (600) MMMs loading on CO_2/CH_4 selectivity (c) Variation of permeance with operating temperature on MXene/PEG (600) MMMs with 25 wt% for CO_2/N_2 separation. Variation in selectivity is also represented (d) Variation of permeance with operating temperature on MXene/PEG (600) MMMs with 25 wt% for CO_2/CH_4 separation; Variation in selectivity is also represented [45]. Copyright 2022. Reproduced with permission from Elsevier.

decrease with time which was more significant for hydrogen than ammonia (Figure 15.18 (e)) [112]. In another work, the preparation and analysis of the gas separation performance of the aqueous solution of PEG loading (15–30 wt%) with MXene was explored (Figure 15.19 (a, b)) [45]. When PEG (600) was added to MXene, the mixture's maximum selectivity for CO_2/N_2 was 32.18 and CO_2/CH_4 was 27.87, respectively. The maximum CO_2 permeance measured at 30 weight percent was 2006.6 GPU. The findings of a study on the gas-transport capabilities exhibited by MXene/PEG (600) MMMs and gas permeation at various permeating temperatures (varying from 25 to 55°C) are displayed in Figure 15.19 (c, d) [45]. At 55°C and 1 bar, the greatest permeance for CO_2 was 2987.25 GPU. Maximum selectivity for the resulting combination is 32.18 for CO_2/N_2 and 27.87 for CO_2/CH_4. Long term stability is a vital parameter deciding the usefulness of an MXene membrane for commercial application. Such studies conducted on MXene membrane with 2 μm thickness for 700 hours (Figure 15.20 (a)) [117] and under a continuous 60-hour gas separation test for MXene/PEG (600) MMMs (Figure 15.20 (b)) are represented [45]. As evident from

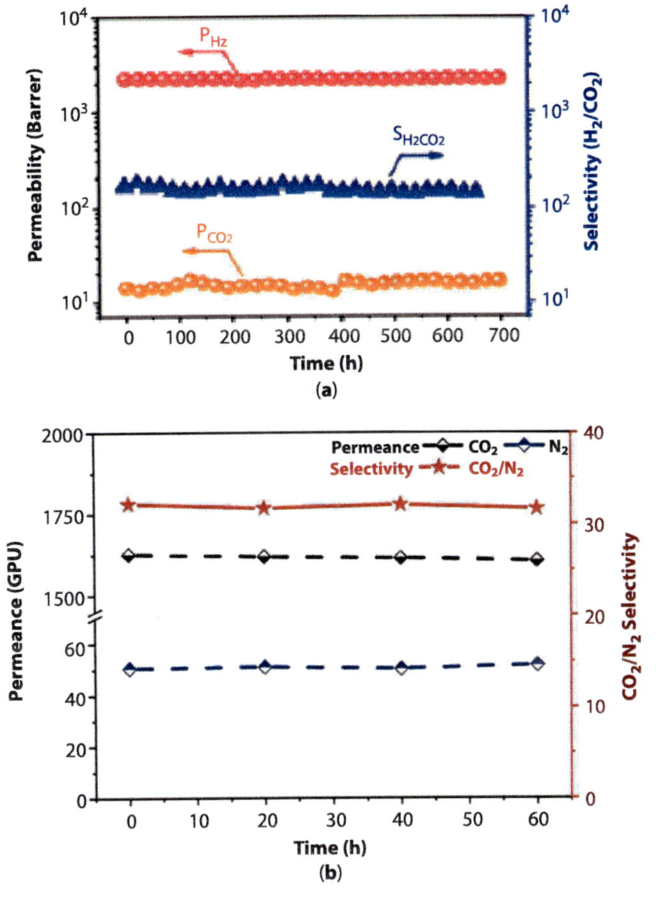

Figure 15.20 Stability test on (a) 2 μm thick membrane MXene towards H_2/CO_2 mixture. Adapted from Ref. [117]. Copyright 2018. Springer Nature, (b) MXene/PEG (600) MMMs for gas separation, with single gas measured at 1 bar at 25 °C for CO_2/N_2 separation [45]. Copyright 2022. Reproduced with permission from Elsevier.

the study, MXene membranes demonstrated exceptional stability and endurance making it viable as a commercial alternative.

15.4.4 Reported Works on Gas Permeation Using MXene-Based Membranes

Theoretically, MXene-based membranes for gas separation capabilities and transportation mechanism were carefully investigated [113–116] and minimal expected works has been done experimentally. Shen *et al.* [25] separated H_2 gas and showed improved H_2/CO_2 selectivity by using well stacked pure MXene nanofilms that are around 20 nm thick. The maximum permeance of 1484 GPU for H_2 gas and selectivity of 27 for H_2/CO_2 was reported using this membrane. The digital micrograph and SEM is shown in Figure 15.21 (a, b) [25]. Additionally, to improve the stacking behavior and (increasing) interlayer spacing for CO_2 gas separation, borate and PEI molecules were added to MXene nanosheets. Maximum permeance up to 350 GPU was exhibited for CO_2 gas by functionalized MXene and a selectivity of was achieved for CO_2/CH_4 (Figure 15.21 (c)) [25]. In addition, during long-term operation test, both pure and functionalized MXene membranes were shown to be extremely

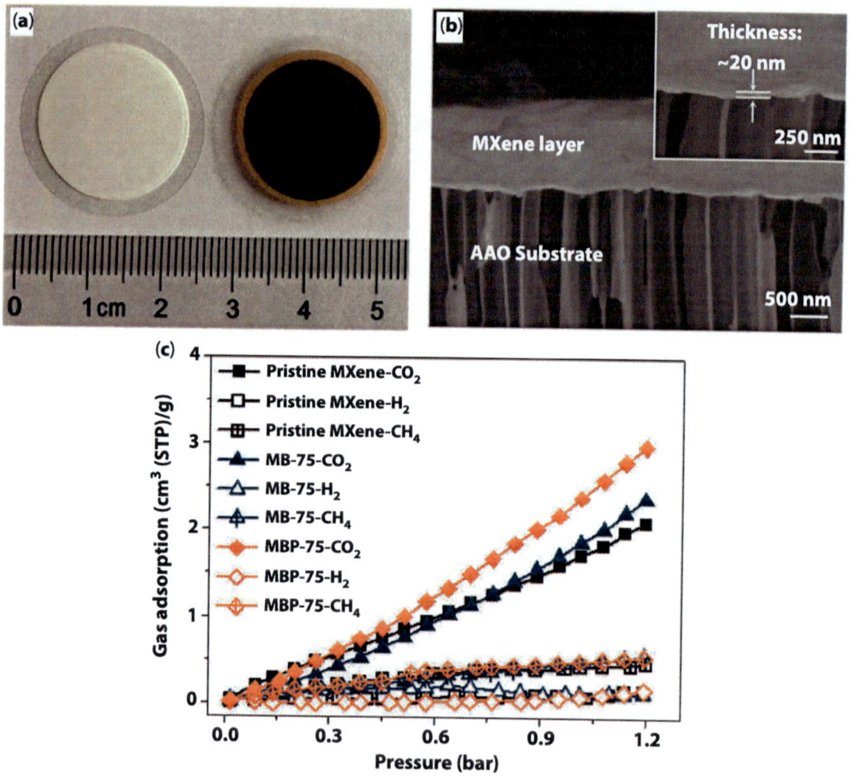

Figure 15.21 Construction and performance of a lamellar membrane based on MXene of 20 nm thick; (a) photographic images of AAO substrate and MXene nanofilms onto substrate; (b) SEM images (cross-sectional) of MXene/AAO substrate nanofilms, a higher magnification of the interface is given in the inset; (c) investigation of the differences between gas adsorption behaviour of pristine and modified MXene membranes [25]. Copyright 2018. Reproduced with permission from John Wiley & Sons.

stable up to 100 hours. For a specific gas separation, Ding *et al.* [117] successfully synthesized 2D $Ti_3C_2T_x$ with high reproducibility Excellent H_2/CO_2 selectivity > 160 and H_2 gas permeance greater than 2200 Barrer (1 Barrer = 3.348 mol $m^{-1}S^{-1}Pa^{-1}$) were demonstrated by these laminates. The MXene laminates that are put together have regular sub nanometer channels, which provide the membrane its exceptional separation performance (Figure 15.22 (a-c) [117]. Additionally, these membranes demonstrated excellent stability for 700 hours while continuously separating an H_2/CO_2 combination. Additionally, the simulation study of these membranes for gas flow was investigated, which supported the experimental findings. Table 15.2 and Table 15.3, gives the gas separation of behaviour of MXene and various other composites-based membranes for H_2/CO_2 separation and N_2/CO_2 respectively. As evident from Table 15.2, the selectivity of MXene based membranes ranges from 5 to 615. It is worth noting that the maximum selectivity is exhibited by MXene/Al_2O_3/Ni^{2+} at a much lower pressure of 249 GPU in comparison with other samples. GO is showing higher values

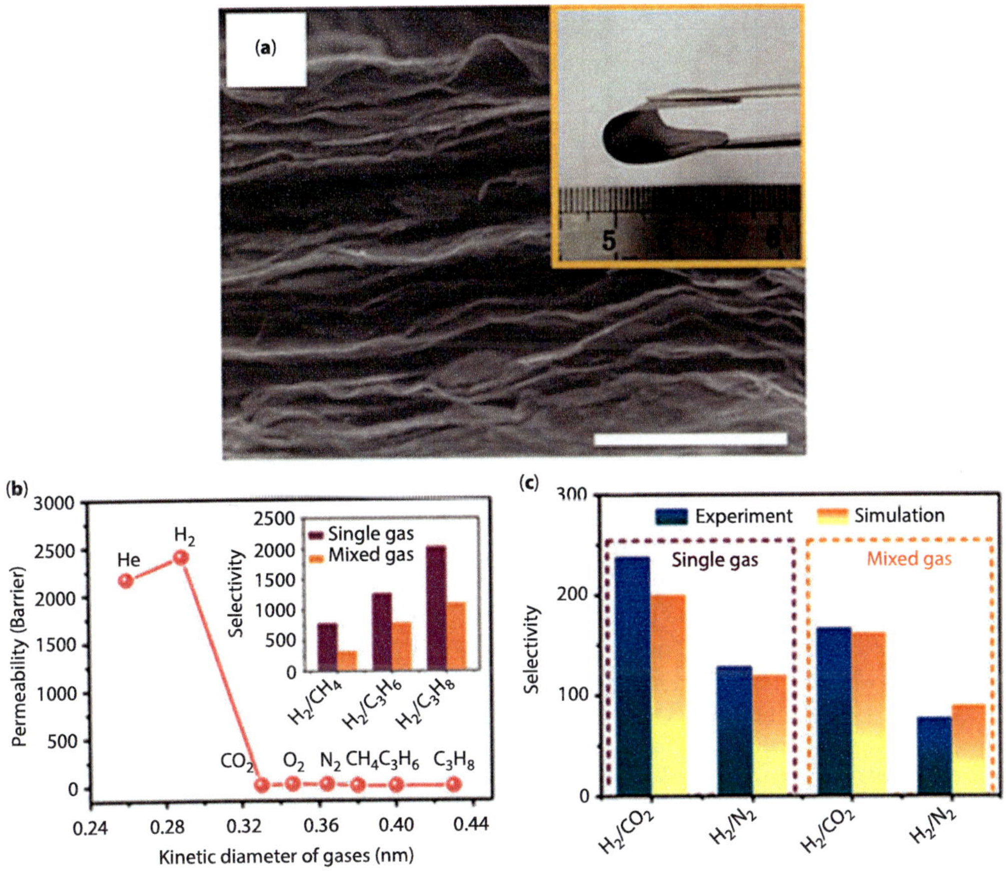

Figure 15.22 (a) SEM analyses (cross-sectional) of MXene laminates (The photograph of the prepared membrane is given in the inset); (b) Influence of gas kinetic diameter upon various gases' permeabilities. In both single- and equimolar mixed gas permeation tests, the selectivity of H_2 gas in relation to other gases is provided in the inset. (c) Comparison on the selectivities of H_2/N_2 and H_2/CO_2 using computational and experimental methods. Adapted from Ref. [117]. Copyright 2018, Springer Nature.

Table 15.2 MXenes' permeability and selectivity for H_2/CO_2 gas separation.

Membrane material	Temperature °C	H_2 Permeance (1 GPU = 3.35×10^{-10} mol $m^{-2}s^{-1}pa^{-1}$) /Barrer	H_2/CO_2 selectivity	Ref.
MXene	25	3202.3 Ba	62	[117]
MXene	25	2226.6 Ba	162	[117]
MXene	25	1473.3 Ba	174	[117]
MXene	25	1302.7 Ba	200	[117]
MXene	40	4 GPU	5	[118]
MXene hollow fiber membrane	25	70.6 GPU	30.3	[119]
MXene-Borate (MB 75)	25	322 GPU	19.13	[25]
Ultrathin (20 nm) MXene-PEBA	25	1584 GPU	27	[25]
MXene-Borate PEI (MBP 75)	25	350 GPU	28.23	[25]
PBI hollow fiber membranes	250	200–400 GPU	20	[120]
PBI hollow fiber membranes	95	90 GPU	20	[121]
MXene-PDMS	25	1612 GPU	25	[25]
MXene/Al_2O_3/Ni^{2+}	25	249 GPU	615	[122]
MXene/Al_2O_3	25	158 GPU	215	[123]

of selectivity compared to other MXene membranes. This shows that the work on MXene for selectivity of H_2/CO_2 has to be focused on having a combination of ceramic and metal nanoparticle intercalation for improved performance. The filler used and its concentration have shown a substantial effect on the selectivity of MXene membranes, and is represented in Table 15.3. The commonly used fillers being PEBAX and ZIF. Selectivity towards CO_2 is better for composite membranes over pristine MXene. The CO_2 permeability is seen to improve with the addition of PEBAX. Due to their remarkable gas separation efficiency, these MXene membranes have a lot of promise for usage in CO_2 capture and H_2 purification in the future. To improve the separation effectiveness of these laminates, it is also feasible to alter the feed concentration, temperature, and pressure. From the description above, it is evident that this is because of its 2D structure and flexible physicochemical properties. For the development of high-performance molecular sieving membrane, MXene appears to have promising application. However, the effect of temperature and humidity has to be

Table 15.3 MXenes' permeability and selectivity for CO_2/N_2 gas separation.

Membrane material	Temperature °C	Pressure MPa	CO_2 permeability (barrer)	CO_2/N_2 selectivity	Ref.
PEBAX/m-MXene	25	0.4	86	105	[124]
PEBAX/0.5 MXene	25	0.4	70.2 – (Pure gas) 85.5 – (Mixed gas)	93.2 90	[124]
MXene/PEBAX	25	0.2	43	72	[125]
Pristine MXene	25	-	6.8	44	[126]
1 wt% MXene on PEBAX	30	0.2	148	63	[127]
10 wt% MXene on PEBAX	30	0.2	584	59	[127]

tackled in order make it suitable for commercialization. The separation performance can therefore still be improved.

15.5 Conclusion and Future Work

Industries have employed membrane technology extensively for pervaporation, gas separation, desalination, and water filtration. This process has recently drawn a lot of interest since it will allow for the creation of gas molecules that are clean and distinct. One of the intriguing compounds that could be employed in membrane separation technologies is MXene. In this chapter, the creation of a composite membrane based on MXene has been comprehensively discussed. Various categories including crosslinked MXene nanosheet, hybrid membranes, mixed matrix, 2D scaffolds, thin films and nanolaminates are discussed by focusing on the construction techniques. The basic mechanism of pervaporation is discussed, keeping focus on the theory, factors and parameters that control the pervaporation. Further, recent progress made in pervaporation using MXene based membranes are discussed in detail. The major challenges posed by the membranes in this application is also discussed. Subsequently, the mechanism, kinetics, and parameters influencing the process are explored in detail in relation to the gas separation component of MXene-based membranes. Recent advancements in the direction of gas separation using these composite membranes are discussed. As evident from the reports, MXenes are having huge potential in the areas of pervaporation and gas separation. In order to improve gas flow, it is necessary to treat the grain boundary, flaws, contaminants, and molecular residues. In order to be useful for pervaporation, it's important to keep the membrane's swelling under control. We need to work on channeling the MXene sheets for improving selectivity. Since the application requires long term exposure to temperature and humidity, the prepared membranes should be robust and inert to changing environmental conditions.

Acknowledgement

The authors thank Vellore Institute of Technology, Vellore for providing 'VIT SEED Grant–RGEMS fund (SG20220118) for carrying out this research work.

References

1. Yuan, S., Li, X., Zhu, J., Zhang, G., Van Puyvelde, P., Van der Bruggen, B., Covalent organic frameworks for membrane separation. *Chem. Soc. Rev.*, 48, 10, 2665–2681, 2019.
2. Liu, G., Jin, W., Xu, N., Two-dimensional-material membranes: A new family of high-performance separation membranes. *Angew. Chem.-Int. Ed.*, 55, 43, 13384–13397, 2016.
3. Mi, B., Graphene oxide membranes for ionic and molecular sieving. *Science*, 343, 6172, 740–742, 2014.
4. Li, J., Li, X., Van der Bruggen, B., An MXene-based membrane for molecular separation. Environ. *Sci. Nano.*, 7, 5, 1289–1304, 2020.
5. Zornoza, B., Seoane, B., Zamaro, J.M., Téllez, C., Coronas, J., Combination of MOFs and zeolites for mixed-matrix membranes. *Chem. Phys. Chem.*, 12, 15, 2781–2785, 2011.
6. Varoon, K., Zhang, X., Elyassi, B., Brewer, D.D., Gettel, M., Kumar, S., Lee, J.A., Maheshwari, S., Mittal, A., Sung, C.Y., Cococcioni, M., Francis, L.F., Mccormick, A.V., Mkhoyan, K.A., Tsapatsis, M., Dispersible exfoliated zeolite nanosheets and their application as a selective membrane. *Science*, 334, 6052, 72–75, 2011.
7. Ahmed, Z., Rehman, F., Ali, U., Ali, A., Iqbal, M., Thebo, K.H., Recent advances in MXene-based separation membranes. *ChemBioEng Rev.*, 8, 2, 110–120, 2021.
8. Peng, Y., Li, Y., Ban, Y., Jin, H., Jiao, W., Liu, X., Yang, W., Metal-organic framework nanosheets as building blocks for molecular sieving membranes. *Science*, 346, 6215, 1356–1359, 2014.
9. Kim, H.W., Yoon, H.W., Yoon, S.M., Yoo, B.M., Ahn, B.K., Cho, Y.H., Shin, H.J., Yang, H., Paik, U., Kwon, S., Choi, J.Y., Park, H.B., Selective gas transport through few-layered graphene and graphene oxide membranes. *Science*, 342, 6154, 91–95, 2013.
10. Wang, Z., Tu, Q., Zheng, S., Urban, J.J., Li, S., & Mi, B., Understanding the aqueous stability and filtration capability of MoS2 membranes. *Nano Lett,. 17*, 12, 7289–7298, 2017.
11. Naguib, M., Kurtoglu, M., Presser, V., Lu, J., Niu, J., Heon, M., Hultman, L., Gogotsi, Y., Barsoum, M.W., Two-dimensional nanocrystals produced by exfoliation of Ti_3AlC_2. *Adv. Mater*, 23, 37, 4248–4253, 2011.
12. Naguib, M., Mashtalir, O., Carle, J., Presser, V., Lu, J., Hultman, L., Gogotsi, Y., Barsoum, M.W., Two-dimensional transition metal carbides. *ACS Nano*, 6, 2, 1322–1331, 2012.
13. Ying, Y., Liu, D., Ma, J., Tong, M., Zhang, W., Huang, H., Yang, Q., Zhong, C., A GO-assisted method for the preparation of ultrathin covalent organic framework membranes for gas separation. *J. Mater. Chem. A*, 4, 35, 13444–13449, 2016.
14. Arshadi, F., Mohammad, M., Hosseini, E., Ahmadi, H., Asadnia, M., Orooji, Y., Korayem, A.H., Noorbakhsh, A., Razmjou, A., The effect of D-spacing on the ion selectivity performance of MXene membrane. *J. Membr. Sci.*, 639, 119752, 2021.
15. Kessler, F.K., Zheng, Y., Schwarz, D., Merschjann, C., Schnick, W., Wang, X., Bojdys, M.J., Functional carbon nitride materials—Design strategies for electrochemical devices. *Nat. Rev. Mater*, 2, 6, 1–17, 2017.
16. Mashtalir, O., Cook, K.M., Mochalin, V.N., Crowe, M., Barsoum, M.W., Gogotsi, Y., Dye adsorption and decomposition on two-dimensional titanium carbide in aqueous media. *J. Mater. Chem. A*, 2, 35, 14334–14338, 2014.

17. Ling, Z., Ren, C.E., Zhao, M.Q., Yang, J., Giammarco, J.M., Qiu, J., Barsoum, M.W., Gogotsi, Y., Flexible and conductive MXene films and nanocomposites with high capacitance. *Proc. Natl. Acad. Sci. U. S. A.*, 111, 47, 16676–16681, 2014.
18. Ying, Y., Liu, Y., Wang, X., Mao, Y., Cao, W., Hu, P., Peng, X., Two-dimensional titanium carbide for efficiently reductive removal of highly toxic chromium (VI) from water. *ACS Appl. Mater. Interfaces*, 7, 3, 1795–1803, 2015.
19. Huang, H., Ying, Y., Peng, X., Graphene oxide nanosheet: an emerging star material for novel separation membranes. *J. Mater. Chem. A*, 2, 34, 13772–13782, 2014.
20. Qu, K., Huang, K., Xu, Z., Recent progress in the design and fabrication of MXene-based membranes. *Front. Chem. Sci. Eng.*, 15, 4, 820–836, 2021.
21. Ding, M., Xu, H., Chen, W., Yang, G., Kong, Q., Ng, D., Lin, T., Xie, Z., 2D laminar maleic acid-crosslinked MXene membrane with tunable nanochannels for efficient and stable pervaporation desalination. *J. Membr. Sci.*, 600, 117871, 2020.
22. Liu, G., Shen, J., Ji, Y., Liu, Q., Liu, G., Yang, J., Jin, W., Two-dimensional Ti_2CT_x MXene membranes with integrated and ordered nanochannels for efficient solvent dehydration. *J. Mater. Chem. A*, 7, 19, 12095–12104, 2019.
23. Ding, L., Li, L.B., Liu, Y.C., Wu, Y., Lu, Z., Deng, J.J., Wei, Y.Y., Caro, J., Wang, H.H., Effective ion sieving with $Ti_3C_2T_x$ MXene membranes for production of drinking water from seawater. *Nat. Sustain.*, 3, 4, 296–302, 2020.
24. Lu, Z., Wei, Y.Y., Deng, J.J., Ding, L., Li, Z.K., Wang, H.H., Self-crosslinked MXene ($Ti_3C_2T_x$) membranes with good antiswelling property for monovalent metal ion exclusion. *ACS Nano*, 13, 9, 10535–10544, 2019.
25. Shen, J., Liu, G., Ji, Y., Liu, Q., Cheng, L., Guan, K., Zhang, M., Liu, G., Xiong, J., Yang, J., Jin, W., 2D MXene nanofilms with tunable gas transport channels. *Adv. Funct. Mater.*, 28, 31, 1801511, 2018.
26. Ding, L., Wei, Y., Wang, Y., Chen, H., Caro, J., Wang, H., A two-dimensional lamellar membrane: MXene nanosheet stacks. *Angew. Chem. Int. Ed.*, 56, 7, 1825–1829, 2017.
27. Xing, Y., Akonkwa, G., Liu, Z., Ye, H., Han, K., Crumpled two-dimensional $Ti_3C_2T_x$ MXene lamellar membranes for solvent permeation and separation. *ACS Appl. Nano Mater.*, 3, 2, 1526–1534, 2020.
28. Pandey, R.P., Rasool, K., Madhavan, V.E., Aïssa, B., Gogotsi, Y., Mahmoud, K.A., Ultrahigh-flux and fouling-resistant membranes based on layered silver/MXene ($Ti_3C_2T_x$) nanosheets. *J. Mater. Chem. A*, 6, 8, 3522–3533, 2018.
29. Wu, X., Cui, X., Wu, W., Wang, J., Li, Y., Jiang, Z., Elucidating ultrafast molecular permeation through well-defined 2D nanochannels of lamellar membranes. *Angew. Chem. Int. Ed.*, 131, 51, 18695–18700, 2019.
30. Shamsabadi, A.A., Isfahani, A.P., Salestan, S.K., Rahimpour, A., Ghalei, B., Sivaniah, E., Soroush, M., Pushing rubbery polymer membranes to be economic for CO_2 separation: Embedment with $Ti_3C_2T_x$ MXene nanosheets. *ACS Appl. Mater. Interfaces*, 12, 3, 3984–3992, 2019.
31. Hao, L., Zhang, H., Wu, X., Zhang, J., Wang, J., Li, Y., Novel thin-film nanocomposite membranes filled with multi-functional $Ti_3C_2T_x$ nanosheets for task-specific solvent transport. *Compos. Part A Appl. Sci. Manuf.*, 100, 139–149, 2017.
32. Liu, G., Cheng, L., Chen, G., Liang, F., Liu, G., Jin, W., Pebax-based membrane filled with two-dimensional mxene nanosheets for efficient co_2 capture. *Chem. Asian J.*, 15, 15, 2364–2370, 2020.
33. Li, S., Dai, J., Geng, X., Li, J., Li, P., Lei, J., Wang, L., He, J., Highly selective sodium alginate mixed-matrix membrane incorporating multi-layered MXene for ethanol dehydration. *Sep. Purif. Technol.*, 235, 116206, 2020.

34. Xu, Z., Liu, G.Z., Ye, H., Jin, W.Q., Cui, Z.F., Two-dimensional MXene incorporated chitosan mixed-matrix membranes for efficient solvent dehydration. *J. Membr. Sci.*, 563, 625–632, 2018.
35. Han, R., Xie, Y., Ma, X., Crosslinked P84 copolyimide/MXene mixed matrix membrane with excellent solvent resistance and permselectivity. *Chin. J. Chem. Eng.*, 27, 4, 877–883, 2019.
36. Cao, X., Jing, W., Xing, W., Fan, Y., Kong, Y., Dong, J., Fabrication of a visible-light response mesoporous TiO_2 membrane with superior water permeability via a weak alkaline sol–gel process. *Chem. Commun.*, 47, 12, 3457–3459, 2011.
37. Zou, D., Qiu, M., Chen, X., Fan, Y., One-step preparation of high-performance bilayer α-alumina ultrafiltration membranes via co-sintering process. *J. Membr. Sci.*, 524, 141–150, 2017.
38. Xia, C., Xu, Z., Yu, J., Sun, Y., Jing, W., Fabrication of microporous GO-TiO_2 membrane via an improved weak alkaline sol–gel method. *J. Membr. Sci.*, 561, 10–18, 2018.
39. Yu, J., Zhang, Y., Chen, J., Cui, L., Jing, W., Solvothermal-induced assembly of 2D-2D rGO-TiO_2 nanocomposite for the construction of nanochannel membrane. *J. Membr. Sci.*, 600, 117870, 2020.
40. Xu, Z., Sun, Y., Zhuang, Y., Jing, W., Ye, H., Cui, Z., Assembly of 2D MXene nanosheets and TiO_2 nanoparticles for fabricating mesoporous TiO_2-MXene membranes. *J. Membr. Sci.*, 564, 35–43, 2018.
41. Sun, Y., Xu, Z., Zhuang, Y., Liu, G., Jin, W., Liu, G., Jing, W., Tunable dextran retention of MXene-TiO_2 mesoporous membranes by adjusting the 2D MXene content. *2D Mater.*, 5, 4, 045003, 2018.
42. Fauzan, N.A.B., Mukhtar, H., Nasir, R., Mohshim, D.F.B., Arasu, N., Man, Z., Mannan, H.A., Composite amine mixed matrix membranes for high-pressure CO_2-CH_4 separation: Synthesis, characterization and performance evaluation. *R. Soc Open Sci.*, 7, 9, 200795, 2020.
43. Maleski, K., Mochalin, V.N., Gogotsi, Y., *Dispersions of two-dimensional titanium carbide MXene in organic solvents*, vol. 1632-1640, Chem. Mater., 29(4, 2017.
44. Ren, C.E., Alhabeb, M., Byles, B.W., Zhao, M.Q., Anasori, B., Pomerantseva, E., Mahmoud, K.A., Gogotsi, Y., Voltage-gated ions sieving through 2D MXene $Ti_3C_2T_x$ membranes. *ACS Appl. Nano Mater.*, 1, 7, 3644–3652, 2018.
45. Luo, W., Niu, Z., Mu, P., Li, J., MXene/poly (ethylene glycol) mixed matrix membranes with excellent permeance for highly efficient separation of CO_2/N_2 and CO_2/CH_4. *Colloids Surf. A Physicochem. Eng. Asp.*, 640, 128481, 2022.
46. Akther, N., Phuntsho, S., Chen, Y., Ghaffour, N., Shon, H.K., Recent advances in nanomaterial-modified polyamide thin-film composite membranes for forward osmosis processes. *J. Membr. Sci. Res.*, 584, 20–45, 2019.
47. Karahan, H.E., Goh, K., Zhang, C., Yang, E., Yıldırım, C., Chuah, C.Y., Ahunbay, M.G., Lee, J., Tantekin-Ersolmaz, Ş.B., Chen, Y., Bae, T.H., MXene materials for designing advanced separation membranes. *J. Adv. Mater.*, 32, 29, 1906697, 2020.
48. Sun, Y., Zhu, J., Bai, L., Li, Q., Zhang, X., Tong, W., Xie, Y., Sandwich-like carbon-anchored ultrathin TiO_2 nanosheets realizing ultrafast lithium storage. *Inorg. Chem. Front.*, 1, 1, 58–64, 2014.
49. Ren, C.E., Hatzell, K.B., Alhabeb, M., Ling, Z., Mahmoud, K.A., Gogotsi, Y., Charge-and size-selective ion sieving through $Ti_3C_2T_x$ MXene membranes. *J. Phys. Chem. Lett.*, 6, 20, 4026–4031, 2015.
50. Xu, W.L., Fang, C., Zhou, F., Song, Z., Liu, Q., Qiao, R., Yu, M., Self-assembly: A facile way of forming ultrathin, high-performance graphene oxide membranes for water purification. *Nano Lett.*, 17, 5, 2928–2933, 2017.
51. Akuzum, B., Maleski, K., Anasori, B., Lelyukh, P., Alvarez, N.J., Kumbur, E.C., Gogotsi, Y., Rheological characteristics of 2D titanium carbide (MXene) dispersions: A guide for processing MXenes. *ACS Nano*, 12, 3, 2685–2694, 2018.

52. Liu, G., Jin, W., Xu, N., Graphene-based membranes. *Chem. Soc Rev.*, 44, 15, 5016–5030, 2015.
53. Manshad, S., Sazegar, M.R., Nawawi, M.G.M., bin Hassan, H., Fabrication of nanohybrid polyetherimide/graphene oxide membranes: Biofuel dehydration by pervaporation process. *RSC Adv.*, 6, 106, 103888–103894, 2016.
54. Naik, P.V., Wee, L.H., Meledina, M., Turner, S., Li, Y., Van Tendeloo, G., Martens, J.A., Vankelecom, I.F., PDMS membranes containing ZIF-coated mesoporous silica spheres for efficient ethanol recovery via pervaporation. *J. Mater. Chem. A*, 4, 33, 12790–12798, 2016.
55. Mixa, A. and Staudt, C., Membrane-based separation of phenol/water mixtures using ionically and covalently cross-linked ethylene-methacrylic acid copolymers. *Int. J. Chem. Eng.*, 2008, 319392, 2008.
56. Baker, R.W., Wijmans, J.G., Huang, Y., Permeability, permeance and selectivity: A preferred way of reporting pervaporation performance data. *J. Membr. Sci.*, 348, 1-2, 346–352, 2010.
57. Baudot, A. and Marin, M., Pervaporation of aroma compounds: Comparison of membrane performances with vapour-liquid equilibria and engineering aspects of process improvement. *Food Bioprod. Process.*, 75, 2, 117–142, 1997.
58. Beaumelle, D., Marin, M., Gibert, H., Pervaporation with organophilic membranes: State of the art. *Food Bioprod. Process.*, 71, 77–89, 1993.
59. Oliveira, T.A., Scarpello, J.T., Livingston, A.G., Pervaporation-biological oxidation hybrid process for removal of volatile organic compounds from wastewaters. *J. Membr. Sci.*, 195, 1, 75–88, 2002.
60. Ganapathi-Desai, S. and Sikdar, S.K., A polymer-ceramic composite membrane for recovering volatile organic compounds from wastewaters by pervaporation. *J. Clean. Prod.*, 2, 3, 140–148, 2000.
61. Feng, X. and Huang, R.Y., Estimation of activation energy for permeation in pervaporation processes. *J. Membr. Sci.*, 118, 1, 127–131, 1996.
62. Hillaire, A. and Favre, E., Isothermal and nonisothermal permeation of an organic vapor through a dense polymer membrane. *Ind. Eng. Chem. Res.*, 38, 1, 211–217, 1999.
63. Wijmans, J.G., Athayde, A.L., Daniels, R., Ly, J.H., Kamaruddin, H.D., Pinnau, I., The role of boundary layers in the removal of volatile organic compounds from water by pervaporation. *J. Membr. Sci.*, 109, 1, 135–146, 1996.
64. Smitha, B., Suhanya, D., Sridhar, S., Ramakrishna, M., Separation of organic–organic mixtures by pervaporation—A review. *J. Membr. Sci.*, 241, 1, 1–21, 2004.
65. Izak, P., Hovorka, Š., Bartovský, T., Bartovska, L., Crespo, J.G., Swelling of polymeric membranes in room temperature ionic liquids. *J. Membr. Sci.*, 296, 1-2, 131–138, 2007.
66. Tarleton, E.S., Robinson, J.P., Salman, M., Solvent-induced swelling of membranes—Mseasurements and influence in nanofiltration. *J. Membr. Sci.*, 280, 1-2, 442–451, 2006.
67. Mulder, M., *Basic Principles of Membrane Technology*, Second Edition, pp. 280–415, Kluwer Academic Publishers, Boston, 1997.
68. Jiang, L.Y., Wang, Y., Chung, T.S., Qiao, X.Y., Lai, J.Y., Polyimides membranes for pervaporation and biofuels separation. *Prog. Polym. Sci.*, 34, 11, 1135–1160, 2009.
69. Shi, G.M., Chen, H., Jean, Y.C., Chung, T.S., Sorption, swelling, and free volume of polybenzimidazole (PBI) and PBI/zeolitic imidazolate framework (ZIF-8) nano-composite membranes for pervaporation. *Polym. J.*, 54, 2, 774–783, 2013.
70. Fathizadeh, M., Aroujalian, A., Raisi, A., Fotouhi, M., Preparation and characterization of thin film nanocomposite membrane for pervaporative dehydration of aqueous alcohol solutions. *Desalination*, 314, 20–27, 2013.
71. Niemistö, J., Pasanen, A., Hirvelä, K., Myllykoski, L., Muurinen, E., Keiski, R.L., Pilot study of bioethanol dehydration with polyvinyl alcohol membranes. *J. Membr. Sci.*, 447, 119–127, 2013.

72. Dobrak, A., Figoli, A., Chovau, S., Galiano, F., Simone, S., Vankelecom, I.F.J., Drioli, E., Van der Bruggen, B., Performance of PDMS membranes in pervaporation: Effect of silicalite fillers and comparison with SBS membranes. *J. Colloid Interface Sci.*, 346, 1, 254–264, 2010.
73. Simone, S., Figoli, A., Santoro, S., Galiano, F., Alfadul, S.M., Al-Harbi, O.A., Drioli, E., Preparation and characterization of ECTFE solvent resistant membranes and their application in pervaporation of toluene/water mixtures. *Sep. Purif. Technol.*, 90, 147–161, 2012.
74. Pandey, R.P., Rasheed, P.A., Gomez, T., Azam, R.S., Mahmoud, K.A., A fouling-resistant mixed-matrix nanofiltration membrane based on covalently cross-linked $Ti_3C_2T_X$ (MXene)/cellulose acetate. *J. Membr. Sci.*, 607, 118139, 2020.
75. Hilal, N. and Johnson, D., The use of atomic force microscopy in membrane characterization, in: *Basic Aspects Of Membrane Science and Engineering*, pp. 337–354, Elsevier Inc, Amsterdam, 2010.
76. Schossig-Tiedemann, M. and Paul, D., Improved preparation of membrane surfaces for field-emission scanning electron microscopy. *J. Membr. Sci.*, 187, 1-2, 85–91, 2001.
77. Wang, J., Chen, P., Shi, B., Guo, W., Jaroniec, M., Qiao, S.Z., A regularly channeled lamellar membrane for unparalleled water and organics permeation. *Angew. Chem. Int. Ed.*, 57, 23, 6814–6818, 2018.
78. Chao, W.C., Huang, S.H., An, Q., Liaw, D.J., Huang, Y.C., Lee, K.R., Lai, J.Y., Novel interfacially-polymerized polyamide thin-film composite membranes: studies on characterization, pervaporation, and positron annihilation spectroscopy. *Polym. J.*, 52, 11, 2414–2421, 2011.
79. Chao, W.C., Huang, S.H., Wei, S.W., Huang, Y.H., Liao, K.S., Lai, C.L., Tsai, H.A., Lee, K.R., Positron annihilation spectroscopic study on fine-structure characterization and ethanol dehydration by pervaporation for interfacially polymerized composite membranes. *J. Membr. Sci.*, 429, 34–43, 2013.
80. Jean, Y.C., Hung, W.S., Lo, C.H., Chen, H., Liu, G., Chakka, L., Cheng, M.L., Nanda, D., Tung, K.L., Huang, S.H., Lee, K.R., Lai, J.Y., Sun, Y.M., Hu, C.C., Yu, C.C., Applications of positron annihilation spectroscopy to polymeric membranes. *Desalination*, 234, 1-3, 89–98, 2008.
81. Satyanarayana, S.V., Subrahmanyam, V.S., Verma, H.C., Sharma, A., Bhattacharya, P.K., Application of positron annihilation: Study of pervaporation dense membranes. *Polym. J.*, 47, 4, 1300–1307, 2006.
82. Yeom, C.K. and Lee, K.H., A study on desorption resistance in pervaporation of single component through dense membranes. *J. Appl. Polym. Sci.*, 63, 2, 221–232, 1997.
83. Hasanoğlu, A., Salt, Y., Keleşer, S., Özkan, S., Dinçer, S., Pervaporation separation of organics from multicomponent aqueous mixtures. *Chem. Eng. Process.*, 46, 4, 300–306, 2007.
84. Hyder, M.N., Huang, R.Y.M., Chen, P., Effect of selective layer thickness on pervaporation of composite poly (vinyl alcohol)–poly (sulfone) membranes. *J. Membr. Sci.*, 318, 1-2, 387–396, 2008.
85. Raisi, A. and Aroujalian, A., Aroma compound recovery by hydrophobic pervaporation: The effect of membrane thickness and coupling phenomena. *Sep. Purif. Technol.*, 82, 53–62, 2011.
86. Mulder, M.H.V. and Smolders, C.A., On the mechanism of separation of ethanol/water mixtures by pervaporation I. Calculations of concentration profiles. *J. Membr. Sci.*, 17, 3, 289–307, 1984.
87. Burshe, M.C., Sawant, S.B., Pangarkar, V.G., Dehydration of glycerine-water mixtures by pervaporation. *J. Am. Oil Chem. Soc*, 76, 2, 209–214, 1999.
88. Domingues, L., Recasens, F., Larrayoz, M., Studies of a pervaporation reactor: Kinetics and equilibrium shift in benzyl alcohol acetylation. *Chem. Eng. Sci.*, 54, 10, 1461–1465, 1999.
89. Ma, J., Zhang, M., Lu, L., Yin, X., Chen, J., Jiang, Z., Intensifying esterification reaction between lactic acid and ethanol by pervaporation dehydration using chitosan–TEOS hybrid membranes. *Chem. Eng. J.*, 155, 3, 800–809, 2009.

90. Wu, X., Hao, L., Zhang, J., Zhang, X., Wang, J., Liu, J., Polymer-$Ti_3C_2T_x$ composite membranes to overcome the trade-off in solvent resistant nanofiltration for alcohol-based system. *J. Membr. Sci.*, 515, 175–188, 2016.
91. Cao, K., Jiang, Z., Zhang, X., Zhang, Y., Zhao, J., Xing, R., Yang, S., Gao, C., Pan, F., Highly water-selective hybrid membrane by incorporating g-C_3N_4 nanosheets into polymer matrix. *J. Membr. Sci.*, 490, 72–83, 2015.
92. Liu, G., Shen, J., Ji, Y., Liu, Q., Liu, G., Yang, J., Jin, W., Two-dimensional Ti_2CT_x MXene membranes with integrated and ordered nanochannels for efficient solvent dehydration. *J. Mater. Chem. A*, 7, 19, 12095–12104, 2019.
93. Liu, G., Shen, J., Liu, Q., Liu, G., Xiong, J., Yang, J., Jin, W., Ultrathin two-dimensional MXene membrane for pervaporation desalination. *J. Membr. Sci.*, 548, 548–558, 2018.
94. Chang, J.H., Yoo, J.K., Ahn, S.H., Lee, K.H., Ko, S.M., Simulation of pervaporation process for ethanol dehydration by using pilot test results. *Korean J. Chem. Eng.*, 15, 1, 28–36, 1998.
95. Guo, Z. and Hu, C., Pervaporation of organic liquid/water mixtures through a novel silicone copolymer membrane. *Chin. Sci. Bull.*, 43, 6, 487–490, 1998.
96. Wolinska-Grabczyk, A., Muszynski, J., Jankowski, A., Applications of polyurethane-based membranes in pervaporation separations. *Chem. Pap.*, 54, 6A, 389–392, 2001.
97. Guo, W.F., Chung, T.-S., Matsuura., T., Pervaporation study on the dehydration of aqueous butanol solutions: A comparison of flux vs. permeance, separation factor vs. selectivity. *J. Membr. Sci.*, 245, 1-2, 199–210, 2004.
98. Castro-Muñoz, R., MXene: A two-dimensional material in selective water separation via pervaporation. *Arab. J. Chem.*, 15, 1, 103524, 2022.
99. Xu, Z., Liu, G., Ye, H., Jin, W., Cui, Z., Two-dimensional MXene incorporated chitosan mixed-matrix membranes for efficient solvent dehydration. *J. Membr. Sci.*, 563, 625–632, 2018.
100. Kang, Z., Fan, L., Sun, D., Recent advances and challenges of metal–organic framework membranes for gas separation. *J. Mater. Chem.*, 5, 21, 10073–10091, 2017.
101. Park, H.B., Graphene-based membranes–a new opportunity for CO_2 separation. *Carbon Manage.*, 5, 3, 251–253, 2014.
102. Tomé, L.C. and Marrucho, I.M., Ionic liquid-based materials: A platform to design engineered CO_2 separation membranes. *Chem. Soc Rev.*, 45, 10, 2785–2824, 2016.
103. Smith, Z.P. and Freeman, B.D., Graphene oxide: A new platform for high-performance gas- and liquid-separation membranes. *Angew. Chem., Int. Ed.*, 53, 39, 10286–10288, 2014.
104. Bernardo, P., Drioli, E., Golemme, G., Membrane gas separation: A review/state of the art. *Ind. Eng. Chem. Res.*, 48, 10, 4638–4663, 2009.
105. Fan, Y., Wei, L., Meng, X., Zhang, W., Yang, N., Jin, Y., Wang, X., Zhao, M., Liu, S., An unprecedented high-temperature-tolerance 2D laminar MXene membrane for ultrafast hydrogen sieving. *J. Membr. Sci.*, 569, 117–123, 2019.
106. Perry, J.D., Nagai, K., Koros, W.J., Polymer membranes for hydrogen separations. *MRS Bull.*, 31, 10, 745–749, 2006.
107. Jin, Y., Fan, Y., Meng, X., Zhang, W., Meng, B., Yang, N., Liu, S., Theoretical and experimental insights into the mechanism for gas separation through nanochannels in 2D laminar MXene membranes. *Processes*, 7, 751, 2019.
108. Javaid, A., Membranes for solubility-based gas separation applications. *Chem. Eng. J.*, 112, 1-3, 219–226, 2005.
109. Ismail, A.F., Khulbe, K.C., Matsuura, T., Fundamentals of gas permeation through membranes, in: *Gas Separation Membranes*, pp. 11–35, Springer Cham, Switzerland, 2015.
110. Stern, S.A., Polymers for gas separations: the next decade. *J. Membr. Sci.*, 94, 1, 1–65, 1994.
111. Dai, Y., Johnson, J.R., Karvan, O., Sholl, D.S., Koros, W.J., Ultem®/ZIF-8 mixed matrix hollow fiber membranes for CO_2/N_2 separations. *J. Membr. Sci.*, 401, 76–82, 2012.

112. Petukhov, D.I., Kan, A.S., Chumakov, A.P., Konovalov, O.V., Valeev, R.G., Eliseev, A.A., MXene-based gas separation membranes with sorption type selectivity. *J. Membr. Sci.*, 621, 118994, 2021.
113. Xu, B., Xiang, H., Wei, Q., Liu, J.Q., Xia, Y.D., Yin, J., Liu, Z.G., Two-dimensional graphene-like C_2N: An experimentally available porous membrane for hydrogen purification. *Phys. Chem. Chem. Phys.*, 17, 23, 15115–15118, 2015.
114. Zhu, L., Xue, Q., Li, X., Wu, T., Jin, Y., Xing, W., C_2N: An excellent two-dimensional monolayer membrane for He separation. *J. Mater. Chem. A*, 3, 42, 21351–21356, 2015.
115. Hu, Q., Sun, D., Wu, Q., Wang, H., Wang, L., Liu, B., Zhou, A., He, J., MXene: A new family of promising hydrogen storage medium. *J. Phys. Chem. A*, 117, 51, 14253–14260, 2013.
116. Li, L., Zhang, T., Duan, Y., Wei, Y., Dong, C., Ding, L., Qiao, Z., Wang., H., Selective gas diffusion in two-dimensional MXene lamellar membranes: Insights from molecular dynamics simulations. *J. Mater. Chem. A*, 6, 11734–11742, 2018.
117. Ding, L., Wei, Y., Li, L., Zhang, T., Wang, H., Xue, J., Ding, L.X., Wang, S., Caro, J., Gogotsi, Y., MXene molecular sieving membranes for highly efficient gas separation. *Nat. Commun.*, 9, 1, 1–7, 2018.
118. Lin, H., Gong, K., Hykys, P., Chen, D., Ying, W., Sofer, Z., Yan, Y., Li, Z., Peng, X., Nanoconfined deep eutectic solvent in laminated MXene for efficient CO_2 separation. *Chem. Eng. J.*, 405, 126961, 2021.
119. Qu, K., Dai, L., Xia, Y., Wang, Y., Zhang, D., Wu, Y., Yao, Z., Huang, K., Guo, X., Xu, Z., Self-crosslinked MXene hollow fiber membranes for H_2/CO_2 separation. *J. Membr. Sci.*, 638, 119669, 2021.
120. Singh, R.P., Dahe, G.J., Dudeck, K.W., Welch, C.F., Berchtold, K.A., High temperature polybenzimidazole hollow fiber membranes for hydrogen separation and carbon dioxide capture from synthesis gas. *Energy Proc.*, 63, 153–159, 2014.
121. Singh, R.P., Dahe, G.J., Dudeck, K.W., Berchtold, K.A., Macrovoid-free high performance polybenzimidazole hollow fiber membranes for elevated temperature H_2/CO_2 separations. *Int. J. Hydrog. Energy*, 45, 51, 27331–27345, 2020.
122. Fan, Y., Li, J., Wang, S., Meng, X., Jin, Y., Yang, N., Meng, B., Li, J., Liu, S., Nickel (II) ion-intercalated MXene membranes for enhanced H_2/CO_2 separation. *Front. Chem. Sci. Eng.*, 15, 4, 882–891, 2021.
123. Li, H., Song, Z., Zhang, X., Huang, Y., Li, S., Mao, Y., Ploehn, H.J., Bao, Y., Yu, M., Ultrathin, molecular-sieving graphene oxide membranes for selective hydrogen separation. *Science*, 342, 6154, 95–98, 2013.
124. Guan, W., Yang, X., Dong, C., Yan, X., Zheng, W., Xi, Y., Ruan, X., Dai, Y., He, G., Prestructured MXene fillers with uniform channels to enhance CO_2 selective permeation in mixed matrix membranes. *J. Appl. Polym. Sci.*, 138, 8, 49895, 2021.
125. Yang, X., Zheng, W., Xi, Y., Guan, W., Yan, X., Ruan, X., Ma, C., Dai, Y., He, G., Constructing low-resistance and high-selectivity transport multi-channels in mixed matrix membranes for efficient CO_2 separation. *J. Membr. Sci.*, 624, 119046, 2021.
126. Hong, X., Lu, Z., Zhao, Y., Lyu, L., Ding, L., Wei, Y., Wang, H., Fast fabrication of freestanding MXene-ZIF-8 dual-layered membranes for H_2/CO_2 separation. *J. Membr. Sci.*, 642, 119982, 2022.
127. Shi, F., Sun, J., Wang, J., Liu, M., Yan, Z., Zhu, B., Li, Y., Cao, X., MXene versus graphene oxide: Investigation on the effects of 2D nanosheets in mixed matrix membranes for CO_2 separation. *J. Membr. Sci.*, 620, 118850, 2021.

Index